Lecture Notes in Computer Science   13224

More information about this series at https://link.springer.com/bookseries/558

Juan Luis Jiménez Laredo · J. Ignacio Hidalgo ·
Kehinde Oluwatoyin Babaagba (Eds.)

Applications of Evolutionary Computation

25th European Conference, EvoApplications 2022
Held as Part of EvoStar 2022
Madrid, Spain, April 20–22, 2022
Proceedings

 Springer

Editors
Juan Luis Jiménez Laredo ⓘD
Université Le Havre Normandie
Le Havre, France

J. Ignacio Hidalgo ⓘD
Universidad Complutense de Madrid
Madrid, Spain

Kehinde Oluwatoyin Babaagba ⓘD
Edinburgh Napier University
Edinburgh, UK

ISSN 0302-9743   ISSN 1611-3349 (electronic)
Lecture Notes in Computer Science
ISBN 978-3-031-02461-0   ISBN 978-3-031-02462-7 (eBook)
https://doi.org/10.1007/978-3-031-02462-7

This Springer imprint is published by the registered company Springer Nature Switzerland AG
The registered company address is: Gewerbestrasse 11, 6330 Cham, Switzerland

Preface

This volume contains the proceedings of EvoApplications 2022, the International Conference on the Applications of Evolutionary Computation. The conference is part of Evo*, the leading event on bio-inspired computation in Europe, and was held in Madrid, Spain as a hybrid event, between Wednesday, April 20 and Friday, April 22, 2022.

EvoApplications, formerly known as EvoWorkshops, aims to bring together high-quality research with a focus on applied domains of bio-inspired computing. At the same time, under the Evo* umbrella, EuroGP focused on the technique of genetic programming, EvoCOP targeted evolutionary computation in combinatorial optimization, and EvoMUSART was dedicated to evolved and bio-inspired music, sound, art, and design. The proceedings for all of these co-located events are available in the LNCS series.

EvoApplications 2022 received 67 high-quality submissions distributed among the main session on Applications of Evolutionary Computation, the EvoApps + EuroGP special joint track on Evolutionary Machine Learning, and 10 additional special sessions chaired by leading experts on the different areas: Analysis of Evolutionary Computation Methods: Theory, Empirics, and Real-World Applications; Applications of Bio-inspired Techniques on Social Networks; Applications of Nature-inspired Computing for Sustainability and Development; Evolutionary Computation in Edge, Fog, and Cloud Computing; Evolutionary Computation in Image Analysis, Signal Processing, and Pattern Recognition; Machine Learning and AI in Digital Healthcare and Personalized Medicine; Evolutionary Robotics; Parallel and Distributed Systems; Resilient Bio-Inspired Algorithms; and Soft Computing Applied to Games. We selected 33 of these papers for full oral presentation, while a further 13 works were presented in short oral presentations and as posters. All accepted contributions, regardless of the presentation format, appear as full papers in this volume.

Obviously, an event of this kind would not be possible without the contribution of a large number of people:

- We express our gratitude to the authors for submitting their works and to the members of the Program Committee for devoting selfless effort in the review process.
- We would also like to thank Nuno Lourenço (University of Coimbra, Portugal) for his dedicated work with the submission and registration system and Sérgio Rebelo and Tiago Martins (University of Coimbra, Portugal) for their important graphic design work.
- We are grateful to Francisco Chicano (University of Málaga, Spain) and João Correia (University of Coimbra, Portugal) for their impressive work in managing and maintaining the Evo* website and handling the publicity, respectively.
- We credit the invited keynote speakers, Gabriela Ochoa (University of Stirling, UK) and Pedro Larrañaga (Technical University of Madrid, Spain), for their fascinating and inspiring presentations.

- We would like to express our gratitude to the Steering Committee of EvoApplications for helping with the organization of the conference.
- We are grateful to the support provided by SPECIES, the Society for the Promotion of Evolutionary Computation in Europe and its Surroundings, for the coordination and financial administration.
- Special thanks to J. Ignacio Hidalgo as local organizer and to the University Complutense of Madrid for organizing and providing an enriching conference venue. Federico Divina from the Pablo de Olavide University of Seville was also involved in the local organization during these years of the COVID-19 pandemic. Despite the fact that the conference could not finally be held in Seville, his contribution was a great help.

Finally, we express our continued appreciation to Anna I. Esparcia-Alcázar, from SPECIES, Europe, whose considerable efforts in managing and coordinating Evo* helped towards building a unique, vibrant, and friendly atmosphere.

April 2022

Juan Luis Jiménez Laredo
Kehinde Oluwatoyin Babaagba
Carola Doerr
Giovanni Iacca
Valentino Santucci
Diego Oliva
Mahshid Helali Moghadam
Pablo Mesejo
Stephen Smith
Kyrre Glette
Carlos Cotta
Penousal Machado
Antonio Mora
Alberto P. Tonda
J. Ignacio Hidalgo Perez
Thomas Bartz-Beielstein
Christine Zarges
Doina Bucur
Fabio Caraffini
Seyed Jalaleddin Mousavirad
Juan Julián Merelo-Guervós
Harith Al-Sahaf
Marta Vallejo
Frank Veenstra
Gustavo Olague
Wolfgang Banzhaf
Pablo García Sánchez

Organization

EvoApplications Conference Chair

Juan Luis Jiménez Laredo   Université Le Havre Normandie, France

EvoApplications Conference Co-chair

J. Ignacio Hidalgo   Universidad Complutense de Madrid, Spain

EvoApplications Publication Chair

Kehinde Oluwatoyin Babaagba   Edinburgh Napier University, UK

Applications of Bio-inspired Techniques on Social Networks Chairs

Giovanni Iacca   University of Trento, Italy
Doina Bucur   University of Twente, The Netherlands

Analysis of Evolutionary Computation Methods: Theory, Empirics, and Real-World Applications Chairs

Thomas Bartz-Beielstein   IDEA, TH Koeln, Germany
Carola Doerr   CNRS, Sorbonne University, France
Christine Zarges   Aberystwyth University, UK

Applications of Nature-inspired Computing for Sustainability and Development Chairs

Valentino Santucci   University for Foreigners of Perugia, Italy
Fabio Caraffini   De Montfort University, UK

Evolutionary Computation in Image Analysis, Signal Processing, and Pattern Recognition Chairs

Pablo Mesejo   Universidad de Granada, Spain
Harith Al-Sahaf   Victoria University of Wellington, New Zealand

Evolutionary Computation in Edge, Fog, and Cloud Computing Chairs

Diego Oliva   Universidad de Guadalajara, Mexico
Seyed Jalaleddin Mousavirad   Hakim Sabzevari University, Iran
Mahshid Helali Moghadam   RISE SICS, Sweden

Machine Learning and AI in Digital Healthcare and Personalized Medicine Chairs

Stephen Smith   University of York, York, UK
Marta Vallejo   Heriot-Watt University, UK

Resilient Bio-inspired Algorithms Chairs

Carlos Cotta   Universidad de Málaga, Spain
Gustavo Olague   CICESE, México

Soft Computing Applied to Games Chairs

Alberto P. Tonda   INRAE, France
Antonio M. Mora   Universidad de Granada, Spain
Pablo García-Sánchez   Universidad de Granada, Spain

Parallel and Distributed Systems Chairs

Juan Julián Merelo-Guervós   Universidad de Granada, Spain
Juan Luis Jiménez Laredo   Université Le Havre Normandie, France

Evolutionary Robotics Chairs

Kyrre Glette   Universitetet i Oslo, Norway
Frank Veenstra   Universitetet i Oslo, Norway

Evolutionary Machine Learning Chairs

Penousal Machado   University of Coimbra, Portugal
Wolfgang Banzhaf   Michigan State University, USA

EvoApplications Steering Committee

Stefano Cagnoni   University of Parma, Italy
Pedro A. Castillo   Universidad de Granada, Spain

Anna I. Esparcia-Alcázar	SPECIES, Europe
Mario Giacobinni	Universitá degli Studi di Torino, Italy
Paul Kaufmann	Johannes Gutenberg-Universität Mainz, Germany
Antonio M. Mora	Universidad de Granada, Spain
Günther Raidl	Technische Universität Wien, Austria
Franz Rothlauf	Mainz University, Germany
Kevin Sim	Edinburgh Napier University, UK
Giovanni Squillero	Politecnico di Torino, Italy
Cecilia di Chio	University of Southampton, UK
(Honorary Member)	

Program Committee

Ahmed Elsaid	University of Puerto Rico at Mayagüez, USA
Aladdin Ayesh	De Montfort University, UK
Alberto P. Tonda	INRAE, France
Ales Zamuda	University of Maribor, Slovenia
Alex Freitas	University of Kent, UK
Amir Dehsarvi	University of York, UK
Anabela Simões	Coimbra Institute of Engineering, Portugal
Anca Andreica	Babes-Bolyai University, Romania
Anders Christensen	University of Southern Denmark, Denmark
Andrea Tettamanzi	University of Nice Sophia Antipolis, France
Andres Faina	IT University of Copenhagen, Denmark
Anil Yaman	Korea Advanced Institute of Science and Technology, South Korea
Anna Paszynska	Jagiellonian University, Poland
Anthony Clark	Pomona College, USA
Antonio Mora-García	University of Granada, Spain
Antonio Cordoba	University of Seville, Spain
Antonio Della Cioppa	University of Salerno, Italy
Antonio González-Pardo	Universidad Rey Juan Carlos, Spain
Antonio J. Fernández Leiva	Universidad de Málaga, Spain
Arkadiusz Poteralski	Silesian University of Technology, Poland
Bernabe Dorronsoro	University of Cadiz, Spain
Carlos Cotta	Universidad de Málaga, Spain
Carlotta Orsenigo	University of Milan, Italy
Carola Doerr	CNRS, Sorbonne University, France
Chien-Chung Shen	University of Delaware, USA
Christine Zarges	Aberystwyth University, UK
Clara Pizzuti	ICAR-CNR, Italy
Daniel Hernandez	Instituto Tecnológico de Tijuana, México
David Megias	Universitat Oberta de Catalunya, Spain

David Pelta	University of Granada, Spain
Diego Oliva	Universidad de Guadalajara, Mexico
Diego Perez Liebana	Queen Mary University of London, UK
Doina Bucur	University of Twente, The Netherlands
Edoardo Fadda	Politecnico di Torino, Italy
Enrico Schumann	University of Basel, Switzerland
Ernesto Tarantino	ICAR-CNR, Italy
Evelyne Lutton	INRAE, France
Fabio Caraffini	De Montfort University, UK
Fabio D'Andreagiovanni	CNRS, Sorbonne University, France
Federico Liberatore	Cardiff University, UK
Federico Divina	Pablo de Olavide University, Spain
Feijoo Colomine	Universidad Nacional Experimental del Tàchira, Venezuela
Fernando Lobo	University of Algarve, Portugal
Ferrante Neri	University of Nottingham, UK
Francesco Fontanella	Università degli Studi di Cassino e del Lazio Meridionale, Italy
Francisco Luna	Universidad de Málaga, Spain
Francisco Chicano	University of Málaga, Spain
Francisco Fernandez De Vega	Universidad de Extremadura, Spain
Frank Veenstra	Universitetet i Oslo, Norway
Gabriel Luque	University of Málaga, Spain
Geoff Nitschke	University of Cape Town, South Africa
Giovanni Iacca	University of Trento, Italy
Giulio Biondi	University of Florence, Italy
Gregoire Danoy	University of Luxembourg, Luxembourg
Guenter Rudolph	TU Dortmund University, Germany
Guillermo Gomez-Trenado	Universidad de Granada, Spain
Gurhan Kucuk	Yeditepe University, Turkey
Gustavo Olague	CICESE, México
Harith Al-Sahaf	Victoria University of Wellington, New Zealand
Heiko Hamann	University of Luebeck, Germany
J. Ignacio Hidalgo	Universidad Complutense de Madrid, Spain
Illya Bakurov	Universidade NOVA de Lisboa, Portugal
Jacopo Aleotti	University of Parma, Italy
James Foster	University of Idaho, USA
János Botzheim	Eötvös Loránd University, Hungary
Jaroslaw Was	AGH University of Science and Technology, Poland
Jaume Bacardit	Newcastle University, UK
Jesús Mayor	Universidad Politécnica de Madrid, Spain

João Correia	University of Coimbra, Portugal
João Macedo	University of Coimbra, Portugal
Joel Lehman	University of Central Florida, USA
Jörg Bremer	University of Oldenburg, Germany
Jorge Novo Buján	Universidade da Coruña, Spain
Jose Santos	University of A Coruña, Spain
José Carlos Ribeiro	Polytechnic Institute of Leiria, Portugal
José Manuel Colmenar	Universidad Rey Juan Carlos, Spain
Juanlu Jimenez	Université du Havre Normandie, France
Karlo Knezevic	University of Zagreb, Croatia
Kehinde Oluwatoyin Babaagba	Edinburgh Napier University, UK
Kenji Leibnitz	National Institute of Information and Communications Technology, Japan
Kevin Sim	Edinburgh Napier University, UK
Krzysztof Michalak	Wrocław University of Economics, Poland
Kyrre Glette	Universitetet i Oslo, Norway
Laura Dipietro	Highland Instruments, USA
Maciej Smołka	AGH University of Science and Technology, Poland
Mahshid Helali Moghadam	RISE SICS, Sweden
Marco Tomassini	University of Lausanne, Switzerland
Marco Villani	University of Modena and Reggio Emilia, Italy
Marco Baioletti	Universitá degli Studi di Perugia, Italy
Marcos Ortega Hortas	Universidade da Coruña, Spain
Mario Giacobini	University of Torino, Italy
Mengjie Zhang	Victoria University of Wellington, New Zealand
Michael Lones	Heriot-Watt University, UK
Michael Guckert	Technische Hochschule Mittelhessen, Germany
Mohamad Alissa	Edinburgh Napier University, UK
Mohamed Wiem Mkaouer	Rochester Institute of Technology, USA
Nuno Lourenço	University of Coimbra, Portugal
Oscar Castillo	Tijuana Institute of Technology, México
Oscar Cordon	University of Granada, Spain
Pablo García-Sánchez	University of Granada, Spain
Pablo Mesejo	University of Granada, Spain
Paolo Mengoni	Hong Kong Baptist University, China
Pedro A. Castillo-Valdivieso	University of Granada, Spain
Penousal Machado	University of Coimbra, Portugal
Petr Pošík	Czech Technical University in Prague, Czech Republic
Philip Bontrager	New York University, USA

Rafael Villanueva   Instituto Universitario de Matematica
                                 Multidisciplinar, Spain
Raneem Qaddoura   Al Hussein Technical University, Jordan
Renato Tinós   Universidade de São Paulo, Brazil
Rolf Hoffmann   TU Darmstadt, Germany
Salem Mohammed   Mustapha Stambouli University, Algeria
Sara Silva   Universidade de Lisboa, Portugal
Sebastian Risi   IT University of Copenhagen, Denmark
Sebastián Ventura   University of Cordoba, Spain
Sevil Sen   University of York, UK
Seyed Jalaleddin Mousavirad   Hakim Sabzevari University, Iran
Shamik Sural   IIT, Kharagpur, India
Simon Wells   Edinburgh Napier University, UK
Stefano Cagnoni   University of Parma, Italy
Stefano Coniglio   University of Southampton, UK
Stephen Smith   University of York, UK
Thomas Bartz-Beielstein   IDEA, TH Koeln, Germany
Thomas Farrenkopf   Technische Hochschule Mittelhessen, Germany
Tiago Baptista   University of Coimbra, Portugal
Tobias Glasmachers   Institut für Neuroinformatik, Germany
Travis Desell   Rochester Institute of Technology, USA
Valentino Santucci   University for Foreigners of Perugia, Italy
Waclaw Kus   Silesian University of Technology, Poland
Will Browne   Queensland University of Technology, Australia
Wolfgang Banzhaf   Michigan State University, USA
Yanan Sun   Sichuan University, China
Yaochu Jin   Universität Bielefeld, Germany
Ying-Ping Chen   National Yang Ming Chiao Tung University,
                                 Taiwan
Yoann Pigné   LITIS, Université Le Havre Normandie, France

Contents

Resilient Bio-inspired Algorithms

Evolutionary Robotics

Analysis of Evolutionary Computation Methods: Theory, Empirics, and Real-world Applications

Applications of Evolutionary Computation

An Enhanced Opposition-Based Evolutionary Feature Selection Approach

Ruba Abu Khurma[1], Ibrahim Aljarah[1], Pedro A. Castillo[2(✉)], and Khair Eddin Sabri[1]

[1] The University of Jordan, Amman, Jordan
[2] ETSIIT-CITIC, University of Granada, Granada, Spain
pacv@ugr.es

Abstract. This paper proposes an enhanced feature selection (FS) approach to improve the classification tasks, taking into account data dimensionality as a significant criterion of the dataset. High dimensionality may cause serious problems in classification that degrade the performance of the classifier. Among these problems: generating complex models (overfitting), increasing the learning time, and including redundant and irrelevant features in the learning model. FS is a data mining technique to minimize the number of dimensions (features) by getting rid of redundant and irrelevant features. Meanwhile, FS tries to maximize the classification performance. As FS is an optimization problem, meta-heuristic optimization algorithms can take place to achieve superior results in solving such problems. This paper proposes the Moth Flame Optimization (MFO) algorithm to tackle the FS problem. A new initialization method called opposition-based is proposed. Furthermore, a new update strategy is proposed to alleviate the local minima. The comparative results find that the proposed approach improves the MFO performance and outperforms other similar approaches.

Keywords: Moth Flame Optimization · MFO · Feature selection · Opposition-based · Flame

1 Introduction

Feature selection (FS) is a data mining technique to reduce the dimensionality of the dataset by excluding noisy features (redundant, irrelevant). It compromises two processes: search and evaluation [7].

In the evaluation, the generated feature subset is evaluated based on the features themselves and their interaction (filter approach) or based on a learning algorithm (wrapper approach). Wrappers are commonly known that they are more accurate and slower compared with filters [12].

In the search process, a searching algorithm is adopted to find the best feature subset. This can be achieved by deterministic algorithms such as gradient-based algorithms to find the maximum best or the minimum best. These methods are time-consuming because they need to generate all the possible feature subsets. In

© Springer Nature Switzerland AG 2022
J. L. Jiménez Laredo et al. (Eds.): EvoApplications 2022, LNCS 13224, pp. 3–14, 2022.
https://doi.org/10.1007/978-3-031-02462-7_1

the meta-heuristic search algorithms, random feature subsets are generated and evaluated. These algorithms achieved promising results in solving FS problems in various domains [6].

Moth Flame Optimization (MFO) [14] is a well-known swarm intelligence algorithm that has been applied in many applications. It is based on initializing a set of random solutions (population) in a random feature space. Each random feature subset represents a binary solution, and it is described using a 1-d array. The values of the solution either 'zero' which means the corresponding feature is not selected or 'one' which means the corresponding feature is selected. These random feature subsets are evaluated using a fitness function to give it a fitness value (e.g. accuracy). The MFO iterates until a stopping condition is satisfied. Then the near (best) solution or called global best (flame) is returned. The flame represents the best moth that has the minimum number of features with maximum performance.

In this paper, two enhancements are proposed to improve the MFO algorithm. The initialization strategy in the original algorithm depends on the uniform random distribution to generate random feature subsets. However, the proposed strategy depends on the opposition-based method, to enhance the update strategy of MFO, a flame reset method is proposed to alleviate the local minima [2].

The paper is sectioned as follows: Sect. 2 presents the basis of the MFO algorithm. Section 3, explains the proposed approach. In Sect. 4, the results of the experiments are presented and discussed. Finally, Sect. 5 the conclusion is given.

2   Moth Flame Optimization

Moth Flame Optimization (MFO) is a recent SI algorithm that was developed in [14]. The natural inspiration of MFO came from the movements of moths at night. These moths move in a straight line and maintain a stable angle to the moon. This movement strategy is called transfer orientation. The transfer orientation is not useful if the source light is close because the moth will try to maintain the same angle with it, forcing it to move in a spiral way and eventually die.

The spiral movement of moths around the flames is formulated in Eq. 1, which describes the movement of moths in spiral path around a candle where Mo_i is the i_{th} moth, Fl_j is the j_{th} flame, and Sp is the function of spiral path.

Equation 2 shows the logarithmic function used to formulate the spiral movement of moths, where Ds_i is the distance between the i_{th} moth and the j_{th} flame as shown in Eq. 3, b is a constant value that determines the shape of the logarithmic spiral, and t is a random number in $[-1, 1]$. The parameter $t = -1$ represents the closest position of a moth to a flame where $t = 1$ represents the farthest position between a moth and a flame. To increase exploitation, the t parameter is selected in the range $[r, 1]$ where r is decreased linearly across iterations from -1 to -2.

$$Moi = Sp(Mo_i, Fl_j) \tag{1}$$

$$Sp(Mo_i, Fl_j) = Ds_i \times e^{bt} \times cos(2\Pi) + Fl_j \tag{2}$$

$$Ds_i = |Mo_i - Fl_j| \tag{3}$$

Equation 4 shows the gradual decrease of the number of flames across the iterations, where Ct is the current number of iterations, Mfl is the maximum number of flames, and Mt is the maximum number of iterations.

$$FlameNumber = round(Mfl - Ct \times (Mfl - 1)/Mt) \tag{4}$$

2.1 Binary Moth Flame Optimization

The original MFO was designed to solve global optimization problems [13]. In such cases, the individual elements are real values. Thus, the optimizer's task is to verify that the upper and lower limits of each element are not exceeded in the initialization and update processes. However, for binary optimization problems, the case is different because each individual's elements are restricted to either "0" or "1". To enable the MFO algorithm to optimize in a binary feature space, the MFO needs to integrate some operators with it. The most common binary operator used to convert continuous optimizers into binary is the transfer function (TF) [14]. The main reason for using TFs is that they are easy to implement without affecting the essence of the algorithm. In this paper, the used TF is the sigmoid function which was originally used in [3] to generate the binary PSO (BPSO). In the MFO algorithm, the first term of Eq. 2 represents the step vector which is redefined in Eq. 5. The function of the sigmoid is to determine a probability value in the range [0, 1] for each element of the solution. Equation 6 shows the formula of the sigmoid function. Each moth updates its position based on Eq. 7 which takes the output of Eq. 6 as its input.

Algorithm 1 shows the BMFO algorithm.

$$\Delta Mo = Ds_i \times e^{bt} \times cos(2\Pi) \tag{5}$$

$$Trf(\Delta Mo_t) = 1/(1 + e^{\Delta Mo_t}) \tag{6}$$

$$Mo_i^d(t+1) = \begin{cases} 0, & \text{if } rand < Trf(\Delta Mo_{t+1}) \\ 1, & \text{if } rand \geqslant Trf(\Delta Mo_{t+1}) \end{cases} \tag{7}$$

2.2 Binary Moth Flame Optimization for Feature Selection

For an FS problem, there are two important issues to consider to establish a correct optimization process: the representation of the individual and the evaluation of it. Usually, the individual in the FS problem is represented using a 1-d array. The value of each element can be assigned to two values, either "0" or

---

Algorithm 1. The pseudo-code of BMFO

Input: Mt, n (# moths), d (# dimensions)
Output: near optimal moth
Initialization process for the moths

  while $Ct \leq Mt$ **do**
    modify the number of flames using Eq.4
    FMo = Fitness(Mo);
    if $Ct == 1$ **then**
      $Fl = sort(Mo)$;
      $FFl = sort(FMo)$;
    else
      $Fl = sort(Mo_{Ct-1}, Mo_{Ct})$;
      $FFl = sort(FMo_{Ct-1}, FMo_{Ct})$;
    end if
    for i = 1: n **do**
      for j = 1: d **do**
        Modify r and t;
        Compute Ds by Eq.3 based on the corresponding moth;
        Modify the step vector of a moth ΔMo using Eq.5.
        Compute the probabilities by Eq.6.
        Modify the position of a moth by Eq.7
      end for
    end for
    $Ct = Ct + 1$;
  end while

"1". The value "0" indicates that the feature is not selected while the value "1" indicates that the feature is selected. The evaluation of the individual in an FS problem depends on using a fitness function that maximizes the performance of a classifier and minimizes the number of selected features simultaneously. The fitness function is formulated in Eq. 8 where Err is the error rate, $|Sf|$ is the number of selected features in the reduced data set, and $|Cf|$ is the number of features in the original data set, and $\alpha \in [0, 1]$, $\beta = (1 - \alpha)$ are two parameters that indicate the significance of classification and the number of selected feature according to [10].

$$Fitness = \alpha \times Err + \beta \times \frac{|Sf|}{|Cf|} \tag{8}$$

3 The Proposed Approach

3.1 Initialization Using Opposition-Based Method

Initialization is an important step in any optimization algorithm in which the candidate solutions are firstly defined. It affects the speed of convergence and determines the final solution quality. Usually, the population is initialized using randomly based methods. However, using intelligent methods such as the opposition-based method may help. In the opposition-method, a set of random solutions simultaneously with their opposite positions are defined. Therefore, this paper studies and analyzes integrating OBL with MFO. Here, a population of size 2M is initialized using a uniform random distribution and OBL strategy. Finally, the fittest solutions (M) out of (2M) are used in the initial population. The pseudo-code of the proposed method is shown in Algorithm 2.

Algorithm 2. Opposition-based population initialization.

Input:Nil
Output:X a set of M moths
$X=$ randomly generated M moths
   for i = 1: M **do**
      for j = 1: d **do**
         $X_{ij}=u_j+l_j$-X_{ij}
         $OX = X \cup \overline{X}$
         OX consists of 2M moths
         Compute fitness of moths in OX using Eq.
         Sort OX with fitness values.
         $X = top(OX/2)$ moths
         Return X
      end for
   end for
   $l = l + 1;$

3.2 Retiring Flame

If the best solution (flame) falls into local minima, then the search process will be bounded by a specific area of the search space. This will hit the classification results. This study proposes a new method that retires flame under such conditions and uses an Enhanced Binary Moth Flame Optimization (EBMFO). By resetting flame, the EBMFO can jump from local minima, and achieves better balance between the global search and the local search. This produces superior classification results. Figure 1a illustrates that moths move towards the flame after a specific searching time. Figure 1b shows that after three iterations, the flame is considered stuck in local minima. Therefore, flame changes its fitness value including classification accuracy and the selected number of features to zero (retires). Figure 1c shows that other moths in the swarm converge toward the reset flame. Figure 1d shows new flame is generated in the area with a lower number of features and higher classification results.

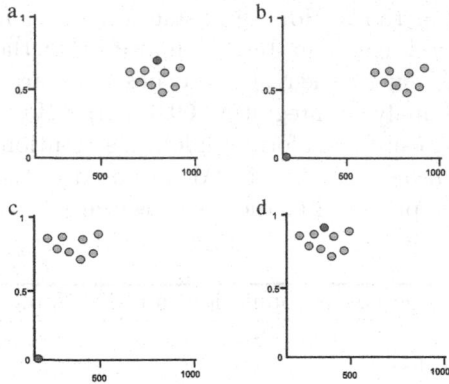

Fig. 1. (a) flame is trapped in a local minima. (b) flame is rest to zero. (c) Moths movement after resetting of flame. (d) Moths congregated towards the updated flame value, improving the individual position.

4   Experimental Setup and Results

In this paper, twelve popular datasets from UCI [1] are used to validate the proposed approach. Table 1 describes the used datasets. The results in table show the average outcomes from consecutive twenty runs. Three well-known FS algorithms are used for comparison with the proposed approach: Binary Grey Wolves (BGWO), Binary Cuckoo Search (BCS), and Binary Bat (BA) algorithms [4,5,8,9].

Table 1. Datasets description

#	Name	# features	# samples	# classes
D1	Abalone	8	3842	11
D2	Glass	9	214	6
D3	Iris	4	150	3
D4	Letter	16	20000	26
D5	Shuttle	9	58000	7
D6	Spambase	57	4601	2
D7	Tae	5	151	3
D8	Vehicle	18	846	4
D9	Waveform	21	5000	3
D10	Wine	13	178	3
D11	Wisconsin	9	683	2
D12	Yeast	8	1484	9

The algorithms are implemented using Python programming language in EvoloPy-FS framework [11]. The maximum number of iterations is 100 and the population size is 10. The evaluation measurements are fitness values, classification accuracy, the number of selected features, and computational time. In this work, the k-nearest neighbor classifier (K-NN) (K = 5 [10]) is used to evaluate the proposed FS wrapper methods. Each dataset is splited into training and testing parts with ratios of 80% and 20% respectively. The parameters are adopted in the experiments as follows: for the GWO, α is in the range [2,0]. For the BA, Qmin Frequency minimum is 0, Qmax Frequency maximum is 2, a Loudness is 0.5, r Pulse rate is 0.5. For the CS, pa is 0.25 and β is 3/2. The studying of the behavior of the proposed EBMFO is studied by comparing it with the standard BMFO in terms of fitness value, accuracy, number of selected features, and computation time. Then, EBMFO is compared with other wrapper algorithms: DGWO, BCS, and BA in terms of the aforementioned evaluation measures.

Inspecting Table 2, it can be found that EBMFO outperforms BMFO in terms of accuracy across 75%. BMFO obtains higher accuracy than EBMFO across three datasets only: Glass, Spambase, and Waveform. Table 3 shows that EBMFO achieves the least number of selected features across 83% of the datasets. BMFO obtains the least number of selected features across only two datasets: Glass and Spambase. Table 4, shows the average fitness values of the BMFO and EBMFO. The fitness value represents the combination between the accuracy and number of selected features, so it gives a more representative evaluation for both approaches. As in Table 2, EBMFO outperforms BMFO across 75% of the datasets. The results of computation time in Table 5 show that there is no difference between BMFO and EBMFO in terms of running time.

Table 2. Average accuracy results of BMFO and EBMFO.

Dataset #	BMFO		EBMFO	
	Avg	Std	Avg	Std
D1	0.650	0.009	**0.675**	0.001
D2	**0.976**	0.005	0.961	0.018
D3	0.967	0.000	**0.991**	0.000
D4	0.964	0.003	**0.977**	0.001
D5	0.973	0.091	**0.985**	0.078
D6	**0.921**	0.087	0.914	0.054
D7	0.659	0.127	**0.660**	0.023
D8	0.756	0.101	**0.777**	0.096
D9	**0.828**	0.259	0.805	0.141
D10	0.589	0.089	**0.596**	0.047
D11	0.982	0.013	**0.985**	0.003
D12	0.645	0.057	**0.699**	0.031

Table 3. Average number of selected features between BMFO and EBMFO

Dataset #	BMFO		EBMFO	
	Avg	Std	Avg	Std
D1	5.758	0.581	**5.000**	0.457
D2	3.733	0.377	**2.782**	0.259
D3	**2.004**	0.012	2.885	0.001
D4	11.091	0.434	**10.005**	0.369
D5	3.871	0.451	**2.112**	0.369
D6	**40.981**	0.569	44.963	0.985
D7	2.001	0.002	**1.822**	0.091
D8	11.569	0.774	**10.833**	0.403
D9	14.887	0.556	**11.299**	0.896
D10	6.892	0.891	**4.281**	0.999
D11	5.221	0.221	**5.011**	0.336
D12	7.631	0.114	**4.332**	0.998

Table 4. Average fitness value of BMFO and EBMFO

Dataset #	BMFO		EBMFO	
	Avg	Std	Avg	Std
D1	0.350	0.002	**0.325**	0.001
D2	**0.024**	0.012	0.039	0.008
D3	0.033	0.000	**0.009**	0.000
D4	0.036	0.015	**0.023**	0.079
D5	0.027	0.084	**0.015**	0.066
D6	**0.079**	0.147	0.086	0.096
D7	0.341	0.087	**0.340**	0.560
D8	0.244	0.189	**0.223**	0.257
D9	**0.172**	0.036	0.195	0.022
D10	0.411	0.001	**0.404**	0.000
D11	0.018	0.000	**0.015**	0.000
D12	0.355	0.044	**0.301**	0.031

The proposed EBMFO is further compared with three wrapper-based approaches. As seen in Table 6, EBMFO achieves the best accuracy across five of the datasets, then BCS gets the highest accuracy across three datasets and finally, BGWO and BBA are the best across only two datasets. From Table 7, it can be seen that EBMFO achieves the fewest number of selected features across six datasets, then BBA across three datasets, then BGWO, and BCS across two

Table 5. Average computational time of BMFO and EBMFO.

Dataset #	BMFO		EBMFO	
	Avg	Std	Avg	Std
D1	**26.358**	2.474	28.221	5.456
D2	9.668	4.176	**8.221**	3.887
D3	**6.412**	3.299	6.956	2.541
D4	**1542.028**	42.896	1755.546	40.563
D5	649.223	23.691	**647.236**	20.159
D6	**190.523**	4.123	194.220	6.889
D7	7.896	2.569	**6.998**	1.563
D8	**10.569**	2.563	10.891	1.336
D9	132.147	20.964	**120.631**	14.552
D10	**0.923**	0.231	9.685	0.154
D11	9.166	0.569	**8.578**	0.412
D12	15.569	4.569	**14.235**	2.110

Table 6. The average accuracy of EBMFO and other approaches.

Dataset #	EBMFO		BGWO		BCS		BBA	
	Avg	Std	Avg	Std	Avg	Std	Avg	Std
D1	**0.675**	0.001	0.670	0.000	0.553	0.002	0.569	0.123
D2	0.961	0.018	0.852	0.045	0.975	0.067	0.951	0.128
D3	**0.991**	0.000	**0.991**	0.125	**0.991**	0.010	**0.991**	0.037
D4	0.974	0.001	0.973	0.015	0.965	0.291	**0.977**	0.204
D5	0.985	0.078	0.985	0.003	0.985	0.029	0.985	0.056
D6	**0.914**	0.054	0.743	0.043	0.911	0.026	0.671	0.044
D7	0.660	0.023	**0.678**	0.016	0.673	0.122	0.659	0.023
D8	0.777	0.096	0.765	0.097	**0.785**	0.136	0.774	0.208
D9	**0.805**	0.414	0.801	0.399	0.800	0.158	0.788	0.054
D10	0.560	0.047	0.551	0.036	**0.583**	0.210	0.555	0.026
D11	**0.985**	0.003	0.984	0.223	0.976	0.125	0.945	0.000
D12	**0.699**	0.031	0.687	0.541	0.663	0.002	0.651	0.149

and one datasets, respectively. Table 8, shows that in terms of fitness values, the EBMFO outperforms other approaches. In Table 9, it is seen that BBA achieves the best running time in six datasets the EBMFO achieves the best running time in five datasets. On the other hand, BGWO and BCS achieve the best running time across one dataset only.

Table 7. The average number of selected features of EBMFO and other approaches

Dataset #	EBMFO		BGWO		BCS		BBA	
	Avg	Std	Avg	Std	Avg	Std	Avg	Std
D1	**5.000**	0.457	5.876	0.154	5.44	0.126	5.100	0.123
D2	2.782	0.259	**2.369**	0.156	2.611	0.111	2.440	0.012
D3	2.885	0.001	2.571	0.012	2.913	0.015	**2.390**	0.057
D4	10.005	0.369	10.981	0.444	11.213	0.432	**9.754**	5.778
D5	**2.112**	0.369	3.051	0.127	2.335	0.321	2.569	0.054
D6	**44.963**	0.985	45.332	0.841	48.771	0.456	46.225	0.329
D7	1.822	0.091	2.220	0.085	3.115	0.044	**1.543**	0.021
D8	10.833	0.403	**9.110**	0.128	11.225	.587	10.456	0.112
D9	**11.299**	0.896	11.885	0.459	12.891	0.669	14.550	0.774
D10	4.281	0.999	5.332	0.952	**4.0231**	0.113	6.112	0.944
D11	**5.011**	0.336	8.254	0.569	6.226	0.125	5.489	0.129
D12	**4.332**	0.998	5.112	0.475	7.221	0.421	4.563	0.336

Table 8. The average fitness value of EBMFO and other approaches

Dataset #	EBMFO		BGWO		BCS		BBA	
	Avg	Std	Avg	Std	Avg	Std	Avg	Std
D1	**0.325**	0.001	0.329	0.000	0.445	0.220	0.421	0.115
D2	0.039	0.008	0.032	0.145	**0.022**	0.001	0.043	0.013
D3	**0.009**	0.000	**0.009**	0.002	**0.009**	0.140	**0.009**	0.018
D4	0.023	0.079	0.025	0.011	0.033	0.000	**0.021**	0.019
D5	**0.015**	0.066	**0.015**	0.054	**0.015**	0.112	**0.015**	0.202
D6	**0.086**	0.096	0.251	0.010	0.089	0.001	0.325	0.116
D7	0.340	0.560	**0.288**	0.443	0.399	0.112	0.420	0.557
D8	0.223	0.257	0.235	0.152	**0.214**	0.087	0.226	0.263
D9	**0.195**	0.022	0.198	0.000	0.197	0.003	0.207	0.025
D10	0.404	0.000	0.511	0.036	**0.339**	0.011	0.425	0.116
D11	**0.015**	0.000	0.010	0.006	0.0124	0.102	0.022	0.114
D12	**0.301**	0.031	0.329	0.106	0.333	0.058	0.327	0.041

Table 9. The average computational time of EBMFO and other approaches.

Dataset name	EBMFO		BGWO		BCS		BBA	
	Avg	Std	Avg	Std	Avg	Std	Avg	Std
D1	28.221	5.456	28.258	6.114	27.119	3.981	**25.113**	4.213
D2	**8.221**	3.887	10.231	1.225	9.116	2.556	8.561	3.459
D3	6.956	2.541	7.113	0.451	9.563	1.260	**4.369**	1.314
D4	1755.546	40.563	1520.331	35.269	1436.991	20.891	**1123.201**	22.598
D5	647.236	20.159	**622.150**	14.369	653.587	22.256	645.111	24.013
D6	**194.220**	6.889	199.115	11.236	184.220	10.256	195.259	5.369
D7	**6.998**	1.563	7.569	2.569	6.999	1.263	8.433	2.255
D8	10.891	1.236	10.459	2.569	**10.125**	2.119	**10.991**	1.000
D9	**120.631**	14.552	129.006	17.236	125.222	11.012	122.369	13.669
D10	**9.685**	0.154	11.236	0.169	14.259	0.336	10.259	0.774
D11	8.578	0.412	8.971	0.593	10.269	0.336	**7.115**	0.221
D12	14.235	2.110	15.987	0.552	14.226	2.569	**13.201**	3.456

5 Conclusions

This paper proposes new enhancement methods to increase the performance of the BMFO algorithm when used as a wrapper-based algorithm. OBL strategy is used to enhance the initialization method by defining a set of solutions and their opposite solutions, then selecting the fittest solutions that match the population size. The second enhancement strategy is using a retiring flame. This can be applied by resetting the best solution (flame) after a specific search time, e.g. the resulted fitness value of the flame is the same for three iterations. This is applied when the flame falls in local minima. Therefore resetting it will help alleviating local minima and enhancing the balance between global search and local search. Twelve UCI datasets are used and four evaluation measures are applied including fitness value, accuracy, the number of selected features, and running time. Three wrapper-based algorithms are used for comparison, including BGWO, BCS, and BBA. The EBMFO outperforms BMFO across 75% in terms of accuracy and it outperforms it across 83% in terms of number of selected features. On the other hand, there is no difference between BMFO and EBMFO in terms of running time. EBMFO shows also promising performance when compared with other algorithms.

For future works, we plan to apply this enhanced version of MFO in applications and we look forward to use the proposed enhancement methods with other optimizers.

Acknowledgments. This work is supported by the Ministerio español de Economía y Competitividad under project PID2020-115570GB-C22 (DemocratAI::UGR).

References

1. Asuncion, A., Newman, D.: UCI machine learning repository (2007)
2. Chuang, L.-Y., Chang, H.-W., Chung-Jui, T., Yang, C.-H.: Improved binary PSO for feature selection using gene expression data. Comput. Biol. Chem. **32**(1), 29–38 (2008)
3. Kennedy, J., Eberhart, R.C.: A discrete binary version of the particle swarm algorithm. In: 1997 IEEE International Conference on Systems, Man, and Cybernetics, 1997. Computational Cybernetics and Simulation, vol. 5, pp. 4104–4108. IEEE (1997)
4. Khurma, R., Castillo, P., Sharieh, A., Aljarah, I.: Feature selection using binary moth flame optimization with time varying flames strategies. In: Proceedings of the 12th International Joint Conference on Computational Intelligence - Volume 1: ECTA, pp. 17–27. INSTICC, SciTePress (2020)
5. Khurma, R., Castillo, P., Sharieh, A., Aljarah, I.: New fitness functions in binary Harris hawks optimization for gene selection in microarray datasets. In: Proceedings of the 12th International Joint Conference on Computational Intelligence - Volume 1: ECTA, pp. 139–146. INSTICC, SciTePress (2020)
6. Abu Khurma, R., Aljarah, I.: A review of multiobjective evolutionary algorithms for data clustering problems. In: Aljarah, I., Faris, H., Mirjalili, S. (eds.) Evolutionary Data Clustering: Algorithms and Applications. AIS, pp. 177–199. Springer, Singapore (2021). https://doi.org/10.1007/978-981-33-4191-3_8
7. Khurma, R.A., Aljarah, I., Sharieh, A.: Improved moth flame optimization based on Harris hawks for genesselection. J. Theoret. Appl. Inf. Technol. **98**, 3794–3807 (2005)
8. Khurma, R.B., Aljarah, I., Sharieh, A.: An efficient moth flame optimization algorithm using chaotic maps for feature selection in the medical applications. In: Proceedings of the 9th International Conference on Pattern Recognition Applications and Methods - Volume 1: ICPRAM, pp. 175–182. INSTICC, SciTePress (2020)
9. Khurma, R.A., Aljarah, I., Sharieh, A.: Rank based moth flame optimisation for feature selection in the medical application. In: 2020 IEEE Congress on Evolutionary Computation (CEC), pp. 1–8. IEEE (2020)
10. Khurma, R.A., Aljarah, I., Sharieh, A.: A simultaneous moth flame optimizer feature selection approach based on levy flight and selection operators for medical diagnosis. Arabian J. Sci. Eng. **46**(9), 8415–8440 (2021). https://doi.org/10.1007/s13369-021-05478-x
11. Khurma, R.A., Aljarah, I., Sharieh, A., Mirjalili, S.: EvoloPy-FS: an open-source nature-inspired optimization framework in python for feature selection. In: Mirjalili, S., Faris, H., Aljarah, I. (eds.) Evolutionary Machine Learning Techniques. AIS, pp. 131–173. Springer, Singapore (2020). https://doi.org/10.1007/978-981-32-9990-0_8
12. Abu Khurmaa, R., Aljarah, I., Sharieh, A.: An intelligent feature selection approach based on moth flame optimization for medical diagnosis. Neural Comput. Appl. **33**(12), 7165–7204 (2020). https://doi.org/10.1007/s00521-020-05483-5
13. Mirjalili, S.: Moth-flame optimization algorithm: a novel nature-inspired heuristic paradigm. Knowl.-Based Syst. **89**, 228–249 (2015)
14. Mirjalili, S., Lewis, A.: S-shaped versus v-shaped transfer functions for binary particle swarm optimization. Swarm Evol. Comput. **9**, 1–14 (2013)

A Methodology for Determining Ion Channels from Membrane Potential Neuronal Recordings

Juan Luis Jiménez Laredo[1](✉) (iD), Loïs Naudin[2] (iD), Nathalie Corson[2] (iD), and Carlos M. Fernandes[3] (iD)

[1] RI2C-LITIS, University of Le Havre Normandy, Le Havre, France
juanlu.jimenez@univ-lehavre.fr
[2] LMAH, University of Le Havre Normandy, Le Havre, France
nathalie.corson@univ-lehavre.fr
[3] LARSyS, University of Lisbon, Lisbon, Portugal
cfernandes@laseeb.org

Abstract. Using differential evolution and statistical analysis, this paper investigates a methodology that is capable of determining the ion channels in a neuron from membrane potential data obtained by the current-clamp method. These data provide the aggregated electrical response of the neuron under stimulation by integrating the individual responses of the different ion channels involved. The proposed methodology aims at determining which are these ion channels based on the hypothesis that each ion channel provides a specific signature in the aggregated response that we are able to detect. In order to assess the methodology, we propose a benchmark of synthetic data where the types of ion channels are predefined in advance. Results show that the methodology is able to determine the correct ion channels in three out of the four data sets. Furthermore, we obtain some hints for future enhancements on the method.

Keywords: Neuroscience · Differential evolution · Conductance-based models · Benchmarking · Non-spiking neurons · C. elegans

1 Introduction

In 1952, the English scientists Alan Hodgkin and Andrew Huxley concluded a series of experiments that aimed to describe the flow of electric current crossing the cell membrane of the squid giant axon [13]. Hodgkin and Huxley would use a set of nonlinear differential equations to describe the electrophysiological dynamics of the neuronal membrane under stimulation using an equivalent electrical circuit representation of the neuron. Conductance-based models (a.k.a. Hodgkin-Huxley type models) were just born. Both scientists would receive the *Nobel prize in Physiology or Medicine* for this contribution to the field of neurosciences.

© Springer Nature Switzerland AG 2022
J. L. Jiménez Laredo et al. (Eds.): EvoApplications 2022, LNCS 13224, pp. 15–29, 2022.
https://doi.org/10.1007/978-3-031-02462-7_2

Ever since, conductance-based models remain the most powerful mathematical tool for describing the electrophysiological characteristics of the cell membrane of excitable cells as every single parameter in the model has a biophysical meaning: the lipid bilayer is represented as a capacitance, ion pumps are represented as current sources and voltage-gated ion channels as a set of nonlinear conductances.

On this basis, the modeling of a neuron can be carried out in several stages: first, thanks to experimental techniques such as the voltage-clamp or current-clamp methods [22], the cell membrane is electrostimulated and electrophyisiological data are collected. Then, a conductance-based model is built taking into account the ion channels present in the neuron [13–17]. Finally, the parameters of the model are adjusted, typically using optimization methods [2–4], so that the dynamics of the model fit the observed experimental data. While the last two stages belong to the field of modeling, the first belongs to the field of electrophysiology and determines the type of data available.

Whole-cell membrane potential data, such as the ones used in this paper, register the membrane potential activity of a neuron under stimulation as shown in Fig. 1. Due to the relative simplicity of the experimental procedure, whole-cell recordings are abundant in the literature and, sometimes, they are the only source of data available for certain neurons [22]. Hence, the relevance of studying this type of data from the perspective of modeling. The limitation of whole-cell techniques, however, is that recordings provide the aggregated electrical response of the neuron without discerning the types of ion channels present in the cell. In that sense, determining the type of ion channels is crucial if the aim is to obtain a model that is biologically equivalent to the targeted neuron.

In this paper, we aim at investigating an in-silico methodology firstly proposed in [23] that is designed to determine the ion channels composing the cell from whole-cell membrane potential recordings. Our working hypothesis is that *each ion channel provides a specific signature in the membrane potential record that can be detected*. We establish first a list of plausible models for a target neuron. Each of these models is composed of a different combination of ion channels and is optimized using differential evolution [23, 29] to fit the model to the experimental data. The list of parameterized models is then ranked according to their cost function following a statistical procedure that involves the Wittkowski and the Wilcoxon statistical tests. The outcome of the methodology is that a model is finally selected, determining the more likely set of ion channels intervening in the neuron.

In order to assess the methodology, we propose a benchmark of synthetic data where the types of ion channels are predefined in advance. These synthetic data are based on whole-cell current-clamp recordings obtained by Liu et al. [22] for the RIM neuron of the nematode *C. elegans* and an ulterior modeling of the same neuron [23, 24]. Specifically, the benchmark consists in four models of the RIM neuron with known parameters and a different combination of some of the following ion channel induced currents: transient potassium current ($I_{K,t}$), leak current (I_L), persistent calcium current ($I_{Ca,p}$) or inward rectifier potassium current (I_{Kir}). The aim is to test whether the methodology is capable of determining the exact ionic composition of each of these models.

The remainder of the paper is organized as follows: Sect. 2 presents a mathematical description of conductance-based models. Section 3 describes the benchmark. Section 4 details the proposed methodology. Results are analyzed in Sect. 5. Finally, conclusions are drawn in Sect. 6.

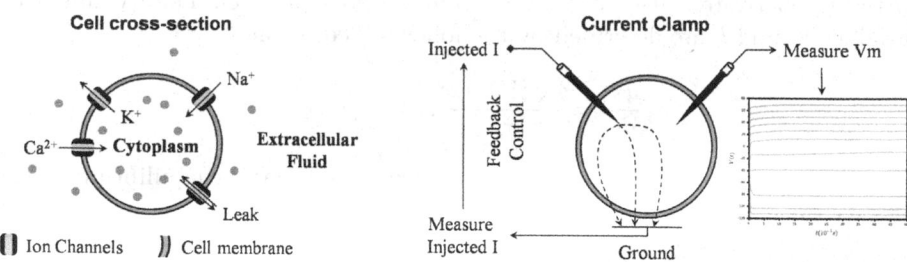

Fig. 1. On the left, cross-section scheme of a neuron with different types of ion channels allowing the exchange of ions between the cytoplasm and the extracellular fluid. The difference in ionic concentrations between the outside and the inside of the cell determines the membrane potential. On the right, scheme of the whole-cell current-clamp technique that allows to electro-stimulate the cell and to perform whole-cell recordings, i.e., registering changes in the membrane potential as a consequence of stimulating the cell with a current I. Changes in the membrane potential respond to the collective activity of the different ion channels allowing the flow of ions across the membrane as a result of the external stimulation I. Membrane potential records do not provide information about the ion channels composing the cell.

2   Conductance-Based Model Description

Conductance-based neuron models are grounded in a series of seminal works initiated by Hodgkin and Huxley in the 1950s [13–17]. In general terms, this framework describes the neuronal dynamics in terms of activation and inactivation of voltage-gated conductances where the dynamics of the membrane potential V is described by a general equation of the form:

$$C\frac{dV}{dt} = -\sum_{ion} I_{ion} + I \tag{1}$$

where C is the membrane capacitance, dV/dt is the time derivative of the membrane potential, $\sum_{ion} I_{ion}$ is the total current flowing across the cell membrane, and I is an applied current. The specific current flowing across every ion channel I_{ion} takes the form:

$$I_{ion} = g_{ion}m_{ion}^{a}h_{ion}^{b}(V - E_{ion})$$

where m (*resp.* h) denotes the probability for an activation (*resp.* inactivation) gate to be in the open state; a and b are the number of activation and inactivation gates, respectively; g_{ion} is the maximal conductance associated with ion; and E_{ion} is the reverse potential. In this paper, a is always assumed equal to 1 and the value of b depends on features of the ionic currents. Channels that do not have inactivation gates ($b = 0$) induce a *persistent* current (noted $I_{ion,p}$) while channels that do inactivate ($b = 1$) induce a *transient* current (noted $I_{ion,t}$).

Channels can be *deactivated* ($m = 0$), partially activated ($0 < m < 1$), or fully activated ($m = 1$); likewise, they can be completely inactivated ($h = 0$), partially inactivated ($0 < h < 1$), or *deinactivated* ($h = 1$). The dynamics of variables m and h are described by the following equation

$$\frac{dx}{dt} = \frac{x_\infty(V) - x}{\tau_x}, \quad x \in \{m, h\} \tag{2}$$

where τ_x is the *constant* time for which x reaches its respective equilibrium value x_∞. We express x_∞ by a Boltzmann sigmoid function [6,20]:

$$x_\infty(V) = \frac{1}{1 + \exp\left(\frac{V_{1/2}^x - V}{k_x}\right)}, \quad x \in \{m, h\} \tag{3}$$

where $V_{1/2}^x$ satisfies $x_\infty(V_{1/2}^x) = 1/2$ and k_x is the slope factor with $k_m > 0$ and $k_h < 0$ as to represent activation and inactivation respectively, *i.e.*, smaller values of $|k_x|$ lead to a sharper x_∞.

3   Defining a Benchmark with Known Types of Ion Channels

In the previous section, we have seen that the dynamics of the membrane potential are described by Eq. 1. In order to fully characterize such an equation, it is necessary to determine all the unknown I_{ion} currents, i.e., the different types of ion channels of the targeted neuron. This seems to be in conflict with the nature of whole-cell recordings as this type of data does not provide information about the ionic currents in the recorded response. If the aim is to validate a methodology that determines the ion channels from whole-cell membrane potential recordings, we will need a benchmark where the ion channels are known in advance and then, check whether the obtained results match these predefined channels.

The benchmark proposed in this paper targets the non-spiking RIM neuron of the nematode *C. elegans* based on the current-clamp recordings obtained by Liu et al. [22]. Nevertheless, instead of Liu's raw membrane potentials, we will use four synthetic models of the neuron described in [23] and shown in Table 1 and Fig. 2. The advantage of using these synthetic data for benchmarking is that we can know the ionic composition of each model beforehand so that we can indistinguishably assess whether the method succeed in determining the ion channels.

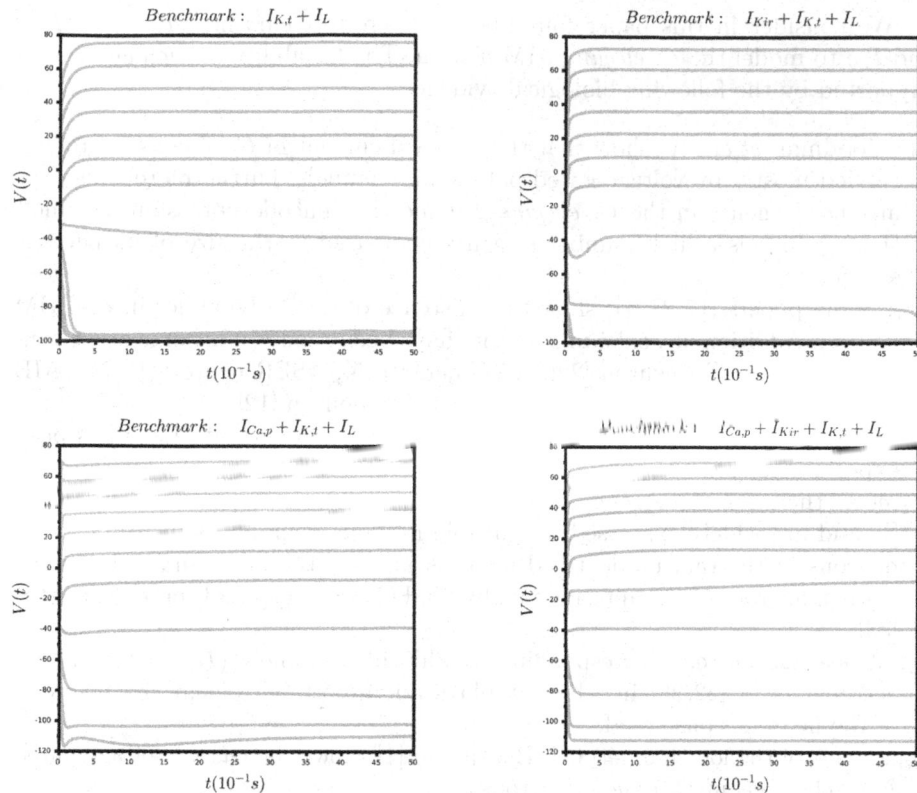

Fig. 2. The four synthetic solutions used as a benchmark in this paper are obtained from [23] and describe the evolution of the membrane potential of the non-spiking *C. elegans* RIM neuron. Specific parameters for the solutions are in Table 3 of the Appendix A. Each figure represents a feasible combination of ion channels that may include: transient potassium current ($I_{K,t}$), leak current (I_L), persistent calcium current ($I_{Ca,p}$) or inward rectifier potassium (I_{Kir}). Traces correspond to a series of current injections, in spans of 5 s, starting from 15 pA and increasing to 35 pA by 5 pA increments.

Table 1. List of some biologically plausible models for the RIM neuron as described in [23]. These models are used as a benchmark in this paper. See Table 3 in the Appendix A for further details.

Feasible RIM models
$I_{K,t} + I_L$
$I_{Kir} + I_{K,t} + I_L$
$I_{Ca,p} + I_{K,t} + I_L$
$I_{Ca,p} + I_{Kir} + I_{K,t} + I_L$

We consider in this paper four plausible types of currents (I_K, I_{Ca}, I_{Kir} and I_L) to model the *C. elegans* RIM neuron. The feasibility of such currents is supported by the following biological evidences:

I_K: Goodman *et al.* [10] show that the outward current in *C. elegans* neurons is carried mostly by voltage-gated potassium channels. Furthermore, there are around 70 genes in the *C. elegans* genome that encode potassium channels [1,28] which is a high number of genes compared to the size of its nervous system.

I_{Ca}: Some papers [11,27,34] show the existence of Ca^{2+} dynamics in the RIM neuron. Additionally, calcium currents have been reported in some other neurons such as AWA neuron [22], AWC neuron [5], ASER neuron [10,21], AIB neuron [5,11], AVA neuron [11,27,34] or RIA neuron [12].

I_{Kir}: The presence of an inwardly rectifying potassium current (I_{Kir}) has been experimentally confirmed in AWA neuron [22] and HSN neuron [9]. Furthermore, there are three genes that encode inward rectifier potassium channels [1] and are widely expressed in multiple neurons including sensory phasmid neurons in the tail, unidentified neurons in both the head and tail of male worms, neurons of the anterior ganglia, and different types of sensory neurons [33].

I_L: A leakage current corresponding to chloride channels (I_L) is taken into account as in [22,26] in which it plays an important role in the behavior of the neurons considered.

I_{Na}: (Lack of the ion channel) C.I. Bargmann [1] shows the lack of voltage-gated Na^+ channels in *C. elegans* neurons.

4   Methodology and Experimental Setup

In the previous section, we defined a list of plausible models that describe the dynamics of the RIM neuron (see Table 1). Using the four feasible solutions presented in Fig. 2 as a benchmark, this section proposes a methodology based on parameter estimation to determine the ionic composition of each of these solutions.

Parameter estimation is carried out through differential evolution (DE) [23, 29][1] by minimizing the root mean square error ($RMSE$) between the *benchmark* voltages $V_{bench}(I, t)$ depicted in Fig. 2 and the voltages $V_{est}^\theta(I, t)$ estimated by differential evolution:

$$RMSE(\theta) = \sqrt{\frac{\sum_t \sum_I \left(V_{bench}(I, t) - V_{est}^\theta(I, t)\right)^2}{N}} \quad (4)$$

where $t \in [0, 50ds]$ corresponds to the biological real time with a sampling period of $\Delta t = 0.004ds$; $N = 10000$ is the number of data points in the measurement

---

[1] DE Parameters and parameter ranges are in Appendix B.

record; I corresponds to successive step values of current injections starting from $-15\,\text{pA}$ and increasing to $35\,\text{pA}$ by intervals of $5\,\text{pA}$; and θ is the vector containing the parameters of a specific model. For instance, the parameter vector of the $I_{K,t} + I_L$-model is expressed by:

$$\theta_{I_{K,t}+I_L} = [g_K \; g_L \; E_K \; E_L \; V_{1/2}^{m_K} \; V_{1/2}^{h_K} k_{m_K} \; k_{h_K} \; \tau_{m_K} \; \tau_{h_K} \; m_K^0 \; h_K^0 \; C]$$

The respective θ-vectors of the benchmark solutions are provided in Table 3. We have implemented DE on Scilab as described in [23]. The code of the algorithm is available at https://github.com/juanluck/evoapps2022 published under GNU Public License v3.0.

To determine the ion channels in the benchmark, each of the four benchmark solutions is tested against the four possible models, resulting in 16 possible benchmark/model combinations. In the ideal case, the proposed methodology should be able to discern the model that corresponds to each solution. Specifically, the methodology consists of the following three stages:

- **Step 1.** Parameter estimation:
  Procedure: We have conducted 30 independent runs of the differential evolution algorithm for each of the 16 benchmark/model combinations.
  Input: 4 benchmark solutions × 4 possible models per solution.
  Output: 30 possible parameterizations per combination benchmark/model (480 parameterizations in total).
- **Step 2.** Ranking the models:
  Procedure: We conduct a Wittkowski statistical test (generalized Friedman rank sum test) [32] in order to produce a ranked list of the four models for every benchmark solution based on the RMSE cost function results.
  Input: 4 models per benchmark solution × 30 parameterizations per model.
  Output: The 4 models tested for each benchmark solution are ranked from best to worst.
- **Step 3.** Selection of the best model:
  Procedure: Preferentially the best ranked model is selected. However, in the case that there are no statistical differences between the best ranked model and the other models, we use the peak performance measure [8] as criterion to select the best model, *i.e.* the parameterization with the smallest cost function prevails over the rest. This step applies pairwise comparisons using paired Wilcoxon signed-rank test [31] corrected for multiple comparisons with the Holm method [18].
  Input: A list of ranked models per benchmark solution.
  Output: 1 model is selected per benchmark solution.

The Wittkowski test uses the marginal likelihood principle to obtain consistent estimates for rank scores (*i.e.* determine which models consistently score better than others). Meanwhile, Wilcoxon signed-rank is a non-parametric test used when distributions cannot be assumed to be normally distributed. The latter test complements the former by determining whether the differences between two models with different ranks are additionally statistically different or not.

5   Experimental Results

Following the methodology proposed in Sect. 4, 30 independent runs of the differential evolution algorithm were conducted for every combination of benchmark solution and model. With the aim of determining the ion channels intervening in the benchmark, results are ranked using the Wittkowski test in Table 2 and complemented with pairwise comparisons between the different models using paired Wilcoxon signed-rank test (Fig. 3).

Table 2. Results of the mean rank of each model obtained using the Wittkowski test. The diagonal cells in gray color stand for the correct choice of model-benchmark. The selected models appear in **underlined bold** font and have been selected according to the criteria established in Sect. 4. In parentheses the results of the best cost function (minimum root mean square error (RMSE)) for the different models. The $*$ stands for the models that are significantly different to the one selected (after pairwise comparisons using paired Wilcoxon signed-rank test).

Model \ Benchmark	I_K+I_L	$I_{Kir}+I_K+I_L$	$I_{Ca}+I_K+I_L$	$I_{Ca}+I_{Kir}+I_K+I_L$
I_K+I_L	$2.6^{(0.011)*}$	$3.9^{(0.139)*}$	$3.8^{(0.239)*}$	$3.9^{(0.242)*}$
$I_{Kir}+I_K+I_L$	$3.4^{(0.014)*}$	$\underline{\mathbf{1.7}}^{(0.003)}$	$3.0^{(0.161)*}$	$3.0^{(0.170)*}$
$I_{Ca}+I_K+I_L$	$\mathbf{2.0}^{(0.009)}$	$2.1^{(0.019)}$	$\underline{\mathbf{1.6}}^{(0.003)}$	$2.0^{(0.039)*}$
$I_{Ca}+I_{Kir}+I_KI_L$	$2.0^{(0.013)}$	$2.3^{(0.009)}$	$1.6^{(0.014)}$	$\underline{\mathbf{1.2}}^{(0.004)}$

At a first glance, Table 2 shows that the proposed methodology is capable of detecting the right model (i.e. ionic composition) in 3 out of the 4 recordings. This result reinforces the initial working hypothesis which assumes that the membrane potential carries the signature of the different intervening ion channels. However, the proposed methodology only works partially since it fails to detect the $I_{K,t}+I_L$ solution of the benchmark, which is the simplest from those under investigation, and select instead the more complex $I_{Ca,p}+I_{K,t}+I_L$ model. A straightforward conclusion is that the latter model, being richer in terms of ionic currents, is capable of emulating the behavior of the former (Fig. 4 offers a glimpse of the accuracy with which the $I_{Ca,p}+I_{K,t}+I_L$ model can emulate the $I_{K,t}+I_L$ response). Despite this setback, we find that the result for this benchmark solution is valuable since it offers hints about how to enhance the methodology: we should take into account not only the goodness of the solutions but also the complexity of the model. This appreciation is confirmed in the analysis of the rest of the benchmark solutions:

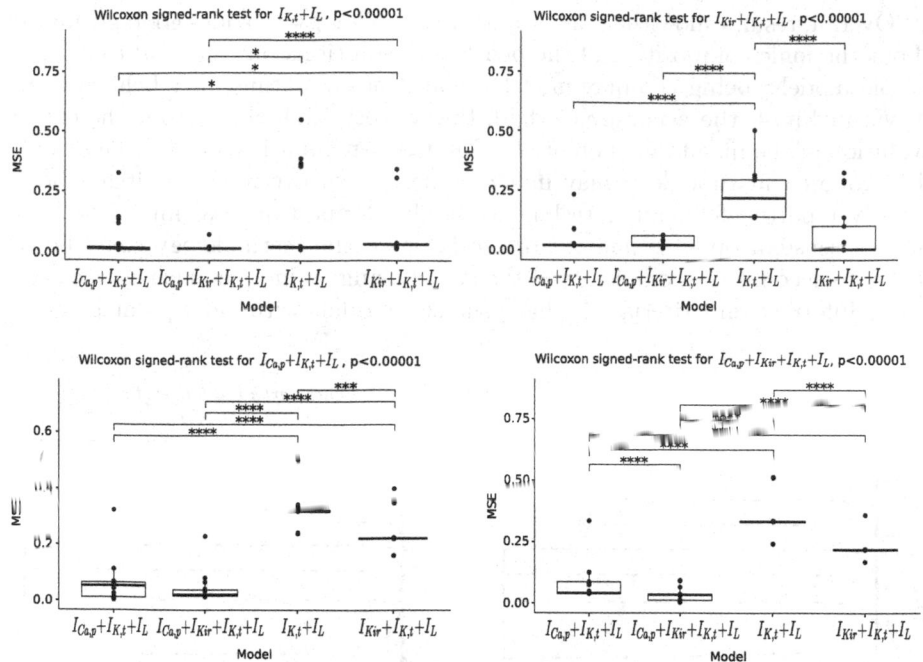

Fig. 3. Pairwise comparisons using paired Wilcoxon signed-rank test and corrected for multiple comparisons with the Holm method. Connected distributions (with stars (*) on top) are significantly different.

- $I_{K,t} + I_L$: As explained above, the methodology fails to determine the ionic currents present in this solution.
- $I_{Kir} + I_{K,t} + I_L$: The methodology succeeds in determining the ionic currents involved in this solution. With a ranking value of 1.7, the $I_{Kir}+I_{K,t}+I_L$ model prevails over the others with respective rankings 2.1, 2.3 and 3.9. Additionally, the peak performance measure (i.e. best found parameterization) is also found for the same model.
- $I_{Ca,p} + I_{K,t} + I_L$: In this benchmark solution, two models are ranked equally with a score of 1.6 and no statistical differences between them. As established by the Step 3 of the methodology, in this case, the model should be selected according to the peak performance measure. Taking into account such metric, the right model is selected. Nonetheless, as in the case of the first benchmark solution, a more complex model (i.e. $I_{Ca,p} + I_{Kir} + I_{K,t} + I_L$) can consistently emulate the response of the targeted benchmark which is simpler in its ionic composition.
- $I_{Ca,p}+I_{Kir}+I_{K,t}+I_L$: The more robust solution is found for this benchmark. Not only the rank of the correct model (1.2) prevails over the rest but also the peak performance measure is the best for this model. Additionally, the RMSE distribution presents significant statistical differences with respect to the rest of models distributions.

Overall, results show that the proposed methodology provides good estimates about the ionic composition of the benchmark solutions but may fail to discern simple models; being complex models capable of emulating their behavior. As shown in Fig. 4, the accuracy of the fitting is very high throughout the entire evolution of the membrane potential for all the benchmark solutions. Therefore, the proposed methodology may induce some sort of overfitting as it has been already reported for similar techniques in the domain of viral infections [25]. In that sense, a possible line of work to improve the methodology could be to take into account the complexity of the model using techniques such as, *e.g.*, the Ikaike Information Criteria [25] that penalizes models with more parameters.

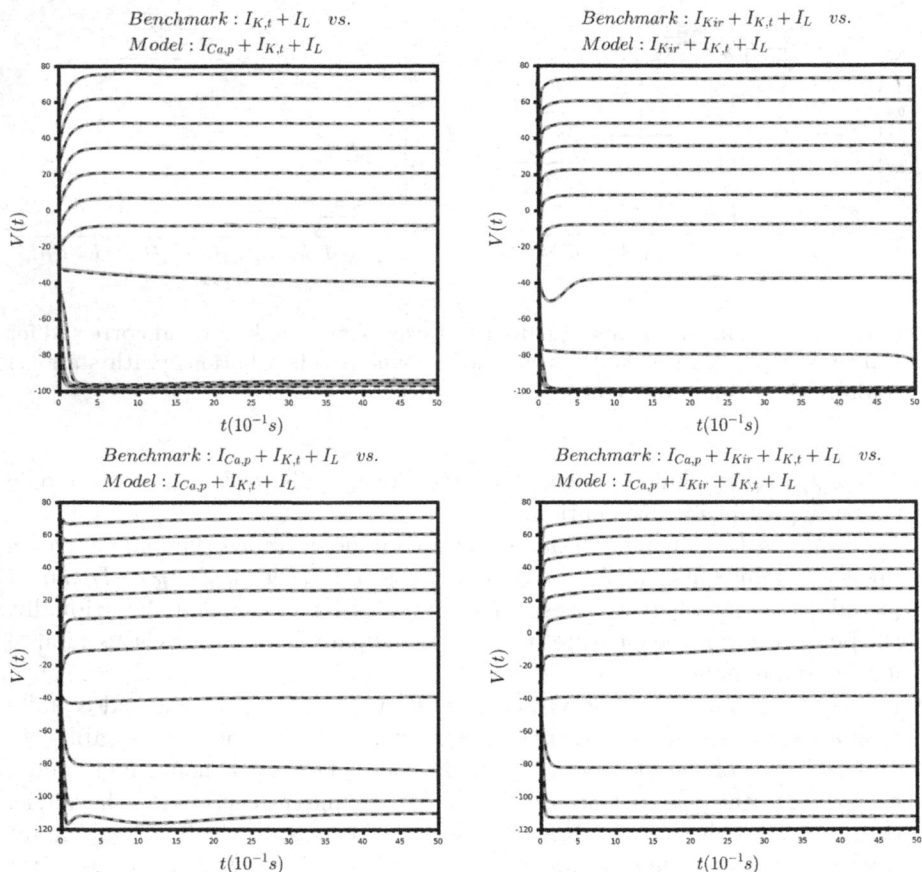

Fig. 4. Evolution of the membrane potential for a series of current injection starting at $-15\,\text{pA}$ and increasing to $35\,\text{pA}$ by $5\,\text{pA}$ increments. Benchmark data (represented in cyan) and outcomes of the selected models (represented in black doted lines) overlap for the same values of current injections. (Color figure online)

6   Conclusions

In this paper, we have investigated a methodology to determine the ion channels of a neuron from whole-cell membrane potential recordings. This type of data provide the aggregated response of the neuron under stimulation without specifying the ion channels composing the cell. The working hypothesis under investigation is whether these data carry information from the different intervening ion channels and whether we can detect those signatures or not. We propose a benchmark where the ion channels are predefined in advance in order to assess whether the methodology works or not and, thus, to confirm or refute such hypothesis. Specifically, the benchmark is defined as four synthetic models of the non-spiking RIM neuron of the nematode *C. elegans*, each using different combinations of ionic currents.

Results show that the methodology is able to determine the correct channels in three out of the four data sets, which only confirms partially the hypothesis under investigation. Certainly, membrane potential recordings carry some signature of the ion channels that have intervened in creating the signal. Nonetheless, we cannot affirm at this stage of the research that such information is enough to unequivocally identify those channels. Despite being a partial conclusion, it is valuable because the methodology may allow to formulate some biological hypotheses about the nature of certain neurons. In that sense, computational modeling reveals itself as a valuable asset to investigate and formulate hypotheses not easily testable through direct experimentation.

In the future, we aim at improving the proposed methodology thanks to the hints obtained in this paper. In general, it seems that complex models (i.e. models with a higher number of ion channels and therefore a higher dimensionality in the solution space) can emulate the response of simpler neurons. Therefore, the complexity of the model should be a factor to be taken into account. Finally, in order to extend the method to the study of spiking neurons, the fundamental step would be to define feature-based errors functions [7] more adapted than the RMSE for quantitatively capturing features relative to spiking patterns, as in [19,30]. This line of research can be promising as spiking neural networks are the most widely studied case in the literature.

Appendix A   Mathematical Description of the Models

Table 3. Characterization of all the mathematical models used in this paper as described in [23].

$I_{K,t} + I_L$-model

$$\begin{cases} C\dot{V} = -g_K m_K h_K (V - E_K) - g_L (V - E_L) + I \\ \tau_{m_K} \dot{m}_K = m_{K\infty}(V) - m_K \\ \tau_{h_K} \dot{h}_K = h_{K\infty}(V) - h_K \end{cases}$$

$\theta = [g_K \ g_L \ E_K \ E_L \ V_{1/2}^{m_K} \ V_{1/2}^{h_K} \ k_{m_K} \ k_{h_K} \ \tau_{m_K} \ \tau_{h_K} \ m_K^0 \ h_K^0 \ C]$

$\theta_{bench} = [13.09 \ 0.36 \ -99.94 \ -19.76 \ -88.30 \ -89.50 \ 30 \ -11.09 \ 14.51 \ 1.07 \ 0.97 \ 0.005 \ 0.008]$

$I_{Kir} + I_{K,t} + I_L$-model

$$\begin{cases} C\dot{V} = g_{Kir} h_{Kir\infty}(V)(V - E_K) - g_K m_K h_K (V - E_K) - g_L (V - E_L) + I \\ \tau_{m_K} \dot{m}_K = m_{K\infty}(V) - m_K \\ \tau_{h_K} \dot{h}_K = h_{K\infty}(V) - h_K \end{cases}$$

$\theta = [g_{Kir} \ g_K \ g_L \ E_K \ E_L \ V_{1/2}^{Kir} \ V_{1/2}^{m_K} \ V_{1/2}^{h_K} \ k_{Kir} \ k_{m_K} \ k_{h_K} \ \tau_{m_K} \ \tau_{h_K} \ m_K^0 \ h_K^0 \ C]$

$\theta_{bench} = [2.86 \ 29.79 \ 0.41 \ -99.99 \ -10.74 \ -89.86 \ -89.19 \ -89.61 \ -19.30$

$\qquad 28.38 \ -1.94 \ 14.95 \ 1.00 \ 0.65 \ 0.001 \ 0.04]$

$I_{Ca,p} + I_{K,t} + I_L$-model

$$\begin{cases} C\dot{V} = -g_{Ca} m_{Ca}(V - E_{Ca}) - g_K m_K h_K (V - E_K) - g_L (V - E_L) + I \\ \tau_{m_{Ca}} \dot{m}_{Ca} = m_{Ca\infty}(V) - m_{Ca} \\ \tau_{m_K} \dot{m}_K = m_{K\infty}(V) - m_K \\ \tau_{h_K} \dot{h}_K = h_{K\infty}(V) - h_K \end{cases}$$

$\theta = [g_{Ca} \ g_K \ g_L \ E_{Ca} \ E_K \ E_L \ V_{1/2}^{m_{Ca}} \ V_{1/2}^{m_K} \ V_{1/2}^{h_K} \ k_{m_{Ca}} \ k_{m_K} \ k_{h_K} \ \tau_{m_{Ca}} \ \tau_{m_K} \ \tau_{h_K} \ m_{Ca}^0 \ m_K^0 \ h_K^0 \ C]$

$\theta_{bench} = [0.30 \ 7.81 \ 0.19 \ 42.09 \ -99.99 \ -66.90 \ -2.00 \ -52.10 \ -90 \ 29.99$

$\qquad 22.65 \ -2.05 \ 0.21 \ 2.37 \ 15 \ 0.25 \ 0.99 \ 0.001 \ 0.03]$

$I_{Ca,p} + I_{Kir} + I_{K,t} + I_L$-model

$$\begin{cases} C\dot{V} = -g_{Ca} m_{Ca}(V - E_{Ca}) - g_{Kir} h_{Kir\infty}(V)(V - E_K) - g_K m_K h_K (V - E_K) - g_L (V - E_L) + I \\ \tau_{m_{Ca}} \dot{m}_{Ca} = m_{Ca\infty}(V) - m_{Ca} \\ \tau_{m_K} \dot{m}_K = m_{K\infty}(V) - m_K \\ \tau_{h_K} \dot{h}_K = h_{K\infty}(V) - h_K \end{cases}$$

$\theta = [g_{Ca} \ g_{Kir} \ g_K \ g_L \ E_{Ca} \ E_K \ E_L \ V_{1/2}^{m_{Ca}} \ V_{1/2}^{Kir} \ V_{1/2}^{m_K} \ V_{1/2}^{h_K}$

$\qquad k_{m_{Ca}} \ k_{Kir} \ k_{m_K} \ k_{h_K} \ \tau_{m_{Ca}} \ \tau_{m_K} \ \tau_{h_K} \ m_{Ca}^0 \ m_K^0 \ h_K^0 \ C]$

$\theta_{bench} = [0.24 \ 0.33 \ 0.12 \ 0.28 \ 105.28 \ -100 \ -81.33 \ -21.03 \ -89.99 \ -17.70 \ -21.28 \ 28.80$

$\qquad -1.20 \ 1.18 \ -4.64 \ 0.16 \ 0.20 \ 5.08 \ 0.34 \ 0.79 \ 0.13 \ 0.02]$

Appendix B Experimental Setup and Parameter Ranges

See Tables 4 and 5.

Table 4. Differential evolution parameters as in [23].

DE parameters	
Generations	1000
F	0.5
CR	0.9
NP	140

Table 5. Parameter ranges have been obtained from the literature [20] and are biologically relevant.

Parameters	Min value	Max value
$g_K, g_{Kir}, g_L, g_{Ca}$	$0\,\mu S$	$50\,\mu S$
E_K	$-100\,mV$	$0\,mV$
E_L	$-90\,mV$	$30\,mV$
E_{Ca}	$20\,mV$	$150\,mV$
$V_{1/2}^m, V_{1/2}^h, V_{1/2}^{Kir}$	$-90\,mV$	$0\,mV$
k_m	$0\,mV$	$30\,mV$
k_h, k_{Kir}	$-30\,mV$	$0\,mV$
τ_m, τ_h	$0\,ds$	$15\,ds$
x_m^0, x_h^0	0	1
C	0	10

References

1. Bargmann, C.I.: Neurobiology of the caenorhabditis elegans genome. Science **282**(5396), 2028–2033 (1998)
2. Buhry, L., Grassia, F., Giremus, A., Grivel, E., Renaud, S., Saïghi, S.: Automated parameter estimation of the hodgkin-huxley model using the differential evolution algorithm: application to neuromimetic analog integrated circuits. Neural Comput. **23**(10), 2599–2625 (2011)
3. Buhry, L., Pace, M., Saïghi, S.: Global parameter estimation of an hodgkin-huxley formalism using membrane voltage recordings: application to neuro-mimetic analog integrated circuits. Neurocomputing **81**, 75–85 (2012)
4. Buhry, L., Saighi, S., Giremus, A., Grivel, E., Renaud, S.: Parameter estimation of the hodgkin-huxley model using metaheuristics: application to neuromimetic analog integrated circuits. In: 2008 IEEE Biomedical Circuits and Systems Conference, pp. 173–176. IEEE (2008)

5. Chalasani, S.H., et al.: Dissecting a circuit for olfactory behaviour in caenorhabditis elegans. Nature **450**(7166), 63 (2007)
6. Dayan, P., Abbott, L.F.: Theoretical Neuroscience: Computational and Mathematical Modeling of Neural Systems. MIT Press, Cambridge (2001)
7. Druckmann, S., Banitt, Y., Gidon, A.A., Schürmann, F., Markram, H., Segev, I.: A novel multiple objective optimization framework for constraining conductance-based neuron models by experimental data. Front. Neurosci. **1**, 1 (2007)
8. Eiben, A.E., Smith, J.E.: Introduction to Evolutionary Computing. NCS, Springer, Heidelberg (2015). https://doi.org/10.1007/978-3-662-44874-8
9. Emtage, L., Aziz-Zaman, S., Padovan-Merhar, O., Horvitz, H.R., Fang-Yen, C., Ringstad, N.: Irk-1 potassium channels mediate peptidergic inhibition of caenorhabditis elegans serotonin neurons via a go signaling pathway. J. Neurosci. **32**(46), 16285–16295 (2012)
10. Goodman, M.B., Hall, D.H., Avery, L., Lockery, S.R.: Active currents regulate sensitivity and dynamic range in C. elegans neurons. Neuron **20**(4), 763–772 (1998)
11. Gordus, A., Pokala, N., Levy, S., Flavell, S.W., Bargmann, C.I.: Feedback from network states generates variability in a probabilistic olfactory circuit. Cell **161**(2), 215–227 (2015)
12. Hendricks, M., Ha, H., Maffey, N., Zhang, Y.: Compartmentalized calcium dynamics in a C. elegans interneuron encode head movement. Nature **487**(7405), 99–103 (2012)
13. Hodgkin, A.L., Huxley, A.F.: A quantitative description of membrane current and its application to conduction and excitation in nerve. J. Physiol. **117**(4), 500–544 (1952)
14. Hodgkin, A.L., Huxley, A.F., Katz, B.: Measurement of current-voltage relations in the membrane of the giant axon of Loligo. J. Physiol. **116**(4), 424–448 (1952)
15. Hodgkin, A.L., Huxley, A.F.: The components of membrane conductance in the giant axon of Loligo. J. Physiol. **116**(4), 473–496 (1952)
16. Hodgkin, A.L., Huxley, A.F.: Currents carried by sodium and potassium ions through the membrane of the giant axon of Loligo. J. Physiol. **116**(4), 449–472 (1952)
17. Hodgkin, A.L., Huxley, A.F.: The dual effect of membrane potential on sodium conductance in the giant axon of Loligo. J. Physiol. **116**(4), 497–506 (1952)
18. Holm, S.: A simple sequentially rejective multiple test procedure. Scand. J. Stat. **6**, 65–70 (1979)
19. Iavarone, E., et al.: Experimentally-constrained biophysical models of tonic and burst firing modes in thalamocortical neurons. PLoS Comput. Biol. **15**(5), e1006753 (2019)
20. Izhikevich, E.M.: Dynamical Systems in Neuroscience. MIT Press, Cambridge (2007)
21. Kuramochi, M., Doi, M.: A computational model based on multi-regional calcium imaging represents the spatio-temporal dynamics in a caenorhabditis elegans sensory neuron. PLoS ONE **12**(1), e0168415 (2017)
22. Liu, Q., Kidd, P.B., Dobosiewicz, M., Bargmann, C.I.: C. elegans awa olfactory neurons fire calcium-mediated all-or-none action potentials. Cell **175**(1), 57–70 (2018)
23. Naudin, L., Corson, N., Aziz-Alaoui, M., Jiménez Laredo, J.L., Démare, T.: On the modeling of the three types of non-spiking neurons of the caenorhabditis elegans. Int. J. Neural Syst. **31**, S012906572050063X (2020)

24. Naudin, L., Laredo, J.L.J., Liu, Q., Corson, N.: Systematic generation of bio-physically detailed models with generalization capability for non-spiking neurons. hal-03474984 (2021)
25. Nguyen, V.K., Hernandez-Vargas, E.A.: Parameter estimation in mathematical models of viral infections using R. In: Yamauchi, Y. (ed.) Influenza Virus. MMB, vol. 1836, pp. 531–549. Springer, New York (2018). https://doi.org/10.1007/978-1-4939-8678-1_25
26. Nicoletti, M., Loppini, A., Chiodo, L., Folli, V., Ruocco, G., Filippi, S.: Biophysical modeling of C. elegans neurons: single ion currents and whole-cell dynamics of AWCon and RMD. PLoS ONE 14(7), e0218738 (2019)
27. Piggott, B.J., Liu, J., Feng, Z., Wescott, S.A., Xu, X.S.: The neural circuits and synaptic mechanisms underlying motor initiation in C. elegans. Cell 147(4), 922–933 (2011)
28. Salkoff, L.B., et al.: Potassium channels in c. elegans. WormBook (2005)
29. Storn, R., Price, K.: Differential evolution-a simple and efficient heuristic for global optimization over continuous spaces. J. Global Optim. 11(4), 341–359 (1997)
30. Venkadesh, S., et al.: Evolving simple models of diverse intrinsic dynamics in hip-pocampal neuron types. Front. Neuroinform. 12, 8 (2018)
31. Wilcoxon, F.: Individual comparisons by ranking methods. In: Kotz, S., Johnson, N.L. (eds.) Breakthroughs in Statistics, pp. 196–202. Springer, New York (1992). https://doi.org/10.1007/978-1-4612-4380-9_16
32. Wittkowski, K.M.: Friedman-type statistics and consistent multiple comparisons for unbalanced designs with missing data. J. Am. Stat. Assoc. 83(404), 1163–1170 (1988)
33. Wojtovich, A.P., DiStefano, P., Sherman, T., Brookes, P.S., Nehrke, K.: Mitochon-drial ATP-sensitive potassium channel activity and hypoxic preconditioning are independent of an inwardly rectifying potassium channel subunit in caenorhabditis elegans. FEBS Lett. 586(4), 428–434 (2012)
34. Zheng, M., Cao, P., Yang, J., Xu, X.S., Feng, Z.: Calcium imaging of multiple neurons in freely behaving C. elegans. J. Neurosci. Methods 206(1), 78–82 (2012)

Swarm Optimised Few-View Binary Tomography

Mohammad Majid al-Rifaie[1]([⊠]) and Tim Blackwell[2]

[1] Univeristy of Greenwich, London SE10 9LS, UK
m.alrifaie@gre.ac.uk
[2] Goldsmiths, University of London, London SE14 6NW, UK
t.blackwell@gold.ac.uk

Abstract. This paper considers a swarm optimisation approach to few-view tomographic reconstruction. DFOMAX, a high diversity swarm optimiser, demonstrably reconstructs binary images to a high fidelity, outperforming a leading algebraic technique, differential evolution and particle swarm optimisation on four standard phantoms. The paper considers the effectiveness of optimisers that have been developed for optimal low dimensional performance and concludes that trial solution clamping on the walls of the feasible search space is important for good performance.

Keywords: Swarm optimisation · Binary tomography · High dimensional optimisation

1 Introduction

Tomographic reconstruction (TR), which is the determination of the internal structure of an opaque object from projected images cast by penetrating radiation, is at the heart of all medical imaging and has widespread application, for example: data compression and data security [24], image processing [35], electron microscopy [14], crystal structure [9], angiography [15], nondestructive testing of homogeneous objects [22], seismic tomography [30], astronomy [12] and geometric, combinatorial and recreational mathematics [19].

Projection data is typically noisy due to the inherent randomness of radiation, detector characteristics and, in medical applications, patient movement, and is usually too sparse for a complete reconstruction. The number of projections should be kept to an absolute minimum in medical contexts due to the damaging effects of radiation. The *few-view* situation is particularly important where the risk is too high, for example in imaging of children.

The standard tomographic numerical reconstruction technique has been, until recently, filtered backprojection (FBP). This algorithm only requires a single iteration but it depends on a large number of projections and is not suitable for few-view imaging [21]. Algebraic Reconstruction Techniques (ART) [24] have effectively replaced FBP in the last few years. ART is an iterative algorithms

© Springer Nature Switzerland AG 2022
J. L. Jiménez Laredo et al. (Eds.): EvoApplications 2022, LNCS 13224, pp. 30–45, 2022.
https://doi.org/10.1007/978-3-031-02462-7_3

based on Kaczmarz's method for solving linear system of equations [35]. ART is applicable to the few-view scenario but can introduce artefacts due to overfitting and has not been proven in large patient populations [20].

Compressed sensing (CS) holds potential for few-view imaging: exact reconstruction is possible if the data can be transformed to a sparse representation [36]. However the sparse representation must be known, and the method involves replacing a non-convex problem with a tractable convex minimisation [13]. Iterative statistical methods have also been applied to TR. These algorithms maximise the likelihood of parameters of an underlying statistical model. MLEM, and an accelerated version known as OSEM, also suffer from overfitting, but noise amplification can be reduced with MAP regularisation. Deep learning (DL), despite its success in natural language and computer vision applications, is yet to improve upon traditional analytical methods [28].

Optimisation techniques, and in particular population based algorithms, are complementary to analytical methods and hold promise for few-view problems due to their resilience, relative lack of assumptions and ability to succeed where analytical methods fail. Several metaheuristic algorithms have been applied to TR, including harmony search [32], tabu search [26], simulated annealing [25], memetics [16] and evolutionary algorithms [8]. Swarm algorithms have also been trialled in binary reconstruction [29], geophysical reconstruction [37], electrical capacitance and impedance tomography [23,38] and surface reconstruction from 3D data [18]. An algorithm based on the movement of particles over a single image, a pixel-swarm, has been developed for binary reconstruction [4].

The few-view TR problem is underdetermined, which indicates the lack of unique solution. An optimisation might find an image of low loss but there is no guarantee that this image is medically feasible. ART and other least-squares methods tend to produce low norm, diffuse solutions with small pixel values. Swarm algorithms make no assumption (such as low norm or sparsity) about the nature of the solution and do not require convexity.

This paper reports on the application of swarm algorithms to few-view binary TR. After an account of binary tomography and swarm algorithms, the feasibility of high dimensional optimisation without any specific coping mechanism such as subspace optimisation dimensional search is considered. Four swarm algorithms are trialled on four standard phantoms and a suggestive mechanism for the effectiveness of wall-clamping is proposed.

2   Binary Tomographic Reconstruction

Incident radiation is typically modelled by a projection matrix $A \in \mathbb{R}_{\geq 0}^{m \times n}$ where m is the total number of rays (projections) and n is the number of pixels in the reconstructed image. Suppose that $b \in \mathbb{R}^m$ is a vector of detector values. Then the continuous/discrete reconstruction problem can be stated as:

$$\text{find } x \begin{cases} \in \mathbb{R}^n \\ \in \{0, 1, \ldots, k-1\}^n, k > 1 \end{cases} \quad \text{such that } Ax = b.$$

In the binary problem, $k = 2$ i.e. $x \in \{0, 1\}^n$.

The equation $Ax = b$ cannot be inverted if $m < n$: an approximate solution y must be found. This trial solution is forward projected:

$$Ay = c$$

with a *reconstruction* error

$$e_1(y) = ||b - c||_1 \qquad (1)$$

An iterative scheme will produce a sequence of candidate solutions of decreasing error but, due to underdetermination, low reconstruction error does not imply faithfulness to the original object x^*. The proximity of y to x^* can be measured:

$$e_2 = ||y - x^*||_1 \qquad (2)$$

In cases where x^* is known, this *reproduction* error provides a test of the ability of an algorithm to find a faithful reconstruction.

3   Swarm Optimisation

An optimisation swarm, for real-space problems $\arg \min f(x)$, where $f : X \to \mathbb{R}$ and $X \subset \mathbb{R}^n$ is the feasible search space, is a population of interacting 'particles'. Each particle position is a possible solution; particles move under each others' influence in an attempt to improve the best found position. Particle interactions might be mediated by current or historical positions of particles in a spatial or social neighbourhood. Two swarm optimisers and differential evolution, a real-space population algorithm that has much in common with swarms, are described below.

PSO. In particle swarm optimisation, particles i in a canonical PSO swarm [27,33,34] of M particles are a triple (x_i, v_i, p_i), representing position, velocity and personal best (pbest), p_i, of the best position they have achieved in the run, as measured by the objective function f. Dynamical variables are updated by the rule

$$\begin{aligned} v_i(t+1) &= wv_i(t) + cu_1 \circ (n_i(t+1) - x_i(t)) \\ &+ cu_2 \circ (p_i(t+1) - x_i(t)) \\ x_i(t+1) &= x_i(t) + v_i(t+1) \end{aligned} \qquad (3)$$

where $u_{1,2} \sim U(0,1)$ are uniform random variables in $[0,1]^D$ and $\circ$ is the Hadamard (entry-wise) product, n_i is the pbest of the best neighbour in i's social network (an arbitrary choice is made in the case of a tie). The inertial weight, w, and acceleration coefficients c, are two arbitrary (but constrained) positive real parameters chosen to balance convergence and exploration and t labels iteration. The pbests may be determined synchronously at the start of the iteration or asynchronously on a particle-by-particle basis.

Two social networks are common in PSO implementations: a global network where particles have access to all pbests (GPSO) and a local ring (LPSO) network where particles can only access 'left' and 'right' neighbours. LPSO has slower information transport; this property inhibits convergence and favours early exploration. LPSO is generally better at more complex multi-modal problems [10].

DE. Differential evolution has many variants. We specify the DE/best/1 version, which is considered competitive and robust [17].

Iterations begin with a determination of the current position, g of the best particle. Then, for each particle i, indices j and k are selected such that $i \neq j \neq k$. A random component $r \in \{1, 2 \ldots n\}$ is also selected. Component d of particle i at x_i is updated:

$$\text{if } u \cdot U(0, 1) \cdot \eta_u \ldots \text{ or}$$
$$y_d = g_d + F(x_{jd} - x_{kd})$$
$$\text{else}$$
$$y_d = x_{id} \tag{4}$$

where y is a trial position and the parameters $C_R \in [0, 1]$ and $F \in [0, 2]$ are known as the 'cross-over rate' and the 'differential weight'. Then, after each component of y has been set, i is conditionally moved:

$$x(t + 1) = \arg\min{}^*(f(y), f(x(t)))$$

DFO Dispersive flies optimisation [1], is a slim PSO variant without memory and velocity whose exploration and exploitation behaviour is studied in [2]. Updates are based on instantaneous, rather than historical, position. In addition, it incorporates component-wise particle jumps [11]. The best overall position $g(t + 1)$ and best ring neighbours $n_i(t + 1)$ are determined (with arbitrary choices in the case of ties). Component d of all particles i other than the swarm best (written x_{id}) updates according to

$$\text{if } u \sim U(0, 1) < \Delta$$
$$x_{id}(t + 1) \sim U(X_d)$$
$$\text{else}$$
$$x_{id}(t + 1) = n_{id}(t + 1) + \phi u_1(g_d(t + 1) - x_{id}(t)) \tag{5}$$

where Δ is a preset jump probability, $U(X_d)$ is the uniform distribution along axis d of the search space X, $u_1 \sim U(0, 1)$ and $\phi \in [0, \sqrt{3}]$. ϕ is invariably set to 1 and Δ to 0.001 in published studies (e.g. [3,5-7,31]). The upper bound on ϕ is derived from a convergence analysis for stochastic difference equations [11]). DFOMAX will henceforth denote DFO with $\phi = \sqrt{3}$. DFOMAX is expected to have the maximum diversity since the particle update rule places the swarm on the edge of divergence.

DFO uses both global and local strategies and formally, with its reliance on instantaneous position, abandonment of particle memory and retention of a static communication network, interpolates between PSO and DE. Formal comparison does not of course necessarily imply intermediate performance.

4   Constrained Search in High Dimensions

The TR problem is high dimensional: a modest 32×32 image has 1024 pixels, therefore, the search space has 1024 dimensions. There is a question whether the algorithms that have been developed for low dimensions (typically $n = 30$), such as the three algorithms specified above, will adapt to the high dimensional problem. There are further issues regarding boundaries and placement of global optima.

The feasible search space for image reconstruction is $X = [0, 255]^n$. A particle might fly outside X and the algorithm must specify if any action is to be taken. Particles may continue to move in $\overline{X} = \mathbb{R}^n \setminus X$ and are either evaluated (if f is defined in $\overline{X}$) or remain unevaluated. Alternatively, particles might be clamped to the boundary ∂X i.e. $x_{id} = \max(\min(255, x_{id}), 0)$. Furthermore, algorithms might behave differently for problems with the global minimum in the interior of X or on ∂X.

The five swarm algorithms defined above (DE, DFO, DFOMAX, GPSO and LPSO) were tested on the Sphere problem, $f(x) = x \cdot x$, with $X = [0, 255]^n$ and with the global optimum at 128^n (interior) and at 255^n (boundary) in a range of dimensions, n, and for clamped and freely moving particles (which are not evaluated outside X. This problem, despite its unimodality and symmetry is far from trivial in high dimensions and with optimum on bounds.

A swarm size of $M = 100$ was chosen for DE, DFO, DFOMAX and G/LPSO. Particles were initialised in X with the uniform distribution and G/LPSO velocities were set to zero. The DFO jump probability Δ was set to 0.001; G/LPSO was run with $w = 0.729844$ and $c = 1.49618$ and the DE/best/1 parameters F and C_R were both set to 0.5.

Tables 1, 2, 3 and 4 report on median errors after 30 runs of 10^5 function evaluations for each algorithm under different boundary conditions and placement of global optimum.

Table 1 shows median errors for the sphere function with optimum in the centre of X and no boundary action where particles can move freely and are evaluated everywhere. All algorithms struggle in higher dimensions where the swarms have barely improved upon their initialised best value: $\mathbb{E}f(x) = n \int_0^{255} x^2 \frac{dx}{255} = 21675n \approx 2 \times 10^7$ ($n = 1000$). DFOMAX is very poor in lower dimensions, presumably because the ϕ-parameter promotes a very high diversity. Placing the optimum at the corner of X (Table 2) does not change the picture; errors are even higher in higher dimensions indicating that this is a harder problem for free movement boundary conditions.

Table 1. Sphere: Particles move freely and are evaluated everywhere, $x^* = 128^n$

Dimension	DE	DFO	DFOmax	GPSO	LPSO
50	1.18e−17	7.53e−12	8.65e+02	3.63e−09	1.40e+00
100	2.26e−05	5.14e−04	3.46e+04	3.04e−01	1.83e+03
200	1.19e+02	1.21e+01	2.91e+05	4.87e+03	9.03e+04
300	5.76e+03	5.80e+02	6.62e+05	5.15e+04	4.70e+05
400	3.51e+04	4.53e+03	1.12e+06	2.41e+05	1.09e+06
500	8.78e+04	1.79e+04	1.54e+06	1.62e+06	1.63e+06
600	1.62e+05	4.52e+04	2.12e+06	2.08e+06	2.14e+06
700	2.65e+05	9.31e+04	2.61e+06	2.48e+06	2.53e+06
800	3.64e+05	1.64e+05	3.15e+06	2.94e+06	3.02e+06
900	5.01e+05	2.58e+05	3.71e+06	3.43e+06	3.45e+06
1000	6.47e+05	3.80e+05	4.27e+06	3.84e+06	3.83e+06

Table 2. Sphere: Particles move freely and are evaluated everywhere, $x^* = 255^n$

Dimension	DE	DFO	DFOMAX	GPSO	LPSO
50	1.81e−17	3.43e−11	1.67e+03	6.67e−09	7.13e+00
100	1.21e−04	2.25e−03	1.19e+05	1.47e+00	8.72e+03
200	8.03e+02	5.81e+01	1.08e+06	3.82e+04	4.10e+05
300	6.13e+04	2.59e+03	2.54e+06	2.95e+05	1.55e+06
400	2.53e+05	2.19e+04	4.21e+06	8.70e+05	3.18e+06
500	6.58e+05	8.36e+04	5.94e+06	1.67e+06	5.04e+06
600	1.17e+06	2.17e+05	7.68e+06	2.65e+06	7.28e+06
700	1.88e+06	4.38e+05	9.78e+06	3.70e+06	9.45e+06
800	2.69e+06	7.65e+05	1.17e+07	5.00e+06	1.21e+07
900	3.53e+06	1.18e+06	1.37e+07	6.54e+06	1.49e+07
1000	4.52e+06	1.70e+06	1.59e+07	8.63e+06	1.70e+07

Tables 3 and 4 show the corresponding experiments with clamped swarms i.e. particles straying outside X are immediately projected onto ∂X. Clamping has no effect when the optimum is in the middle of the search space, but drastically improves DFO, DFOMAX and LPSO performance for high dimensions when the optimum is placed at the corner of X. Setting ϕ to its maximum value is advantageous at $n = 1000$; in this case the larger diversity is aiding search.

The Sphere problem offers some clues to the relationship between large n performance, optimum placement and boundary conditions but is different in several important respects from the reconstruction problem $f_{TR}(x) = ||b − Ax||_1$.

f_{TR} has an infinity of solutions lying on the $(n − m)$ dimensional hyperplane $\{x : Ax = b\}$, a rather extreme multimodality. A found exact solution might not equate to the original image, x^*, i.e. zero reconstruction error (e_1) does

Table 3. Sphere: Particles are clamped to $X = [0, 255]^D$, $x^* = 128^n$

Dimension	DE	DFO	DFOMAX	GPSO	LPSO
50	6.29e−18	9.37e−12	1.01e+02	2.59e−09	9.09e−01
100	2.01e−05	4.54e−04	9.40e+03	1.64e+04	1.00e+03
200	1.55e+02	1.21e+01	1.43e+05	7.05e+04	3.72e+04
300	1.01e+04	5.32e+02	4.41e+05	1.89e+05	1.57e+05
400	5.22e+04	4.40e+03	8.40e+05	3.37e+05	3.53e+05
500	1.31e+05	1.66e+04	1.28e+06	5.31e+05	5.81e+05
600	2.33e+05	4.38e+04	1.77e+06	7.66e+05	8.31e+05
700	3.49e+05	8.89e+04	2.27e+06	1.01e+06	1.09e+06
800	5.40e+05	1.52e+05	2.77e+06	1.29e+06	1.36e+06
900	6.51e+05	2.37e+05	3.28e+06	1.54e+06	1.65e+06
1000	8.53e+05	3.49e+05	3.82e+06	1.87e+06	1.90e+06

Table 4. Sphere: Particles are clamped to $X = [0, 255]^D$, $x^* = 255^n$

Dimension	DE	DFO	DFOMAX	GPSO	LPSO
50	3.23e−27	0.00e+00	0.00e+00	0.00e+00	0.00e+00
100	4.80e−25	0.00e+00	0.00e+00	2.60e+05	0.00e+00
200	3.64e−11	0.00e+00	0.00e+00	9.75e+05	0.00e+00
300	4.96e−06	0.00e+00	0.00e+00	1.85e+06	0.00e+00
400	2.68e−03	0.00e+00	0.00e+00	2.96e+06	0.00e+00
500	2.53e−01	0.00e+00	0.00e+00	4.00e+06	0.00e+00
600	4.34e+00	0.00e+00	0.00e+00	4.97e+06	0.00e+00
700	3.26e+04	0.00e+00	0.00e+00	6.08e+06	0.00e+00
800	6.53e+04	0.00e+00	0.00e+00	7.51e+06	0.00e+00
900	1.31e+05	2.29e−08	0.00e+00	8.49e+06	0.00e+00
1000	2.00e+05	5.69e+01	0.00e+00	9.88e+06	0.00e+00

not ensure zero reproduction error (e_2). Level sets of f_{TR} are flat in $n - m$ dimensions; the Sphere's level sets are curved in n dimensions. High-n curvature renders update unlikely because the interior of the level set has a much smaller volume than the exterior. However level set flatness will remain problematic in few-view TR because the topography remains curved in $m \approx n$-dimensions.

A dummy TR problem was devised in order to test if the Sphere results might generalise. The five algorithms were trialled on a uniform phantom $x^* = 255^{32 \times 32}$ under free movement and clamping. The solution hyperplane intersects with $X = [0, 255]^{1024}$ at a single point (a corner) so the problem is unimodal in X. Tables 5 and 6 report on $e_{1,2}$ under the above experiment settings and for $m = 6, 8, 16, 32$. Free movement performance is poor for all algorithms and for

Table 5. Uniform 32×32 phantom: particles move freely and are evaluated everywhere

e_1	DE	DFO	DFOMAX	GPSO	LPSO
$m = 6$	27512	36946	229072	61075	194348
$m = 8$	52050	62997	331839	107265	285517
$m = 16$	188214	181379	781586	348052	674266
$m = 32$	472357	381192	1618504	847332	1444595

e_2	DE	DFO	DFOMAX	GPSO	LPSO
$m = 6$	287601	423121	529438	139996	150539
$m = 8$	281366	388361	509529	141577	150056
$m = 16$	245387	318266	472701	137075	153196
$m = 32$	211071	228996	433595	144072	151037

Table 6. Uniform 32×32 phantom, clamped particles

e_1	DE	DFO	DFOMAX	GPSO	LPSO
$m = 6$	25000	0	0	434558	30689
$m = 8$	33865	0	0	586305	39757
$m = 16$	73463	0	0	1163191	83098
$m = 32$	137865	0	0	2316471	176054

e_2	DE	DFO	DFOMAX	GPSO	LPSO
$m = 6$	4612	0	0	77392	5865
$m = 8$	4737	0	0	77775	5610
$m = 16$	5006	0	0	77902	5865
$m = 32$	4804	0	0	77647	6120

all numbers of projections m. The reproduction error even exceeds the maximum value of $261120 = 32 \times 32 \times 255$ in some instances, indicating that best positions lie outside X. Clamping tells a different story: both errors are improved and DFO and DFOMAX find the corner optimum corresponding to the original phantom. We see that clamping can play an important role in TR, rendering even high-n problems tractable.

5   Reconstructions

Four standard binary TR test phantoms, as depicted in the leftmost column of Fig. 1, were chosen for the algorithm trials. Two sizes, 32×32 and 64×64 and 6, 8, 16 and 32 projections (few-view scenarios) were tested. Phantom imaging was

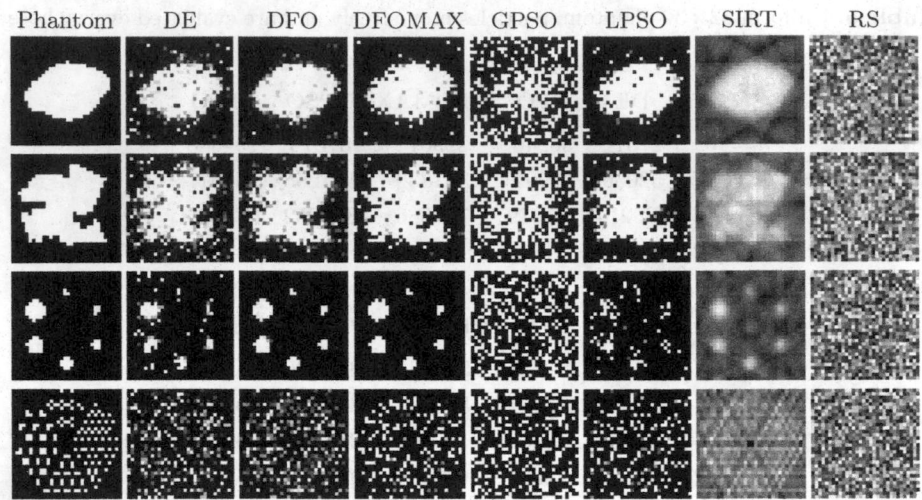

Fig. 1. Phantoms $1-4$ (top to bottom) and sample reconstructions by the algorithms, with phantom size of 32×32 and 6 projections.

conducted by the ASTRA tomography toolbox[1] using parallel geometry with the number of rays set to 32 and 64 for the the 32×32 and 64×64 phantoms respectively.

Six algorithms were chosen, including the five swarm (DE is considered here as a pseudo-swarm) algorithms from the previous section and SIRT, an algebraic reconstruction algorithm from the toolbox. A previous examination (paper under review) with the same experiment set-up confirmed that SIRT was the superior algebraic algorithm. In addition, random search (RS) was used as a control because the swarm algorithms rely on extensive sampling. All data from DE, DFO, G/LPSO, SIRT and RS experiments are taken from a recent paper (under review).

Particles were clamped to $[0, 255]$ in each dimension and swarms and RS were run for 100,000 function evaluations. Swarm parameter settings were identical to those of Sect. 4 experiments. All algorithms with randomisation were run 30 times on each problem.

6   Results

Figure 2 depicts algorithm reconstructions for each of the eight problems. Reconstruction improves in all cases with increased projections and for the smaller phantom. The extreme few-view case ($64 \times 64, m = 6$) is very challenging. SIRT produces blurred images and swarm reconstructions are sharper but contain pepper noise. GPSO is particularly bad, with reconstructions that are hardly better than random search. Otherwise, DFO and DFOMAX appear to be the best performers.

---

[1] https://www.astra-toolbox.com.

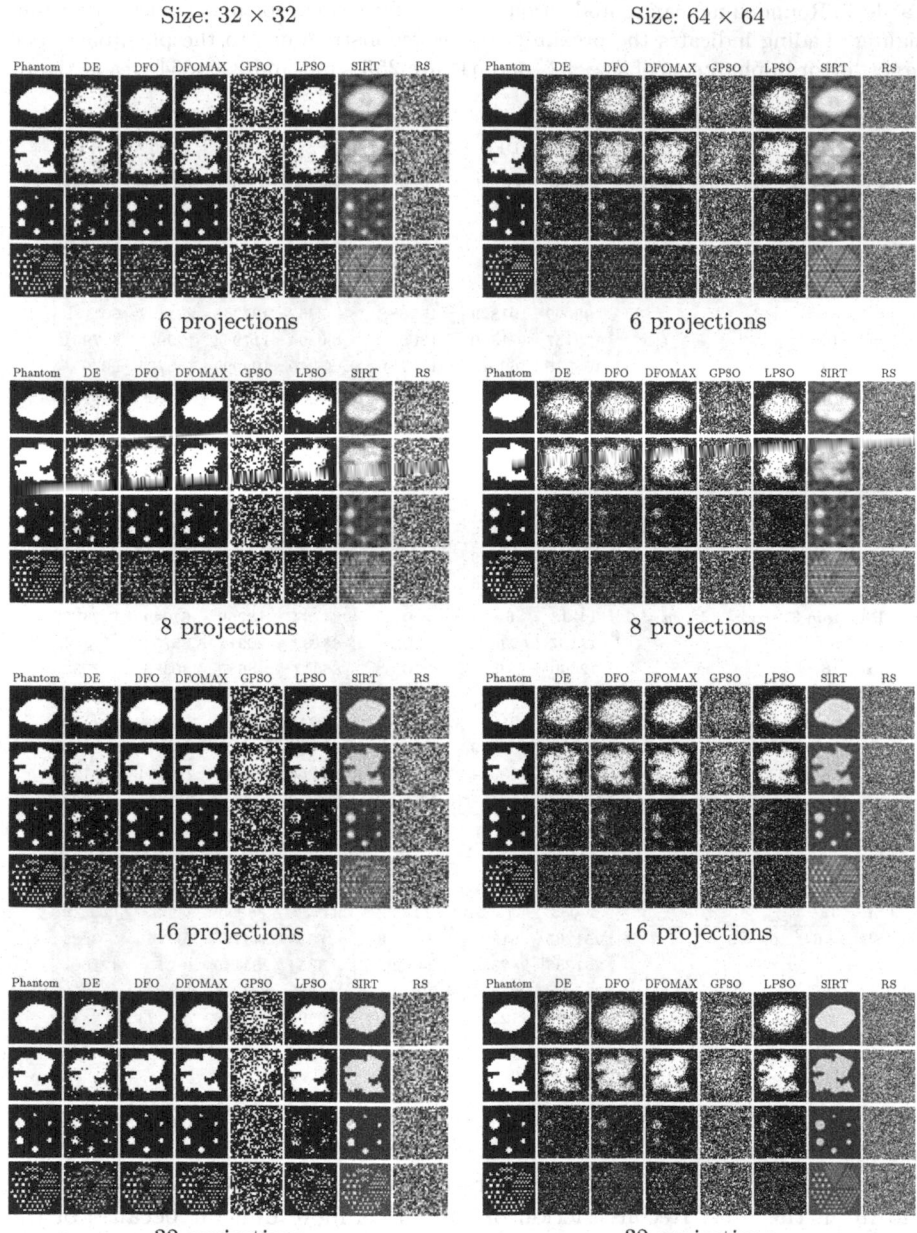

Fig. 2. Reconstructed phantoms

Table 7. Rounded median reproduction error, e_2, for each problem and each algorithm. Lighter shading indicates the proximity of the reconstructions to the phantoms. The largest error in phantoms of sizes 32^2 and 64^2 are 255×32^2 and 255×64^2 respectively.

	DE	DFO	DFOMAX	GPSO	LPSO	SIRT	RS
Phantom 1, size = 32^2, $m = 6$	24360	11392	9462	90896	17006	52254	123172
$m = 8$	18306	4250	0	88231	12444	53452	122473
$m = 16$	10800	0	0	87185	8898	42305	123530
$m = 32$	8646	0	0	87993	8540	27998	122754
Size = 64^2, $m = 6$	210940	207107	169545	435764	154123	246054	506638
$m = 8$	193099	191829	151688	432352	137348	241197	507789
$m = 16$	172757	171379	130625	430594	117964	179971	507951
$m = 32$	163515	160018	117159	430219	109999	150775	507091
Phantom 2, size = 32^2, $m = 6$	39667	27905	27878	93150	28901	75948	122240
$m = 8$	35209	22588	22280	92225	22997	71291	122429
$m = 16$	17269	785	0	90415	8816	52369	122056
$m = 32$	11701	0	0	87261	7921	30212	122628
Size = 64^2, $m = 6$	251333	243877	210000	438343	190386	308809	507088
$m = 8$	237651	230159	196678	436555	170015	280526	506604
$m = 16$	189399	188739	151242	428432	120831	241244	504912
$m = 32$	171271	171116	128279	430698	109442	181168	504911
Phantom 3, size = 32^2, $m = 6$	14213	697	0	84787	17259	61049	122377
$m = 8$	17187	2302	0	88102	20912	65759	121866
$m = 16$	12000	0	0	86317	18689	40385	122053
$m = 32$	10536	0	0	85807	18903	31715	121947
Size = 64^2, $m = 6$	173616	161874	126078	425212	156488	243334	504370
$m = 8$	181518	166219	134297	423937	159573	297473	504085
$m = 16$	178099	158374	121136	421260	160890	219134	505120
$m = 32$	175900	155551	117687	424702	149612	141821	504338
Phantom 4, size = 32^2, $m = 6$	54862	48549	48357	93915	52286	106043	123660
$m = 8$	58459	51857	54111	94987	56592	101792	122876
$m = 16$	48886	37014	37679	95115	44677	90408	123858
$m = 32$	40083	17595	11636	94222	38095	60032	123279
Size = 64^2, $m = 6$	251265	231391	226598	437839	244786	369251	472486
$m = 8$	261236	242350	244098	443354	253149	402474	471607
$m = 16$	252098	225371	222921	442305	244071	378158	473273
$m = 32$	242980	212200	202150	440890	231698	250803	473312

Table 7 confirms these visual findings. DFO and DFOMAX find exact reproductions for phantoms 1, 2 and 3 of size $n = 32^2$ at 32 projections and phantoms 1 and 3 at $m = 16$. DFOMAX recovers the 32^2 phantom 3 at all projections, and in all the runs. Reconstruction of the 64^2 is more difficult because of the extreme few-view scenario.

Wilcoxon rank tests of algorithm performance (e_1 and e_2) were conducted at $\alpha = 0.05$. The results of the 32 problems is reported in Table 8. SIRT dominates the rankings for e_1 but produces significantly worse reproductions (e_2) than all algorithms apart from GPSO and RS. DFOMAX is the top reproduction performer, beating DE, GPSO and SIRT in each problem; DFOMAX also returned better reconstruction errors (e_1) than DE and GPSO. Setting DFO's

Table 8. Algorithms comparison based on e_1 and e_2. The numbers indicate statistically significant wins for the algorithm in the left hand column versus the algorithm in the top row. Total number of problems are 32: 4 phantoms (1, 2, 3, 4) × 2 sizes (32^2, 64^2) × 4 projections (6, 8, 16, 32)

e_1	DE	DFO	DFOMAX	GPSO	LPSO	SIRT	RS
DE	NA	0	0	32	4	0	32
DFO	32	NA	6	32	23	5	32
DFOMAX	32	25	NA	32	24	9	32
GPSO	0	0	0	NA	0	0	32
LPSO	24	8	7	32	NA	0	32
SIRT	32	27	23	32	32	NA	32
RS	0	0	0	0	0	0	NA
e_2	DE	DFO	DFOMAX	GPSO	LPSO	SIRT	RS
DE	NA	0	0	32	4	30	32
DFO	29	NA	1	32	18	30	32
DFOMAX	32	26	NA	32	22	32	32
GPSO	0	0	0	NA	0	2	32
LPSO	27	11	8	32	NA	31	32
SIRT	2	2	0	30	1	NA	32
RS	0	0	0	0	0	0	NA

ϕ to $\sqrt{3}$ and thereby ensuring maximum diversity without explosion improves DFO in this setting (beating DFO 25 times and 26 times for reconstruction and reproduction error respectively).

The results of the preliminary trials on optimum-on-corner-Sphere and the uniform phantom are confirmed for these phantoms. It seems that clamping ameliorates the curse of high dimension in cases where the desired optimum is on the boundary of the search space. The results for DFO and DFOMAX are very promising considering the few-view conditions and the high dimensionality of the problems. The apparent feasibility of a conversion of reconstruction into an optimisation problem is surprising where the swarm is moving in the space of all possible 32^2 or 64^2 binary images.

7 Discussion

The results of all trials indicate the importance of imposing clamping in cases where the desired optimum lies on a corner of the search space. This finding can be supported by considering a model algorithm optimising Sphere.

The model algorithm is assumed to produce trial points with spherical symmetry. In particular, trials are generated in a ball of radius r with probability $\frac{1}{2}$ about a centre x. For the Sphere problem, x lies on a hyperspherical level set of radius R. All points within the level set have lower function value; a trial point is accepted if it lies in the n-ball of radius R which we can assume is centred at O. If P_u is the relative volume of intersection of the trial ball $B_n(x, r)$ with the ball of points with lower function value $B_n(O, R)$, i.e.

$$P_u = \frac{\text{vol}(B_n(x, r) \cap B_n(O, R))}{\text{vol}(B_n(x, r))}$$

then the probability of updating in a single trial is $\frac{P_u}{2}$. Figure 3 illustrates three possibilities for three different search radii. The geometry is depicted in Fig. 4.

The update region $B_n(x, r) \cap B_n(O, R)$ in the leftmost diagram of Fig. 3 is an n-dimensional lens. The length of the major axis, a, of this lens is $R \sin \phi$ (Fig. 4). a is larger than the minor axes for $\phi \in (0, \pi)$. From Fig. 4, $\sin \phi = \rho\sqrt{1 - \frac{\rho^2}{4}}$, and, with $\rho = \frac{r}{R}$,

$$P_u < \frac{(R \sin \phi)^n}{r^n} = \left(1 - \frac{\rho^2}{4}\right)^{\frac{n}{2}} \quad (\rho < 1)$$

For $\rho \geq 1$, middle and rightmost diagrams of Fig. 3, $B_n(x, r) \cap B_n(O, R)$ is covered by a ball of radius R, so

$$P_u \leq \rho^{-n} \quad (\rho \geq 1)$$

Hence $\lim_{n \to \infty} P_u = 0$ for all scenarios in which $\rho > 0$. The pathology of high dimensions is manifest as a varnishing update probability for spherically symmetric search on the Sphere function; for example, in 1000 dimensions and $\rho = \frac{1}{2}$, the chances of finding a better trial position are less than 10^{-14}.

The above conclusions is valid for unconstrained search. Clamping at a wall significantly increases the update probability. Figure 5 depicts wall clamping for optimisation of the Sphere with optimum at a corner. A trial landing in region A will be projected onto the edge of the search box and into the update region. Furthermore, a particle that has been repositioned on the wall will have a significantly greater chance of updating (because region A is a half-space) at the next iteration. The effective augmentation of the update region by clamping is a possible mechanism for the evident improved performance of all swarm algorithms for corner problems (although it does not explain the superiority of this pair over PSO and DE).

 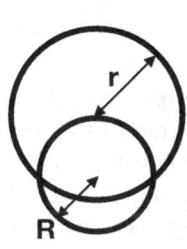 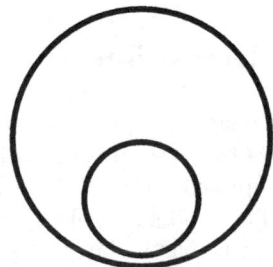

Fig. 3. Three configurations of the update region $B_n(x, r) \cap B_n(O, R)$

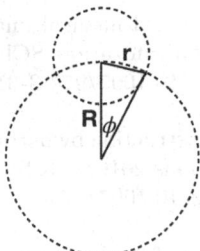

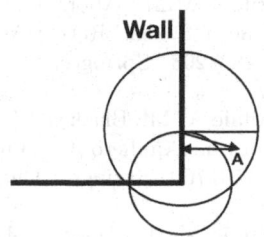

Fig. 4. Geometry of the lens-shaped update region

Fig. 5. Advantageous effect of clamping

6   Conclusions

This paper establishes that swarm search algorithms developed for optimisation in low dimensions ($n \sim 30$) can transfer to reconstruction problems in higher dimension ($n = 1024 - 4096$) providing that particles are clamped to the walls of the search space. Experiments with four standard phantoms under few-view conditions demonstrate good reconstructions when compared to the original phantom for two swarm optimisers: DFO and DFOMAX. The latter is a diversity boosted version of the former and offers superior performance. The higher diversity possibly encourages motion outside the search space; particles are clamped back to the search space wall, effectively increasing update probability. Theoretical arguments based on an idealised optimiser support this picture.

DFOMAX reconstructions are comparable to SIRT, the best algebraic technique for the problems under consideration. DFOMAX can achieve sharper and higher fidelity reconstructions in the few-view regime. Further tuning of DFO to this problem set (swarm size and jump probability remain fixed at optimal settings for low dimensional problems) is important in the light of these findings. An extension of these trials to discrete tomography is also of interest.

Acknowledgement. The authors would like to thank Darren Wise for his support in facilitating access to the HPC machines at the University of Greenwich.

References

1. al-Rifaie, M.M.: Dispersive flies optimisation. In: M. Ganzha, L: Maciaszek, M.P. (ed.) Proceedings of the 2014 Federated Conference on Computer Science and Information Systems. Annals of Computer Science and Information Systems, vol. 2, pp. 529–538. IEEE (2014). https://doi.org/10.15439/2014F142
2. al-Rifaie, M.M.: Investigating knowledge-based exploration-exploitation balance in a minimalist swarm optimiser. In: IEEE Congress on Evolutionary Computation. CEC 2021. IEEE (2021)

3. al-Rifaie, M.M., Aber, A.: Dispersive flies optimisation and medical imaging. In: Fidanova, S. (ed.) Recent Advances in Computational Optimization. SCI, vol. 610, pp. 183–203. Springer, Cham (2016). https://doi.org/10.1007/978-3-319-21133-6_11

4. al-Rifaie, M.M., Blackwell, T.: Binary tomography reconstruction by particle aggregation. In: Squillero, G., Burelli, P. (eds.) EvoApplications 2016. LNCS, vol. 9597, pp. 754–769. Springer, Cham (2016). https://doi.org/10.1007/978-3-319-31204-0_48

5. al-Rifaie, M.M., Cavazza, M.: Evolutionary optimisation of beer organoleptic properties: a simulation framework. Foods 11(3), 351 (2022). https://doi.org/10.3390/foods11030351

6. al-Rifaie, M.M., Ursyn, A., Zimmer, R., Javid, M.A.J.: On symmetry, aesthetics and quantifying symmetrical complexity. In: Correia, J., Ciesielski, V., Liapis, A. (eds.) EvoMUSART 2017. LNCS, vol. 10198, pp. 17–32. Springer, Cham (2017). https://doi.org/10.1007/978-3-319-55750-2_2

7. Aparajeya, P., Leymarie, F.F., al-Rifaie, M.M.: Swarm-based identification of animation key points from 2D-medialness maps. In: Ekárt, A., Liapis, A., Castro Pena, M.L. (eds.) EvoMUSART 2019. LNCS, vol. 11453, pp. 69–83. Springer, Cham (2019). https://doi.org/10.1007/978-3-030-16667-0_5

8. Batenburg, K.J., Kosters, W.A.: Solving nonograms by combining relaxations. Pattern Recogn. 42(8), 1672–1683 (2009)

9. Batenburg, K.J., Palenstijn, W.J.: On the reconstruction of crystals through discrete tomography. In: Klette, R., Žunić, J. (eds.) IWCIA 2004. LNCS, vol. 3322, pp. 23–37. Springer, Heidelberg (2004). https://doi.org/10.1007/978-3-540-30503-3_2

10. Blackwell, T., Kennedy, J.: Impact of communication topology in particle swarm optimization. IEEE Trans. Evol. Comput. 23(4), 689–702 (2019)

11. Blackwell, T.: A study of collapse in bare bones particle swarm optimization. IEEE Trans. Evol. Comput. 16(3), 354–372 (2011)

12. Butala, M., Hewett, R., Frazin, R., Kamalabadi, F.: Dynamic three-dimensional tomography of the solar corona. Sol. Phys. 262(2), 495–509 (2010)

13. Candes, E.J., Romberg, J.K., Tao, T.: Stable signal recovery from incomplete and inaccurate measurements. Commun. Pure Appl. Math. 59(8), 1207–1223 (2006)

14. Carazo, J.M., Sorzano, C.O., Rietzel, E., Schröder, R., Marabini, R.: Discrete tomography in electron microscopy. In: Herman, G.T., Kuba, A. (eds.) Discrete Tomography. ANHA, pp. 405–416. Birkhäuser Boston, Boston, MA (1999). https://doi.org/10.1007/978-1-4612-1568-4_18

15. Carvalho, B.M., Herman, G.T., Matej, S., Salzberg, C., Vardi, E.: Binary tomography for triplane cardiography. In: Kuba, A., Šáamal, M., Todd-Pokropek, A. (eds.) IPMI 1999. LNCS, vol. 1613, pp. 29–41. Springer, Heidelberg (1999). https://doi.org/10.1007/3-540-48714-X_3

16. Cipolla, M., Bosco, G.L., Millonzi, F., Valenti, C.: An island strategy for memetic discrete tomography reconstruction. Inf. Sci. 257, 357–368 (2014)

17. Das, S., Suganthan, P.N.: Differential evolution: a survey of the state-of-the-art. IEEE Trans. Evol. Comput. 15(1), 4–31 (2011). https://doi.org/10.1109/TEVC.2010.2059031

18. Gálvez, A., Iglesias, A.: Particle swarm optimization for non-uniform rational b-spline surface reconstruction from clouds of 3D data points. Inf. Sci. 192, 174–192 (2012)

19. Gardner, R.J.: Geometric Tomography, vol. 1. Cambridge University Press, Cambridge (1995)

20. Geyer, L.L., et al.: State of the art: iterative CT reconstruction techniques. Radiology **276**(2), 339–357 (2015)
21. Giussani, A., Hoeschen, C.: Imaging in Nuclear Medicine. Springer, Cham (2013). https://doi.org/10.1007/978-3-642-31415-5
22. Hampel, U.: High resolution gamma ray tomography scanner for flow measurement and non-destructive testing applications. Rev. Sci. Instrum. **78**(10), 103704 (2007)
23. Hu, G., Chen, M., He, W., Zhai, J.: Clustering-based particle swarm optimization for electrical impedance imaging. In: Tan, Y., Shi, Y., Chai, Y., Wang, G. (eds.) ICSI 2011. LNCS, vol. 6728, pp. 165–171. Springer, Heidelberg (2011). https://doi.org/10.1007/978-3-642-21515-5_20
24. Irving, R., Jerrum, M.: Three-dimensional data security problems. SIAM J. Comput. **23**, 170–184 (1994)
25. Jarray, F., Tlig, G., Dakhli, A.: Reconstructing hv-convex images by tabu research approach. In: International Conference on Metaheuristics and Nature Inspired Computing, p. 3 (2010)
26. Jarray, F., Tlig, G.: A simulated annealing for reconstructing hv-convex binary matrices. Electron. Not. Discr. Math. **36**, 447–454 (2010)
27. Kennedy, J.: Small worlds and mega-minds: effects of neighborhood topology on particle swarm performance. In: Proceedings of the 1999, Congress of Evolutionary Computation, vol. 3, pp. 1931–1938. IEEE Press (1999)
28. Lucas, A., Iliadis, M., Molina, R., Katsaggelos, A.K.: Using deep neural networks for inverse problems in imaging: beyond analytical methods. IEEE Signal Process. Mag. **35**(1), 20–36 (2018)
29. Miklós, P.: Particle swarm optimization approach to discrete tomography reconstruction problems of binary matrices. In: 2014 IEEE 12th International Symposium on Intelligent Systems and Informatics (SISY), pp. 321–324. IEEE (2014)
30. Nolet, G., et al.: A breviary of seismic tomography. Imaging the Interior (2008)
31. Oroojeni, H., al-Rifaie, M.M., Nicolaou, M.A.: Deep neuroevolution: Training deep neural networks for false alarm detection in intensive care units. In: European Association for Signal Processing (EUSIPCO) 2018, pp. 1157–1161. IEEE (2018). https://doi.org/10.23919/EUSIPCO.2018.8552944
32. Ouaddah, A., Boughaci, D.: Improving reconstructed images using hybridization between local search and harmony search meta-heuristics. In: Proceedings of the Companion Publication of the 2014 Annual Conference on Genetic and Evolutionary Computation, pp. 1475–1476. ACM (2014)
33. Poli, R., Kennedy, J., Blackwell, T.: Particle swarm optimization: An overview. Swarm Intell. **1**, 33–57 (2007)
34. Shi, Y., Eberhart, R.: A modified particle swarm optimizer. In: Congress on Evolutionary Computation, pp. 69–73 (1998)
35. Shliferstein, A.R., Chien, Y.: Some properties of image-processing operations on projection sets obtained from digital pictures. IEEE Trans. Comput. **26**(10), 958–970 (1977)
36. Tao, T.: Compressed sensing or: the equation ax= b, revisited. Mahler Lecture Series (2009)
37. Tronicke, J., Paasche, H., Böniger, U.: Crosshole traveltime tomography using particle swarm optimization: a near-surface field example. Geophysics **77**(1), R19–R32 (2012)
38. Wang, P., Lin, J., Wang, M.: An image reconstruction algorithm for electrical capacitance tomography based on simulated annealing particle swarm optimization. J. Appl. Res. Technol. **13**(2), 197–204 (2015)

Comparing Basin Hopping with Differential Evolution and Particle Swarm Optimization

Marco Baioletti[1], Alfredo Milani[1], Valentino Santucci[2(✉)], and Marco Tomassini[3]

[1] Department of Mathematics and Computer Science, University of Perugia, 06123 Perugia, Italy
{marco.baioletti,alfredo.milani}@unipg.it
[2] University for Foreigners of Perugia, 06123 Perugia, Italy
valentino.santucci@unistrapg.it
[3] Faculty of Economics, Department of Information Systems, University of Lausanne, 1015 Lausanne, Switzerland
marco.tomassini@unil.ch

Abstract. Using a well known benchmarking and profiling environment, we compare the performances of three simple and easy to use metaheuristics for global optimization: Differential Evolution, Basin Hopping and Particle Swarm Optimization. The comparison was done on a test set of 24 functions featuring many characteristics found on real-world problems and on four different space dimensions. Our results statistically show that there is no clear winner overall. The three methods perform well in general and the actual differences are related to the different groups of functions in the benchmark with Basin Hopping being the most robust technique, and Differential Evolution and Particle Swarm Optimization excelling on highly multi-modal functions.

Keywords: Global continuous optimization · Metaheuristics · Benchmarking

1 Introduction

Global function optimization deals with the mathematical problem of maximising or minimizing a function, possibly subject to some constraints. It is a fundamental technique that is very often needed in many fields of science, technology, and economics. Given a function f, the global minimization problem is to find the minimum value m of f over a domain $\mathcal{X}$ and can be stated as follows:

$$\min_{\mathbf{x}} \{ f(\mathbf{x}) \ : \ \mathbf{x} \in \mathcal{X} \}$$

Usually, one also wants to know the argument, i.e., the point, or set of points, $\mathbf{x}$ that provide the minimum value m of the function:

J. L. Jiménez Laredo et al. (Eds.): EvoApplications 2022, LNCS 13224, pp. 46–60, 2022.
https://doi.org/10.1007/978-3-031-02462-7_4

$$\underset{\mathbf{x}}{\operatorname{argmin}} \{\mathbf{x} \in \mathcal{X} \; : \; f(\mathbf{x}) = m\}$$

Here $\mathbf{x}$ is a real column vector of scalar variables $[x_1, x_2, \ldots, x_n]^T$, $\forall x_i \in \mathbb{R}$ and $\mathcal{X} \subseteq \mathbb{R}^n$ is the feasible set to which any solution $\mathbf{x}$ must belong. Maximization is obtained by replacing $f(\mathbf{x})$ with $-f(\mathbf{x})$. In this form, the definition also implicitly covers constrained optimization problems with a proper definition of $\mathcal{X}$. However, here we will only consider so-called "box constraints" that limit each variable to a segment of the real line, thus restricting the search space to the hyperrectangle $[l_1, u_1] \times, \ldots, \times [l_i, u_j] \times, \ldots, \times [l_n, u_n]$, where l_i and u_i are the lower and upper bounds of variable x_i.

Because of the importance of the problem, many algorithms have been devised to solve it by a variety of techniques, for a good recent presentation see, e.g., [8]. Most of these algorithms are very efficient for convex functions, for which a locally optimal solution is also globally optimal. However, many important applications give rise to problem formulations that are neither linear nor convex. Instead, they are often non-linear, highly multimodal, discontinuous, or even non-differentiable. In particular, it is often the case that no analytical form for the function is known and the function value is provided by a simulation or a measurement, a situation that is aptly called a "black box" and which requires algorithms that only rely on function values. To approach those more general problems, in the last few decades new heuristic methods have been introduced for global function optimization that are, in some sense, inspired by natural phenomena, the most well known being Evolution Strategies (ES) [2], Differential Evolution (DE) [14], Simulated Annealing (SA) [7], and Particle Swarm Optimization (PSO) [6]. Some methods are trajectory-based, such as simulated annealing, while others are population-based. These approaches, often dubbed *metaheuristics*, have variable and sometimes unknown convergence behavior and give no global optimality guarantee, but they all have the valuable common feature of searching the space globally by being potentially able to escape from local optima. Another advantage of metaheuristics for practitioners is that they do not require a large amount of mathematical knowledge and are generally easy to understand and to implement. However, it must be acknowledged that, to get good results, all metaheuristics require some parameters to be tuned, which is usually done by trial and error or by applying some rules of thumb. There also exist other, more mathematically based global optimization methods, either deterministic or stochastic, or both (see, e.g. [8,9]) but they are more difficult to understand, program, and parametrize for the non-specialist. In the present study we shall not consider them further and limit our investigation to metaheuristics; however, a comparison with nature-inspired techniques based on a custom benchmark function set can be found in a recent publication by Sergeyev et al. [12]. Another previous study comparing ES, PSO, Artificial Bee Colony, and the Bees Algorithm has appeared in [11]. However, it uses a different benchmark suite and does not include Basin Hopping.

A less well known trajectory-based heuristic for global optimization is *Basin Hopping* (BH). Basin Hopping has its origins in computational physical chemistry, where it has been successfully used for years to determine minimum energy configurations of atomic clusters and biological macromolecules [15,16]. Although over the years the technique has tended to become somewhat specialized for the above tasks by taking advantage of known chemical and physical constraints, it is also a very simple but powerful general global optimization method and it is from this point of view that it will be studied here. To our knowledge, Basin Hopping has never been systematically compared to other nature-inspired algorithms before. Thus, our goal in this study is to find out how BH compares to two well established metaheuristics: Differential Evolution and Particle Swarm Optimization. General ES techniques, such as Covariance Matrix Adaptation ES, are known to perform very well on the type of functions of interest here but we did not include them in the comparison because they are more sophisticated and in general require more expertise on the part of the user. Moreover, there are several similar methods in existence and it is difficult to choose a particular one if one is not familiar with them (see, e.g., [2]). As a comparison testbed, we used the widespread BBOB benchmark test suite [5] provided in the *IOHprofiler* environment [4].

The article is structured as follows. In the next section we give an introduction to Basin Hopping, Differential Evolution, and Particle Swarm Optimization, with an emphasis on BH which is the less well known methodology. This is followed by a description of the benchmarking environment, including the function test set. The following sections describes the results obtained and discusses them. Finally, we draw our conclusions.

2   The Metaheuristics Studied

For the sake of completeness, in this section we briefly describe the metaheuristics that have been compared with an emphasis on Basin Hopping which is probably less well known among practitioners.

2.1   Basin Hopping

The outline of the BH algorithm is deceptively simple, see pseudocode 1, where solutions s, x, y, z are to be understood as n-dimensional vectors.

Algorithm 1. Basin Hopping

$s \leftarrow$ generate initial solution
$x \leftarrow$ minimize(f, s)
while termination condition not met **do**
    $y \leftarrow$ perturb(x)
    $z \leftarrow$ minimize(f, y)
    $x \leftarrow$ acceptance(x, z)
end while
return $x, f(x)$

The algorithm starts by generating an initial solution s either randomly or heuristically. Unless the fitness landscape is flat, this solution must belong to the basin of attraction of some local optimum whose coordinates x are found by using a local search procedure starting at s. After that the algorithm iterates three stages. First, the current solution x is perturbed by some kind of coordinates change yielding the new solution y. Next, starting at y, a local minimizer finds the new local minimum z. There are two possibilities: either z is different from x or it is the same. In the first case, the algorithm has successfully jumped out of the basin of attraction of x. Otherwise, the perturbation has been insufficient and the point y belongs to the original basin of attraction, causing the search to find the same minimum again. Finally, the acceptance phase consists in deciding whether the new solution z is accepted as the starting point of the next cycle. If $f(z) = f(x)$ the search resumes by trying another perturbation from this point. Otherwise, it is accepted either unconditionally or subject to some condition. For example, it could be accepted only if $f(z) < f(x)$. In the original Basin Hopping algorithm the acceptance of the new solution was done conditionally using a Monte Carlo test as in simulated annealing [16]. The new solution z is always accepted if it is better than x. Otherwise, if it is worse than x, it is accepted with probability $e^{-\beta(f_z - f_x)}$, where β is a parameter inversely related to a simulated temperature. As in other metaheuristic approaches the termination condition is met after a predetermined number of iterations, when a given time has elapsed, or when the solution doesn't change within a given precision during a given number of iterations.

So, the three basic components of basin hopping are the minimization algorithm, the perturbation technique, and the acceptance criterion. A good synergy between these components is key for the efficiency of the search. Mathematical minimization techniques for well-behaved functions have been developed over several decades and are efficient and reliable. For this part, in our custom Python implementation of BH, we use a state-of-the art algorithm called "L-BFGS-B" in the SciPy library, which is an extension of the Broyden-Fletcher-Goldfarb-Shanno (BFGS) algorithm [13]. The latter approximates the inverse Hessian by using first-derivative, i.e., gradient information. However, given that metaheuristics are often used in a black-box context in which only the function value at a given point is known, it is important to note that BH can also be used for non-continuous or non-differentiable functions since the local search phase can be performed with any working minimization algorithm. For example, one could use the BFGS algorithm above without providing derivatives, which will then approximated by finite differences, or the Nelder-Mead algorithm [10], or any other derivative-free method.

The perturbation technique is difficult to get right in a general way: if perturbations are too small with respect to the typical basin size of the problem at hand then the search will often fall back into the starting basin obviously causing a loss of efficiency. If, on the other hand, jumps are too long then there is the risk that the search degenerates into a random walk in solution space, hardly an efficient technique. On top of this, each particular function has its own landscape which is in principle unknown a priori unless one samples the function space beforehand or during the search. We have seen above that the acceptance phase can also be implemented in various ways that impose a different intensification/diversification ratio thus influencing the speed of the search and its convergence.

2.2   Differential Evolution

Differential Evolution (DE) is a population-based metaheuristic for function optimization that was introduced by Storn and Price [14]. In the basic DE each individual **x** in the population is varied by recombining three randomly chosen but distinct individuals **a**, **b**, and **c** to produce individual **z** according to the formula $\mathbf{z} = \mathbf{a} + F(\mathbf{b} - \mathbf{c})$, where F is a weight parameter usually chosen in $[0, 2]$. A random coordinate direction j in the n-dimensional space is then chosen and the new candidate individual **x**′ is constructed using binary crossover between **x** and **z**, with crossover probability CR as follows:

$$x_i' = \begin{cases} z_i \text{ if } i = j \text{ or with probability } CR \\ x_i \text{ otherwise} \end{cases} \tag{1}$$

Finally, the new solution **x**′ replaces **x** if $f(\mathbf{x}') \leq f(\mathbf{x})$ assuming function minimization. Pseudocode 2 below summarizes the algorithm.

Algorithm 2. Differential Evolution

    initialize the individuals in the population P with random positions
    while termination condition not met **do**
        for each individual $\mathbf{x}_i \in P$ **do**
            choose three different individuals $\mathbf{a},\mathbf{b},\mathbf{c} \neq \mathbf{x}_i$ at random
            combine $\mathbf{a},\mathbf{b}$, and $\mathbf{c}$ to produce $\mathbf{z}$
            build candidate solution $\mathbf{x}'$ by binary crossover between $\mathbf{x}_i$ and $\mathbf{z}$
            if $f(\mathbf{x}') \leq f(\mathbf{x}_i)$ **then**
                replace $\mathbf{x}_i$ with $\mathbf{x}'$ in the population P
            else
                keep solution $\mathbf{x}_i$ in the population P
            end if
        end for
    end while
    return best solution

The above pseudocode describes the standard DE algorithm but several more advanced variants have been developed both for the intermediate design of **z** as well as for crossover.

2.3   Particle Swarm Optimization

Inspired by animal behavior such as flocks of birds or a swarm of insects, Eberhart and Kennedy [6] proposed an optimization method called *Particle Swarm
Optimization*. In this approach, a number of particles simultaneously explore
a problem's search space with the goal of finding the globally optimum configuration. Here we describe the canonical version of the algorithm. PSO is a
population-based metaheuristic in which the position $\mathbf{x}_i$ of each particle i corresponds to a possible solution to the problem with objective value $f(\mathbf{x}_i)$. In each
iteration of the search algorithm the particles move as a function of their veloc-
ity $\mathbf{v}_i$ within the specified continuous search space. Two quantities, $\mathbf{x}_i^{best}(t)$ and
$\mathbf{B}(t)$, have to be defined and updated in each iteration. The first one, $\mathbf{x}_i^{best}(t)$,
which is often called *particle-best*, corresponds to the best fitness point visited
by particle i since the beginning of the search. The second quantity, $\mathbf{B}(t)$, called
global-best, is the best fitness point reached by the population as a whole up to
time step t.

The particles' movement in PSO is determined by three contributions. In
the first place, there is a term accounting for the "inertia" of the particles: this
term tends to keep them on their present trajectory. Second, they are attracted
towards $\mathbf{B}(t)$, the global best. And third, they are also attracted towards their
best fitness point $\mathbf{x}_i^{best}(t)$. The movement of a particle from one iteration to the
next is described by the following equations for particle i's new velocity $\mathbf{v}_i(t+1)$
and new position $\mathbf{x}_i(t+1)$:

$$\mathbf{v}_i(t+1) = \omega\mathbf{v}_i(t) + c_1 r_1(t+1)[\mathbf{x}_i^{best}(t) - \mathbf{x}_i(t)]$$
$$+ c_2 r_2(t+1)[\mathbf{B}(t) - \mathbf{x}_i(t)]$$
$$\mathbf{x}_i(t+1) = \mathbf{x}_i(t) + \mathbf{v}_i(t+1)$$

where ω, c_1 and c_2 are constants to be specified, and r_1 and r_2 are pseudo-random
numbers uniformly distributed in the interval $[0, 1]$. The c_1 parameter reflects the
individual's own "perception," and c_2 takes into account the group's behavior.
A common choice for these parameters is $c_1 \approx c_2 \approx 2$. The ω parameter is the
inertia constant, whose value is in general chosen as being slightly less than one.

In the initialization phase of the algorithm the particles are distributed in a
uniform manner in the search domain and are given zero initial velocity. In the
algorithm loop, which stops after a given termination condition is met, at each
iteration n candidate solutions are generated, one per particle, and the set of
solutions is used to construct the next generation according to the dynamical
update equations above.

3   The Benchmarking Environment

To meaningfully compare the performances of the algorithms described in the
previous section we have used a rich and widespread testing environment called
IOHprofiler [4] which, besides providing different collections of test functions,

also possesses on-line tools for the visualisation and statistical evaluation of the results[1]. To put this into perspective it is useful to recall some established facts. It has been proved that the performance of any black box optimization method averaged over all possible discrete functions is the same (see the no free lunch theorems [17]). This result has been extended to the continuous scenario, e.g., see [1]. Thus, one might argue that there is no point in comparing algorithms on a given finite and usually small function set because even if an algorithm, say A_1, is more efficient on that set, there will always be other functions on which it is beaten by another algorithm A_2. However, most possible functions are essentially random and do not appear in practical problems. Therefore, it is interesting to benchmark an algorithm on a test set which contains functions that are representative of problems that appear in real-world applications. Still, the results cannot straightforwardly be generalized to other functions but can nevertheless provide useful indications for practitioners.

To this end, we have used the real-parameter optimization benchmarking set called BBOB which comprises 24 noiseless scalable real-parameter single-objective test functions (fully described in [5]). The functions are designed with the goal of exposing the typical difficulties encountered in practice. They include separable and non-separable functions, functions with conditioning, multi-modal functions of various kinds, the role of symmetry and deception. To generate these function features various transformations are applied such as random shifting of the global optimum position, and linear and non-linear transformations of the search space. The functions $\{f_1, f_2, \ldots, f_{24}\}$ are classified into five groups, emphasizing different characteristics as follows:

- separable functions (f_1 to f_5);
- functions with low or moderate conditioning (f_6 to f_9);
- unimodal functions with high conditioning (f_{10} to f_{14});
- multi-modal with adequate global structure (f_{15} to f_{19});
- multi-modal with weak global structure (f_{20} to f_{24}).

The global optima (minimization is assumed) are sought in the search domain defined by the closed compact $[-5, 5]^D$ where D is the space dimension and the typical target value for all functions is $f_{opt} + 10^{-8}$, where f_{opt} is the known optimum of the function at hand and 10^{-8} is the allowed tolerance.

4   Experimental Setup

For the tests we used publicly available versions of the DE and PSO algorithm while the basic BH implementation is our own. Furthermore, since all these algorithms have several parameters that have to be chosen in advance, in the following we describe the values used in this study. The DE implementation comes from the **nevergrad**[2] library with its default parametrization, i.e., the crossover rate is

---

[1] See https://iohprofiler.github.io/.
[2] See https://github.com/FacebookResearch/Nevergrad.

set to $CR = 0.5$, the mutation operator used is "curr-to-best" with $F = 0.8$ and the population size is 30. The initial population is randomly sampled from the function domain. For the PSO algorithm, we used the RealSpacePSO implementation of **nevergrad**, again with the default parametrization: $\omega = 0.5/\log(2)$, $c_1 = c_2 = 0.5 + \log(2)$ and the population size is 40. It would be useful and interesting to vary some of the important metaheuristic variables to see their influence on the results. However, for this first basic study we couldn't afford the computing time needed to do so. This aspect should certainly be investigated more fully in future work.

Finally, we have implemented the BH algorithm as it was described in Sect. 2.1 and by considering the standard setting which adopts the sharp acceptance criterion and a random perturbation strength sampled from the interval $[-0.5, +0.5]$.

Each metaheuristic has been executed 15 times on the first 15 instances of each function in the BBOB benchmark, using the values $5, 10, 20, 40$ for the dimension D. Hence, the overall number of runs for each metaheuristic was $15 \times 24 \times 4 \times 15 = 21,600$. Each run had a fixed budget of $200,000$ objective function evaluations to ensure that a fair comparison is performed, regardless of the computation time needed by each metaheuristic.

5   Experimental Results

In this section we report and discuss the results of running the three metaheuristics previously described under the *IOHprofiler* environment. To give an overall view of the results we report the scores, i.e. the ratio between the final objective value obtained by an execution and the best objective function value obtained by the three metaheuristics on the instance of the benchmark's test function at hand. In order to improve the visual plotting, we report the logarithm of the score defined as $log(fitness/ best\_fitness)$, where $best\_fitness$ is the best objective value found on the current instance in 45 executions, 15 for each algorithm.

In Fig. 1 we report a boxplot of the scores of each metaheuristic aggregated on the D value. Results depend slightly on problem dimension but BH seems to be consistently better and shows less variation with respect to DE and PSO over the four dimensionalities tested. Figure 2 again shows aggregated results for the three algorithms, this time by groups of functions in the benchmark as defined in [5] (see also Sect. 3 above). The idea here is to try to understand whether there are statistical differences between the approaches according to function group. This appears to be the case for BH, which shows better and more robust performance than either DE and PSO in the first three groups of functions, i.e. separable functions and functions with low and high conditioning. On the other hand, the results are not dramatically different on the multi-modal functions, with or without a clear global structure.

In the following Figs. 3 and 4 we show the average scores for each function in the test set for $D = 40$, the hardest set of problems. Function numbers and names are drawn from reference [5] which provides all kind of details. It is difficult to

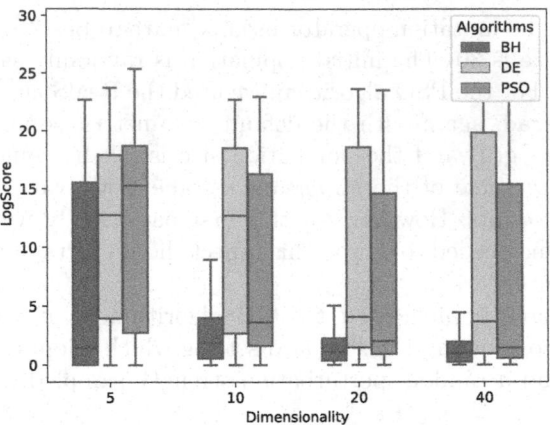

Fig. 1. Scores for BH, DE, and PSO averaged over all test functions for space dimensions $D = 5, 10, 20, 40$. See text for the interpretation of *logScore* on the ordinate axis.

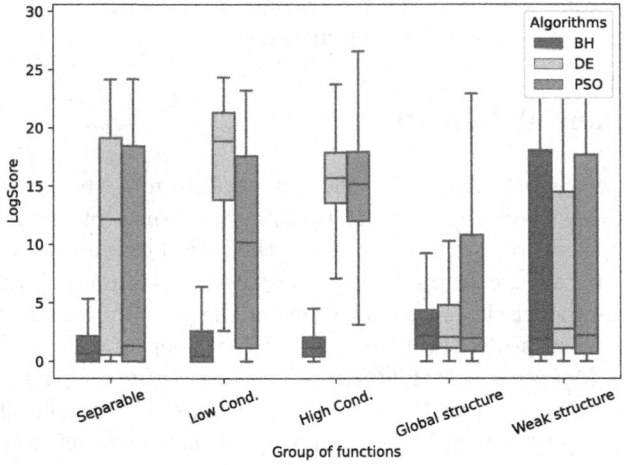

Fig. 2. Scores for BH, DE, and PSO aggregated and averaged according to the function groups as defined in the text.

draw general conclusions from the plots. However, we see that DE seems to have some trouble minimizing the relatively easy f_1 (sphere function), f_2 (ellipsoidal function), and f_5 (linear slope), while highly multimodal functions such as f_3 (Rastrigin) or f_4 (Böuche-Rastrigin) are easily optimized. In the same vein, the functions with conditioning (f_6 to f_{14}) seem also to be harder for DE and, to a lesser extent, for PSO, confirming the aggregated results previously shown in Fig. 2. On all these functions, BH seems to be the more stable method, although not always the best. In the last group of functions (multi-modal with weak global structure), f_{21} (Gallagher's Gaussian 101-me peaks function) is the function that

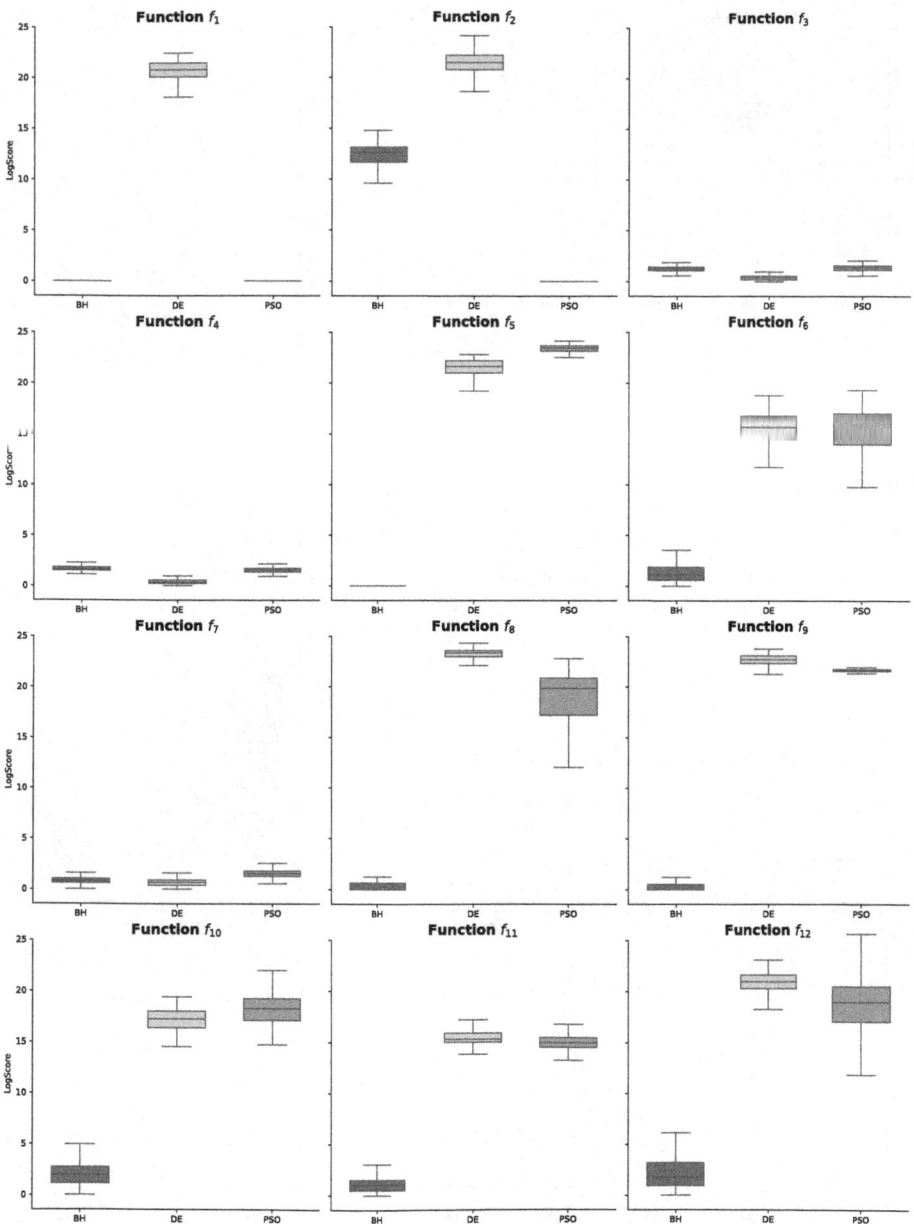

Fig. 3. Box-plots of the log-scores for the three metaheuristics for the functions from f_1 to f_{12} in the benchmark test set for dimension $D = 40$

shows the highest fluctuation in the results for all algorithms. This is a difficult highly multimodal function for which PSO obtains by far the best results.

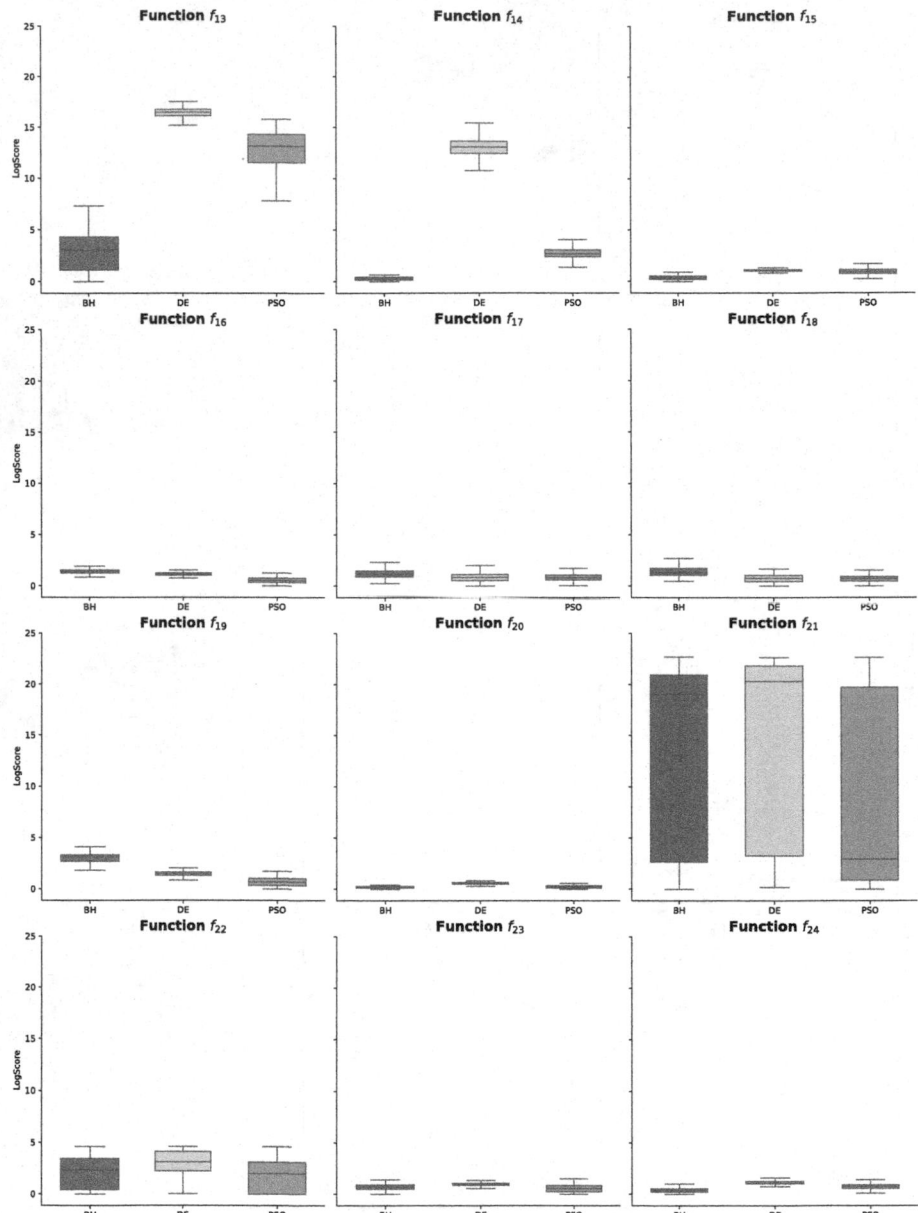

Fig. 4. Box-plots of the log-scores for the three metaheuristics for the functions from f_{13} to f_{24} in the benchmark test set for dimension $D = 40$

In order to summarize the results in an easier to see and more synthetic way, we refer the reader to Table 1. In the table the first four columns contain a comparison between BH and DE for each of the 24 functions in the test set and for the four space dimensions studied. The second four columns do the same for the pair BH/PSO. For each function a Mann-Whitney test has been performed and a p-value derived [3]. According to the p-value –a standard significance level of 0.05 is considered–, differences can be statistically significant or not. In the last case the corresponding entry in the table is "=". If the differences are statistically significant and BH is better than either DE or PSO a black up-pointing triangle is drawn; otherwise, a white down-pointing triangle is drawn.

Table 1. Results of a Mann-Whitney test for the significance of the differences between the two pairs of algorithms BH vs DE and BH vs PSO. If BH wins the entry is a black up-pointing triangle. If DE or PSO win the entry is a white down-pointing triangle. The entry "=" means that there is no statistically significant difference.

Function	BH vs DE				BH vs PSO			
	5D	10D	20D	40D	5D	10D	20D	40D
f_1	▲	▲	▲	▲	=	=	=	=
f_2	▲	▲	▲	▲	=	▽	▽	▽
f_3	=	▽	▽	▽	▲	▲	▲	▲
f_4	▽	▽	▽	▽	▽	▲	▲	▽
f_5	▲	▲	▲	▲	▲	▲	▲	▲
f_6	▲	▲	▲	▲	▽	▽	▲	▲
f_7	▽	=	▲	▽	▲	▲	▲	▲
f_8	▲	▲	▲	▲	▲	▲	▲	▲
f_9	▲	▲	▲	▲	▲	▲	▲	▲
f_{10}	▲	▲	▲	▲	▲	▲	▲	▲
f_{11}	▲	▲	▲	▲	▲	▲	▲	▲
f_{12}	▲	▲	▲	▲	▲	▲	▲	▲
f_{13}	▲	▲	▲	▲	▲	▲	▲	▲
f_{14}	▲	▲	▲	▲	▲	▲	▲	▲
f_{15}	▲	▲	▲	▲	▲	▲	▲	▲
f_{16}	▽	▽	▽	▽	▽	▽	▽	▽
f_{17}	▽	▽	▽	▽	▽	▽	▽	▽
f_{18}	▽	▽	▽	▽	▽	▽	▽	▽
f_{19}	▲	▲	▽	▽	▲	▲	▽	▽
f_{20}	▽	▽	▲	▲	=	▲	▲	▲
f_{21}	▽	▽	=	▲	▽	▽	▽	▽
f_{22}	▽	▽	=	▲	=	▽	▽	▽
f_{23}	▲	▲	▲	▲	=	▽	▽	▽
f_{24}	▲	▲	▲	▲	▲	▲	▲	▲
Wilcoxon	=	▲	▲	▲	▲	=	▲	=

From the table, one sees that BH is better overall for the group of unimodal functions with high conditioning (f_{10} to f_{14}) and for the group $f_6 - f_9$. On the other groups results are mixed with DE being statistically better than BH on multimodal functions $f_{15} - f_{19}$ and PSO prevailing on $f_{20} - f_{23}$, while BH is always better on f_{24}. Again, these are only statistical results and more study is needed to really understand the causes for the different behavior of the three metaheuristics on functions having different structure.

To conclude our study, we show here the convergence curves for a couple of functions of dimension $D = 40$. The curves are averaged over all runs for a given function and dimension and report the behavior of the objective value as a function of time, i.e. the number of function evaluations. The shaded areas around the curves show the standard deviations. Figure 5 depicts such a plot for function f_{18} (Shaffer's function F7 moderately ill-conditioned). In agreement with the corresponding entries in Table 1, both DE and PSO converge faster and get better average results than BH. Figure 6 shows the convergence curves for function f_{24} (Lunaceck bi-Rastrigin function) and we now see that BH obtains the best results, also in agreement with Table 1. In both cases BH starts with higher variability but it recovers over time. These two examples are only meant to be illustrative: no general behavior can be drawn at this stage.

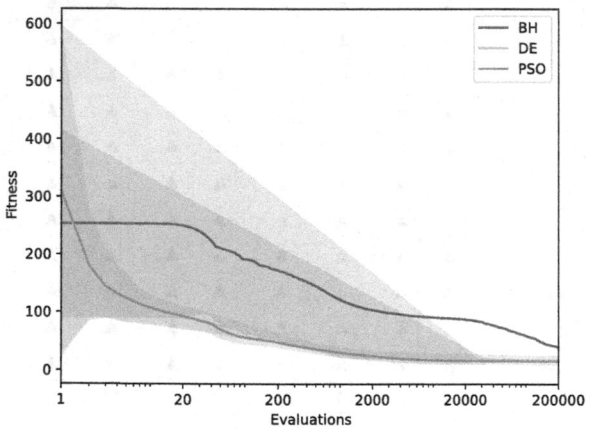

Fig. 5. Convergence curves for function f_{18} and $D = 40$.

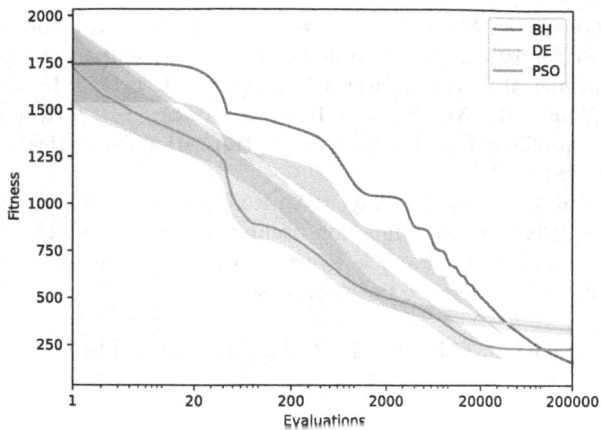

Fig. 6. Convergence curves for function f_{24} and $D = 40$.

6   Conclusions

In this contribution we started from the hypothesis that, when confronted with difficult numerical optimization problems, researchers that are not well versed in specialized mathematical optimization algorithms, may wish to use metaheuristics which do not provide global optimization guarantees but are more intuitive and easy to use. Therefore, among the most well known metaheuristics we singled out PSO, DE, and BH as being widely available and relatively easy to parameterize, and compared them on a standard recognized function test set that is representative of real-world function features commonly encountered in optimization. The picture that emerges from the comparison, using the benchmarking profiler tools and standard statistics is one in which, averaged over all functions tested, no algorithm is really superior to the others. Depending on the group of functions, one or the other offers the best results. Overall, it can be said that BH is probably the more robust of the three metaheuristics, getting consistently good results on most functions while PSO and DE appear well suited for highly multimodal functions in the last two groups. Besides, BH offers the added advantage of being really simple to use and is thus advisable for preliminary pilot studies before embarking on more complex approaches. Future studies include investigating the effect of metaheuristic parameters on the results and the inclusion of some real-world functions arising from important applications such as machine learning.

References

1. Alabert, A., Berti, A., Caballero, R., Ferrante, M.: No-free-lunch theorems in the continuum. Theoret. Comput. Sci. **600**, 98–106 (2015)
2. Bäck, T., Foussette, C., Krause, P.: Contemporary Evolution Strategies. Springer, Heidelberg (2013). https://doi.org/10.1007/978-3-642-40137-4

3. Derrac, J., García, S., Molina, D., Herrera, F.: A practical tutorial on the use of nonparametric statistical tests as a methodology for comparing evolutionary and swarm intelligence algorithms. Swarm Evol. Comput. 1(1), 3–18 (2011)
4. Doerr, C., Wang, H., Ye, F., van Rijn, S., Bäck, T.: IOHprofiler: a benchmarking and profiling tool for iterative optimization heuristics. arXiv preprint arXiv:1810.05281 (2018)
5. Hansen, N., Finck, S., Ros, R., Auger, A.: Real-parameter black-box optimization benchmarking 2009: Noiseless functions definitions. Ph.D. thesis, INRIA (2009)
6. Kennedy, J., Eberhart, R.: Particle swarm optimization. In: Proceedings of ICNN 1995-International Conference on Neural Networks, vol. 4, pp. 1942–1948. IEEE (1995)
7. Kirkpatrick, S., Gelatt, C.D., Vecchi, M.P.: Optimization by simulated annealing. Science 220(4598), 671–680 (1983)
8. Kochenderfer, M.J., Wheeler, T.A.: Algorithms for Optimization. MIT Press, Cambridge (2019)
9. Liberti, L.: Introduction to global optimization. Ecole Polytechnique (2008)
10. Nelder, J.A., Mead, R.: A simplex method for function minimization. Comput. J. 7(4), 308–313 (1965)
11. Pham, D.T., Castellani, M.: Benchmarking and comparison of nature-inspired population-based continuous optimisation algorithms. Soft. Comput. 18(5), 871–903 (2013). https://doi.org/10.1007/s00500-013-1104-9
12. Sergeyev, Y.D., Kvasov, D.E., Mukhametzhanov, M.S.: On the efficiency of nature-inspired metaheuristics in expensive global optimization with limited budget. Sci. Rep. 8(1), 1–9 (2018)
13. Shanno, D.F.: On Broyden-Fletcher-Goldfarb-Shanno method. J. Optim. Theory Appl. 46, 87–94 (1985)
14. Storn, R., Price, K.: Differential evolution-a simple and efficient heuristic for global optimization over continuous spaces. J. Global Optim. 11(4), 341–359 (1997)
15. Wales, D.J., Doye, J.P.: Global optimization by basin-hopping and the lowest energy structures of Lennard-Jones clusters containing up to 110 atoms. J. Phys. Chem. A 101(28), 5111–5116 (1997)
16. Wales, D.J., Scheraga, H.A.: Global optimization of clusters, crystals, and biomolecules. Science 285(5432), 1368–1372 (1999)
17. Wolpert, D.H., Macready, W.G.: No free lunch theorems for optimization. IEEE Trans. Evol. Comput. 1(1), 67–82 (1997)

Combining the Properties of Random Forest with Grammatical Evolution to Construct Ensemble Models

Daniel Parra[1](✉)[ID], Alberto Gutiérrez[1,2][ID], Jose-Manuel Velasco[1][ID], Oscar Garnica[1][ID], and J. Ignacio Hidalgo[1][ID]

[1] Complutense University, 28040 Madrid, Spain
{dparra02,albegu02,mvelascc,ogarnica,hidalgo}@ucm.es
[2] Biztools S.L., Madrid, Spain
alberto.gutierrez@biztools.es

Abstract. Random Forest algorithm is a prediction technique where a set of tree predictors are combined to construct an ensemble model. If a set of conditions are satisfied, we can affirm that random forest avoids overfitting and converges. On the other hand, grammatical evolution, the popular variant of genetic programming where solutions are built following a grammar, has been successfully applied to a plethora of different problems. Among them, symbolic regression is one of the hits of grammatical evolution. Although encoded in codons and decoded by a grammar, solutions in grammatical evolution are trees that represent mathematical expressions. In this paper, we investigate the convenience of combining the best of both approaches, and we propose Random Structured Grammatical Evolution as an adaptation of Random Forest to a symbolic regression problem. Using structured Grammatical Evolution, a set of weak predictors are built and combined on an ensemble model for prediction.

Keywords: Grammatical evolution · Differential evolution · Symbolic regression

1 Introduction

Several statistical techniques exist to find the relationship between an independent variable and one or more explanatory variables. If, a priori, we think there is a linear relationship, linear regression can be applied. However, as we can assume, this technique can apply only to a subset of problems. If we do not start from any prior model and want to do a broad search for possible mathematical formulations that fit our dataset, we face a symbolic regression problem.

Genetic programming (GP) [9] has historically been a bio-inspired technique especially gifted for symbolic regression, and today we have available several software packages that employ GP for symbolic regression [2,8,12]. In GP, mathematical functions are expressed as trees in which the nodes can be operators,

© Springer Nature Switzerland AG 2022
J. L. Jiménez Laredo et al. (Eds.): EvoApplications 2022, LNCS 13224, pp. 61–76, 2022.
https://doi.org/10.1007/978-3-031-02462-7_5

constants, or variables, and the search consists of recombination of these trees following an evolutionary strategy. There is a popular variant of GP where solutions are built following a grammar, Grammatical Evolution (GE) [14], which has been successfully applied to a plethora of different problems [15]. Among them, symbolic regression is also one of the main hits of this technique. In GE, again, solutions are trees that represent mathematical expressions but they are encoded in codons and decoded by a grammar.

In addition to the bio-inspired techniques, we can find in the literature other physics-inspired techniques [17,18], techniques based on machine learning [1,13] and on Bayesian methods [7]. All of these recent articles can give us an idea of the importance that researchers give to symbolic regression at this moment because of its interpretability trait versus the deep learning approach.

On the other hand, in the field of linear regression, several techniques have been used to avoid the overfitting of the model to the training data by using ensembles of predictors [20]. Instead of finding the best model that describes the data, in ensemble learning, several different models are developed whose predictions are averaged to get the final solution. Within this line of research we can find: boosting [16], bagging [3] and random forests [4].

In this paper, we investigate the convenience of combining the best of both, and we propose a kind of *Random Grammatical Evolution* as an adaptation of Random Forest to symbolic regression problems. Random Grammatical Evolution builds a set of weak predictors combined on an ensemble model for prediction. GE has several drawbacks, the most prominent are redundancy and the locality of the representation. This project uses structured-GE (SGE), a modification of GE that attempts to deal with these limitations [11].

In [6] several evolutionary and machine learning algorithms were tested for solving the glucose prediction problem for different time horizons, and GE was shown as one of the best tools for this kind of problem if faced as a symbolic regression problem. In [19], authors presented a methodology to automatize the decision of the insulin bolus presenting a new algorithm, Random-GE, which combines the properties of Grammatical Evolution and Random Forests. Finally, in [10] SGE was also applied for generating predictive models for the glucose levels of diabetic patients. SGE was able to evolve models that can predict glucose more accurately when compared with GE. The authors also claimed that models were more robust since the behavior in training and test was similar and with a small variance. Inspired by those works and trying to extrapolate those results to other kinds of problems, we start here by implementing a new version of Random GE that uses SGE.

The main contributions of this work are:

- The development of an adaptation of Structured Grammatical Evolution inspired in the Random Forest Algorithm.
- A standard SGE run has been compared with three different RSGE configurations.

- Two different instances of two types of symbolic regression problems have been used to compare the results at different levels of uncertainty in the data.

The rest of this paper is structured as follows: Sect. 2 describes how the proposed Random Grammatical Evolution proceeds. Section 2.1 shows the specific adaptation of the SGE grammars to the problem of symbolic regression. Section 3 presents the experimental setup of the paper, whereas Sect. 4 shows the results of applying Random Grammatical Evolution to two datasets. Finally, Sect. 5 closes the paper by giving some concluding remarks on the work carried out.

2   Methodology

Random Forests (RF) algorithm was proposed by Breiman [4] as a combination of previous ideas from the same author and others. He defines RF as "*a combination of tree predictors such that each tree depends on the values of a random vector sampled independently and with the same distribution for all trees in the forest.*" [4] proves that RF always converges and avoids overfitting when it satisfies the following conditions:

1. For each tree, a random vector of features is generated independent of the past random vectors but with the same distribution.
2. A large number of trees.
3. A voting system.

In [6], we observed that RF obtained good results solving a glucose prediction problem for some of the subjects of the study, although performed worst than GE (as applied in [5]) for some of the datasets. Although encoded in codons and decoded by a grammar, GE solutions are trees representing mathematical expressions. So, combining the best of both approaches, we proposed a first implementation of *Random-GE* as an adaptation of RF to the use of GE to symbolic regression with good results for this particular problem.

In this paper, we investigate the extension of this work by exploring the performance of structured GE (see Sect. 2.1 and [10]) for two kinds of symbolic regression problems.

2.1   Structured Grammatical Evolution

As one of the different variants of genetic programming, we can find Grammatical evolution (GE), having as one of its most significant characteristics how it obtains the phenotype from the genotype. In order to make this process, GE uses a set of rules closely related to the problem to be solved, in other words, a grammar. It allows to specify the interaction between the different variables (if desired) and establish restrictions (if needed). To obtain the phenotype in GE, taking the grammar provided in Fig. 1, we will take <*start*> as the starting point, then we have to choose among the possible alternatives offered by the grammar for the

Table 1. Non-terminal and genotype

Non-terminals	<start>	<expr>	<op>	<var>	<const>
Genotype	[0]	[0, 2, 3]	[1]	[2]	[0]

given non-terminal. This decision is made using the modulo operation, taking the remainder of the division between the first integer of the genotype and the number of options proposed by the grammar. In this case *<start>* only has one option to derive, *<expr>*, so it rewrites and consumes the first integer of the genotype. Then the next integer is read, and the same process is repeated for the leftmost non-terminal in the derivation. This process is repeated from left to right until the phenotype is complete.

Grammatical evolution has been spreading in a wide variety of fields, but it is not free of some drawbacks. Some of the most referenced are those related to the locality and the redundancy of its representations. A representation is said to have a high locality if a small modification in the genotype also implies a small modification in the phenotype. Due to this property, it is possible to perform a correct sampling of the search space, but with a low locality, it is possible that the search performed is closer to a random search. On the other hand, a representation is considered redundant if there are several different genotypes that produce the same phenotype. This is not necessarily a drawback, but it should be considered for some problems.

In order to deal with these limitations, Structured Grammatical Evolution was proposed by Lourenço et al. [10]. The main feature of SGE is the use of a one-to-one mapping between the genotype and the non-terminals. In the structured representation of SGE, a gene is linked directly to a non-terminal of the grammar, each gene is composed by a list of integers used to select the expansion option. The objective is to guarantee that a change in a single genotypic position will not affect the derivation options of the other non-terminals. In SGE the input grammar must be pre-processed, if the grammars allow infinite recursion in some of the rules, it restricts the maximum number of times it can be performed.

To understand better how SGE decodes, let's see an example of how the phenotype is obtained from the genotype. The grammar used is shown in Fig. 1. In this case, the set of non-terminals would be:

{*<start>*, *<expr>*, *<op>*, *<var>*, *<const>*},

so the genotype will be composed of 5 genes, one for each non-terminal (Table 1). Following the grammar, we can see the different options for each of the non-terminals. For this example we have a grammar with a total of four possible operators (*<op>*), three variables (*<var>*) and three constants (*<const>*).

The non-terminals are solved from left to right using the grammar to obtain the phenotype. The steps of this process are:

1. Take the first integer associated with the non-terminal that we are solving.
2. Check in the grammar the option that represents the integer for that non-terminal.

```
<start> ::= <expr> #(0)

<expr> ::= <expr> <op> <expr> #(0)
        |(<expr> <op> <expr>) #(1)
        |<var> #(2)
        |<const> #(3)

<op> ::= + #(0)
      |-       #(1)
      |*       #(2)
      |\ eb_div_\eb #(3)

<var> ::= x₀ #(0)
       | x₁ #(1)
       | x₂ #(2)
```

$$<var> ::= x_0 \ \#(0)$$
$$| \ x_1 \ \#(1)$$
$$| \ x_2 \ \#(2)$$

```
<const> ::= 1.0 #(0)
         | 0.1 #(1)
         | 10 #(2)
```

Fig. 1. One of the grammars used in this work.

3. Replace the non-terminal by the chosen value and consume the integer of the genotype.

The process is repeated until the complete phenotype is obtained. One-to-one correspondence between genes and non-terminals removes the need to use modulo operation in genotype-phenotype mapping, reducing redundancy levels. For this example the first element to treat is $<start>$ and in the gene associated to this non-terminal appears the integer 0. Checking in the grammar, 0 corresponds to $<expr>$, we replace $<start>$ by that value and consume the 0 of the gene. This time the element to be treated is $<expr>$, looking up the position associated to the non-terminal in the genotype, the option to be consulted in the grammar is 0, which represents the option $<expr><op><expr>$. We replace it and consume the 0, leaving us with [2,3]. Following the same steps we continue solving from left to right. We take the $<expr>$ and the integer 2, this time the selected option is $<var>$. After replacing $<expr>$ with $<var>$ and eliminating the integer used, the gene associated with the non-terminal $<expr>$ is [3]. The option selected for $<var>$ is indicated by the 2 of the gene, which according to the grammar, represents x_2, we replace it and consume the integer. The next non-terminal we must solve for is $<op>$, the value associated with 1 in this case is subtraction (-), so we substitute and consume. The process would keep repeating until the phenotype is complete, for this example we obtained: x_2 - 1.0.

Table 2 shows how the phenotype is obtained step by step.

Table 2. Obtaining the phenotype step by step.

Derivation step	Integers left
$<start>$	$[[0], [0, 2, 3], [1], [2], [0]]$
$<expr>$	$[[], [0, 2, 3], [1], [2], [0]]$
$<expr> <op> <expr>$	$[[], [2, 3], [1], [2], [0]]$
$<var> <op> <expr>$	$[[], [3], [1], [2], [0]]$
$x_2 <op> <expr>$	$[[], [3], [1], [], [0]]$
$x_2 - <expr>$	$[[], [3], [], [], [0]]$
$x_2 - <const>$	$[[], [], [], [], [0]]$
$x_2 - 1.0$	$[[], [], [], [], []]$

2.2   Random Structured Grammatical Evolution for Symbolic Regression Problems

In this work we proposed the use of structured GE for generating trees representing solutions of symbolic regression problems. The objective of the algorithms is:

Given a set of n data points, $(\vec{x}_i, f(\vec{x}_i))_{i=1...n}$, find a mathematical expression of $f(\vec{x}_i)$ that fit the best to the data set.

In Random-SGE, we generate a (high) number (r) of solutions running a SGE algorithm during a reduced number of generations (g) with a reduced number (p) of individuals. For each one of the training datasets, we obtain r Random-GE solutions by running the algorithm r times. The ensemble model will be obtained after the following steps:

1. Generate a population P of p random solutions (individuals) following the grammar.
2. Run SGE for g generations starting with population P.
3. Repeat Steps 1 and 2 r times.
4. Select the best individual of each run in order to obtain a set of r models.

$$\widehat{M_R} = \{m_1, m_2, \cdots, m_{r-1}, m_r\}. \tag{1}$$

5. Order $\widehat{M_R}$ by fitness, F_i, and assign an ordinal i to each model m_i^*, being $i = 1$ the ordinal of the best model in training m_1^*, and $i = r$ the ordinal of the model with the worst fitness in the training phase m_r^*.

$$\begin{aligned}
\{m_1^*, F_1^*\} &= \{m_i, F_i\} \setminus F_i = MIN(F_1, F_2, \cdots F_{r-1}, F_r) \\
\{m_i^*, F_i^*\} &= \{m_i, F_i\} \setminus F_i = i^{th}(F_1, F_2, \cdots F_{r-1}, F_r) \\
\{m_r^*, F_r^*\} &= \{m_i, F_i\} \setminus F_i = MAX(F_1, F_2, \cdots F_{r-1}, F_r)
\end{aligned} \tag{2}$$

6. Name the ordered set of models as

$$\widehat{M_R^*} = \{m_1^*, m_2^*, \cdots, m_{r-1}^*, m_r^*\}. \tag{3}$$

and being

$$F_1^* < F_2^* < \cdots < F_{r-1}^* < F_r^* \tag{4}$$

7. Define a Range $[L, H]$, with $L, H \in [0.1]$ and $L < H$
8. Compute two ordinals l and h:

$$l = \text{Int}(r \cdot L)$$
$$h = \text{Int}(r \cdot H)$$

9. Generate an ensemble model using $\widehat{M_s}^{\text{ensemble}}$ using the ordinals L and H and (5).

$$\widehat{M_s}^{\text{ensemble}}(\vec{x}) = \sum_{i=l}^{h} w_i \times \widehat{m_i^*}(\vec{x}) \tag{5}$$

where w_i is the weight of model i in the predicted value and:

$$\sum_{i=l}^{h} w_i = 1 \tag{6}$$

In this way, we obtain r models for the prediction, satisfying the conditions 1 and 2 regarding the RF convergence. Condition 3 is accomplished by using bagging in the selection of the ensemble model. Bagging is a method, also proposed by Breiman [3], where the ensemble model prediction is generated by averaging the prediction of a set of models. The use of bagging as a selection method is mainly a first approximation, and it would be interesting to carry out a study testing other techniques. After the selection of the s models, we construct the final expression by generating an ensemble model from this set.

3   Experimental Setup

This section explains the experimentation performed to test random structured grammatical evolution. As explained above, and as a first test of the concept, the algorithms are faced with two kinds of symbolic regression problems (see Sect. 3.1). Three different configurations of RSGE are implemented and compared with a configuration of a *long* configuration of SGE. Next, the benchmark problems and the configurations of the algorithms are detailed.

3.1   Study Problems

Two different kinds of problems were chosen for testing our proposal. On the one hand, two instances of the family of Vladislavleva functions were selected, polynomial functions commonly used to measure the performance of the algorithms on SR problems. Vladislavleva family problems were selected in order to test the performance of the algorithm when seeking for well-defined and known functions. The first problem is Vladislavleva-F5 (VF5 in the experiments), a function of three variables $\vec{x} = (x_2, x_1, x_0)$ presented in (7). Figure 2a shows a graphical representation of the function.

$$f_{v_5}(x_2, x_1, x_0) = 30 \cdot \frac{(x_2 - 1) \cdot (x_0 - 1)}{x_1^2 \cdot (x_2 - 10)} \tag{7}$$

The second function is Vladislavleva-F8 (VF8 in the experiments), a function of two variables $\vec{x} = (x_1, x_0)$ represented by (8). Figure 2b depicts the function.

$$f_{v_8}(x_1, x_0) = \frac{(x_1 - 3)^4 + (x_0 - 3)^3 - (x_0 - 3)}{(x_1 - 2)^4 + 10} \tag{8}$$

In these cases, to obtain the sets of train and test, 1000 points have been selected and distributed in a proportion of 70% for train and 30 for test.

On the other hand, the time series of the blood glucose level of a person with diabetes is used as test bench. With this time series, two prediction problems are generated, one for 30-min prediction horizon and another for 60-min, and they are solved as a regression problem using the training dataset. With glucose problems, the intention is to analyze the performance of RSGE in an environment with greater uncertainty, where we do not known the exact mathematical expression of the solution and that is more similar to real-life problem. The objective is to predict the glucose value of a person for a specific time horizon based on the current and historical values of glucose, ingested food (as the amount of carbohydrates) and administered insulin (basal and boluses). The time horizons we work with are 30 and 60 min. Figures 2c and 2d represent the time series of glucose divided into training and test cases, respectively. In this case, the data sets have been selected from the data of one subject, approximately 8.3 h for training and another 8.6 h for test, the data sets have been adapted to the 30 and 60 min horizons.

Also is interesting to compare the results of the vladislavleva functions with those obtained with glucose in order to observe how it handles problems with different uncertainty in the data.

It is important to note that we are using the very similar grammar for all the problems, i.e., this fundamental part of the algorithm is not tunned, especially for the glucose problem. Our goal is precisely to test RSGE in fairly conditions. Solutions with more complex grammar and better results can be found in the literature.

3.2   Configuration of the Algorithms

For the experimental phase, different combinations for RSGE have been tested, varying the number of runs, generations, and size of the population. In addition, a longer run of SGE is performed. For each one of the RSGE and SGE configurations, 400000 evaluations are performed. Table 3 shows the configurations of the different combinations tested. For this project, a total of 30 runs have

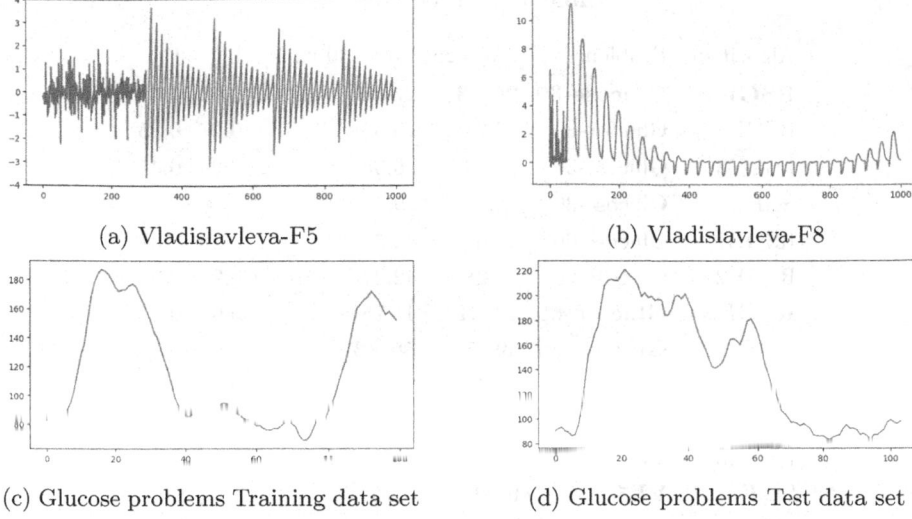

(a) Vladislavleva-F5

(b) Vladislavleva-F8

(c) Glucose problems Training data set

(d) Glucose problems Test data set

Fig. 2. Study problems.

been performed for each of the algorithms presented in this table. Depending on the type of algorithm the number of runs, the population and the number of generations have been modified.

As explained in Sect. 2.2, in RSGE, the ensemble model uses two additional parameters, L and H, that were initially fixed to 0.25 and 0.75, respectively. The objective of these parameters is to select the models that contribute to the ensemble. With the selected values, the models with higher and lower fitness during the training phase are discarded for the ensemble model, trying to avoid the contribution not only of the worst models but also overfitted models. This decision was revised, and we also analyze several combinations of L and H in Sect. 4.

Table 3. Configuration of the different SGE algorithms.

Algorithm	Runs	Generations	Population
RSGE1	40	250	40
RSGE2	20	250	80
RSGE3	32	250	50
SGE	1	1000	400

4   Results

In Table 4, we can observe the results obtained with the different algorithm configurations for the selected problems. Table shows the best fitness (root mean

Table 4. Table of results.

Algorithm	Problem	Average	Best (30 runs)	L	H
RSGE1	**Glucose-30**	**26.68**	**25.59**	0.0	0.0
RSGE2	Glucose-30	27.21	26.53	0.2	0.35
RSGE3	Glucose-30	26.70	26.33	0.2	0.45
SGE	Glucose-30	33.10	29.45	–	–
RSGE1	Glucose-60	45.05	42.75	0.25	0.35
RSGE2	Glucose-60	45.28	42.27	0.25	0.3
RSGE3	**Glucose-60**	**44.81**	41.78	0.0	0.05
SGE	**Glucose-60**	49.05	**38.28**	–	–
RSGE1	VF5	0.37	0.19	0.0	0.05
RSGE2	VF5	0.36	0.22	0.0	0.1
RSGE3	VF5	0.39	0.23	0.0	0.0
SGE	**VF5**	**0.31**	**0.17**	–	–
RSGE1	**VF8**	1.24	**0.90**	0.0	0.0
RSGE2	VF8	1.23	1.05	0.0	0.0
RSGE3	**VF8**	**1.18**	0.97	0.0	0.05
SGE	VF8	1.58	1.05	–	–

squared error, RMSE), the average fitness for 30 runs and, in the case of RSGE, the values of L and H with the best result.

Figures 3 and 4 represent an analysis of the convergence of the algorithms under two perspectives, from the view of SGE (1000 generations) Fig. 3, and from the short view of RSGE (250 generations), Fig. 4. Figures represent the average fitness of the best individuals for each generation, and the shadows represent the standard deviation. Each figure is divided into four sub-figures, one for each problem studied, and collects information from the three RSGE configurations and the SGE. As expected, the higher the number of individuals in the population, the better the fitness. However, for some of the benchmarks, the selected configuration of SGE seems to suffer from premature convergence. At first sight, it seems that the best results are obtained with SGE, but it must be taken into account that these results are for training, and that is why the model may be over-fitting. Regarding RSGE for the vladislavleva family problems, the results show a similar convergence for the different configurations in the case of VF5, as shown in Figs. 3a and 4a. In the case of VF8 , we find a more staggered convergence, as observed in 3b and 4b.

The results of the glucose problem with a time horizon of 30 min can be seen in Fig. 5 for training set and Fig. 6, for test set. Each figure is divided into four sub-figures, one for each RSGE configuration and SGE. For each RSGE case, the real value, the mean of the solutions, and a band covering the space enclosed between the best and worst solution of the chosen set are displayed.

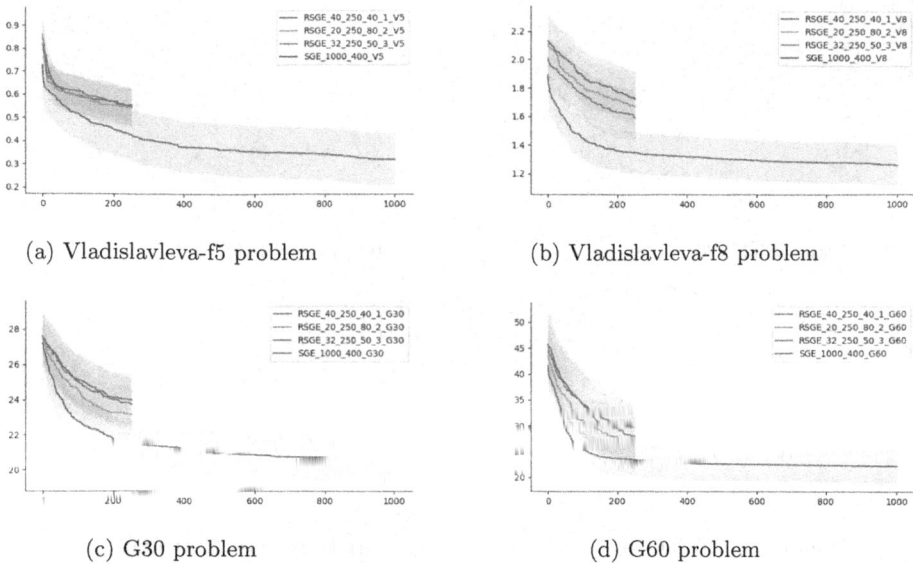

(a) Vladislavleva-f5 problem

(b) Vladislavleva-f8 problem

(c) G30 problem

(d) G60 problem

Fig. 3. Analysis of the convergence of the different algorithms for the four problems, view of 1000 generations.

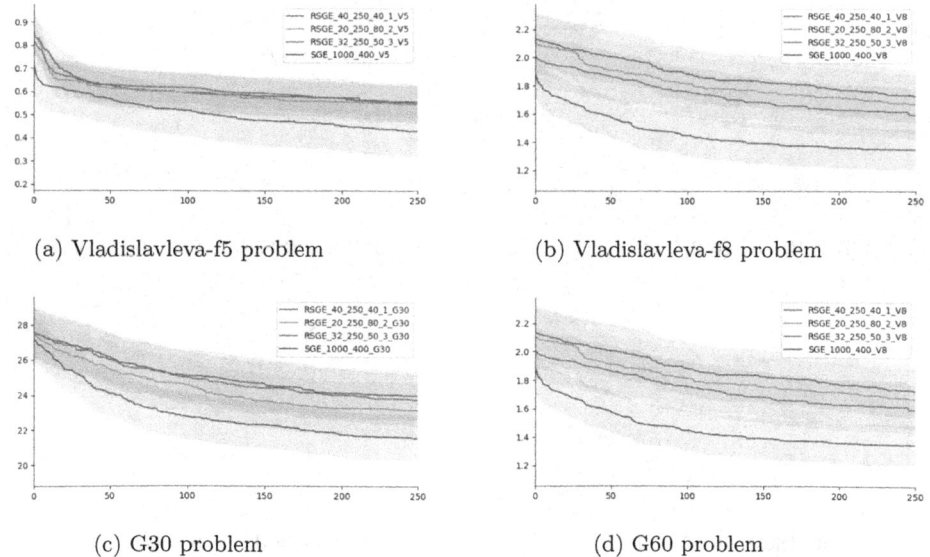

(a) Vladislavleva-f5 problem

(b) Vladislavleva-f8 problem

(c) G30 problem

(d) G60 problem

Fig. 4. Analysis of the convergence of the different algorithms for the four problems, view of 250 generations.

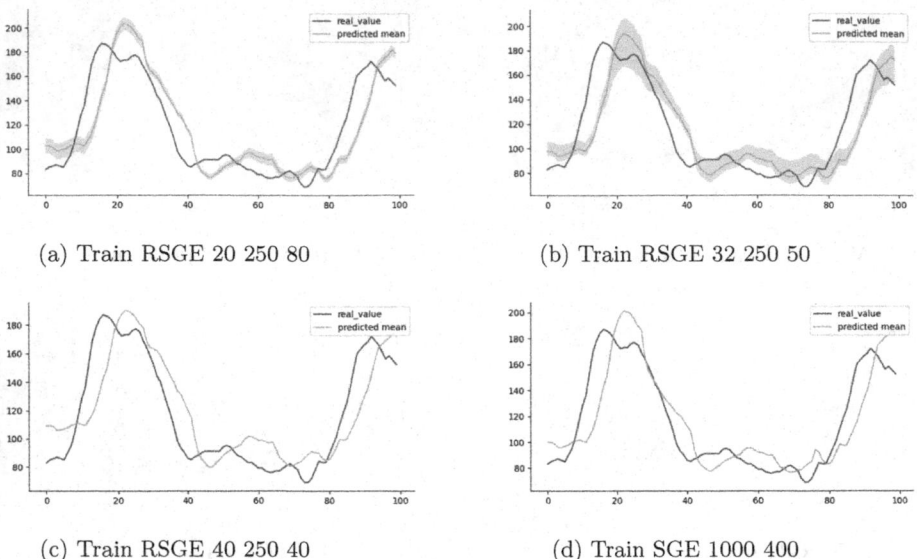

(a) Train RSGE 20 250 80

(b) Train RSGE 32 250 50

(c) Train RSGE 40 250 40

(d) Train SGE 1000 400

Fig. 5. Training of the best solution for the G30 problem obtained by the four different algorithm configurations.

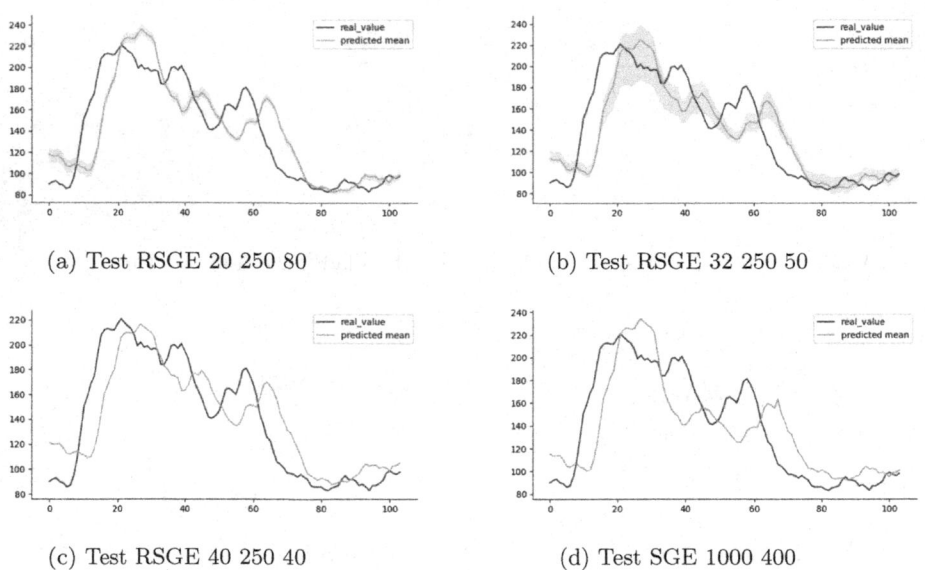

(a) Test RSGE 20 250 80

(b) Test RSGE 32 250 50

(c) Test RSGE 40 250 40

(d) Test SGE 1000 400

Fig. 6. Test of the best solution for the G30 problem obtained by the four different algorithm configurations.

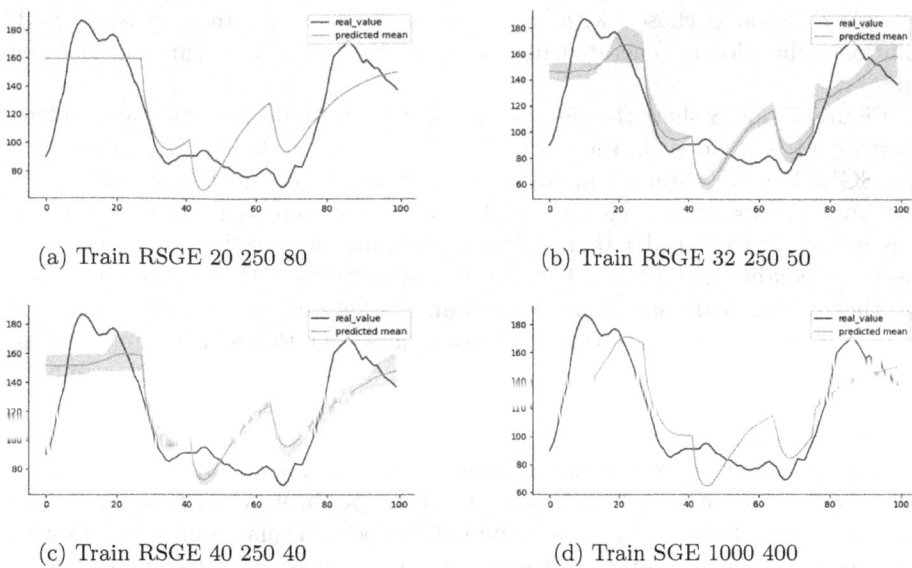

(a) Train RSGE 20 250 80

(b) Train RSGE 32 250 50

(c) Train RSGE 40 250 40

(d) Train SGE 1000 400

Fig. 7. Training of the best solution for the G60 problem obtained by the four different algorithm configurations.

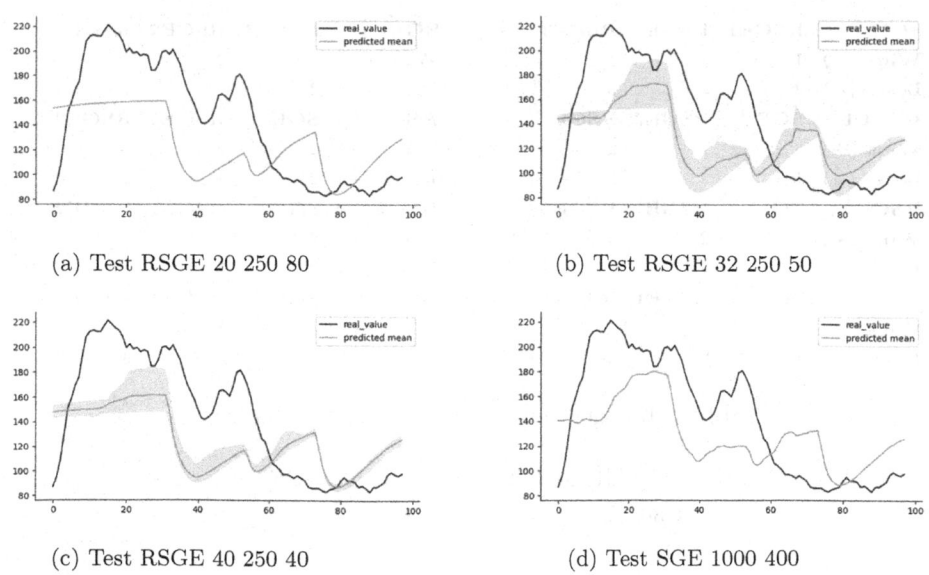

(a) Test RSGE 20 250 80

(b) Test RSGE 32 250 50

(c) Test RSGE 40 250 40

(d) Test SGE 1000 400

Fig. 8. Test of the best solution for the G60 problem obtained by the four different algorithm configurations.

In the cases where this band does not appear for the RSGE runs, as in Fig. 6c, a single solution is chosen with these values of L and H. In the case of SGE solutions, the plot is generated from the results of the best run and the real value.

Figures 7 and 8 show the results for the 60 min prediction glucose problem. Each figure is divided into four sub-figures, one for each RSGE configuration and SGE. Results obtained in training and test are worse than in the rest of the problems for RSGE. This is mainly due to the complexity of the problem. It is necessary to consider that a simple grammar was used to compare under the same conditions and in a fair way the algorithms with the four instances. In other works in the literature, high-quality solutions to this problem can be found with a higher customization of grammars, but this article aim is not to find the best solution for this problem rather to test what happens with symbolic regression problems of different complexity. Although RSGE was tested only with four different datasets, and it is difficult to test statistical significance with such amount of problems, we present in Tables 5, 6, and 7, a ranking analysis of the solutions, we can observe that RSGE3 tends to perform better than the others configurations. However, the best individuals appear in some configurations with only one run, which means that it was just a random search solution. The effect of the L and H parameters should be investigated in future works.

Table 5. Pairwise comparisons (Average of 30 runs).

SGE	RSGE1	RSGE2	RSGE3
Wins (+)	1	1	1
Loses (−)	3	3	3
RSGE1	**SGE**	**RSGE2**	**RSGE3**
Wins (+)	3	2	2
Loses (−)	1	2	2
RSGE3	**SGE**	**RSGE1**	**RSGE2**
Wins (+)	3	2	3
Loses (−)	1	2	1
RSGE2	**SGE**	**RSGE1**	**RSGE3**
Wins (+)	3	2	1
Loses (−)	1	2	3

Table 6. Pairwise comparisons (Best solution found).

SGE	RSGE1	RSGE2	RSGE3
Wins (+)	2	2	2
Loses (−)	2	2	2
RSGE1	**SGE**	**RSGE2**	**RSGE3**
Wins (+)	2	3	3
Loses (−)	2	1	1
RSGE3	**SGE**	**RSGE1**	**RSGE2**
Wins (+)	2	1	3
Loses (−)	2	3	1
RSGE2	**SGE**	**RSGE1**	**RSGE3**
Wins (+)	2	1	1
Loses (−)	2	3	3

Table 7. Ranking comparison (Average of 30 runs).

	RSGE1	RSGE2	RSGE3	SGE
G30	1	3	2	4
G60	2	3	1	4
V-F5	3	4	2	1
V-F8	3	2	1	4
Avg	2.25	3	1.5	3.25

5 Conclusions

In this paper, we have presented an adaptation of structured Grammatical Evolution inspired in the random Forest Algorithm, namely *Random structured Grammatical Evolution*, Two different instances of two kinds of symbolic regression problems were used as test cases:

- Two instances of the family of functions known as Vladislavleva functions (instances 5 and 8).
- A glucose time series prediction problems with two prediction horizons (30 and 60 min).

Three different configurations of RSGE were compared against a run of a standard SGE which can use the same computational effort, i e the same number of evaluations (100.000), Experimental results have shown that RSGE is capable of outperforming SGE for some problems; in particular, the configuration of RSGE using 40 runs, 250 generations and 40 individuals. This configuration obtains the best results for two of the four cases, in terms of the best solution and at least one configuration of RSGE outperformed SGE in three of the four problems.

From the results obtained, it seems that RSGE could performs better on those problems with more uncertainty in the data. However, this first approximation of RSGE needs further analysis and validation with a whole set of problems under different properties in terms of landscape complexity. Another viable work proposal is to use new SGE configurations for the comparison. At the time of writing this paper, we are investigating the performance of RSGE compared to GE, Genetic Programming and Random Forest for more instances of the problems discussed in this paper. We are working with all Vladislavleva functions, other symbolic regression instances of real problems and different instances of glucose data from 25 different subjects.

Acknowledgments. Work financed by the Community of Madrid and co-financed by the EU Structural Funds through the Community of Madrid projects B2017/BMD3773 (GenObIA-CM) and Y2018/NMT-4668 (Micro-Stress - MAP-CM). Also financed by the PhD project IND2020/TIC-17435 and Spanish Ministry of Economy and Competitiveness with number RTI2018-095180-B-I00.

References

1. Al-Roomi, A.R., El-Hawary, M.E.: Universal functions originator. Appl. Soft Comput. **94**, 106417 (2020)
2. Ashok, D., Scott, J., Wetzel, S.J., Panju, M., Ganesh, V.: Logic guided genetic algorithms. CoRR abs/2010.11328 (2020)
3. Breiman, L.: Bagging predictors. Mach. Learn. **24**(2), 123–140 (1996)
4. Breiman, L.: Random forests. Mach. Learn. **45**(1), 5–32 (2001)
5. Hidalgo, J.I., Maqueda, E., Risco-Martin, J.L., Cuesta-Infante, A., Colmenar, J.M., Nobel, J.: glucmodel: a monitoring and modeling system for chronic diseases applied to diabetes. J. Biomed. Inform. **48**, 183–192 (2014)

6. Hidalgo, J.I., Colmenar, J.M., Kronberger, G., Winkler, S.M., Garnica, O., Lanchares, J.: Data based prediction of blood glucose concentrations using evolutionary methods. J. Med. Syst. **41**(9), 142 (2017)
7. Jin, Y., Fu, W., Kang, J., Guo, J., Guo, J.: Bayesian symbolic regression (2020)
8. Kommenda, M., Kronberger, G., Wagner, S., Winkler, S., Affenzeller, M.: On the architecture and implementation of tree-based genetic programming in heuristiclab. In: Proceedings of the 14th Annual Conference Companion on Genetic and Evolutionary Computation, New York, NY, USA , pp. 101–108. GECCO 2012, ACM (2012)
9. Koza, J.R.: Genetic Programming: On the Programming of Computers by Means of Natural Selection. The MIT Press, Cambridge (1992)
10. Lourenço, N., Colmenar, J.M., Hidalgo, J.I., Garnica, O.: Structured grammatical evolution for glucose prediction in diabetic patients. In: Proceedings of the Genetic and Evolutionary Computation Conference, pp. 1250–1257. ACM (2019)
11. Lourenço, N., Pereira, F.B., Costa, E.: Unveiling the properties of structured grammatical evolution. Genet. Program Evol. Mach. **17**(3), 251–289 (2016). https://doi.org/10.1007/s10710-015-9262-4
12. Oliveira, L.O.V.B., Martins, J.F.B.S., Miranda, L.F., Pappa, G.L.: Analysing symbolic regression benchmarks under a meta-learning approach. In: Proceedings of the Genetic and Evolutionary Computation Conference Companion. GECCO 2018, New York, NY, USA, pp. 1342–1349. Association for Computing Machinery (2018)
13. Petersen, B.K., Larma, M.L., Mundhenk, T.N., Santiago, C.P., Kim, S.K., Kim, J.T.: Deep symbolic regression: recovering mathematical expressions from data via risk-seeking policy gradients (2021)
14. Ryan, C., Nicolau, M., O'Neill, M.: Genetic algorithms using grammatical evolution. In: Foster, J.A., Lutton, E., Miller, J., Ryan, C., Tettamanzi, A. (eds.) EuroGP 2002. LNCS, vol. 2278, pp. 278–287. Springer, Heidelberg (2002). https://doi.org/10.1007/3-540-45984-7_27
15. Ryan, Conor, O'Neill, Michael, Collins, J.J. (eds.): Handbook of Grammatical Evolution. Springer, Cham (2018). https://doi.org/10.1007/978-3-319-78717-6
16. Schapire, R.E., Freund, Y.: Boosting: Foundations and Algorithms. The MIT Press, Cambridge (2012)
17. Schmidt, M., Lipson, H.: Distilling free-form natural laws from experimental data. Science **324**(5923), 81–85 (2009)
18. Udrescu, S.M., Tegmark, M.: Ai feynman: a physics-inspired method for symbolic regression. Sci. Adv. **6**(16), 2631 (2020)
19. Velasco, J.M., Garnica, O., Lanchares, J., Botella, M., Hidalgo, J.I.: Combining data augmentation, EDAS and grammatical evolution for blood glucose forecasting. Memetic Comput. **10**(3), 267–277 (2018)
20. Zhou, Z.H.: Ensemble Learning, pp. 411–416. Springer, Boston (2015). https://doi.org/10.1007/978-1-4899-7488-4_293

EvoCC: An Open-Source Classification-Based Nature-Inspired Optimization Clustering Framework in Python

Anh T. Dang[1], Raneem Qaddoura[2($\boxtimes$)], Ala' M. Al-Zoubi[3,4], Hossam Faris[2,3,4], and Pedro A. Castillo[3]

[1] Hanoi University of Science and Technology, Hanoi 550000, Vietnam
[2] Al Hussein Technical University, Amman 11831, Jordan
{raneem.qaddoura,hossam.faris}@htu.edu.jo, inv.htaris@ugr.es
[3] Research Centre for Information and Communications Technologies of the University of Granada (CITIC-UGR), University of Granada, Granada, Spain
alzoubi@correo.ugr.es, pacv@ugr.es
[4] The University of Jordan, Amman 11942, Jordan
hossam.faris@ju.edu.jo
http://evo-ml.com/

Abstract. EvoCC framework is an open-source, free, and cross-platform framework implemented in Python which combines clustering, classification, and evolutionary computation methods. It optimizes the classification process by generating a classification model for each group generated by a clustering process where the clustering process is optimized by evolutionary optimization techniques. It includes the most well-known and recent nature-inspired metaheuristic optimization algorithms, well-known datasets, different fitness functions, and distance measures, and several well-known and highly-used classifiers. The aim is to provide the practitioners and researchers with a user-friendly and customizable implementation of classification-based nature-inspired optimization clustering algorithms that can be used by experienced and non-experienced users for the classification process in different domains. The current implementation of the framework includes eleven classification algorithms and five evaluation measures. It also utilizes the implementation of the EvoCluster framework which has ten metaheuristic optimizers, thirty datasets, five objective functions, more than twenty distance measures, and ten different ways for detecting the number of clusters (k value). The source code of EvoCC is publicly available at (https://evo-ml.com/evocc/)

Keywords: EvoCC · EvoCluster · Clustering · Classification · Evolutionary computation

© Springer Nature Switzerland AG 2022
J. L. Jiménez Laredo et al. (Eds.): EvoApplications 2022, LNCS 13224, pp. 77–92, 2022.
https://doi.org/10.1007/978-3-031-02462-7_6

1 Introduction

Data have been used and exploited in the past decade to understand and solve various problems in domains like engineering, medical agriculture security, and so on [18]. Therefore, comprehending the outcome of such data and predicting their pattern can be done magnificently by using Machine Learning Methods (MLM). The most common types of MLM are classification and clustering, each of which operates according to certain standards [10]. Classification is using an algorithm to group information into categories. This can be used for different purposes; pattern recognition for fraud detection or simple categorizing of objects [16]. Clustering, on the other hand, breaks down large amounts of data into smaller similar data groups. This can be used in many particular instances like clustering different cars into two or three categories [10]. In addition, clustering does not consider learning by knowing the actual groups of some seen data while classification considers learning by knowing beforehand the actual grouping of data to predict the grouping of other unseen data [20].

However, utilizing the classification or clustering models without tuning their parameters can lead sometimes to negative performance. The solution to such a dilemma can be achieved through evolutionary computation algorithms that select and tune these parameters [45]. The evolutionary algorithms are designed to solve complex problems quickly, simultaneously, and for a variety of tasks. Thus, we can define evolutionary computation as a technique that improves upon itself by generating new possible solutions and then either keeping or discarding them based on the determined fitness [7]. Such a technique uses heuristics to find solutions, which can sometimes provide exact results to a problem. Further, the algorithms can easily be adjusted to meet the needs of a particular problem. It is possible to change almost any aspect of an algorithm and customize it.

Recently, researchers implement various types of frameworks that obtain the MLM components to handle many problems. Many frameworks have been proposed in the literature, namely, Scikit-learn [26], OpenCV [3], Dlib [15,37], Pandas [21], XGBoost [6], TensorFlow [1], and Keras [13]. These frameworks can be adjusted and improved easily due to their structure and design. The term framework refers to an actual or conceptual structure intended to serve as a guide or support for the implementation of different elements that can be expanded depending on its structure [46]. Moreover, frameworks are often layered structures that describe what the components can or should include, and how they interact. For example, classification frameworks can obtain elements such as parameters selection, models, evaluation and splitting criteria, produced figures, results designed, and different characters of datasets [46]. Many features can be taken advantage of in the frameworks related to classification, including, performance-optimizing, concise and clear defining, user-friendly and easy to understand, prompt model development and deployment, update-able, and community-supported.

In this study, we propose an open-source framework, named EvoCC, that includes both the clustering and classification techniques while optimizing the clustering learning process through the Evolutionary Computation optimization approach. The framework uses the evolutionary clustering approach to categorize data into different groups and then generate a classification model for each group of data to classify the data into different classification categories. The process of generating a classification model for each clustering group adds an advantage over generating one classification model over the whole data by considering the similarity of data for the group through evolutionary clustering to produce an optimized classification model and thus generating a better prediction process for the whole data. The aim is to optimize the classification process to generate an accurate prediction of categories through the evolutionary clustering process.

The main contribution of this work is the proposal and development of the open-source framework EvoCC which has three important features:

- Automatic generation of groups of data points that have similar characteristics through the clustering process which is optimized by the evolutionary computation process.
- Ability to classify similar groups of data into different categories by generating a specialized model for each group.
- Providing researchers and practitioners a flexible open-source framework where they can select different evolutionary optimization algorithms, Optimization parameters, fitness functions, distance measures for clustering, classifiers, and classifiers' parameters. The framework also supports different well-known datasets and allows users to add a dataset of their choice.

The remainder of this paper is organized as follows: Sect. 2 discusses related works including the available frameworks in the fields of clustering, classification, and evolutionary computation. Section 3 represents the methodology followed for constructing and implementing the proposed framework. Section 4 discusses in detail the proposed framework including the parameters and datasets that are provided, the evolutionary clustering and classification processes, the measures which are used to evaluate the classification process, and the results management process included in the framework. Section 5 shows a sample of the results which are generated by experimenting with the framework on certain conditions. Finally, this paper is concluded in Sect. 6.

2   Related Works

The process of classifying can be described as supervised learning that learns from the data to make new observations [27]. Using classification, we can determine what particular data point belongs to. These are sometimes referred to as labels, categories, or targets. It consists of the process of estimating the output variables (Y) from input variables (X) [5].

Various frameworks have been proposed in the literature to facilitate different machine learning tasks for researchers, including, Elki [25], Scikit-learn [26] and Weka [11]. Frameworks which are oriented to the classification tasks are very common in the literature due to the popularity of the classification methods and the problems that need to be solved by such tasks. For example, the aforementioned frameworks (Elki, Scikit-learn, and Weka) can also operate classification tasks. Also, the OpenNN framework [24] applies machine learning techniques to solve a wide range of predictive analytics and data mining problems across several domains. Problems related to chemistry, energy, and engineering have been resolved by this framework. OpenNN is known for its high performance, due to being developed in C++. There are several sophisticated algorithms and utilities in the framework to accomplish forecasting, classification, regression, etc. Moreover, the Shogun framework is an open-source machine learning library that utilizes a number of data structures and machine learning algorithms [39]. In contrast to other popular machine learning frameworks, Shogun employs kernel machines to solve regression and classification problems.

As for the clustering frameworks, a number of them have been developed in recent years. For example, ClustEval [44] and clusterNOR [23] are other frameworks designed particularly for clustering. ClustEval framework contains about twenty algorithms with more than ten evaluation measures, while clusterNOR on the other hand, considers a parallel framework with nine clustering algorithms. There are also frameworks that are specific to certain areas: TimeClust [19] is a tool for clustering time series of gene expression. In [38], a framework named clusterExperiment was developed for clustering RNA-Seq data.

Most of these frameworks and libraries contain only traditional and basic classification and clustering algorithms, however, they don't take into account the nature-inspired metaheuristic algorithms. Several frameworks and libraries have been developed to support nature-inspired metaheuristic algorithms for use in a variety of applications. A recent Python framework called EvoloPy [8] implements a set of the latest and well-known optimization algorithms. Moreover, two extended versions of this framework are implemented; EvoCluster framework [34,35] works as an evolutionary clustering framework while Evolopy-FS [14] operates the feature selection method as the main criterion. Further, NiaPy [41] is also a python microframework that considers building optimization algorithms for solving different types of problems. There are several other well-known frameworks, including HeuristicLab [42], ParadisEO [4], EO [12], and DEAP [9]. Whereas, certain frameworks specialize in particular domains: GEATbx [28] is a MATLAB framework dedicated to Genetic Algorithms and Genetic Programming. Another framework (GAlib [43] that supports parallel environments for the genetic algorithm is implemented using a C++ library. Such frameworks utilize general optimization methods without considering classification or clustering techniques.

Many frameworks combine these methods, for instance, Spark MLlib that developed by Apache [22]. The framework can apply different tasks such as feature extraction, regression, Optimization, dimensional reduction, classification, and clustering. Another well-known framework that used these techniques together is Scikit-learn [17,26]. This Python library can handle various tasks, namely, classification, clustering, feature and model selection, regression, and preprocessing. Moreover, the aforementioned EvoCluster framework uses several evolutionary optimization techniques to optimize the clustering process.

None of the aforementioned studies considered combining clustering, classification, and evolutionary computation methods in one framework. In this study, we propose such a framework, which is named EvoCC. It optimizes the classification process by generating a classification model for each group generated by a clustering process where the clustering process is optimized by evolutionary optimization techniques.

3   Methodology

The methodology of the EvoCC framework can be represented in Fig. 1. As shown in the figure, the dataset is divided into training and testing portions. The labels of the training part are excluded to perform the clustering task. The training part is clustered using the EvoCluster framework, which performs an optimization of the clusters' centroids to generate the intended clusters.

The instances of each cluster are then combined with their actual labels to perform the classification task. Each cluster is classified using a classification algorithm to generate a specific model for the cluster.

The other part of the methodology considers the testing part of the dataset. The framework performs the labeling operation based on the generated clusters. The labeling operation is performed for each instance in the testing part by measuring the distance between each instance and each centroid, then selecting the model that corresponds to the centroid which has the minimum distance with the instance.

The selected model for an instance is used to apply the classification task for the instance to predict a label that corresponds to that instance. The predicted labels and the actual labels are then evaluated using different evaluation measures to assess the performance of the experiment.

4   Framework Overview

This section describes in detail the components of the proposed framework including the parameters that the user should select, the datasets that are available in the framework, the clustering and classification part of the framework, the evaluation measures that are used to assess the performance of the classification, and the type of files and figures that the framework provides to show the results.

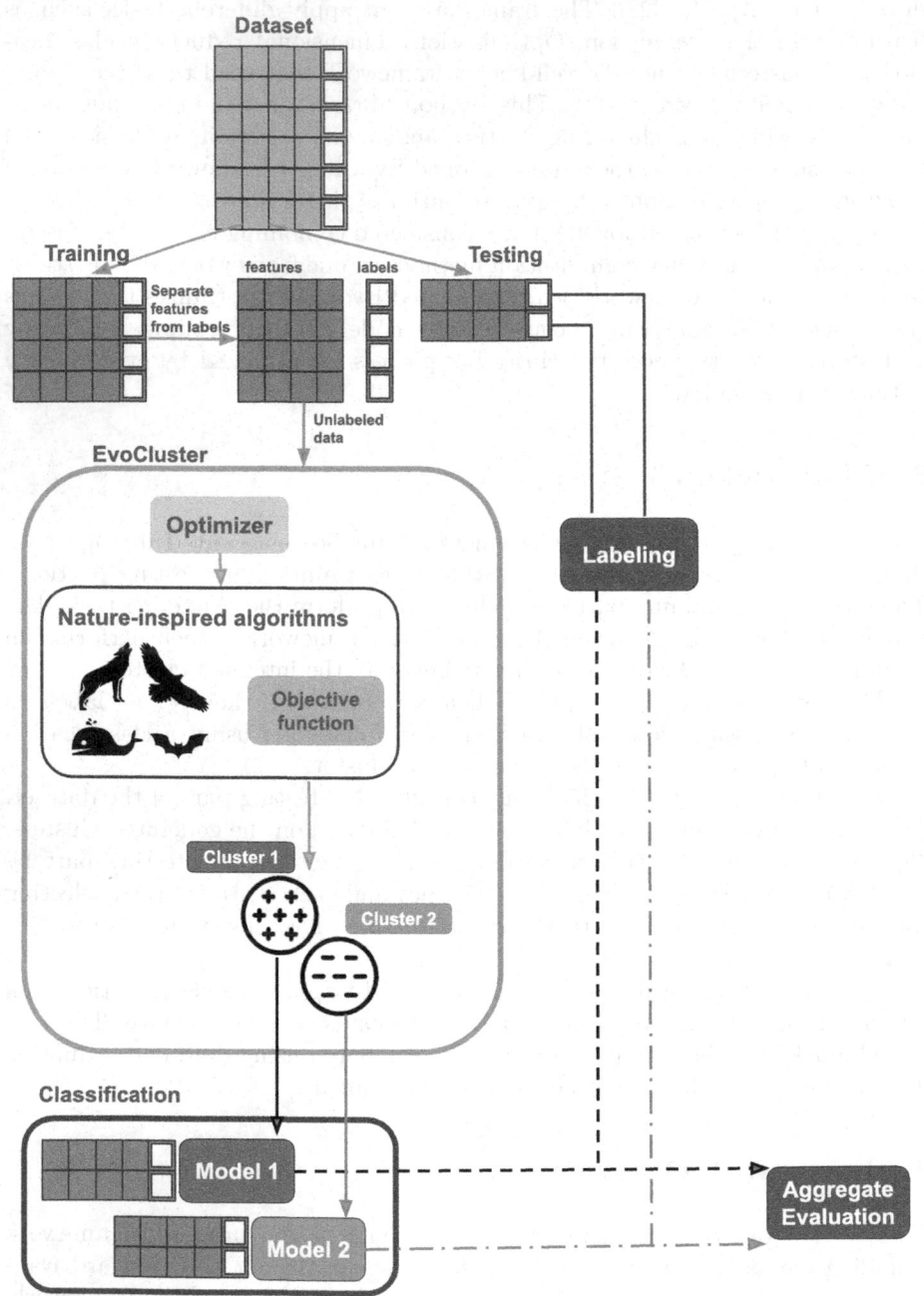

Fig. 1. EvoCC methodology

4.1 Parameters

The parameters are divided into two types: the ones required by EvoCluster and others that are specific to the proposed framework. The following parameters are requested by the EvoCluster framework and need to be selected by the user:

- Optimizers: the user selects one of the available optimizers which include Salp Swarm Algorithm (*SSA*), Particle Swarm Optimization (*PSO*), Genetic Algorithm (*GA*), Bat Algorithm (*BAT*), Firefly Algorithm (*FFA*), Gray Wolf Optimizer (*GWO*), Whale Optimization Algorithm (*WOA*), Multi-Verse Optimizer (*MVO*), Moth Flame Optimizer (*MFO*), and Cuckoo Search Algorithm (*CS*).
- Objective functions: Sum of squared error (*SSE*), Total Within Cluster Variance (*TWCV*), Silhouette Coefficient (*SC*), Davies-Bouldin (*DB*), and Dunn Index (*DI*) can be selected to measure the convergence of the solution constructed by the optimizer.
- Datasets: the user can select different datasets from the ones available in the framework, which are further discussed in Sect. 4.2.
- Optimizers' parameters: the population size and the number of iterations are two important parameters that control the advancement of the evolutionary computations.
- Auto cluster: The auto cluster parameter accepts a Boolean value of whether a specific number of clusters for each dataset is specified explicitly by the user or it can be auto-detected by the framework.
- Number of clusters: This parameter is dependent on the previous one; if the user selects a *True* value for auto cluster parameter, one of the following strings can be considered as a value for this parameter: *elbow, gap* analysis, *SC*, Calinski-Harabasz (*CH*), *DB*, Bayesian Information Criterion score (*BIC*), *min, max, median,* and *majority,* as specified by the EvoCluster framework. In contrast, if the user selects a *False* value, the user much specify the list of the number of clusters for this parameter. The *supervised* value that are allowed for EvoCluster is not applicable for the EvoCC framework, since the classification is the task specified by the Framework.
- metric: the metric parameter indicated the type of distance measure used for the clustering process if applicable for the selected objective function which includes any option allowed by *scipy.spatial.distance* package [40]

The other parameters that are specific to the proposed framework are as follows:

- Number of runs: The parameter indicates the number of independent runs that the user selects. When the user is experimenting with the framework or applying it for research, choosing more than one run is usually the case.
- Classifiers: the list of classifiers are selected as values to this parameter to be used for the classification process, which is further discussed in Sect. 4.4.

– Classifiers' parameters: For each selected classifier of the previous parameter, a dictionary of parameters can be selected by the user according to the documentation provided by scikit learn[1] for each classifier.

4.2 Datasets

The datasets that are available at the time of writing this paper are gathered from scikit learn (See footnote 1), UCI machine learning repository[2], School of Computing at University of Eastern Finland[3], ELKI[4], KEEL[5], and Naftali Harris Blog[6]. Table 1 shows the datasets that are available which have a different number of instances, features, and labels. Other datasets can be added to the framework in the future and they can also be added by the user when experimenting with the framework.

4.3 Clustering with EvoCluster

The clustering task is performed using the EvoCluster framework, which utilizes nature-inspired algorithms to optimize the clustering task according to the selected objective function. An individual represents a solution containing the position of the centroids for the clusters. A population of individuals (s) is generated for each iteration, which consists of the features (f) of each centroid for (k) clusters. This is observed in Fig. 2.

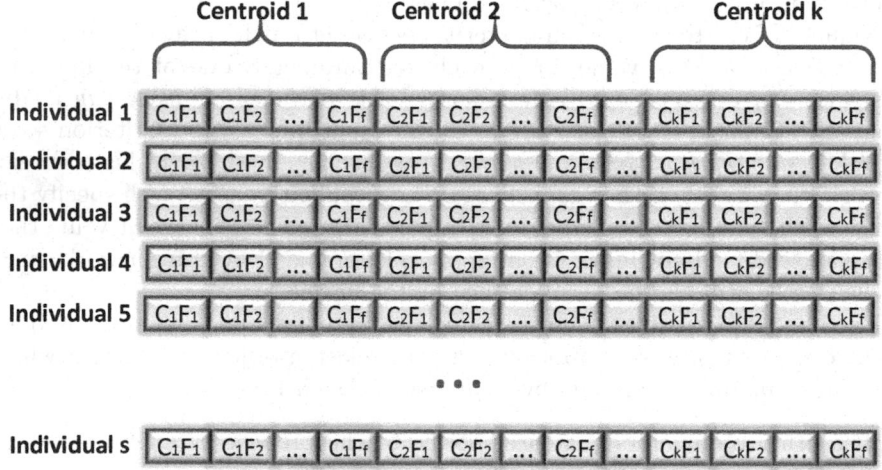

Fig. 2. The EvoCluster Population of individuals at a specific iteration [34, 35]

---

[1] http://scikit-learn.org/stable/datasets/index.html.
[2] https://archive.ics.uci.edu/ml/.
[3] http://cs.uef.fi/sipu/datasets/.
[4] https://elki-project.github.io/datasets/.
[5] https://sci2s.ugr.es/keel/datasets.php.
[6] https://www.naftaliharris.com/blog/visualizing-K-means-clustering/.

4.4   Classification

The classification process is performed by generating a model for each cluster that is specified by the clustering process. The model is generated by applying the selected classifier to the training data. Then, each instance in the testing data is attached to a specific cluster that has the closest distance between the instance and the cluster's centroid. According to the model generated for the cluster, the instances that are attached to the cluster are classified by the corresponding model and evaluated by the evaluation measures.

The list of classifiers that are available at the time of writing this paper is as follows:

Table 1. Data sets properties which show the name, number of labels, number of instances, number of features, data set type, and source

ID	Data set	k	#Points	#Features	Type	Source
1	Aggregation	7	788	2	Artificial	University of Eastern Finland (See footnote 3)
2	Aniso	3	1500	2	Artificial	scikit learn (See footnote 1)
3	Appendicitis	2	106	7	Real	KEEL (See footnote 5)
4	Balance	3	625	4	Real	UCI (See footnote 2)
5	Backnote	2	1372	4	Real	UCI (See footnote 2)
6	Blobs	3	1500	2	Artificial	scikit learn (See footnote 1)
7	Blood	2	748	4	Real	UCI (See footnote 2)
8	Circles	2	1500	2	Artificial	scikit learn (See footnote 1)
9	Diagnosis II	2	120	6	Real	UCI (See footnote 2)
10	Ecoli	5	327	7	Real	UCI (See footnote 2)
11	Flame	2	240	2	Artificial	University of Eastern Finland (See footnote 3)
12	Glass	6	214	9	Real	UCI (See footnote 2)
13	Heart	2	270	13	Real	UCI (See footnote 2)
14	Iris	3	150	4	Real	UCI (See footnote 2)
15	Iris 2D	3	150	2	Real	UCI (See footnote 2)
16	Ionosphere	2	351	344	Real	UCI (See footnote 2)
17	Jain	2	373	2	Artificial	University of Eastern Finland (See footnote 3)
18	Liver	2	345	7	Real	UCI (See footnote 2)
19	Moons	2	1500	2	Artificial	scikit learn (See footnote 1)
20	Mouse	3	490	2	Artificial	ELKI (See footnote 4)
21	Pathbased	3	300	2	Artificial	University of Eastern Finland (See footnote 3)
22	Seeds	3	210	7	Real	UCI (See footnote 2)
23	Smiley	4	500	2	Artificial	naftaliharris (See footnote 6)
24	Sonar	2	208	60	Real	UCI (See footnote 2)
25	Varied	3	1500	2	Artificial	scikit learn (See footnote 1)
26	Vary Density	3	150	2	Artificial	ELKI (See footnote 4)
27	Vertebral2	2	310	6	Real	UCI (See footnote 2)
28	Vertebral3	3	310	6	Real	UCI (See footnote 2)
29	WDBC	2	569	30	Real	UCI (See footnote 2)
30	Wine	3	178	13	Real	UCI (See footnote 2)

- Support Vector Machine (SVM): It classifies the instances of the dataset by mapping the data into a higher dimension. It fits the boundary of the training instances using the support vectors by generating non-linear kernels [2].
- Linear SVM: It is an SVM that generates a linear classifier.
- SGD Classifier: It generated a regularized linear model with stochastic gradient descent (SGD) method where the gradient of the loss works for each instance to update the model with decreasing strength schedule.
- k-Nearest Neighbor (KNN): The KNN finds the predicted label of an instance I according to the nearest k instances ($k - NN(I) = \{nn_1, nn_2...nn_k\}$) by calculating the distance between the instance I and each other instance in $k - NN(I)$ [32,36].
- Naive Bayes (NB): It is a probabilistic method based on Bayes' Theorem where features are independent of each other.
- Decision Tree (DT): The model is a tree where the internal nodes at each level represent the possible values of a feature in the dataset and the leaf nodes represent the labels. The model is used to find a predicted label of an unseen instance, by following the route aligning the values of the instance with the values of the route at each level, reaching the label at the leaf note which is considered as the predicted label.
- Multi-Layer Perceptron (MLP): It is a network that has multiple hidden layers by adjusting the weights of the connected hidden, input, and output layers where the weights help in calculating the values for the nodes at each layer [30].
- Adaboost: It is an ensemble method taking a vote of the predicted labels of multiple classifiers by applying a weight to each training instance at each iteration and retraining the data on the weighted instances.

4.5   Evaluation Measures

The Accuracy, Precision, Recall, F1-score, and G-mean and measures are considered for evaluating the classification task of the framework which are represented by Eqs. 1–5: [29]

$$Accuracy = \frac{TP_{class1} + TP_{class2} + ... + TP_{classL}}{N} \qquad (1)$$

$$Precision_{class} = \frac{TP_{class}}{TP_{class} + FP_{class}} \qquad (2)$$

$$Recall_{class} = \frac{TP_{class}}{TP_{class} + FN_{class}} \qquad (3)$$

$$f1\text{-}score_{class} = \frac{2 \times Precision_{class} \times Recall_{class}}{Precision_{class} + Recall_{class}} \qquad (4)$$

$$G\text{-}mean = \sqrt[L]{Recall_{class1} \times Recall_{class2} \times ... \times Recall_{classL}} \qquad (5)$$

where TP_{class}, TN_{class}, FP_{class}, and FN_{class} are the true positive, the true negative, the false positive, and the false negative of a class while L is the number of classes and N is the number of instances.

4.6 Results Management

There are four different forms of results that are generated upon completion of running the framework. Two CSV files for the numeric results and two different types of plots as graphical representations. These files are listed as follows:

- Average results file: the file consists of the different measures considered in the framework including the average values for all the runs for the Accuracy, G-mean, Precision, Recall, and F1-measure in addition to the True Positive (TP), True Negative (TN), False Positive (FP), and False Negative (FN). It also contains the average execution time and the value of the objective function for each iteration. The values are presented for each combination of optimizer, objective function, classifier, and data set.
- Detailed results file: the file consists of the aforementioned values for each run.
- Convergence curve plot: The average values of the objective function for different iterations for each combination of classifier, dataset, and objective function are presented by a convergence plot. The convergence plot is a multiple line graph representing the values as a line for each optimizer. The convergence curve shows the tendency of the optimizer in finding better solutions throughout iterations [33].
- Box plot: The box plot shows the different values for each evaluation measure for several runs of the framework. The values are presented for each combination of the optimizer, dataset, evaluation measure, and objective function. Each box represents the interquartile range, best value, and worst value for a classifier as the box, the upper whiskers, and the lower whiskers, respectively [31].

5 Experiments and Visualizations

This section shows an example of the figures and results file that the framework provides. The examples present running the framework 10 times using a population size of 50 and iterations of 100. It also presents the clustering by CPSO, CGA, CGWO, and CMVO clustering Evolutionary Algorithms with an SSE objective function and a classification of SVM, MLP, NB, and DT with the default parameters for different datasets including Aggregation, Flame, Iris, and Seeds. These settings are summarized in Table 2. In addition, the results that are generated from running the experiments can be observed in Figs. 3, 4, and 5 showing the box plots, average results file, and detailed results file, respectively. Since selected parameters for the classifiers are not specified, the box plot charts are having some large standard deviation, and are represented here to show the capabilities of the framework only.

Table 2. The settings applied for an example results

Settings	Values
Number of Runs	10
Population Size	50
Iterations	100
Evolutionary Algorithm	CPSO, CGA, CGWO, and CMVO
Objective Function	SSE
Classifiers	SVM, NB, DT, and MLP
Datasets	Aggregation, Flame, Iris, and Seeds
Distance Measure	Euclidean

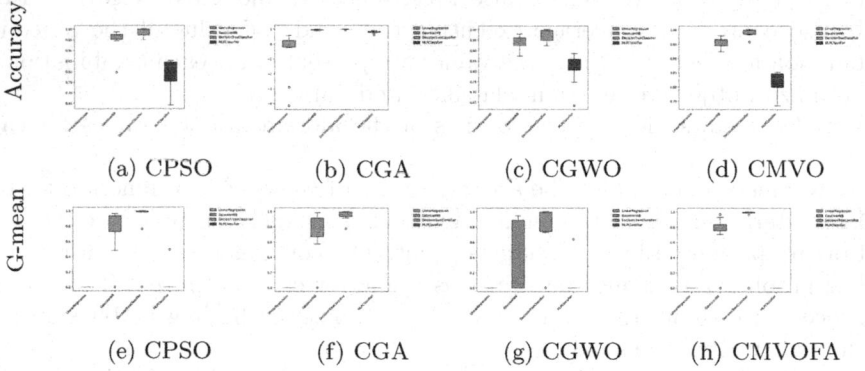

(a) CPSO   (b) CGA   (c) CGWO   (d) CMVO

(e) CPSO   (f) CGA   (g) CGWO   (h) CMVOFA

Fig. 3. Box plot of Accuracy and G-mean with CPSO, CGA, CGWO for SVM, MLP, NB, and DT using SSE objective function for the Aggregation data set

Dataset	classifier	classifier_	Optimizer	objfname	k	Execution	Accuracy	g-mean	f1_score	precision	recall	iters
aggregation	LinearRegression	{}	PSO	SSE	7	0.008203	nan	0	[0.088235	[0.088235	[0.088235	[14.757, 13.513, 13.3229999999
aggregation	GaussianNB	{}	PSO	SSE	7	0.012198	0.950211	0.755833	[0.959585	[0.948325	[0.981970	[14.757, 13.513, 13.3229999999
aggregation	DecisionTreeClassifier	{}	PSO	SSE	7	0.007062	0.983966	0.872315	[0.989935	[0.992698	[0.987524	[14.757, 13.513, 13.3229999999
aggregation	MLPClassifier	{}	PSO	SSE	7	0.489279	0.794093	0.049184	[0.867881	[0.834399	[0.920956	[14.757, 13.513, 13.3229999999
aggregation	LinearRegression	{}	GA	SSE	7	0.006093	-0.61193	0	[0.811651	[0.857128	[0.782008	[14.757, 14.323000000000002,
aggregation	GaussianNB	{}	GA	SSE	7	0.010688	0.939662	0.713495	[0.964645	[0.968253	[0.964731	[14.757, 14.323000000000002,
aggregation	DecisionTreeClassifier	{}	GA	SSE	7	0.005544	0.986498	0.953418	[0.992240	[0.994642	[0.990039	[14.757, 14.323000000000002,
aggregation	MLPClassifier	{}	GA	SSE	7	0.465114	0.801688	0	[0.881188	[0.836186	[0.937964	[14.757, 14.323000000000002,
aggregation	LinearRegression	{}	GWO	SSE	7	0.005666	nan	0	[0.807776	[0.847339	[0.786181	[15.809000000000001, 14.8939
aggregation	GaussianNB	{}	GWO	SSE	7	0.010979	0.933333	0.50044	[0.967450	[0.978521	[0.960167	[15.809000000000001, 14.8939
aggregation	DecisionTreeClassifier	{}	GWO	SSE	7	0.005907	0.972152	0.745146	[0.988402	[0.991612	[0.985540	[15.809000000000001, 14.8939
aggregation	MLPClassifier	{}	GWO	SSE	7	0.492049	0.81097	0.063825	[0.907243	[0.875398	[0.945292	[15.809000000000001, 14.8939
aggregation	LinearRegression	{}	MVO	SSE	7	0.006292	nan	0	[0.805365	[0.829030	[0.796185	[14.516, 13.809000000000001,
aggregation	GaussianNB	{}	MVO	SSE	7	0.011144	0.952743	0.734591	[0.980448	[0.991428	[0.971795	[14.516, 13.809000000000001,
aggregation	DecisionTreeClassifier	{}	MVO	SSE	7	0.005809	0.989451	0.892904	[0.991117	[0.994642	[0.987865	[14.516, 13.809000000000001,
aggregation	MLPClassifier	{}	MVO	SSE	7	0.522283	0.813502	0	[0.916236	[0.895355	[0.942319	[14.516, 13.809000000000001,
flame	LinearRegression	{}	PSO	SSE	2	0.002728	0.170104	0.174802	[0.790866	[0.655443	[1.0, 0.054	[7.374, 6.534000000000001, 6.:
flame	GaussianNB	{}	PSO	SSE	2	0.003391	0.943056	0.916489	[0.957687	[0.921395	[0.998039	[7.374, 6.534000000000001, 6.:
flame	DecisionTreeClassifier	{}	PSO	SSE	2	0.002631	0.969444	0.962295	[0.976434	[0.973837	[0.980008	[7.374, 6.534000000000001, 6.:
flame	MLPClassifier	{}	PSO	SSE	2	0.171026	0.741667	0.578064	[0.804321	[0.812749	[0.839004	[7.374, 6.534000000000001, 6.:
flame	LinearRegression	{}	GA	SSE	2	0.003126	0.117634	0.200048	[0.792193	[0.657167	[1.0, 0.061	[7.129, 6.992999999999999, 6.:
flame	GaussianNB	{}	GA	SSE	2	0.00315	0.938889	0.908613	[0.955148	[0.916860	[0.998039	[7.129, 6.992999999999999, 6.:
flame	DecisionTreeClassifier	{}	GA	SSE	2	0.002557	0.9625	0.955384	[0.970671	[0.966718	[0.975608	[7.129, 6.992999999999999, 6.:
flame	MLPClassifier	{}	GA	SSE	2	0.220471	0.748611	0.603426	[0.810333	[0.813526	[0.837851	[7.129, 6.992999999999999, 6.:
flame	LinearRegression	{}	GWO	SSE	2	0.002436	0.193825	0.175311	[0.790845	[0.655329	[1.0, 0.053	[7.769000000000001, 6.983, 6.:
flame	GaussianNB	{}	GWO	SSE	2	0.003015	0.938889	0.909959	[0.954771	[0.916157	[0.998039	[7.769000000000001, 6.983, 6.:
flame	DecisionTreeClassifier	{}	GWO	SSE	2	0.002435	0.970833	0.966327	[0.977273	[0.977268	[0.978047	[7.769000000000001, 6.983, 6.:
flame	MLPClassifier	{}	GWO	SSE	2	0.296702	0.786111	0.595143	[0.852061	[0.815508	[0.907667	[7.769000000000001, 6.983, 6.:

Fig. 4. The average results file for 10 runs of the setting presented in Table 2

Dataset	classifier	classifier_	Optimizer	objfname	k	Execution	confusion	Accuracy	g-mean	f1_score	precision	recall	iters
aggregatic	DecisionTr{}		GA	SSE	7	0.004349	[[[202, 1],	0.987342	0.972705	0.985507	0.971428	[1.0, 1.0, 1	['13.25', '13.25', '13.25', '13.25', '13.25', '13.0'
aggregatic	DecisionTr{}		GA	SSE	7	0.006072	[[[205, 0],	0.995781	0.995475	0.984126	[1.0, 1.0, 1	0.96875,	['15.59', '13.39', '13.39', '13.39', '13.39', '13.1'
aggregatic	DecisionTr{}		GA	SSE	7	0.007091	[[[194, 0],	0.987342	0.965128	[1.0, 1.0, 1	[1.0, 1.0, 1	[1.0, 1.0, 1	['14.34', '14.34', '14.34', '13.64', '13.64', '13.6'
aggregatic	DecisionTr{}		GA	SSE	7	0.005713	[[[197, 1],	0.991561	0.982278	0.987341	0.975, 1.0	[1.0, 1.0, 1	['14.21', '14.21', '14.21', '14.21', '12.37', '12.3'
aggregatic	DecisionTr{}		GA	SSE	7	0.004566	[[[192, 0],		1	1	[1.0, 1.0, 1	[1.0, 1.0, 1	['16.99', '15.11', '15.11', '12.84', '12.84', '12.8'
aggregatic	DecisionTr{}		GA	SSE	7	0.003643	[[[192, 0],	0.945148	0.776164	0.988764	[1.0, 1.0, 1	0.977777	['15.59', '15.59', '13.78', '13.78', '13.26', '12.5'
aggregatic	DecisionTr{}		GA	SSE	7	0.006905	[[[205, 0],	0.983122	0.919459	[1.0, 1.0, 1	[1.0, 1.0, 1	[1.0, 1.0, 1	['14.89', '14.89', '14.89', '14.89', '14.89', '14.3'
aggregatic	DecisionTr{}		GA	SSE	7	0.00481	[[[206, 0],	0.983122	0.929624	[1.0, 1.0, 1	[1.0, 1.0, 1	[1.0, 1.0, 1	['15.38', '15.12', '15.12', '13.7', '13.64', '13.64'
aggregatic	DecisionTr{}		GA	SSE	7	0.005015	[[[196, 0],	0.995781	0.996479	0.987654	[1.0, 1.0, 1	0.975609	['12.82', '12.82', '12.82', '12.82', '12.82', '12.8'
aggregatic	DecisionTr{}		GA	SSE	7	0.00728	[[[191, 0],	0.995781	0.996865	0.989010	[1.0, 1.0, 1	0.978260	['14.51', '14.51', '12.93', '12.93', '12.93', '11.7'
flame	DecisionTr{}		GA	SSE	2	0.002875	[[[23, 2], [	0.972222	0.959166	0.979166	0.959183	[1.0, 0.92]	['7.93', '7.57', '7.57', '7.57', '7.33', '7.33'
flame	DecisionTr{}		GA	SSE	2	0.002788	[[[24, 1], [	0.944444	0.94801	0.956521	0.977777	0.936170	['7.31', '7.31', '7.31', '6.08', '6.08', '6.08'
flame	DecisionTr{}		GA	SSE	2	0.002193	[[[24, 2], [	0.972222	0.960769	0.978723	0.958333	[1.0, 0.923	['6.84', '6.84', '6.84', '6.65', '6.65', '6.65'
flame	DecisionTr{}		GA	SSE	2	0.002929	[[[22, 2], [	0.972222	0.957427	0.979591	[0.96, 1.0]	[1.0, 0.916	['7.03', '7.03', '7.03', '6.85', '6.85', '6.85'
flame	DecisionTr{}		GA	SSE	2	0.00293	[[[31, 0], [	0.958333	0.96272	0.962025	[1.0, 0.911	0.926829	['8.32', '7.32', '6.91', '6.91', '6.91', '6.63'
flame	DecisionTr{}		GA	SSE	2	0.002216	[[[27, 2], [	0.958333	0.953615	0.965517	0.954545	0.976744	['7.08', '7.08', '7.08', '7.08', '6.46', '6.46'
flame	DecisionTr{}		GA	SSE	2	0.0025	[[[26, 1], [	0.972222	0.970342	0.977777	0.977777	0.977777	['5.88', '5.88', '5.88', '5.88', '5.88', '5.88'
flame	DecisionTr{}		GA	SSE	2	0.002374	[[[21, 0], [	0.972222	0.980196	0.98, 0.95	[1.0, 0.913	0.960784	['6.52', '6.52', '6.52', '6.33', '6.33', '6.2', '
flame	DecisionTr{}		GA	SSE	2	0.002175	[[[19, 3], [	0.958333	0.92932	0.970873	0.943396	[1.0, 0.863	['6.68', '6.68', '6.68', '6.49', '6.49', '6.49'
flame	DecisionTr{}		GA	SSE	2	0.002587	[[[24, 3], [	0.944444	0.932275	0.956521	0.936170	0.977777	['7.7', '7.7', '7.0', '7.0', '7.0', '6.56', '6.48', '6.48
iris	DecisionTr{}		GA	SSE	3	0.00405	[[[28, 0], [	0.822222	0.789609	[1.0, 0.666	[1.0, 0.727	[1.0, 0.615	['10.64', '9.26', '9.05', '8.97', '8.14', '8.14', '8.0
iris	DecisionTr{}		GA	SSE	3	0.003279	[[[28, 0], [	0.866667	0.881177	[1.0, 0.745	[1.0, 0.6, 1	[1.0, 1.0, 0	['18.38', '18.38', '18.08', '9.39', '5.75', '5.75', '
iris	DecisionTr{}		GA	SSE	3	0.003848	[[[25, 0], [	0.977778	0.973672	[1.0, 0.96	[1.0, 0.923	[1.0, 1.0, 0	['23.25', '16.16', '15.9', '15.9', '8.18', '8.18', '8.
iris	DecisionTr{}		GA	SSE	3	0.003399	[[[33, 0], [	0.888889	0.894998	[1.0, 0.838	[1.0, 0.928	[1.0, 0.764	

Fig. 5. The detailed results file for 10 runs of the setting presented in Table 2

6　Conclusion and Future Works

This paper proposed the EvoCC framework as an open-source, free, and cross-platform classification-based nature-inspired optimization clustering algorithms framework. It optimizes the classification process by generating a classification model for each group generated by a clustering process where the clustering process is optimized by evolutionary optimization techniques. The current implementation of the framework considers eleven classification algorithms, five evaluation measures, and different options in which the EvoCluster framework provides. The framework is designed in a user-friendly implementation so that practitioners and researchers can customize it by implementing classifiers and evaluation measures. This paper represents several automatically-generated visual figures in the form of box plots and convergence curve plots. It also represents the average and detailed results files which show the experimental results of a selected sample applied using specific sample parameters which are required by the framework. For future work, we plan to include more classification algorithms and evaluation measures in the framework. We are also looking into experimenting with the framework with other types of datasets that are used in certain applications.

Acknowledgment. This work is supported by the Ministerio español de Economía y Competitividad under project PID2020-115570GB-C22 (DemocratAI::UGR).

References

1. Abadi, M.: Tensorflow: learning functions at scale. In: Proceedings of the 21st ACM SIGPLAN International Conference on Functional Programming, p. 1 (2016)

2. Aljarah, I., et al.: Intelligent detection of hate speech in Arabic social network: a machine learning approach. J. Inf. Sci., 0165551520917651 (2020)
3. Bradski, G.: The opencv library. Dr. Dobb's J. Softw. Tools Prof. Programm. **25**(11), 120–123 (2000)
4. Cahon, S., Melab, N., Talbi, E.-G.: Paradiseo: a framework for the reusable design of parallel and distributed metaheuristics. J. Heuristics **10**(3), 357–380 (2004)
5. Cervantes, J., Garcia-Lamont, F., Rodríguez-Mazahua, L., Lopez, A.: A comprehensive survey on support vector machine classification: applications, challenges and trends. Neurocomputing **408**, 189–215 (2020)
6. Chen, T., He, T., Benesty, M., Khotilovich, V., Tang, Y., Cho, H., et al.: Xgboost: extreme gradient boosting. R package version 0.4-2 **1**(4), 1–4 (2015)
7. Črepinšek, M., Liu, S.-H., Mernik, M.: Exploration and exploitation in evolutionary algorithms: a survey. ACM Comput. Surv. (CSUR) **45**(3), 1–33 (2013)
8. Faris, H., Aljarah, I., Mirjalili, S., Castillo, P.A., Guervós, J.J.M.: Evolopy: an open-source nature-inspired optimization framework in python. In: IJCCI (ECTA), pp. 171–177 (2016)
9. Fortin, F.-A., De Rainville, F.-M., Gardner Gardner, M.-A., Parizeau, M., Gagné, C.: Deap: evolutionary algorithms made easy. J. Mach. Learn. Res. **13**(1), 2171–2175 (2012)
10. Goodfellow, I., Bengio, Y., Courville, A.: Machine learning basics. Deep Learn. **1**(7), 98–164 (2016)
11. Hall, M., Frank, E., Holmes, G., Pfahringer, B., Reutemann, P., Witten, I.H.: The weka data mining software: an update. ACM SIGKDD Expl. Newsletter, **11**(1), 10–18 (2009)
12. Keijzer, M., Merelo, J.J., Romero, G., Schoenauer, M.: Evolving objects: a general purpose evolutionary computation library. In: Collet, P., Fonlupt, C., Hao, J.-K., Lutton, E., Schoenauer, M. (eds.) EA 2001. LNCS, vol. 2310, pp. 231–242. Springer, Heidelberg (2002). https://doi.org/10.1007/3-540-46033-0_19
13. Ketkar, N.: Introduction to keras. In: Deep learning with Python, pp. 97–111. Springer (2017)
14. Khurma, R.A., Aljarah, I., Sharieh, A., Mirjalili, S.: EvoloPy-FS: an open-source nature-inspired optimization framework in python for feature selection. In: Mirjalili, S., Faris, H., Aljarah, I. (eds.) Evolutionary Machine Learning Techniques. AIS, pp. 131–173. Springer, Singapore (2020). https://doi.org/10.1007/978-981-32-9990-0_8
15. King, D.E.: Dlib-ml: a machine learning toolkit. J. Mach. Learn. Res. **10**, 1755–1758 (2009)
16. Kotsiantis, S.B., Zaharakis, I., Pintelas, P., et al.: Supervised machine learning: a review of classification techniques. Emerging Artif. Intell. Appl. Comput. Eng. **160**(1), 3–24 (2007)
17. Porcu, V.: Scikit-learn. In: Python for Data Mining Quick Syntax Reference, pp. 235–253. Apress, Berkeley, CA (2018). https://doi.org/10.1007/978-1-4842-4113-4_11
18. Liu, Y., Liu, S., Wang, Y., Lombardi, F., Han, J.: A survey of stochastic computing neural networks for machine learning applications. IEEE Trans. Neural Networks Learn. Syst. (2020)
19. Magni, P., Ferrazzi, F., Sacchi, L., Bellazzi, R.: Timeclust: a clustering tool for gene expression time series. Bioinformatics **24**(3), 430–432 (2008)
20. Mahesh, B.: Machine learning algorithms-a review. Int. J. Sci. Res. (IJSR) **9**, 381–386 (2020)

21. McKinney, W., et al.: pandas: a foundational python library for data analysis and statistics. Python High Performance Sci. Comput. **14**(9), 1–9 (2011)
22. Meng, X., et al.: Mllib: machine learning in apache spark. J. Mach. Learn. Res. **17**(1), 1235–1241 (2016)
23. Mhembere, D., Zheng, D., Priebe, C.E., Vogelstein, J.T., Burns, R.: Clusternor: a numa-optimized clustering framework. arXiv preprint arXiv:1902.09527 (2019)
24. NN Open. An open source neural networks c++ library. http://opennn.cimne.com/: 04(10), pp. 2008 (2016)
25. Palop, J.J., Mucke, L., Roberson, E.D.: Quantifying biomarkers of cognitive dysfunction and neuronal network hyperexcitability in mouse models of alzheimer's disease: depletion of calcium-dependent proteins and inhibitory hippocampal remodeling. In: Alzheimer's Disease and Frontotemporal Dementia, pp. 245–262. Springer (2010)
26. Pedregosa, F., et al.: Scikit-learn: Machine learning in python. J. Mach. Learn. Res. **12**, 2825–2830 (2011)
27. Phyu, T.N.: Survey of classification techniques in data mining. In: Proceedings of the International Multiconference of Engineers and Computer Scientists, vol. 1 (2009)
28. Pohlheim, H.: Geatbx®-the genetic and evolutionary algorithm toolbox for matlab® (2007). http://www.geatbx.com/. Accessed 24 June 2012
29. Qaddoura, R., Al-Zoubi, A.M., Almomani, I., Faris, H.: A multi-stage classification approach for iot intrusion detection based on clustering with oversampling. Appl. Sci. **11**(7), 3022 (2021)
30. Qaddoura, R., Al-Zoubi, M., Faris, H., Almomani, I., et al.: A multi-layer classification approach for intrusion detection in iot networks based on deep learning. Sensors **21**(9), 2987 (2021)
31. Qaddoura, R., Aljarah, I., Faris, H., Almomani, I.: A classification approach based on evolutionary clustering and its application for ransomware detection. In: Evolutionary Data Clustering: Algorithms and Applications, p. 237 (2021)
32. Qaddoura, R., Faris, H., Aljarah, I.: An efficient clustering algorithm based on the k-nearest neighbors with an indexing ratio. Int. J. Mach. Learn. Cybern. **11**(3), 675–714 (2020)
33. Qaddoura, R., Faris, H., Aljarah, I.: An efficient evolutionary algorithm with a nearest neighbor search technique for clustering analysis. J. Ambient. Intell. Humaniz. Comput. **12**(8), 8387–8412 (2020). https://doi.org/10.1007/s12652-020-02570-2
34. Qaddoura, R., Faris, H., Aljarah, I., Castillo, P.A.: EvoCluster: an open-source nature-inspired optimization clustering framework in python. In: Castillo, P.A., Jiménez Laredo, J.L., Fernández de Vega, F. (eds.) EvoApplications 2020. LNCS, vol. 12104, pp. 20–36. Springer, Cham (2020). https://doi.org/10.1007/978-3-030-43722-0_2
35. Qaddoura, R., Faris, H., Aljarah, I., Castillo, P.A.: Evocluster: an open-source nature-inspired optimization clustering framework. SN Comput. Sci. **2**(3), 1–12 (2021)
36. Qaddoura, R., Faris, H., Aljarah, I., Guervós, J.J.M., Castillo, P.A.: Empirical evaluation of distance measures for nearest point with indexing ratio clustering algorithm. In: IJCCI, pp. 430–438 (2020)
37. Rehurek, R., Sojka, P., et al.: Gensim-statistical semantics in python. Retrieved from genism. org (2011)

38. Risso, D., et al.: Clusterexperiment and rsec: a bioconductor package and framework for clustering of single-cell and other large gene expression datasets. PLoS Comput. Biology **14**(9), e1006378 (2018)
39. Sonnenburg, S., et al.: The shogun machine learning toolbox. J. Mach. Learn. Res. **11**, 1799–1802 (2010)
40. Virtanen, P., et al.: SciPy 1. 0: Fundamental Algorithms for Scientific Computing in Python. Nature Methods **17**, 261–272 (2020)
41. Vrbančič, G., Brezočnik, L., Mlakar, U., Fister, D., Fister, I.: Niapy: python microframework for building nature-inspired algorithms. J. Open Source Softw. **3**(23), 613 (2018)
42. Wagner, S., et al.: Architecture and design of the heuristiclab optimization environment. In: Advanced Methods and Applications in Computational Intelligence, pp. 197–261. Springer (2014). https://doi.org/10.1007/978-3-319-01436-4_10
43. Wall, M.: Galib: A c++ library of genetic algorithm components. Mech. Eng. Department, Massachusetts Institute of Technology **87**, 54 (1996)
44. Wiwie, C., Baumbach, J., Röttger, R.: Comparing the performance of biomedical clustering methods. Nat. Methods **12**(11), 1033–1038 (2015)
45. Yu, X., Gen, M.: Introduction to evolutionary algorithms. Springer Science & Business Media (2010)
46. Zöller, M.-A., Huber, M.F.: Benchmark and survey of automated machine learning frameworks. J. Artif. Intell. Res. **70**, 409–472 (2021)

Evolution of Acoustic Logic Gates in Granular Metamaterials

Atoosa Parsa[1]([envelope])[iD], Dong Wang[2][iD], Corey S. O'Hern[2][iD],
Mark D. Shattuck[3][iD], Rebecca Kramer-Bottiglio[2][iD], and Josh Bongard[1][iD]

[1] Department of Computer Science, University of Vermont,
Burlington, VT 05405, USA
{atoosa.parsa,josh.bongard}@uvm.edu
[2] Department of Mechanical Engineering and Materials Science, Yale University,
New Haven, CT 06520, USA
{dong.wang,corey.ohern,rebecca kramer}@yale.edu
[3] Benjamin Levich Inst. & Physics Dept., City College of New York,
New York, NY 10031, USA
shattuck@ccny.cuny.edu

Abstract. Granular metamaterials are a promising choice for the realization of mechanical computing devices. As preliminary evidence of this, we demonstrate here how to embed Boolean logic gates (AND and XOR) into a granular metamaterial by evolving where particular grains are placed in the material. Our results confirm the existence of gradients of increasing "AND-ness" and "XOR-ness" within the space of possible materials that can be followed by evolutionary search. We measure the computational functionality of a material by probing how it transforms bits encoded as vibrations with zero or non-zero amplitude. We compared the evolution of materials built from mass-contrasting particles and materials built from stiffness-contrasting particles, and found that the latter were more evolvable. We believe this work may pave the way toward evolutionary design of increasingly sophisticated, programmable, and computationally dense metamaterials with certain advantages over more traditional computational substrates.

Keywords: Granular metamaterials · Mechanical computing · Inverse design problem

1 Introduction

The concept of mechanical computing can be traced back to the second century BC when the earliest known analogue computer, the Antikythera mechanism, was invented [4]. Since then many other mechanical devices were invented for applications other than astronomical calculations such as basic mathematical operations [18], solving arbitrary equations [5], and even differentiation and

This material is based upon work supported by the National Science Foundation under the DMREF program (award number: 2118810).

J. L. Jiménez Laredo et al. (Eds.): EvoApplications 2022, LNCS 13224, pp. 93–109, 2022.
https://doi.org/10.1007/978-3-031-02462-7_7

integration [21]. With the invention of operational amplifiers in the early twentieth century, electronic analog computers became feasible. Soon after, digital computers emerged and rapidly became the dominant form for computation [9]. Although they impose a more abstract form of system representation, digital computers rapidly outpaced their mechanical counterparts. Moreover, higher precision and the capacity for miniaturization helped make the digital computing paradigm more desirable. But, recently, rapid advances in the chemical, biological and materials sciences have opened new opportunities for embedding computation directly into physical substrates [20].

Metamaterials are one such promising class of substrate that has surfaced in recent years. Metamaterials are engineered composite materials that exhibit properties different from their constituent materials, and from material properties observed in nature [7]. Granular metamaterials (GMMs) are a specific class of metamaterials consisting of discrete particles. GMMs exhibit increased plasticity compared to continuous metamaterials because they can be dynamically programmed by reconfiguring the material's physical structure or changing particles' properties using external stimuli [19]. In this paper, we investigate the potential of granular metamaterials as a physical substrate for mechanical computation.

Starting with logic gates as the basic computational blocks upon which more complex units can be built, we focus our work on evolving GMMs that act as acoustic logic gates: vibrations with near-zero and non-zero amplitudes arriving at and leaving the material are treated as incoming and outgoing bits. There are several advantages for this type of computation compared to the conventional approach of designing logic gates using electrical transistors. First, by moving to a mechanical substrate, we can avoid analogue to digital conversion, also thereby bypassing all of the limitations of abstract representations and discretizations necessary for a digital computing system [21]. Second, outsourcing computation to the physical substrate provides opportunities for conflating computational, mechanical, energetic, sensing and actuation properties into the same material. This could lead to robots built from continua of materials rather than modular components. This in turn could allow these machines to better exploit the natural dynamics of the materials, leading to better energy efficiency and higher robustness and stability. Finally, our approach affords a bottom-up design view point for computer architecture where the exact form of computation is not predetermined [11] and useful, non-intuitive exploitations of the material itself can be found by evolutionary search [16].

Some work on embedding mechanical computation into materials has been conducted. In [15], a universal logic gate is implemented as a nonlinear mass-spring-damper model. In [13] a soft bistable building block is designed and used in the implementation of soft mechanical diodes and logic gates. [6] utilizes a bistable spring embedded in a unit cell to implement simple logic gates. In [17] connected origami units are used to program the behavior of a mechanical bit and thus produce logic gates. [8] and [1] present examples of acoustic gate design in a 1D chain of elastic particles. Despite these advances, in none of the aforementioned works is the material automatically optimized for the desired computational function. Instead, the building blocks are hand-designed based on human intuition. In contrast, we here propose using evolutionary algorithms for automatically optimizing materials to exhibit desired computation. Moreover, the abovementioned works involved continuous metamaterials or 1D particle chains, while we here investigate the computational potential of 2D granular metamaterials. This type of material can exhibit different responses to different environmental stimuli by reconfiguring their physical structure and changing the material properties of individual particles. Thus, we anticipate a greater potential for extending our work to more complex computational operations.

Granular metamaterials have many parameters that affect the response of the system. For example, the position, size, shape, stiffness and mass of each particle can affect the eigenfrequencies of the system and consequently the propagation of acoustic waves through the material. With so many design parameters, deciding on the optimal micro-structure to achieve a desired macro-behavior (i.e. a logic gate) is a non-trivial optimization problem. For this reason, evolutionary algorithms have already proven useful for designing metamaterials that exhibit mechanical properties [10, 12] rather than the computational properties we study here. In this paper, we apply evolutionary algorithms to the design of 2D granular metamaterials to act as acoustic logic gates, where the input and output signals are acoustic waves. The granular assembly is designed such that it passes or filters the propagation of certain waves and thus acts as a Boolean logic gate.

The remainder of the paper is organized as follows: first, we formally define the problem and introduce the simulator used for optimization. Then, simulation results are presented for two different cases: designing an AND gate and designing an XOR gate. Finally, the results are discussed and the paper is concluded with some future possible directions of work.

2 Problem Statement

Our material is comprised of a two dimensional assembly of two types of circular particles placed on a hexagonal lattice. As mentioned in the introduction, the goal is to reconfigure this granular material to act as a logic gate where the inputs and the output receive and emit acoustic waves respectively. The setup is shown in Fig. 1. We choose two particles on one side of the material to serve as the input ports and one particle on the other side to serve as the output port. Each input particle i receives a sinusoidal wave with amplitude A_i and frequency ω applied in the x direction (the particle is vibrated left and right). When this signal is applied to a particle, it causes a displacement from its initial position x_i^0.

In order to represent a logic gate in this substrate, a representation for the bits must be chosen. One option is to use the amplitude of the displacement signal $(x_i(t))$ as the bit abstraction. In this case, applying a sinusoidal wave with amplitude zero $(A_i = 0)$ to an input port denotes the presentation of a zero at that port. We fixed the non-zero amplitude to 1×10^{-2}, which is 10%

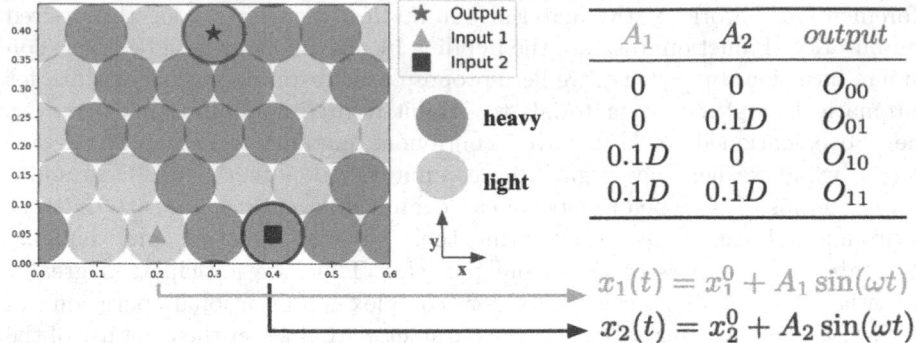

Fig. 1. Problem formulation: a logic gate with two inputs (green and blue) and one output (red) embedded in a 2D granular assembly composed of heavy (dark blue) and light (light blue) particles. A_1 and A_2 denote the amplitudes of oscillations applied in the x direction to the input ports (D = particle diameter). ω denotes the input frequency. The truth table indicates how 'bits' are supplied to the input ports, and the signal obtained at the output port that will be used to determine how much of a desired logical function the material encodes. (Color figure online)

of the diameter of a particle. This is at present an arbitrary design choice. The frequency of the applied signal (ω) is also fixed to 7, chosen based on the frequency spectrum of a typical random configuration of light and heavy particles, which will be discussed in the next section. The truth table in Fig. 1 shows our bit representation. The output signal $O_{ij}(i, j \in 0, 1)$ denotes the amplitude of the displacement of the particle at the output port. The three particles chosen to represent the three ports is also currently an arbitrary design choice.

Using zero and non-zero amplitudes directly to interpret bits arriving at the output port however is problematic: excitation of both input ports biases the system to produce larger amplitudes at the output port compared to excitation of just one input port, biasing any material configuration toward linear functions. Thus we instead use low and high gain at the output port to represent bits. In each of the four input cases, the gain of the system is defined as the amplitude of the fast Fourier transform ($\hat{f}$) at the driving frequency (ω) at the output divided by the sum of the amplitudes of the fast Fourier transform at the driving frequency (ω) in the inputs:

$$G_{ij} = \frac{\hat{f}(O_{ij})}{\hat{f}(\text{in}_i) + \hat{f}(\text{in}_j)} \quad i, j \in 0, 1 \tag{1}$$

In our experiments, for each material we measure the gain for each of the four input cases. In order for the material to act as a logic gate, the relative magnitude of the gain in each case must be consistent with desired functionality of the gate. For example, for an AND gate when both input ports are driven with a sinusoidal wave, we expect to see a high amplitude of oscillation at the output and therefore we expect a high gain (G_{11}). But in the other three cases

$(00, 01, 10)$ we expect a low gain. Based on this, we can measure the similarity of any material's functionality to a desired logic gate by taking the distance between the four expected output bits and the four gain values.

3 Simulation Setup

In this section, we first present some details of our 2D granular metamaterial simulator. Then, we provide details of the evolutionary algorithm used for optimization.

3.1 2D Granular Simulator

We model a simplified granular metamaterial inspired by [10]. The system is 2D and composed of frictionless circular disks with fixed and equal diameters. The particles can be assigned differing masses and/or stiffnesses. The particles are placed on a 5 by 6 hexagonal lattice, resulting in 30 particles available for optimization. The system has a periodic boundary condition in the x direction and a fixed boundary condition in the y direction. There is no gravity in the system and the only forces acting on the particles are the result of a purely repulsive linear spring potential between the disks which can be formalized as Lennard-Jones potential. This system is simulated using Discrete Element Method (DEM). At each simulation time step, repulsive forces are calculated for those particles in contact with other particles, based on their distances to particles with which they are in contact. Then the accelerations, velocities and positions of each particle are updated using Verlet integration. Before probing the bulk properties of the material, a post-processing step is taken to ensure that the system is at equilibrium: the sum of the total forces between particles is near zero. This ensures that the particle packing is statistically stable. This is done by calculating the total force acting on each particle and updating their positions using the steepest-descent method to reduce total force. In the experiments where we have particles with different stiffnesses, we need to find the stable initial positions for each configuration separately. This will increase the total simulation time of our optimizations. In those cases Fast Inertial Relaxation Engine (FIRE) was used in order to reduce computational effort.

As computational metamaterials must selectively amplify or extinguish certain input waves to perform logical functions in the frequency domain, it is useful to take a closer look at their frequency spectrum. One useful property of granular metamaterials is the existence of band gaps in their vibrational density of states [2]: a contiguous range of input frequencies extinguished by the material. To locate a material's band gap, its mass-weighted dynamical matrix is calculated using the Hessian of the total potential energy. The eigenvalues of this matrix are the eigenfrequencies of the system and the eigenvectors are the modes. If the eigenfrequency spectrum is plotted by sorting the frequencies in increasing order, gaps in the spectrum become visible. The widest gap is denoted as the *band gap* (An example is shown in Fig. 2c). If the granular system is excited at a frequency within the band gap, the signal will not propagate through the material. On the other hand if the signal is outside the band gap, the system will be excited at one of its resonant frequencies and the output will be magnified. For this reason, we will choose input frequencies near the low and high cut-off frequencies of a typical material's band gap to facilitate the evolution of computational metamaterials capable of selectively amplifying or suppressing input waves, as explained in the next section.

3.2   Optimization Method

For optimization we use Age-Fitness Pareto Optimization (AFPO) [14]. AFPO is a multi-objective, multi-deme evolutionary algorithm that periodically injects new random individuals into the population and temporarily reduces selection pressure on their resulting lineages, thereby achieving diversity maintenance without requiring additional hyperparameter tuning. In all experiments we employed a direct encoding scheme for the genome: length-30 binary vectors indicated which particles were light or heavy in the mass-contrasting experiments, and which particles were soft or stiff in the stiffness-contrasting experiments. Two way tournament selection was employed to select which individuals produced offspring. Offspring were mutated by flipping each bit with probability 0.05. Crossover was not employed because there is no evidence that combining parts of two materials preserves any of their respective bulk behaviors. In all the experiments (unless mentioned otherwise) a population size of 50 was used, and each evolutionary trial was conducted for 200 generations. Three replicates were performed for each experiment. Each replicate began with a different random initial population. The fitness function for each experiment will be introduced in the subsequent sections.

4   Results and Discussion

In the first experiment, the goal was to evolve a particle configuration that maximizes the band gap. This evolved band gap was used to choose input frequencies for the subsequent experiments in which AND and XOR gates are evolutionarily embedded into two different metamaterials: those with mass-differing particles and those with stiffness-differing particles. Source code for all of the experiments is available in our GitHub repository.

4.1   Evolution of an Acoustic Band Gap

Vibrational frequency band gaps can be used to shield materials from vibrations and other perturbations. As mentioned before, vibrations with frequencies within the band gap do not propagate into the material. To block vibrations over a range of frequencies, wider band gaps are necessary. We can also tune the location of the central region of the band gap to block perturbations over different frequency ranges. A granular metamaterial's band gap can be altered by changing the number of particles embedded within it, as well as the particles' masses, positions, stiffnesses and shapes. Here, we focus on particle arrangements and define the optimization problem as finding the placement of a fixed number of heavy and light particles on a hexagonal lattice to maximize the band gap. In this section, we assume that we have 9 light and 21 heavy particles (30 particles in total), an arbitrary design chose at present. The genome encodes the positions of the light particles on the lattice. The result of three evolutionary trials is shown in Fig. 2.

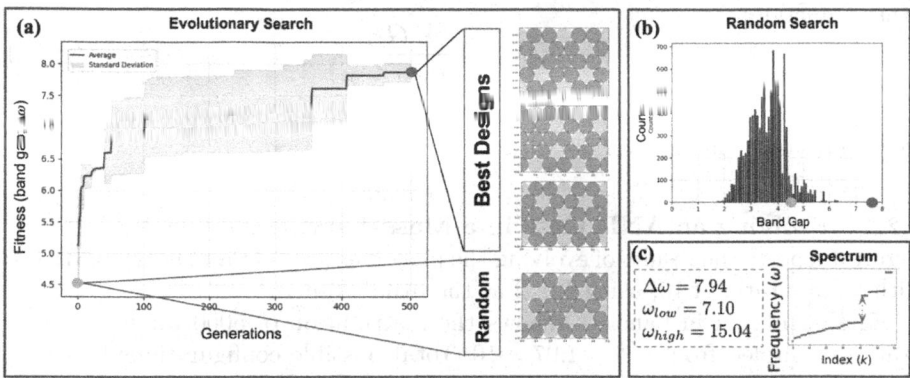

Fig. 2. Evolving the placement of 9 light and 21 heavy particles on a hexagonal lattice to maximize the acoustic band gap. (a): The fitness curve (solid blue) reports evolutionary progress as the fitness of the best individual from each generation averaged over 3 runs. The three most fit solutions from three independent runs are shown on the right. (b): The histogram shows the distribution of 10000 randomly generated samples. (c): The plot shows the frequency spectrum of the best solutions along with the width, start and end points of the band gap on the left. (Color figure online)

This problem admits $C_9(30) = \binom{30}{9} = 143071150$ possible materials. To judge the quality of optimization, we generated 10000 random configurations and calculated the band gap for each of them (Fig. 2b). The mean value of this distribution lies near 4. At the end of 500 generations (with a population size of 30), AFPO was able to find configurations with a band gap of $\Delta\omega = 7.94$. It's worth mentioning that in this section, we chose to increase the number of generations, because probing the evolutionary progress showed a continued improvement. Interestingly, the optimal designs with the highest band gaps are symmetric and show an ordered arrangement, which is consistent with our knowledge from materials sciences. Band gaps are known to occur in crystalline mixtures with regular patterns [3]. Moreover, because of the periodic boundary condition in the x direction, the three best solutions are the same configuration, just shifted different distances horizontally.

4.2   Evolving an AND Gate

In this experiment, particle configurations are evolved on a hexagonal lattice to act as much like an acoustic AND gate as possible. As we mentioned in the simulation section, we can measure the gain of the system (the relative amplitude of output oscillations to the input oscillations) for each of the four possible input cases $(00, 01, 10, 11)$. G_{00} is trivial: if the input is 00—no displacement is applied to either of the input particles—the output particle will yield no displacement either. For the other three cases, a significant gain should only be observed when both inputs are activated (high G_{11}). To achieve this, we defined the following fitness function:

$$f_{\text{“AND-ness”}} = \frac{G_{11}}{(G_{10} + G_{01})/2} \tag{2}$$

The next two sections present the results of evolving mass-varying and stiffness-varying granular metamaterials with this fitness function.

4.2.1   Evolving an AND Gate in a Mass-Varying Material

Figure 3 reports the results of evolving the placement of light and heavy particles, with a mass ratio of 10, using Eq. 2 as the fitness function.

The histogram in panel (a) shows the distribution of 5000 random configurations sampled from $2^{30} = 1.07 \times 10^{9}$ total possible configurations based on the measure of “AND-ness” (Eq. 2). The mean AND-ness for a random configuration is 0.715. The best configuration found by random search has a fitness of 3.52. Optimization was able to find a configuration with a fitness of 8.21 after 200 generations. Figure 5 reports the fittest designs from the three independent evolutionary trials. We notice that these configurations are not intuitive or symmetric, which makes it much harder to obtain from scratch in material design. Figure 4 illustrates how one of these best designs approximates the behavior of an AND gate: there is significant gain in the signal at the output port, at the driving frequency, only when both input ports are excited at that frequency.

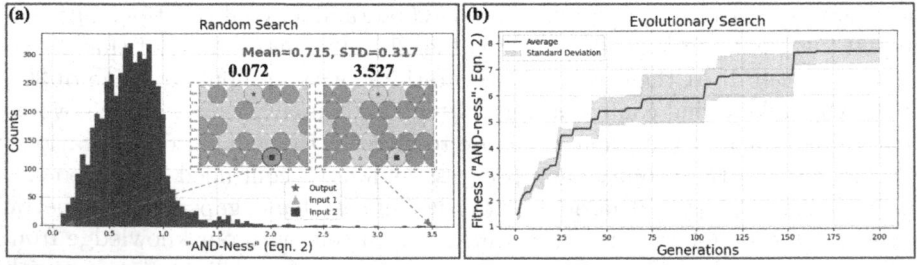

Fig. 3. Designing an AND gate in a mass-contrasting assembly of particles. (a): The histogram shows the distribution of the AND-ness in 5000 random configurations. (b): The plot shows the progress of optimization during 200 generations.

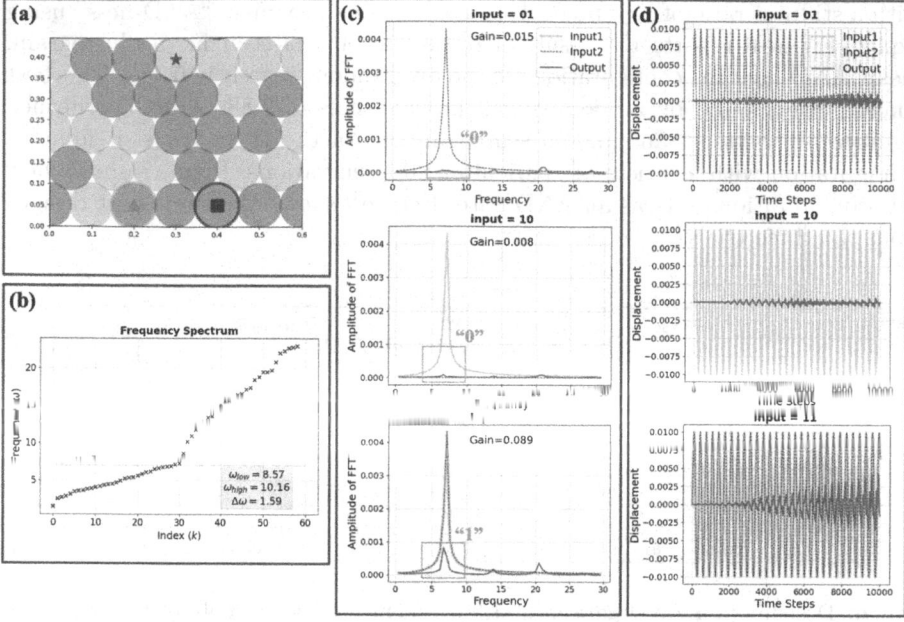

Fig. 4. (a): One of the best designs for an AND gate with mass-contrasting particles. (b): its band gap characteristics. (c): amplitudes at the driving frequency (7), and all other frequencies, at the input and output ports, for three of the four input cases, in the frequency domain. (d): the same signals, shown in the time domain. The orange rectangle in each of the plots in panel (c) highlights the behavior of the output port. The 00 → 0 case is not shown as it is trivial and always holds, regardless of material, because no energy can enter the material except through the input ports. (Color figure online)

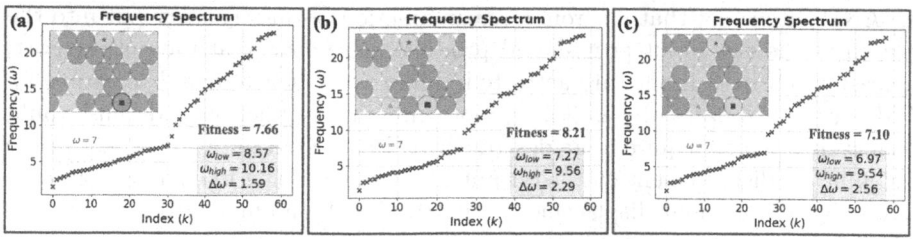

Fig. 5. The fittest AND gate designs from the three evolutionary trials, using mass-contrasting particles. Their frequency spectra, band gap features, and fitness values (Eq. 2) are also shown.

4.2.2   Evolving an AND Gate in a Stiffness-Varying Material

To investigate how different materials facilitate or obstruct the ability to evolve computational abilities into them, we evolved materials composed of particles with the same mass but differing stiffnesses: AFPO places stiff and soft particles, with a stiffness ratio of 10, and evolves materials to maximize "AND-ness" using Eq. 2. Figure 6 reports the result of optimization. As seen in the histogram, mean AND-ness for configurations found by random search is 0.760. The best configuration found by random search has a fitness of 7.88. AFPO performed significantly better than random search: in one of the three trials, it found a configuration with a fitness of 10.61 after 200 generations. Figure 7 shows how this configurations acts as an AND gate. Figure 8 shows the three best designs from the three trials.

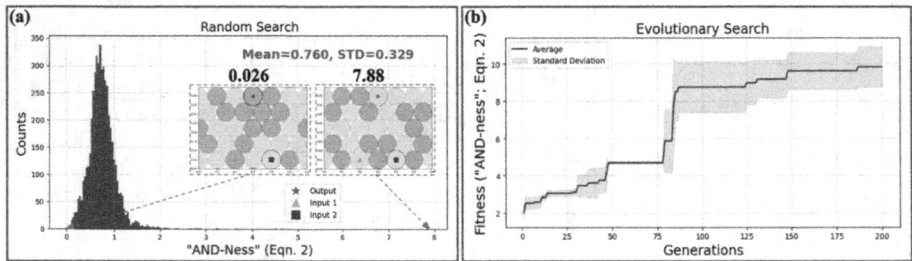

Fig. 6. Designing an AND gate in a stiffness-contrasting assembly of particles. (a): The histogram shows the distribution of "AND-ness" in 5000 random configurations, as well as the best and worst material found. Light and dark green colors indicate soft and stiff particles, respectively. (b): The plot shows the progress of optimization during 200 generations. (Color figure online)

4.3   Evolving an XOR Gate

An AND gate is not that far from a linear function, as more energy put into the system at the two input port should produce more energy at the output port, at least when both inputs are activated. Thus we next attempted to evolve an XOR gate into materials, as it is a more non-linear function and thus would intuitively seem to require more design effort. In an XOR gate, we expect to see a significant displacement at the output if only one of the input ports is being driven by a sinusoidal displacement (the 01 and 10 input cases). In order to achieve this, we defined an "XOR-ness" fitness function as follows:

$$f_{\text{"XOR-ness"}} = \frac{(G_{10} + G_{01})/2}{G_{11}} \tag{3}$$

such that increasing values denote materials that act increasingly like an XOR gate. As before, we investigated evolving materials with mass-contrasting particles and materials with stiffness-contrasting particles against this fitness function.

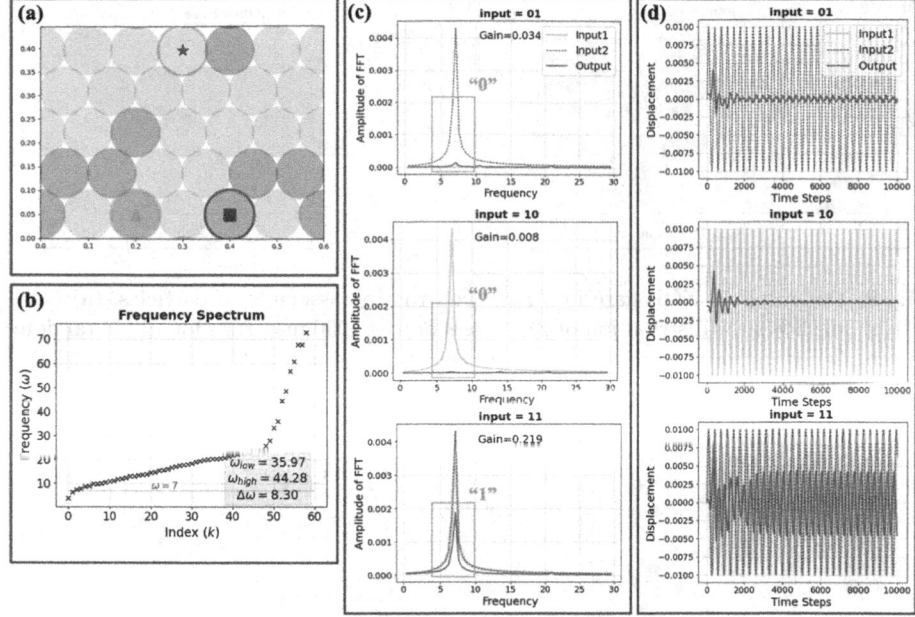

Fig. 7. One of the best designs for an AND gate with stiffness-contrasting particles. (a) and (b) show the configuration and its band gap respectively. (c) shows the response of the material at the driving frequency, and the other frequencies, for three of the four input cases. (d) shows the displacements of the input ports and the output port over time.

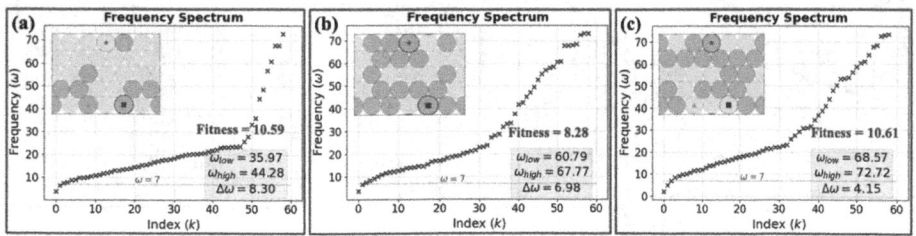

Fig. 8. The most "AND-like" stiffness-varying materials from the three evolutionary trials. The band gap for each configuration is shown below each of them.

4.3.1   Evolving an XOR Gate in a Mass-Varying Material

We performed three evolutionary trials that optimize the placement of heavy and light particles (mass ratio = 10) into materials such that they maximize "XOR-ness" (Eq. 3). Figure 9 shows the result of optimization. There we see that mean XOR-ness in materials found by random search is 1.775, while the best material had an XOR-ness of 26.59. Evolutionary search performed significantly better: it

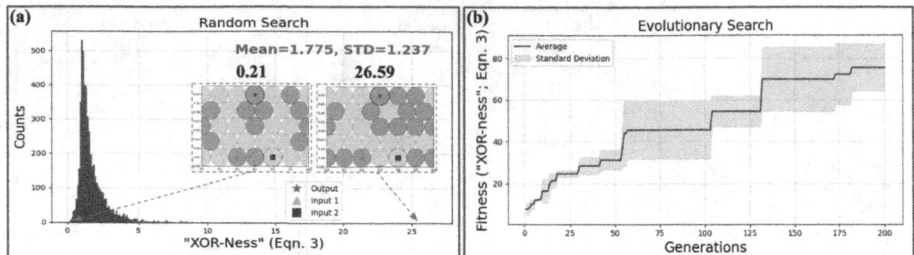

Fig. 9. Designing an XOR gate in a mass-contrasting assembly of particles. (a): The histogram shows the distribution of XOR-ness across 5000 materials found via random search. (b): Progress of evolutionary optimization.

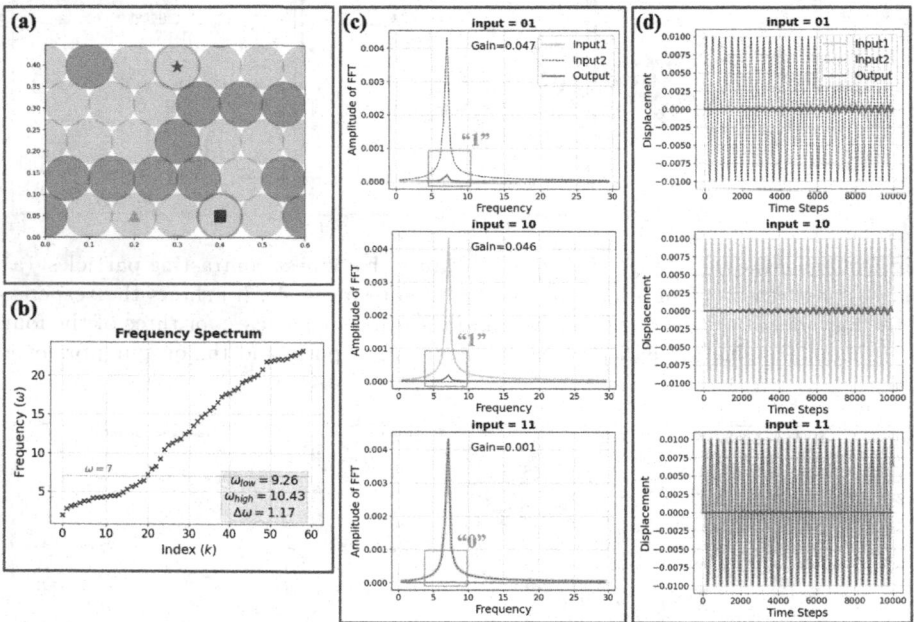

Fig. 10. One of the best designs for an XOR gate found for materials with mass-contrasting particles. (a) and (b) report the evolved configuration and its band gap features. (c) reports the material's response at the driving frequency (7) and all other frequencies. (d) shows material's behavior over time when presented with three of the four input cases.

found a material with an XOR-ness of 91.08. Figure 11 shows that material and the best materials found in the other two trials. Figure 10 reports the detailed behavior of one of these materials.

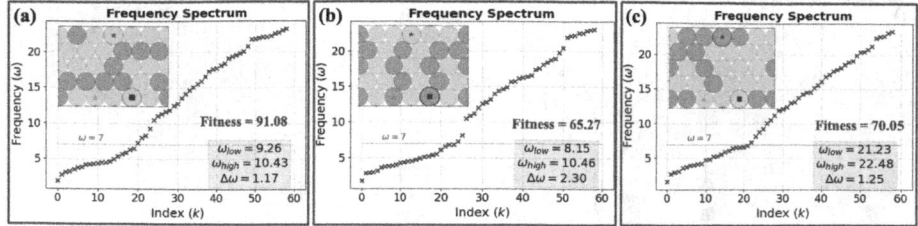

Fig. 11. The most XOR-like mass-varying materials found by each of the three evolutionary trials. The frequency spectrum of each configuration along with its fitness values is shown below each one.

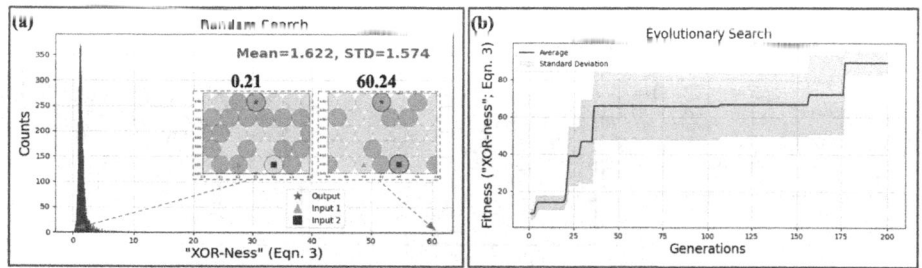

Fig. 12. Evolving XOR behavior into a stiffness-contrasting assembly of particles. (a): The distribution of XOR-ness (as defined in Eq. 3) in 5000 materials found by random search. (b): Average performance of the three evolutionary trials.

4.3.2   Evolving an XOR Gate in a Stiffness-Varying Material

As with the AND gate experiments, we investigated whether different kinds of materials facilitate or obstruct evolving XOR gates into them. To do so, we performed another three evolutionary trials using the XOR-ness fitness function (Eq. 3) on materials in which 30 soft or stiff particles could be placed into the material. Figure 12 reports the relative performance of random and evolutionary search. Although the mean fitness of materials found by random search is only 1.622, the best material achieved an XOR-ness of 60.24. Evolutionary search, on the other hand, again performed significantly better than random search: it found a particle assembly with a fitness of 94.08. Figure 13 shows one of the three best solutions from evolutionary optimization. Figure 13c shows that this material does indeed act as an XOR gate: the amplitude of oscillations at the output is significantly higher when only one of the input ports is activated. Figure 14 shows the best solutions from the three trials.

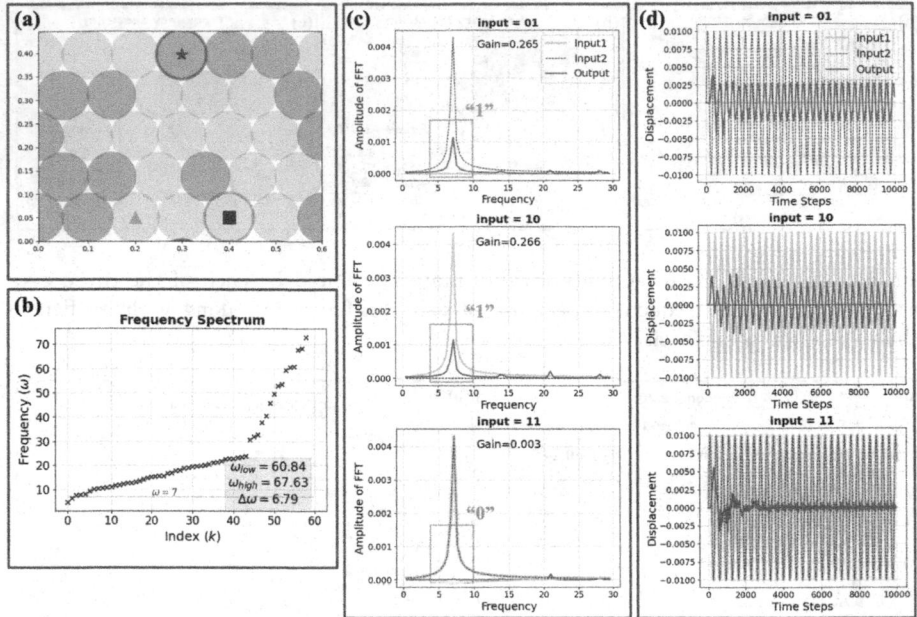

Fig. 13. One of the most XOR-like evolved materials, composed from stiffness-contrasting particles. (a) and (b) report the configuration and its band gap features. (c) reports how the material responds at the driving frequency (7) and all other frequencies. (d) shows how the three ports displace, over time, during three of the four input cases.

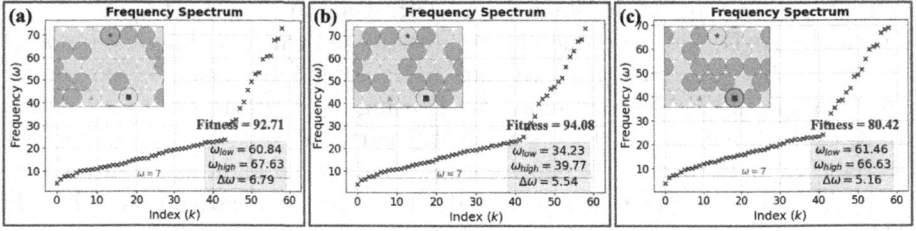

Fig. 14. The most XOR-like stiffness-varying materials, produced by each of the three evolutionary trials. The frequency spectrum of each configuration along with its fitness is shown below each one.

5   Conclusion and Future Work

Here we demonstrated the potential for granular metamaterials to act as physical substrates for computation expressed as amplification or extinction of acoustic waves. The significant performance advantage obtained by evolutionary search over random search, for two different logical operations and two different kinds of granular materials, indicates that such materials can embed computation, but finding ones that do is non-trivial, even for simple Boolean operations. Moreover, as the computation becomes more challenging, the efficacy of evolutionary over random search increases: evolutionary search is 10 times better at finding materials that act like AND gates compared to random search (Figs. 3, 6); it is 100 times better at finding materials that act like XOR gates (Figs. 9, 12).

Moreover, we noticed that evolutionary search can embed computation into some materials better than others. For example, the most AND-like mass-varying material shows a much weaker '1' signal at its output port (Fig. 4c) compared to the '1' output by the most AND-like stiffness varying material (Fig. 7c). Similarly, the '1's output by the most XOR-like mass-varying material (Fig. 10c) are weaker than the '1's output by the most XOR-like stiffness-varying material (Fig. 13c).

The non-intuitive nature of embedding computation into granular metamaterial is also evidenced by the lack of obvious common patterns across the evolved materials that best embody the logic gates: each has unique ratios of light/heavy or soft/stiff particles, geometric patterns, and there is no obvious regularity or symmetry (Figs. 5, 8, 11 and 14). This emphasizes the utility of automated design in this domain. Designing a configuration of particles to behave as a logic gate is a rather difficult task to accomplish without the aid of computer optimization.

Future work is planned in which the design space is expanded by expanding lattice resolution, and subjugating particle shape and input/output port placement to evolutionary optimization as well. Analytic efforts will focus on attempting to understand how vibrating particles encode computation by training machine learning methods to seek common patterns across successful designs not visible to human inspection. We will also investigate whether successful designs compute sub-functions in different regions of the material and then combine them downstream, or do not need recourse to such divide-and-conquer strategies.

We will also explore verifying our simulation results in physical hardware. It is possible that, given the discrete nature of granular metamaterials compared to continuous media, crossing the reality gap may prove easier for former compared with the latter. Also, because different designs are currently just different placements of two types of particles on a predefined grid, we expect the fabrication process to be cheaper, faster and easier as well.

Granular metamaterials, unlike continuous media, afford the possibility of serving not just as computational substrates but as reconfigurable computational substrates: it may be possible to build physical GMMs from particles that dynamically change stiffness in response to external stimuli such as temperature. This may allow for the packing of more computational ability within the same dynamic material. Creating increasingly computationally dense GMMs will also be investigated by providing waves with increasingly complex and diverse shape, at more input ports, with summed oscillatory components that drive different computations in the same material at the same time. This may in time show that granular metamaterials, or other emerging exotic materials, may be competitive with or possibly superior to current electronic devices as vehicles for computation.

Acknowledgment. The authors would like to thank Qikai Wu for providing the simulator for mass-contrasting assembly of circular particles. The computational resources provided by the Vermont Advanced Computing Core are also gratefully acknowledged.

References

1. Bilal, O.R., Foehr, A., Daraio, C.: Bistable metamaterial for switching and cascading elastic vibrations. Proc. Natl. Acad. Sci. **114**(18), 4603–4606 (2017)
2. Deymier, P.A.: Acoustic Metamaterials and Phononic Crystals, vol. 173. Springer Science & Business Media (2013)
3. Fornleitner, J., Kahl, G., Likos, C.N.: Tailoring the phonon band structure in binary colloidal mixtures. Phys. Rev. E **81**(6), 060401 (2010)
4. Freeth, T., et al.: Decoding the ancient greek astronomical calculator known as the antikythera mechanism. Nature **444**(7119), 587–591 (2006)
5. Hartree, D.R.: The bush differential analyser and its applications. Nature **146**(3697), 319–323 (1940)
6. Ion, A., Wall, L., Kovacs, R., Baudisch, P.: Digital mechanical metamaterials. In: Proceedings of the 2017 CHI Conference on Human Factors in Computing Systems, pp. 977–988 (2017)
7. Kadic, M., Milton, G.W., van Hecke, M., Wegener, M.: 3d metamaterials. Nature Rev. Phys. **1**(3), 198–210 (2019)
8. Li, F., Anzel, P., Yang, J., Kevrekidis, P.G., Daraio, C.: Granular acoustic switches and logic elements. Nat. Commun. **5**(1), 1–6 (2014)
9. MacLennan, B.J.: A review of analog computing. Department of Electrical Engineering & Computer Science, University of Tennessee, Technical report UT-CS-07-601 (September) (2007)
10. Miskin, M.Z., Jaeger, H.M.: Adapting granular materials through artificial evolution. Nat. Mater. **12**(4), 326–331 (2013)
11. Nakajima, K.: Physical reservoir computing-an introductory perspective. Jpn. J. Appl. Phys. **59**(6), 060501 (2020)
12. O'Hern, C.S., Shattuck, M.D.: Highly evolved grains. Nat. Mater. **12**(4), 287–288 (2013)
13. Raney, J.R., Nadkarni, N., Daraio, C., Kochmann, D.M., Lewis, J.A., Bertoldi, K.: Stable propagation of mechanical signals in soft media using stored elastic energy. Proc. Natl. Acad. Sci. **113**(35), 9722–9727 (2016)
14. Schmidt, M., Lipson, H.: Age-fitness pareto optimization. In: Riolo, R., McConaghy, T., Vladislavleva, E. (eds.) Genetic Programming Theory and Practice VIII, pp. 129–146. Springer, New York (2011). https://doi.org/10.1007/978-1-4419-7747-2_8
15. Serra-Garcia, M.: Turing-complete mechanical processor via automated nonlinear system design. Phys. Rev. E **100**(4), 042202 (2019)
16. Thompson, A.: An evolved circuit, intrinsic in silicon, entwined with physics. In: Higuchi, T., Iwata, M., Liu, W. (eds.) ICES 1996. LNCS, vol. 1259, pp. 390–405. Springer, Heidelberg (1997). https://doi.org/10.1007/3-540-63173-9_61
17. Treml, B., Gillman, A., Buskohl, P., Vaia, R.: Origami mechanologic. Proc. Natl. Acad. Sci. **115**(27), 6916–6921 (2018)

18. Tympas, A.: The delights of the slide rule. In: Calculation and Computation in the Pre-electronic Era, pp. 7–38. Springer, London (2017)
19. Wu, Q., Cui, C., Bertrand, T., Shattuck, M.D., O'Hern, C.S.: Active acoustic switches using two-dimensional granular crystals. Phys. Rev. E **99**(6), 062901 (2019)
20. Yasuda, H., Buskohl, P.R., Gillman, A., Murphey, T.D., Stepney, S., Vaia, R.A., Raney, J.R.: Mechanical computing. Nature **598**(7879), 39–48 (2021)
21. Zangeneh-Nejad, F., Sounas, D.L., Alù, A., Fleury, R.: Analogue computing with metamaterials. Nat. Rev. Mater. **6**(3), 207–225 (2021)

Public-Private Partnership: Evolutionary Algorithms as a Solution to Information Asymmetry

Simone Pellegrino[1]ⓘ, Massimo Rebuglio[2][✉]ⓘ, and Giovanni Squillero[2]ⓘ

[1] University of Turin, Turin, Italy
`simone.pellegrino@unito.it`
[2] Politecnico di Torino, Turin, Italy
{`massimo.rebuglio,giovanni.squillero`}`@polito.it`

Abstract. In a free market, the creation of hospitals, schools, sports and public residential facilities, requires the expertise—and possibly the capital—of the private sector. The traditional contract, in which the public administration pays private operators to make or maintain buildings and services, is flanked by public private partnership, in which the private operator is usually delegated to carry out the entire process receiving a fixed fee. For years, governments and administrations have been incentivized to use this kind of contract, assuming that it would increase the building qualities and reduce the risk of higher expenses. Empirical evidence refutes this assumption, and this can be caused by to the so-called moral hazard of the private operator. One of the main problem in public private partnership is the difficulty to define an optimal risk allocation, as there no formulas exist to simulate the performance of the contract in advance. In this paper, Evolutionary Algorithms are used to compute an optimal specifications document, while, at the same time, foreseeing an optimal effort in work. Experimental results clearly demonstrate the feasibility of this approach, also helping the public administration to check if their knowledge is sufficient to structure an efficient specifications document.

Keywords: Public private partnerships · Information asymmetry · Evolutionary algorithms · Stochastic optimization

1 Introduction

The success of the public private partnership (PPP), is based on the quality of risks attribution; in essence, those who make mistakes, pay. For example, in a construction project, delays in obtaining the necessary authorizations should be paid by the Public Administration (PA), and delays in the works by the private operator (OP). The real world is much less clear-cut: if the work includes

---

Authors are listed in alphabetical order.

© Springer Nature Switzerland AG 2022
J. L. Jiménez Laredo et al. (Eds.): EvoApplications 2022, LNCS 13224, pp. 110–123, 2022.
https://doi.org/10.1007/978-3-031-02462-7_8

hundreds of processes, each of them strongly correlated with the other, and in presence of exogenous factors not attributable to any actor, the risk analysis is anything but trivial. The interest in mitigating the risk is asymmetrical, since the administrations, in front of the law or public opinion, are always responsible for the provision of certain services: a dishonest OP who takes part to a PPP regardless of the risks knows these dynamics and can take advantage of them. The attribution of risks, in the European Union, follows a standardized model called *risk matrix*, and contributes to the definition of the public sector comparator (PSC), useful for determining the convenience, or inconvenience, of a work in partnership than that achieved with traditional methods. Like other parameters of the PSC, it is often seen as a necessary formality, or even worse as an element to be arranged to make hands meet. Even if the attribution is well executed, it is not necessarily well formalized: the *mitigation tools*, or who pays what, and the eventual penalties and rewards, can be written in discursive form, dispersed in the specifications[1]. Despite the attempts to guide the administrations, the European PPP remains an inconvenient tool for the public part, so that in 2018 the European Court of Auditors suggested not to use it until better protocols are defined. Among the main critical issues, the following is cited: *"The private sector was ineffective between public partners risks, while the high rates of remuneration (up to 14 %) of the private partner's risk capital did not always reflect the risks incurred."* [12].

In this paper we try to adopt a completely new strategy, which simulates a real iteration between an OP, who will act in order to earn as much as possible, and a public administration, which will act in order to maximize the social value of the work. Therefore the public administration does not use a top-down approach, but rather can step into the OP shoes, i.e., calibrate its choices according to the expected reactions of the counterpart.

We tested the model on various test cases and on a real case taken from the motorway sector, showing that in each case the algorithm is capable of optimizing.

2   The Problem

The main elements for defining risk allocation and mitigation tools, which are the parameters that this text seeks to optimize, are the allocation of unexpected costs deriving from external factors and when and how to impose a penalty. Allocation and penalties must be consistent with the fee paid, which is also subject to optimization. Over the years, governments have tried to define standardized processes to define some of these parameters. United Kingdom provides an official and standardized tool to calculate whether the PPP is more or less

---

[1] Drafting of this text various process managers of a large Italian administration that regularly uses the PPP were interviewed, and in all cases the matrix was presented as insignificant. In addition, more than 50,000 cases (not necessarily PPPs) present in the Lombardy Region Sintel aggregation center were automatically analyzed, noting that the matrix between the attachments was present in only about fifty cases.

convenient than a traditional contract, those use is mandatory[2]. Italy asks the contracting authority to identify the risk, through brainstorming and statistical analysis, quantifying each risk with its average value weighted with respect to the probability that it will occur. None of these approaches has proved to be fully effective, but they have forced contracting authorities to find more and more data and make more and more reflections, of which this model will make extensive use.

3   Proposed Approach

As previously discussed, as a characteristic of the PPP, it is not possible to exercise precise control over the actions of the OP, nor over how much he will spend to carry out the work in a workmanlike manner [1,7,8]. We therefore focused on what is possible to know. First of all, the OP's goal is known: in the worst case, to make as much money as possible [8]. Knowing this, we can know how to maximize a fitness function for him. Also known are the basic dynamics foreseen by any procurement code [9], such as the penalties for unattained results and the respective key performance indicators (KPIs). From the benchmarks, literature and experience of contracting authorities it is possible to estimate the impact of the operator's actions on KPI, and the impact of a choice on a task on the costs of subsequent tasks [2]. Finally, the concept of moral hazard is known [4], so, for each action, literature and experience of the contracting authorities can identify a percentage of information asymmetry, which consists of the part of the damage caused by the operator that can be presented as due to exogenous factors.

By combining these elements with the goal of the administration, which is to maximize social value, we have set up an adversarial optimization. Note that while the operator optimize ex-post, as he can observe the course of events and adapt his behavior accordingly, the administration must prepare a complete specification before the start of the works. Consequently, the optimization of the public administration is stochastic, since must take into account exogenous random factors that can alter costs and behaviors.

3.1   The Model

The model tries to reflect the progress of a PPP as closely as possible, representing in the genotypes the parameters that PA and OP have to choose. The

---

[2] The tool manages the risk by associating a *sensitivity multiplier* and a *optimism variance* to some preset items (construction/start-up costs, project duration, operating costs excluding labor costs, labor costs, transaction transaction costs, costs for closing the relationship, proceeds for providing the service). The optimism bias is a fixed percentage of worsening of the parameter in question, justified by the fact that the conditions are always too optimistic. The sensitivity multiplier represents the possibility of the costs to fluctuate: each parameter is varied individually with discrete steps, showing the impact on the PSC.

simulation of real world economic parameters via EA has been exploited in literature [10].

Let J be the number of activities[3] and T the number of periods. Let F be the matrix of ideal flows, with $F_{j,t}$ equal to the flow of activity j in period t. The flows represent the estimated expense to perform an activity in the absence of externalities and exogenous factors.

The operator can alter the expected flows, spending less or more. This variation is represented by a matrix of *effort costs* E, with each Ej, t representing the percentage variation with respect to the flows $F_{j,t}$. The variation can determine a significant decrease —or increase— in quality in achieving the result of the activity, but it must in any case guarantee its conclusion within the foreseen time frame. A negative effort cost represents an activity not performed in a workmanlike manner, which, usually determines the increase in the cost of subsequent tasks and/or the worsening of one or more KPIs.

There are two risk factors: endogenous, or externalities, deriving from the actions of the OP, and exogenous, deriving from the environment. Each risk can be determined in whole or in part by the actions of the OP and in whole or in part by exogenous factors. Ideally, the costs of an externality are always paid by the OP, while the exogenous risk is divided by agreement between the OP and the administration. Let R be the number of identified risks, and $R^\%$ the vector for sharing the risk from external factors, with $R_r^\%$ the percentage charged by the administration for the risk r. However, the administration is not always able to accurately determine the cause of the realization of a risk. Let $A^\%$ be the vector of the information asymmetry. Each $A_r^\%$, relating to the r risk, is equal to the percentage of externalities that the OP can propose to the administration as an exogenous factor. Let ε be the matrix of exogenous factors, with $\varepsilon_{r,t}$ the realization of the random variable characterizing the exogenous risk factors r at time t.

Let fee represents a predetermined fee by the administration towards the OP, to be paid every period.

KPIs are the administration's tool to monitor the good progress of the process. Let K be the number of KPIs, and $f_k(E)$ a function that taken as input the effort matrix returns a matrix of KPIs, of size equal to K for the contract periods[4].

---

[3] An activity is any action performed by the OP, with any level of detail. Ex: the design of the building, construction of walls, maintenance of the electrical system, pruning of green areas ...

[4] The ideal KPI function is the linear application of the identity matrix for the effort, which provides the administration with the real cash flow of the operator for each activity and for each period in order to guarantee a work of art realization. Note that the total flow of the operator for a certain task is influenced not only by the need to develop a work of art, but also by any externalities or exogenous factors discussed below. The KPI does not monitor *how much the operator spends*, but *how much the operator spends in order for the work to be good*. An increase in flows to face an unexpected event or to calm the results of a previous saving are not measured by the KPI.

The administration defines *pnlt* as the monetary penalty matrix, with relative *th* thresholds. For each $f_k(E)_{k,t} < th_k$, a $pnlt_k$ penalty is imposed, if kpi k is active at year t.

Any variation of the flows envisaged by the administration through non-zero effort can produce an externality, or an increase or decrease in costs for carrying out the subsequent activities in a workmanlike manner. Usually the externality is positive, i.e. directly correlated to the effort cost. Let $f_x(E)$ be a function that, taking the effort matrix as input, returns the matrix of externalities, of dimensions equal to E, with $f_x(E)_{j,t}$ the value of the externality undergoing activity j at time t^5.

An operator failure scenario would require repeating the contract, administrative costs, reputational damage, alienation of officials. So let $f_d(x)$ be an increasing monotone function that represents the higher costs that the administration has to face when the OP accrues a certain liability x.

Inflation and interest on borrowed money require the discounting of the flows. Inflation affects the cost of goods and penalizes those who have cash in hand[6]. With respect to interest, it is good to remember how a standard PPP works: an operator builds a structure, typically by borrowing from a bank; when he has finished, he receives an annual management fee, and with a part of this he repays us the bank. To cope with major maintenance, whether planned or not, this scheme can be much less clear-cut: the operator can ask for credit in several moments. Assuming maximum flexibility on credit, the OP can go into debt at any time and repay money at any time, as long as he pays the interest on it. To calculate this dynamic, the nominal discount rate t_{sn} is used[7] In most cases, interest rates are referred to as *real*, i.e. net of inflation. The nominal rate can be calculated from the real rate with the formula:

$$t_{sn} = (1 + t_i) \cdot (1 + t_{sr}) - 1 \qquad (1)$$

So, the values of F and *pnlt* are to be considered discounted at the rate of t_i already; the *fee* does not vary with inflation; the resulting flows are discounted at the t_{sn} discount rate.

3.2   Data

The problem data are J, T, K, R such as number of activities, periods, KPIs and risks; the F matrix of the expected flows; the random variables that make ε; the functions f_x for calculating externalities, f_k for calculating KPIs and f_d for calculating bankruptcy costs; the discount rate t_s and the inflation rate t_i.

---

[5] The externality functions must be modeled or customized by domain experts, and improve system predictions the more accurate they are.

[6] Example: if the operator has to buy goods for 100 euros but waits a year, he will pay around 102 euros (with an inflation rate of 2%).

[7] Example, if the operator in year 1 has a liability of €100 (which he asks the bank) and in year 2 an asset of €100 (which he pays to the bank), the total flow will not be zero. Assuming a $ts_n = 5\%$, the discounted flow will be about -5 (that is, the bank demands another 5 euros).

3.3 Adversarial Optimization

The model tries to simulate the real interaction between an OP, who will act in order to earn as much as possible (EA1), and a public administration, which will act in such a way as to maximize the social value of the work, within its own economic and financial constraints (EA2).

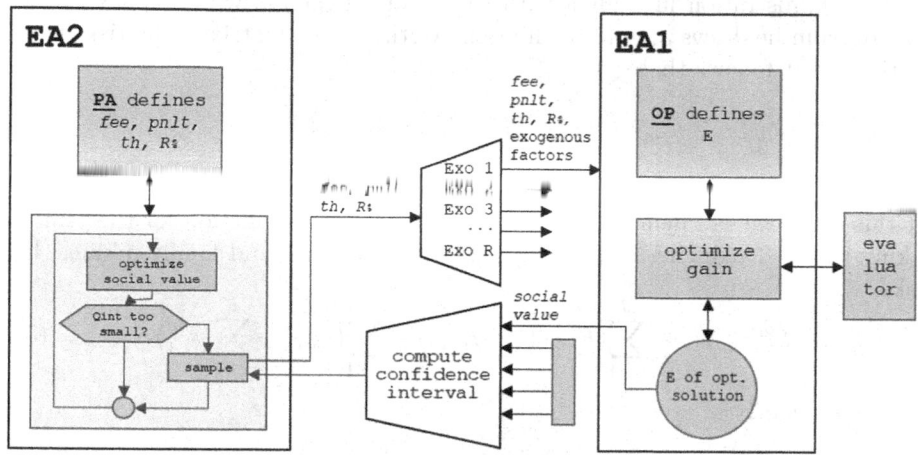

Fig. 1. Algorithm scheme. Red arrow represent confidence intervals. (Color figure online)

3.4 Operator (EA1)

The operator is interested in maximizing his earnings, and consequently will tend to reduce the effort of construction and operation as much as possible. Limits to the reduction are the administration penalties and positive externalities[8].

Genotype. The EA1 (OP) genotype expresses the E percentage effort cost matrix through a number of floating-point genes of equal size, $J \cdot T$, with $E_{j,t} > 0$ $\forall\ j, t$. Stricter limits can be set.

Fitness. Let p be the matrix of the penalties imposed. If in the year t, for the KPI k, the OP does not reach the established service level, th_k, a penalty equal to $pnlt_k$ is imposed:

$$p_{k,t} = \begin{cases} pnlt_{k,t} & \text{if } af_k(E)_{k,t} < th_k \wedge \text{ KPI k is active in time t} \\ 0 & \text{otherwise} \end{cases} \tag{2}$$

---

[8] Remember that the externality is positive when the coefficient between effort costs and externality is positive: consequently a positive externality causes damage compared to a reduction costs.

Let ε^* be the matrix of the risks that can be propined to the administration:

$$\varepsilon_{r,t}^* = \varepsilon_{r,t} + f_x(E)_{r,t} \cdot A_r^\% \tag{3}$$

The first addend represents the exogenous factors, and the second the externalities. Both addends can be both positive and negative, and if the exogenous risk plus the propinable risk result in a negative sum, the OP requests the support of the administration in dealing with the greater expense, while if they give a positive sum he draws a profit for himself. With r^{op} the matrix of the risks paid by the OP, it follows that

$$r_{r,t}^{op} = \begin{cases} \varepsilon_{r,t}^* \cdot (1 - R_r^\%) & \text{if } \varepsilon_{r,t}^* < 0 \\ \varepsilon_{r,t}^* & \text{otherwise} \end{cases} \tag{4}$$

At this point we can define a contribution to the fitness function for the year t, taking the *fee*, reducing it by fixed costs and penalties, and finally adding the risks.

$$L_t^{op} = fee - \sum_{j=1}^{J}(F_{j,t} \cdot (1 + E_{j,t})) - \sum_{k=1}^{K}(p_{k,t}) + \sum_{r=1}^{K}(r_{r,t}^{op}) \tag{5}$$

and consequently

$$L_{op} = \sum_{t=1}^{T}(L_t^{op} \cdot \frac{1}{(1 + t_{sn})^t}) \tag{6}$$

3.5 Public Administration (EA2)

Genotype. The EA2 (PA) genotype expresses: the concession *fee*, with a floating-point gene; the matrix of monetary penalties *pnlt* and thresholds *th*, both with a number of floating-point genes equal to K; the risk attribution vector $R^\%$, with a number of floating point genes equal to R.

The limits for *th* are to be calibrated with respect to the codomain of the function $f_k(E)$, of which they represent a yardstick. The genes of *pnlt* and *fee* are limited according to the user's common sense, and always greater than zero. The genes of $R^\%$ are limited between zero and one.

Fitness. The administration is interested in optimizing the *social value*, i.e., maximizing the effort exerted by the operator to carry out the work in a workmanlike manner, and minimizing the expense. In the model described, the social value is equal to the expected cash flows multiplied by the effort, less the *fee*. Therefore, some corrective measures must be defined, first of all with respect to risk sharing. When the administration has to pay for an exogenous risk (real or served up), it loses money that it could invest in other works. The social value is therefore reduced by what the r^{pa} matrix indicates, specular of r^{op}, with

$$r_{r,t}^{pa} = \begin{cases} \varepsilon_{r,t}^* \cdot R_r^\% & \text{if } \varepsilon_{r,t}^* < 0 \\ 0 & \text{otherwise} \end{cases} \tag{7}$$

On the other hand, it was not considered useful to calculate a gain to the administration for the fines imposed. The algorithm thus interprets the sanctions as defined by law and common sense, or as a tool to direct the work towards the expected quality.

We can now define a contribution to the fitness function for the year t as

$$L_t^{pa} = -fee + \sum_{j=0}^{J}(F_{j,t} \cdot (1 + E_{j,t})) + \sum_{r=0}^{r}(r_{k,t}^{pa}) \qquad (8)$$

Note that until now the administration is not in the least interested in the good performance of the OP's checkouts, but simply aims to provide the best possible service by paying for it as little as possible. Contexts of this type would present very high penalties for any disservice, no premiums or of a value lower than the effort to obtain them, a concession fee reduced to a simulation minimum, the assumption of a zero percentage of risk. We then complete the fitness function with the penalty function for failure f_d:

$$L_{pa} = \sum_{t=0}^{T}(L_t^{pa} \cdot \frac{1}{(1 + t_{sn})^t}) + f_d(L_{op}) \qquad (9)$$

4 Experimental Evaluation

The algorithm was implemented in Python3 and DEAP Framework [6] and is available on Github[9] under EUPL 1.2 license. The simulation parameters are configured by extending an apposite class. They have the same names as those described in the model. Dynamics of simulation follows what has been described above: the public administration generates a certain number of solutions, which represent fee, $pnlt$, th and $R_\%$, and for each compute a certain number of samplings, with different exogenous factors. For each sample, the values of the exogenous factors and the specifications are passed to the simulation of the OP, who maximize his gain choosing an optimal effort E. With the results obtained, the public administration can proceeds in her steps and optimizes social value [3–5].

The inner optimization, which simulates the choices of the OP, are performed with a full EA, with tournament selection and polynomial mutation. The following section contains some preliminary simulations to verify the feasibility of the approach.

4.1 Stochastic Optimization

The outer optimization, which simulates the choices of the PA, it's much harder, since L_{pa} it does not return numerical values, but confidence intervals. Two confidence intervals centered in c_0 and c_1, with $c_1 > c_0$, of dimensions s_0 e s_1,

---

[9] https://github.com/rebuglio/evolutionary-psc-v4.

Table 1. Test case results

Test	Mean SV first-best		Optimization output				
	After 1 min	At end	i	Fee	th_i	$pnlt_i$	R_i
1a	-770	-29	0	374	-0.054	206	67
			1		-0.005	224	
1b	-318	-4	0	371	-0.050	437	17
			1		-0.005	95	
2	-1000	-403	0	534	0.074	498	36
			1		0.198	242	
3	-255	-31	0	306	-0.176	382	20
			1		-0.199	316	
4	-206 ± 1768	-352 ± 215	0	149	-0.149	402	97
			1		-0.113	260	

are comparable if $c_1 - c_0 > \frac{s_0 + s_1}{2}$. In case of overlapping intervals it is necessary to provide new samplings. A global index of comparability of the population finesses is defined as follow:

$$Q_{int} = \frac{\text{comparable pairs}}{\text{total pairs}} \tag{10}$$

The PA optimization algorithm is stopped when Q_{int} is below a certain pre-determined level Q_{int}^{th}, generating new samples. Also in this case, tournament selection and polynomial mutation are adopted. The entire process is represent in Fig. 1.

4.2 Analysis

The examples in this section, unless otherwise specified, have the following common data. Inflation rate $t_i = 0$, real discount rate $t_{sr} = 0.05$. Two activities: j_1 construction, first period, and j_2 operation, lasting five periods; consequently $T = 6$. One risk, $R = 1$. The OP can vary each of his efforts $E_{j,t}$ up to 20% more or less. There is joint and several liability with penalty function for failure $f_d(x) = 5x$ with $x < 0$. Two KPIs, the first determined equally by construction and operation $f_k(E)_{k=1,t} = E_{j=1,t}/2 + E_{j=2,t}/2$, and the second direct observable of the conduction $f_k(E)_{k=2,t} = E_{j=2,t}$. Flows defined as follows:

t	1	2	3	4	5	6
build	1000					
operate		250	250	250	250	250

Test 1. The first test is in the absence of exogenous factors, $\varepsilon_{r,j} = 0$, externality, $f_x(E)_{j,t} = 0 \ \forall \ j,t$ and asymmetry, $A_r^\% = 0 \ \forall \ r$. We run the test twice with different random seed.

The algorithm has consistently recognized the need to introduce penalties and to reduce the concession fee to a reliable value, approximately equal to the value of the flows that the OP must face. The simulation was performed twice with different random suits, the results are in Table 1.

Test 2. Now let's set up a harder default-penalty function to manage, $f_d(x) = x + 1000$ con $x < 0$. Also in this case the algorithm manages to optimize. Optimization result is shown in Table 1.

Test 3. Based on first test data, now introducing positive externalities for the first two years of operation equal to one third of the effort cost under construction:

$$f_x(E)_{j,t} = \begin{cases} \frac{1000 * E_{j=1,t=1}}{3} & \text{con } t \in [2,3], j = 2 \\ 0 & \text{otherwise} \end{cases} \tag{11}$$

The strong externality has given its effects: the social value is better, even in the face of penalties with rather low thresholds. Optimization result is shown in Table 1.

Test 4. Now, based on the data from Test 3, let's add a noise factor as well. As can be seen from Fig. 2, the size of the intervals decreases, and at the same time the average increases. Optimization result is shown in Table 1.

4.3   Real World Case

The motorway sector follows the model discussed with regard to externalities [2,11]. The a99 motorway is an example taken from Martiniello, a member of the UTFP[10], which reports the data in anonymized form of a real motorway concession [9]. In the following example, the amounts and risks are taken— with some modifications—from Martiniello, while the externality functions, the asymmetry, and the choice of KPIs are arbitrary.

Data and Settings. We consider an inflation rate $t_i = 2.5\%$, a real discount rate $t_{sr} = 5\%$, and then a nominal discount rate

$$t_{sn} = (1 + t_i) \cdot (1 + t_{sr}) - 1 = 7.625\% \tag{12}$$

---

[10] Italian authority for supporting the drafting of PPPs.

Table 2. A99 highway discounted flows.

	2003 k€	2004 k€	2005–2008 k€	2009 k€	2010 k€	2011–2032 k€
J0	34.884	–	–	–	–	–
J1	113.374	110.609	–	–	–	–
J2	7.412	7.231	27.204	–	–	–
J3	229.800	224.195	843.416	–	–	–
J4	21.179	20.662	77.732	–	–	–
J5	1.491	1.455	5.472	–	–	–
J6	–	–	–	–	–	326.215
J7	–	–	–	–	3.975	66.642
J8	–	–	–	948	925	15.500
J9	–	–	–	9.562	9.329	156.401
J10	–	–	–	11.372	11.095	186.006
J11	–	–	–	4.362	4.256	71.353
J12	–	–	–	3.793	3.701	62.044

Given the considerable complexity of simulation, we group the "affine periods" as an unique period[11]. For this purpose, vector L of lengths of each group of affine periods is defined:

$$L = 1, 1, 4, 1, 1, 22 \tag{13}$$

and consequently a vector of multipliers M_g is defined equal to the sum of the inflation indices of the periods in the group. For example, for the group of periods 2005–2008 (from the third to the sixth year), the multiplier is:

$$M_{2005-2008} = \sum_{t=3}^{6} \frac{1}{(1+t_i)^y} = 3.67 \tag{14}$$

by repeating the operation for each group we obtain:

$$M = [1.000, 0.975, 3.670, 0.862, 0.841, 14.104] \tag{15}$$

Using this multipliers, we obtain the grouped and discounted raw psc shown in Table 2. OP can vary each of his efforts $E_{j,t}$ from -20% up to $+20\%$. The bankruptcy costs functions is defined as $f_d(x) = 2x$. Exogenous factors, externalities functions and asymmetries are shown in Table 3[12].

---

[11] Affine periods are the periods that share the same raw psc, the same exogenous factors and the same externalities, according to the intuitive logic that under the same conditions, OP and PA will do the same actions. This compression is lossy, the model allows for partial or total compression as the user needs.

[12] Exogenous factor is reporting, according to Martiniello, as a percentage of same raw psc value (i.e. J1, J2...). In this way we don't need other actualization or discounting.

Table 3. A99 exogenous factors and externalities.

	Exogenous factors ε							Externalities f_x	Asymmetry $A_\%$	Active
R0	Value	% on	J0	0	0	20,00	60,00	80,00		0.20	2003–2003
	Occurrency	%		0	0,24	0,31	0,25	0,20			
R1	Value	% on	J1	3,00	0	5,00	8,00	12,00		0.30	2003–2004
	Occurrency	%		0,05	0,20	0,40	0,20	0,15			
R2	Value	% on	J2	0	0	8,00	12,00	15,00		0.10	2003–2008
	Occurrency	%		0	0,22	0,20	0,28	0,30			
R3	Value	% on	J3	−10,00	0	10,00	20,00	30,00	$-E_{0,0}*0.05$	0.40	2003–2008
	Occurrency	%		0,01	0,10	0,44	0,38	0,07			
R4	Value	% on	J3	−5,00	0	7,00	12,00	15,00	$-E_{0,0}*0.05$	0.50	2003–2008
	Occurrency	%		0,05	0,20	0,25	0,20	0,30			
R5	Value	% on	J5	0	0	4,00	7,00	10,00		0.10	2003–2008
	Occurrency	%		0	0,25	0,35	0,25	0,15			
R6	Value	% on	J6	0	0	0,00	4,00	12,00	$-E_{3,0}*0.10$	0.20	2011–2032
	Occurrency	%		0	0,30	0,35	0,25	0,10			
R7	Value	% on	J7	0	0	10,00	20,00	30,00	$-E_{3,0}*0.10$	0.30	2011–2032
	Occurrency	%		0	0,35	0,30	0,25	0,10			
R8	Value	% on	J7	0	0	10,00	20,00	30,00		0.30	2011–2032
	Occurrency	%		0	0,35	0,26	0,27	0,12			

Results. The simulation was run in three different configurations: simulation 1 as described, simulation 2 with total information asymmetry, simulation 3 with absence of information asymmetry. Trends of the best individuals fitnesses are respectively reported in Fig. 2, 3, 4 and 5. Optimization result is shown in Table 4.

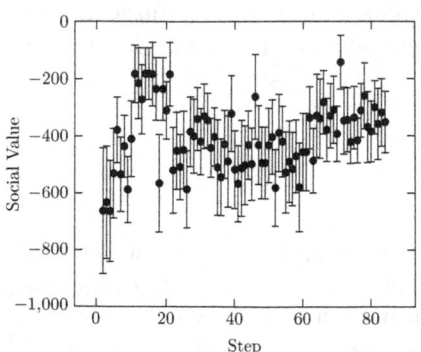

Fig. 2. Sym test 4, trends of first-best conf. int.

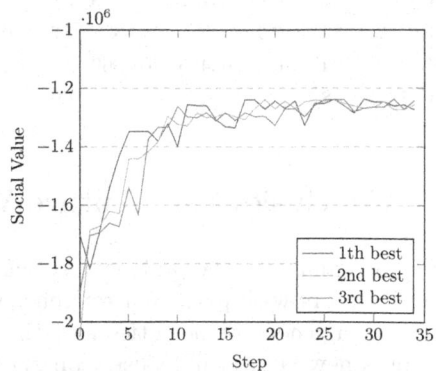

Fig. 3. Sym 1, best three trends.

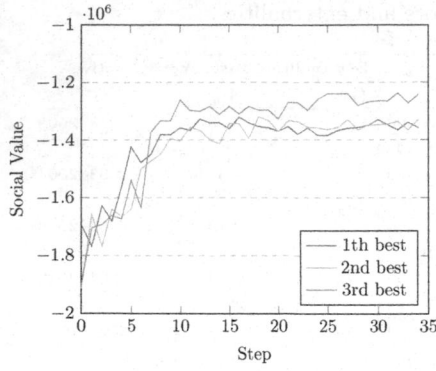

Fig. 4. Sym 2, best three trends.   **Fig. 5.** Sym 3, best three trends.

Table 4. A99 optimization output.

i	Sym 1			Sym 2			Sym 3		
	Best SV −1211036 ± 24031			−1357461 ± 35712			−1298873 ± 19867		
	Fee k€ 704.975			762.752			775.300		
	Ri **Thi**		**Pnlti**	**Ri** **Thi**		**Pnlti**	**Ri** **Thi**		**Pnlti**
	%		k€	%		k€	%		k€
1	77	−0.11881565	393.659	93	0.07983518	168.400	74	0.08584949	315.309
2	4	0.04202931	2.779	12	0.11218364	45.717	5	−0.03578963	209.046
3	56	0.04563693	423.960	74	0.01201033	93.668	69	−0.04796428	68.179
4	79	−0.02966594	210.129	13	−0.06665999	452.266	9	−0.04004140	487.925
5	80	0.00396262	489.459	8	−0.01997641	92.883	40	−0.19989360	116.775
6	72	0.12899980	444.327	77	0.15985616	383.401	91	0.15461932	313.755
7	64	−0.13563455	401.369	1	−0.19887337	405.815	1	−0.07271345	423.954
8	98			69			34		

5   Conclusions and Future Work

We have analyzed the problems associated with the writing of an efficient PPP specification, with particular reference to the allocation of unexpected expenses, the management of penalties and the determination of the fixed fee amount. Using a new approach, focusing on known data and relationships, we have modeled an adverse optimization, in which the public administration can calibrate specifications according to the expected reactions of the counterpart. We have implemented the model in python language, making use of evolutionary algorithms, and we have tested it on various test cases at and on a real highway concession. In all cases, the algorithm was able to optimize the parameters of the specifications, showing the feasibility of this approach. In the future, we intend to greatly the operator optimization (EA1), trying to replace the evolutionary algorithm with a more efficient deterministic optimizer, in order to reduce simulation times. Another important development will be the test of the algorithm on new real cases, supported by domain experts.

References

1. Amatucci, F., Pezzani, F., Vecchi, V. (eds.): Le scelte di finanziamento degli enti locali. EGEA, Milano (2009). oCLC: 606734566
2. Bardaka, E., Zhang, Z., Labi, S., Sinha, K.C., Mannering, F.: Statistical assessment of the cost effectiveness of highway pavement warranty contracts. J. Infrastruct. Syst. 22(3), 04016017 (2016)
3. Deb, K., Agrawal, R.B., et al.: Simulated binary crossover for continuous search space. Complex Syst. 9(2), 115–148 (1995)
4. Dembe, A.E., Boden, L.I.: Moral hazard. a question of morality? NEW SOLUTIONS: J. Environ. Occupational Health Policy 10(3), 257–279 (2000)
5. Fogel, D.B., Back, T., Michalewicz, Z. (eds.): Evolutionary computation. Institute of Physics Publishing, Bristol; Philadelphia (2000). oCLC: ocm44807816
6. Fortin, F.A., De Rainville, F.M., Gardner, M.A., Parizeau, M., Gagné, C.: DEAP: evolutionary algorithms made easy. J. Mach. Learn. Res. 13, 2171–2175 (2012)
7. Menezes, F., Wooders, M., Iossa, E., Martimort, D., Martimort, D., et al.: The simple microeconomics of public-private partnerships. J. Pub. Econ. Theory 17(1), 4–48 (2015)
8. Martimort, D., Pouyet, J.: To build or not to build: normative and positive theories of public-private partnerships. Int. J. Ind. Organ. 26(2), 393–411 (2008)
9. Martiniello, U.: The calculation of the public sector comparator in the road transport sector. RIREA, Roma (2005). oCLC: 1260488508
10. Pellegrino, S., Perboli, G., Squillero, G.: Balancing the equity-efficiency trade-off in personal income taxation: an evolutionary approach. Econ. Politica 36(1), 37–64 (2018)
11. Singh, R.: Public-private partnerships vs. traditional contracts for highways. Indian Econ. Rev. 53(1–2), 29–63 (2018). https://doi.org/10.1007/s41775-018-0032-0
12. European Court of Auditors: Public private partnerships in the EU: Widespread shortcomings and limited benefits (2018). https://www.eca.europa.eu/Lists/ECADocuments/SR1809/SRPPPIT.pdf

The Asteroid Routing Problem:
A Benchmark for Expensive Black-Box
Permutation Optimization

Manuel López-Ibáñez[(⊠)] [ID], Francisco Chicano[ID], and Rodrigo Gil-Merino

ITIS Software, Universidad de Málaga, Málaga, Spain
{manuel.lopez-ibanez,chicano,gilmerino}@uma.es, chicano@lcc.uma.es

Abstract. Inspired by the recent 11th Global Trajectory Optimisation Competition, this paper presents the asteroid routing problem (ARP) as a realistic benchmark of algorithms for expensive bound-constrained black-box optimization in permutation space. Given a set of asteroids' orbits and a departure epoch, the goal of the ARP is to find the optimal sequence for visiting the asteroids, starting from Earth's orbit, in order to minimize both the cost, measured as the sum of the magnitude of velocity changes required to complete the trip, and the time, measured as the time elapsed from the departure epoch until visiting the last asteroid. We provide open-source code for generating instances of arbitrary sizes and evaluating solutions to the problem. As a preliminary analysis, we compare the results of two methods for expensive black-box optimization in permutation spaces, namely, Combinatorial Efficient Global Optimization (CEGO), a Bayesian optimizer based on Gaussian processes, and Unbalanced Mallows Model (UMM), an estimation-of-distribution algorithm based on probabilistic Mallows models. We investigate the best permutation representation for each algorithm, either rank-based or order-based. Moreover, we analyze the effect of providing a good initial solution, generated by a greedy nearest neighbor heuristic, on the performance of the algorithms. The results suggest directions for improvements in the algorithms being compared.

Keywords: Spacecraft trajectory optimization · Unbalanced Mallows Model · Combinatorial Efficient Global Optimization · Estimation of distribution algorithms · Bayesian optimization

1 Introduction

Several space programs from different countries focus on small bodies orbiting around the Sun and the Earth. Apart from the search for answers about the origin and evolution of the Solar System, there are other practical reasons that space agencies consider. One of these reasons is related to the present decrease of mineral resources on Earth. These resources, heavily used in technological devices like mobile phones and computers, include silicon, quartz, boronite and

© Springer Nature Switzerland AG 2022
J. L. Jiménez Laredo et al. (Eds.): EvoApplications 2022, LNCS 13224, pp. 124–140, 2022.
https://doi.org/10.1007/978-3-031-02462-7_9

others, and are hard to mine for: the density of these minerals is usually high, and they fell to inner strata during Earth formation, which translates into an extra difficulty when mining for these materials. As an alternative to Earth mining for techno-minerals, space agencies have proposed asteroids and near Earth objects (NEOs) [3,4] as near-term mining targets. These objects contain varying amounts of rare earth metals, minerals and water at surface levels, consequently much easier to mine for than on Earth. A clear drawback to these plans is the transport from and to the asteroids, which will imply high consumption of energy. A solution to reduce the energy consumption is to build human settlements on large asteroids and/or on orbit, together with space solar power stations that could fuel spacecrafts and those settlements. This idea firstly appeared on a science-fiction novel and later it was popularized as a "gedankenexperiment" by Dyson in the 60's. Nowadays, the idea of a swarm of space solar power stations is called "the Dyson sphere". In its last edition, the 11th Global Trajectory Optimisation Competition (GTOC11, https://gtoc11.nudt.edu.cn) proposed to computationally implement a Dyson ring, which is a set of space stations (12 in this case) orbiting the Sun in a circular orbit, to mine close to Earth passing-by asteroids. The Dyson ring construction was proposed as an optimisation problem in the context of orbital mechanics. The first part of the optimization problem consists in finding a plan (trajectory) to visit a selected subset of asteroids with a spacecraft launched from Earth. This spacecraft does not return to Earth. The second part of the problem consists in activating a device, called asteroid transfer device or ATD, located in the visited asteroids to guide them to the space stations in the Dyson ring. Once the Dyson ring is formed, regular visits to the space stations in the Dyson ring are supposed to bring the minerals in the asteroids to Earth. We are interested in this paper in solving the first optimization problem, for which we provide a formulation as a challenging expensive black-box permutation problem that we call Asteroid Routing Problem (ARP) that may be used to compare and benchmark optimization approaches.

In recent years, there has been growing interest in tackling expensive black-box permutation problems [1,6,9,11]. In permutation problems, the decision space is the space of permutations of a given length n, usually denoted with S_n. Black-box optimization, as considered in this paper, tackles an optimization problem without requiring an explicit analytical model of its objective function and constraints. Finally, in expensive optimization, each function evaluation is often the result of a costly simulation or physical experiment, thus the total budget of evaluations allowed before returning the best solution found is severely restricted, typically no more than 1000 [8,9]. The combination of these three characteristics gives rise to a challenging family of optimization problems. Typical approaches include estimation-of-distribution algorithms [6] and combinatorial Bayesian optimizers [11] or combinations of both [9].

A potential criticism of recent works is that the benchmark problems considered—e.g., the travelling salesman problem, the linear ordering problem, etc.—are taken from the classical combinatorial optimization literature and they are neither truly expensive nor black-box. The ARP proposed here is much more expensive to evaluate than classical combinatorial optimization problems, even for short permutations. Moreover, there is no explicit analytical model of the problem from the routing perspective, thus classical techniques such as fast neighborhood evaluation or gradient calculation are not possible. In our formal definition of the ARP, we aim to preserve the most challenging features of the problem, without requiring expert knowledge of the astrophysical details. Benchmark generation and evaluation code is made publicly available. We also provide preliminary results using state-of-the-art optimization methods. Our hope is that this work will raise interest in the field of expensive black-box combinatorial optimisation and its applications to trajectory optimisation.

This paper is structured as follows. Section 2 gives a brief introduction to basic astrodynamical concepts required to understand the problem as an optimization benchmark. We do not explain in detail the underlying calculations because (1) these are standard formulas available in many physics textbooks, (2) they are implemented by multiple open-source software libraries and (3) they are not directly available to any of the optimization algorithms considered here. The optimization model of the proposed Asteroid Routing Problem is given in Sect. 3. The optimization algorithms evaluated in this work are described in Sect. 4. We provide an experimental analysis of these algorithms on the ARP in Sect. 5 and conclude in Sect. 6 with a summary of findings and suggestions for further research motivated by these findings as well as possible extensions of the ARP benchmark.

2   Background

In this section we will briefly revise the basic theory behind the astrodynamics of the problem. We will start describing the two-body problem and its solution leading to the Keplerian orbits. Then, we will describe one type of maneuver used by the spacecrafts to change their trajectory in space and we will end this section presenting the Lambert problem to determine the impulses required to reach a point in space and time from any other point.

2.1   Two-Body Problem

The dynamics of two bodies in space subject to the gravitational force is determined by Newton's second law of dynamics and gravitation law. Combined, they lead to the following vector equation:

$$\frac{d^2\mathbf{r}}{dt^2} = -\frac{\mu}{r^3}\mathbf{r}, \tag{1}$$

where $\mathbf{r}$ is the position vector of the second object respect to the first one, $r = |\mathbf{r}|$ is the module of the position vector, t is time, and μ is usually called *gravitational parameter*, which for our solar system is typically assumed to be a constant due to the mass of the Sun being much larger than the mass of any other object.

Despite the dynamics of the two bodies happening in 3D space, the conservation of the angular momentum forces the movement to happen in a plane and the analytical solutions of the problem are conic curves (also called Keplerian orbits when they refer to the two-body problem): ellipse (and circumference as a particular case), hyperbola and parabola[1]. Both, the hyperbola and parabola are open curves, which means that the object following one of them will come from outside the solar system and after approaching the Sun will disappear outside the solar system without coming back again. Solar system objects, like planets and asteroids, follow an elliptic (or circular) orbit around the Sun.

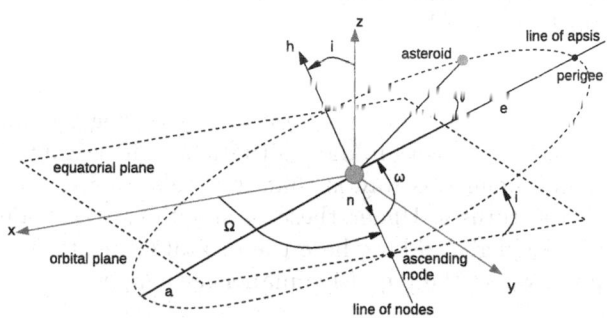

Fig. 1. The five parameters to describe an elliptic orbit.

When we fix a reference frame in space, elliptic orbits can be characterized by five *orbital elements*: a (semi-major axis), e (eccentricity), i (inclination), ω (argument of periapsis[2]) and Ω (longitude of the ascending node) (see Fig. 1). The position of the object in the orbit is characterized by a sixth parameter which changes with time: the true anomaly, ν. The term *epoch* is used in this context to refer to a specific *date* and *time*. The movement of an object, i.e., its position and velocity vectors, in an elliptical orbit is completely determined by the six orbital elements described above in a given epoch.

When there are more than two bodies, as it happens in our solar system, the movement equations cannot be solved analytically, except for a few particular cases. In that non-analytical case, the solution of the two-body problem is still useful. The presence of additional bodies can be considered in most of the cases a minor perturbation in the Hamiltonian describing the two-body problem and, as a consequence, the orbital elements are slowly changing with time [5, Chapter 11]. In our formulation of the problem, however, we do not consider these perturbations, and we assume that all objects follow a Keplerian orbit.

---

[1] The trajectory of the second object could also be linear, but this is not of practical interest, because this would end with the object destroyed in the Sun.

[2] The periapsis is the nearest point of an elliptic orbit to the object in the focus. When the object is the Sun, the periapsis is also called perihelion.

2.2   Maneuvers in Space

Artificial spacecrafts also follow a Keplerian orbit around the Sun when they are far from the gravitational influence of the Earth or any other solar system object. They usually have an engine that allows them to change the orbit using operations called *maneuvers*. There are two main kinds of maneuvers: *continuous* and *impulsive*. In a continuous maneuver the engine of the spacecraft is on during a long period of time (e.g., days) and generates a force whose direction and magnitude can be controlled. We will only consider impulsive maneuvers here.

Impulsive maneuvers cover a wide range of different space maneuvers, from small orbital adjustments when two spacecrafts encounter each other, to large orbital corrections between a low-Earth orbit and an interplanetary transit. To carry out an impulsive maneuver, usually small engine firings take place, changing only the direction and magnitude of the velocity vector instantaneously. This means that the position vector does not change during the impulse, which, although an idealisation, it works in most situations. The advantage of treating the impulsive maneuver in this way is that it avoids solving the equations of motion including the thrusts. Under these assumptions, the trajectory of the spacecraft is still a Keplerian orbit where the orbital elements change instantaneously at the point where the impulsive maneuver happens.

2.3   Lambert Problem

At the heart of Astrodynamics, a key problem is to determine the Keplerian orbit connecting two positions in a given period of time Δt. This is the Lambert's problem, also known as the Gauss' problem [7].

We can use the solution of Lambert's problem to determine the impulsive maneuvers required for a spacecraft to do a rendezvous with an asteroid. The transit orbit and impulsive maneuvers can be completely characterized by two times: the epoch at which the transit starts τ and the transit time t. These two times determine the two points P_1 and P_2 in space we need to join. P_1 is the point at which the spacecraft is at τ. P_2 is the point at which the asteroid is at $\tau + t$. Lambert's problem is solved to go from P_1 to P_2 in time t. Two impulsive maneuvers are used to move from P_1 to P_2. A first impulsive maneuver $\Delta \mathbf{v}_1$ at P_1 inserts the spacecraft in the transit orbit determined by solving Lambert's problem. A second impulsive maneuver $\Delta \mathbf{v}_2$ at P_2 inserts the spacecraft in the same orbit as the asteroid (rendezvous). Let's denote with s the orbit of the spacecraft (six orbital parameters in an epoch) and with a the orbit of the asteroid[3]. Then, $\Delta \mathbf{v}_1$ and $\Delta \mathbf{v}_2$ are determined by s, a, τ and t:

$$(\Delta \mathbf{v}_1, \Delta \mathbf{v}_2) = \text{Lambert}(s, a, \tau, t) \tag{2}$$

---

[3] We previously used a to mean the semi-major axis of an orbit, from now on we will use it to denote the complete orbit of an asteroid.

3   Asteroid Routing Problem

The *Asteroid Routing Problem* consists in finding a route followed by a spacecraft launched from Earth to visit a given set of n asteroids $A = \{a_1, a_2, \ldots, a_n\}$ that minimizes both the sum of velocity impulses required by the route (related to consumption) and the total time required to visit all of them. A solution to this problem is a pair $(\pi, \mathbf{t})$, where $\pi \in S_n$ is a permutation representing the order in which the asteroids are visited and $\mathbf{t} \in \mathbb{R}_{\geq 0}^{2n}$ is a vector of $2n$ non-negative real numbers representing parking and transit times to reach each asteroid.

We will consider a start epoch τ_0, determined by the instance, when the spacecraft is on Earth. The spacecraft remains on Earth during a time t_1 after it. Then, it is launched at epoch $\tau_0 + t_1$ to reach the first asteroid in the sequence, $a_{\pi(1)}$, using a transit orbit computed by solving Lambert's problem between the Earth and the asteroid. In the launch, the spacecraft experiments an impulse $\Delta\mathbf{v}_1$ that inserts the spacecraft in a transit orbit to reach asteroid $a_{\pi(1)}$ in time t_2 after the launch epoch. Once the asteroid is reached, a new impulse $\Delta\mathbf{v}_2$ is used to make the spacecraft to follow the orbit of asteroid $a_{\pi(1)}$.

In order to visit the i-th asteroid, $a_{\pi(i)}$, for $i > 1$, the spacecraft remains during time t_{2i-1} in the orbit of asteroid $a_{\pi(i-1)}$ and receives an impulse $\Delta\mathbf{v}_{2i-1}$ to reach asteroid $a_{\pi(i)}$ in time t_{2i}. Then, a new impulse $\Delta\mathbf{v}_{2i}$ changes the orbit of the spacecraft to coincide with the one of $a_{\pi(i)}$.

For the transit between asteroids $a_{\pi(i-1)}$ and $a_{\pi(i)}$, the impulses $\Delta\mathbf{v}_{2i-1}$ and $\Delta\mathbf{v}_{2i}$ are computed using Eq. (2). All times at odd positions of the time vector, t_{2i-1}, are parking times, because the spacecraft is waiting in an asteroid (or Earth's) orbit, while the times at even positions of the time vector, t_{2i} are transit times, because the spacecraft is traveling through a transit orbit to reach the next asteroid.

The two objectives of the problem to minimize are the sum of all the velocity impulses, which is related to energy consumption, and the sum of all parking and transit times, which gives the elapsed time from launch to arriving at the last asteroid. More formally, the two objectives are:

$$\Delta v = \sum_{i=1}^{2n} |\Delta\mathbf{v}_i| \quad \text{and} \quad T = \sum_{i=1}^{2n} t_i \ , \tag{3}$$

where $\Delta\mathbf{v}_i$ is computed as follows:

$$(\Delta\mathbf{v}_{2i-1}, \Delta\mathbf{v}_{2i}) = \text{Lambert}\left(a_{\pi(i-1)}, a_{\pi(i)}, \tau_{i-1}, t_{2i}\right) \ , \tag{4}$$

where $\tau_{i-1} = \tau_0 + \sum_{j=1}^{2i-1} t_j$ is the launch epoch from orbit $a_{\pi(i-1)}$ and we consider that $a_{\pi(0)}$ is Earth's orbit.

An instance of the problem is completely determined by the orbital parameters at a given epoch of Earth and all asteroids, the starting epoch τ_0 and the gravitational parameter μ.

In this paper, we will focus on a scalarized variant of the above bi-objective problem that simply aggregates the two objectives in Eq. (3) as follows:

$$f(\pi, t) = \Delta v + \frac{2 \text{ km/s}}{30 \text{ days}} \cdot T, \tag{5}$$

where the constant in front of T is based on preliminary experiments.

4 Optimization Algorithms

In order to use the ARP as a benchmark problem for expensive black-box combinatorial optimization, we tackle the problem in a hierarchical manner. The *inner* optimization problem decides the values of the vector of times given a permutation of the asteroids, while the *outer* problem aims to find an optimal permutation of the asteroids. Since our focus here is the outer problem, we always tackle the inner problem using the deterministic SLSQP (Sequential Least Squares Programming) algorithm, thus ensuring that we obtain the same objective function value given the same permutation. For the outer problem, we evaluate and compare two algorithms for expensive black-box optimization problems, namely, Unbalanced Mallows Model (UMM) [6] and Combinatorial Efficient Global Optimization (CEGO) [11]. In addition, we study the effect that a good initial solution has on the behavior of the black-box algorithms. This initial solution is found by means of a gray-box greedy nearest neighbor heuristic. All these algorithms are described below.

4.1 Sequential Least Squares Programming (SLSQP)

The inner optimization of the vector of times is done using the implementation of SLSQP provided by SciPy [10] without an explicit gradient, which is estimated by the SLSQP algorithm using 2-point finite differences with an absolute step size of 1.49e-08. Each transfer is optimized independently, that is, SLSQP solves a sequence of n problems with two numerical decision variables (x_0, x_1), which correspond respectively to the parking and transit times to visit a particular asteroid. Parking times are bounded by $[0, 730]$, while transit times are bounded by $[1, 730]$, both measured in days.

4.2 Greedy Nearest Neighbor Heuristic

A reasonably good permutation of the asteroids can be generated by iteratively visiting the nearest asteroid, in Euclidean distance, to the last-visited asteroid, after calculating the positions of all unvisited asteroids at the arrival time at the last-visited asteroid. This heuristic is presented in Algorithm 1. The algorithm keeps track of the orbit of the spacecraft s and its arrival time (epoch) τ at the last-visited asteroid, which are initialized to the Earth's orbit $a_{\pi(0)}$ and the initial epoch τ_0, and a set of unvisited asteroids U. At each iteration i, an asteroid $a_{\pi(i)}$ is chosen from U according to the minimum value of $d(s, a, \tau)$, which calculates the Euclidean distance between the Cartesian positions of the orbits of the spacecraft s and asteroid a when both are considered at time τ. After the asteroid $a_{\pi(i)}$ is chosen, we solve the inner problem—using SLSQP as described above—to find the parking time t_{2i-1} at asteroid $a_{\pi(i-1)}$ and transit time t_{2i} to asteroid $a_{\pi(i)}$. The algorithm finishes when all asteroids are visited, which results in a permutation π of the asteroids and the corresponding vector $\mathbf{t}$ of parking and transit times.

Algorithm 1. Greedy Nearest Neighbor Heuristic.

1: $s := a_{\pi(0)}$ // Earth's orbit
2: $\tau := \tau_0$ // Epoch of the spacecraft
3: $U := \{1, \dots, n\}$
4: **for** $i := 1$ **to** $n - 1$ **do**
5:     $\pi(i) := \arg\min_{j \in U} d(s, a_j, \tau)$ // Euclidean distance between orbits at epoch τ
6:     $(t_{2i-1}, t_{2i}) := \mathsf{SLSQP}(s, a_{\pi(i)})$ // Solve inner problem
7:     $\tau := \tau + t_{2i-1} + t_{2i}$
8:     $U := U \setminus \{\pi(i)\}$
9:     $s := a_{\pi(i)}$
10: **return** $(\pi, \mathbf{t})$

4.3   Unbalanced Mallows Model (UMM)

The UMM algorithm [6] is an estimation-of-distribution algorithm based on the Mallows model and unbalanced Borda learning. The Mallows model for permutation spaces is defined by a reference permutation σ_0 and a dispersion parameter θ, which are analogous to the mean and variance of the Gaussian distribution. At each step of UMM, the permutations evaluated so far together with their objective values are used to calculate a *weighted mean* permutation $\hat{\sigma}_0$. From this $\hat{\sigma}_0$ and a value of the dispersion parameter θ, a new permutation is sampled at each iteration from the Mallows model. The dispersion parameter θ is set by UMM in such a way that, at the first iteration, the expected Kendall's-τ distance of newly sampled permutations from the reference permutation is at half of the expected distance of an uniform sample. In subsequent iterations, the parameter θ is adjusted such that the expected Kendall's-τ distance linearly decreases until the expected distance is 1 at the last iteration.

Empirical results of UMM on expensive black-box variants of the linear ordering problem and the permutation flowshop scheduling problem showed results [6] comparable to the Combinatorial Efficient Global Optimization (CEGO), with the additional benefit of being computationally much less expensive than CEGO.

4.4   Combinatorial Efficient Global Optimization (CEGO)

CEGO extends classical Bayesian optimizers based on Gaussian processes to unconstrained black-box combinatorial optimization problems [11]. A Gaussian process may be used as a surrogate of the continuous landscape of an expensive optimization problem. In order to model the landscape of a permutation space, CEGO uses a distance measure for permutation spaces, such as the Kendall's-τ distance, to *interpolate* between solutions. CEGO uses a genetic algorithm (GA) to explore the landscape of the surrogate model. The GA optimizes the expected improvement criterion, which takes into account both the expected mean and variance of a solution. The solution returned by the GA is then evaluated on the true objective function and this information is used to update the Gaussian process surrogate model.

5   Experimental Study

In this section we describe the experiments performed and the results obtained. Our goal is to answer two research questions:

RQ1: How do the algorithms perform on the problem in a black-box setting (without any problem knowledge)?
RQ2: How do they perform when some a priori information about the problem, in the form of a good initial solution, is introduced?

In Subsect. 5.1 the methodology used to answer both research questions is presented. Then, the experimental results to answer RQ1 and RQ2 are presented and described in Subsect. 5.2 and 5.3, respectively.

5.1   Experimental Methodology

We prepared an instance generator that, given an instance size n and seed for the random number generator, randomly selects n asteroids from the 83 453 asteroids provided by the GTOC11 competition to create an instance of the ARP. In this manner, we generated two instances (seeds 42 and 73) of each size $n = \{10, 15, 20, 25, 30\}$. In the remainder of the paper, instances are named n_seed, that is, instance 10_73 is an instance of size 10 generated with seed 73.

In our first set of experiments, designed to answer RQ1, both UMM and CEGO start the search from a set of 10 initial solutions generated by a *max-min-distance sequential design*, that is, solutions are added iteratively to a set of existing solutions by maximizing the minimum distance to solutions already in the set. The first solution in the set is generated uniformly at random. Generating the max-min-distance design only takes into account the distance between permutations and, thus, it does not require any evaluation of the objective function. These 10 solutions are then evaluated on the objective function function (Eq. 5) and they become the initial population of either CEGO or UMM. In order to answer RQ2, we initialize the max-min-distance design with the heuristic solution returned by the greedy nearest neighbor heuristic described above.

Both UMM and CEGO stop after evaluating 400 permutations using the objective function (Eq. (5)). Each evaluation requires solving the inner problem by optimizing the vector of times **t** using a sequence of n runs of SLQSP. Each run i of SLSQP optimizes the pair of parking and transit times (t_{2i-1}, t_{2i}), starting from the initial solution $(0, 30)$ and performing a maximum of 1000 iterations.

For each algorithm, we consider two representations, namely, whether the permutations are represented as the *order* or as the *ranking* of the asteroids in the visiting sequence. The *order*-based representation is the one used in the description of the problem in Sect. 3, i.e., $\pi(i) = j$ denotes that asteroid j is visited at step i of the route. Its *ranking*-based counterpart is given by its inverse π^{-1}, where $\pi(i) = j \Leftrightarrow \pi^{-1}(j) = i$. This distinction is important when calculating distances between permutations, learning a probability distribution, as done by UMM, or a surrogate model, as done by CEGO [6].

The GA used by CEGO to explore the surrogate model uses a budget of 10^4 evaluations of the surrogate model, population size of 20 individuals, cycle crossover for permutations with crossover rate of 0.5, swap mutation (i.e., exchanging two randomly selected elements) with mutation rate of $1/n$, and tournament selection of size 2 with probability of 0.9. The population used by the GA is different from the solutions evaluated by CEGO because the GA explores the surrogate model, which does not require evaluations of the expensive problem.

We use here the original implementation of GECO v2.4.2 (https://cran.r-project.org/package=CEGO). UMM is implemented in Python based on the original code [6]. We have implemented the proposed ARP benchmark in Python using the software package `poliastro` (v0.16) [2] for astrophysical calculations.

Each experiment was repeated 30 times with different random seeds. The experiments were run in the Picasso supercomputing facility of the University of Málaga with 126 SD530 servers with Intel Xeon Gold 6230R (26 cores each) at 2.10 GHz, 200 GB of RAM and an InfiniBand HDR100 network.

5.2 Results of the Black-Box Setting

In this section we compare UMM and CEGO (combined with SLSQP) in a black-box setting to solve the 10 instances of the ARP. Table 1 shows the results of CEGO and UMM using the two representations, rank-based or order-based. As a baseline, we show the results obtained by a simple *random search* that iteratively evaluates 400 random permutations. We can also see in Fig. 2 the evolution of the search for two particular instances. Each plot shows as a solid line the mean, over 30 runs, of the best objective value found up to a given number of function evaluations. The 95% confidence interval around the mean is shown as a shaded region. We applied the non-paired Wilcoxon sumrank test to compare the different algorithm configurations in each instance. The p-values are shown in Table 2. We assume that the differences are statistically significant when the p-value is below 0.01 (confidence level).

Results show that, whenever there are significant differences (see columns two and three of Table 2), rank-based UMM outperforms order-based UMM in terms of the objective function, whereas order-based CEGO outperforms rank-based CEGO. This result matches what we expect given the assumptions made by each algorithm. That is, UMM internally assumes that the permutation π it

Table 1. Objective value of the best solution found by each algorithm in a single run. Mean (and standard deviation) over 30 independent runs. RS denotes random search.

| | CEGO | | UMM | | |
Instance	Order	Rank	Order	Rank	RS
10_42	379.3 (17.5)	390.1 (16.4)	413.6 (15.2)	388.7 (20.8)	419.3 (14.1)
10_73	346.0 (17.7)	361.0 (14.3)	374.5 (16.6)	361.4 (18.9)	375.2 (15.3)
15_42	575.1 (29.2)	610.1 (27.4)	626.9 (26.1)	594.6 (37.7)	636.2 (18.5)
15_73	582.4 (33.1)	591.6 (29.1)	626.2 (24.7)	595.3 (32.6)	636.2 (25.2)
20_42	806.2 (45.2)	833.2 (38.7)	868.9 (26.9)	841.6 (43.7)	877.8 (26.4)
20_73	839.2 (49.8)	838.0 (43.0)	883.0 (28.4)	876.8 (32.7)	897.2 (26.3)
25_42	1048.8 (51.2)	1074.5 (56.5)	1121.2 (28.3)	1076.9 (48.5)	1122.1 (20.8)
25_73	1069.1 (49.5)	1096.8 (50.1)	1131.4 (35.8)	1115.5 (45.5)	1139.7 (25.6)
30_42	1272.2 (65.8)	1334.4 (66.6)	1372.0 (24.3)	1347.3 (35.6)	1357.3 (38.4)
30_73	1320.9 (63.2)	1325.1 (66.4)	1385.9 (27.5)	1360.6 (51.1)	1398.6 (24.9)

generates represents a ranking ($\pi(3) = 6$ means that object 3 has position 6) and, thus, it makes sense that it works better when the fitness function is aligned with this assumption [6]. CEGO, on the other hand, assumes that the permutation it generates represents an order ($\pi(3) = 6$ means that the third object is 6), and that explains the good results of CEGO when the objective function interprets the permutation as an order.

If we compare both algorithms, we observe that CEGO is usually better than UMM using both representations (with only two exceptions for the ranking representation in instances 10_42 and 15_42). The best CEGO variant is clearly better than the best UMM variant, although not always with statistical significance (see column four in Table 2).

Given these results, we select rank-based UMM and order-based CEGO for the next step in our analysis.

5.3 Results of the Informed Setting

In this section, we evaluate the effect of providing a good initial solution to the UMM-rank and CEGO-order variants selected above. This initial solution is generated using the greedy nearest neighbor heuristic as the initial point of the max-min-distance sequential design, as explained above, that creates the 10 initial solutions of both UMM and CEGO.

Figure 3 shows the evolution of the algorithms. In addition to the results provided by CEGO and UMM with and without the initial greedy solution, we plot the objective function value of the greedy solution as a horizontal line. Table 3 provides a summary of the results of the two best variants of CEGO and UMM together with the result obtained by the Greedy Nearest Neighbor Heuristic. A first observation is that UMM only reaches the fitness level of the greedy solution after 10 evaluations. The reason is that we evaluate first the 9 random solutions generated by the max-min-distance design to show how much the greedy solution improves over them. Moreover, without knowledge of the

Table 2. p-values of a Wilcoxon sumrank test of all the hypotheses checked in during the experimental study for each instance of the problem. Columns two and three correspond to a test comparing the two representations (rank and order) for CEGO and UMM in a black-box setting. The fourth column compares the best black-box UMM (using ranking) with the best black-box CEGO (using order). The final two columns compare UMM-ranking and CEGO-order in a black-box versus informed settings (including the greedy initialization).

	Rank vs. Order		CEGO-order	Greedy vs. BB	
Instance	CEGO	UMM	vs. UMM-rank	UMM	CEGO
10_42	6.46e−03	2.00e−06	4.48e−02	2.65e−02	1.18e−01
10_73	2.37e−03	1.33e−02	7.07e−04	8.34e−07	1.65e−02
15_42	3.77e−05	6.12e−04	5.14e−02	2.95e−11	1.72e−12
15_73	3.35e−01	2.75e−04	1.50e−01	6.52e−09	3.69e−10
20_42	1.42e−02	8.01e−03	2.74e−03	2.41e−05	3.25e−07
20_73	7.75e−01	4.67e−01	2.22e−03	2.84e−11	2.31e−11
25_42	3.71e−02	7.74e−05	3.45e−02	3.02e−11	4.04e−11
25_73	4.46e−02	1.77e−01	5.43e−04	2.26e−11	2.84e−11
30_42	5.43e−04	4.11e−03	1.30e−05	2.95e−11	3.02e−11
30_73	6.12e−01	3.71e−02	1.83e−02	2.78e−11	3.00e−11

Table 3. Objective value of the best solution found by each algorithm in a single run. For CEGO and UMM, the values are the mean (and standard deviation) over 30 independent runs.

Inst	GreedyNN	CEGO		UMM	
		Order	Ord + Greedy	Rank	Rank + Greedy
10_42	391.3	379.3 (17.5)	374.9 (6.9)	388.7 (20.8)	382.4 (1.1)
10_73	398.3	346.0 (17.7)	355.9 (18.3)	361.4 (18.9)	385.0 (6.4)
15_42	508.1	575.1 (29.2)	497.2 (1.1)	594.6 (37.7)	501.7 (3.8)
15_73	576.4	582.4 (33.1)	525.6 (4.7)	595.3 (32.6)	545.0 (9.3)
20_42	841.7	806.2 (45.2)	737.0 (31.0)	841.6 (43.7)	797.6 (28.3)
20_73	691.5	839.2 (49.8)	661.8 (7.8)	876.8 (32.7)	684.7 (6.1)
25_42	946.3	1048.8 (51.2)	881.5 (23.3)	1076.9 (48.5)	921.7 (15.2)
25_73	918.3	1069.1 (49.5)	873.6 (15.3)	1115.5 (45.5)	911.9 (10.0)
30_42	1131.7	1272.2 (65.8)	1084.6 (13.4)	1347.3 (35.6)	1115.4 (11.0)
30_73	1024.7	1320.9 (63.2)	967.7 (17.9)	1360.6 (51.1)	1017.3 (6.9)

greedy solution, CEGO or UMM are only able to match the greedy solution under 400 evaluations on a few small instances ($n \leq 20$). Unsurprisingly, when starting from the greedy solution, both algorithms outperform their *uninformed* counterparts. However, UMM struggles to find any improving solutions, whereas

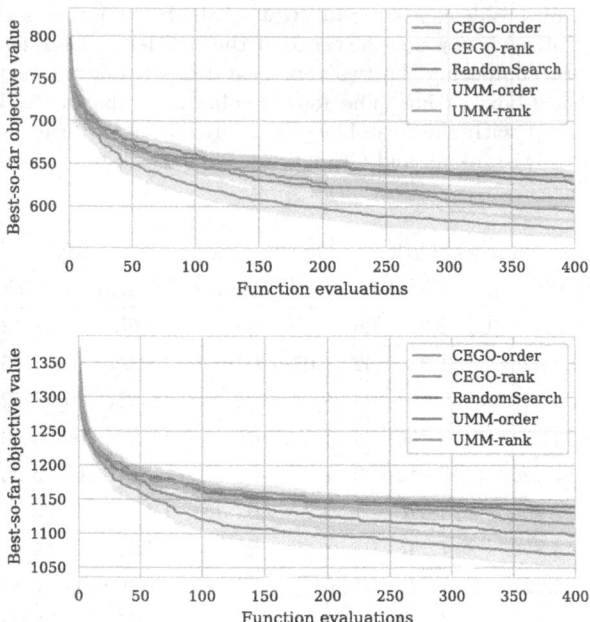

Fig. 2. Convergence of the best-so-far objective value over the number of evaluations. Each line is the mean value over 30 runs and the shaded area is the 95% confidence interval around the mean. The instances shown are 15_42 (top) and 25_73 (bottom).

CEGO is typically able to further improve the greedy solution, although not by much. In the case of UMM, this behavior is explained by the fact that the sampling of new permutations is strongly biased by the relative objective values of the permutations already evaluated. Since the greedy solution is so much better than any other initial permutation, the algorithm remains stuck at the initial permutation. In the case of CEGO, starting from a good initial solution provides a good starting point around which the surrogate model can be built. However, in several instances, after a quick initial improvement of the greedy solution, CEGO appears stuck and unable to further find any improvements. This behavior of CEGO was already reported for other problems [6] and it is attributed to the inability of the surrogate model to accurately estimate the underlying objective function beyond a certain point, thus leading to a blind search.

We also report the mean runtime of CEGO, UMM and the greedy heuristic in Table 4. The runtime of the greedy algorithm is relatively fast, taking around 6 s to generate and evaluate a single solution. UMM requires between 24 and 69 min per run on average, depending on the instance size, where each run involves 400 solution evaluations. The runtime per evaluation is larger than in GreedyNN due to the update of and sampling from the probabilistic model. CEGO also evaluates 400 solutions per run, however, it is more than 100 times slower than

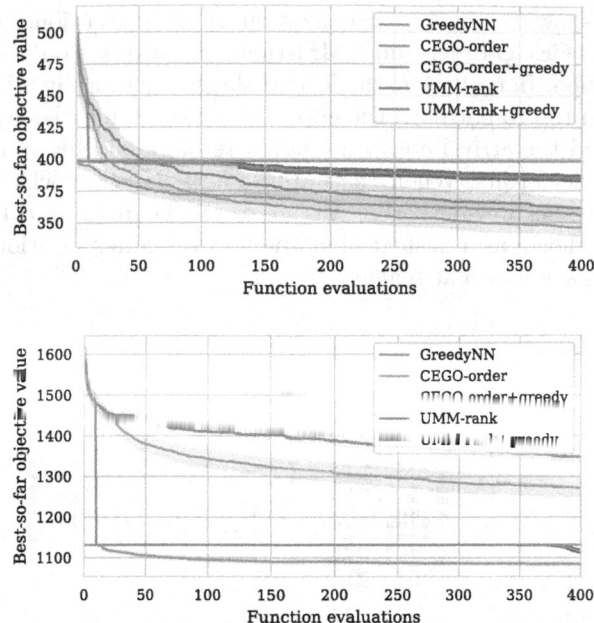

Fig. 3. Convergence of the best-so-far objective value over the number of evaluations. Each line is the mean value over 30 runs and the shaded area is the 95% confidence interval around the mean. The instances shown are 10_73 (top) and 30_42 (bottom).

UMM, requiring almost 2 CPU-days per run on average for the largest instances tested here. Although part of this difference is due to implementation choices, the main difference is the fact that building the Gaussian Process model in CEGO is significantly more expensive than building the probabilistic model in UMM.

Table 4. Mean CPU-time (minutes) of a single run of each algorithm.

		CEGO		UMM	
Instance	GreedyNN	Order	Ord + Greedy	Rank	Rank + Greedy
10_42	0.1	2365.4	2147.8	26.0	26.2
10_73	0.1	2372.9	2365.3	24.2	25.7
15_42	0.1	2445.9	2394.1	34.5	32.4
15_73	0.1	2375.6	2404.2	35.5	36.2
20_42	0.2	2411.4	2348.4	47.8	48.7
20_73	0.2	2223.9	2387.4	48.2	50.8
25_42	0.2	2338.1	2452.5	59.4	58.9
25_73	0.2	2397.3	2657.6	47.3	60.3
30_42	0.2	2375.9	2553.7	61.9	69.1
30_73	0.2	2138.3	2611.0	56.0	67.4

Finally, we show in Fig. 4 a visualization of three solutions found by the greedy heuristic, CEGO-order and UMM-rank, the two last ones starting from the greedy solution. In each plot, the legend shows the order in which asteroids are visited (from top to bottom), the epoch at which an impulse was applied to change orbit, and the arrival epoch at each asteroid. Solid lines indicate either parking or transfer orbits. When the impulse and arrival epochs coincide, the parking time at the previous orbit was zero and no parking orbit is shown. Earth's orbit is shown as a dashed blue line and the Sun's position as a yellow point. Distances are shown at scale.

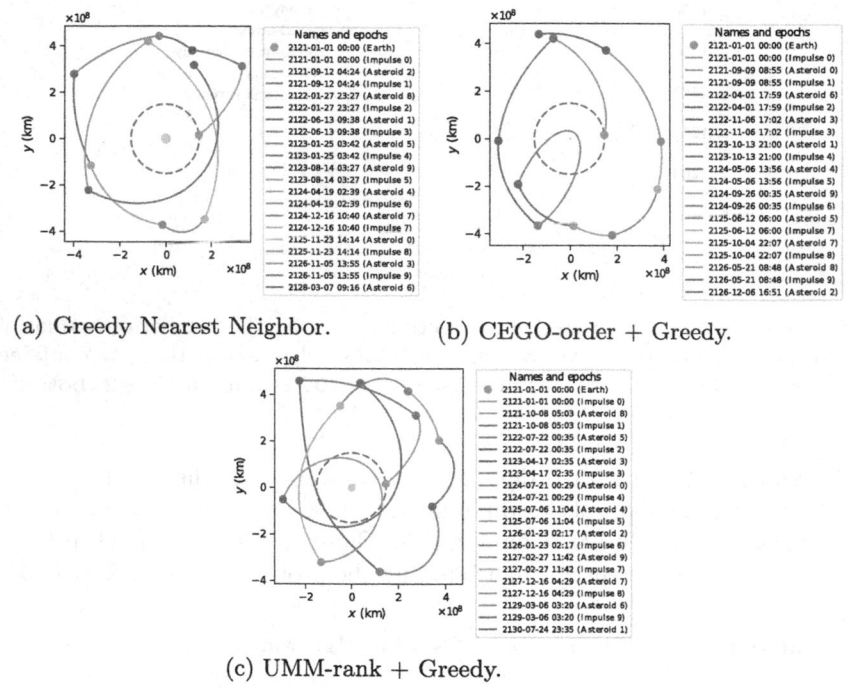

(a) Greedy Nearest Neighbor.   (b) CEGO-order + Greedy.

(c) UMM-rank + Greedy.

Fig. 4. Example solutions for instance 10_73.

6   Conclusions

We have proposed in this paper a realistic benchmark for evaluating algorithms designed to tackle expensive black-box optimization problem in permutation space. The results highlight already some limitations of the evaluated algorithms. First, although CEGO consistently obtains better results than UMM as long as the right choice of representation is made, this comes at a cost of 100 times more computation time. Second, without any additional information, very rarely either algorithm is able to match the quality of the solution found by a greedy approach, suggesting that their results are far from optimal. Finally, only CEGO makes some use of the information provided by a good initial solution, whereas UMM is very rarely able to improve over it.

Future work should evaluate other black-box algorithms for permutation problems in the ARP. Improvements to CEGO and UMM to make better use of good initial solutions would also be welcome. In fact, it is possible to generate multiple greedy solutions that may be used as starting points by the black-box optimizers. Finally, we plan to change the optimizer of the inner problem from the current deterministic SLSQP to a stochastic one, which changes the context of the outer problem from a deterministic permutation problem to a stochastic one. Under such context, a black-box optimizer for the outer problem must decide whether it is worth evaluating again already seen permutation in the hope that a new evaluation may further improve the objective function.

Reproducibility. Datasets and source code for reproducing the experiments reported are available from https://doi.org/10.5281/zenodo.5725837.

Acknowledgement. This work is partially funded by the Universidad de Málaga, Consejería de Economía y Conocimiento de la Junta de Andalucía and FEDER (grant UMA18-FEDERJA-003) and MCIN/AEI/10.13039/501100011033 (grant PID 2020-116727RB-I00). Thanks to the Supercomputing and Bioinnovation Center (SCBI) of the University of Málaga for their provision of computational resources and support. M. López-Ibáñez is a "Beatriz Galindo" Senior Distinguished Researcher (BEAGAL 18/00053) funded by the Spanish Ministry of Science and Innovation (MICINN).

References

1. Bartz-Beielstein, T., Zaefferer, M.: Model-based methods for continuous and discrete global optimization. Appl. Soft Comput. **55**, 154–167 (2017). https://doi.org/10.1016/j.asoc.2017.01.039
2. Cano Rodríguez, J.L., et al.: poliastro: astrodynamics in Python. Zenodo (2015). https://doi.org/10.5281/zenodo.174
3. European Space Agency: Hera Mission. https://wwww.esa.int/Safety_Security/Hera/Hera (2019). Accessed 22 Nov 2021
4. European Space Agency: Science & Exploration: Asteroids. https://www.esa.int/Science_Exploration/Human_and_Robotic_Exploration/Exploration/Asteroids (2021). Accessed 22 Nov 2021
5. Goldstein, H.: Classical Mechanics. Addison-Wesley (1980)
6. Irurozki, E., López-Ibáñez, M.: Unbalanced mallows models for optimizing expensive black-box permutation problems. In: Proceedings of GECCO (2021). https://doi.org/10.1145/3449639.3459366
7. Izzo, D.: Revisiting Lambert's problem. arXiv 1403.2705 [astro-ph.EP] (2014)
8. Knowles, J., Corne, D., Reynolds, A.: Noisy multiobjective optimization on a budget of 250 evaluations. In: Ehrgott, M., Fonseca, C.M., Gandibleux, X., Hao, J.-K., Sevaux, M. (eds.) EMO 2009. LNCS, vol. 5467, pp. 36–50. Springer, Heidelberg (2009). https://doi.org/10.1007/978-3-642-01020-0_8

9. Cáceres, L.P., López-Ibáñez, M., Stützle, T.: Ant colony optimization on a limited budget of evaluations. Swarm Intell. 103–124 (2015). https://doi.org/10.1007/s11721-015-0106-x
10. Virtanen, P., et al.: SciPy 1.0: fundamental algorithms for scientific computing in Python. Nat. Meth. **17**, 261–272 (2020). https://doi.org/10.1038/s41592-019-0686-2
11. Zaefferer, M., Stork, J., Friese, M., Fischbach, A., Naujoks, B., Bartz-Beielstein, T.: Efficient global optimization for combinatorial problems. In: Proceedings of GECCO, pp. 871–878 (2014). https://doi.org/10.1145/2576768.2598282

On the Difficulty of Evolving Permutation Codes

Luca Mariot[1]([⊠]), Stjepan Picek[1], Domagoj Jakobovic[2], Marko Djurasevic[2], and Alberto Leporati[3]

[1] Digital Security Group, Radboud University,
Postbus 9010, 6500 GL Nijmegen, The Netherlands
{luca.mariot,stjepan.picek}@ru.nl
[2] Faculty of Electrical Engineering and Computing,
University of Zagreb, Unska 3, Zagreb, Croatia
{domagoj.jakobovic,marko.djurasevic}@fer.hr
[3] DISCo, Università degli Studi di Milano-Bicocca,
Viale Sarca 336/14, 20126 Milan, Italy
alberto.leporati@unimib.it

Abstract. Combinatorial designs provide an interesting source of optimization problems. Among them, permutation codes are particularly interesting given their applications in powerline communications, flash memories, and block ciphers. This paper addresses the design of permutation codes by evolutionary algorithms (EA) by developing an iterative approach. Starting from a single random permutation, new permutations satisfying the minimum distance constraint are incrementally added to the code by using a permutation-based EA. We investigate our approach against four different fitness functions targeting the minimum distance requirement at different levels of detail and with two different policies concerning code expansion and pruning. We compare the results achieved by our EA approach to those of a simple random search, remarking that neither method scales well with the problem size.

Keywords: Permutation codes · Evolutionary algorithms · Incremental construction · Powerline communications · Flash memories · Block ciphers

1 Introduction

Permutation codes (also called permutation arrays) are a particular kind of error-correcting codes where the codewords are permutations. In particular, a permutation code $PA(n, d)$ is a set of permutations of length n such that any two permutations in it disagree in at least d positions.

These combinatorial objects have several applications, for example, as error-correcting codes in *powerline communications* [2]. The basic approach for powerline transmission is to encode the data by small voltage variations, with the

© Springer Nature Switzerland AG 2022
J. L. Jiménez Laredo et al. (Eds.): EvoApplications 2022, LNCS 13224, pp. 141–156, 2022.
https://doi.org/10.1007/978-3-031-02462-7_10

requirement of keeping the power output as constant as possible. Further, transmission over powerlines is affected not only by white Gaussian noise but also by impulse and narrow-band noise due to electrical interference and magnetic fields. Using permutation codes in a modulation scheme, as suggested by Han Vinck [10], provides a good trade-off between power variation and correcting errors introduced by these kinds of noise. A second domain where permutation codes have been extensively applied is *flash memories*, particularly in the so-called *rank-modulation scheme* [11]. In traditional designs, the cells in a flash memory encode the information using different charge levels, allowing them to store a set of discrete values. On the other hand, rank-modulation encodes the data in the cells with a permutation that specifies the relative ranks of the charges instead of directly using their absolute values. Using a permutation code in this scheme improves the writing speed and the correction of errors introduced by charge leakage, which becomes progressively frequent in aging memories.

Finally, permutation codes have also been applied to a smaller extent in cryptography, specifically for the design of *block ciphers* [24]. In the *Substitution-Permutation Network* (SPN) paradigm for block ciphers, the plaintext is encrypted by iteratively applying several times a *round function*. The round function, in turn, consists of a *confusion layer*, which aims at making the relationship between the ciphertext and the symmetric key as complicated as possible, and a *diffusion layer*, whose goal is to spread the statistical structure of the plaintext over the ciphertext. The resulting block is mixed with a *round key* to get the corresponding ciphertext, which is then given as input to the next application of the round function. The diffusion layer is usually implemented through a *Maximum Distance Separable (MDS) matrix* as it happens, for example, in AES [6]. An alternative approach is to use a set of different permutations coming from a permutation code $PA(n, d)$. By dynamically choosing a different permutation from the code at each round, two different input blocks are guaranteed to result in output blocks at Hamming distance of at least d, thereby implementing a *multi-permutation* as defined by Vaudenay [25].

Despite their simple definition, the construction of permutation codes is far from being a trivial problem. Indeed, finding the largest permutation code is a particular instance of the *sphere-packing problem* studied in coding theory [5], and of the MAX-CLIQUE problem in graph theory, which is known to be **NP**−complete [12]. In particular, one of the main open questions in this research field is to determine the largest permutation code for a given length n and minimum distance d, i.e., the maximum number of permutations that can partake in a $PA(n, d)$. Such a number is usually denoted as $M(n, d)$, and its exact value is known only for a few specific cases. Generally, one resorts to coding-theoretic results to provide lower and upper bounds on $M(n, d)$. Apart from algebraic constructions, for which the reader may find a survey in [3], a few heuristic algorithms have also been developed to construct large permutation codes [18,21], mostly based on branch and bound and iterative clique search approaches. As far as we know, up to now, there have been no attempts in the literature to employ Evolutionary Algorithms (EA) to address this problem, although some authors used EA in the past to evolve other kinds of combinatorial designs, such as *orthogonal Latin squares* [16], *orthogonal arrays* [17] and *disjunct matrices* [14].

This paper investigates the suitability of EA to optimize permutation codes. We do so by from the previous permutations. The process is repeated until either a given fitness budget expires or an upper bound on $M(n, d)$ is reached (since this means that the code cannot be expanded further). We evaluate our incremental EA approach under four fitness functions and two *update policies*. The first update policy expands the code as soon as a suitable permutation is found by the EA. The second policy also removes some rows at random from the current code after a certain amount of fitness evaluations with no improvement has elapsed. The number of removed rows is decreased over time, similarly to the cooling schedule used in simulated annealing. For the sake of comparison, we also adopt a baseline random search (RS) method and perform experiments over 15 problem instances. The results show that both EA and RS cannot scale well on this optimization problem, with the largest codes found that lie far from the best-known lower bounds [20].

2   Preliminaries

We denote by $[n]$ the set $\{1, \cdots, n\}$ of the first $n \in \mathbb{N}$ positive integers. Next, S_n denotes the *symmetric group* of order n, i.e., the set of all permutations over $[n]$. Given a permutation $\pi \in S_n$, we encode it as a vector $\pi = (p_1, \cdots, p_n)$ of length n, where each coordinate $\pi[i]$ specifies the value of the permutation when evaluated on $i \in [n]$. Further, given two permutations $\pi, \sigma \in S_n$, the *Hamming distance* $d_H(\pi, \sigma)$ is the number of coordinates where π and σ differ.

Definition 1. *Let $n \in \mathbb{N}$ and $d \leq n$. A permutation code (also permutation array, PA) of length n and minimum distance d, denoted by $PA(n, d)$, is a subset P of the symmetric group S_n such that $d_H(\pi, \sigma) \geq d$ for every pair of distinct permutations $\pi, \sigma \in P$.*

Using the error-correcting codes terminology, the permutations in a $PA(n, d)$ are also called *codewords*. If P is composed of m codewords, one can represent it through a $m \times n$ matrix where each row corresponds to one of the permutations in P. The ordering of the rows in such a matrix is irrelevant since it does not change the pairwise Hamming distances of the permutations. In the following, we will mostly use this matrix-based notation to represent PA, although the set-theoretic notation will also be useful to describe the operations performed by our evolutionary algorithms to search for such arrays.

One of the main problems is determining the largest number of codewords that can partake in a permutation code. Following the notation from [3], given n and d we denote by $M(n, d)$ the maximum number of rows in a $PA(n, d)$. The values of $M(n, d)$ are generally unknown, but several theoretical bounds exist. Two well-known results in this direction, originating from coding-theoretical considerations, are the *Gilbert-Varshamov lower bound* and the *sphere-packing upper bound*, which we summarize below for permutation codes.

Theorem 1. *Let $n, d \in \mathbb{N}$ with $d \leq n$. Then, the following inequalities hold for the maximum number of codewords in a permutation code:*

$$\frac{n!}{\sum_{k=0}^{d-1} \binom{n}{k} D_k} \leq M(n,d) \leq \frac{n!}{\sum_{k=0}^{\lfloor \frac{d-1}{2} \rfloor} \binom{n}{k} D_k} \,, \tag{1}$$

where D_k is the number of derangements *of k elements (i.e., the number of permutations of length k without fixed points), which for all $k \in \mathbb{N}$ equals:*

$$D_k = k! \sum_{i=0}^{k} \frac{(-1)^i}{i!} \,. \tag{2}$$

The Gilbert-Varshamov and sphere-packing bounds are rather crude, but in practice, they are helpful to decide whether a specific instance of n and d is suitable to construct a permutation code large enough for a specific application. Of course, tighter bounds have been proved in the related literature of permutation codes, either by combinatorial arguments or by providing concrete constructions. The latter case usually occurs for specific values of the minimum distance. For instance, if $d = n$, then it is rather easy to constructively prove that $M(n, n) = n$, by considering any *Latin square* of order n as an example of permutation code reaching this bound. Indeed, a Latin square of order n is a $n \times n$ array such that each number in $[n]$ appears exactly once in each row and each column. Thus, each row of the square is a permutation, and any two rows differ in all coordinates since there cannot be any repeated number in any column. A simple construction for a Latin square of order n is to take all *cyclic shifts* of the identity permutation $(1, 2, \cdots, n)$, which proves the existence of a $PA(n, n)$ for every $n \in \mathbb{N}$.

Latin squares also provide a construction for a better lower bound on $M(n, d)$ when $d = n - 1$. In particular, two Latin squares are called *orthogonal* if their superposition yields all ordered pairs in the Cartesian product $[n] \times [n]$, and a set of k *mutually orthogonal Latin squares* (k-MOLS) is a family of k Latin squares of order n that are pairwise orthogonal. Colbourn et al. [4] showed how to construct a $PA(kn, n - 1)$ by using a set of k-MOLS of order n, thereby proving that $kn \leq M(n, n - 1)$. We emphasize that determining the maximum size of a MOLS family for a given n is also a long-standing open problem in design theory, but several results are known for specific cases [22].

It is also easy to determine the maximum number of rows in a permutation array for low minimum distances. Indeed, two permutations cannot differ in only one position since this would imply that both vectors have a repeated value (thus, not making it a permutation). Therefore, the minimum distance is always at least 2, i.e., when two permutations differ by a single *transposition*, or swap. This means that the largest $PA(n, 2)$ corresponds to the symmetric group S_n itself, hence $M(n, 2) = n!$. Additionally, any two distinct permutations in the *alternating group* A_n (i.e., the set of *even* permutations of length n) are always at a minimum distance of 3. Since the alternating group is exactly half of the size of the symmetric group, it follows that $M(n, 3) = n!/2$. We conclude this section by summarizing the above results as follows:

Theorem 2. *Let $n, d \in \mathbb{N}$ with $d \leq n$. Then:*

- $M(n, 1) = 1$; $M(n, 2) = n!$; $M(n, 3) = n!/2$;
- $M(n, n) = n$; $kn \leq M(n, n - 1)$, *if there exists a set of k-MOLS of order n.*

Tables reporting more refined lower and upper bounds for various values of n and d may be found in [20].

3 Incremental Construction with EA

From an intuitive point of view, it seems natural to cast the search of a permutation code as a combinatorial optimization problem. Given the length n of the permutations, the (minimum) distance d and the number m of desired permutations in the array, one needs to find a set of m elements from S_n such that the Hamming distance between any two permutations in it is at least d. Therefore, disregarding the bounds on m induced by the distance parameter, the size of the resulting search space $\mathcal{S}_{m,n}$ is $|\mathcal{S}_{m,n}| = \binom{n!}{m}$ since we need to pick m elements from a set of size $n!$. Exhaustively searching for a solution would be already prohibitive for very small values of n and m: for example, there are only $|S_5| = 120$ permutations of length $n = 5$. However, visiting all subsets of S_5 of size $m = 12$ would imply a search space of $\binom{120}{12} \approx 1.05 \cdot 10^{17}$ elements, which clearly cannot be explored in a reasonable amount of time. Consequently, it seems interesting to address this optimization problem with evolutionary algorithms.

3.1 Evolving Subsets of Permutations

Given n, m and d, a straightforward option is to set up an EA that searches for permutations codes by directly evolving a set of m permutations. A candidate solution in the population is represented as a matrix A of size $m \times n$, where each row is a permutation of the set $[n]$. Then, this candidate solution would be evaluated through a fitness function that measures how close is A from being a $PA(n, d)$. This could be accomplished, e.g., by counting the number of pairs of rows in A that are at Hamming distance at least d and maximizing such fitness. In this approach, one could use common operators for permutation-based chromosomes. For crossover, these include among others *partially mapped crossover* [9] and *cycle crossover* [19]. For mutation, the most natural solution is to apply a simple swap operator that randomly exchanges two values in a permutation [1], but other methods have been proposed such as, e.g., the *inversion operator* [8] and the *scramble operator* [23]. Still, when evolving permutation codes, one deals with sets of permutations. Thus, a possible solution for this problem would be to apply the variation operators in a *row-wise* manner. For example, given two $m \times n$ arrays A and B, define an offspring array C by first applying a permutation-based crossover to the first row of A and B, then to the second one, and so on until C is completed. Although straightforward, this optimization approach suffers from several drawbacks:

- As the aim is constructing a permutation array in an "all-at-once" fashion, an EA would directly explore m-subsets of the symmetric group S_n, which results in a very large search space already for small values of m and n.
- The fitness function would need to consider the Hamming distance of each pair of rows in the arrays. Hence, the computational complexity required to evaluate a single candidate solution would be quadratic in the number of permutations of the array, as there are $\binom{m}{2} = \frac{m(m-1)}{2} = \mathcal{O}(m^2)$ pairwise Hamming distances to compute in an $m \times n$ array.
- This optimization approach relies on the number of rows m composing the desired permutation to define the problem instance. This implies that one would need to check in advance if m rows are attainable by a permutation code of length n and distance d.

3.2   Iterative Approach

Given the problems featured by the "all-at-once" method, we chose to follow an *iterative* optimization approach, greatly reducing the search space handled at each step by the evolutionary algorithm. Given $n, d \in \mathbb{N}$, the idea is to start from an empty set and add a random permutation $p_1 \in S_n$ of length n: trivially, P forms a $PA(n, d)$ with $m = 1$ rows. Then, an EA evolves a single permutation $p_2 \in S_n$, until it finds one whose Hamming distance from p_1 is at least d. When it is found, p_2 is added to P, thereby expanding the permutation code to $m = 2$ rows. The process is repeated by evolving a new permutation until a general termination criterion is met, such as reaching a theoretical upper bound for $M(n, d)$ or a specified number of fitness evaluations. By construction, the obtained array will be a permutation code $PA(n, d)$ with a certain number of rows m. At each stage, the EA only explores the set S_n of all permutations of length n instead of the whole set of m-subsets of permutations.

More formally, given a $PA(n, d)$ $P = \{p_1, \cdots, p_m\}$ with m rows, the decoding of a candidate chromosome $p_{m+1} \in S_n$ results in the following phenotype: $P' = P \cup \{p_{m+1}\}$. Clearly, since P already satisfies the properties of a permutation code of minimum distance d, the fitness of the candidate solution P' encoded by p_{m+1} is evaluated only by taking into account the m Hamming distances $d_H(p, p_{m+1})$, with p ranging over P. This is a much more efficient fitness function since its computational complexity scales linearly with the number of rows in P. Further, suppose that $\chi : S_n \times S_n \to S_n$ is a crossover operator for single permutations. Then, given two parent permutations $p_1, p_2 \in S_n$, the phenotype C for the offspring child candidate solution is defined as $C = P \cup \{\chi(c_1, c_2)\}$, that is, crossover is limited only on the new row. Accordingly, the same approach is adopted for mutation by applying the corresponding operator $\mu : S_n \to S_n$ only on the new permutation optimized by the EA.

Algorithm 1 reports the pseudocode for the incremental EA informally introduced above. The input parameters are the length of the permutations n, the required minimum distance d, the fitness budget fb, the target number of rows M (which specifies, for example, a known upper bound for $M(n, d)$), the size

Algorithm 1. INCREMENTAL-EA-PA$(n, d, fb, M, popsize, \theta)$

```
P ← {}
π ← GEN-RAND-PERMUTATION(n)
P ← P ∪ {π}
eval ← 0
while |P| < M AND eval < fb do
    pop ← INIT-POPULATION(n, popsize)
    EVALUATE-FITNESS(pop, P, n, d, ev)
    best ← UPDATE-BEST-IND(pop)
    while eval < fb AND (NOT IS-PA(n, d, P, best.π)) do
        pop ← UPDATE-POP-EA(n, d, P, pop, θ, eval)
        best ← UPDATE-BEST-IND(pop)
    if IS-PA(n, d, P, best.π) then
        P ← P ∪ {best.π}
return P
```

of the EA population $popsize$, and a vector θ specifying the parameters for the underlying evolutionary algorithm.

The subroutines GEN-RAND-PERMUTATION and INIT-POPULATION respectively generate at random a single permutation and a population of $popsize$ candidate permutations. An individual ind in the population is assumed to be a record composed of two items, namely $ind.\pi$ (the vector specifying the permutation) and $ind.fit$ (the fitness value of the permutation). EVALUATE-FITNESS computes the underlying fitness function for each individual in the population, while UPDATE-BEST-IND returns a pointer to the best individual in the current population. The specific structures for these two subroutines depend on the details of the fitness function, which we will address in the next section. IS-PA is a predicate returning true if and only if the union of a $PA(n, d)$ and a new permutation is still a $PA(n, d)$, and it is used to determine when to exit from the inner while loop of the EA. When an optimal solution is found, IS-PA$(n, d, P, best.\pi))$ returns true, and the permutation code P is extended by adjoining to it the permutation of the best individual in the population.

The actual EA is implemented by the UPDATE-POP-EA subroutine. In particular, depending on the underlying EA, the population might be updated completely, as in a generational approach (possibly coupled with elitism) or only partially, with only a few new offspring individuals entering into the population at each step. In our experiments, we adopted a steady-state genetic algorithm (GA) with tournament selection. This means that each time UPDATE-POP-EA is invoked, t individuals are drawn at random from the population, and the two with the best fitness values are selected for crossover. The resulting offspring then undergoes mutation with probability p_μ, that replaces the worst individual in the tournament. Algorithm 2 gives the pseudocode for our steady-state GA implementing the UPDATE-POP-EA subroutine. The parameters vector θ is replaced by the pair (t, p_μ), whose components respectively specify the tournament size and the mutation probability. Notice also that $eval$, which is a counter used to keep track of the number of fitness evaluations performed by the algorithm, is assumed to be a global variable: in fact, it is used in the invariants for the while loops in Algorithm 1.

Algorithm 2. UPDATE-POP-EA$(n, d, P, pop, (t, p_\mu), eval)$

$tourn \leftarrow$ RANDOM-SELECT(t, pop)
$(p_1, p_2) \leftarrow$ SELECT-BEST$(tourn)$
$c \leftarrow$ CROSSOVER(p_1, p_2)
$c \leftarrow$ MUTATION(c, p_μ)
$c.fit \leftarrow$ FITNESS(n, d, P, c)
$eval \leftarrow eval + 1$
$worst \leftarrow$ SELECT-WORST$(tourn)$
REPLACE$(worst, c)$
return pop

The subroutines RANDOM-SELECT, SELECT-BEST, and SELECT-WORST respectively return a random subset of t individuals from the population, the two best individuals and the worst one in the tournament concerning their fitness values. The offspring chromosome is created from p_1 and p_2 by first applying CROSSOVER and then MUTATION. After evaluating the fitness function – and increasing the counter of fitness evaluations – the subroutine REPLACE changes the worst individual in the tournament to the newly created offspring.

3.3   Fitness Functions

We defined four fitness functions to be optimized by the iterative EA described in the previous section, which we describe below. In what follows, we assume that the goal is to compute the fitness of a permutation $p \in S_n$ when adjoined to a $PA(n, d)$ of m rows, $P = \{p_1, \cdots, p_m\}$.

The first fitness function directly sums the Hamming distances of each pair (p, p_i) of permutations, but only if they are at least equal to the required minimum distance d:

$$fit_1(p) = \sum_{p_i \in P} \delta_i \cdot d_H(p, p_i), \text{ where } \delta_i = \begin{cases} 1, & \text{if } d_H(p, p_i) \geq d, \\ 0, & \text{otherwise} \end{cases} . \qquad (3)$$

Note that this fitness function completely neglects the permutation pairs' information at Hamming distance lower than d. For this reason, the second fitness function has the same form of fit_1, but also takes into account the invalid pairs by discounting them through an exponential factor:

$$fit_2(p) = \sum_{p_i \in P} \delta_i' \cdot d_H(p, p_i), \text{ where } \delta_i' = \begin{cases} 1, & \text{if } d_H(p, p_i) \geq d, \\ 2^{d_H(p, p_i)-d}, & \text{otherwise} \end{cases} . \qquad (4)$$

Indeed, when $d_H(p, p_i) < d$ the factor δ_i' is a number between 0 and 1, which decreases as the difference $d_H(p, p_i) - d$ gets smaller. In this way, the more a pair (p, p_i) is closer to the required minimum distance d, the more it contributes to the fitness function.

The third fitness function considered in our experiments corresponds to the minimum Hamming distance between p and each permutation in P, that is,

$$fit_3(p) = \min_{p_i \in P} \{d_H(p, p_i)\} . \qquad (5)$$

Hence, the permutation p is an optimal solution as soon as $fit_1(p) \geq d$, since this is precisely the characterizing property of a $PA(n, d)$. Although straightforward, this fitness function suffers from a limited range of possible values (especially for small values of d), making many candidate solutions very similar. This may hamper, in turn, the EA's ability to exploit specific regions of the search space.

The three fitness functions described up to now are all meant to be maximized as an optimization objective. On the contrary, the fourth fitness function is based on counting the number of pairs that do not meet the minimum distance requirement, clearly with the objective of minimizing them:

$$fit_4(p) = |\{(p, p_i) : p_i \in P, \ d_H(p, p_i) < d\}| \ . \tag{6}$$

4   Experimental Evaluation

A problem instance for the permutation array problem is defined by the length of the permutation n and the (minimum) distance d. To evaluate the suitability of the incremental EA on this problem, one possibility is to compare the maximum number of rows obtained by it for a $PA(n, d)$ and the corresponding lower/upper bounds known in the literature. As far as we are aware, Smith and Montemanni [20] report the most up-to-date table that reports such bounds for $6 \leq n \leq 18$ and $4 \leq d \leq 18$. Since d must always be less than or equal to n, the total number of problem instances to be tested for a complete comparison is $\sum_{i=3}^{15} i = 117$, which might be unfeasible depending on how much time a single run of the EA takes. Therefore, it makes sense to perform the experiments on a subset of instances, limiting the size of the permutations to $n = 10$ and minimum distance $n - 2 \leq d \leq n$. In this way, we get a total of 15 problem instances to test. These instances also have practical relevance in the design of modulation schemes for powerline communications (see, e.g., [7], where PA of length at most 8 are considered for this task) and for the design of block ciphers, where $n = 8$ is a popular permutation size in the diffusion layers of lightweight block ciphers [15]. The instances where $n = d$ correspond to the problem of finding a Latin square of order n. Furthermore, although the size of the symmetric group S_n, for $6 \leq n \leq 10$, is sufficiently limited to be completely explored, recall that the unfeasibility of the exhaustive search approach stems from the fact that we are trying to construct *subsets* of permutations. This already yields a search space of size $\binom{6!}{120} \approx 3.07 \cdot 10^{140}$, for the $PA(6, 4)$ instance, and thus it cannot be exhaustively explored. For each considered combination of n and d, Table 1 reports the size of the corresponding search space computed as $\binom{n!}{M(n,d)}$ and the corresponding best value known for $M(n, d)$ taken from [20]. The search space size decreases as the minimum distance approaches the length, with $n = d$ giving the smallest sizes – although still not amenable to exhaustive search.

4.1   Experimental Settings

In our experiments, we evaluated our evolutionary approach to construct PA along three different components: namely, the *fitness functions* described in

Table 1. Approximate search space size $\mathcal{S}_{n,d}$ and code size bound $M(n,d)$ for each considered problem instance. Bold values represent non-tight lower bounds.

$d \backslash n$		6	7	8	9	10
$n-2$	$\mathcal{S}_{n,d}$	$3.07 \cdot 10^{140}$	$2.31 \cdot 10^{277}$	$1.81 \cdot 10^{843}$	$1.20 \cdot 10^{1\,658}$	$3.83 \cdot 10^{2\,978}$
	$M(n,d)$	120	**77**	336	504	720
$n-1$	$\mathcal{S}_{n,d}$	$3.41 \cdot 10^{36}$	$1.91 \cdot 10^{106}$	$1.10 \cdot 10^{184}$	$3.26 \cdot 10^{297}$	$1.61 \cdot 10^{453}$
	$M(n,d)$	18	42	56	72	49
n	$\mathcal{S}_{n,d}$	$1.89 \cdot 10^{15}$	$1.63 \cdot 10^{23}$	$1.73 \cdot 10^{34}$	$3.01 \cdot 10^{45}$	$1.10 \cdot 10^{60}$
	$M(n,d)$	6	7	8	9	10

Sect. 3.3, the underlying *search algorithm*, and the adopted *update policy*. The search algorithm refers to the particular procedure used to select a suitable permutation to be added in the current code, i.e., the content of the while loop at lines 11–12 in Algorithm 1. In this case, we adopted a permutation-based genetic algorithm (which we will refer to as EA in the following) and a simple random search (RS) as a baseline method for comparison. In particular, the RS works by drawing at random a new permutation at each iteration of the while loop, which is subsequently added only if it is at a minimum distance d from all previous permutations. On the other hand, the EA follows the UPDATE-POP-EA steady-state procedure described in Algorithm 2.

The update policy is the strategy by which the algorithm constructs the permutation code. The iterative approach laid out in Sect. 3 is based on the INCREMENTAL-EA-PA procedure (Algorithm 1), which only expands the code when a new permutation at minimum distance d from all the current ones is found. However, this update policy might easily get stuck in local optima. Intuitively, the size achievable by a permutation array constructed incrementally highly depends on the initial permutations chosen. Therefore, if the search algorithm makes a few "wrong choices" initially, it might end up with a relatively small list of permutations that cannot be further expanded.

For this reason, we also experimented with a *random reset* update policy: if a new permutation satisfying the minimum distance requirement is not found within a given number of fitness evaluations in the inner while loop of Algorithm 1, then some previous permutations – chosen at random – are removed from the current code. Also, the number of permutations to be removed is chosen randomly, but the maximum value is modeled after the cooling policy as employed in simulated annealing [13] Initially, the maximum number of codewords to remove can be as high as one-third of the current PA size ($|P|$) but is then decreased at every subsequent random reset, in order to favor the exploration of the search space at the beginning of the optimization process and its exploitation in the later stages. The actual number of permutations to be removed is a random value in $\{1, \cdots, r\}$, where r is set as $r = \frac{1}{3} \times |P| \times e^{-evals/10^6}$. The condition to invoke this reset is defined as the number of successive evaluations without increasing the PA size, e.g., the number of unsuccessful attempts to add a new permutation to the current

PA. In our experiments, this number was defined as $\max(n!, 10^5)$; the reasoning behind this is that $n!$ evaluations of the exhaustive search would be enough to find out whether any new permutation can be added to the current PA. Consequently, we use this number as the stagnation detection threshold. Note that the random reset policy can be used with any search algorithm (i.e., any type of population update) and any fitness function.

In what follows, we will denote by EA1 and EA2 the incremental EA equipped respectively with the plain update policy as in Algorithm 1 (where new permutations are only added) and with the above random reset update policy. Likewise, RS1 and RS2 will denote the analogous variants of the RS baseline algorithm concerning the update policy.

Prior to the experiments on the selected problem instances, a short tuning phase was performed on problem instance $PA(7, 5)$ to estimate the appropriate parameter values for the population size and the mutation rate p_μ of the GA. Based on those results, the population size was set to 1 000 individuals, and the mutation rate was kept at 30%. In all the experiments, the total number of evaluations (the fitness budget fb) was set to 10^7, and each experiment was executed in 30 repetitions. Concerning the variation operators, we employed the permutation-based crossovers and mutations implemented in the ECF framework[1], by choosing them uniformly at random at each evaluation.

4.2 Results

Figures 1 and 2 display the results obtained in our experiments with all search methods and across all considered problem instances, except those where $n = d$. In fact, in all those cases, each search variant managed to construct a $PA(n, n)$, or equivalently a Latin square of order n. For each problem instance (n, d), the corresponding boxplot shows the four fitness functions against the largest code size achieved by the corresponding variant of a search method. The legend for the four considered combinations of search method and update policy is reported on top of each figure.

First, one can see from the plots that all considered methods, independently from the fitness function, the search method, and the update policy, cannot scale very well concerning the problem size. Indeed, optimal solutions reaching known values for $M(n, d)$ are consistently found only in the $n = 6$ case, with the exception of a single $PA(7, 6)$ of size 42 found by EA2 with fitness function fit_3. Contrarily, for $n \geq 7$, all considered variants find PA that are significantly smaller than the best-known bounds reported in [20]. Our methods are always able to outperform the Gilbert-Varshamov bound, which is, however, quite loose as reported in Sect. 2.

As expected, the random reset update policy generally achieves better results than the plain one. This effect is particularly evident in the $n = 6$ problem instances, with the combinations adopting the plain update policy obtaining

---

[1] Framework available at http://ecf.zemris.fer.hr/.

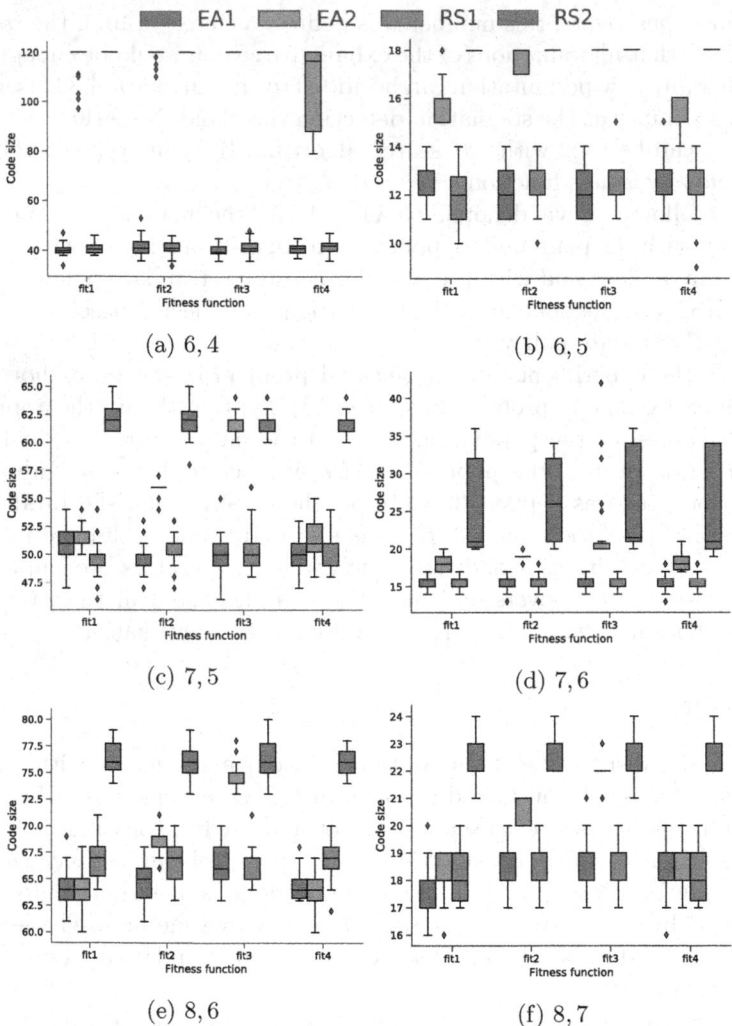

Fig. 1. Largest code size achieved by all methods across the problem instances with $n = 6, 7, 8$ and $d = n - 2, n - 1$.

considerably smaller codes than those using random resets, which instead find almost always an optimal solution. Surprisingly, by comparing the results concerning the update policies, there is no significant difference between the code sizes obtained by EA and RS. A second surprising remark concerns the fitness functions: while we expected fit_3 to be the worst-performing one in Sect. 3.3, it generally achieved larger code sizes than the other three.

As for fitness functions, the most interesting remark is that the best performing one is also the simplest, namely fit_3, that measures the minimum distance of the new candidate solution from all permutations in the current code.

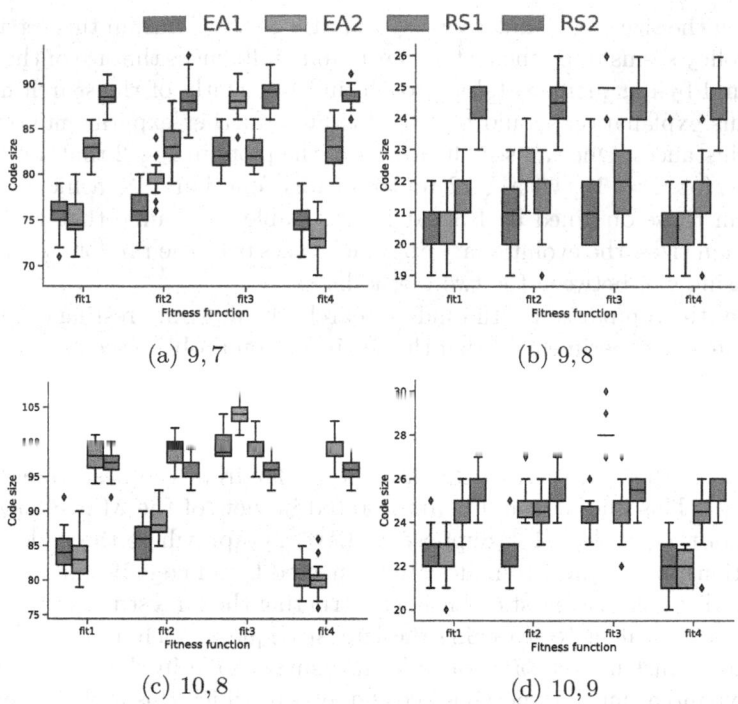

Fig. 2. Largest code size achieved by all methods across the problem instances with $n = 9, 10$ and $d = n - 2, n - 1$.

This is a fairly straightforward translation of the property characterizing a permutation code into an objective function to be maximized, and we hypothesized that it could underperform due to its limited range of values. The reason why fit_3 achieves the largest code sizes over all considered instances might reside in the size of the "local" search space, i.e., in the set of all permutations of size n that the EA searches at each stage of the incremental construction. Indeed, we targeted relatively small permutations, where the symmetric group is composed of at most $10! \approx 3 \cdot 10^6$ in the largest considered instance. Using finer-grained fitness functions such as fit_1, fit_2, and fit_4 might have hampered the EA search process, investing many fitness evaluations in optimizing much more information – e.g., the discounted invalid distances in Eq. (4) – than what was needed. It could be interesting to perform experiments on larger instances where the symmetric group S_n is not amenable to exhaustive search to see if this trend continues or if the additional information exploited by the other fitness functions gives an advantage over fit_3.

The relatively small size of the local search space of permutations might also be related to the substantial equivalence of the EA and RS performances. It could be the case that the results are quite similar because it does not make any difference how the local permutation is selected to incrementally expand the code, given the size of the underlying symmetric group. Thus, in this setting, the update policy seems to be the factor that mainly influences the size of the largest codes found by our incremental approach, independently of the search method. Again, this explanation should be tested against further experiments on larger problem instances. One can see already from the plots in Fig. 2 that a difference does arise for $n = 10$, with EA2 under fitness function fit_3 achieving larger codes than those obtained by RS2. It is reasonable to assume that with larger permutation sizes, the evolutionary approach takes over the random search, with an increasing gap between the two methods.

Besides the comparison with random search, the most interesting observation arising from our experiments is that this optimization problem seems to be exceptionally difficult for evolutionary algorithms. Only for the smallest instances of $PA(6, 4)$ and $PA(6, 5)$ could we obtain optimal solutions concerning the code size (neglecting the outlier found for $PA(7, 6)$). In all other cases, the largest code found (either with EA or RS) always lies far from the best lower bounds for $M(n, d)$. This finding could be interpreted in view of the MAX-CLIQUE formulation of the problem [18]. Suppose we have a graph where the nodes are the permutations in S_n, and two nodes are connected by an edge if and only if their Hamming distance is at least d. Then, constructing the largest permutation code $PA(n, d)$ is equivalent to searching the largest clique in such a graph. With the incremental construction, one starts from a single node in the graph and then tries to expand as much as possible the clique(s) to which this node belongs. Our evolutionary algorithm, on the other hand, does not take into account the *topology* of this graph, which involves both the region where the initial permutation is located and its *neighborhood*, i.e., the set of its adjacent nodes. This problem might also be worsened because EAs are population-based methods. Hence, by starting with a set of candidate solutions generated at random, one might waste many fitness evaluations to "move" the population close to the neighborhood of the clique constructed up to that point. One strategy to cope with this issue could be to experiment with smaller population sizes or to integrate the EA with a local search step that also considers the graph representation.

5   Conclusions and Future Work

This paper addresses the optimization problem of constructing permutation codes using EA, which, as far as we know, has not been addressed before. The main question in this domain concerns finding the largest code size for a given permutation length n and a minimum distance d. We have developed an incremental construction approach, starting from a single random permutation chosen at random and then using an EA to iteratively expand the code. We evaluated our method with four fitness functions using two different update policies and in comparison to a baseline random search algorithm. Most importantly, the results of our experiments show that this optimization problem is particularly difficult for evolutionary techniques, with the largest codes found by our EA lying far from the best-known lower bounds in most of the considered problem instances.

Further interesting findings include the fact that the simplest fitness function performed the best and that the update policy seems to be crucial for finding large codes rather than the underlying search method.

In future work, we plan to improve our incremental EA approach by following the directions outlined above: experimenting with larger problem instances and including a local search optimization step. We also envision investigating a concept closely related to equidistant permutation codes, where the Hamming distance between codewords must be exactly equal to d. Equidistant permutation codes are thus a subset of permutation codes, making our iterative procedure applicable. However, since equidistant permutation codes are more rare, we believe this problem to be even harder.

References

1. Banzhaf, W.: The "molecular" traveling salesman. Biol. Cybern. **64**(1), 7–14 (1990)
2. Chu, W., Colbourn, C.J., Dukes, P.: Constructions for permutation codes in powerline communications. Des. Codes Cryptogr. **32**(1–3), 51–64 (2004)
3. Colbourn, C.J., Dinitz, J.H.: Combinatorial designs. In: Rosen, K.H., Michaels, J.G., Gross, J.L., Grossman, J.W., Shier, D.R. (eds.) Handbook of Discrete and Combinatorial Mathematics. CRC Press (1999)
4. Colbourn, C.J., Kløve, T., Ling, A.C.H.: Permutation arrays for powerline communication and mutually orthogonal Latin squares. IEEE Trans. Inf. Theory **50**(6), 1289–1291 (2004)
5. Conway, J.H., Sloane, N.J.A.: Sphere Packings, Lattices and Groups, Grundlehren der mathematischen Wissenschaften, vol. 290. Springer (1988). https://doi.org/10.1007/978-1-4757-6568-7
6. Daemen, J., Rijmen, V.: The Design of Rijndael - The Advanced Encryption Standard. AES), Second Edition. Springer (2020). https://doi.org/10.1007/978-3-662-04722-4
7. Ferreira, H.C., Vinck, A.H.: Interference cancellation with permutation trellis codes. In: IEEE 52nd Vehicular Technology Conference Fall 2000, vol. 5, pp. 2401–2407. IEEE (2000)
8. Fogel, D.B.: Applying evolutionary programming to selected traveling salesman problems. Cybern. Syst. **24**(1), 27–36 (1993)
9. Goldberg, D.E., Jr., R.L.: Alleles, loci and the traveling salesman problem. In: Grefenstette, J.J. (ed.) Proceedings of the 1st International Conference on Genetic Algorithms, Pittsburgh, PA, USA, July 1985, pp. 154–159. Lawrence Erlbaum Associates (1985)
10. Han Vinck, A.: Coded modulation for powerline communications. AEU Int. J. Eletron. Commun. **54**(1), 45–49 (2000)
11. Jiang, A., Mateescu, R., Schwartz, M., Bruck, J.: Rank modulation for flash memories. In: Kschischang, F.R., Yang, E. (eds.) 2008 IEEE International Symposium on Information Theory, ISIT 2008, Toronto, ON, Canada, 6–11 July 2008, pp. 1731–1735. IEEE (2008)
12. Karp, R.M.: Reducibility among combinatorial problems. In: Miller, R.E., Thatcher, J.W. (eds.) Proceedings of a symposium on the complexity of computer computations, held 20–22 March 1972, at the IBM Thomas J. Watson Research Center, Yorktown Heights, New York, USA, pp. 85–103. The IBM Research Symposia Series, Plenum Press, New York (1972)

13. Kirkpatrick, S., Gelatt, C.D., Vecchi, M.P.: Optimization by simulated annealing. Science **220**(4598), 671–680 (1983)
14. Knezevic, K., Picek, S., Mariot, L., Jakobovic, D., Leporati, A.: The design of (almost) disjunct matrices by evolutionary algorithms. In: Fagan, D., Martín-Vide, C., O'Neill, M., Vega-Rodríguez, M.A. (eds.) Theory and Practice of Natural Computing - 7th International Conference, TPNC 2018, Dublin, Ireland, December 12-14, 2018, Proceedings. Lecture Notes in Computer Science, vol. 11324, pp. 152–163. Springer (2018). https://doi.org/10.1007/978-3-030-04070-3_12
15. Liu, M., Sim, S.M.: Lightweight MDS generalized circulant matrices. In: Peyrin, T. (ed.) Fast Software Encryption - 23rd International Conference, FSE 2016, Bochum, Germany, 20–23 March 2016, Revised Selected Papers. Lecture Notes in Computer Science, vol. 9783, pp. 101–120. Springer (2016). https://doi.org/10. 1007/978-3-662-52993-5_6
16. Mariot, L., Picek, S., Jakobovic, D., Leporati, A.: Evolutionary algorithms for the design of orthogonal Latin squares based on cellular automata. In: Bosman, P.A.N. (ed.) Proceedings of the Genetic and Evolutionary Computation Conference, GECCO 2017, Berlin, Germany, 15–19 July 2017, pp. 306–313. ACM (2017)
17. Mariot, L., Picek, S., Jakobovic, D., Leporati, A.: Evolutionary search of binary orthogonal arrays. In: Auger, A., Fonseca, C.M., Lourenço, N., Machado, P., Paquete, L., Whitley, L.D. (eds.) Parallel Problem Solving from Nature - PPSN XV - 15th International Conference, Coimbra, Portugal, 8–12 September 2018, Proceedings, Part I. Lecture Notes in Computer Science, vol. 11101, pp. 121–133. Springer (2018). https://doi.org/10.1007/978-3-319-99253-2_10
18. Montemanni, R., Barta, J., Smith, D.H.: Graph colouring and branch and bound approaches for permutation code algorithms. In: Rocha, Á., Correia, A.M.R., Adeli, H., Reis, L.P., Teixeira, M.M. (eds.) New Advances in Information Systems and Technologies - Volume 1 [WorldCIST'16, Recife, Pernambuco, Brazil, March 22–24, 2016]. Advances in Intelligent Systems and Computing, vol. 444, pp. 223–232. Springer (2016). https://doi.org/10.1007/978-3-319-31232-3_21
19. Oliver, I.M., Smith, D.J., Holland, J.R.C.: A study of permutation crossover operators on the traveling salesman problem. In: Grefenstette, J.J. (ed.) Proceedings of the 2nd International Conference on Genetic Algorithms, Cambridge, MA, USA, July 1987, pp. 224–230. Lawrence Erlbaum Associates (1987)
20. Smith, D.H., Montemanni, R.: A new table of permutation codes. Des. Codes Cryptogr. **63**(2), 241–253 (2012)
21. Smith, D.H., Montemanni, R.: Permutation codes with specified packing radius. Des. Codes Cryptogr. **69**(1), 95–106 (2013)
22. Stinson, D.R.: Combinatorial designs - constructions and analysis. Springer (2004). https://doi.org/10.1007/b97564
23. Syswerda, G., Palmucci, J.: The application of genetic algorithms to resource scheduling. In: Belew, R.K., Booker, L.B. (eds.) Proceedings of the 4th International Conference on Genetic Algorithms, San Diego, CA, USA, July 1991, pp. 502–508. Morgan Kaufmann (1991)
24. De la Torre, D., Colbourn, C., Ling, A.: An application of permutation arrays to block ciphers. Congressus Numerantium, pp. 5–8 (2000)
25. Vaudenay, S.: On the need for multipermutations: cryptanalysis of MD4 and SAFER. In: Preneel, B. (ed.) Fast Software Encryption: Second International Workshop. Leuven, Belgium, 14–16 December 1994, Proceedings. Lecture Notes in Computer Science, vol. 1008, pp. 286–297. Springer (1994). https://doi.org/10. 1007/3-540-60590-8_22

Improving the Convergence and Diversity in Differential Evolution Through a Stock Market Criterion

Mario A. Navarro[1], Alfonso Ramos-Michel[1], Angel Gaspar[1], Diego Oliva[1(✉)],
Salvador Hinojosa[2], Seyed Jalaleddin Mousavirad[3],
and Marco Pérez-Cisneros[1]

[1] Universidad de Guadalajara, CUCEI, 44430 Guadalajara, Mexico
{mario.navarro,angel.gaspar}@alumnos.udg.mx,
alfonso.rmichel@academicos.udg.mx, {diego.oliva,marco.perez}@cucei.udg.mx
[2] School of Engineering and Science, Tecnologico de Monterrey,
45201 Zapopan, Mexico
salvador.hinojosa@tec.mx
[3] Computer Engineering Department, Hakim Sabzevari University, Sabzevar, Iran

Abstract. Most of the Evolutionary Algorithms (EA) use a population of candidate solutions to explore the search space following specific rules during an iterative process. These algorithms are designed expecting a good balance between exploration and exploitation during the search process. Besides, the diversity of the population is crucial to properly explore the search space. This article introduces an improved version of the Differential Evolution (DE) algorithm, which employs the moving average (MA) to determine when the population should diversify or intensify by using additional operators. The MA is one of the most used stock market indicators, providing recommendations for selling or buying stocks based on historical data. Here, the MA of the historical fitness and dimension-wise diversity is analyzed to determine if the DE continues operating normally or should diversify or intensify the search using additional operators. An exhaustive benchmark involving 37 optimization functions with different complexity levels confirmed the effectiveness of the proposed approach.

Keywords: Moving average · Diversity · Differential evolution · Evolutionary algorithms · Global optimization

1 Introduction

Several implementations aim to minimize or maximize objective functions from optimization problems, such as time, energy, power, errors, etc. The complexity of such problems depends on different aspects, including the number of decision variables, which define the search space. The related literature uses evolutionary algorithms (EA) to explore the search space and find the optimal solution [7,19].

© Springer Nature Switzerland AG 2022
J. L. Jiménez Laredo et al. (Eds.): EvoApplications 2022, LNCS 13224, pp. 157–172, 2022.
https://doi.org/10.1007/978-3-031-02462-7_11

The Differential Evolution (DE) is an EA that performs a stochastic search using operators such as crossover, selection, and mutation to guide the population towards optimal solutions in the search space [20]. Although EAs work on different domains, they can be stuck in sub-optimal solutions. Therefore the use of diversity can guide the search to avoid premature convergence [14]. Diversity is a characteristic of individual distribution used in many population-based algorithms to obtain relevant information about the exploration of the search space [13,16].

The moving average (MA) is one of the most used indicators in stock and futures markets [12]. The MA uses the historical price in a predetermined period to decide whether to buy or sell. In this article, we use MA to analyze the evolution of the diversity during the execution of DE and, if needed, apply other operators. One addressed operator here is the Opposition-Based Learning (OBL) [22], which computes an opposite position of a current solution, compares their quality and discards the worst. The MA has been previously used in EA, some examples are [1,2], where the authors propose versions of DE with self-adaptation of the mutation and crossover factor by using the exponential MA. While other papers explore the application of the exponential weighted MA in multi-objective approaches [11].

This article introduces an improved version of the DE algorithm based on the MA to decide whether to apply the original DE algorithm operator, the OBL, to intensify DE's exploration or change its parameters to intensify the exploitation. Moreover, the MA of the diversity is used as a metric to avoid the entire population falling into sub-optimal solutions. Using the MA, the proposed approach (MADE) has different cases that permit the modification of the search operator based on the evolution of the MA of fitness during the iterations. The MADE has been tested over benchmark functions with different complexity and multiple dimensions.

The rest of the paper is organized as follows: Sect. 2 presents the preliminaries of the proposal. In Sect. 3 the proposed MADE algorithm is detailed. Meanwhile, Sect. 4 discusses the experiments and results. Finally, the conclusions and future work are presented in Sect. 5.

2   Background

2.1   Differential Evolution

DE uses a population P of N_p individuals, described by Eq. (1):

$$P^G = \left[X_1^G, X_2^G, ..., X_{N_p}^G \right] \tag{1}$$

where G is the iteration number and N_p is the population size.

The population is randomly initialized, and a mutation operator is applied to generate a mutant vector V_i^G for each X_i^G in the current population. One of the most common mutation strategies is shown in Eq. (2).

$$DE/best/1 : V_i^G = X_{best}^G + F\left(X_{r_1}^G - X_{r_2}^G\right) \tag{2}$$

where $r_1 \neq r_2 \in [1, N_p]$ are different random integers, and they are also not equal to the index i, F is a real and constant factor $\in [0, 2]$ which controls the amplification of the differential variation of $\left(X_{r_1}^G - X_{r_2}^G\right)$ on Eq. (2), and X_{best}^G represents the global best individual at generation G.

After mutation, a crossover operation is applied. For each individual X_i^G, the trial vector U_i^G is generated by using Eq. (3).

$$U_{i,j}^G = \begin{cases} V_{i,j}^G, rand \leq CR \ or \ j = j_{rand} \\ X_{i,j}^G, otherwise \end{cases} \tag{3}$$

Here the crossover rate is $CR \in [0, 1]$, and it is an important parameter that controls the crossover probability [17]. j_{rand} is a random integer generated between $range[1, D]$, which guarantees that the trial vector gets at least one component from the mutant vector V_i^G.

The trial vector U_j^G is then compared with its target vector X_i^G. The better one will be selected to survive for the next generation. Finally, mutation, crossover, and selection are repeated until the termination criterion is met and a final solution is returned.

2.2 Moving Average

A moving average is a time series constructed by taking averages of several sequential values of another time series [8]. The most commonly used moving average is the simple moving average (SMA) which considers the mean of the previous s data points in time series data as shown in Eq. (4).

$$SMA = \frac{1}{s} \sum_{i=0}^{s-1} P_{M-i} \tag{4}$$

Where P_M represents the value of the data point at time M and s represents the number of data points used in the calculation.

2.3 Population Diversity

The most important factor affecting the performance of optimization algorithms is a balance between exploration and exploitation. In this paper we apply the Dimension-wise Population Position Diversity [3,7] to orient the search and achieve that balance, this measure of diversity is defined in Eq. (5) and Eq. (6).

$$Div_j = \frac{1}{N_p} \sum_{i=1}^{N_p} \left| median(x^j) - x_i^j \right| \tag{5}$$

$$Div = \frac{1}{d} \sum_{j=1}^{d} Div_j \tag{6}$$

where x_i^j is the position of the individual i in dimension j, $median(x^j)$ is the median position of the whole population in dimension j, N_p is the candidate solution number and d is the number of dimensions. The variable Div_j expresses the measure between each solution and each specific dimension concerning the median of the entire population. Then, Div expresses the average value of the diversity of all dimensions.

2.4   Opposition-Based Learning

Opposition-based learning (OBL) [22] considers the current solution's position and its opposite location at the same time in order to achieve a better approximation for the current candidate solution. Equation (7) shows the opposite candidate solution in one-dimensional search space and Eq. (8) for a d-dimensional space.

$$W^o = a + b - W \tag{7}$$

$$W_q^o = a_q + b_q - W_q \tag{8}$$

where $W_1, W_2, ..., W_d$ is a point in d-dimensional space, and $W_q \in [a_q, b_q]; q = \{1, 2, .., d\}$ [15].

3   Proposed Approach

This paper proposes an improved algorithm called Moving Average Differential Evolution (MADE) based on the standard DE algorithm combined with the Moving Average (MA). MADE collects historical data over the iterations to analyze the MAs of the historical fitness and dimension-wise diversity [18]. The analysis performed at each iteration determines if the population should diversify or intensify the search using additional operators. MADE uses Opposition Based Learning (OBL) [22] to enhance the exploration of the search space, and a modified DE operator to promote exploitation with the mutation strategy $DE/best/1$ [23]. In Fig. 1 the flowchart of the MADE shows the general approach. First, the initial population is generated. Then, the algorithm uses the original DE algorithm during the first $periodMA$ iterations to generate historical data. Afterward, there is enough information to calculate the MA considering the last samples of the fitness history and the diversity history for the selected period defined as $periodMA$ and represented as P_M in Eq. (4).

MADE stores the historical data in four vectors, getting the information regarding each iteration defined as it; $histFit$ stores the historical values of fitness as $histDiv$ stores the historical data of Diversity values. The MA of the fitness history called $MAfit$ and the moving average of diversity called $MAdiv$ are calculated from their respective historical data vectors. The values of each vector on the current iteration it are compared, and three cases are defined:

Case 1 ($histFit(it) = MAfit(it)$ or $histDiv(it) = MAdiv(it)$) Stagnation. The OBL operator is selected to promote exploration and diversify the population.

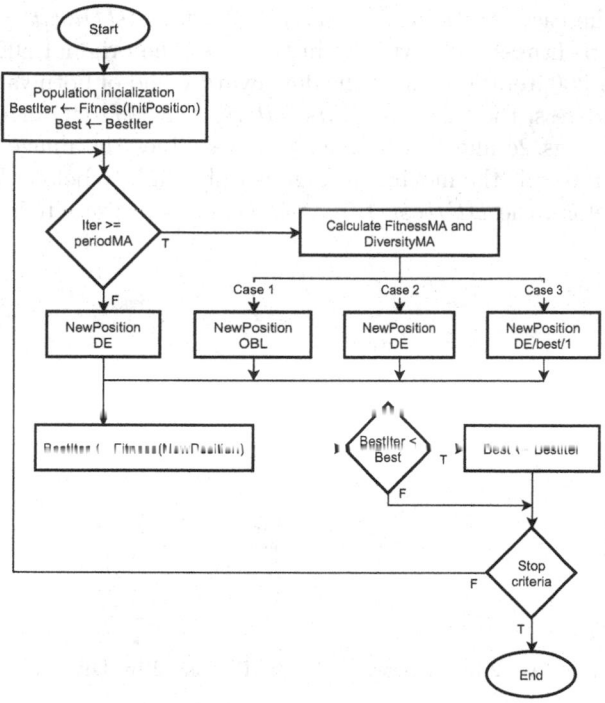

Fig. 1. Flow chart of the proposed MADE.

Case 2 $(histFit(it) < MAfit(it)$ *or* $histDiv(it) < MAdiv(it))$ Downward trend. The original search strategy of the DE algorithm is executed by mutation and crossover.

Case 3 $(histFit(it) > MAfit(it)$ *or* $histDiv(it) > MAdiv(it))$ Upward trend. The mutation strategy $DE/best/1$ is used to attract the population towards the best solution granting faster convergence speed and a higher accuracy while reducing the population diversity.

The algorithm executes the operator selection process until it reaches the stop criteria. Although the possible logical combinations exceed the three presented cases, these combinations were not estimated considering the nature of the behavior of the algorithms and the variables used. Figure 2 shows the considered cases in the methodology after selecting a sample of each case by choosing random functions from the test set; Figs. 2a, 2c, and 2e, correspond to the fitness history depicted as dark blue while the MA is orange.

Also, the plots in Figs. 2b, 2d, and 2f show the historical diversity in green color while its MA is plotted in yellow. At Fig. 2a it is observed that after 325 iterations the condition $(histFit(it) = MAfit(it))$ is true and the OBL operator is selected; in the same way Fig. 2b shows the instant in which the condition $(histDiv(it) = MAdiv(it))$ is true to execute the OBL operator, the Figs. 2c and 2d show the cases $histFit(it) < MAfit(it)$ and $histDiv(it) < MAdiv(it))$ corresponding to fitness and diversity, in this case the original DE algorithm is selected for the 200 iteration due to the downward trend of both variables during the iterative process, the third case $(histFit(it) > MAfit(it)$ or $histDiv(it) > MAdiv(it))$ in Figs. 2e and 2f where in both variables the image is zoomed in on the area of interest, the moving average is only slightly below the fitness and diversity; therefore, the $DE/best/1$ mutation strategy is selected.

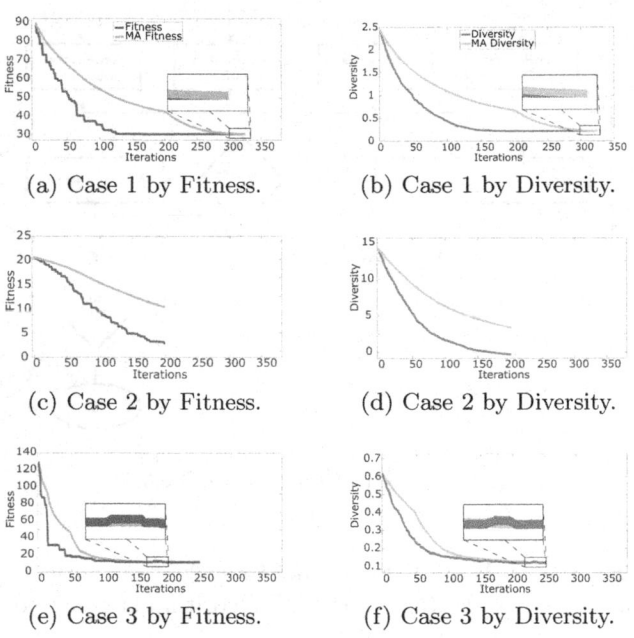

(a) Case 1 by Fitness.   (b) Case 1 by Diversity.

(c) Case 2 by Fitness.   (d) Case 2 by Diversity.

(e) Case 3 by Fitness.   (f) Case 3 by Diversity.

Fig. 2. Different cases of the MA for the fitness and diversity. (Color figure onine)

4   Experiments and Results

In the experiments, it is considered a set of benchmark functions: 5 unimodal functions, 28 multimodal functions, and 4 composite functions [6][1] considering 30 and 50 dimensions. The list of the benchmark functions is presented in Table 1.

The comparative study presents 6 different algorithms: the original DE version [20], three improved variants (JADE [25], BeSD [4] and L-SHADE [21]), and two classical algorithms well known in the state-of-the-art such as PSO [10] and GA [5].

All the algorithms are configured with 50,000 access functions as stop criteria. The algorithms with the fixed population parameter were configured with $Np = 30$. For the enhanced versions of DE, the parameters are as follows:

---

[1] Some functions can also be found at https://www.sfu.ca/~ssurjano/optimization.html.

Table 1. Function reference table

Function reference	Function name	Limits	Global minimum
F1	Ackley	$[-30, 30]^n$	0
F2	Dixon	$[-10, 10]^n$	0
F3	Griewank	$[-600, 600]^n$	0
F4	Infinity	$[-1, 1]^n$	0
F5	Levy	$[-10, 10]^n$	0
F6	Mishra1	$[0, 1]^n$	2
F7	Mishra2	$[0, 1]^n$	2
F8	Mishra11	$[-10, 10]^n$	0
F9	MultiModal	$[-10, 10]^n$	0
F10	Penalty1	$[-50, 50]^n$	0
F11	Penalty2	$[-50, 50]^n$	0
F12	Perm1	$[-n, n]^n$	0
F13	Perm2	$[-n, n]^n$	0
F14	Plateau	$[-5.12, 5.12]^n$	30
F15	Powell	$[-4, 5]^n$	0
F16	Qing	$[-500, 500]^n$	0
F17	Quartic	$[-1.28, 1.28]^n$	0
F18	Quintic	$[-10, 10]^n$	0
F19	Rastringin	$[-5.12, 5.12]^n$	0
F20	Rosenbrock	$[-5, 10]^n$	0
F21	Schwefel21	$[-100, 100]^n$	0
F22	Schwefel22	$[-100, 100]^n$	0
F23	Schwefel26	$[-500, 500]^n$	0
F24	Step	$[-100, 100]^n$	0
F25	Styblinski-Tank	$[-5, 5]^n$	$-39.1659n$
F26	Trid	$[-n^2, n^2]^n$	$-n(n+4)(n-1)/6$
F27	Vincent	$[0.25, 10]^n$	$-n$
F28	Zakharov	$[-5, 10]^n$	0
F29	Rothyp	$[-65.536, 65.536]^n$	0
F30	Schwefel2	$[-100, 100]^n$	0
F31	Sphere	$[-5, 5]^n$	0
F32	Sum Squares	$[-10, 10]^n$	0
F33	Sum of Different Powers	$[-1, 1]^n$	0
F34	fx15	$[-100, 100]^n$	0
F35	fx16	$[-100, 100]^n$	$n - 1$
F36	fx17	$[-100, 100]^n$	$(1.1n) - 1$
F37	fx18	$[-100, 100]^n$	$n - 1$

- JADE: $uF = 0.6$ and $uCR = 0.5$ [25].
- L-SHADE: the historical memory size $H = 6$, $pvalue = 0.11$, external archive size $|A| = Npop * 2.6$ [21].
- BeSD: $N = Np$, where N is Size of the pattern matrix and $1 \leq K \leq 5$, where K denotes the size control value of S [4].

Besides, all the algorithms performed 30 independent runs over each benchmark problem. Internal parameters of the other algorithms (GA and PSO and DE) were set following the standard versions and the references of their original publications. For the MADE, the low and high dimension scaling factors were established as $beta_{min} = 0.1$ and $beta_{max} = 0.9$, respectively. The crossover probability was set to $CR = 0.2$ according to DE standard version. To calculate the MA, the typical values used are 20, 50, 100, and 200 periods. However, in the related literature, a fixed period of 200 is suggested since a long MA should prevent losses from false trading signals when the markets are ranging [9]. The MADE was tested with different periods in this work, obtaining the best results with a 200-period long-term MA. Moreover, a non-parametric pairwise Wilcoxon rank test is conducted to validate the results [24] with 5% of the significance level.

4.1 Experiments over 30 Dimensions

The experiments are conducted over benchmark functions in 30 dimensions. The average (Avg) of the fitness and standard deviation (Std. Dev) are used. Table 2 presents the results of the Avg and the Std. Dev for the seven algorithms. The results in boldface represent the optimal values found by each method. In terms of the Avg, the proposed MADE achieved the global optimal in 65% of the functions. The rest of the algorithms obtain the optimal solutions: JADE in 14%, BeSD in 5%, L-SHADE in 30%, DE in 11%, PSO in 11%, and GA in 16% of the functions. For the Std. Dev the MADE obtains the lower value in 59% of the cases. These results mean that the MADE is more stable and accurate than the other methods.

The p-values from the Wilcoxon test are obtained by comparing the best value of each run between the proposed method against every other algorithm; the null hypothesis is rejected for all values lower than (0.05), which means that the methods compared are significantly different from each other. In a simplified form, to identify the outcomes between the algorithms, we use the symbols ▲, ▼, and ►, representing the performance of the competing algorithms concerning the proposed algorithm. In the Table 2 the symbol ▲ means that the Wilcoxon test refutes the null hypothesis and is significantly different from MADE, ▼ shows that the null hypothesis is not rejected for MADE, and ► specifies that the analysis did not identify any difference between MADE and its competitors. The cases where there are not statistically significant differences with JADE are in the functions *F2, F6, F7, F14, F24, F25* and *F27*, with BeSD in *F6, F7, F14, F23* and *F27*, with L-SHADE in the functions *F2, F6, F7, F14, F23-F25, F27, F35* and *F37*, with the DE algorithm in functions *F6, F7, F14, F23* and *F25*, with PSO for functions *F9, F14* and *F24*, and with GA in *F6, F7, F9, F14* and

F27. Finally, for the algorithms where the null hypothesis it is not satisfied are listed below:

- JADE in functions *F8*, *F10*, *F16*, *F19*, *F20*, *F26* and *F36*.
- BeSD in functions *F17*, *F26*, *F28* and *F36*.
- L-SHADE in functions *F10*, *F12*, *F15-F21*, *F26*, *F28* and *F36*.
- DE in functions *F8*.
- PSO in functions *F8*, *F16*, *F18* and *F36*.
- GA in functions *F8*, *F17*, *F19* and *F26*.

Figures 3 and 4 present the convergence curves for a single run in 30 dimensions[2]. The plots show the evolution of the global best regarding the number of function evaluations. All the graphs are in log scale for better representation except for *F14*, *F23*, *F25* and *F27*. From these figures, it is possible to observe the MADE behavior (light blue line), in which it can reach the minimum values in a reduced number of function evaluations. Besides, from some problems as *F10* to *F13*, it is possible to see the MA's influence due to the drastic change of the fitness. The results of Figs. 3 and 4 confirm the comparison presented in Table 2.

4.2   Experiments over 50 Dimensions

For the second series of experiments, there are used 50 dimensions for the benchmark functions. Table 3 shows the Avg and the Std results. Dev. for the MADE, JADE, BeSD, L-SHADE, DE, PSO, and GA. From this table, it is possible to analyze that the MADE obtains the global optimal in 62% of the functions, JADE gets the optimal value in 14%, BeSD in 3% of the functions, L-SHADE obtain the optimal value in 27% while DE and PSO reach the optimal of the functions in 8% and 3% respectively. Finally, the GA reaches its best values in 16% of the functions. In terms of stability, the MADE has a low Std. Dev in 54% of the problems. These facts provide evidence that the accuracy of the MADE is not affected by the number of dimensions.

For 50 dimensions, we also performed the Wilcoxon test. From Table 3 the MADE in most experiments refutes the null hypothesis. The algorithms where it is not satisfied are listed below:

- JADE in functions *F8*, *F10*, *F11*, *F15*, *F16*, *F18*, *F19*, *F26*, *F27* and *F36*.
- BeSD in functions *F12*, *F15*, *F17*, *F21*, *F26*, *F28* and *F36*.
- L-SHADE in functions *F10-F12*, *F15-F18*, *F20*, *F21*, *F26-F28* and *F36*.
- DE on functions *F8*.

---

[2] The convergence curves for the 50 dimensions also have advantages for MADE. However, they were not considered due to the space limitation.

Table 2. Average and Standard deviation for 30 dimensions

Function	MADE Avg	MADE Std Dev	JADE Avg	JADE Std Dev	BeSD Avg	BeSD Std Dev	L-SHADE Avg	L-SHADE Std Dev	DE Avg	DE Std Dev	PSO Avg	PSO Std Dev	GA Avg	GA Std Dev
F1	1.39E-12	3.66E-13	5.73E+00	1.63E+00	3.98E-01	1.84E-01	1.89E-06	1.27E-06	3.06E+00	4.38E-01	2.09E-05	5.69E-05	5.87E-01	3.80E-01
F2	6.72E-01	2.66E-02	6.67E-01	6.03E-04	1.83E+00	9.84E-01	6.67E-01	3.88E-10	2.49E-01	5.59E+00	7.67E+00	1.17E+02	4.66E+00	2.71E+00
F3	0.00E+00	0.00E+00	6.98E-04	1.88E-03	8.20E+01	1.23E-01	1.21E-10	8.68E-11	1.05E+00	1.48E-02	1.15E+00	1.02E+02	1.36E+00	8.30E-02
F4	1.19E-55	2.66E-55	3.00E-21	1.54E-20	2.57E-11	4.56E-11	9.19E-33	2.82E-32	1.36E-07	6.23E-08	2.19E-17	4.68E-17	3.45E-14	6.92E-14
F5	1.71E-24	1.65E-24	4.65E-02	1.38E-01	2.98E-02	1.20E-02	3.58E-12	4.58E-12	7.79E-01	1.45E-01	1.09E-01	2.28E-01	5.11E-01	7.68E-01
F6	2.00E+00	0.00E+00	2.00E+00	0.00E+00	NaN	NaN	2.00E+00	0.00E+00	2.00E+00	0.00E+00	1.22E+01	1.79E-01	2.00E+00	0.00E+00
F7	2.00E+00	0.00E+00	2.00E+00	0.00E+00	NaN	NaN	2.00E+00	0.00E+00	2.00E+00	0.00E+00	1.54E+01	3.56E+01	2.00E+00	0.00E+00
F8	7.95E-06	3.12E-05	0.00E+00	0.00E+00	3.27E-04	1.39E-04	1.07E-05	3.54E-05	0.00E+00	0.00E+00	1.53E-30	7.13E-30	2.23E-11	8.44E-11
F9	0.00E+00	0.00E+00	9.75E-49	5.34E-48	4.33E+00	2.15E-48	9.00E-79	4.57E-78	4.21E-65	1.39E-64	0.00E+00	0.00E+00	0.00E+00	0.00E+00
F10	7.17E-01	9.71E-01	3.14E-02	2.05E-02	5.00E-06	2.13E-05	7.95E-01	3.75E-01	1.25E+06	3.15E-05	8.23E+01	3.47E-01	8.28E+01	3.66E+00
F11	9.14E+01	4.20E+00	1.36E-02	2.01E-01	4.39E-04	1.82E-04	1.09E-02	2.92E+00	1.60E+05	3.43E-04	9.92E+01	4.60E+00	1.00E+02	6.03E+00
F12	1.85E-80	3.42E-80	5.99E-82	4.16E-81	1.71E-82	2.30E-82	5.80E-80	1.30E-81	5.46E-81	6.69E-81	1.33E-81	4.02E-81	1.15E-82	2.01E-82
F13	4.76E+01	1.03E+01	4.42E+01	4.66E-01	2.51E-03	5.91E-03	5.94E-01	1.02E-02	5.88E-38	3.22E-39	8.26E+09	4.54E-10	5.38E+02	6.46E+02
F14	3.00E+01	0.00E+00	3.00E-01	0.00E+00	3.00E-01	0.00E+00	3.00E-01	0.00E+00	3.00E+01	0.00E+00	3.02E+01	9.13E-01	3.00E-01	0.00E+00
F15	3.81E-02	1.76E-02	1.06E+00	4.12E+00	3.00E-01	1.44E-01	2.96E-06	2.89E-06	3.91E+02	6.14E-01	1.78E+02	1.17E-02	2.69E-01	1.50E-01
F16	1.14E+00	1.97E+00	7.60E+00	6.59E+00	1.04E-03	4.78E-02	5.62E+00	1.20E-01	2.07E+05	8.63E-04	2.23E-05	5.86E-05	8.07E+01	5.94E-01
F17	9.97E-09	5.82E-01	1.12E+01	5.64E+01	9.07E+00	4.43E-01	9.26E+00	3.80E-01	1.21E+01	6.18E-01	1.06E+01	5.92E-01	9.33E+00	3.46E-01
F18	1.89E-04	9.22E-04	2.34E+00	1.20E+00	8.35E+00	4.67E+00	2.97E-03	5.34E-03	5.55E+01	6.22E+00	7.91E-05	2.85E-04	4.48E-01	5.09E-01
F19	3.74E-01	6.18E+00	1.77E+01	2.97E+00	6.93E-01	6.40E+00	1.14E-01	3.06E+00	4.20E+01	3.56E+00	5.65E+01	2.48E-01	8.34E+00	2.25E+00
F20	2.79E+01	1.03E+01	2.61E+01	1.82E+01	1.06E+02	5.36E-01	2.10E-01	6.82E-01	3.62E+02	5.57E-01	2.27E+04	3.21E+04	1.09E+02	4.61E+01
F21	1.43E-01	4.38E-02	6.15E+00	1.47E+01	2.81E+00	7.28E-01	5.45E-03	2.74E-03	3.95E+01	2.89E+00	6.85E+00	2.44E+00	3.16E+00	5.71E+01
F22	2.49E-13	1.22E-13	7.21E-01	4.85E+00	5.64E+00	1.74E+00	1.24E-01	1.79E-01	4.58E+00	6.60E-01	1.83E+02	7.75E-01	2.45E+00	6.22E+00
F23	-1.25E+04	1.62E+02	-9.74E+03	6.80E+02	NaN	NaN	-1.20E-04	2.02E-02	-1.24E+04	4.60E+01	-9.28E+03	7.94E+02	-1.08E+04	3.26E+02
F24	0.00E+00	0.00E+00	0.00E+00	0.00E+00	3.50E+00	1.98E+00	0.00E+00	0.00E+00	8.30E+00	1.29E-00	0.00E+00	1.29E+00	1.70E+00	2.27E+01
F25	-1.17E+03	5.97E-14	-1.16E+03	1.82E+01	-9.63E-02	1.88E-01	-1.17E-03	9.59E-07	-1.15E-03	8.29E+00	-1.07E+04	3.29E-01	-1.10E+03	5.59E+03
F26	1.33E-04	1.42E+04	-4.43E-03	4.84E+01	2.87E-02	6.30E-02	-4.93E-03	3.43E-02	6.84E+05	1.46E+05	4.07E+04	7.19E-04	4.31E+03	4.47E+01
F27	-2.96E-01	2.73E-01	-2.99E-01	3.33E-02	NaN	NaN	-2.98E-01	8.46E-02	-2.62E+00	7.00E-01	-2.78E-01	1.48E-00	-3.00E+01	7.18E-04
F28	8.58E-01	1.72E-01	4.29E+00	8.54E+01	2.63E-01	1.25E-01	4.20E-04	4.56E-04	3.82E+02	5.07E-01	1.76E+02	9.49E+01	1.10E+02	4.47E+01
F29	2.02E-22	1.52E-22	4.05E-05	2.12E-04	5.32E-06	2.38E+00	3.40E-10	3.64E-10	2.36E+01	5.14E-01	3.44E+03	4.84E-03	1.51E+00	3.75E+00
F30	6.67E-21	5.51E-21	2.06E+03	1.05E+02	1.61E+02	8.05E-01	2.30E-08	1.92E-08	5.03E+02	9.81E-01	1.78E+02	2.22E-05	1.89E+02	3.58E+02
F31	1.06E-25	6.48E-26	1.98E+08	8.27E+08	2.44E-03	1.37E-03	1.19E-13	9.87E-14	1.42E-02	3.84E-03	5.71E-13	8.45E-13	1.84E-04	1.75E+00
F32	6.03E-24	4.51E-24	8.13E-08	2.65E-07	1.11E-01	4.20E-02	1.05E-11	8.31E-12	5.34E-01	1.40E-01	5.33E+01	1.11E-02	6.60E-02	1.69E-01
F33	3.31E-78	1.60E-77	1.39E-16	7.60E-16	5.03E-18	8.19E-18	1.43E-31	5.92E-31	2.08E-10	2.89E-10	9.21E-25	3.83E-24	4.41E-11	1.97E-10
F34	4.31E-14	2.55E-14	8.98E-01	4.89E+00	1.77E+00	7.61E-06	1.25E-05	7.61E-06	8.11E+00	1.78E+00	6.69E+02	2.55E-03	1.64E+01	1.47E-01
F35	2.90E+01	1.37E-14	3.32E+01	3.11E+01	3.27E+01	5.36E+00	2.90E-01	1.99E-07	7.88E+01	1.01E+01	6.87E+01	1.63E-01	3.79E+01	5.82E+00
F36	1.99E+02	2.85E+01	1.60E+02	1.04E+02	9.21E+01	2.01E-01	3.20E-01	5.49E-05	1.27E+03	2.29E+02	9.75E+01	3.10E-01	4.71E+02	1.46E+02
F37	2.90E+01	2.53E-13	3.41E+01	5.23E+01	3.28E+01	2.41E+01	2.90E-01	8.18E-05	4.85E+01	5.73E+01	1.17E+02	1.24E+02	4.04E+01	7.35E+00

Table 3. Average and Standard deviation for 50 dimensions

Function	MADE Avg	MADE Std. Dev	JADE Avg	JADE Std. Dev	BeSD Avg	BeSD Std. Dev	L-SHADE Avg	L-SHADE Std. Dev	DE Avg	DE Std. Dev	PSO Avg	PSO Std. Dev	GA Avg	GA Std. Dev
F1	7.78E-09	1.66E-09	1.11E+01	3.27E+00	2.55E+00	2.96E+01	4.17E+03	1.26E+03	1.45E+01	1.34E+01	9.20E-01	7.55E-01	1.76E+00	2.59E-01
F2	6.73E-01	3.45E-02	6.89E+01	6.57E+02	1.74E+01	4.12E+00	6.68E+01	2.51E+03	6.57E+04	1.36E+04	7.88E+03	2.96E+04	2.16E+01	1.06E+01
F3	7.13E-15	1.19E-14	4.10E+03	6.96E+03	1.24E+00	7.28E-02	9.33E+04	5.49E+04	4.04E+01	5.36E+01	1.61E-02	2.25E-02	6.64E-01	1.87E-01
F4	5.14E-34	6.87E-34	2.49E+08	9.89E+08	9.50E+09	8.14E+09	1.76E-18	2.69E-18	7.47E-02	2.00E-02	3.41E-08	9.24E-08	2.36E-11	2.62E-11
F5	2.07E-16	1.55E-16	4.90E+01	6.78E+01	8.25E+01	2.64E+01	2.42E+05	1.62E-05	6.61E+01	8.05E+01	5.71E+00	5.37E-06	2.47E+00	1.31E+00
F6	2.00E+00	0.00E+00	2.00E+00	0.00E+00	NaN	NaN	2.00E+00	5.34E-08	2.00E+00	0.00E+00	5.52E+06	7.85E+06	2.00E+00	0.00E+00
F7	2.00E+00	0.00E+00	2.00E+00	0.00E+00	NaN	NaN	2.00E+00	1.99E+08	2.00E+00	0.00E+00	8.75E+06	7.98E+06	2.00E+00	0.00E+00
F8	8.30E-05	4.55E-04	0.00E+00	0.00E+00	6.89E+04	2.14E+04	6.71E+03	2.45E-03	0.00E+00	0.00E+00	8.28E-06	2.73E-05	9.77E-10	2.21E-09
F9	0.00E+00	0.00E+00	1.34E-61	7.32E-61	5.19E-61	2.79E-60	8.85E-89	4.85E-88	1.42E-30	3.70E-30	0.00E+00	0.00E+00	0.00E+00	0.00E+00
F10	5.40E+02	1.75E+02	1.45E-04	8.09E+03	5.05E+06	1.45E-06	3.22E+02	3.26E-02	4.83E+07	8.23E+06	9.14E+04	1.53E-04	2.23E+02	2.00E+02
F11	2.31E+02	1.56E+02	1.24E+03	6.02E+02	3.45E+05	7.75E-04	3.66E+02	4.17E-02	2.84E+07	7.97E+06	9.30E+03	2.62E+03	2.33E+02	2.65E+01
F12	2.70E+162	6.55E-04	2.48e+164	6.55E-04	2.05e+163	6.55E+04	1.13e+162	6.55E+04	1.22E+164	6.55E-04	8.29E+162	6.55E+04	3.05E+164	6.55E+04
F13	2.30E+01	5.17E-01	7.59e+104	4.16e+105	8.39e+47	4.59e+48	8.30E+02	1.14E-03	2.27E+155	6.55E-04	8.24E+116	5.06E-117	5.57E+70	2.95E+71
F14	3.00E+01	0.00E+00	3.00E+01	0.00E+00	3.00E+01	1.83E-01	3.00E+01	0.00E+00	3.79E+01	1.30E-01	8.23E+01	3.50E+00	3.06E+01	7.28E-01
F15	2.40E+01	3.25E-01	2.32E+03	1.44E+03	2.74E+00	8.33E-01	3.57E+03	1.57E-03	6.71E+03	1.17E-03	9.07E+02	7.16E-02	3.12E+00	1.63E+00
F16	4.26E+02	2.29E+02	1.45E+02	2.08E+02	8.66E+04	6.02E+04	6.57E+01	7.79E-01	6.59E+09	1.57E-02	8.46E+02	5.75E-02	3.11E+03	2.30E+03
F17	1.97E+01	7.41E-01	2.17E+01	1.47E+00	1.74E+01	7.32E-01	1.77E+01	7.69E-01	3.09E+01	1.48E-01	8.25E+01	1.62E-00	1.83E+01	7.06E-01
F18	6.94E-01	9.40E-01	6.54E+00	4.12E+00	3.54E+01	1.55E-01	1.76E+00	7.94E-01	6.92E+03	1.90E-03	9.49E+00	2.18E+00	1.94E+00	3.95E+00
F19	1.52E+02	1.37E+01	6.01E+01	6.78E+00	1.79E+02	1.61E+01	1.11E+02	1.09E-01	2.82E+02	9.73E-01	8.66E+02	4.53E-01	4.15E+01	9.40E+00
F20	5.62E+01	2.04E+01	7.68E-01	4.03E+00	2.05E+03	6.49E-02	4.60E+01	5.53E-01	2.39E+04	3.65E-03	2.06E+05	9.84E-04	2.33E+02	7.09E+01
F21	4.57E+00	1.32E+00	5.03E+01	3.03E+01	5.35E+00	7.30E+01	5.36E+01	1.39E-01	7.91E+01	3.49E-01	2.97E+01	3.90E+01	1.51E+01	3.01E+00
F22	9.48E-09	3.20E-09	4.68E+01	1.79E+01	4.30E+01	1.14E-01	6.46E+00	2.35E-00	3.78E+17	9.78E-17	2.21E+02	1.58E-02	3.61E+01	3.29E+01
F23	-1.99E-04	7.45E+02	-1.33E+04	1.44E+03	NaN	NaN	-1.42E+04	3.65E-02	-1.67E+04	2.89E-12	-3.34E+04	1.26E-03	-1.65E+04	5.90E+02
F24	0.00E+00	0.00E+00	3.33E+02	1.83E+01	4.12E+01	1.14E+01	0.00E+00	0.00E+00	4.74E+03	4.95E-02	2.27E+00	3.44E+00	1.15E+01	4.45E+00
F25	-1.95E-03	6.36E+00	-1.85E+03	4.14E+01	-1.41E+03	4.72E+01	-1.95E+03	8.96E-00	-1.27E+03	4.13E-00	-7.73E+03	4.19E+01	-1.77E+03	4.17E+01
F26	3.79E+05	3.65E+05	-2.77E+03	1.31E+04	2.62E+04	7.79E+03	-5.73E+03	3.07E+03	2.28E+07	2.74E-05	1.10E+06	9.87E+05	1.03E+05	4.41E+04
F27	-4.47E+01	1.13E+00	-4.90E+01	1.69E+01	NaN	NaN	-4.43E+01	8.47E-01	-3.27E+01	1.20E-01	-3.33E+01	2.90E+00	-5.00E-01	3.89E-02
F28	4.24E+01	5.46E+01	8.93E+02	1.18E+02	1.89E+02	5.85E+01	5.98E+00	5.16E-00	9.71E+02	6.75E-00	9.42E+02	1.87E-02	3.95E+02	1.22E+02
F29	1.63E-14	1.11E-14	6.25E+03	3.27E+02	1.97E+02	6.12E+01	5.39E+03	2.85E+03	2.78E+04	3.44E-01	1.73E+04	2.04E+04	1.23E+02	1.98E+02
F30	8.99E-13	8.50E-13	5.74E+03	1.61E+02	1.08E+04	3.40E+03	2.09E+00	1.65E-00	8.58E+05	1.01E-01	2.66E+06	2.09E-06	1.01E+04	1.42E+04
F31	5.63E-18	3.46E-18	8.89E+05	4.87E+02	6.07E+02	2.14E+02	1.41E+06	1.07E-06	1.21E+01	1.79E-01	1.18E-05	2.75E-05	3.19E-03	1.71E-03
F32	3.80E-16	2.27E-16	1.68E+05	8.76E+05	4.93E+00	1.69E+00	1.10E+04	5.52E-05	6.46E+02	6.80E-01	3.47E+02	3.51E-02	1.47E+00	3.55E+00
F33	4.39E-60	1.43E-59	7.72E-17	3.23E-16	4.56E-17	1.20E-16	5.94E-24	1.44E-23	2.01E-03	1.22E-03	1.88E-15	4.84E-15	1.31E-10	5.00E-10
F34	1.14E-09	2.66E-10	2.43E+01	8.10E+01	4.01E+01	8.80E+00	5.30E+02	1.95E-02	5.45E+03	7.51E-02	8.37E+03	1.02E+04	2.41E+00	1.11E+00
F35	4.90E-13	2.01E-13	9.59E+01	1.16E+01	7.77E+01	1.07E+01	4.93E+01	1.41E+00	6.82E+02	4.46E-01	1.63E+02	2.40E+01	1.04E+02	1.53E+00
F36	9.44E+02	1.63E+02	3.66E+02	2.56E+02	3.91E+02	1.13E+02	6.08E+01	1.92E+00	2.33E+07	7.25E-06	8.39E+02	1.64E-02	2.02E+03	6.89E+02
F37	4.94E+01	2.39E+00	1.02E+02	2.37E+01	8.00E+01	1.38E+01	4.97E+01	1.70E+00	6.49E+02	2.60E-01	6.48E+02	3.86E-02	1.07E+02	1.57E+01

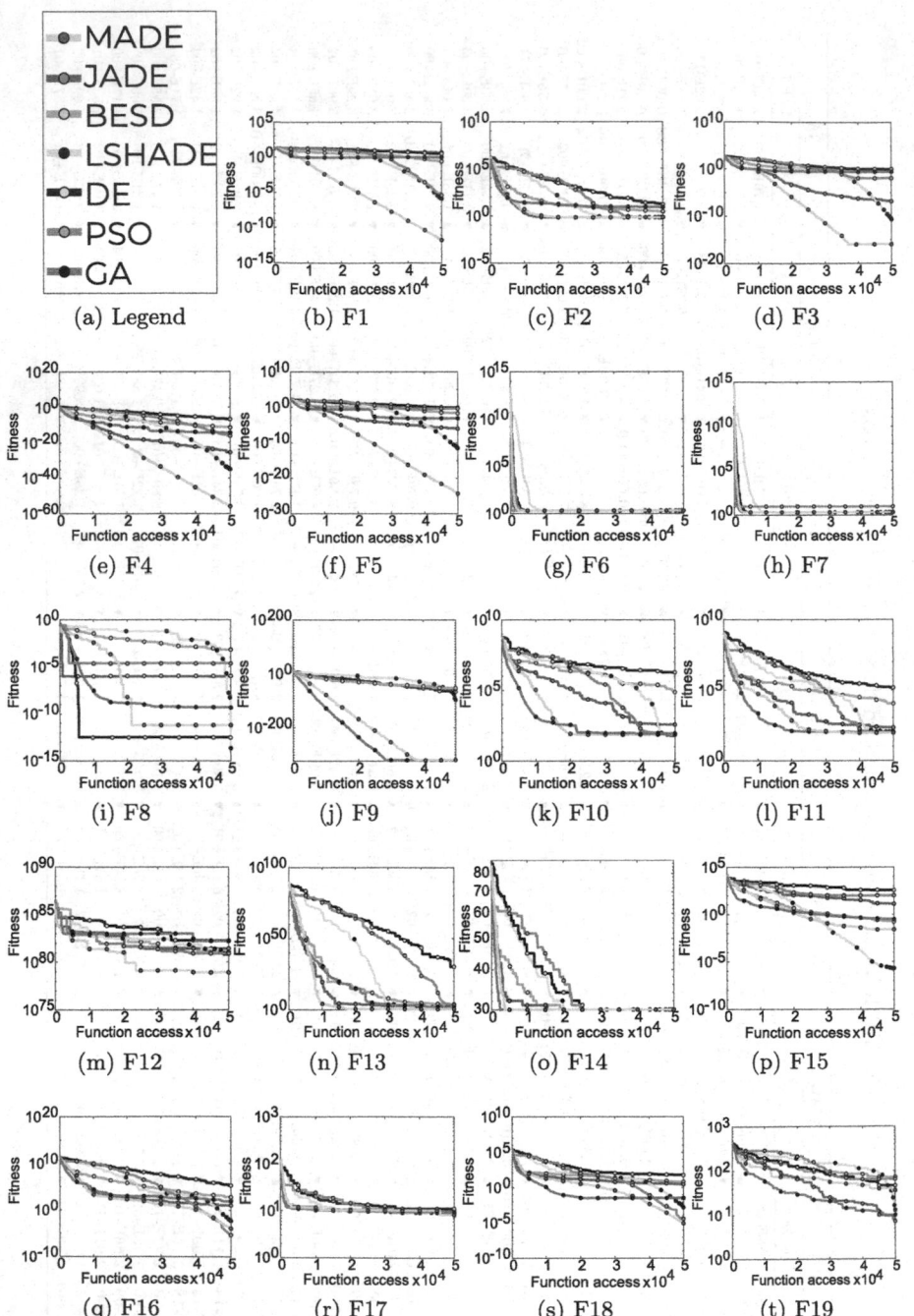

Fig. 3. Convergence curves for functions F1 to F24 in 30 dimensions.

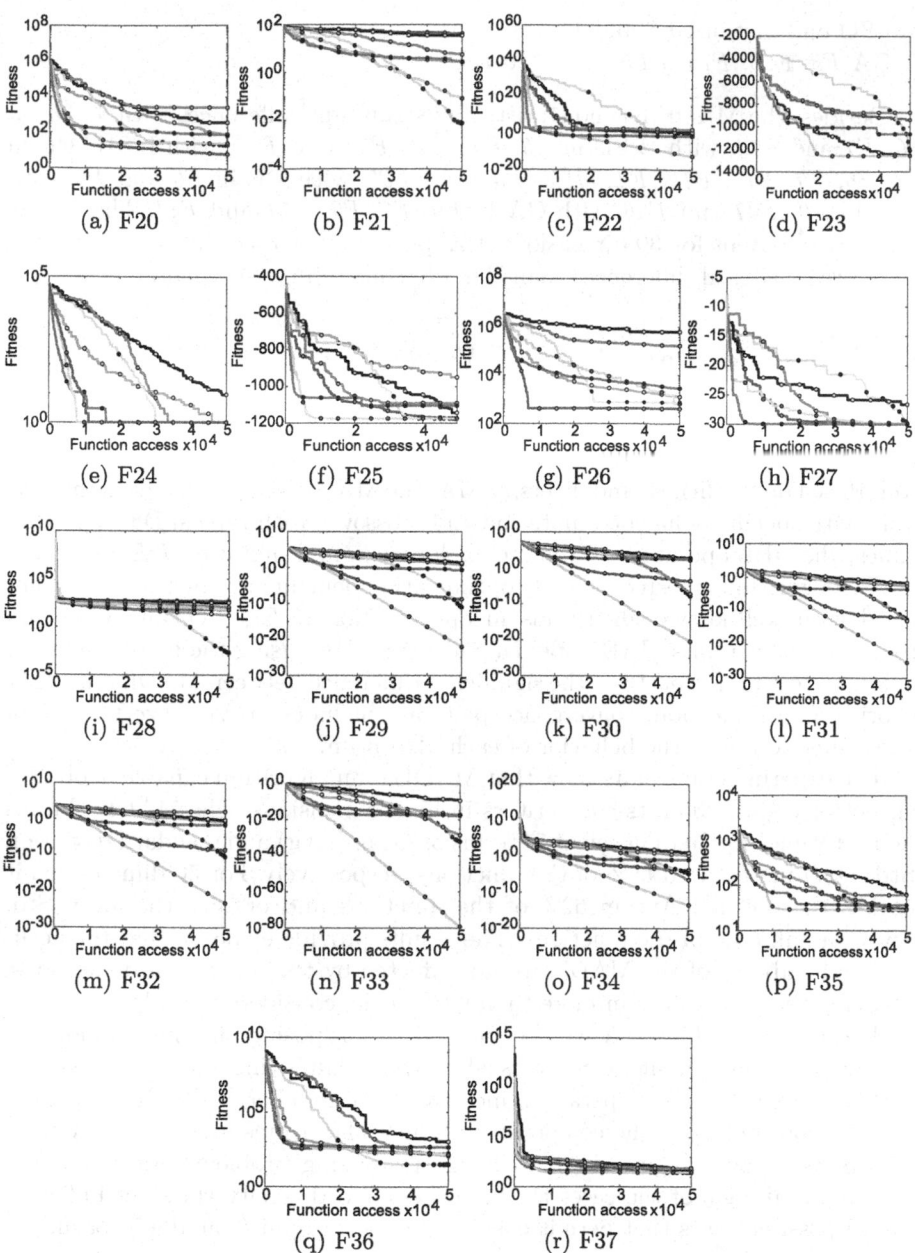

(a) F20   (b) F21   (c) F22   (d) F23

(e) F24   (f) F25   (g) F26   (h) F27

(i) F28   (j) F29   (k) F30   (l) F31

(m) F32   (n) F33   (o) F34   (p) F35

(q) F36   (r) F37

Fig. 4. Convergence curves for functions F25 to F37 in 30 dimensions

- PSO on functions *F8* and *F12*.
- GA *F8*, *F10*, *F15*, *F17*, *F19*, *F26-F28*.

Besides, the MADE has no statistically significant differences with JADE in *F6*, *F7* and *F14*, with BeSD in *F6*, *F7*, *F14*, *F23* and *F27*, with L-SHADE in *F2*, *F6*, *F7*, *F14*, *F19*, *F24*, *F25*, *F35* and *F37*, with DE in *F6* and *F7*, with PSO in *F9*, *F27* and *F36*, with GA in *F6*, *F7*, *F9*, *F11* and *F14*. The results are more apparent for 30 dimensions; this permits us to see that the algorithm has more statistical differences than the other high-dimension methods.

5   Conclusions and Future Work

This paper proposes an improved DE-based algorithm called MADE. It uses the moving average calculation (MA) to trade-off between exploration and exploitation. Based on the fitness and diversity MA, the MADE selects three search operators with specific behaviors: mutation and crossover with typical DE algorithm values, the OBL operator to promote exploration, and mutation $DE/best/1$.

The algorithm is tested using 37 optimization benchmark functions and compared with well-known algorithms in optimization (DE, PSO, and GA) and improved DE variants (JADE, BeSD, and L-SHADE) using the non-parametric Wilcoxon test to demonstrate the significant difference between MADE and other algorithms. In addition, convergence plots are included to visualize the use of access functions and the behavior of each algorithm.

The experimental results show that MADE is an algorithm capable of obtaining better results than its competitors in high dimensions. The MADE obtains the best values for 30 dimensions in terms of mean (Avg) and standard deviation (Std. Dev) in 65% and 59% of the functions, respectively. For 50 dimensions, it reaches the optimal Avg in 62% of the functions and obtains the lower Std. Dev values in 54% of the functions. The results provide evidence that the accuracy and stability of the MADE are not affected by the number of dimensions, achieving the best values in more than 62% of the considered problems.

The future of this work will be focused on proposing the methodology to improve the classical metaheuristics algorithms employing the MAs selection method, integrating new operators, increasing the number of dimensions of the optimization problems, and competing with new algorithms. Besides, the MADE will be tested over engineering and image processing problems. Moreover, the combinatorial logic of the cases for the fitness and diversity variables indicates several possible cases that permit designing a more sophisticated approach.

References

1. Aalto, J., Lampinen, J.: A mutation adaptation mechanism for differential evolution algorithm. In: 2013 IEEE Congress on Evolutionary Computation, pp. 55–62. IEEE (2013)

2. Aalto, J., Lampinen, J.: A mutation and crossover adaptation mechanism for differential evolution algorithm. In: 2014 IEEE Congress on Evolutionary Computation (CEC), pp. 451–458. IEEE (2014)
3. Cheng, S., Shi, Y., Qin, Q., Zhang, Q., Bai, R.: Population diversity maintenance in brain storm optimization algorithm. J. Artif. Intell. Soft Comput. Res. **4**(2), 83–97 (2014)
4. Civicioglu, P., Besdok, E.: Bezier search differential evolution algorithm for numerical function optimization: a comparative study with crmlsp, mvo, wa, shade and lshade. Expert Syst. Appl. **165**, 113875 (2021)
5. Goldberg, D.E., Holland, J.H.: Genetic algorithms and machine learning. Mach. Learn. **3**, 95–99 (1988). https://doi.org/10.1023/A:1022602019183
6. Hussain, K., Salleh, M.N.M., Cheng, S., Naseem, R.: Common benchmark functions for metaheuristic evaluation: a review. JOIV: Int. J. Inf. Visualization **1**(4–2), 218–223 (2017)
7. Hussain, K., Salleh, M.N.M., Cheng, S., Shi, Y.: On the exploration and exploitation in popular swarm-based metaheuristic algorithms. Neural Comput. Appl. **31**(11), 7665–7683 (2018). https://doi.org/10.1007/s00521-018-3592-0
8. Hyndman, R.J.: Moving averages. In: Lovric, M. (ed.) International Encyclopedia of Statistical Science, pp. 866–869. Springer, Heidelberg (2011). https://doi.org/10.1007/978-3-642-04898-2_380
9. Indera, N., Yassin, I., Zabidi, A., Rizman, Z.: Non-linear autoregressive with exogeneous input (narx) bitcoin price prediction model using pso-optimized parameters and moving average technical indicators. J. Fund. Appl. Sci. **9**(3S), 791–808 (2017)
10. Kennedy, J., Eberhart, R.: Particle swarm optimization. In: Proceedings of ICNN 1995-International Conference on Neural Networks, vol. 4, pp. 1942–1948. IEEE (1995)
11. Kukkonen, S., Coello, C.A.C.: Applying exponential weighting moving average control parameter adaptation technique with generalized differential evolution. In: 2016 IEEE Congress on Evolutionary Computation (CEC), pp. 4755–4762. IEEE (2016)
12. Metghalchi, M., Marcucci, J., Chang, Y.H.: Are moving average trading rules profitable? Evidence from the European stock markets. Appl. Econ. **44**(12), 1539–1559 (2012)
13. Oliva, D., Martins, M.S.: A Bayesian based hyper-heuristic approach for global optimization. In: 2019 IEEE Congress on Evolutionary Computation (CEC), pp. 1766–1773. IEEE (2019)
14. Oliva, D., et al.: Balancing the influence of evolutionary operators for global optimization. In: 2020 IEEE Congress on Evolutionary Computation (CEC), pp. 1–8. IEEE (2020)
15. Roy, P.K., Sur, A., Pradhan, D.K.: Optimal short-term hydro-thermal scheduling using quasi-oppositional teaching learning based optimization. Eng. Appl. Artif. Intell. **26**(10), 2516–2524 (2013)
16. Scoczynski, M., et al.: A selection hyperheuristic guided by Thompson sampling for numerical optimization. In: Proceedings of the Genetic and Evolutionary Computation Conference Companion, pp. 1394–1402 (2021)
17. Shen, X., Zou, D., Duan, N., Zhang, Q.: An efficient fitness-based differential evolution algorithm and a constraint handling technique for dynamic economic emission dispatch. Energy **186**, 115801 (2019)
18. Shi, Y., Eberhart, R.C.: Population diversity of particle swarms. In: 2008 IEEE Congress on Evolutionary Computation (IEEE World Congress on Computational Intelligence), pp. 1063–1067. IEEE (2008)

19. Slowik, A., Kwasnicka, H.: Evolutionary algorithms and their applications to engineering problems. Neural Comput. Appl. **32**(16), 12363–12379 (2020). https://doi.org/10.1007/s00521-020-04832-8
20. Storn, R., Price, K.: Differential evolution-a simple and efficient heuristic for global optimization over continuous spaces. J. Global Optim. **11**(4), 341–359 (1997)
21. Tanabe, R., Fukunaga, A.S.: Improving the search performance of shade using linear population size reduction. In: 2014 IEEE congress on evolutionary computation (CEC), pp. 1658–1665. IEEE (2014)
22. Tizhoosh, H.R.: Opposition-based learning: a new scheme for machine intelligence. In: International Conference on Computational Intelligence for Modelling, Control and Automation and International Conference on Intelligent Agents, Web Technologies and Internet Commerce (CIMCA-IAWTIC 2006), vol. 1, pp. 695–701. IEEE (2005)
23. Wang, S., Li, Y., Yang, H.: Self-adaptive mutation differential evolution algorithm based on particle swarm optimization. Appl. Soft Comput. **81**, 105496 (2019)
24. Wilcoxon, F.: Individual comparisons by ranking methods. In: Kotz, S., Johnson, N.L. (eds.) Breakthroughs in Statistics. Springer Series in Statistics (Perspectives in Statistics), pp. 196–202. Springer, New York (1992). https://doi.org/10.1007/978-1-4612-4380-9_16
25. Zhang, J., Sanderson, A.C.: Jade: self-adaptive differential evolution with fast and reliable convergence performance. In: 2007 IEEE Congress on Evolutionary Computation, pp. 2251–2258. IEEE (2007)

Search-Based Third-Party Library Migration at the Method-Level

Niranjana Deshpande[1]([⊠]), Mohamed Wiem Mkaouer[1], Ali Ouni[2],
and Naveen Sharma[1]

[1] Rochester Institute of Technology, New York, USA
{nd7896,mwmvse,nxsvse}@rit.edu
[2] Ecole de Technologie Superieure, Montréal, Canada
ali.ouni@etsmtl.ca

Abstract. In software development, third-party libraries are commonly
used to reduce implementation efforts and errors, while delivering high-
quality, reliable and secure software. To support software evolution,
newer libraries are continuously released to offer added features, and
critical updates such as bug and vulnerability fixes. As a result, old
(source) libraries and their methods must be replaced with their newer,
updated counterparts (target libraries) during the library migration process. This is known to be a time-consuming and error-prone process as
developers need to analyze both the source and target library's Application Programming Interface (API) documentation and implementation
to replace every source API with target API(s). Recent studies have
utilized various techniques to recommend the appropriate target library
for replacement, but do not provide generalizable guidelines on how to
replace each source API with target library APIs. To address this limitation, our work leverages evolutionary search algorithms to recommend
APIs by (1) formulating API migration as a combinatorial optimization
problem, and (2) using genetic algorithms (GA) to recommend suitable
APIs during migration based on the method signature and documentation similarity, and co-occurrence. We conduct an empirical study on 9
popular library migrations from 57,447 open-source Java projects and
demonstrate that GA can recommend multiple APIs for replacement
with up to 100% precision for certain library migrations.

Keywords: Genetic Algorithm · Search based software engineering ·
Library migration · API Migration

1 Introduction

In modern software development, third-party library adoption is strongly
encouraged to reduce errors and implementation effort, and to avoid redundancy [5,7,13]. To support software evolution, newer libraries are often released
with added features, bug fixes or improved security. These libraries offer their

© Springer Nature Switzerland AG 2022
J. L. Jiménez Laredo et al. (Eds.): EvoApplications 2022, LNCS 13224, pp. 173–190, 2022.
https://doi.org/10.1007/978-3-031-02462-7_12

services through their Application Programming Interfaces (APIs) that represent a collection of public objects and methods that can be called once a library is deployed [2–4,13,16,21]. As a result, existing libraries need to be periodically replaced with more recent, updated libraries to maintain high-quality, reliable and secure software. This process of replacing old and obsolete libraries with other newer, up-to-date libraries is known as *library migration*.

Migrating from a *source* library to a new *target* library consists of locating all source library methods currently used in software and replacing them using target library methods. To do so, developers must carefully study the documentation and implementation of each target method to find a replacement for every source method. Identifying target method(s) is highly time-consuming and error-prone as source and target libraries may be designed differently *e.g.*, in naming conventions, separation of concerns, the number of methods etc. [6] An added challenge during library migration is that numerous combinations of source and target methods (or *mappings*) must be evaluated to select the right replacement. For example, if we consider that 10 source methods must be replaced using a target library with 100 methods, then the total number of possible source-target mappings is $100^{10} = 1e^{20}$. Hence, manually evaluating all possible mappings becomes increasingly tedious as the number of source and target methods increases [6,10,23]. Thus, automated tools and techniques are needed to support efficient developer decision-making for library migration at the API level ('API migration').

Various approaches have been proposed to support automated library migration [5,8,10,15,17,18], however, these techniques focus on specialized training routines and can only recommend one target method to replace each source method (*i.e.*, one-to-one mappings). That is, existing techniques cannot generalize to different types of library migrations, *e.g.*, when more than one target method needs to be recommended or when data for newer migrations is unavailable. Recent studies have also used evolutionary algorithms to recommend the most suitable library to replace a deprecated one [16]. These studies recommend the best library to use without providing support at the method level, *i.e.*, they do not suggest what method(s) from the recommended library can be used to replace each method from the deprecated library.

To address the above-mentioned challenges, we formulate fine-grained library migration as a combinatorial optimization problem and leverage the Genetic Algorithm (GA) to find suitable source-target method mappings. That is, we find suitable replacing methods from the target library for one or more source methods. We determine the similarity between source and target methods based on a combination of three measurements. First, we measure the closeness of two methods according to their signatures, and then we measure their documentation similarity using word vector embeddings obtained from a state-of-the-art neural network - the Universal Sentence Encoder (USE) [9]. Finally, we leverage data from existing projects to recommend APIs that have successfully been used before by developers in Open Source Software (OSS) projects. To evaluate our approach, we conduct an empirical study using 9 popular library migrations collected from a corpus of 57,447 open-source Java projects [6], and compare GA performance to two baselines - random search and stochastic hill-climbing. Our results demonstrate that GA outperforms both baselines to recommend API mappings with high precision. In summary, our main contributions are:

- We formulate API migration as an optimization problem, and leverage Genetic Algorithm (GA) to support library migration at the method level. Our approach can recommend multiple methods for replacement when replacing old libraries.
- We conduct an empirical study using a popular dataset containing 9 Java library migrations collected from 57,447 open-source Java projects. Our results demonstrate that GA outperforms random search and hill-climbing to recommend source-target mappings with up to 100% precision for certain migrations.
- We provide a replication package of our artifacts for reproducibility.[1]

2 Background and Motivation

2.1 Background

For ease of reference, we first define the terminology used in this paper as follows:

- **Library:** A library consists of objects and functions that can be accessed using an Application Programming Interface (API) [20].
- **Library Migration:** This refers to the process of replacing a *source* library with a *target* library by removing all functional dependencies of the source library from a software [16,19].
- **Migration Rule:** A migration rule *source* → *target* denotes that a *source* library is replaced by a *target* library. For example, the rule *easymock* → *mockito* denotes that all *easymock* APIs are replaced using *mockito* APIs [6].
- **Migration Mapping:** This is a set of mappings which contains one or many methods or APIs from the source library that are replaced using one or many APIs from the target library [5].

2.2 Motivating Example

To illustrate the challenges associated with API migration, we consider real-world examples from the dataset used in our experiments [5]. When replacing *source* library methods with those from a *target* library, developers need to manually inspect each method and its implementation in the target library carefully to pick suitable replacements without modifying source code behavior. Furthermore, due to differences in library design, naming conventions, separation of concerns etc. [6], a developer does not know beforehand which target methods should be used. For example, in Fig. 3, the source methods `create()` and

---

[1] https://github.com/niranjanadeshpande/search-based-api-migration.

toJson(T) are replaced by the target API writeValueAsString(Object)[2]. In this example, two source methods are replaced by one target method, and their input parameter types and method names are entirely different. Thus, developers need to study each source and target method combination, leading to a *combinatorial explosion*. For example, consider that a developer must replace 5 methods from the source library with 5 methods from a given target library containing around 500 methods. The total number of source-target mappings to be examined when replacing all 5 methods is $500^5 = 3.1e^{13}$; thus, evaluating each source-target mapping combination is highly time-consuming and labor-intensive.

```
HttpServletRequest request = createMock(HttpServletRequest.class);
expect(request.getHeaders("Authorization")).andReturn(
when(request.getHeaders("Authorization")).thenReturn(
```

Fig. 1. A many-to-many mapping where expect(T) and andReturn(T) are replaced using when(T) and thenReturn(T).

```
String value = (String) grid.get("auto_start_hub");
String value = grid.getAsJsonObject().get("auto_start_hub").toString();
```

Fig. 2. A one-to-many mapping where get(String) is replaced by the getAsJsonObject(), get(String) and toString() for the rule *gson* → *json*.

```
new GsonBuilder().create().toJson(getChanges(Integer.parseInt(params[0]))));
objectMapper.writeValueAsString(changes));
```

Fig. 3. A many-to-one mapping where the source methods create(), toJson(T) are replaced by the writeValueAsString(Object) method.

```
ForeignKeyConstraint key = EasyMock.createMock(ForeignKeyConstraint.class);
ForeignKeyConstraint key = mock(ForeignKeyConstraint.class);
```

Fig. 4. A one-to-one mapping where one source method createMock(Class) is replaced by one target method mock(Class) from *mockito*.

---

[2] http://github.com/ybonnel/CodeStory/commit/32d3dbb0246e08d71651e20eb7f9c
7a9bc8c0956.

As seen from Figs. 1, 2, 3 and 4, another critical challenge during API migration is posed by the *cardinalities* of mappings. That is, each source method may need to be replaced by multiple target methods. To make the situation worse, multiple source methods may need to be replaced by one target method. This increases the number of potential source-target mappings to be studied to a number much greater than $3.1e^{13}$. Additionally, when evaluating potential source-target method mappings, developers must also take into account the number and types of input parameters, returned values from each method call and the order in which each method in the source fragment is invoked, thus making API migration a challenging, laborious and error-prone process. Next, we provide concrete examples of different cardinalities of source-target mappings:

- *Many-to-Many Mapping*: When many source methods are replaced by many target methods, as in Fig. 1, it is a *many-to-many* mapping. The expect(T) and andReturn(T) methods are replaced using when(T) and thenReturn(T).
- *One-to-Many Mapping*: In Fig. 2 one source method get(String) is replaced by multiple target methods getAsJsonObject(), get(String) and toString() when migrating from *gson* to *json*. This is a challenging API mapping to recommend because: (a) the variable returned by getAsJsonObject().get("auto_start_hub") can be typecast using the toString() method or prefixing (String) (b) there is no equivalent getAsJsonObject() method used in the source mapping. In this scenario, developers will spend relatively more time and effort to find the right target methods and ensure that their result is similar to the source methods.
- *Many-to-One Mapping*: In Fig. 3, multiple source methods create(),toJson(T) are replaced using a single target method writeValueAsString(Object). This migration is not easy to recommend as the create(), toJson(T) and writeValueAsString(Object) all have different names and input parameter types. In this case, semantic similarity would not be sufficient to find correct mappings and other measures such as their co-occurrence in existing mined data may be more accurate.
- *One-to-One Mapping*: Fig. 4 shows a *one-to-one* migration for the rule *easymock* → *mockito*, where the source method createMock(Class) is replaced by one target method mock(Class). In this instance, the method names are similar and have the same return type and input parameters; so finding this mapping is a relatively simple task.

We observe from these examples that API migration is a challenging and complex task due to the large number of potential source-target API mappings to be evaluated. This is due to several factors such as mapping cardinalities, differences in method names, return types, input parameters, and API documentation (if it exists). Current approaches [5,10] that address API migration are designed to address one-to-one method mappings, so they cannot accurately recommend *one-to-many, many-to-one* and *many-to-many* mappings. Furthermore, these approaches do not easily generalize to migration rules or programming languages that have not been used in train-

Table 1. The 9 popular migration rules used in this study.

Source → Target
commons-logging → slf4j
slf4j → log4j
easymock → mockito
google-collect → guava
gson → jackson
testng → junit
json → gson
commons-lang → slf4j
json-simple → gson

ing. To address these limitations, we formulate API migration as a search-based problem and use evolutionary algorithms to identify source-target mappings of different cardinalities without the need for specialized training routines. Additionally, unless source and target libraries share similar designs or specialized functionalities, the number of source-target mappings to be manually evaluated poses a significant challenge to developers. Our approach aims to ease this burden by searching for and recommending a subset consisting of the most suitable method mappings based on their similarities.

3   Search-Based API Migration

In this work, the identification of suitable source-target method mappings for each migration rule is formulated as a 0/1 knapsack problem [11]. Each *knapsack* has an associated *capacity* that dictates how many *items* can be selected. Every *item* is associated with a *profit* and *weight*, based on which an item is either selected (*i.e.,* encoded as 1) or not (*i.e.,* encoded as 0). The goal of the knapsack problem is to maximize overall *profit* without exceeding its *capacity*. In this work, we create a *knapsack* for each migration rule by randomly sampling source methods to be replaced and potential target methods to be considered for replacement. Each potential target method has a weight of +1 and a *profit* that is equivalent to its fitness or similarity score described in Sect. 3.2. Since developers do not know beforehand which target methods should be used for replacement or how many will be required, the *capacity* value is equal to the number of source methods that need to be replaced. The capacity parameter is critical because it controls the number of target methods that can be recommended for selection, thus impacting performance. Without a constraint on the number of recommended methods, a knapsack's profit will be maximized by setting capacity to a high value, resulting in more target method recommendations; however, the time spent by developers to examine all recommended mappings increases considerably as the number of recommended target methods increases.

Thus, to ensure that accurate source-target mappings are recommended while minimizing the number of recommendations, a penalty is imposed on solutions that exceed the capacity constraint by recommending too many target methods. If the number of recommended methods exceeds capacity, a penalty equal to the difference between the number of recommended methods and knapsack capacity is imposed.

3.1 Solution Representation

In Fig. 5, we demonstrate how source-target method mappings (mappings) are represented as a *chromosome* and encoded using a binary bit string used in GA. Each chromosome contains $m \times n$ *genes* corresponding to possible source-target method mappings, where m and n are the number of source and target API methods respectively. In Fig. 5, we evaluate three possible target methods (from *mockito*, on the right) to replace two source methods (from *easymock*, on the left), leading to chromosome containing 6 genes. Search algorithms select the most suitable target method (highlighted in green, on the right) to replace each source method (on the left), and set its corresponding gene value to 1 while others are set to 0: *e.g.*, `static <T> T createMock(Class<T>)` is replaced by `static <T> T mock(Class<T>)`, so its corresponding gene/index in the chromosome has a value of one. The candidate mappings shown in Fig. 5 are one-to-one and one-to-many, so one source method is replaced by one or more target methods. In case of multiple source methods in a code fragment such as in many-to-one, many-to-many mappings, each source method is represented separately as a one-to-many mapping. For example, in Fig. 1, the source code fragment is split into two separate methods: `expect(T)` and `andReturn(T)`. Furthermore, each source method is mapped to many target methods, *e.g.*, both `when(T)` and `thenReturn(T)` from the target library can be recommended for `expect(T)`, and their corresponding gene value is set to 1 in the chromosome.

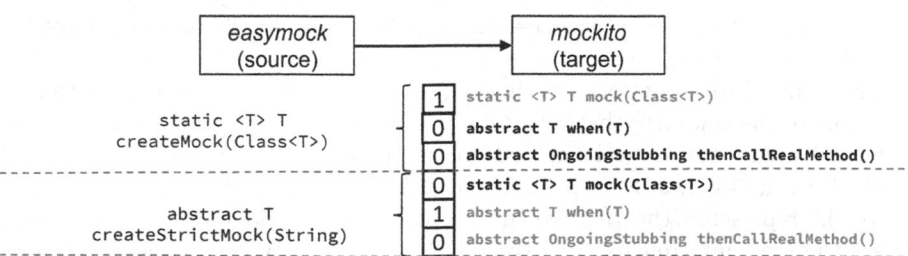

Fig. 5. An example of source-target method mappings for the rule *easymock* → *mockito* represented as a *chromosome*. (Color figure online)

3.2 Calculating the Fitness Function

When applying GA, the *fitness function* is crucial to evaluate and compare potential source-target mappings to select the best solution. In general, solutions with a higher fitness potentially contain more accurate mappings. In this work, we calculate solution fitness by aggregating the similarity between the source and target method signatures, their documentation and co-occurrence probability. Using these similarities allows us to identify and leverage information about potential source-target mappings *e.g.,* how many times a particular source and target API method have been used previously or if they offer similar functionalities. We describe each similarity measure as follows:

1. **Co-Occurrence**: Co-occurrence probability $coOc(s, t)$ is defined as the number of times a source *and* target method pair is observed *together* in previous *migration diffs*. A source code migration diff represents the set of removed and added lines of code that we use to identify which source methods were removed and which target methods were added. Code diffs can be identified using modern development tools and frameworks, such as GitHub. The intuition behind using this metric is to leverage the "wisdom of the crowd" and assign higher fitness scores to source-target mappings that have been found in various pre-existing programs.

2. **Method Similarity**: The intuition behind this metric is: two methods delivering the same functionality tend to be similar in terms of their naming, return type, and input parameters. To calculate the similarity between a source and target method pair, we split both the source and target methods into return types, method names and a list of input parameter types to calculate their similarity using the formula by Nguyen et al. [14]. Consider a source method s named n_s that returns a variable of type rt_s using a list of input parameters p_s, and similarly, a corresponding target method t with parameters rt_t, n_t and p_t. The similarity sim between the method pair s and t is calculated as follows:

$$sim(s,t) = 0.5 * sqSim(n_s, n_t) + 0.25 * sqSim(p_s, p_t) + 0.25 * strSim(rt_s, rt_t)$$

where the $sqSim$ function uses the longest common sub-sequence algorithm to calculate the similarity between two word sequences, while the $strSim$ function computes token-level similarity between two strings. Similarly to Nguyen et al. work [14], we use 0.5, 0.25 and 0.25 to weigh each component. Each weight represents the relative importance of a component: method names are the most informative, and the return and input types are assigned equal weights.

3. **Documentation Similarity**: Method documentation is a textual description of how a specific method works written by a developer. Intuitively, two methods with identical behavior have similar documentation. To compare source and target documentation, we use a state-of-the-art neural network, the Universal Sentence Encoder (USE) [9], that is trained on a wide variety of natural language data sources specifically to identify similarities between

sentences. In this work, we use it to calculate documentation similarity [9] by generating word embeddings that are used to calculate cosine similarity. Let s and t be the source and target methods, respectively, whose documentation is an input to the USE network. We obtain the source and target documentation embeddings W_s and W_t and calculate cosine documentation similarity as:

$$simDoc(s,t) = (W_s \cdot W_t)/(\|W_s\| \times \|W_t\|)$$

Our last step involves aggregating the $coOc$, sim and $simDoc$ for each source-target method pair s, t to calculate the *fitness* of a solution. First we normalize the *co-occurrence*, while the *method* and *documentation* similarities are already normalized. We then normalize this aggregate fitness using the total number of possible source-target mappings being considered as follows:

$$fitness(s,t) = (coOc(s,t) + sim(s,t) + simDoc(s,t))/(3 \times m \times n)$$

3.3 Genetic Algorithm Operators and Parameters

The Genetic Algorithm (GA) iteratively evolves a set of candidate solutions until a termination criteria is met using the *crossover* and *mutation* operators. Crossover and mutation are recombination operators used to evolve existing solutions, *i.e.,* they *generate* various candidate solutions in different ways to obtain a diverse set of *offspring* solutions. The offspring population candidates compete amongst each other to generate potentially better solutions for the next iteration. We use the crossover and mutation operators similarly to [12]. In this work, we use half-uniform crossover with a probability of 1.0, *i.e.,* the entire parent population generates new offspring. Similarly to mutate solutions, we use the bit flip mutation operator with a probability of 0.01. Additionally, for our termination condition, we specify a maximum number of iterations that GA can run and return the solution with the highest fitness score. Additionally, the number of candidate solutions at each iteration (population size) is also specified beforehand. In this work, we set the population size to 100 and the maximum number of iterations to 250. Additionally, as we do not know how many methods will be required for replacement beforehand, the *capacity* for each migration rule is set to the number of source methods being replaced. Note that this is a soft constraint that can be exceeded, but solutions with the number of methods not exceeding the capacity value are preferred. Each pair of candidate solutions in a population is compared using the linear dominance comparator that uses an weighted linear aggregate function to compare both the penalty and fitness score. To ensure robust search performance, we use tournament selection with elitism to select candidates with the most suitable method mappings.

4 Experimental Evaluation

In this section, we evaluate the efficacy of our approach when recommending correct source-target method mappings. In that regard, we compare GA performance to that of random search and stochastic hill-climbing in terms of the precision and recall achieved during recommendation.

4.1 Dataset Used

To evaluate our approach, we use the dataset collected by Alrubaye et al. [5] containing a set of manually curated mappings belonging to 9 popular library migrations collected from 57,447 open-source Java projects. Each of these Java projects contains API migrations from atleast one of the 9 migration rules, thus providing a large and diverse dataset. Table 1 lists the source and target libraries used in our study. We randomly sample source and target methods from this dataset and pre-process different types of mappings to create a bit string representation as described in Sect. 3. Randomly sampling different datasets allows us to test our approach on a wide range of API migration scenarios containing different number and cardinalities of mappings. We implemented GA and random search using the MOEA framework [1], while hill-climbing was implemented using Python. To ensure statistically meaningful results, we run each search algorithm 30 times and report average results.

4.2 Metrics Used

Our goal is to reduce developer effort by accurately recommending source-target mappings while minimizing the number of recommended target APIs. To evaluate the efficacy of our approach, we evaluate our results using the following:

- **Precision**: Precision is the proportion of positive mappings that were correctly identified compared to all positively identified mappings. It is calculated as *precision = (True Positives/True Positives+False Positives)*.
- **Recall**: Recall is the fraction of positive mappings that were correctly identified compared to the total number of correct mappings, and is calculated as *recall = (True Positives/True Positives+False Negatives)* .

4.3 Results

Table 2 shows the precision and recall obtained using GA, random search and hill-climbing averaged over 30 runs for different migration rules. Additionally, we use the Wilcoxon signed rank test ($p < 0.05$) and observe that our results are statistically significant. In general, we observe that GA has a higher precision than both random search and hill-climbing for all migration rules, indicating that it can identify correct source-target method mappings. We observe that GA achieves the highest values of precision and recall compared to both baselines for rules *google − collect → guava* and *json → gson*. These libraries contain a majority of one-to-one mappings and well-defined similarity scores resulting in good performance. While GA also achieves perfect precision when migrating between *commons − lang → slf4j − api* and *json − simple → gson*, it recommends mappings extremely conservatively leading to poor recall. Similarly,

Table 2. Precision and recall values for GA, random search and hill-climbing obtained using 100 individuals and 50000 function evaluations.

Migration rule (↓)	Precision			Recall		
	GA	RS	HC	GA	RS	HC
logging → slf4j	**0.7166**	0.3148	0.3039	0.0855	0.3621	**0.4950**
slf4j-api → log4j	**0.4639**	0.2019	0.1944	0.1216	0.3807	**0.4912**
easymock → mockito	**0.0331**	0.0339	0.0331	0.0622	0.4177	**0.5000**
google-collect → guava	**0.4444**	0.0685	0.0721	**0.6666**	0.2444	0.4777
gson → jackson	**0.5000**	0.0478	0.0486	0.2500	0.3444	**0.5444**
testng → junit	**0.1417**	0.0769	0.0831	0.0500	0.3615	**0.5038**
json → gson	**0.6666**	0.4500	0.2152	**0.6666**	0.4388	0.5333
commons-lang → slf4j-api	**1.0000**	0.1281	0.1252	0.1666	0.2944	**0.4944**
json-simple → gson	**1.0000**	0.2541	0.2216	0.1666	0.3055	**0.4925**

GA achieves high precision and poor recall when migrating between *logging →
slf4j*, *slf4j − api → log4j*, *gson → jackson* because these libraries contain a
majority of *many-to-one*, *many-to-many and one-to-many* mappings that makes
the search process difficult. Overall, the worst performance is observed for the
rules *easymock → mockito* and *testng → junit* where the search space is complex due to multiple local optima and close to zero similarity scores. We discuss
the different factors impacting GA performance in greater detail as follows:

The Effect of Capacity: The capacity value is critical to GA performance as
it determines which and how many target methods are recommended. However,
the value of capacity is difficult to determine exactly because we cannot know
how many mappings exist between two libraries unless all possible combinations
are explored. Due to these reasons, we use a case study to conduct an experiment to determine the effect of different capacity values on GA performance.
In this regard, we use the rule *commons − lang → slf4j − api* as a case study
and evaluate the effect of increasing capacity on precision and recall in Fig. 6.
We observe that precision values are highest for low capacity values, in fact, we
obtain perfect precision up to a capacity value of 4. This is expected because
GA recommends methods conservatively leading to almost no false positives.
As the capacity values increase, more correct and incorrect source-target methods are recommended. This increases the likelihood of selecting wrong mappings
resulting in more false positives and lower precision values. Moreover, as capacity
increases more correct methods are also identified leading to fewer false negatives
thereby increasing recall. From the graph, we observe that a balance between
precision and recall is achieved when the capacity is 12, which is also the number
of correct source-target mappings that exist in the ground truth. That is, GA
balances precision and recall when the capacity is set to the number of source
methods to be replaced (which we know beforehand). Thus, in subsequent experiments, we set the capacity for each migration rule to be equal to the number of
source methods to be replaced.

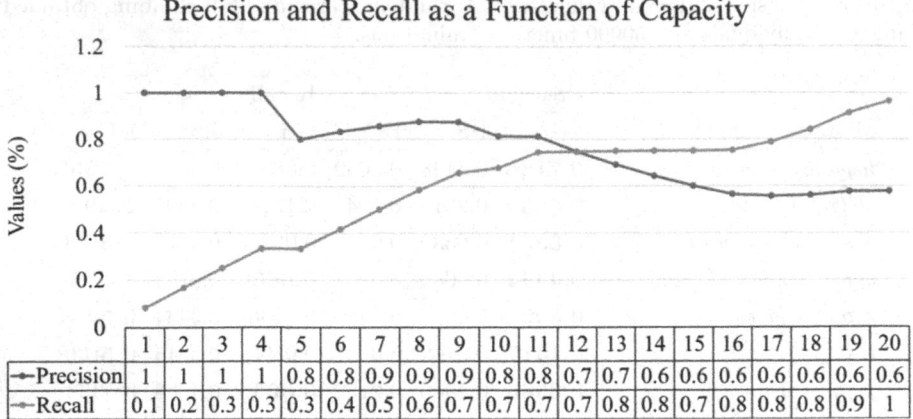

Fig. 6. This graph demonstrates the effect of increasing the capacity values for the rule $commons - lang \rightarrow slf4j - api$. As the capacity increases from 1 to 20, we observe a decline in precision values and an increase in recall values. This is because more source-target mappings are recommended for higher capacities.

Next, we study the impact of capacity on different migration rules. When the capacity value is well-specified, GA achieves high precision and recall as demonstrated in Table 2 for the rules $google - collect \rightarrow guava$ and $json \rightarrow gson$. This also indicates that the source and target libraries involved in these migrations deliver the same functionality and thus have many similar methods leading to several correct potential mappings $i.e.,$ they have equivalent functionality. This is also demonstrated in Figs. 7(c), 7(d), where GA consistently achieves relatively higher recall scores as the capacity value is equal to the number of mappings that must be found. When the capacity value is relatively high, GA evolves solutions containing a large number of mappings which increases the probability of finding the wrong mappings (more false positives), resulting in low precision. This is reflected in GA performance for rules $logging \rightarrow slf4j$, $slf4j - api \rightarrow log4j$, $gson \rightarrow jackson$ where it obtains high values of precision but does not select all correct mappings. When the capacity value is relatively low, GA becomes $picky$, and selects very few source-target method mappings. This increases precision because selected methods are correct, but also results in poor recall because a large number of correct mappings are not selected, resulting in low recall $e.g.,$ $commons-lang \rightarrow slf4j-api$ and $json-simple \rightarrow gson$. This is illustrated further in Figs. 7(a) and 7(b) where GA consistently achieves perfect precision and poor recall because only a few methods are correctly recommended compared to the actual number of correct mappings that can be found. In an extreme scenario, we also observe that GA is extremely picky when recommending mappings between $easymock \rightarrow mockito$ because it must evaluate a large number of target methods to recommend very few mappings (1% of all potential mappings). In this scenario, GA is unable to find a solution and instead randomly selects target methods to fulfill the capacity constraints.

The Effect of Various Cardinalities: In general, it is difficult to find replacement methods for *many-to-one, one-to-many and many-to-many* mappings. It is relatively easier to recommend for one-to-one mappings, because only one target method (usually with the highest similarity score) needs to be recommended for each source method. Thus, the search process is simplified because it must only find m target methods with the highest similarity. This is reflected in both the precision and recall values when migrating between $google - collect \rightarrow guava, testng \rightarrow junit$ where the majority of mappings in these datasets are one-to-one, so the number of recommended methods is close to the number of actually correct source-target method mappings that exist. This leads to fewer false positives, thus resulting in high values of precision. Furthermore, GA exhibits perfect precision but poor recall when migrating between $commons - lang \rightarrow slf4j - api$ and $json - simple \rightarrow gson$ because it selects very few methods. These selected methods are correct and therefore lead to almost no false positives, however, GA cannot accurately identify all correct mappings and thus returns a high number of false negatives that lead to a low recall. This occurs because of a majority of one-to-many, many-to-one or many-to-many mappings where several methods are 'helper' functions that do not contain the main programming logic but are used to perform repetitive tasks *e.g.,* basic error handling or typecasting using the `toString()` method. As a result, there are many correct source-target mappings with similar fitness scores and GA is unable to differentiate between them, resulting in poor recall.

The Effect of Libraries' Characteristics: Additionally, we find that two characteristics impact search performance: the number of methods in source and target libraries (dataset size) and the number of potential source-target mappings with a zero similarity score. The migrations $logging \rightarrow slf4j$ and $testng \rightarrow junit$ contain various methods that are close in signature and documentation, *i.e.,* they contain the *least* number of source-target mappings with zero similarity. This results in a difficult search space with multiple *local optima*, thus making it hard for the fitness to discriminate wrong mappings. In contrast, the migration rules $easymock \rightarrow mockito, google - collect \rightarrow guava, gson \rightarrow jackson, json \rightarrow gson$, and $commons - lang \rightarrow slf4j - api$ contain a large number of source-target mappings that have a fitness of zero (distant method signatures and different documentations). That is, most of the correct source-target mappings have a non-zero similarity score. Moreover, the recall value for the rule $gson \rightarrow jackson$ is relatively higher compared to $easymock \rightarrow mockito$ and $testng \rightarrow junit$. This because the dataset for this rule contains the largest number of source-target methods with zero similarity, leading to fewer local optima.

The Effect of Different Similarity Scores: In general, each scoring measure (MS, DS and CO) captures a different dimension that can be useful for selecting correct source-target mappings. However, there are some exceptions: for example, some libraries do not have proper documentation for each method. In these scenarios, documentation similarity (DS) is zero and not only it does not provide the intended result, but it also biases the search: in our approach, if the documentation similarity is calculated as zero, it may drive the overall fitness

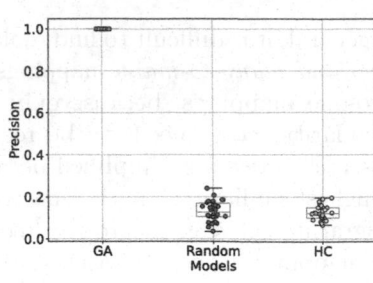

(a) Precision for $commons-lang \rightarrow slf4j-api$

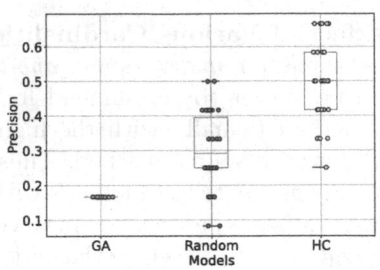

(b) Recall for $commons-lang \rightarrow slf4j-api$

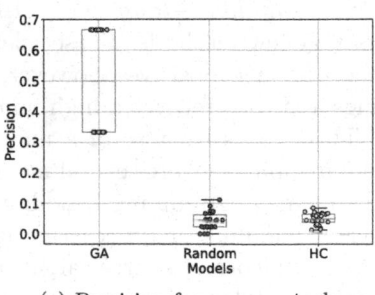

(c) Precision for $gson \rightarrow jackson$

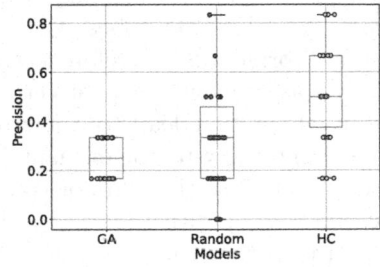

(d) Recall for $gson \rightarrow jackson$

Fig. 7. Precision and recall values over 30 runs demonstrate that GA outperforms random search and hill-climbing to recommend API mappings with high precision.

closer to zero and adversely impact search performance. Thus, search algorithms may not select correct mappings due to poor documentation scores.

Combining the MS, DS and CO results in better recommendations as compared to using them separately, especially if one of them cannot be properly calculated (*e.g.,* in case of poor documentation). As a result, during our manual validation, we found that GA can also discover other mappings that are correct but have not been seen previously (absent from ground truth). For example, in Fig. 8(b), the correct mapping from `public abstract void warn(java.lang.Object, java.lang.Throwable)` to `public abstract void warn(java.lang.String)` has been returned by our algorithm, while being absent from the existing dataset. Another example in Fig. 8(a) shows how the methods `public static <T> org.easymock.IExpectationSetters<T> expectLastCall()` and `public abstract T should()` have signature and documentation similarities as zero, but a non-zero co-occurrence score, thus leading to it being correctly recommended.

```
- public static <T> org.easymock.IExpectationSetters<T>
expectLastCall();
+ public abstract org.mockito.stubbing.OngoingStubbing<T>
then(org.mockito.stubbing.Answer<?>);
+public abstract T should();
```

(a) A one-to-many mapping. The method similarity score for *expectLast-Call()* to *should()* is zero as their signatures are zero, as is the documentation score. However, as this mapping has been used before in our dataset, it has a co-occurrence of 1.

```
- public abstract void warn(java.lang.Object,
java.lang.Throwable);
+ public abstract void warn(java.lang.String);
```

(b) A discovered one-to-one mapping selected by GA. In this case, GA selects this mapping because the method similarity before normalization is calculated as 0.5 due to the same return type and method name.

Fig. 8. Examples of correct mappings found by our algorithm.

4.4 Discussion and Limitations

- In our experiments, we observed how tuning the search favors precision over recall. We make these decisions because we prefer to recommend fewer but correct mappings, rather than many incorrect ones that force developers to do more verification and testing. To find a better tradeoff between fitness and capacity, we will pose API migration as a multi-objective problem.

- We also observed that documentation similarity may be zero for certain specific migration rules because (1) selected method mappings may not have documentation, and (2) documentation often contains code examples that we did not leverage. In order to utilize the USE network, we removed code examples because it is pre-trained on a variety of general-purpose NLP datasets that do not include code. In the future, we will leverage code examples when calculating documentation similarity to retain useful information about API usage.

- In our experiment, we used a maximum number of function evaluations as a termination condition and set the population size to 100. However, GA may perform better with different parameter values or a different termination criteria. As we multi-objectivize the problem, we will also explore the use of different termination criteria, performance indicators and rigorous statistical testing for better convergence towards the pareto front.

- A limitation of this work is that source-target method mappings for *one-to-many* and *many-to-many* mappings may not be recommended in order. In future work, we plan to address this limitation by ranking recommended target methods for each source API.

5 Related Work

In this section, we discuss the similarities and differences between our approach and current techniques. Several related works [16,19,20] address migration at the library level, *i.e.,* they recommend which library can be used to replace a source library. Our work complements these studies by taking their output (recommended library) as input, and recommends suitable methods from the target library to replace methods from the retired library.

Other works [8,15,17,18] identify API mappings when migrating libraries or projects from one programming language or framework to another. The approach presented by Bui et al. [8] models method level migrations using mined data from parallel corpora, *i.e.,* source code of the same software implemented in both source and target languages. Similarly, the approaches presented in [15,18] analyze the method calls surrounding an API and hypothesize that the use of surrounding methods remains consistent across both source and target languages. However, these approaches require that both source and target libraries are sufficiently similar in their design that often may not be the case. Our approach does not make this assumption or require specialized training data to recommend suitable method mappings. Instead, GA only requires a fitness function to recommend the right mappings. In this work, we leverage semantic similarity between source and target APIs and co-occurrence probability to recommend method mappings.

Other related works [21–23] recommend method-level migrations for source and target libraries that are different versions of the same software library. These works use semantic similarity to score possible source-target method mappings and/or use a differencing based approach to find and store all changes between two versions of a library and use them during migration. In addition to a list of changes, these approaches may also utilize mined data from existing projects. However, these approaches may lead to inaccurate recommendations when major changes are made to one version of the library *e.g.,* naming conventions are changed, or if mined data is unavailable. To address these challenges, our approach leverages word vector representations that are robust to changes in library design, and utilize three measures of similarity for accurate recommendation of source-target mappings. Moreover, our approach can recommend mappings for source and target libraries that are not different versions of the same library.

Collie et al. [10] uses a program synthesis technique to identify semantically equivalent source-target API mappings using compiler-specific low-level intermediate representations. However, this approach is only applicable for some specific languages, is difficult to understand and validate, and also requires significant training data to learn to recommend mappings. Another approach by Alrubaye et al. [5] uses manually validated source-target mappings to predict whether a given source-target method pair is valid or invalid using a decision tree classifier. However, manually generating labels for each mapping can become a labor-intensive and error-prone process as the size of the dataset increases. Furthermore, this approach cannot recommend mappings for migration rules that are not included in the training data. Different from these works, our approach does not require labeled data or compiler specific representations and can be applied to recommend mappings between different programming languages and libraries.

6   Conclusion

In this work, we formulate API Migration as a combinatorial optimization problem and demonstrate the benefits of using a Genetic Algorithm (GA) to recommend source-target method mappings with high precision. Our approach aims to find similar source and target methods using their co-occurrence, signature and documentation similarities. To evaluate the efficacy of our approach, we conducted an empirical study using a popular dataset containing 9 popular migrations collected from 57,447 open-source Java projects. Our results demonstrate that GA recommends potentially correct source-target mappings with high precision while outperforming random search and hill-climbing. Our study also reveals many challenges related to the complex nature of identifying method replacements.

References

1. The MOEA framework. http://moeaframework.org/index.html
2. AlOmar, E.A., AlRubaye, H., Mkaouer, M.W., Ouni, A., Kessentini, M.: Refactoring practices in the context of modern code review: an industrial case study at Xerox. In: 2021 IEEE/ACM 43rd International Conference on Software Engineering: Software Engineering in Practice (ICSE-SEIP), pp. 348–357. IEEE (2021)
3. Alrubaye, H., Alshoaibi, D., Alomar, E., Mkaouer, M.W., Ouni, A.: How does library migration impact software quality and comprehension? An empirical study. In: International Conference on Software and Software Reuse, pp. 245–260. Springer (2020). https://doi.org/10.1007/978-3-030-64694-3_15
4. Alrubaye, H., Mkaouer, M.W.: Automating the detection of third-party java library migration at the function level. In: CASCON, pp. 60–71 (2018)
5. Alrubaye, H., Mkaouer, M.W., Khokhlov, I., Reznik, L., Ouni, A., Mcgoff, J.: Learning to recommend third-party library migration opportunities at the API level. Appl. Soft Comput. **90**, 106140 (2020)
6. Alrubaye, H., Mkaouer, M.W., Ouni, A.: On the use of information retrieval to automate the detection of third-party Java library migration at the method level. In: 2019 International Conference on Program Comprehension (2019)
7. Alrubaye, H., Mkaouer, M.W., Peruma, A.: Variability in library evolution: an exploratory study on open-source java libraries. In: Software Engineering for Variability Intensive Systems, pp. 295–320. Auerbach Publications (2019)
8. Bui, N.: Towards zero knowledge learning for cross language API mappings. In: International Conference on Software Engineering: Companion Proceedings (2019)
9. Cer, D., et al.: Universal sentence encoder (2018)
10. Collie, B., Ginsbach, P., Woodruff, J., Rajan, A., O'Boyle, M.F.: M3: semantic API migrations. In: International Conference on Automated Software Engineering (ASE) (2020)
11. Cormen, T.H., Leiserson, C.E., Rivest, R.L., Stein, C.: Introduction to Algorithms, Third Edition. The MIT Press, 3rd edn. (2009)
12. Eiben, A.E., Smith, J.E.: Introduction to Evolutionary Computing. Springer (2015). https://doi.org/10.1007/978-3-662-05094-1
13. Kula, R.G., German, D.M., Ouni, A., Ishio, T., Inoue, K.: Do developers update their library dependencies? Empirical Softw, Engg (2018)

14. Nguyen, H.A., Nguyen, T.T., Wilson, G., Nguyen, A.T., Kim, M., Nguyen, T.N.: A graph-based approach to API usage adaptation. In: International Conference on Object Oriented Programming Systems Languages and Applications (2010)
15. Nguyen, T.D., Nguyen, A.T., Phan, H.D., Nguyen, T.N.: Exploring API embedding for API usages and applications. In: International Conference on Software Engineering (2017)
16. Ouni, A., Kula, R.G., Kessentini, M., Ishio, T., German, D.M., Inoue, K.: Search-based software library recommendation using multi-objective optimization. Inf. Softw. Technol. **83**, 55–75 (2017)
17. Pandita, R., Jetley, R.P., Sudarsan, S.D., Williams, L.: Discovering likely mappings between APIs using text mining. In: 2015 International Working Conference on Source Code Analysis and Manipulation (2015)
18. Phan, H.D., Nguyen, A.T., Nguyen, T.D., Nguyen, T.N.: Statistical migration of API usages. In: International Conference on Software Engineering Companion (ICSE-C) (2017)
19. Teyton, C., Falleri, J.R., Palyart, M., Blanc, X.: A study of library migrations in Java. J. Softw. Evol. Process **26**(11), 1030–1052 (2014)
20. Teyton, C., Falleri, J.R., Blanc, X.: Mining library migration graphs. In: 19th Working Conference on Reverse Engineering (2012)
21. Wu, W., Guéhéneuc, Y.G., Antoniol, G., Kim, M.: AURA: a hybrid approach to identify framework evolution. In: 2010 International Conference on Software Engineering (2010)
22. Xing, Z., Stroulia, E.: API-evolution support with Diff-CatchUp. IEEE Trans. Softw. Eng. **33**, 818–836 (2007)
23. Xu, S., Dong, Z., Meng, N.: Meditor: inference and application of API migration edits. In: International Conference on Program Comprehension (2019)

Multi-objective Optimization of Extreme Learning Machine for Remaining Useful Life Prediction

Hyunho Mo and Giovanni Iacca(✉)

Department of Information Engineering and Computer Science,
University of Trento, Trento, Italy
giovanni.iacca@unitn.it

Abstract Given that physics-based models can be difficult to derive, data-driven models have been widely used for remaining useful life (RUL) prediction, which is a key element for predictive maintenance. In industrial applications, although the models have to be trained in a short time with limited computational resources, recent research using back propagation neural networks (BPNNs) has focused only on minimizing the RUL prediction error, without considering the time needed for training. Driven by this motivation, here we consider a simple and fast neural network, named extreme learning machine (ELM), and we optimize it for the specific case of RUL prediction. In particular, we propose to apply both single-objective and multi-objective optimization to search for the best ELM architectures in terms of a trade-off between RUL prediction error and training time, the latter being determined by the number of trainable parameters. We perform a comparative analysis on a recent benchmark dataset, the N-CMAPSS, in which we compare the proposed methods with other algorithms based on BPNNs. The results show that while the optimized ELMs perform slightly worse than the BPNNs in terms of RUL prediction error, they require a significantly shorter (up to 2 orders of magnitude) training time.

Keywords: Evolutionary algorithm · Multi-objective optimization · Extreme learning machine · Remaining useful life · N-CMAPSS

1 Introduction

With the advent of Industry 4.0, the remaining useful life (RUL) prediction of industrial components has become one of the mainstream elements in predictive maintenance (PdM) research [1]. Maintenance of industrial components is in fact closely related to costs and reliability, since industry stakeholders can cut their loss by reducing any unplanned downtime. Moreover, the performance of the industrial equipment—as well as the quality of products—can be improved by offering timely maintenance before failures occur. A paradigmatic example is given by the airline industry, where it is crucial to predict the RUL of aircraft engines accurately and timely, in order to guarantee the aircraft operation

© Springer Nature Switzerland AG 2022
J. L. Jiménez Laredo et al. (Eds.): EvoApplications 2022, LNCS 13224, pp. 191–206, 2022.
https://doi.org/10.1007/978-3-031-02462-7_13

safety, and make appropriate maintenance decisions. By estimating the RUL of the aircraft engines, airlines can improve maintenance schedules and indeed avoid major disasters. The RUL prediction is thus an essential requirement for minimizing maintenance cost, as well as guaranteeing safety [2].

Recently, many ML-based methods using neural networks have been introduced to handle the above task. One of the earliest approaches, discussed in [3], is to employ a multi-layer perceptron (MLP) for the RUL prediction of aircraft engines. In the same work, the authors also propose to use a convolutional neural network (CNN), that is a widely used network used especially for computer vision tasks. Instead of using traditional back propagation-neural networks (BPNNs), such as a MLP or a CNN, Yang et al. [4] propose to exploit extreme learning machines (ELM) [5], a kind of neural network that requires a much shorter training time compared to (even shallow) BPNNs. As an alternative approach, recurrent neural networks (RNNs), including long short term memory (LSTM), have been proven able to predict the RUL by directly recognizing temporal patterns of the data, instead of extracting their convolutional features [4]. More recently, deeper neural networks have been proposed to improve the RUL prediction accuracy [6]. Other deep learning (DL)-based RUL prediction methods consider a combination of an autoencoder (AE) with a CNN [7], or with a RNN [8]. An attention-based DL framework has been proposed in [9]. Another recent work [10] applied evolutionary computation to optimize deep networks tailored to the RUL prediction task. In this case the authors propose to use a genetic algorithm (GA) to optimize the architecture of a multi-head CNN-LSTM, which was handcrafted in their previous work [11].

Most of the methods proposed so far in the literature mainly focus on achieving a minimization of the RUL prediction error on a well-established benchmark called the commercial modular aero-propulsion system simulation (CMAPSS) dataset [12], which is the *de facto* standard benchmark for RUL prediction. The CMAPSS dataset has been widely used to develop and evaluate the RUL prediction models after it became publicly available on the NASA's data repository in 2008. However, one shortcoming of this dataset is that the data are solely based on MATLAB simulations, without considering real flight conditions, so that each time series is rather short (just a few hundred samples). In 2021, the NASA's data repository released the new CMAPSS (N-CMAPSS) [13] dataset, that contains data acquired under real flight conditions. One major difference with respect to the previous dataset is that each time series consists of millions of samples, thus the total size of the dataset is significantly larger. Therefore, this new realistic dataset provides a chance to develop reliable algorithms for RUL prediction in a real-world context. Moreover, while the previous dataset was small enough to allow researchers to focus only on the minimization of the RUL prediction error, without consider the training time, the N-CMAPSS dataset, due to its much larger amount of data, requires algorithms that are faster to train, without compromising the RUL prediction error.

It should be noted that, although reducing the training time (which correlates to the number of trainable parameters) is a crucial objective in industrial contexts that do not normally have access to expensive computing infrastructures, this aspect has not been considered so far in the literature, also because of the aforementioned limitations of the CMAPSS dataset.

To achieve the multiple (and conflicting) objectives of reducing the RUL prediction error while also minimizing the training time, here we propose to use an ELM and we optimize its architecture and parameters using both a single-objective optimization (SOO) and a multi-objective optimization (MOO) algorithm. For the SOO case, we use a custom GA with two different fitness function formulations (including or not a penalty on the neural network complexity): we use its results as baseline. For the MOO case, we use the well-known non-dominated sorting genetic algorithm II (NSGA-II) [14], which has been recently applied successfully also for neural architecture search [15]. The choice of using an ELM is motivated by the experimental results presented in [16], that are based on the CMAPSS dataset, where the authors show that ELM-based models can provide comparable performance to BPNNs in terms of RUL prediction error, while enabling a considerable saving in terms of training time. The main limitation of the study presented in [16], however, is that it only considers for the comparative analysis a small set of manually designed ELM architectures. On the other hand, exploring the search space of those networks with an automatic search process can potentially provide further performance improvements: our goal is to fill this gap. Moreover, to take into account a proper assessment of the training time, we perform our experiments on the N-CMAPSS dataset.

Another important advantage of our proposal is that it is based on a fully data-driven approach that is able to predict the RUL directly, by leveraging the relation between the degradation pattern of the raw sensor data and the RUL, thus without using any physics-based model of the degradation process (as recently done in [17]). This is also made possible by the large amount of data available in the N-CMAPSS dataset: in fact, these sensor data allow to model the RUL prediction problem as a regression task on a multivariate time series.

To summarize, the main contributions of this work can be identified in the following elements:

- We achieve a successful trade-off between two conflicting objectives, namely the RUL prediction error and the number of trainable parameters.
- To the best of our knowledge, this is the first use case of multi-objective neural architecture search applied to RUL prediction (and, in particular, on the N-CMAPSS dataset).

The rest of the paper is organized as follows: in Sect. 2, the background concepts on ELMs are introduced. The details of the proposed methods are presented in Sect. 3. Then, Sect. 4 discusses the details of our experiments and Sect. 5 presents the numerical results. Finally, Sect. 6 provides the conclusions of this work.

2   Background

An ELM is a fast training algorithm for *single-hidden layer feed-forward neural networks* (SLFNs). While BPNNs tune their parameters iteratively with (usually, time-consuming) gradient-based computations, ELMs merely use an analytically determined non-iterative solution as the output weights, together with randomly initialized input weights.

An ELM can be formally described as follows. Let N be the number of labeled samples, where each sample is a pair made of a d-dimensional input vector and the corresponding c-dimensional label, which are denoted by $\boldsymbol{x}_i$ and $\boldsymbol{t}_i$ respectively. A given set of training samples can then be written as $(\boldsymbol{x}_i, \boldsymbol{t}_i)$, $i \in [1, N]$ with $\boldsymbol{x}_i \in \mathbb{R}^d$ and $\boldsymbol{t}_i \in \mathbb{R}^c$. In our experiments on the N-CMAPSS dataset, d is the number of monitoring signals, and $c = 1$ (i.e., the label is a real number representing a RUL value).

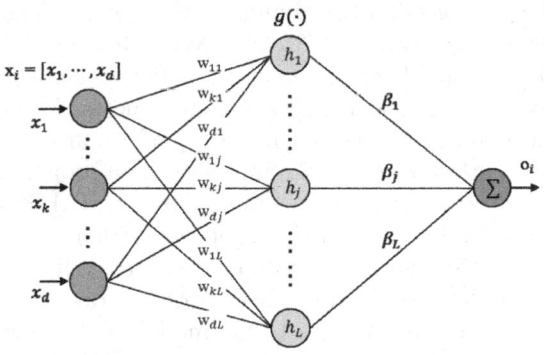

Fig. 1. Illustration of ELM with the structure of a SLFN.

The notation used for describing the ELM and the structure of a SLFN are visualized in Fig. 1. For a given input sample $\boldsymbol{x}_i$, the output of a SLFN o_i with L hidden neurons and activation function $g(\cdot)$ is defined by:

$$o_i = \sum_{j=1}^{L} \beta_j g(\boldsymbol{w}_j \cdot \boldsymbol{x}_i + b_j). \tag{1}$$

where $\boldsymbol{w}_j = [w_{1j}, \ldots, w_{dj}]$ is a vector of weights on the connections between the d input neurons (which are assumed to be linear) and each j-th hidden neuron, β_j is the weight on the connection between the j-th hidden neuron and the output neuron, and b_j denotes the bias for the j-th hidden neuron. The computation for all the N equations (one for each of the N samples) can be written compactly as:

$$\boldsymbol{H} \cdot \boldsymbol{\beta} = \begin{bmatrix} g(\boldsymbol{w}_1 \cdot \boldsymbol{x}_1 + b_1) & \cdots & g(\boldsymbol{w}_L \cdot \boldsymbol{x}_1 + b_L) \\ \vdots & \ddots & \vdots \\ g(\boldsymbol{w}_1 \cdot \boldsymbol{x}_N + b_1) & \cdots & g(\boldsymbol{w}_L \cdot \boldsymbol{x}_N + b_L) \end{bmatrix} \begin{bmatrix} \beta_1 \\ \vdots \\ \beta_L \end{bmatrix} \tag{2}$$

where H is the hidden layer output matrix (of size $N \times L$), and β (of size L) consists of the weights of all the connections between the hidden neurons and the output neuron.

To train the SLFN defined above is equivalent to find a least square solution $\hat{\beta}$ to the linear system $H \cdot \beta = T$ where $T = [t_1, \cdots, t_N]^\top$. Therefore, the mathematical formulation of the training procedure can be expressed as:

$$\|H\hat{\beta} - T\| = \min_{\beta}\|H\beta - T\| \tag{3}$$

As discussed in [5], the smallest norm least squares solution of the above equation is determined by:

$$\hat{\beta} = H^\dagger T \tag{4}$$

where $H^\dagger$ denotes the Moore-Penrose generalized inverse of the matrix H. The pseudo-inverse can be calculated as $(H^\top H)^{-1}H^\top$. Moreover, an L2 regularization term αI (with $\alpha \in \mathbb{R}$ being an arbitrarily small value) is added to prevent the inverse term from being singular (i.e., $H^\top H$ is replaced by $H^\top H + \alpha I$). The solution to Eq. (3) is then defined by:

$$\hat{\beta} = (H^\top H + \alpha I)^{-1}H^\top T. \tag{5}$$

This last equation describes the ELM training algorithm for SLFNs[1].

As pointed out in the seminal paper on ELMs [5], the training process described by Eq. (5) is *extremely* fast (hence their name). Moreover, it has been shown that ELMs can even achieve a better generalization performance than back-propagation training algorithms [18,19] and that, compared to BPNNs, they obtain comparable results in terms of prediction accuracy on a variety of regression tasks [16,20]. Considering these advantages, ELMs then appear an appropriate tool for RUL prediction tasks in industrial contexts that require a fast learning process and a stable generalization performance.

On the other hand, the computational complexity of the ELM training algorithm derives from the size of the matrix H: more specifically, it is $\mathcal{O}(NL^2 + L^3)$. In other words, the complexity is cubic w.r.t. the number of hidden neurons, which, in turn, is the same as the number of trainable parameters (i.e., the β values), see also the empiric characterization of the training time shown in Fig. 2. Yet, increasing the number of hidden neurons L does not always contribute to decreasing the RUL prediction error, and may lead to overfitting.

Finding the optimal value of L, as well as of the other parameters of an ELM, is therefore a crucial element for the ELM performance. However, this task that cannot be easily achieved by manual design or by empiric considerations. On the contrary, using evolutionary search to explore this parameter space and to discover optimized ELMs automatically (in terms of both the RUL prediction error and the number of trainable parameters) seems to be a more viable solution.

---

[1] We should note that, strictly speaking, "ELM" refers to the training algorithm only. However in the literature (as well as in the rest of this paper) "ELM" is generically used to refer to both the training algorithm and the neural network itself.

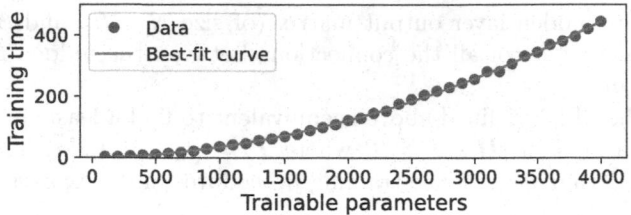

Fig. 2. Correlation between the number of trainable parameters and the training time of the ELM models. The best fit curve shows a super-quadratic dependency between the training time and the number of parameters, in line with the cubic complexity derived analytically, $\mathcal{O}(NL^2 + L^3)$.

3   Methods

We present now the details of the proposed methods: Sect. 3.1 describes the individual encoding, while Sect. 3.2 describes the SOO and MOO evolutionary algorithms.

3.1   Individual Encoding

Given the baseline structure of the ELM model described in Sect. 2, we consider the optimization of the following integer parameters:

- number of hidden neurons with hyperbolic tangent activation (n_{tanh});
- number of hidden neurons with sigmoid activation (n_{sigm});
- L2 regularization parameter (r);

In preliminary experiments, we observed that the RUL prediction error as well as the computational complexity are largely affected by the number of hidden neurons. Furthermore, using different activation function $g(\cdot)$ for the hidden neurons also produces different results. Therefore, we encode the number of hidden neurons with two different activation functions into n_{tanh} and n_{sigm}, respectively. The remaining parameter, r, refers to the order of magnitude of the L2 regularization parameter α described in Sect. 2, i.e., $\alpha = 10^{-r}$.

Overall, the lower and upper bounds for each parameter considered in our experiments are set as follows: $[1, 200]$ (multiplied by a fixed value of 10) for both n_{tanh} and n_{sigm}, and $[2, 6]$ for r. These values have been chosen empirically. In particular, we use a discretization on the number of hidden nodes to reduce the search space yet allowing ELMs of up to $2,000$ tanh hidden neurons and $2,000$ sigmoid hidden neurons. This upper bound is chosen to allow to train each ELM generated during the evolutionary search within a maximum training time of 500 seconds, so that the overall runtime of the search remains affordable. Concerning r, its lower bound corresponds to $\alpha = 10^{-2}$, which we find being the largest value that does not too much affect the ELM performance; the upper bound, on the other hand, corresponds to $\alpha = 10^{-6}$, which we set as the smallest value that makes the inverse term in Eq. (4) non-singular.

3.2   Optimization Algorithms

In order to optimize the ELM parameters described in Sect. 3.1, we consider first a SOO approach based on a GA aimed at minimizing simply the RUL prediction error. However, as we will see in Sect. 5, in this case the discovered ELMs tend to have a large number of trainable parameters. To avoid this, we consider a second SOO approach where the number of trainable parameters is included as a penalty factor into the fitness evaluation. Finally, we consider a MOO approach to look explicitly for the best trade-off solutions in terms of RUL prediction error and number of trainable parameters. In all the three approaches, the fitness of each individual is calculated by generating an ELM (the phenotype) associated to the corresponding genotype, i.e., a vector containing the three parameters introduced in Sect. 3.1. Moreover, given that the N-CMAPSS dataset consists of a training set D_{train} and a test set D_{test}, we further split D_{train} into training purpose data, E_{train}, and validation purpose data, E_{val} (i.e., $D_{train} = E_{train} \cup E_{val}$). The fitness is calculated on the latter.

Single-Objective Optimization. First, we initialize the population by generating n_{pop} individuals at random. In the main loop of the evolutionary search, we use crossover and mutation as genetic operators. As for crossover, we implement a specialized one-point crossover: at first, the population is sorted according to the fitness of its individuals. Then we mate the individual in the $(2i)$-th position with the one in the $(2i + 1)$-th position following a crossover probability p_{cx}, where $i \in [0, n_{pop}/2-1]$. This allows us not only to exploit the best individuals, but also to explore the areas of the search space that are distant from the region in which the best individuals lie. Then, we apply uniform mutation following a mutation probability p_{mut}, according to which each gene can be mutated to another value uniformly drawn from its bounds specified in Sect. 3.1. The expected number of mutations is determined by a p_{gene} parameter which is set to 0.4, so that we have, on average, 1.2 mutated genes out of 3. This enables us to have a relatively fast search process by having on average at least one gene mutation, while avoiding disruptive mutations. After that, we create the population for the next generation with implicit elitism, i.e., the worst parent involved in the crossover is replaced by its offspring if the latter has a better fitness. We set both p_{cx} and p_{mut} to 0.5.

We stop the above process after a fixed number of generations n_{gen}, after which the algorithm returns the best individual found during the evolutionary search based on the specific fitness. We set n_{pop} to 28 and n_{gen} to 30. We have empirically found that these settings allow enough evaluations to observe convergence on the fitness across generations.

We consider two different scenarios in terms of SOO, that we refer to respectively as SOO-ELM(1) and SOO-ELM(2). In the first case, the fitness of each individual is defined as the root mean square error (RMSE) of its phenotype on E_{val}, after training it on E_{train}. We indicate this validation RMSE with $RMSE_{val}$. As said, the limitation of this approach is that the size of the discovered ELM tends to be very large in order to decrease $RMSE_{val}$. In other words,

n_{tanh} and n_{sigm} tend to converge to their upper bounds throughout the evolutionary process. To overcome this limitation, we consider a second SOO approach in which the fitness formulation includes the number of trainable parameters of the ELM. This is obtained by calculating the fitness as $RMSE_{val} + \tau L$, where τ is a constant weight. This way, we can prevent the survival of the unnecessarily large ELMs by penalizing their fitness with τL.

Multi-objective Optimization. As we discussed earlier, minimizing $RMSE_{val}$ and L are conflicting objectives in the architecture search of the ELMs. While the SOO-ELM(2) approach discussed above somehow goes in the direction of compromising those two objectives, the best model still largely depends on a human decision, since it depends on how the value of τ is parametrized.

To tackle this limitation, we propose to use a MOO algorithm, NSGA-II, to search explicitly for a set of trade-off ELMs. We refer to this method as MOO-ELM. We follow the procedure of the original NSGA-II algorithm. At the beginning of the evolutionary run, a population of n_{pop} individuals is randomly initialized. In the main loop after the initial generation, an offspring population of the same size is created by using tournament selection, crossover and mutation. The tournament selection primarily checks the dominance across the individuals in the population. Then, the secondary criteria, crowding distance, is considered only once all the non-dominated individuals have been considered. Regarding the crossover and the mutation, we use the same strategies specified in the single-objective case, with the same parametrization. The combined population of the parents and the offspring is then sorted according to non-domination. The best non-dominated sets are inserted into the new population, until no more sets can be accommodated. For the next non-dominated set, which would make the new population be larger than the fixed population size n_{pop}, only the solutions largest crowding distance values are inserted in the remaining slots in the new population. When the new population is ready, its offspring population is created and the same process continues to the next generation.

Also in this case, we set n_{pop} to 28 and n_{gen} to 30. After the fixed number of generations, the evolutionary search returns a set of the trade-off solutions that have the top dominance level (i.e., the method returns a Pareto front).

4   Experimental Setup

In this section, we present the details of our experimentation: first, we briefly describe the N-CMAPSS dataset in Sect. 4.1. In the following Sect. 4.2, we describe two BPNNs that are used for the comparison with our methods. Then, the computational setup and data preparation steps are outlined in Sect. 4.3.

4.1   Benchmark Dataset

The proposed methods are evaluated on the N-CMAPSS dataset. Specifically, we only use its sub-dataset DS02, that has been developed for data-driven methods

[13]. It consists of the run-to-failure degradation trajectories of nine turbofan engines with unknown and different initial conditions. The synthetic trajectories were generated with the CMAPSS dynamic model implemented in MATLAB, but a fidelity gap between simulation and reality is mitigated by reflecting real flight conditions recorded on board of a commercial jet. Furthermore, the relation between the degradation and its operation history is considered, to extend the degradation modeling [13]. Among the nine engines, we use 6 units (u_2, u_5, u_{10}, u_{16}, u_{18} and u_{20}) for the training set D_{train}, and the remaining 3 units (u_{11}, u_{14} and u_{15}) for the test set D_{test}. In particular, the u_{14} and u_{15} relate to shorter and lower altitude flights compared to those of the training units, so that the evaluation results on the D_{test} can implicitly reflect the generalization capability of the RUL prediction model.

Table 1 describes each unit in the dataset. The total number of samples (i.e., timestamps) is 5.26M in D_{train} and 1.25M in D_{test}, with a sampling rate of 1Hz. The end-of-life time t_{EOL} points out the counted flight cycles at the end of the engine's lifespan, i.e., t_{EOL} is the same as the initial value of the labeled RUL. There are two distinctive failure modes in the dataset: the abnormal high-pressure turbine (HPT) and the low-pressure turbine (LPT). The combination of the two failure modes for a unit means that the unit is subject to a more complex failure mode than a single-failure mode.

The dataset provides condition monitoring signals that are related to the useful life of the flight engine. Following the setup used in [17], we select the same 20 signals. The multivariate time series from the 20 signals is used as an input for the ELM model, therefore the dimension of the input sample d is 20.

Table 1. Overview of each unit in the DS02 of N-CMAPSS dataset w.r.t the number of samples m_i, the end-of-life time t_{EOL} and the failure modes.

Training set (D_{train})				Test set (D_{test})			
Unit	m_i(M)	t_{EOL}	Failure mode	Unit	m_i(M)	t_{EOL}	Failure mode
u_2	0.85	75	HPT	u_{11}	0.66	59	HPT + LPT
u_5	1.03	89	HPT	u_{14}	0.16	76	HPT + LPT
u_{10}	0.95	82	HPT	u_{15}	0.43	67	HPT + LPT
u_{16}	0.77	63	HPT + LPT				
u_{18}	0.89	71	HPT + LPT				
u_{20}	0.77	66	HPT + LPT				

4.2   Back-Propagation Neural Networks (BPNNs)

To compare the proposed methods to BPNNs, we consider a MLP and a CNN whose architectures were manually designed in [17]. As we discussed in Sect. 1, the previous works on the old CMAPSS dataset have mostly used deep networks with complex architectures. On the other hand, on the new dataset a simple feed-forward neural network (a MLP) and a 1D CNN are still used as state-of-the-art neural networks, considering the great amount of data (in the order of millions of samples) obtained from real flight conditions.

The architecture of the considered MLP has four hidden layers: the first three of them have 200 neurons each, while the last one has 50 neurons. All the hidden nodes use the ReLU activation function. Similarly to the ELM, a 20-dimensional vector for each timestamp is used as an input for the MLP, so the input layer has 20 nodes, while the output layer consists of a single node.

In contrast, the CNN requires time-windowed data as an input to apply 1D convolution in the temporal direction. Therefore, all the given time series are sliced by a time window of length 50 and stride 1. Each input for the CNN then spans 50 timestamps and has a size of 50×20. Regarding the architecture, the CNN is made up of three convolutional layers followed by a fully connected layer. The first two convolutional layers consist of 10 filters of size 10, while the last convolutional layer has only one filter of the same size. The following fully connected layer has 50 neurons. ReLU is used as activation function for all the nodes in the network.

Lastly, we should note that we follow the training setup used in [17]. In particular, stochastic gradient descent (SGD), with mini-batch size set to 1024, is used to compute the gradient, and $AMSgrad$ [21] is used as an optimization algorithm after initializing the weights with $Xavier$ initialization [22]. For the training iterations, the maximum number of epochs is set to 60 and 30 for the MLP and the CNN respectively, whereas the learning rate is set to 0.001. To handle overfitting, early stopping is considered with a patience of 5.

4.3 Computational Setup and Data Preparation

All the neural networks used in our work are implemented in Python. In particular, TensorFlow 2.3 is used to implement the BPNNs. To implement the ELM, we use the high performance toolbox for ELM (HP-ELM)[2] that supports GPU computation. All the experiments have been conducted on the same workstation with an NVIDIA TITAN Xp GPU, so that we can have a reliable comparison between different models w.r.t the training time. Both the SOO and MOO algorithms are implemented using the DEAP library[3]. Our code is available online[4].

Since we employ the neural networks, each time series is normalized to $[-1, 1]$ by a min-max normalization. As we mentioned in Sect. 3.2, we split D_{train} in E_{train} and E_{val}: for that, we randomly choose 80% of the data in D_{train} and assign them to E_{train}, used for training each individual. The remaining 20% are designated as E_{val} for the fitness evaluation. For the experiments on the BPNNs, following the experimental setup used in [17], 90% of the data in D_{train} are used to train the networks, and the remaining 10% are reserved for early stopping.

---

[2] https://github.com/akusok/hpelm.
[3] https://github.com/DEAP/deap.
[4] https://github.com/mohyunho/MOO_ELM.

5   Experimental Results

The aim of our experiments is to evaluate the optimized ELMs found with the proposed methods described in Sect. 3.2, by comparing their results with those obtained by the two BPNNs described in Sect. 4.2. The comparison is mainly based on two metrics: 1) the RMSE on D_{test}, and 2) the number of trainable parameters. To highlight the advantage of MOO-ELM, we additionally compare the training times of the methods under study, which as seen in Fig. 2 in the case of ELM is super-quadratic w.r.t. the number of trainable parameters. To improve the reliability of the results from the GA-based methods, we execute 10 independent runs with different random seeds.

Let us first analyze the convergence of MOO-ELM. For each MOO-ELM run, we collect the hypervolume (HV) [23] across 30 generations, and we normalize it to $[0, 1]$ by a min-max normalization. As shown in Fig. 3, a gradual improvement across the generations of MOO-ELM is observed by an increase in the normalized HV. The monotonic increase of the mean HV indicates that the MOO-ELM algorithm keeps finding better non-dominated solutions across the generations. In addition, the slope of the mean HV and its std. dev. indicate convergence at the end of the generations. This means that the algorithm explores the search space enough within 30 generations, even if it starts from different initial populations.

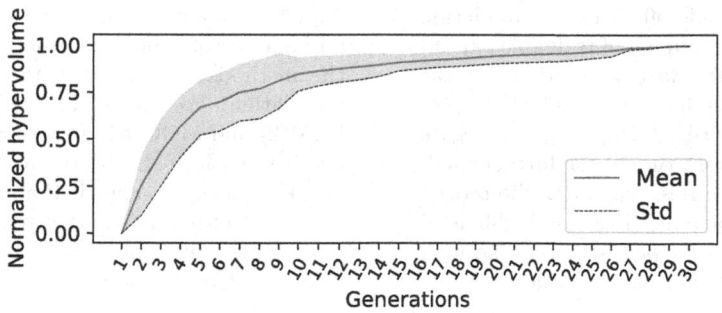

Fig. 3. Normalized hypervolume across generations (mean ± std. dev. across 10 independent runs) for the proposed MOO-ELM approach.

To compare the results obtained by the different methods under study, we aggregate the 10 independent runs in the following way: in the case of the two SOO-ELM approaches, each run returns a single solution that has the best fitness during the evolution. The aggregation is then simply the mean of the test RMSE of the 10 best individuals. On the other hand, each run of NSGA-II returns a number of solutions on the final Pareto front. In our case, we collected 417 non-dominated solutions across the 10 runs. For the sake of comparison, instead of using all of them, we select a fraction of the solutions based on their density in the fitness space, as described in Fig. 4. When we do not have any preference for a certain objective, this strategy can be used to derive a subset of the solutions which are implicitly "preferred" by the MOO algorithm. As shown in Fig. 4, we first place all the solutions from the 10 runs on the fitness space, which is discretized in equally-spaced bins. The density of the solutions can then be measured by counting the number of solutions lying in each bin. As a result, we choose the 28 solutions from the bin with the highest density and use the average of their test RMSE as the final result for MOO-ELM, shown in Table 2.

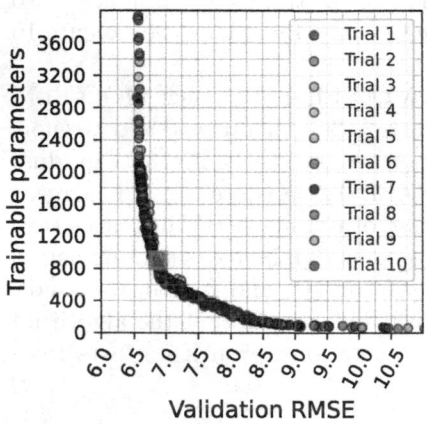

Fig. 4. Trade-off between validation RMSE and number of trainable parameters at the last generation for the 10 independent runs of the proposed MOO-ELM approach (aggregate results across runs). For further analysis, the fitness space is discretized in 20 × 20 bins. The bin highlighted in yellow corresponds to the one with the highest density of solutions. (Color figure online)

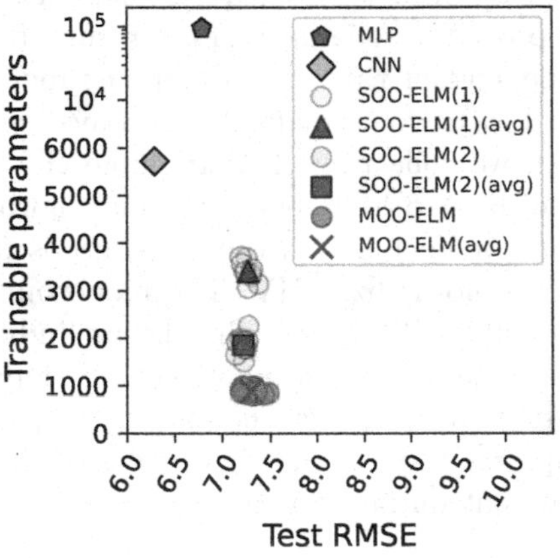

Fig. 5. Trade-off between test RMSE and number of trainable parameters for the methods considered in the experimentation. For SOO-ELM(1), SOO-ELM(2) and MOO-ELM we report the result of each of the 10 available runs, and their average. For MLP and CNN, we report only one solution related to one single run (since their computations are deterministic).

The comparative results of all the considered methods are presented in Table 2. It can be seen that MOO-ELM achieves comparable results to state-of-the-art MLP and CNN models designed by human experts [17]. As mentioned earlier, in the scope of data-driven methods, those deep networks offer indeed state-of-the-art RUL predictions (in terms of test RMSE) on the N-CMAPSS dataset. However, the MLP contains a huge amount of trainable parameter throughout four stacked layers, and training those parameters with an iterative approach requires more than 18 minutes. The CNN shows better prediction accuracy with a lower number of parameters by leveraging parameter sharing, but the training time of this DL architecture is almost twice as big as that of the MLP. Note that the test RMSE values of the BPNNs in Table 2 is different from those reported in [17], which are based on an early version of DS02 that has lower noise level on the sensor readings and a sampling rate of 0.1Hz (instead of 1Hz).

Table 2. Summary of the comparative analysis. For MLP and CNN, we report only one value related to one single run (since their computations are deterministic). For the proposed methods, SOO-ELM(1), SOO-ELM(2) and MOO-ELM, we report mean $\pm$ std. dev. obtained across 10 independent runs. Note that for the proposed methods the worst/median/best architectures are reported in Table 3. The boldface indicates the best result in each column.

Methods	Architecture	Test RMSE (on D_{test})	Trainable parameters	Training time (s)
MLP [17]	4 hidden layers	6.79	94,701	1,081
CNN [17]	3 convolutional layers	**6.29**	5,722	1,969
SOO-ELM(1)	–	7.27 ± 0.05	3,405 ± 202	337 ± 49
SOO-ELM(2)	–	7.21 ± 0.04	1,859 ± 207	110 ± 22
MOO-ELM	–	7.29 ± 0.07	**898 ± 60**	**55 ± 10**

Table 3. Worst, median and best architectures among the best solutions found in each of the 10 runs for each of the three proposed methods, with the corresponding test RMSE and number of trainable parameters. The architectures are denoted by $[n_{tanh}{\cdot}10,\ n_{sigm}{\cdot}10]$ representing the number of tanh and sigmoid hidden neurons, respectively. The L2 regularization parameter r is almost always the same, with a value of 6 for most of the best solutions (5 for the remaining few cases). Thus, we omit its value for brevity.

Methods	Test RMSE			Trainable parameters		
	Worst	Median	Best	Worst	Median	Best
SOO-ELM(1)	[1,840, 1,310]	[1,740, 1,390]	[1,930, 1,760]	[1,930, 1,760]	[1,910, 1,490]	[1,750, 1,310]
SOO-ELM(2)	[1,780, 10]	[1,900, 50]	[1,860, 70]	[1,960, 310]	[1,860, 70]	[1,500, 10]
MOO-ELM	[740, 110]	[790, 80]	[850, 60]	[970, 20]	[840, 20]	[740, 70]

On the contrary, all the optimized ELMs have a considerably smaller number of trainable parameters, which reflects in a much shorter (up to 2 orders of magnitude) training time. The architecture discovered by SOO-ELM(1) tends to have almost the maximum available number of hidden neurons, because it simply uses the validation RMSE as the fitness for its evolutionary search. Yet, in this case the ELM does not suffer from overfitting and its performance does not get worse even if we increase the number of hidden neurons excessively. In other words, the oversized ELM network merely can make a negligible improvement but requires unnecessarily large computational cost for training the redundant parameters. The SOO-ELM(2) approach, on the other hand penalizes those oversized ELM by introducing the penalty factor described in Sect. 3.2. We set the constant weight τ to 10^{-4}, small enough that the penalty does not dominate the overall fitness. We can see that SOO-ELM(2) successfully prevents the use of the redundant neurons, so that it can preserve the RUL prediction accuracy using only almost half of the neurons, w.r.t. the previous method.

Although SOO-ELM(2) uses almost a half of the neurons and requires less than two minutes of training time, a proper value of τ must be determined by empirical considerations. In contrast, MOO-ELM can overcome this problem using NSGA-II for automatically searching a set of trade-off network architectures without considering a tunable parameter such as the τ used before. Compared to SOO-ELM(2), this method achieves almost the same test RMSE but can further halve the number of trainable parameters. Moreover, it only needs, on average, less than one minute to train the best trade-off networks.

Table 4. Execution time for 30 generations of the three proposed methods (mean $\pm$ std. dev. across 10 independent runs). The boldface indicates the best result.

	SOO-ELM(1)	SOO-ELM(2)	MOO-ELM
Execution time (hours)	16.19 ± 1.68	9.67 ± 0.44	$\mathbf{6.43 \pm 0.27}$

In addition, MOO-ELM has a clear advantage in terms of execution time. As shown in Table 4, on average, the evolutionary search process of MOO-ELM is much shorter than that of the two SOO-ELM approaches. For SOO-ELM(1), very large ELM models, that take long time to evaluate, tend to survive throughout the generations, because they can have better validation RMSE even though their improvement may be trivial. In contrast, those individuals hardly survive during a run of MOO-ELM, because the NSGA-II algorithm proceeds in the direction of finding better trade-off solutions.

Finally, the comparative results are visualized in Fig. 5, which easily enables to compare the performance of the different methods in terms of trade-off between the two conflicting objectives. We can observe that the CNN dominates the MLP. Among the compared algorithms, MOO-ELM obtains the best solutions in terms of number trainable parameters (on average, about 900, i.e., less than 16% compared to the CNN). Moreover, compared to the CNN, the models discovered by MOO-ELM have an approximately 97% shorter training time, while their test RMSE is on average only 16% larger.

6   Conclusions

In this paper, we applied evolutionary algorithms to search for optimized ELMs for RUL prediction tasks. We considered three methods, two single-objective based on a custom GA (SOO-ELM) and one multi-objective based on NSGA-II (MOO-ELM), and applied them to automatic design ELMs. Our goal was to achieve a trade-off between two competing objectives: the test RMSE and the number of trainable parameters. We compared our methods to state-of-the-art MLP and CNN models from the literature. The comparative evaluation was based on the experimental results on the N-CMAPSS, one of the most up-to-date benchmarks in the area of RUL prediction. The results show that MOO-ELM can search ELMs that are much smaller in size compared to ELMs obtained by optimizing only the RUL prediction error, but have similar prediction performance. Compared to the MLP and CNN, MOO-ELM performs slightly worse in terms of the test RMSE, but the number of trainable parameters is considerably smaller and the training time is significantly shorter. Hence, our work can be used as an efficient RUL prediction tool for industrial applications that require to compromise training time and prediction accuracy.

One major limitation of ELMs is that they comprise a single hidden layer not work. To overcome this limitation multiple hidden layer ELMs (MELMs) have been proposed recently [24]. In future work, we expect to obtain further improvements by applying our method to optimize the architecture of the MELMs. Another interesting future direction would be to employ recurrent extreme learning machines (RELMs) [25]. Similar to this work, we can attempt to optimize the parameters of the RELMs and their architecture in order to achieve higher accuracy while saving time during the training stage.

References

1. Zhang, W., Yang, D., Wang, H.: Data-driven methods for predictive maintenance of industrial equipment: a survey. IEEE Syst. J. **13**(3), 2213–2227 (2019)
2. Zheng, C., et al.: A data-driven approach for remaining useful life prediction of aircraft engines. In: International Conference on Intelligent Transportation Systems (ITSC), pp. 184–189 (2018)
3. Babu, G.S., Zhao, P., Li, X.L.: Deep convolutional neural network based regression approach for estimation of remaining useful life. In: International Conference on Database Systems for Advanced Applications (DASFAA), pp. 214–228. Springer (2016). https://doi.org/10.1007/978-3-319-32025-0_14
4. Zheng, S., Ristovski, K., Farahat, A., Gupta, C.: Long short-term memory network for remaining useful life estimation. In: International Conference on Prognostics and Health Management (ICPHM), pp. 88–95. IEEE (2017)
5. Huang, G.B., Zhu, Q.Y., Siew, C.K.: Extreme learning machine: a new learning scheme of feedforward neural networks. In: International Joint Conference on Neural Networks (IJCNN), vol. 2, pp. 985–990. IEEE (2004)
6. Li, X., Ding, Q., Sun, J.Q.: Remaining useful life estimation in prognostics using deep convolution neural networks. Reliabil. Eng. Syst. Safety **172**, 1–11 (2018)
7. Ye, Z., Yu, J.: Health condition monitoring of machines based on long short-term memory convolutional autoencoder. Appl. Soft Comput. **107**, 107379 (2021)
8. Cheng, Y., Hu, K., Wu, J., Zhu, H., Shao, X.: Auto-encoder quasi-recurrent neural networks for remaining useful life prediction of engineering systems. IEEE/ASME Trans. Mechatron. 1 (2021)
9. Chen, Z., Wu, M., Zhao, R., Guretno, F., Yan, R., Li, X.: Machine remaining useful life prediction via an attention-based deep learning approach. IEEE Trans. Industr. Electron. **68**(3), 2521–2531 (2021)

10. Mo, H., Custode, L., Iacca, G.: Evolutionary neural architecture search for remaining useful life prediction. Appl. Soft Comput. **108**, 107474 (2021)
11. Mo, H., Lucca, F., Malacarne, J., Iacca, G.: Multi-head CNN-LSTM with prediction error analysis for remaining useful life prediction. In: 2020 27th Conference of Open Innovations Association (FRUCT). IEEE, pp. 164–171 (2020)
12. Saxena, A., Goebel, K., Simon, D., Eklund, N.: Damage propagation modeling for aircraft engine run-to-failure simulation. In: Proceedings of the International Conference on Prognostics and Health Management. IEEE, pp. 1–9 (2008)
13. Arias Chao, M., Kulkarni, C., Goebel, K., Fink, O.: Aircraft engine run-to-failure dataset under real flight conditions for prognostics and diagnostics. Data **6**, 5 (2021)
14. Deb, K., Pratap, A., Agarwal, S., Meyarivan, T.: A fast and elitist multiobjective genetic algorithm: NSGA-II. IEEE Trans. Evol. Comput. **6**(2), 182–197 (2002)
15. Lu, Z., et al.: NSGA-NET: neural architecture search using multi-objective genetic algorithm. In: Genetic and Evolutionary Computation Conference (GECCO), pp. 419–427 (2019)
16. Yang, Z., Baraldi, P., Zio, E.: A comparison between extreme learning machine and artificial neural network for remaining useful life prediction. In: Prognostics and System Health Management Conference (PHM), pp. 1–7 (2016)
17. Chao, M.A., Kulkarni, C., Goebel, K., Fink, O.: Fusing physics-based and deep learning models for prognostics. Reliabil. Eng. Syst. Safety **217**, 107961 (2022)
18. Huang, G.B., Zhu, Q.Y., Mao, K., Siew, C., Saratchandran, P., Sundararajan, N.: Can threshold networks be trained directly? IEEE Trans. Circuits Syst. II Express Briefs **53**, 187–191 (2006)
19. Huang, G.B., Zhu, Q.Y., Siew, C.K.: Extreme learning machine: theory and applications. Neurocomputing **70**(1–3), 489–501 (2006)
20. Popoola, S., Misra, S., Atayero, P.A.: Outdoor path loss predictions based on extreme learning machine. Wirel. Pers. Commun. **99** (2018). https://doi.org/10.1007/s11277-017-5119-x
21. Kingma, D., Ba, J.: Adam: a method for stochastic optimization. In: International Conference on Learning Representations (ICLR) (2014)
22. Glorot, X., Bengio, Y.: Understanding the difficulty of training deep feedforward neural networks. J. Mach. Learn. Res. Proc. Track **9**, 249–256 (2010)
23. Shang, K., Ishibuchi, H., He, L., Pang, L.M.: A survey on the hypervolume indicator in evolutionary multiobjective optimization. IEEE Trans. Evol. Comput. **25**(1), 1–20 (2020)
24. Xiao, D., Li, B., Mao, Y.: A multiple hidden layers extreme learning machine method and its application. Math. Probl. Eng. **2017** (2017). https://doi.org/10.1155/2017/4670187
25. Ertugrul, O.F.: Forecasting electricity load by a novel recurrent extreme learning machines approach. Int. J. Electr. Power Energy Syst. **78**, 429–435 (2016)

Explainable Landscape Analysis in Automated Algorithm Performance Prediction

Risto Trajanov[1], Stefan Dimeski[1], Martin Popovski[1], Peter Korošec[2] ,
and Tome Eftimov[2(✉)]

[1] Faculty of Computer Science and Engineering, Ss. Cyril and Methodius,
University - Skopje, Skopje, North Macedonia
{stefan.dimeski.1,martin.popovski}@students.finki.ukim.mk
[2] Computer Systems Department, Jožef Stefan Institute, Ljubljana, Slovenia
{peter.korosec,tome.eftimov}@ijs.si

Abstract. Predicting the performance of an optimization algorithm on
a new problem instance is crucial in order to select the most appropriate algorithm for solving that problem instance. For this purpose, recent
studies learn a supervised machine learning (ML) model using a set of
problem landscape features linked to the performance achieved by the
optimization algorithm. However, these models are black-box with the
only goal of achieving good predictive performance, without providing
explanations which landscape features contribute the most to the prediction of the performance achieved by the optimization algorithm. In this
study, we investigate the expressiveness of problem landscape features
utilized by different supervised ML models in automated algorithm performance prediction. The experimental results point out that the selection of the supervised ML method is crucial, since different supervised
ML regression models utilize the problem landscape features differently
and there is no common pattern with regard to which landscape features
are the most informative.

Keywords: Exploratory landscape analysis · Algorithm performance
prediction · Machine learning · Feature importance

1 Introduction

Automated algorithm performance prediction plays a crucial part in the automated algorithm selection and configuration tasks [1,5,6,10,14]. The most common practices are to train a supervised machine learning (ML) model using a
set of problem landscape features. The ML model links the characteristics of
the problem instance landscape to the performance achieved by an optimization
algorithm that is run on that instance. However, such ML models are still black-
box with a limited explanations of how each landscape feature of the problem
instance influences the prediction of the end performance result achieved by an
optimization algorithm.

© Springer Nature Switzerland AG 2022
J. L. Jiménez Laredo et al. (Eds.): EvoApplications 2022, LNCS 13224, pp. 207–222, 2022.
https://doi.org/10.1007/978-3-031-02462-7_14

To describe the characteristics of a problem instance, the Exploratory Landscape Analysis (ELA) [16] is used, where for each problem instance a set of landscape features are calculated, known as ELA features. These features are coming from different groups (e.g., statistical, information theory, etc.) and require a selection of a sampling technique and a sample size that will be used for their calculation. They can be split into two groups, cheap and expensive, based on the computational time required to calculate them.

The idea behind the automated algorithm performance prediction is to link the problem instances landscape data to the performance data achieved by an optimization algorithm. For this purpose, the algorithm is run on a set of benchmark problem instances (i.e., in most cases on already defined benchmark suite such as COCO [4]) to collect the performance data. Next, the ELA features are calculated in order to describe the characteristics of the problem instances. Finally, in order to predict the performance of the algorithm on a new problem instance, a supervised ML method is trained using the landscape data as input data and the performance data as a target data. In recent studies, this is done by learning a single ML model using a set of ELA features that works well across all problem instances [8,9,12]. All these studies used classical feature selection methods to select the ELA features that should improve the performance of the ML model. However, the key element missing here is the explainability of the ML performance, or providing explanations which ELA features contribute to the end performance prediction. This kind of analysis is more than needed to understand which ELA features are the most informative ones and can be used to make a good algorithm performance prediction. These empirical insights will also provide new directions for theoretical research. Even more, it has been shown that using different supervised ML methods with the same landscape and performance data provide different results [9]. Therefore, the selection of the ML algorithm depends on the optimization algorithm whose performance data is used as a target.

In this paper, we present a ML pipeline that can be utilized to understand the expressiveness of the ELA features in automated algorithm performance prediction. The main contribution of the paper is that the ELA feature importance changes when different supervised ML methods are utilized, so their expressiveness on the automated algorithm performance prediction is questionable. It depends from the problem instance being solved, the optimization algorithm run on that problem instance, and the supervised ML methods used to learn a predictive model.

2   Related Work

There are two different types of studies where the ELA features are utilized:

- Studies performed only in the landscape space.
- Studies performed to link the landscape data to the performance data.

In the first type of studies, the ELA features are used to describe the problem instances and then these representations are used to perform complementary analysis between different sets of problem instances [13,25]. With this kind of analysis, similar problem instances can be detected. In addition, sensitivity analysis of the ELA features are preformed concerning different sampling techniques and sample sizes, where it has been shown that the ELA features are really sensitive to the sampling techniques and sample sizes that are used for their calculation [19,21]. Different ELA features portfolios have been also investigated in order to see if the information they convey is enough to classify each instance to the problem to which belongs [3,20]. The common thing of all above-mentioned analyses is that all are done only in the landscape space, and the relations with the performance space have not been explored.

The second type of studies involve automated algorithm performance prediction as a regression task. Here, it has been shown that different supervised ML regression models provide different results when they are utilized for the same learning task [9]. So depending on which optimization algorithm is used in the prediction, different supervised ML model should be selected for learning the model. In addition, it has been shown that personalizing the regression models to the problem instance that is being solved can decrease the predictive error [2]. Furthermore, a recent study provides global (across all benchmark problem instances) and local (for a single problem instance) explanations of which ELA features are most important when a supervised ML algorithm is used to predict an optimization algorithm performance [22]. However, these explanations have not been analyzed when different supervised ML methods are used in the predictive task. This analysis is extremely important to investigate if there is some pattern showing which ELA features are the most informative when automated algorithm performance prediction is investigated no matter which supervised ML model is used.

3   Automated Algorithm Performance Prediction

Previous studies have already shown that models trained to predict the target precision reached by an algorithm or its logarithmic value perform differently [8,9]. There are problem instances for which the model trained to predict the target precision works well, however there are also problem instances for which the model trained to predict the logarithmic value of the target precision works better. To decide which model should be used, several previous studies have analyzed different empirical thresholds for switching between the original and the logarithmic model. To analyze the importance of the ELA features in automated algorithm performance prediction, we investigated different supervised ML regression methods (decision trees (DT), random forest (RF), and deep neural network (DNN)) in Single Target Regression (STR) and Multi Target Regression (MTR) learning scenario. Two STR models will be investigated (i.e., one per the target precision and one per its logarithmic value) together with one MTR model that predicts both values (i.e., the target precision and its logarithmic value) simultaneously. This will be explored using the three supervised ML regression methods.

4 Experimental Setup

Next, the experimental setup is explained in more detail providing information about the landscape and performance data. In addition, the utilized supervised ML models together with their hyper-parameters are presented in more detail.

4.1 Data

Landscape Data. The COCO benchmark platform [4], comprised of 24 single-objective continuous optimization problems, is selected to represent the problem space. For the experiments presented here, the problem dimension is set to $D = 5$ and the first 50 instances per problem are included. Because of this experimental design, there are 1,200 problem instances, 50 per each problem. The R package "flacco" [11] has been used for the calculation of the landscape characteristics (i.e. ELA features) of the problem instances. Out of all the calculated ELA features, 99 have been selected. The calculated ELA features are from the following groups: *cm_angle*, *cm_grad*, *disp*, *ela_conv*, *ela_curv*, *ela_distr*, *ela_level*, *ela_local*, *ela_meta*, *ic*, and *nbc*. The calculation of the selected ELA features has been done using the improved latin hypercube sampling method [24] utilizing $50D$ sample points. This process has been repeated 10 times for each problem instance and each ELA feature is actually the median value of its 10 repetitions. The reason for choosing the median over the mean is that the median value is more statistically robust than the mean value. Regarding the computation cost, the selected ELA features are the cheap ones and they do not have any missing values.

Performance Data. One randomly selected configuration for modular CMA-ES has been investigated as performance data. The selected CMA-ES configuration has the following hyper-parameters: Active update = FALSE, Elitism = TRUE, Orthogonal Sampling = TRUE, Sequential selection = FALSE, Threshold Convergence = TRUE, Step Size Adaptation = tpa, Mirrored Sampling = mirrored, Quasi-Gaussian Sampling = halton, Recombination Weights = default, Restart Strategy = BIPOP. [17] contains additional information about the hyper-parameters of the modular CMA-ES. Only one randomly selected configuration has been presented as a proof of concept about the analysis, but in our GitHub repository [23], there are results for another 14 CMA-ES configurations, which makes this analysis a personalized task. Each CMA-ES configuration has been run 10 times on each problem instance in a fixed budget scenario, where the budget has been set to 50,000 function evaluations. In our analysis, the focus of the performance prediction is on the best reached target precision, thus as a final result, the median across all 10 runs of the best reached precision has been selected. The target variables in our case are the target precision and its logarithmic transformation with a slight modification: before base 10 logarithm is performed, one is added to each original target precision reached. The purpose of this modification is for getting a better interpretation of the performance measure that is used for evaluating the regression models.

Table 1. Hyper-parameter values for tree-based regression models.

Algorithm	Hyper-parameters
Decision Tree	$crit =$ "mae",
	$max\_depth = 10$ (MTR) and $max\_depth = 9$ (STR)
Random Forest	$crit =$ "mae"
	$max\_depth = 7$ $(MTR$ and $STR)$
	$n\_estimators = 20$ (MTR) and $n\_estimators = 10$ (STR)

4.2 Regression Models and Their Hyper-parameters

Here, the three regression models (i.e., Decision Tree (DT), Random Forest (RF), and Deep Neural Network (DNN)) together with their hyper-parameters are explained in more detail. The DT and RF have been selected because recent studies [7,9] showed that they provide one of the most promising results. The selected hyper-parameters for both DT and RF are summarized in Table 1. For both models, "mae" has been selected to measure the quality of the split. The maximum depth of the DT in Multi Target Regression (MTR) scenario has been set to 10, while in Single Target Regression (STR) is set to 9. The max_depth of the DT in STR scenario is set to 9 so that the maximum possible sum of the number of leaf nodes of the two DT used in STR is equal to the maximum possible number of leaf nodes of the DT in the MTR scenario, since the trees are binary. In case of RF, the number of trees in the forest in the MTR scenario has been set to 20, while in STR is set to 10. The maximum depth of the tree is set to 7 in both scenarios. Since in-depth hyper-parameter analysis has not been performed for this study, the number of trees in the MTR RF scenario is actually the sum of trees that appear in the STR RF models.

The tested DNN in STR scenario is presented in Fig. 1a, while the DNN for MTR is presented in Fig. 1b. We have decided for similar architectures to test the prediction results, where finding the best neural architecture design has not been the focus of this study. The focus of this study is to investigate the ELA features importance that are utilized by different supervised ML methods used for the same learning task. The designs of the DNNs used in STR and MTR scenarios are very similar, they have the same layer structure (i.e., seven layers) and use the same activation functions in the corresponding layers. In terms of activation functions, the hidden dense layers all use the ReLU activation function, while the output layer uses the linear activation function.

There are several main differences between the DNNs' design for the two scenarios:

- The dense, concatenate, and dropout layers in the STR scenario all have exactly half of the number of inputs and outputs as the corresponding layers in the MTR scenario. The only exception to this is the first layer after the input layer, because it has the same number of inputs in both scenarios, since the same features are used.

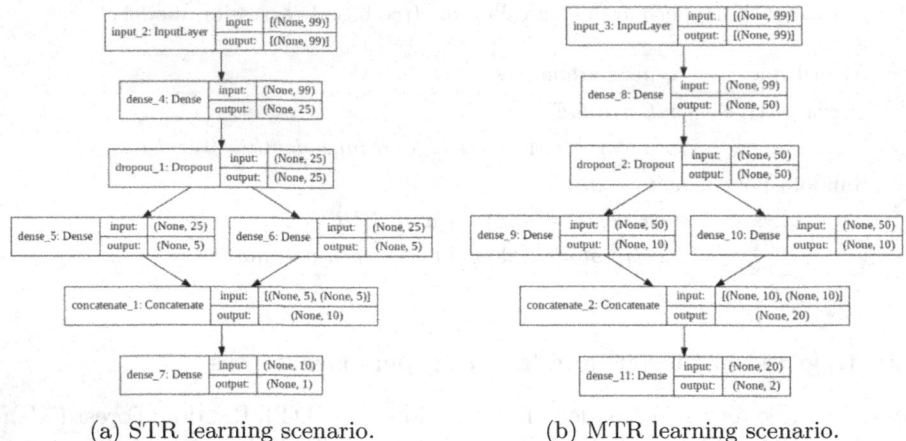

(a) STR learning scenario.   (b) MTR learning scenario.

Fig. 1. DNN architectures.

- The output layer of each of the two DNNs in the STR scenario has one output neuron, while the output layer of the DNN in the MTR scenario has two output neurons.
- The number of parameters is 2,771 for STR and 6,062 for MTR (roughly twice of the number of parameters used in STR).

Both DNNs were trained for 100 epochs each, using the Adam optimizer with *learning_rate* set to 0.001 and *batch_size* set to 10.

4.3 Evaluation

The Single Target Regression (STR) and Multi Target Regression (MTR) models are trained and evaluated using 50-fold cross validation. The data set consisting of 1,200 problem instances has been split into 50 folds such that the first fold contains the first instances of all 24 problems, the second contains the second instances of all 24 problems and so on till the 50th fold which contains the 50th instances of all 24 problems. The learning process was repeated fifty times, each time using one of the folds for testing and the others for training the regression model. This evaluation follows the idea of leave-one problem-instance out, since leaving all instances of one problem out does not provide promising results and does not transfer the knowledge learned by the model. This comes from the fact that the existing benchmark problem suites are not representative enough to generalize a performance of a ML model, which is not the focus of this study. The selected benchmark suite has been used to provide explanations about the expressiveness of the ELA features.

5   Results and Discussion

Tables 2 and 3 present the mean absolute error (MAE) across all 50 folds for each of the 24 benchmark problem separately obtained by DT, RF, and DNN models in STR and MTR learning scenarios. From the results, there is no general conclusion if the MTR models are better than the STR models, so there are benchmark problems for which STR models provide better mean absolute error, and vice versa. In general, the STR models obtained by the DT models provide better MAE across all benchmark problems (i.e., 25.088 and 0.180) than the MTR models (i.e., 27.950 and 0.208) for the target precision and its logarithmic transformation, respectively. In case of the RF models, the MTR model provides better MAE (i.e., 21.381) across all benchmark problems than the STR model learned for the original target precision reached (i.e., 21.559). The opposite is true when the prediction is done for the logarithmic transformation of the original target. The same holds true for the DNN models in the STR and MTR scenarios.

Table 2. Mean absolute error across the 50 folds for each COCO benchmark problem obtained by DT and RF models in STR and MTR scenario.

f	DT				RF			
	Target		log_target		Target		log_target	
	STR	MTR	STR	MTR	STR	MTR	STR	MTR
1	0.074	**0.057**	0.013	**0.012**	1.095	**0.461**	**0.012**	0.013
2	**132.867**	139.609	**0.348**	0.405	106.448	**97.127**	0.346	**0.335**
3	**4.105**	4.198	**0.106**	0.146	2.775	**1.942**	**0.105**	0.117
4	**1.605**	1.824	**0.073**	0.082	**3.236**	3.499	**0.068**	0.101
5	0.000	0.000	0.014	**0.000**	0.577	**0.561**	**0.007**	0.019
6	**6.897**	7.393	**0.034**	0.143	**2.449**	3.699	**0.084**	0.125
7	**105.273**	141.233	**1.041**	1.218	**113.143**	118.557	0.987	**0.964**
8	**0.719**	0.807	**0.074**	0.121	**0.688**	0.839	**0.092**	0.114
9	6.740	**6.671**	**0.069**	0.101	**1.629**	2.972	**0.058**	0.077
10	**137.634**	138.199	0.463	**0.441**	**98.529**	100.714	0.358	**0.340**
11	**39.892**	51.325	**0.301**	0.483	**39.745**	41.054	**0.324**	0.410
12	**139.217**	152.809	**0.343**	0.356	109.999	**106.186**	0.265	**0.256**
13	**4.340**	4.638	0.193	**0.192**	**3.912**	3.976	**0.162**	0.164
14	1.416	**1.269**	**0.050**	0.089	**0.747**	0.968	**0.061**	0.146
15	1.436	**1.086**	0.090	**0.082**	5.527	**2.260**	**0.084**	0.111
16	**1.254**	1.374	0.133	**0.126**	2.571	**1.916**	**0.130**	0.216
17	**3.446**	3.545	**0.174**	0.213	**4.418**	4.960	**0.126**	0.265
18	11.109	**10.196**	0.280	**0.209**	**13.109**	13.503	**0.237**	0.351
19	0.195	**0.055**	0.022	**0.019**	**1.170**	1.627	**0.050**	0.207
20	0.289	**0.231**	**0.036**	0.037	1.091	**0.388**	**0.039**	0.077
21	**0.593**	0.709	**0.152**	0.167	**0.652**	1.684	**0.135**	0.143
22	**0.658**	0.776	**0.166**	0.178	1.423	**0.911**	0.153	**0.150**
23	0.549	**0.459**	0.047	**0.036**	0.418	**0.411**	**0.033**	0.042
24	**1.816**	2.349	**0.102**	0.148	**2.065**	2.929	**0.105**	0.122
Mean	**25.089**	27.951	**0.180**	0.209	21.559	**21.381**	**0.168**	0.203

Table 3. Mean absolute error across the 50 folds for each COCO benchmark problem obtained by DNN models in STR and MTR scenarios.

f	Target		log_target	
	STR	MTR	STR	MTR
1	0.559	**0.292**	**0.048**	0.078
2	**92.219**	93.970	**0.327**	0.332
3	2.228	**1.756**	**0.126**	0.158
4	**2.399**	3.055	**0.087**	0.140
5	0.175	**0.123**	**0.026**	0.051
6	1.199	**0.435**	0.094	**0.092**
7	90.217	**89.238**	0.911	**0.904**
8	1.126	**0.665**	0.116	**0.100**
9	0.644	**0.538**	**0.072**	0.092
10	**88.878**	92.989	**0.307**	0.393
11	5.737	**2.176**	0.231	**0.181**
12	**105.375**	106.930	**0.417**	0.417
13	6.862	**5.478**	**0.228**	0.335
14	**1.938**	2.193	**0.169**	0.451
15	2.468	**1.594**	0.135	**0.130**
16	2.361	**2.009**	**0.156**	0.202
17	**3.100**	3.492	0.176	**0.175**
18	14.605	**14.400**	**0.293**	0.364
19	0.614	**0.309**	**0.037**	0.077
20	0.953	**0.640**	**0.067**	0.119
21	0.609	**0.578**	**0.135**	0.143
22	**0.632**	0.736	**0.148**	0.163
23	**0.493**	0.495	**0.060**	0.085
24	2.339	**2.289**	**0.138**	0.172
Mean	17.822	**17.766**	**0.186**	0.223

Evaluating across the models, the DNN MTR model provides the best MAE (i.e., 17.765) across all benchmark problems for the original target precision, while the RF STR model provides the best MAE for predicting the logarithmic transformation of the target precision reached (i.e., 0.167). The obtained results show us that there is no practical difference in investigating the performance prediction (i.e., original and its logarithmic transformation) in STR and MTR scenario. The only benefit could be the time required to train the models, instead of training two STR models, one MTR model can provide very similar results.

Comparing the models on a single-problem level trained for prediction the target precision, it is obvious that all models in STR (DT, RF, and DNN) obtain large errors on the 2nd, 7th, 10th, 12th, and 18th problem. However for the 11th problem, both DT and RF provide large errors, while the DNN provides an error which is much more smaller than the other two models.

No matter which ML model has been utilized in the STR or MTR learning scenario, the results are similar. The focus of this study is actually to go into more explanations by providing which ELA features are used by different models to provide their predictions. For this purpose, the SHAP method is used to provide explanations on global and local level [15]. The SHAP explanation method computes Shapley values from coalitional game theory with the goal of explaining the prediction of a single instance. It is a method to explain individual predictions. It does this explanation by computing the contribution of each feature to the prediction. The feature values of a data instance play the role of players in a coalition. Shapley values tell us how to fairly distribute the "payout" (i.e., the prediction) among the features. Two levels of explanations are provided by the SHAP method:

- Global explanations - the SHAP values for each problem instance are used to study the impact to the target variable (the impact across all problem instances involved in the learning process). This is done by combining the SHAP values for each problem instance.
- Local explanations - allow us to study the similarities/differences in the importance of the features for different problem instances (we can see how the feature importance changes across different problem instances). This is done through the use of the SHAP values for each problem instance. Each problem instance gets its own set of SHAP values.

In our experiments, the explanations are provided for the models trained on the first fold and a clustering analysis of the models is presented across all folds based on the Shapley values of the ELA features.

Global Explanations. Figure 2 presents the SHAP value impact of the STR and MTR DT models, both for the original target precision and its logarithmic transformation. The plots presented in this figure present the positive and negative relationships with the original target precision or its logarithmic transformation. The dots presented in the plots correspond to all instances from the training data set (i.e., in our case the first fold). The descending order of the ELA features presents their importance starting from the most important one. The colors used are related to the magnitude of the ELA feature value, where higher values are red and lower value are blue. The impact of the ELA feature value to the target variable prediction is its horizontal location.

Using the figure and focusing on the prediction of the original target precision reached by the DT models in the STR and MTR scenario, the models differ in one from 10 most important ELA features. The STR model uses the "ela_local.fun_evals", while the MTR model uses "ela_distr.number_of_peaks". The patterns of the other nine shared features are almost the same in the STR and the MTR scenario. For example, if we have a high value of the "ela_curv.grad_norm.min", it contributes by adding a large value to the original target precision reached. The feature "ela_curv.grad_norm.min", which is most important feature in most of the scenarios, represents the aggregation of minimum values of

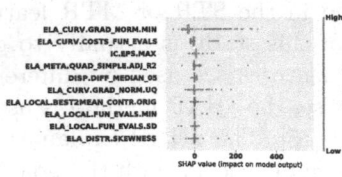

(a) Target precision STR DT.

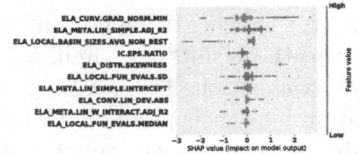

(b) Logarithmic transformation of the target precision STR DT.

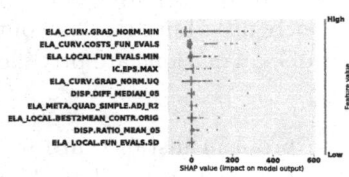

(c) Target precision MTR DT.

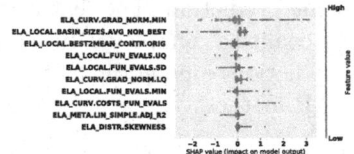

(d) Logarithmic transformation of the target precision MTR DT.

Fig. 2. SHAP value impact of the STR and MTR DT models.

the gradients length in all the runs when searching for the optimal solution. This means that high value "*ela_curv.grad_norm.min*" is an indication of difficulty in solving the benchmark problem because it takes us away from the optimum reached with adding a large value on the target precision (i.e., error). In addition, looking at the "*ela_curv.grad_norm.uq*", which represents aggregation of the upper quartile of the gradients lengths, lower values do not take us away from the target precision reached, while higher values of this ELA feature can decrease or increase the reached target precision, showing that the benchmark problem is difficult to be solved there. In the future, such kind of analysis can allow us to estimate and rank the problem difficulty concerning the values of the ELA features and their Shapley values. Looking at the most important features it seems that they all come from the classical ELA features group, except one that comes from the information content group. Focusing on the logarithmic transformation of the original target precision reached used as a target variable in the STR and MTR scenario by DT models, it seems that most of the features are overlapping. These models also utilize one feature from the nearest-better clustering group. The impact ELA patterns of both models are very similar for each ELA feature separately.

Comparing the STR DT models for both performance targets (i.e., the original target precision and its logarithmic transformation), it is obvious that the ELA features importance is different and there is overlapping in a small number of features.

In the case when RF is used to learn the predictive model, the STR and MTR scenario differ in the ELA features they utilize, for both the original target precision reached and its logarithmic transformation. They differ in five out of the 10 most important ELA features. Similar to the DT models, the most

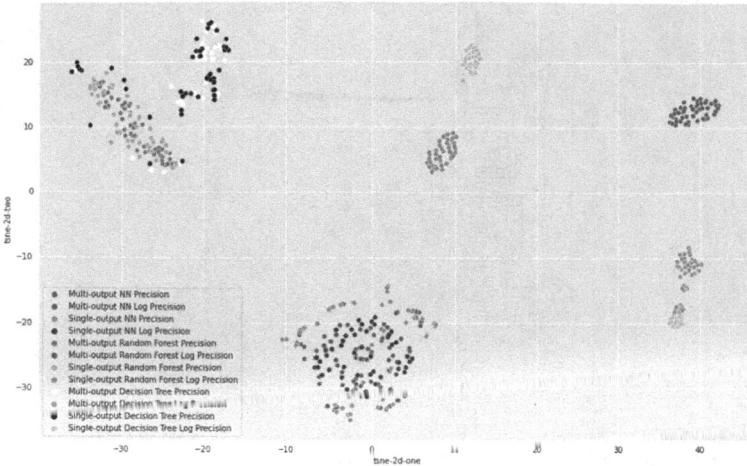

Fig. 3. t-SNE visualization of the STR and MTR models for original target precision and its logarithmic transformation trained per each fold. The models are represented as vectors of 99 Shapley values (i.e., one per each ELA feature).

utilized features are the classical ELA features. The "$ela\_curv.grad\_norm.min$" is the most important ELA feature for both DT and RF models in STR and MTR learning scenario. Comparing the DNN STR and MTR models concerning the target precision, the importance of the ELA features is similar. They are overlapping in nine out of the 10 most important features. In case of the logarithmic transformation prediction, they are overlapping in five out of the 10 most important features. Compared to the RF and DT models, the DNN models utilized also features from the information content and nearest neighbours groups. Figures about the RF and DNN scenarios are available at out GitHub repository.

Figure 2 presents the explanations obtained from the training data from the first fold. To see if these results are consistent across different folds (i.e., if the features importance is consistent within a model across the folds) in case when DT, RF, and DNN are used, we have represented each ML model in each learning scenario as a vector of 99 Shapley values. For this purpose, we averaged the Shapley value for each ELA feature across all problem instances that belong to each training data fold. The vector with the averaged Shapley values is the ML model representation.

Figure 3 presents t-SNE visualization of each ML model in two-dimensional space using their Shapley representation. We have used the default parameters from the t-SNE visualization available from the python package *scikit-learn* [18]. Looking at the figure, we can assume that the models trained by the same ML algorithm across different folds are consistent since their Shapley representations place them close together. This result indicates that no matter which fold is used, the ELA features importance utilized by the models is similar. The only exceptions are the STR DT models trained for predicting the logarithmic

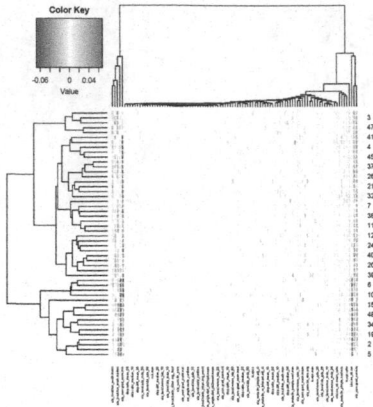

Fig. 4. A heatmap where the DT models trained on the 50 folds for predicting the logarithmic transformation are reordered concerning how they utilized the ELA features.

transformation of the reached target. These results indicate that the same DT model trained on different folds utilized different ELA features (i.e., their importance changed). To check this, Fig. 4 presents a heatmap with dendrogram, where the DT models trained on the 50 folds for predicting the logarithmic transformation are reordered concerning their Shapley values representation (i.e., 99 Shapley values, one per each feature). Looking at the dendrogram presented in the rows, it is obvious that these models are split into two clusters according to the information of the ELA feature importance, which is actually the result presented using the t-SNE visualization. This happens because different folds consist of different instances from the same problems that are obtained with different random transformations. These results indicate that the DT model is less robust to variations that exist between the ELA values across instances from the same problem when predicting the logarithmic transformation.

To compare which ELA features are utilized across different ML models (i.e., DT, RF, and DNN), Fig. 5 presents the intersections between the top 10 most important ELA features utilized by the DT, RF, and DNN models in STR and MTR learning scenario separately. For this purpose, the vectors for each ML model in each scenario used in the t-SNE visualization have been averaged within the scenario (e.g., STR and target precision, etc.). Based on the averaged Shapley values, the top 10 ELA features for each model and each scenario have been selected and used for the intersection analysis. From the figure, it follows that the tree-based models (DT and RF) share more similar ELA features. When we are comparing them to DNN models, the DNN models are most similar to the RF models and only few ELA features are overlapping with the DT models. The "*ela_curv.grad_norm.min*" is the ELA feature that belongs to the 10 most important features for every ML algorithm from our portfolio no matter whether it is the STR or MTR scenario. In the future, the union of these features could be reused as a general feature selection to train a ML model.

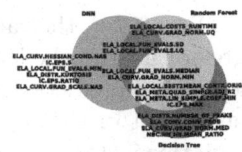

(a) STR target precision.

(b) STR logarithmic transformation.

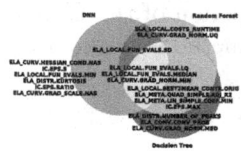

(c) MTR target precision.

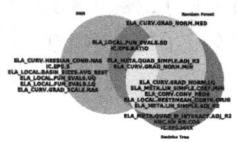

(d) MTR logarithmic transformation.

Fig. 5. Intersection between the top 10 most important ELA features utilized by the DT, RF, and DNN models in STR and MTR learning scenario.

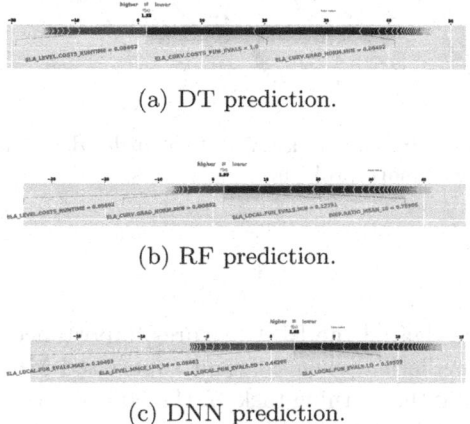

(a) DT prediction.

(b) RF prediction.

(c) DNN prediction.

Fig. 6. SHAP value impact of the STR models to explain the predicted target precision reached on the first instance of the 11th benchmark problem.

Local Explanations. After presenting the global impact of the ELA features concerning different supervised ML regression models, here the focus is on local explainability. To provide local explanations, the first instance of the 11th benchmark problem from the first fold was selected. This was done with the purpose to see what makes the prediction of this problem difficult using DT or RF (i.e., which ELA feature influence here) and why the DNN model has lower error compared to the other two models.

Figure 6 presents the SHAP impact of the 10 most important ELA features that contribute to the original target prediction for the selected instance by the DT, RF, and DNN models in STR scenario. Form the figure, it is obvious that using the most important features in the cases of the DT and RF, the range of the prediction is from -15 to 45. However, the DNN regressor utilizes the features by making the prediction in the range from -5 to 13, so smaller prediction errors are achieved. To go in more detail, Fig. 7 presents the impact of the top most important 10 ELA features utilized by the DNN model on this instance, together with their importance when the DT or RF models are used. From the figure, it follows that the importance of these 10 features is not the same when the DT and RF are used. These results indicate that the selection of the ELA features portfolio is also dependent on which supervised ML method will be used.

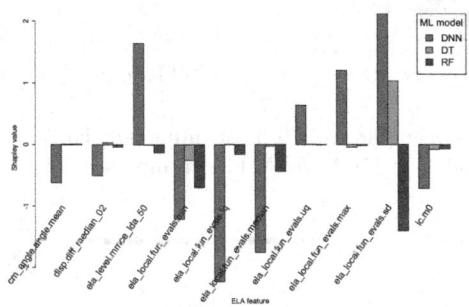

Fig. 7. The top 10 most important ELA feature utilized by the DNN on the first instance of the 11th benchmark problem.

6   Conclusion

In this study, we investigated the ELA features importance in automated algorithm performance prediction when different supervised ML regression methods are utilized. Evaluating the learning task on the 50 instances from the 24 COCO benchmark problems and one CMA-ES configuration by using decision tree, random forest, and deep neural network, it follows that in this learning scenario the most important ELA features are coming from the classical *ela* group. The calculated Shapley values were used to estimate the contribution of each ELA feature to the performance prediction. The experimental results showed that a different set of ELA features are important for different problem instances depending on which supervised ML method is utilized. This indicates that selection of ELA features is very dependent on the ML method that is utilized for learning the predictive model. So, depending on which supervised ML method is used, the impact of the ELA features to the model prediction performance changes.

For future work, we are planning to analyse different families of optimization algorithms to investigate which ELA features are related to their exploration and exploitation capabilities, and further recommend the most appropriate ML method that can be used to perform automated algorithm performance prediction.

Acknowledgments. This work was supported by projects from the Slovenian Research Agency: research core funding No. P2-0098 and projects No. Z2-1867 and N2-0239.

References

1. Blot, A., Marmion, M., Jourdan, L., Hoos, H.H.: Automatic configuration of multi-objective local search algorithms for permutation problems. Evol. Comput. **27**(1), 147–171 (2019). https://doi.org/10.1162/evco_a_00240
2. Eftimov, T., Jankovic, A., Popovski, G., Doerr, C., Korošec, P.: Personalizing performance regression models to black-box optimization problems. arXiv preprint arXiv:2104.10999 (2021)
3. Eftimov, T., Popovski, G., Renau, Q., Korošec, P., Doerr, C.: Linear matrix factorization embeddings for single-objective optimization landscapes. In 2020 IEEE Symposium Series on Computational Intelligence (SSCI), pp. 775–782. IEEE (2020)
4. Hansen, N., Auger, A., Ros, R., Mersmann, O., Tušar, T., Brockhoff, D.: COCO: a platform for comparing continuous optimizers in a black-box setting. Optim. Methods Softw. **36**, 1–31 (2020)
5. Hutter, F., Kotthoff, L., Vanschoren, J. (eds.): Automated Machine Learning. TSSCML, Springer, Cham (2019). https://doi.org/10.1007/978-3-030-05318-5
6. Jankovic, A., Doerr, C.: Landscape-aware fixed-budget performance regression and algorithm selection for modular CMA-ES variants. In: Proceedings of Genetic and Evolutionary Computation Conference (GECCO 2020), pp. 841–849. ACM (2020). https://doi.org/10.1145/3377930.3390183
7. Jankovic, A., Doerr, C.: Landscape-aware fixed-budget performance regression and algorithm selection for modular CMA-ES variants. In: Proceedings of the 2020 Genetic and Evolutionary Computation Conference, pp. 841–849 (2020)
8. Jankovic, A., Eftimov, T., Doerr, C.: Towards feature-based performance regression using trajectory data. In: Castillo, P.A., Jiménez Laredo, J.L. (eds.) EvoApplications 2021. LNCS, vol. 12694, pp. 601–617. Springer, Cham (2021). https://doi.org/10.1007/978-3-030-72699-7_38
9. Jankovic, A., Popovski, G., Eftimov, T., Doerr, C.: The impact of hyper-parameter tuning for landscape-aware performance regression and algorithm selection. arXiv preprint arXiv:2104.09272 (2021)
10. Kerschke, P., Trautmann, H.: Automated algorithm selection on continuous black-box problems by combining exploratory landscape analysis and machine learning. Evol. Comput. **27**(1), 99–127 (2019). https://doi.org/10.1162/evco_a_00236
11. Kerschke, P., Dagefoerde, J., Kerschke, M.P.: Package 'flacco' (2017)
12. Kerschke, P., Hoos, H.H., Neumann, F., Trautmann, H.: Automated algorithm selection: survey and perspectives. Evol. Comput. **27**(1), 3–45 (2019)
13. Lang, R.D., Engelbrecht, A.P.: An exploratory landscape analysis-based benchmark suite. Algorithms **14**(3), 78 (2021)
14. Liefooghe, A., Daolio, F., Vérel, S., Derbel, B., Aguirre, H.E., Tanaka, K.: Landscape-aware performance prediction for evolutionary multiobjective optimization. IEEE Trans. Evol. Comput. **24**(6), 1063–1077 (2020). https://doi.org/10.1109/TEVC.2019.2940828

15. Lundberg, S.M., Lee, S.I.: A unified approach to interpreting model predictions. In: Guyon, I., et al. (eds.) Advances in Neural Information Processing Systems 30, pp. 4765–4774. Curran Associates, Inc. (2017). http://papers.nips.cc/paper/7062-a-unified-approach-to-interpreting-model-predictions.pdf
16. Mersmann, O., Bischl, B., Trautmann, H., Preuss, M., Weihs, C., Rudolph, G.: Exploratory landscape analysis. In: Proceedings of the 13th Annual Conference on Genetic and Evolutionary Computation, pp. 829–836 (2011)
17. de Nobel, J., Vermetten, D., Wang, H., Doerr, C., Bäck, T.: Tuning as a means of assessing the benefits of new ideas in interplay with existing algorithmic modules. CoRR abs/2102.12905 (2021). https://arxiv.org/abs/2102.12905
18. Pedregosa, F., et al.: Scikit-learn: machine learning in Python. J. Mach. Learn. Res. 12, 2825–2830 (2011)
19. Renau, Q., Doerr, C., Dreo, J., Doerr, B.: Exploratory landscape analysis is strongly sensitive to the sampling strategy. In: Bäck, T., et al. (eds.) PPSN 2020. LNCS, vol. 12270, pp. 139–153. Springer, Cham (2020). https://doi.org/10.1007/978-3-030-58115-2_10
20. Renau, Q., Dreo, J., Doerr, C., Doerr, B.: Towards explainable exploratory landscape analysis: extreme feature selection for classifying BBOB functions. In: Castillo, P.A., Jiménez Laredo, J.L. (eds.) EvoApplications 2021. LNCS, vol. 12694, pp. 17–33. Springer, Cham (2021). https://doi.org/10.1007/978-3-030-72699-7_2
21. Škvorc, U., Eftimov, T., Korošec, P.: The effect of sampling methods on the invariance to function transformations when using exploratory landscape analysis. In: 2021 IEEE Congress on Evolutionary Computation (CEC), pp. 1139–1146. IEEE (2021)
22. Trajanov, R., Dimeski, S., Popovski, M., Korošec, P., Eftimov, T.: Explainable landscape-aware optimization performance prediction. arXiv preprint arXiv:2110.11633 (2021)
23. Trajanov, R., Dimeski, S., Popovski, M., Korošec, P., Eftimov, T.: GitHub repository containing all source code and data of the study presented in this paper (2021). https://github.com/risto-trajanov/explainable-landscape-aware-performance-regression
24. Xu, Q., Yang, Y., Liu, Y., Wang, X.: An improved Latin hypercube sampling method to enhance numerical stability considering the correlation of input variables. IEEE Access 5, 15197–15205 (2017)
25. Škvorc, U., Eftimov, T., Korošec, P.: Understanding the problem space in single-objective numerical optimization using exploratory landscape analysis. Appl. Soft Comput. 90, 106138 (2020). https://doi.org/10.1016/j.asoc.2020.106138. https://www.sciencedirect.com/science/article/pii/S1568494620300788

Search Trajectories Networks of Multiobjective Evolutionary Algorithms

Yuri Lavinas[1](✉)⬤, Claus Aranha[1], and Gabriela Ochoa[2]⬤

[1] University of Tsukuba, Tsukuba, Japan
lavinas.yuri.xp@alumni.tsukuba.ac.jp, caranha@cs.tsukuba.ac.jp
[2] University of Stirling, Stirling, UK
gabriela.ochoa@cs.stir.ac.uk

Abstract. Understanding the search dynamics of multiobjective evolutionary algorithms (MOEAs) is still an open problem. This paper extends a recent network-based tool, search trajectory networks (STNs), to model the behavior of MOEAs. Our approach uses the idea of decomposition, where a multiobjective problem is transformed into several single-objective problems. We show that STNs can be used to model and distinguish the search behavior of two popular multiobjective algorithms, MOEA/D and NSGA-II, using 10 continuous benchmark problems with 2 and 3 objectives. Our findings suggest that we can improve our understanding of MOEAs using STNs for algorithm analysis.

Keywords: Algorithm analysis · Search trajectories · Continuous optimization · Visualization · Multi-objective optimization

1 Introduction

Most real-world optimization problems involve multiple conflicting objectives. This has prompted the development of a variety of multiobjective evolutionary algorithms (MOEAs), which can be classified into three broad categories, based on dominance [5], indicators [1] and decomposition [21]. There has been significant progress in improving MOEAs in all categories, and these algorithms are widely used in practice. However, algorithm development and improvement are mostly guided by intuition and empirical performance comparisons. We argue that there is a lack of accessible tools to analyze, contrast and visualize the dynamic behavior of MOEAs.

The main goal of this article is to generalize to multiobjective optimization a recent graph-based modeling tool, search trajectory networks (STNs), which was originally proposed for single objective optimization [14, 15]. Our approach uses decomposition, a key strategy in multiobjective optimization, where the multiobjective problem is transformed into several single-objective problems. This is convenient as it allows us to use the existing tools and features proposed for single objective STNs [14, 15].

© Springer Nature Switzerland AG 2022
J. L. Jiménez Laredo et al. (Eds.): EvoApplications 2022, LNCS 13224, pp. 223–238, 2022.
https://doi.org/10.1007/978-3-031-02462-7_15

This study is exploratory and interpretative in nature, taking the form of a case study using 10 continuous benchmark problems with 2 and 3 objectives and two broadly known MOEAs, MOEA/D [21] and NSGA-II [5]. To the best of our knowledge, this is the first effort to apply STNs for modeling the search behavior of MOEAs.

The paper is organized as follows. Section 2 overviews previous work related to visualization in multiobjective optimization. Section 3 introduces relevant concepts. Our proposal to extend STNs to model MOEAs is described in Sect. 4. The experimental setup and results are presented in Sects. 5 and 6, respectively. Finally, Sect. 7 outlines our main findings and suggestions for future work.

2   Related Work

Most visualization approaches in the literature for multiobjective optimization focus entirely on the objective space, completely ignoring the decision space. The classical visualization shows the true or approximated Pareto front for 2 or 3 objectives in a standard scatter plot. Extensions to this idea for visualizing problems with more than three objectives, using dimensionality reduction techniques, have been proposed [19]. However, by purely focusing on the objective space, the interaction effects from the decision variables are ignored; therefore, almost no information on the structural properties of the problem landscapes can be derived.

Very few visualization techniques in the literature provide a joint view of the decision and objective spaces in continuous multiobjective optimization. Fonseca et al. [7] proposed the cost landscapes, which use dominance ranking to evaluate points in the decision space with respect to global optimal trade-offs. This approach, however, does not capture local optimal sets. Kerschke and Grimme [8] proposed the gradient field maps to explicitly address local optimal sets, with further extensions plotting landscapes with optimal trade-offs [16], and providing an accessible dashboard for visualization [17]. In combinatorial optimization, recent work has adapted the local optima networks model [13] to multiobjective optimization, providing visual insights into the distribution and connectivity pattern of Pareto local optimal solutions [11] and dominance-based hill-climbing [6]. These are insightful visual approaches; however, they concentrate on the structural configuration of local and global optima in the optimization landscapes, rather than on the dynamic (trajectory) behavior of the search process.

3   Preliminaries

3.1   Search Trajectory Networks

The original STN model definitions for single objective optimization can be found in [15]; we rephrase them here for completeness and to guide our proposed extension to multiobjective optimization. To define a network model, we need to specify their nodes and edges. The relevant definitions are given below.

Representative solution. Is a solution to the problem at a given iteration that represents the status of the search process. For population-based algorithms, the solution with the best fitness in the population at a given generation is chosen as the representative solution.

Location. Is a non-empty subset of solutions that results from a mapping process. Each solution in the search space is mapped to one location. Several similar solutions are generally mapped to the same location, as the locations represent a partition of the search space.

Search trajectory. Given a sequence of representative solutions in the order in which they are encountered during the search process, a search trajectory is defined as a sequence of locations formed by replacing each solution with its corresponding location.

Node. Is a location in a search trajectory of the search process being modeled. The set of nodes is denoted by N.

Edges. Edges are directed and connect two consecutive locations in the search trajectory. Edges are weighted with the number of times a transition between two given nodes occurred during the process of sampling and constructing the STN. The set of edges is denoted by E.

Search trajectory network (STN). Is a directed graph $STN = (N, E)$, with node set N, and edge set E as defined above.

The data to construct an STN model is gathered while the studied algorithm is running. Specifically, the required output from a run is a list of steps connecting two adjacent representative solutions in the search process. Each search step is stored as an entry in a log file containing the two consecutive representative solutions being linked with the step; these transitions become the edges of the network model. Once the data logs of a predefined number of runs of a given algorithm-problem pair are gathered, a post-processing maps solutions to locations, aggregates all the locations and transitions, and constructs a network object.

3.2 Multiobjective Optimisation Problems

In Multiobjective Optimization Problems (MOPs), a solution to the problem is evaluated by multiple objective functions, which are possibly conflicting. One key characteristic of MOPs is that a single solution rarely is the optimal solution for all objectives. This leads to the concept of "Pareto dominance": A solution x_j is said to be dominated by x_i if x_i is better than x_j in one objective, and at least equivalent in all others. Otherwise, x_j is non-dominated by x_i. Note that two solutions can be mutually non-dominated (each solution is better in a different objective).

As a consequence of Pareto dominance, the search status of an MOEA cannot be represented by a single representative solution. Therefore, the definitions of location, node, edge, and search trajectory must be extended in a multiobjective context. Our proposed extension of STNs to multiobjective optimization relies on a few operations that are used in dominance-based MOEAs such as NSGA-II [5] and decomposition-based MOEAs such as MOEA/D [21]. Brief descriptions of these operations are given below.

Non-dominated sorting. Sorts a set X of solutions based on their dominance relationship. Initially, we find the subset of solutions, X_{ND}, in X that are non-dominated in relation to all other solutions and give the solutions in X_{ND} rank 0. Then we repeat this procedure on the set $X - X_{ND}$, giving the new non-dominated solution set the rank 1. The procedure keeps repeating until we have ranked all solutions in X.

Decomposition. Breaks down a MOP into a set of single objective subproblems, where each subproblem is a combination of the objectives, characterized by a weight vector. There are several methods to generate a set of weight vectors (decomposition methods). In this work, we use the Uniform Design [20], as it allows us to choose the number of weight vectors to be generated explicitly.

Scalar aggregation function. Takes the objective values of one solution and the weight vector of one subproblem and calculates a scalar value that represents the quality of that solution for that particular subproblem. In a decomposition-based MOEA, these quality values are used to associate one representative solution to each subproblem. There are several scalar aggregation functions (also called scalarization functions), and in this work, we use the Weighted Tchebycheff [4,12], which is less affected by the shape of the Pareto front in a given MOP.

4   STN Extension for the Multiobjective Domain

As discussed in the previous section, STNs are calculated based on the sequence of representative solutions in an algorithm execution. However, each iteration in an MOEA does not contain a single representative solution. In this section, we describe a method to calculate STNs for multiobjective optimization.

The key idea of this method is that we keep track of a small number of decomposition vectors, match a representative solution to a vector, and then merge the vector trajectories of each vector into a single multiobjective STN. The following processes describes the steps to associate a solution to each of the vectors:

1. Initially, we choose the number of decomposition vectors n and generate these vectors using the decomposition technique, using the technique discussed in the previous section.
2. Every iteration, we apply the non-dominated sorting procedure on the solution set and select the set of rank 0 solutions as the set of candidate solutions. If the number of solutions in this set is less than n, we add solutions from rank 1, 2, and so on until we have n or more candidate solutions.
3. Every iteration, we select a representative solution for each vector from the set of candidate solutions, using the scalar aggregation function. In the case of ties, we choose the newest solution, discarding solutions already associated with that vector.

After associating one representative solution for each weight vector, we map the locations of each representative solution as the nodes. To create this mapping, we use a precision parameter to portion the space into hypercubes with length 10^{-03}. In other words, a node in the STN is equivalent to a $0.001 \times 0.001 \times 0.001$ hypercube. Edges are given by the sequence of locations of consecutive iteration. Each decomposition vector is modeled as an STN. Finally, we merge these STNs, establishing the merged STN, that models the search progress of an algorithm in a problem. We merge the STNs as follows:

1. The merged STN model merges the n STNs of each decomposition vector and is obtained by the graph union of the n individual graphs.
2. The merged graph contains the nodes and edges present in at least one of the vectors graphs. Attributes are kept for the nodes and edges, indicating whether they were visited by both algorithms (shared) or by one of them only.

5   Experiments

We conduct the following experimental and exploratory study to investigate whether STNs can discriminate between MOEA/D-DE and NSGA-II. This analysis is conducted by characterizing and visualizing the search behavior of these MOEAs.

In this work, we use the UF benchmark set [23], with 2 and 3 objectives. Both algorithms were widely studied, and collectively these works have shown that MOEA/D performs better than NSGA-II in this group of problems. Thus the UF benchmark set helps show that the STN can discriminate these metaheuristics. For all functions in all benchmark sets, we set the number of dimensions to $D = 10$. The implementation of the test problems available from the *smoof* package [2] was used in all experiments. The UF Benchmark set comprises ten unconstrained test problems with Pareto sets designed to be challenging to existing algorithms [9]. Problems UF1-UF7 are two-objective MOPs, while UF8-UF10 are three-objective problems [23].

5.1   Experimental Parameters

We used the MOEA/D variant using the Differential Evolution (MOEA/D-DE) parameters as they were introduced in the work of Li and Zhang [10] in all tests. Details of these parameters can be found in the documentation of package MOEADr and the original MOEA/D-DE reference [3,4,22]. We used the NSGA-II parameters as they were introduced in the work of Deb et al. [5] in all tests, except for the use of the DE mutation operator, with the same parameters as MOEA/D. Details of these parameters can be found in the documentation of package nsga2R and the original reference [18]. Table 1 summarizes the experimental parameters for both algorithms.

Table 1. Experimental parameter settings.

MOEA/D Parameters	Value
DE mutation	$F = 0.25$
Polynomial mutation	$\eta_m = 20$ $prob = 0.01$
Restricted Update	$nr = 2$ $\delta_p = 0.9$
Neighborhood size	$T = 20\%$ of the pop. size
SLD weight vectors	250

NSGA-II Parameters	Value
Tournament Size	2
DE mutation	$F = 0.25$
Polynomial mutation	$\eta_m = 3$ $prob = 0.1$

Population size	Value
MOEA/D and NSGA-II	250

Experiment Parameters	Value
Repeated runs	3
Computational budget	30000 evals.

For creating the STNs, we use $n = 5$ different vectors, and for each vector, there is one solution at a given iteration. Our idea here was to use a small number of vectors that could capture visually valuable information about the pair algorithm-problem. More work is required to understand the effect of different values for the number of vectors.

5.2   Metrics

MOP Metrics: We use the following criteria to compare the results of the different strategies: (a) final approximation hypervolume (HV) and Inverted Generational Distance (IGD). For calculating the hypervolume, we set the reference point to (1.1, 1.1) for two objective problems and (1.1, 1.1, 1.1) for three objective problems.

STN Metrics: We use seven STN metrics to assess the global structure of the trajectories and bring insight into the behavior of the MOEAs modeled. These metrics are (1) the number of unique nodes and (2) the number of unique edges, and we also calculate (3) the number of shared nodes between vectors, (4) the number of Pareto optimal solutions, (5) the number of components of the network and (6) the mean and (7) maximum values of the in- and out-degree of nodes. These metrics are summarised in Table 2. It is worth noting that additional metrics could also be considered, such as the distances between optimal solutions and centrality metrics for the optimal solutions.

Table 2. Description of STN metrics

Metric	Justification
Number of nodes	Shows unique locations visited.
Number of edges	Shows unique search transitions between nodes.
Number of shared nodes	Shows nodes visited by more than one components.
Number of optimal solutions	Shows what nodes are in the theoretical PF.
Number of Components	Show the distribution of the search progress.
Mean in/out-degree	Shows the mean degree of branching and loops.
Maximum in/out-degree	Shows the maximum degree of branching and loops.

5.3 Reproducibility

For reproducibility purposes, all the code and experimental scripts are available online at https://github.com/yurilavinas/STNMOP/tree/evostar.

6 Results

This section compares the metaheuristic algorithms MOEA/D and NSGA-II in use of the STN model. For that, we compare the (a) STNs visualizations and (b) the performance in terms of MOP metrics, both HV and IGD, (c) and the STN metrics: (1) the number of nodes and (2) edges, (3) number of shared nodes, (4) the number of Pareto optimal solutions, (5) the number of components of the network, (6) the mean and (7) maximum in-degree and (8) the mean and (9) maximum out-degree.

The visualisations in Figs. 1, 2, 3 illustrate the STN by three independent runs of both candidate algorithms on the UF3, UF5 and UF8 benchmark MOPs. Figure 4 shows the number of nodes for each vector traversed by MOEA/D and NSGA-II on these same three MOPs. The first two MOPs are problems with two objectives, while UF8 has three objectives. Table 3 displays the traditional HV (higher is better) and IGD (lower is better) metrics and the STN metrics.

Considering the colors used in the STN visualizations, we use yellow squares to indicate the start of trajectories and black triangles to indicate the end of trajectories. The red color shows the best Pareto optimal solutions, and light grey circles show shared locations visited by more than one vector of the same algorithm. Finally, each vector has its color: light orange for V1; green for V2; purple for V3; light pink for V4; and light blue for V5.

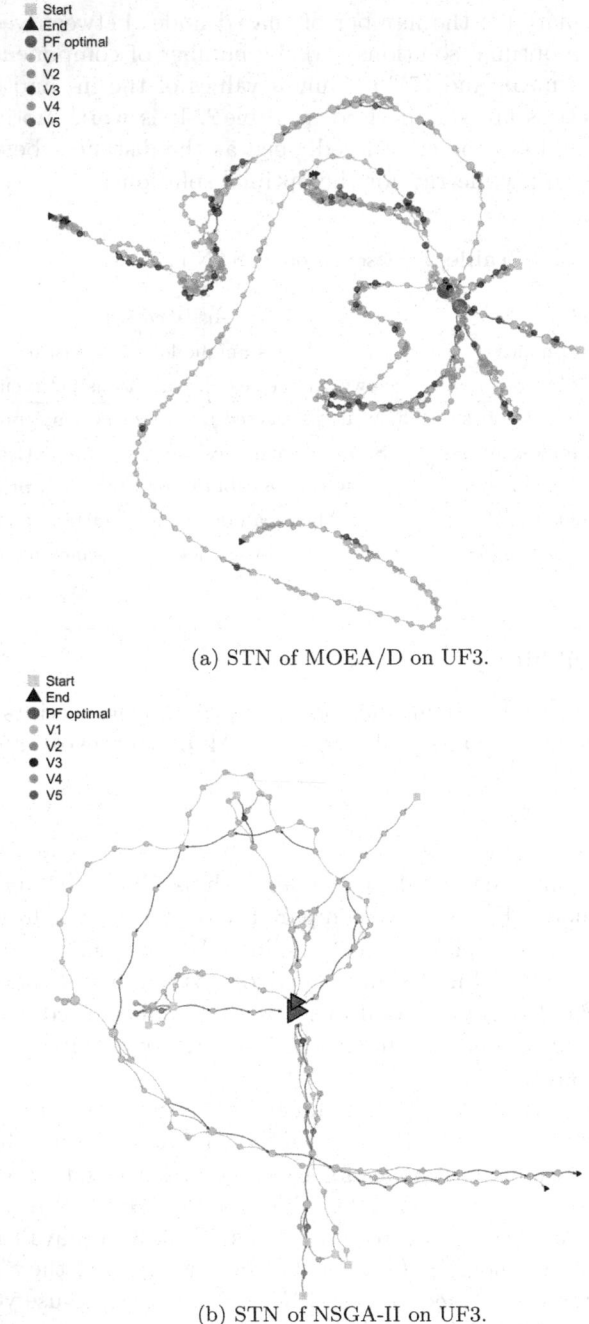

(a) STN of MOEA/D on UF3.

(b) STN of NSGA-II on UF3.

Fig. 1. We can see that MOEA/D (top) is able to find many optimal solutions while NSGA-II (bottom) finds optimal solutions at the end of their trajectory (red triangles). This reflects NSGA-II characteristic of using the non-dominance relationship as update criterion. (Color figure online)

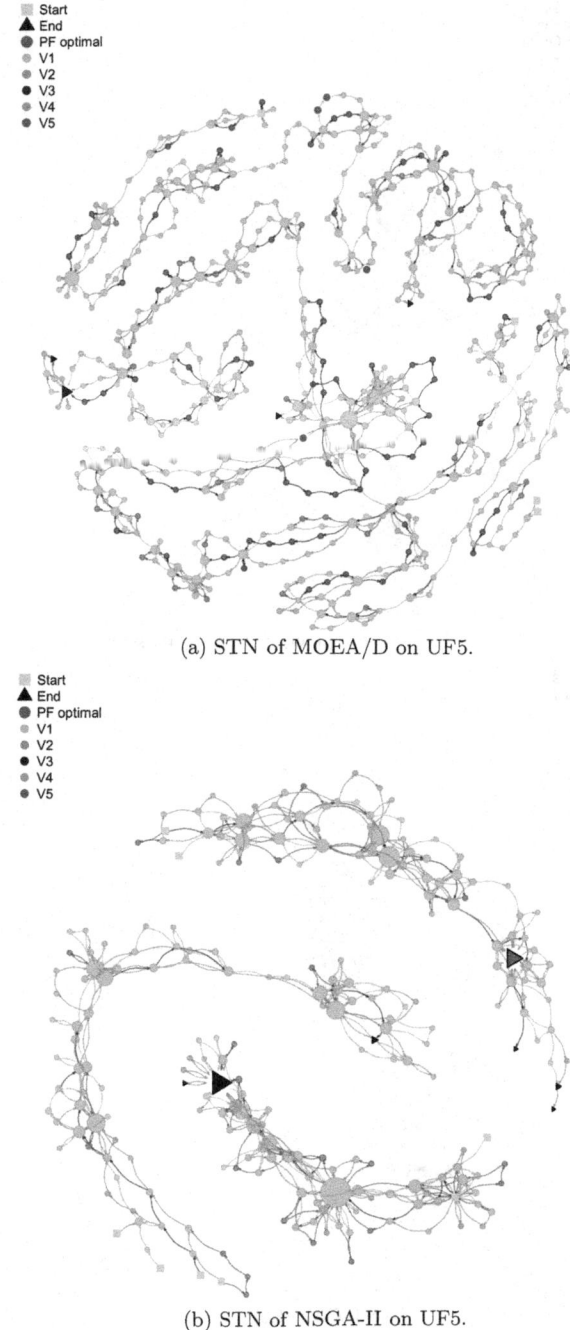

(a) STN of MOEA/D on UF5.

(b) STN of NSGA-II on UF5.

Fig. 2. MOEA/D (top) is able to find many optimal solutions, in two different decomposition vectors and NSGA-II (bottom) only finds optimal solutions in one of the three vectors. (Color figure online)

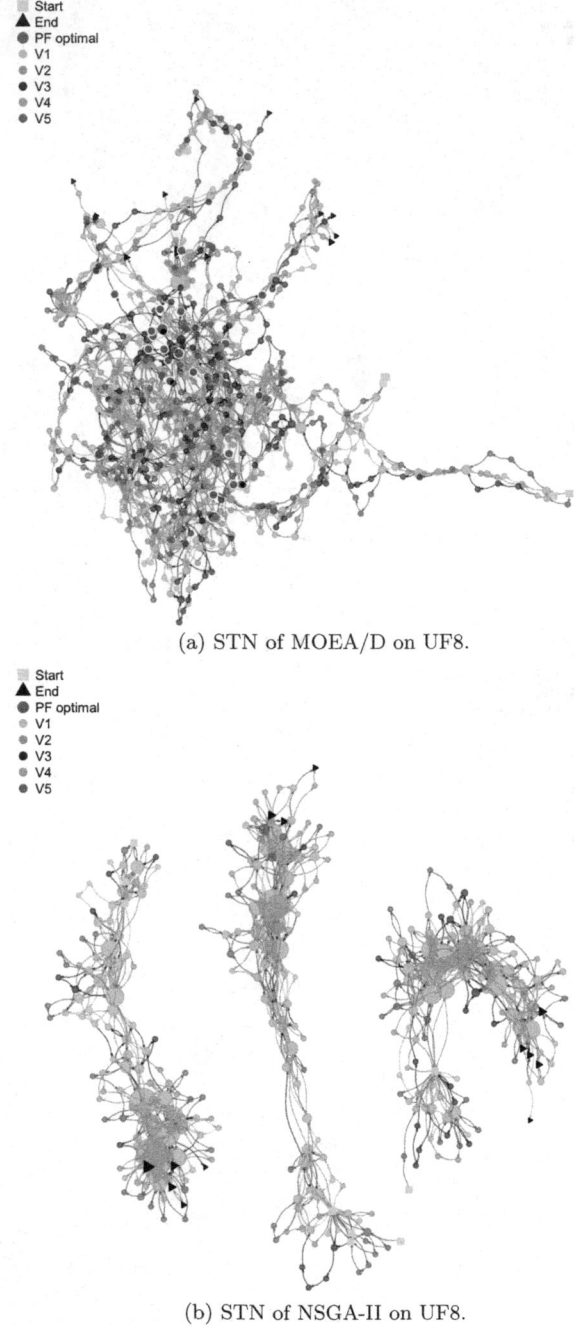

(a) STN of MOEA/D on UF8.

(b) STN of NSGA-II on UF8.

Fig. 3. MOEA/D finds many optimal solutions, around many different areas and NSGA-II (bottom) progresses the search with all vectors sharing nodes, from start to end. (Color figure online)

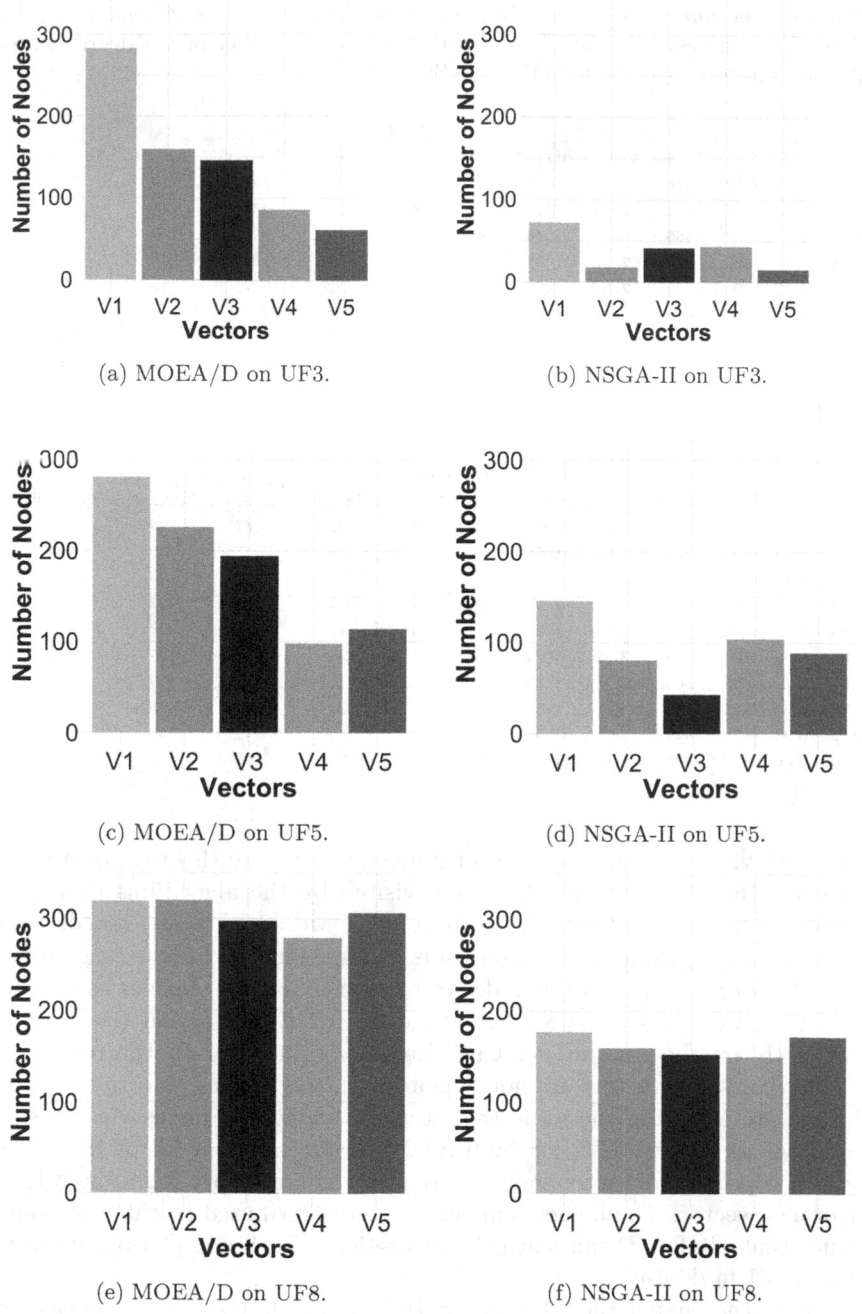

Fig. 4. The number of nodes for each vector found by MOEA/D is higher than those found by NSGA-II. MOEA/D seems to find more nodes via V1 and V2 vectors, showing a more unbalanced distribution while the distribution of nodes found by NSGA-II depends on the problem in question.

Table 3. *HV*, *IGD*, *Nodes*, Ratio of *Edges* given the number of nodes, Ratio of *Shared* Nodes given the number of nodes, Number of *Opt*imal Solutions, *Comp*onents, *Max* number of *in*-degrees, *Mean* number of *in*-degrees, *Max* number of *out*-degrees, *Mean* number of *out*-degrees of MOEA/D and NSGA-II.

								MOEA/D-DE			
	HV	IGD	Nodes	Edges	Shared	Opt.	Comp.	Max in.	Mean in.	Max out.	Mean out.
UF1	0.86	0.01	578	2.64	0.72	0	3	10	3.04	10	3.04
UF2	0.86	0.01	693	2.40	0.65	0	3	10	2.53	10	2.53
UF3	0.57	0.17	465	3.04	0.76	6	1	23	3.77	23	3.77
UF4	0.47	0.04	593	2.44	0.61	0	3	18	2.96	18	2.96
UF5	0.63	0.04	529	2.76	0.70	4	3	13	3.32	13	3.31
UF6	0.64	0.00	646	2.47	0.63	0	3	10	2.72	10	2.72
UF7	0.70	0.01	679	2.35	0.62	0	3	13	2.58	13	2.58
UF8	0.66	0.07	1051	1.61	0.30	24	3	12	1.67	12	1.67
UF9	1.08	0.03	1022	1.62	0.28	1	3	10	1.72	10	1.72
UF10	0.07	0.42	839	1.95	0.38	0	3	13	2.09	13	2.09

								NSGA-II			
	HV	IGD	Nodes	Edges	Shared	Opt.	Comp.	Max in.	Mean in.	Max out.	Mean out.
UF1	0.85	0.02	304	3.79	0.68	0	3	29	5.77	29	5.77
UF2	0.85	0.02	321	3.64	0.58	0	3	31	5.47	30	5.47
UF3	0.41	0.23	126	2.74	0.82	2	1	21	13.93	15	13.93
UF4	0.49	0.03	230	4.99	0.63	0	3	53	7.63	52	7.63
UF5	0.49	0.12	233	4.16	0.72	1	3	32	7.53	32	7.53
UF6	0.62	0.01	333	3.61	0.59	0	3	25	5.27	25	5.27
UF7	0.70	0.01	363	3.35	0.63	0	3	39	4.83	39	4.83
UF8	0.53	0.11	381	4.11	0.58	0	3	26	4.61	26	4.61
UF9	0.93	0.10	421	3.67	0.56	0	3	39	4.17	39	4.17
UF10	0.00	1.34	345	4.28	0.51	0	3	59	5.09	59	5.09

In these visualizations, the size of nodes and the width of edges are proportional to how many times they were visited by the algorithms during the aggregation of runs used to extract the model. Moreover, these visualisations use *force-directed* graph layout algorithms as implemented in R package igraph [?], that focuses on position the nodes to have few crossing of edges as possible.

In Fig. 1, we can see the STN modeled for MOEA/D (a, on the top) and NSGA-II (b, on the bottom) on UF3. Before commenting on the results, we highlight that some vectors are not appearing in the images, and the reason is that they are following the same trajectory, reaching the nodes via the same paths. Looking at the STN for MOEA/D, we can see that all of the Pareto front (PF) optimal solutions, shown in red, in the trajectory found by V1, the light orange vector. The higher number of edges attributed to this component indicates that MOEA/D can reach the theoretical PF after exploring the region related to V1 in depth.

On the other hand, the STN of NSGA-II is much smaller, as we can also see in the number of nodes for each vector. It is no surprise that all the PF solutions found by NSGA-II are also at the end of the search since NSGA-II only accepts new solutions if they dominate old ones. Given the size of the red triangles, NSGA-II tends to come back to these solutions frequently.

Figure 2 shows that the STNs of MOEA/D and NSGA-II are bigger than the STNs modeled for UF3. We can see that the STN modeled for MOEA/D shows four optimal solutions found by different vectors during the search progress. After finding these solutions, MOEA/D continues exploring the problem until the end of the search. Now, NSGA-II also finds few optimal solutions, only one, but tends to come back to this solution frequently. We can see that the network components for both STN have little interaction with which other, suggesting that this problem poses more difficulties to both MOEA/D and NSGA-II than UF3.

Figure 3 illustrates the UF8 results. We can see that MOEA/D and NSGA-II conduct different searches over the three objective problems (as can also be seen in Table 3). Looking first at the results of MOEA/D, we can see that this algorithm progresses with the search with a higher exploration rate since, for all vectors, the number of edges and nodes is higher. Also, we can see that the STN of MOEA/D for the UF8 problem is denser than the NSGA-II for the same problem. MOEA/D performs the best in terms of the number of optimal solutions found in a problem, with 24 solutions in the theoretical Pareto front. By contrast, NSGA-II finds no optimal solution, and we can see that the network components are apart, suggesting that the starting point of the search poses a great impact on the search progress of NSGA-II.

Turning now to the STN metrics for each trajectory, in Fig. 4, we can see that each of the MOEA/D vectors visit more nodes than the vectors of NSGA-II, meaning that MOEA/D finds more unique solutions than NSGA-II. This higher number of nodes, overall vectors, is not exclusive to the MOPs shown by Fig. 4, as can be seen in Table 3. Moreover, these results, combined with the higher number of optimal solutions, lower values of maximum and mean in-/out-degrees, reinforce that MOEA/D can explore better the search space given the higher number of optimal solutions.

In terms of the traditional MOP metrics HV and IGD in Table 3, we can see that MOEA/D achieves higher HV values and lower IGD values, in comparison with NSGA-II for all MOPs. However, in UF7, both algorithms perform the same, and in UF4, NSGA-II performs the best in both metrics. Overall, for this set of MOPs, we consider that MOEA/D presents better performance than NSGA-II.

Table 3 also shows that the NSGA-II algorithm, for most problems, has more than three times more edges than nodes (MOEA/D keeps a lower ratio, with almost the same number of nodes and edges, with a ratio of edges around 2). This higher number of edges and nodes indicates that NSGA-II is visiting some nodes multiple times, which suggests that this metaheuristic might not be exploring the search space in depth during the search progress. A closer look at this Table also shows that the maximum out-degree value differs from the in-degree value in UF2, UF3, and UF4. This result indicates that NSGA-II visits the first node more than once, thus finding it challenging to find the right search trajectory direction to follow at the beginning of the search. Overall, these two results indicate that NSGA-II faces difficulties in some problems, not exploring many new areas of the search space.

In summary, the STN models suggest an explanation for the differences in HV and IGD performance between MOEA/D and NSGA-II. The STNs showed that MOEA/D extensively explores the search space, finding more optimal solutions without re-visiting nodes too many times. On the other hand, the STNs modeled for NSGA-II indicate that this algorithm faces difficulties when exploring new areas of the search space and that its search progress depends on having good starting points. These results show that using STN for discriminating MOEAs is a viable method for analyzing and aggregating helpful information to the traditional practices for comparing such algorithms, such as the HV and IGD. A note of caution is due here since we only analyzed two MOEAs in a handful set of problems, with only two- and three-objectives.

7   Conclusion

The main goal of the current study was to determine and design a methodology for applying the search trajectory networks (STNs) for modeling the search behavior of Multi-Objective Evolutionary Algorithms (MOEAs). The proposed method is based on the simple idea of extracting the STN features from some regions of the problem in question. We create STN models in a benchmark set that the most famous MOEAs, MOEA/D and NSGA-II, have different performances. Finally, we show that STNs can be effectively applied to differentiate visually and quantitatively.

Overall, this study strengthens the idea that characterizing and visualizing the search behavior of MOEAs can provide insightful advances into comprehending how distinct the search behaviors dynamics of such algorithms are and their overall performance. That is because we provided an STN modeling process that can discriminate well different MOEAs, aggregating helpful information to the traditional practices for comparing such algorithms, such as the HV and IGD. Thus, these findings have significant implications for understanding how different MOEAs perform when applied to solve multi-objective problems (MOPs). Moreover, we understand that this work shows excellent results of interest for the whole bio-inspired computation community.

A natural progression of this work is to investigate the generalization of the model to describe the search behavior of MOEAs other than MOEA/D and NSGA-II, especially in real-world and constrained MOPs. Furthermore, an analysis of the importance of features STN used and introduced in this work should be conducted in the context of automated landscape-aware selection and configuration MOEAs.

References

1. Beume, N., Naujoks, B., Emmerich, M.: SMS-EMOA: multiobjective selection based on dominated hypervolume. Eur. J. Oper. Res. 181(3), 1653–1669 (2007)

2. Bossek, J.: smoof: single- and multi-objective optimization test functions. R J. (2017). https://journal.r-project.org/archive/2017/RJ-2017-004/index.html
3. Campelo, F., Aranha, C.: MOEADr: Component-wise MOEA/D implementation (2018). https://cran.R-project.org/package=MOEADr. r package version 1.2.0
4. Campelo, F., Batista, L., Aranha, C.: The MOEADr package: a component-based framework for multiobjective evolutionary algorithms based on decomposition. J. Stat. Softw. (2020). https://arxiv.org/abs/1807.06731
5. Deb, K., Pratap, A., Agarwal, S., Meyarivan, T.: A fast and elitist multiobjective genetic algorithm: NSGA-II. IEEE Trans. Evol. Comput. **6**(2), 182–197 (2002)
6. Fieldsend, J.E., Alyahya, K.: Visualising the landscape of multi-objective problems using local optima networks. In: Proceedings of the Genetic and Evolutionary Computation Conference Companion, GECCO 2019, pp. 1421–1429. Association for Computing Machinery, New York (2019). https://doi.org/10.1145/3319619.3326838
7. Fonseca, C.M., Fleming, P.J.: On the performance assessment and comparison of stochastic multiobjective optimizers. In: Voigt, H.-M., Ebeling, W., Rechenberg, I., Schwefel, H.-P. (eds.) PPSN 1996. LNCS, vol. 1141, pp. 584–593. Springer, Heidelberg (1996). https://doi.org/10.1007/3-540-61723-X_1022
8. Kerschke, P., Grimme, C.: An expedition to multimodal multi-objective optimization landscapes. In: Trautmann, H., et al. (eds.) EMO 2017. LNCS, vol. 10173, pp. 329–343. Springer, Cham (2017). https://doi.org/10.1007/978-3-319-54157-0_23
9. Li, H., Deb, K., Zhang, Q., Suganthan, P., Chen, L.: Comparison between MOEA/D and NSGA-III on a set of novel many and multi-objective benchmark problems with challenging difficulties. Swarm Evol. Comput. **46**, 104–117 (2019)
10. Li, H., Zhang, Q.: Multiobjective optimization problems with complicated Pareto Sets, MOEA/D and NSGA-II. IEEE Trans. Evol. Comput. **13**(2), 284–302 (2009)
11. Liefooghe, A., Derbel, B., Verel, S., López-Ibáñez, M., Aguirre, H., Tanaka, K.: On pareto local optimal solutions networks. In: Auger, A., Fonseca, C.M., Lourenço, N., Machado, P., Paquete, L., Whitley, D. (eds.) PPSN 2018. LNCS, vol. 11102, pp. 232–244. Springer, Cham (2018). https://doi.org/10.1007/978-3-319-99259-4_19
12. Miettinen, K.: Nonlinear multiobjective optimization, volume 12 of international series in operations research and management science (1999)
13. Ochoa, G., Tomassini, M., Verel, S., Verel, C.: A study of NK landscapes' basins and local optima networks. In: Genetic and Evolutionary Computation Conference, GECCO, pp. 555–562. ACM Press, New York (2008)
14. Ochoa, G., Malan, K.M., Blum, C.: Search trajectory networks of population-based algorithms in continuous spaces. In: Castillo, P.A., Jiménez Laredo, J.L., Fernández de Vega, F. (eds.) EvoApplications 2020. LNCS, vol. 12104, pp. 70–85. Springer, Cham (2020). https://doi.org/10.1007/978-3-030-43722-0_5
15. Ochoa, G., Malan, K.M., Blum, C.: Search trajectory networks: a tool for analysing and visualising the behaviour of metaheuristics. Appl. Soft Comput. **109**, 107492 (2021). https://www.sciencedirect.com/science/article/pii/S1568494621004154
16. Schäpermeier, L., Grimme, C., Kerschke, P.: One PLOT to show them all: visualization of efficient sets in multi-objective landscapes. In: Bäck, T., et al. (eds.) PPSN 2020. LNCS, vol. 12270, pp. 154–167. Springer, Cham (2020). https://doi.org/10.1007/978-3-030-58115-2_11
17. Schäpermeier, L., Grimme, C., Kerschke, P.: To boldly show what no one has seen before: a dashboard for visualizing multi-objective landscapes. In: Ishibuchi, H., et al. (eds.) EMO 2021. LNCS, vol. 12654, pp. 632–644. Springer, Cham (2021). https://doi.org/10.1007/978-3-030-72062-9_50

18. Tsou, C.S.V.: Elitist non-dominated sorting genetic algorithm based on r (2013). https://cran.r-project.org/web/packages/nsga2R/nsga2R.pdf
19. Tusar, T., Filipic, B.: Visualization of pareto front approximations in evolutionary multiobjective optimization: a critical review and the prosection method. IEEE Trans. Evol. Comput. **19**(2), 225–245 (2015). https://doi.org/10.1109/tevc.2014.2313407
20. Wang, R., Zhang, T., Guo, B.: An enhanced MOEA/D using uniform directions and a pre-organization procedure. In: 2013 IEEE Congress on Evolutionary Computation, pp. 2390–2397 (2013)
21. Zhang, Q., Li, H.: MOEA/D: a multiobjective evolutionary algorithm based on decomposition. IEEE Trans. Evol. Comput. **11**(6), 712–731 (2007)
22. Zhang, Q., Liu, W., Li, H.: The performance of a new version of MOEA/D on CEC 2009 unconstrained MOP test instances. In: IEEE Congress on Evolutionary Computation 2009, CEC 2009, pp. 203–208. IEEE (2009)
23. Zhang, Q., Zhou, A., Zhao, S., Suganthan, P.N., Liu, W., Tiwari, S.: Multiobjective optimization test instances for the CEC 2009 special session and competition. University of Essex, Colchester, UK and Nanyang technological University, Singapore, special session on performance assessment of multi-objective optimization algorithms, technical report 264 (2008)

EvoMCS: Optimising Energy and Throughput of Mission Critical Services

Miguel Arieiro and Bruno Sousa(✉)

DEI, UC, Pinhal de Marrocos, 3030-290 Coimbra, Portugal
marieiro@student.dei.uc.pt, bmsousa@dei.uc.pt

Abstract. Mission critical services have stringent requirements in terms of reliability, energy-efficiency and performance. Optimisation approaches are required to maximise missions' service time by minimising the consumption of energy without compromising performance like the throughput in scenarios with high density in terms of connected devices. EvoMCS is an evolutionary algorithm approach able to determine the optimal configurations for IEEE 802.11 wireless networks in dense environments with multi objectives, such as minimisation of consumed energy and maximisation of the throughput in each device. EvoMCS is evaluated in wildfire scenarios, with the ns-3 simulator, determining the optimal values for configuration parameters in 802.11n and 802.11ax technologies. The achieved results demonstrate that EvoMCS is able to provide optimal configuration values, that reduces in a factor of three the energy that is consumed by devices.

Keywords: Mission critical services · IEEE 802.11ax · Energy efficiency · Evolutionary algorithm · ns-3

1 Introduction

Mission Critical Services (MCS) have stringent Quality of Service (QoS) requirements, in terms of resilience and delay-sensitivity since they can be associated with people's lives. Wildfires, in Mediterranean countries like Portugal, have a high occurrence probability in summer, requiring strategical plans at national level [1] to coordinate human and technical resources for an efficient fire combat. Public Mobile Radio (PMR) technologies, are outdated and do not fulfil the required capabilities of first responders [8], for instance no support for video or IoT data communications. In this regard, technologies like Long Term Evolution (LTE) and 5G [3] have included several feature to enhance the support of MCS (e.g. PROSE), within the same security levels of PMR technologies [9]. Despite the evolution in wireless technologies, wildfires can damage communication infrastructures. This demonstrates the need for solutions to support fast deployment that can be easily managed and that scale/adapt according to the conditions of wildfire scenario.

© Springer Nature Switzerland AG 2022
J. L. Jiménez Laredo et al. (Eds.): EvoApplications 2022, LNCS 13224, pp. 239–254, 2022.
https://doi.org/10.1007/978-3-031-02462-7_16

Nonetheless the deployment of technologies, like Wifi 6 supporting features for simplified deployments with a high number of end-user devices and efficient power saving mechanisms [10], require careful configuration of diverse parameters to fully support the demanding requirements of MCS. For instance, IEEE 802.11ax is able to enhance throughput and reduce collisions in scenarios with a high number of connected devices, but configuration parameters of physical and medium access layers need to be set properly, for high throughput and efficient power usage.

The research on MCS have been focusing LTE [17] and its evolution (5G and beyond). The efforts mainly consider the performance of services, or functionalities like Device to device communications and do not entail scenarios without a supporting infrastructure, or integrated with the vehicles providing support for first responders in wildfires combat.

EvoMCS, here in proposed is a multi-objective optimization approach [18] that provides optimal values for configuration parameters in IEEE 802.11n and IEEE 802.11ax networks with dense deployments in terms of simultaneous connected users. The achieved results, demonstrated that EvoMCS is able to reduce in a factor of three the energy consumed by devices, when compared with suboptimal configuration profiles (i.e., factory default sets). EvoMCS has the following contributions:

1. Design, validation of a multi-objective Genetic Algorithm to optimise energy-efficiency and throughput in MCS.
2. Validate, through simulation, the feasibility of employing WiFi 6, as a supporting wireless technology to enable communications between teams in the field.
3. Identification of optimised configuration profiles to extend the battery lifetime of end-user devices or to optimise performance of critical applications.
4. The implementation of EvoMCS is released as Open Source[1].

The remaining of this paper is organised as follows: Sect. 2 compares EvoMCS with the related work, while Sect. 3 details the algorithm of EvoMCS, which is evaluated according to the methodology described in Sect. 4. Achieved results are presented in Sect. 5, while Sect. 6 concludes the paper.

2   Related Work

Sabtah et al. [17] propose a parameter selection framework for LTE to enable Mission Critical Networks. The proposed framework tackles the problem of finding the best set of parameters for MCS, considering aspects like eNB deployment, allocated spectrum, achieved throughput and latency. The evaluations conducted via simulation does not consider energy consumption. Contrasting with the work of Sotirios et al. [6], which propose a model to maximise the coverage and minimise the power consumption in LTE networks. The devised model relies on

---

[1] https://github.com/MiguelArieiro/EvoMCS.

Differential Evolution algorithms and consider real data from the city of Ghent, regarding the location of base stations.

Niyato et al. [13] use evolutionary game approaches to optimise the network selection problem in the presence of multiple wireless technologies (IEEE 802.11, WiMAX, etc.). Results demonstrate that evolutionary population is able to reach equilibrium faster in comparison to reinforcement learning, but requires centralised components to gather information of users and services.

EvoMCS employs evolutionary algorithms that have been used in the literature to optimize solutions for diverse problems. Soumaya et al. [20] propose a multi-objective evolutionary algorithm to improve network planning by enhancing the location accuracy of IEEE 802.11 Access Points. In the same line, Alfredo [14] employs a multi-objective evolutionary algorithm to place nodes in wireless sensor networks, tackling the nodes placement and the total energy dissipated with the placement.

As per the analysed works, distinct optimisation approaches have been considered to tackle energy efficiency, service performance, devices locations, etc. EvoMCS enables a multi-objective evolutionary algorithm aiming to optimise energy efficiency and throughput in mission critical scenarios with stringent requirements.

3   EvoMCS: Multi-objective Optimization

This section provides the design and the implementation details of EvoMCS as a multi-objective optimisation approach.

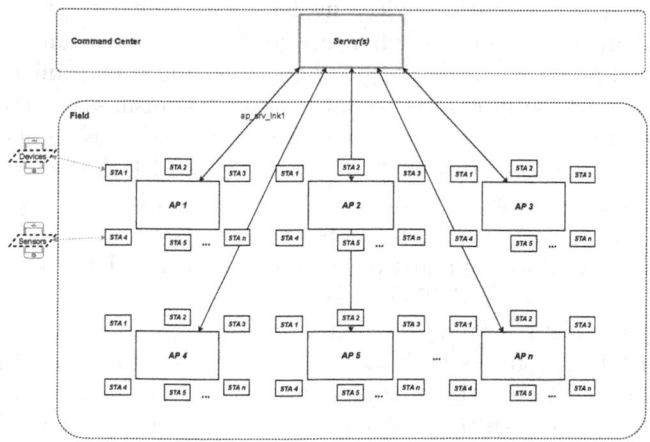

Fig. 1. Evaluation scenario

3.1   Scenario and Technologies

EvoMCS is tailored to IEEE 802.11n, 802.11ax and beyond technologies, which can be employed in wildfire scenarios, considering legal and physical aspects (i.e. water pressure levels) where the distance of firemen to the supporting vehicles cannot exceed 100 m [12]. IEEE 802.11 technologies are known to be deployed rapidly and easily integrated with existing resources (i.e. vehicles), as Fig. 1.

3.2   Evolutionary Algorithm

The individuals of the genetic algorithm's population are codified as per the fields in Table 1 (e.g. .11ax-{1,5,20,0,800,0,0,0,0}, .11n-{0,5,20,0,800,0,n/a,n/a,0}).

Table 1. Information of individuals of the population

Tech	Freq. (GHz)	Channel Band.	moCS	GI	RTS	Ext. Block	OBSS	UDP/ TCP
.11ax	{5,6}	{20, 40, 80, 160}	[0,11]	{800, 1600, 3200}	{0,1}	{0,1}	{0,1}	{0,1}
.11n	{5}	{20, 40, 80, 160}	[0,23]	{400, 800}	{0,1}	n/a	n/a	{0,1}

EvoMCS considers the values of channel bandwidth in MHz, the Modulation Coding Scheme (moCS) as per the values in each technology. The Guard Interval (GI), reported in nanoseconds (ns), corresponds to the parameter that defines the wait time between symbols being transmitted. Higher values (3200) of GI lead to higher delays in the spread propagation, thus with better resilience, while lower values (800) lead to better throughput values but with higher levels of interference, which impacts devices in dense environments. The Request to Send (RTS) parameter aims to enhance transferring performance solving the hidden terminal problem [5] that occurs when a station is transmitting, but the other one in the same service set is not aware of such transmission. RTS allows to inform stations in the same service set regarding the transmission of others, thus reducing the collision probability [10]. The Extended Block allows to perform multiple acknowledgements in a single frame, for bandwidth savings.

The type of application also impacts the energy-efficiency in wireless scenarios. EvoMCS considers applications sending data (i.e., UL-uplink direction) using protocols employed in IoT context, like MQTT (TCP-based) and CoAP (UDP-based) [7]. The reasoning to choose UL data applications is related with the enhancements introduces with WiFi 6 (11.ax) regarding UL transmissions. The utilisation of TCP, with reliability mechanisms can lead to higher levels of energy consumption, since a data packet needs to be acknowledged, or re-transmitted multiple times, if loss is verified.

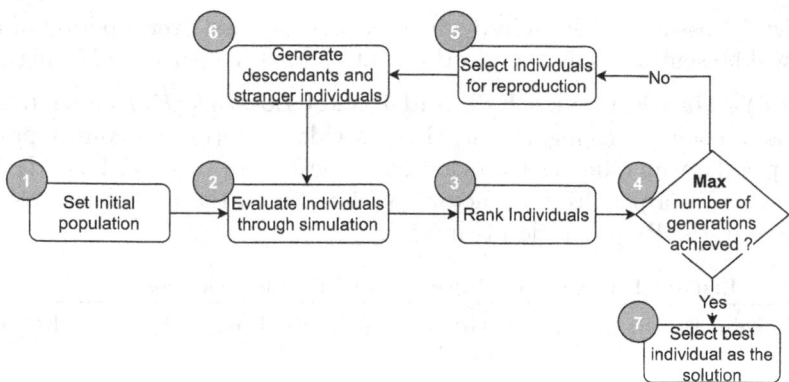

Fig. 2. Evolutionary algorithm

Figure 2 depicts the diverse steps of EvoMCS, relying on a genetic algorithm approach. The first step corresponds to the generation of the initial population, in a random fashion, and considering the possible values for genomes, as per the parameter's values in Table 1. The second step includes the evaluation of the population through the ns-3 simulator [15], that assesses the energy consumed and the achieved throughput of each WiFi device in a specific scenario. The ns-3 simulator assesses the performance of each individual, that correspond to possible WiFi configurations. These individuals are ranked, after the simulation in step 3, considering the devised heuristic. Each generation includes the best individuals of the original population and the descendants and individuals generated randomly (strangers). The number of maximum generations serves as a limit to stop the finding process of best possible configurations, after n generations. At the end, in step 7, the best solution are selected.

3.3   Heuristic for Fitness

Individuals are evaluated considering two approaches for the heuristics.

$H_1(E/T)$ **- Heuristic with Energy/Throughput.** The average energy/throughput ratio metric $H_1(E/T)$ assesses how well a specific individual performs in terms of energy efficiency and throughput in a period of time. Thus, $H_1(E/T)$ determines the required average energy to send a bit of information (Joules/bit) in a time period. H_1 ranks as best individuals the ones with lower values, that is the ones that are able to send more information with less energy.

Listing 1.1. Heuristic $H_1(E/T)$ - Energy/throughput ratio

```
1  def  heuristic (energy, throughput, min_thr=0.0):
2      if (throughput < min_thr):
3          return sys.maxsize
4      return (energy/throughput)
```

$H_1(E/T)$ assumes that an individual has data to send over a period of time, which will be sent according to a certain throughput, as per code Listing 1.1.

$H_2(E/T)$ - Heuristic with E/T and Packet Loss. $H_2(E/T)$ aims to avoid the issue of choosing configurations that provide an optimized consumption of energy per data unit but can provide high packet loss ratios. Thus, H_2 relies on H_1 but adds a penalization factor associated with the packet loss ratio, as illustrated in the Python code Listing 1.2.

Listing 1.2. Heuristic $H2(E/T)$ - E/T with packet loss ratio

```
1   def heuristic (energy, throughput, pktLoss=.0, min_thr=0.0):
2       if (throughput < min_thr):
3           return sys.maxsize
4       return (energy/throughput *(1.0+pktLoss))
```

3.4 Selection Strategy

The selection strategy in EvoMCS, depicted in listing 1.3, is performed in step 5, when the maximum number of generations has not been achieved. First individuals are sorted in reverse order (line 3) as per the fitness value. Secondly, each individual receives a probability to be selected, as per lines 4–5. The probability value considers the position that an individual occupies in the population list, as the best values appear in the first positions, due to the sorting per the heuristic value. Higher values of fitness correspond to the worst individuals, which spend more energy to send a specific amount of data.

Listing 1.3. Selection of individuals

```
1   def rankPop (population, n):
2       pop = copy.deepcopy(population)
3       pop.sort(reverse=True, key=lambda x: x[-1])
4       probs = [(2*i)/(pop_size*(pop_size + 1))
5           for i in range (1, pop_size + 1)]
6       parents = []
7       for _ in range (n):
8           value = random.uniform(0,1)
9           index = 0
10          total = probs [index]
11          while total < value:
12              index += 1
13              total += probs[index]
14          parents.append(pop[index])
15      return parents
```

After setting the probability values, the n individuals are selected based on a random distribution, as per lines 7–14. The n–number of individuals that are selected can be configured, since this is a parameter in the *rankPop()* function.

Step 6 of EvoMCS uses the input of the n–individuals selected by the *rankPop()* function and combines the selected with stranger individuals. This is required to avoid having non–dominated solutions and to avoid the local minimum issues of genetic algorithms [16]. The use of stranger individuals prevents premature convergence to local optima. The stranger individuals are randomly generated, as per the logic of the initial population in step 1.

3.5   Operators to Generate Descendants

The operators to generate descendants in step 6, rely on different approaches: The Single Gene Mutation Operator (SGMO), and the Multi-Gene Probabilistic Mutation Operator (MGMO).

The SGMO only affects a single gene that is selected randomly, considering the total number of genes in an individual. After the random selection of a gene, it is verified the corresponding parameter field, for instance, if it correlates to the MCS parameter. As per the type of field the random values that can be assigned to the gene, must consider the possible values in Table 1, within the specific technology set. For instance, the GI gene can only have value in the set {400,800} when considering IEEE 802.11n.

Listing 1.4. Example code for MGMO

```
1   def mutateProb(original_indiv):
2     global param
3     global mutation_prob
4     indiv = original_indiv
5     indiv [−1] = −1
6     for i in range(len(indiv) − 1):
7       if random.random() < mutation_prob:
8         if (i == 1) or (i == 2):
9           temp=indiv[i]
10          while (temp == indiv[i]):
11            temp = param[i][indiv[i −1]]
12            [random.randint(0,
13            len(param[i][indiv[i −1]]) − 1)]
14          indiv[i] = temp
15    # partial listing for demo purposes
16    return indiv
```

The MGMO can modify multiple genes according to a probability value, as depicted in listing 1.4. MGMO determines a random probability for a gene and if the generated value is below a pre-configured probability that specific gene is modified (line 8), as per the SGMO logic. Lines 10–16 illustrate modification of frequency and channel width parameters.

4   Experimentation

This section depicts the experimentation methodology to assess the performance of the EvoMCS in finding the optimal configurations, and to validate design choices, such as single–SGMO or multiple–MGMO.

4.1 Validation Scenarios

Table 2 illustrates the implemented scenarios. The Basic scenario was employed to validate the workflow of running the EvoMCS algorithm implemented in Python and interacting with ns-3 to validate the configuration of individuals. This workflow involved mainly steps 2 and step 3.

Table 2. Validations scenarios

Scenario	#Access points	#Stations per AP	Duration	Bit rate
Basic	4	4	10s	100 Kbps
Intermediate	9	16	10s	100 Kbps

The Intermediate scenario, is more complex and leads to a total of 144 stations, and is close to the deployments that can be observed in real wildfire combating scenarios [1]. This is the scenario that is employed to evaluate the performance of the optimal profiles determined by EvoMCS.

4.2 Configuration Parameters

Diverse configuration parameters were considered in the experimentation, as documented in this subsection. Such parameters are associated with the energy model of WiFi stations and access points (assumed to be on support vehicles). The WiFi stations follow the *LiIonEnergySource* model, which states the capacity of the battery in mAh and the respective voltage. The technical characteristics of the Samsung Note 10 have been considered for the energy model, in particular with 4300 mAh and 3.85 V. The access points are assumed to operate without energy constraints, they are connected to the power source of the vehicles[2].

The radio energy follows the WiFi model that includes how much energy is spent in the different states. For instance, how much is spent in the transmission–TxCurrentA, in reception–RxCurrentA, sleep mode (SleepCurrentA), and the amount of energy spent in the state transition (SwitchingCurrentA). The values of such parameters have been considered as per related works [2,19].

No mobility has been considered in the scenarios evaluating the EvoMCS to avoid having values affected by the variation of stations' location. For this, the RandomWalk2dMobilityModel was configured with no velocity. Stations and access points are placed in a grid layout, with the access point in the centre. Stations are placed at a distance to the access points around 50 m (recall Fig. 1). The mobility is set to uniform values in the range of [2, 4] m/s in the profiles evaluation, where the aim is to assess the gains with the optimal configurations determined by EvoMCS.

---

[2] Such power source relies normally on diesel generators.

Table 3. GA configuration parameters in EvoMCS

Parameter	Value(s)	Description
popSize	25	Population size
nGeneration	15	16 with the initial population as generation '0'
muteOperat	{SGMO, MGMO}	Mutation operator
muteProb	50%	Mutation probability of a single gene in MGMO

The configuration parameters of the Genetic algorithm of EvoMCS are summarised in Table 3. The population size is set 25, as per the recommendations in the literature, regarding the higher performance values when employing pop ulations with low size [4]. Other relevant configuration parameters include the number of antennas in stations and access points that is set to 4, which with 4 spatial streams for tx-transmission and rx-reception. The threshold for OBSS is set to -82 dBm with the $ns3::ConstantObssPdAlgorithm$.

4.3   Evaluation Metrics

The evaluation metrics employed to evaluate an individual are summarised in Table 4, thus being used to determine the fitness values of H_1 or H_2.

Table 4. Evaluation metrics used in fitness of heuristics

Metric	Description	How is measured
(E) conEne	Energy consumed by stations in Joules	$E = \sum conEne_i/nSta$
(T) goodPut	Throughput at application layer in bits/s	$T = \sum throughput_i/nSta$

The metrics used to assess the performance of the EvoMCS' genetic algorithm are summarised in Table 5.

The nBest–number of best individuals corresponds to percentage of elitism that is configured, in this case 30%. The remaining individuals are the descendants of the population, in a ratio of 50%, while 20% are configured for the

Table 5. GA evaluation metrics in EvoMCS

Metric	Description	How is measured
bestMax	Maximum of best	$max(H(E/T))$
bestMin	Minimum of best	$min(H(E/T))$
avBest	Average of best	$[\sum H_b(E/T)]/nBest$
$avFit_x$	Average of fitness in generation x	$[\sum H_i(E/T)]/nGeneration$
avDiver	Average diversity	$\frac{2.0*nCountDiff}{popSize*(popSize-1)}$

stranger individuals. The avBest–metric is determined based on the b–best individuals (in a ratio of 30%). The nCountDiff–different individuals in a population are determined according to the Hamming distance [11].

4.4   Profiles Validation - Inputs from EvoMCS

The output of EvoMCS is then evaluated in the intermediary scenario (recall Table 2) with mobility activated in the WiFi Stations. The mobility considers velocities varying uniformly in the [2, 4] m/s range. This scenario is also validated according to the Packet Error Ratio (%), the flow mean delay (ms), the direction of the flow either download or upload, and the receiving throughput in kbit/s. The performance gain of each profile is determined in terms of the required average energy to send a bit of information (Joules/bit).

5   Results

5.1   Operators for the EvoMCS in $H_1(E/T)$

The choice of the mutation operators are evaluated considering the $avFit_x$ average of the fitness value, and the $avDiver$–average diversity.

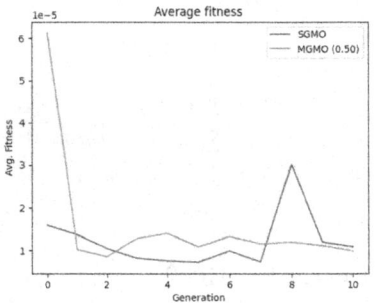 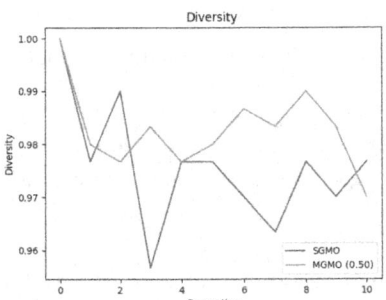

(a) Average fitness per generation   (b) Average diversity per generation

Fig. 3. Operators performance of EvoMCS in $H1(E/T)$

Figure 3a depicts the average fitness in the diverse generations for the SGMO and MGMO with a probability of 50%. The SGMO has the lowest average value for fitness, since the MGMO starts initially with the worst value of fitness. As pictured the variation of the fitness value is higher in the MGMO approach, as opposed to the SGMO, which has average fitness values varying in the range of 7.36×10^{-06} and 3.03×10^{-05}.

Figure 3b depicts the diversity metric, demonstrating a better performance with the MGMO (higher values are more interesting). MGMO is able to introduce more diversity due to the high probability value of 50% to mutate genes in individuals, Nonetheless, the difference between SGMO and MGMO is minimal. The SGMO can provide better results when considering the addition of stranger individuals. Thus, SGMO was the approach that was employed regarding the operator to mutate individuals in the populations of EvoMCS.

5.2 Optimal Configurations

On the experiments performed, EvoMCS was able to identify the configurations in Table 6 as the best and worst profiles for stations in MCS scenarios.

Table 6. Information of individuals of the population

Heu	Profile	Tech	Freq.	Channel Band.	MCS	GI	RTS	Ext. Block	OBSS	App
H_1	Best	.11ax	6	20	3	800	0	0	0	TCP
H_2	Best	.11ax	6	160	4	800	0	0	1	TCP
all	Worst	.11n	5	160	23	1600	1	n/a	n/a	TCP

The values summarised in Table 6 corresponds to the overall configuration that is assessed as optimal, according to $H_1(E/T)$ and $H_2(E/T)$ heuristics.

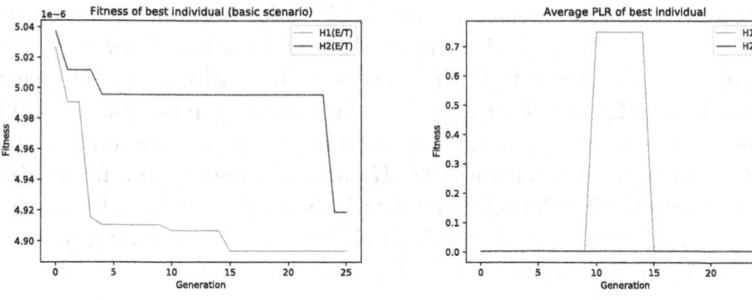

(a) Average fitness per heuristic   (b) Average PLR of the best individual

Fig. 4. Performance differences between $H1(E/T)$ and $H2(E/T)$

The best configuration/profile in both heuristics stands out as being the IEEE 802.11ax with TCP applications in the 6 GHz Frequency. The lowest value of the Guard Interval set to 800 ns for better values of throughput. TCP applications, due to the reliability mechanisms (e.g. acknowledgement, re-transmissions, etc.) is known to introduce more impact on the energy. EvoMCS discards the employment of UDP due to lower values in the *goodPut* metric (recall Table 4), since UDP does not provide reliability mechanisms, it leads to higher packet losses.

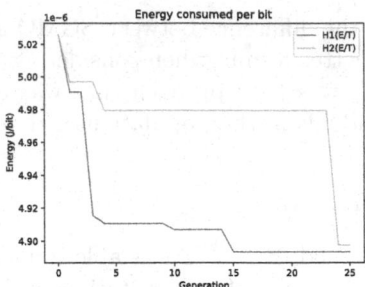

Fig. 5. Consumed energy per bit of H1 and H2 heuristics

$H_2(E/T)$ selects the use of spatial reuse with BSS Colouring (OBSS/PD) and with a channel with higher bandwidth 160 MHz, as opposed to the 20 MHz suggested by $H_1(E/T)$. Figure 4 puts in evidence the difference between the heuristics in terms of fitness and PLR. H_2 by penalizing the individuals with packet loss has higher values in the fitness and has lower PLR ratios, as depicted in Fig. 4b. Another difference relies in the consumed energy to send a data unit, as pictured in Fig. 5.

5.3 Optimal Profiles in Scenarios with Dense-Environments

The results of packet loss are demonstrated in Fig. 6 for both directions of traffic for the H_1, H_2 optimal values and worst profiles identified by EvoMCS.

The download traffic is sent from the servers to the stations, for instance considering the acknowledgements messages in the reliability mechanisms of TCP. The difference in the achieved performance (average of all the stations) is preferable with the best profiles determined by H_1 and H_2, being more noticeable in the downlink direction. The Cumulative Distribution values (CDF) demonstrates more than 75% of nodes have packet losses below 25% in the upload direction, as pictured in Fig. 7b for H_1, while for H_2 this value ascends to 80%, as illustrated

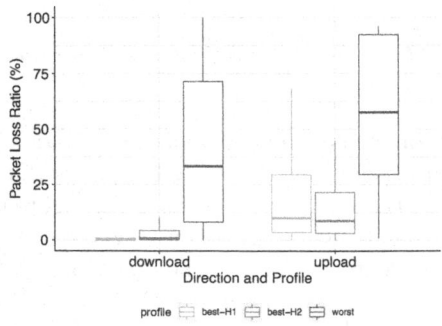

Fig. 6. Packet loss ratio (best and worst)

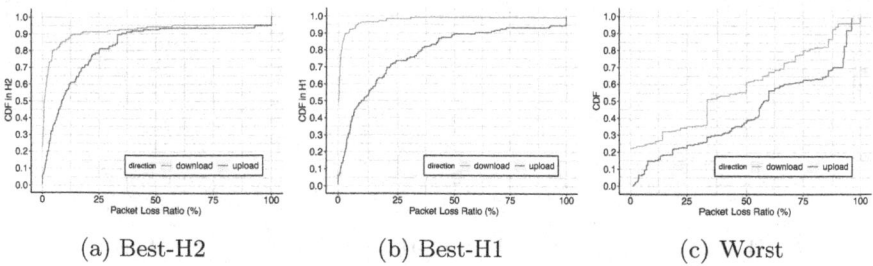

(a) Best-H2   (b) Best-H1   (c) Worst

Fig. 7. CDF for packet loss ratio

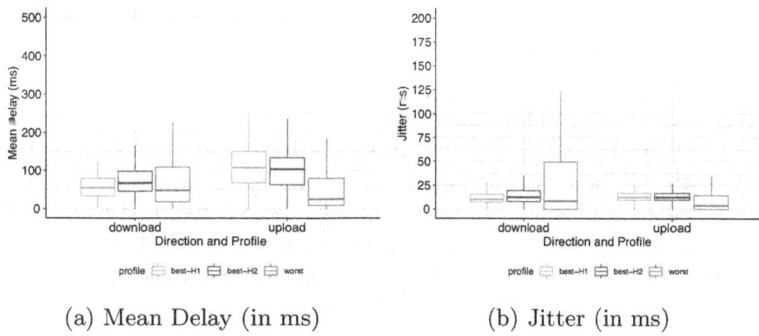

(a) Mean Delay (in ms)   (b) Jitter (in ms)

Fig. 8. Delay and jitter

in Fig. 7a. H_1 and H_2 results contrast with the worst profile, where 75% of the nodes have packet losses close to 75% levels, as pictured in Fig. 7c.

The optimal profiles determined by H_1 and H_2 introduce stable results regarding delay and jitter (i.e., which is the delay variation), as depicted in Fig. 8. The results of the mean delay may seem better with the worst profile, but one should not forget the high packet loss ratios that are observed with this configuration. Indeed, the it is observed higher variation in the delay with the worst configurations as illustrated in Fig. 8b. On the other hand, with more packets delivered H_1 and H_2 have higher values for delay, in particular with the upload direction given that the majority of the traffic is sent in this direction.

The metric of RX throughput measures the amount of data that is received per unit of time. The simulation was configured constant data rates of 1000 kbps. Figure 9 depicts the CDF values for the RX throughput in kbps, where the difference relies in the x-scale with the best H_1 and H_2 profiles provide data rates close to 200 kbps. Indeed, almost 50% of the nodes have a throughput higher than 150 kbps, as illustrated in Fig. 9.

The RX throughput with the worst configuration leads to unacceptable values, where more than 90% of the nodes have a throughput bellow 25 kbps. Considering the higher packet loss ratios, the worst configurations, with the 802.11n

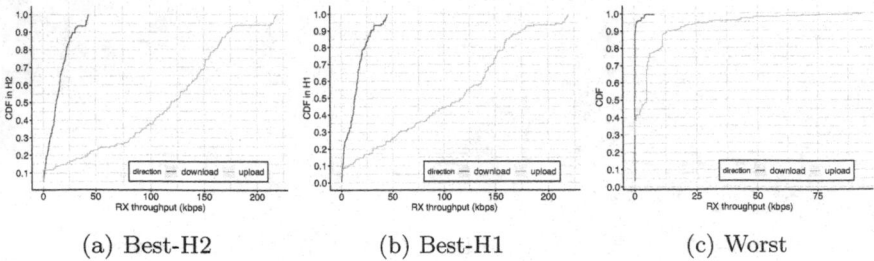

(a) Best-H2   (b) Best-H1   (c) Worst

Fig. 9. CDF for RX throughput

lead to performance values that do not meet the requirements of mission critical services.

The consumed energy plays a key roles in mission critical scenarios, in particular wildfires scenarios, since these missions have a long duration. Figure 10 depicts the ratio of the consumed energy per bit. As demonstrated, the performance of the best profiles lead to lower values in the consumed energy per each bit. The worst configuration is three times worse than the best, which means that the battery of the device will last, at least three times less.

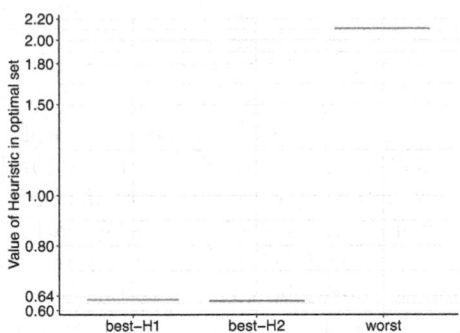

Fig. 10. Heuristic with ratio of consumed energy per bit

The achieved results put in evidence the performance achieved with the best profiles and the worst. The bests, relying on the IEEE 802.11ax, are arguably 3 times better then the worst, since IEEE 802.11ax has more stable results.

6   Conclusions

EvoMCS is a multi-objective approach relying on a evolutionary approach that is able to determine optimal values for the maximisation of throughput and minimisation of consumed energy.

Achieved results demonstrate that EvoMCS is able to determine optimal configurations for IEEE 802.11ax and IEEE 802.11n, that improve in a order of three the ratio of energy consumed and throughput. The H_1 and H_2 heuristics select optimal profiles that are able to provide a better performance.

Our next steps include the optimisation for devices with heterogeneous technologies, integrating 5G and IEEE 802.11 support and heuristics that perform the dynamic selection of best profiles. In addition, EvoMCS will be compared with other multi-objective evolutionary approaches, to include support for other objectives such as latency reduction.

Acknowledgments. This work was supported in part by the European Regional Development Fund (FEDER), through the Regional Operational Programme of Lisbon (POR LISBOA 2020); in part by FCT/MCTES through national funds under the project MH-SDVanet (PTDC/EEI-COM/5284/2020); in part by the national funds through the FCT - Foundation for Science and Technology, I.P., within the scope of the project CISUC - UID/CEC/00326/2020; and in part by the European Social Fund, through the Regional Operational Program Centro 2020.

References

1. ANPC: Diretiva Operacional Nacional n.2 - DECIR - Dispositivo Especial de Combate a Incêndios Rurais, April 2020
2. Bernardo, V.M.H.: Energy efficient multimedia communications in IEEE 802.11 networks. Ph.D. thesis, University of Coimbra (2015). http://hdl.handle.net/10316/28311
3. CEPT ECC: Draft ECC Report 218: Harmonised conditions and spectrum bands for the implementation of future European Broadband Public Protection and Disaster Relief (BB-PPDR) systems, October 2015
4. Chen, T., Tang, K., Chen, G., Yao, X.: A large population size can be unhelpful in evolutionary algorithms. Theoret. Comput. Sci. **436**, 54–70 (2012)
5. Deng, D.J., Chen, K.C., Cheng, R.S.: IEEE 802.11ax: next generation wireless local area networks. In: Proceedings of the 2014 10th International Conference on Heterogeneous Networking for Quality, Reliability, Security and Robustness, QSHINE 2014, vol. 1, pp. 77–82 (2014)
6. Goudos, S.K., Deruyck, M., Plets, D., Martens, L., Joseph, W.: Optimization of power consumption in 4G LTE networks using a novel barebones self-adaptive differential evolution algorithm. Telecommun. Syst. **66**(1), 109–120 (2017)
7. Gündoğan, C., Kietzmann, P., Lenders, M., Petersen, H., Schmidt, T.C., Wählisch, M.: NDN, COAP, and MQTT: a comparative measurement study in the IoT. In: ICN 2018 - Proceedings of the 5th ACM Conference on Information-Centric Networking, pp. 159–171 (2018)
8. International Forum to Advance First Responders Forum: Common Capability Gaps (2021). https://www.internationalresponderforum.org/capability_gaps
9. Prabhu, K., et al.: A reliability-aware, delay guaranteed, and resource efficient placement of service function chains in softwarized 5G networks. IEEE Trans. Cloud Comp. (2020). https://doi.org/10.1109/TCC.2020.3020269
10. Khorov, E., Kiryanov, A., Lyakhov, A., Bianchi, G.: A tutorial on IEEE 802.11ax high efficiency WLANs. IEEE Commun. Surv. Tutor. **21**(1), 197–216 (2019)

11. Kim, Y.-H., Moon, B.-R.: Distance measures in genetic algorithms. In: Deb, K. (ed.) GECCO 2004. LNCS, vol. 3103, pp. 400–401. Springer, Heidelberg (2004). https://doi.org/10.1007/978-3-540-24855-2_43
12. Ministério da Administração Interna - Autoridade Nacional de Emergência e Proteção Civil: Despacho n.o 7316/2016, Junho 2016
13. Niyato, D., Hossain, E.: Dynamics of network selection in heterogeneous wireless networks: an evolutionary game approach. IEEE Trans. Veh. Technol. 58(4), 2008–2017 (2009)
14. Perez, A.J.: M-SPOT: a hybrid multiobjective evolutionary algorithm for node placement in wireless sensor networks. In: 2018 32nd International Conference on Advanced Information Networking and Applications Workshops (WAINA), January 2018, pp. 264–269. IEEE, May 2018
15. Riley, G.F., Henderson, T.R.: The ns-3 network simulator. In: Wehrle, K., Güneş, M., Gross, J. (eds.) Modeling and Tools for Network Simulation, pp. 15–34. Springer, Heidelberg (2010). https://doi.org/10.1007/978-3-642-12331-3_2
16. Rocha, M., Neves, J.: Preventing premature convergence to local optima in genetic algorithms via random offspring generation. In: Imam, I., Kodratoff, Y., El-Dessouki, A., Ali, M. (eds.) IEA/AIE 1999. LNCS (LNAI), vol. 1611, pp. 127–136. Springer, Heidelberg (1999). https://doi.org/10.1007/978-3-540-48765-4_16
17. Sabbah, A., Jarwan, A., Bonin, L., Ibnkahla, M.: A high-level parameter selection framework for irregular LTE-based mission critical networks. In: 2019 IEEE Wireless Communications and Networking Conference (WCNC), April 2019, pp. 1–6. IEEE, April 2019
18. Tian, Y., et al.: Evolutionary large-scale multi-objective optimization: a survey. ACM Comput. Surv. 54(8), 1–34 (2021)
19. Yang, H., Deng, D.J., Chen, K.C.: On energy saving in IEEE 802.11 ax. IEEE Access 6, 47546–47556 (2018)
20. Zirari, S., Abdou, W., Canalda, P., Spies, F.: IEEE 802.11 network planning based on ESBEA evolutionary algorithm to improve location accuracy. In: International Conference on Indoor Positioning and Indoor Navigation, vol. 13 (2012). p. 15th

RWS-L-SHADE: An Effective L-SHADE Algorithm Incorporation Roulette Wheel Selection Strategy for Numerical Optimisation

Seyed Jalaleddin Mousavirad[1]([⊠]), Mahshid Helali Moghadam[2,3],
Mehrdad Saadatmand[3], Ripon Chakrabortty[4], Gerald Schaefer[5],
and Diego Oliva[6]

[1] Computer Engineering Department, Hakim Sabzevari University, Sabzevar, Iran
jalalmoosavirad@gmail.com
[2] RISE Research Institutes of Sweden, Västerås, Sweden
[3] Mälardalen University, Västerås, Sweden
[4] School of Engineering and Information Technology, UNSW Canberra at ADFA,
Canberra, Australia
[5] Department of Computer Science, Loughborough University, Loughborough, UK
[6] Depto. de Innovacion Basada en la Informacion y el Conocimiento,
Universidad de Guadalajara, CUCEI, Guadalajara, Mexico

Abstract. Differential evolution (DE) is widely used for global optimisation problems due to its simplicity and efficiency. L-SHADE is a state-of-the-art variant of DE algorithm that incorporates external archive, success-history-based parameter adaptation, and linear population size reduction. L-SHADE uses a current-to-pbest/1/bin strategy for mutation operator, while all individuals have the same probability to be selected. In this paper, we propose a novel L-SHADE algorithm, RWS-L-SHADE, based on a roulette wheel selection strategy so that better individuals have a higher priority and worse individuals are less likely to be selected. Our extensive experiments on the CEC-2017 benchmark functions and dimensionalities of 30, 50 and 100 indicate that RWS-L-SHADE outperforms L-SHADE.

Keywords: Differential evolution · L-SHADE algorithm · Roulette wheel selection strategy · Optimisation · CEC-2017 benchmark functions

1 Introduction

Global optimisation refers to finding the optimal solution(s) among all feasible solutions for a given problem according to some criteria. While conventional optimisation algorithms such as gradient-based approaches are popular in the literature, they suffer from some difficulties, such as their tendency to get stuck in local

J. L. Jiménez Laredo et al. (Eds.): EvoApplications 2022, LNCS 13224, pp. 255–268, 2022.
https://doi.org/10.1007/978-3-031-02462-7_17

optima and the need to calculate derivative information [13]. Population-based metaheuristic algorithms such as particle swarm optimisation [17] and human mental search (HMS) [6] have been shown to successfully address these problems.

Differential evolution (DE) [18] is a population-based metaheuristic algorithm that, despite its simplicity, has shown considerable efficacy in solving various optimisation problems such as image processing [12,15], pattern recognition [1, 9,14], and food quality [21]. DE is based on three main operators: mutation to generate new individuals based on differences among individuals, crossover to integrate the mutant vector with the parent one, and selection to select a better individual from the new individual and its parent.

In recent years, a wide variety of DE variants have been proposed such as opposition-based DE [16], centre-based DE [10], competition-based DE [7], one-array DE [11], and adaptive differential evolution with optional external archive (JADE) [23]. Among them, success-history based parameter adaptation for DE (SHADE) [19] has shown superior performance compared to other approaches [8]. SHADE employs a current-to-pbest/1/bin strategy for updating an individual, which is based on selecting an individual from the best individuals in the current population and an external archive. SHADE also employs a dynamic approach to select the scaling factor F and crossover probability CR based on earlier values with good performance in past iterations.

L-SHADE [20] improved the SHADE algorithm using linear population size reduction (LPSR) which decreases the population size steadily during the optimisation process. L-SHADE has shown outstanding performance in solving optimisation problems; for example, it ranked first at the CEC-2014 Competition on Real-Parameter Single Objective Optimization [4].

L-SHADE also employs a current-to-pbest/1/bin strategy for updating individuals. Here, two individuals are selected randomly, with all individuals having an equal chance of being elected. In this paper, we propose a novel L-SHADE algorithm, RWS-L-SHADE, based on a roulette wheel selection strategy so that individuals in the mutation step are selected with a likelihood based on their objective function value. In other words, RWS-L-SHADE selects better individuals with a higher probability, and worse performing individuals with a lower probability.

The remainder of the paper is organised as follows. Section 2 briefly describes DE and L-SHADE. Section 3 introduces our RWS-L-SHADE algorithm, while Sect. 4 provides experimental results. Section 5 concludes the paper.

2 Background

2.1 Differential Evolution

Differential evolution (DE) [5] is an effective population-based metaheuristic algorithm that has shown good performance in solving various optimisation problems [2,3]. In the following, we briefly characterise its main components:

1. *Initialisation:* DE starts with a set of randomly-generated individuals.
2. *Mutation:* generates a mutant vector based on differences among distinct individuals as

$$v_i = x_{r1} + F(x_{r2} - x_{r3}), \tag{1}$$

where x_{r1}, x_{r2}, and x_{r3} are three different individuals randomly selected from the current population and F is a scaling factor.

3. *Crossover:* combines the mutant vector with the parent one, typically based on binomial crossover, defined as

$$u_{i,j} = \begin{cases} v_{i,j}, & \text{if } rand(0,1) \leq CR \text{ or } j == j_{rand} \\ x_{i,j}, & \text{otherwise} \end{cases}, \tag{2}$$

where u_i is the trial vector, CR is the crossover rate, and j_{rand} is a random integer number between 1 and the number of dimensions.

4. *Selection:* selects the better individual from the new individual and the old one.

2.2 L-SHADE

Success-History based Adaptive DE with linear population size reduction (L-SHADE) [20] is a powerful variant of DE whose main components are:

1. *External archive:* An external archive (A) is used to retain diversity. Here, parent vectors x_i which are worse than the trial vector u_i are maintained, and the mutation operator employs the union of the external archive and the current population, $(A \cup P)$, to update an individual.

2. *Mutation scheme:* A current-to-pbest/1 scheme is employed, where a new individual is generated as

$$v_i^t = x_i^t + F_i^t(x_{pbest} - x_i^t) + F_i^t(x_{r1} - x_{r2}), \tag{3}$$

where x_{pbest} is a randomly-selected individual from the top $N \times p$ individuals with $p \in [0, 1]$ defining the greediness of the strategy, F_i denotes the scaling parameter for x_i, x_{r1} is an individual randomly selected from the current population, ands x_{r2} is an individual randomly selected from $A \cup P$. The size of the archive is set to the population size.

p is adaptively adjusted so that each individual includes a specific p defined as

$$p_i = rand[p_{min}, 0.2], \tag{4}$$

with $p_{min} = \frac{2}{NP}$, where NP is the population size.

3. *History-based parameter adaptation:* uses a historical memory with H entries to maintain the control parameters CR and F. First, the memory elements M_{CR} and M_F are set to 0.5. A random entry r_i between 1 and H is then selected in each generation, and CR_i and F_i obtained as

$$CR_i = randn_i(M_{CR,r_i}, 0.1) \tag{5}$$

and

$$F_i = randc_i(M_{F,r_i}, 0.1), \tag{6}$$

where $randn_i(\mu, \sigma^2)$ and $randc_i(\mu, \sigma^2)$ generate random numbers with mean μ and variance σ^2 using the normal and Cauchy distribution, respectively. CR_i and F_i values employed by successful individuals are stored in S_{CR} and S_F. The memory elements are then updated as

$$M_{CR,K}^{t+1} = \begin{cases} mean_{WA}(S_{CR}) & \text{if } S_{CR} \neq 0 \\ M_{CR,K}^t & \text{otherwise} \end{cases} \tag{7}$$

and

$$M_{F,K}^{t+1} = \begin{cases} mean_{WL}(S_F) & \text{if } S_F \neq 0 \\ M_{F,K}^t & \text{otherwise} \end{cases}, \tag{8}$$

where K in an index identifying the location in memory between 1 and H and initialised to 1, and $mean_{WA}$ and $mean_{WL}$ are the weighted arithmetic and weighted Lehmar means.

4. *Linear population size reduction (LPSR):* decreases the population size over the course of iterations as

$$NP^{t+1} = round(\frac{P^{min} - P^{int}}{NFE_{\max}} NFE + P^{int}), \tag{9}$$

where NP^{t+1} is the population size in the next iteration, P_{min} is the smallest population size (here $P_{min} = 4$ since the current-to-pbest/1 scheme needs a minimum 4 individuals), P^{int} is the population size in the first iteration, $NFE_{\max}$ is the maximum number of function evaluations, and NFE is the current number of function evaluations.

3   RWS-L-SHADE

As mentioned, L-SHADE employs a current-to-pbest/1 strategy for mutation. In Eq. (3), x_{r1} is a randomly-selected individual from the current population, while x_{r2} is also a randomly-selected individual but selected from the union of the current population and the archive. The selection probabilities for x_{r1} and x_{r2} are

$$Pr_{x_{r1}} = \frac{1}{\text{population size}} \tag{10}$$

and

$$Pr_{x_{r2}} = \frac{1}{\text{population size+Archive size}}, \tag{11}$$

respectively.

It can be seen that the selection probability for all members of the population is the same, i.e. there high and low performing individuals are not distinguished. However, selecting better individuals with a higher probability should be beneficial. To address this issue, our proposed RWS-L-SHADE employs an objective function-proportionate selection strategy. In particular, a roulette wheel selection (RWS) strategy is used in the mutation operator so that individuals will be selected also based on their objective function value in order to take advantage of good characteristics of better individuals.

The employed RWS strategy assigns a selection probability to each individual proportional to its objective function value as

$$Pr_i = \frac{f_i}{\sum_{j=1}^{NP} f_i},$$ (12)

where f_i is the objective function value for i-th individual. M individuals is selected by repeating the RWS strategy M times. Algorithm 1 lists the RWS strategy in form of pseudo-code.

Input : $f = f_1, f_2, ..., f_{NP}$: set of objective functions
Output: i: an index

for $i = 2$ **to** NP **do**
  | $f_i = max(f) - f_i$;
end
for $i = 2$ **to** NP **do**
  | $f_i = \frac{f_i}{\sum_{j=1}^{NP} f_i}$
end
$p(0) = 0$;
$p(1) = f(1)$;
for $i = 2$ **to** NP **do**
  | $p(i) = f(i-1) + f(i)$;
end
$r =$ random number between 0 and 1 ;
for $i = 1$ **to** NP **do**
  | **if** $p(i-1) < r \le p(i)$ **then**
  | | $return\ i$
  | **end**
end

Algorithm 1: Pseudo-code of RWS strategy (for a minimisation problem).

In RWS-L-SHADE, the RWS strategy is employed in two ways:

1. Selection of x_{r_1}: the RWS strategy selects an individual from the current population;
2. Selection of x_{r_2}: the RWS strategy selects an individual from the union of the archive and the current population.

Algorithm 2 gives the RWS-L-SHADE algorithm in form of pseudo-code.

4 Experimental Results

To assess the efficacy of RWS-L-SHADE, and to compare it to L-SHADE, a set of challenging rotated and shifted benchmark functions, introduced in the CEC2017

Input : D: dimensionality of problem; NFE_{max}: maximum number of function evaluations; NP_{max}: maximum population size; NP_{min}: minimum population size; H: Memory size

Output: x^*: the best individual

$M_F = 0.5$, $M_{CR} = 0.5$, $NFE = 0$, $t = 1$, $NP = NP_{max}$;
Randomly generate initial population Pop;
Evaluate objective function value of each individual;
while $NFE < MAX_{NFE}$ **do**
  $S_F = 0, S_R = 0$;
  for $i \leftarrow 1$ **to** NP **do**
    Generate random index $r_i = rand(1, H)$;
    Generate CR_i^t as $randn_i(M_{CR,r_i}, 0.1)$;
    Generate F_i^t as $randc_i(M_{F,r_i}, 0.1)$;
    Generate p_i^t as $rand[p_{min}, 0.2]$;
    Select x_{r1} from current population using Algorithm 1 ;
    Select x_{r2} from combination of current population and archive using Algorithm 1 ;
    Generate trial vector as $v_i^t = x_i^t + F_i^t(x_{best} - x_i^t) + F_i^t(x_{r1} - x_{r2})$;
    for $j \leftarrow 1$ **to** D **do**
      if $rand_j[0, 1] < C_R$ *or* $j == j_{rand}$ **then**
        $u_{i,j}^t = v_{i,j}^t$
      else
        $u_{i,j}^t = x_{i,j}^t$
      end
    end
  end
  for $i \leftarrow 1$ **to** NP **do**
    if $f(u_i^t) \leq f(x_i^t)$ **then**
      $x_i^{t+1} \leftarrow u_i^t$;
    else
      $x_i^{t+1} \leftarrow x_i^t$;
    end
    if $f(u_i^t) < f(x_i^t)$ **then**
      $A \leftarrow x_i^t$;
      $S_{CR} \leftarrow CR_i^t$;
      $S_F \leftarrow F_i^t$;
    end
  end
  When size of archive exceeds $|A|$, randomly delete individuals so that $|A| \leq |P|$;
  if $S_{CR} \neq 0$ *and* $S_F \neq 0$ **then**
    Compute $M_{F,K}^{t+1} = mean_{WL}(S_F)$;
    Compute $M_{CR,K}^{t+1} = mean_{WA}(S_{CR})$;
  end
  Perform LPSR as $NP = round[\frac{NP_{max} - NP_{min}}{MAX_{NFE}}.NFE + NP_{max}]$;
  t=t+1 ;
end

Algorithm 2: Pseudo-code of RWS-L-SHADE algorithm.

Table 1. Summary of CEC2017 benchmark functions [22]. N indicates the number of basic functions to form hybrid and composite functions. The search range is $[+100, -100]^D$ in all cases.

		Optimum
Unimodal functions		
F1	Shifted and Rotated Bent Cigar Function	100
F2	Shifted and Rotated Sum of Different Power Function	200
F3	Shifted and Rotated Zakharov Function	300
Multimodal functions		
F4	Shifted and Rotated Rosenbrock's Function	400
F5	Shifted and Rotated Rastrigin's Function	500
F6	Shifted and Rotated Expanded Scaffer's Function	600
F7	Shifted and Rotated Lunacek Bi_Rastrigin Function	700
F8	Shifted and Rotated Non-Continuous Rastrigin's Function	800
F9	Shifted and Rotated Levy Function	900
F10	Shifted and Rotated Schwefel's Function	1000
Hybrid multimodal functions		
F11	Hybrid Function 1 (N = 3)	1100
F12	Hybrid Function 2 (N = 3)	1200
F13	Hybrid Function 3 (N = 3)	1300
F14	Hybrid Function 4 (N = 4)	1400
F15	Hybrid Function 5 (N = 4)	1500
F16	Hybrid Function 6 (N = 4)	1600
F17	Hybrid Function 7 (N = 5)	1700
F18	Hybrid Function 8 (N = 5)	1800
F19	Hybrid Function 9 (N = 5)	1900
F20	Hybrid Function 10 (N = 6)	2000
Composite functions		
F21	Composition Function 1 (N = 3)	2100
F22	Composition Function 2 (N = 3)	2200
F23	Composition Function 3 (N = 4)	2300
F24	Composition Function 4 (N = 4)	2400
F25	Composition Function 5 (N = 5)	2500
F26	Composition Function 6 (N = 5)	2600
F27	Composition Function 7 (N = 6)	2700
F28	Composition Function 8 (N = 6)	2800
F29	Composition Function 9 (N = 3)	2900
F30	Composition Function 10 (N = 3)	3000

Table 2. Results for $D = 30$. For each function, we report the average difference to the optimal value, standard deviation, and IAR (in bold if IAR $\geq$ 1). The bottom row gives the number of wins (IAR > 1), ties (IAR = 1), and losses (IAR < 1) of RWS-L-SHADE over L-SHADE.

Function		L-SHADE	RWS-L-SHADE	IAR
F1	avg.	3.28E−07	**2.28E−08**	14.39
	std. dev.	8.41E−07	**4.09E−08**	
F2	avg.	**6.40E−01**	1.04E+00	0.62
	std. dev.	**1.04E+00**	1.86E+00	
F3	avg.	5.55E−08	**3.49E−08**	1.59
	std. dev.	1.04E−07	**5.62E−08**	
F4	avg.	5.88E+01	**5.87E+01**	1.01
	std. dev.	1.21E+00	**4.15E−01**	
F5	avg.	1.89E+01	**1.83E+01**	1.03
	std. dev.	**3.47E+00**	3.79E+00	
F6	avg.	2.01E−05	**1.68E−05**	1.20
	std. dev.	**1.39E−05**	1.42E−05	
F7	avg.	4.96E+01	**4.91E+01**	1.01
	std. dev.	3.97E+00	**3.21E+00**	
F8	avg.	**1.64E+01**	1.74E+01	0.94
	std. dev.	2.96E+00	**2.91E+00**	
F9	avg.	**0.00E+00**	**0.00E+00**	1.00
	std. dev.	**0.00E+00**	**0.00E+00**	
F10	avg.	**2.43E+03**	2.53E+03	0.96
	std. dev.	**3.13E+02**	3.24E+02	
F11	avg.	3.24E+01	**2.52E+01**	1.29
	std. dev.	3.05E+01	**2.79E+01**	
F12	avg.	1.26E+03	**1.19E+03**	1.06
	std.dev.	**4.17E+02**	4.97E+02	
F13	avg.	4.84E+01	**3.00E+01**	1.61
	std. dev.	3.20E+01	**1.94E+01**	
F14	avg.	2.59E+01	**2.52E+01**	1.03
	std. dev.	**2.10E+00**	2.33E+00	
F15	avg.	1.21E+01	**1.18E+01**	1.03
	std. dev.	**3.35E+00**	4.30E+00	
F16	avg.	3.50E+02	**2.35E+02**	1.49
	std. dev.	1.42E+02	**1.17E+02**	
F17	avg.	**7.08E+01**	**7.08E+01**	1.00
	std. dev.	1.25E+01	**1.06E+01**	
F18	avg.	**2.88E+01**	2.95E+01	0.98
	std. dev.	**6.17E+00**	6.77E+00	
F19	avg.	1.24E+01	**1.07E+01**	1.16
	std. dev.	2.21E+00	**2.03E+00**	
F20	avg.	1.04E+02	**8.71E+01**	1.19
	std. dev.	4.14E+01	**3.15E+01**	
F21	avg.	**2.17E+02**	2.19E+02	0.99
	std. dev.	**2.93E+00**	4.93E+00	
F22	avg.	**1.00E+02**	**1.00E+02**	1.00
	std. dev.	3.25E−12	**2.09E−12**	
F23	avg.	3.64E+02	**3.63E+02**	1.01
	std. dev.	**4.31E+00**	3.64E+00	
F24	avg.	**4.36E+02**	**4.36E+02**	1.00
	std. dev.	**2.76E+00**	3.25E+00	
F25	avg.	**3.87E+02**	**3.87E+02**	1.00
	std. dev.	**2.02E−02**	2.80E−02	
F26	avg.	1.03E+03	**1.02E+03**	1.01
	std. dev.	6.31E+01	**5.81E+01**	
F27	avg.	**5.03E+02**	**5.03E+02**	1.00
	std. dev.	**4.75E+00**	6.84E+00	
F28	avg.	**3.42E+02**	3.51E+02	0.97
	std. dev.	5.57E+01	**5.17E+01**	
F29	avg.	**4.82E+02**	4.77E+02	1.01
	std. dev.	**1.21E+01**	1.46E+01	
F30	avg.	2.05E+03	**2.04E+03**	1.01
	std. dev.	8.25E+01	**7.51E+01**	
wins/ties/losses for RWS-L-SHADE				18/6/6

Table 3. Results for $D = 50$, laid out in the same fashion as Table 2.

function		L-SHADE	RWS-L-SHADE	IAR
F1	avg.	2.65E+00	**1.74E+00**	1.52
	std. dev.	**1.99E+00**	2.84E+00	
F2	avg.	**3.45E+08**	3.97E+08	0.87
	std. dev.	**1.66E+09**	1.95E+09	
F3	avg.	9.61E−02	**5.66E−02**	1.70
	std. dev.	1.40E−01	**5.11E−02**	
F4	avg.	**1.09E+02**	1.12E+02	0.97
	std. dev.	**4.67E+01**	5.14E+01	
F5	avg.	**4.68E+01**	4.76E+01	0.98
	std. dev.	**7.30E+00**	8.72E+00	
F6	avg.	6.50E−04	**1.03E−04**	6.31
	std. dev.	1.08E−03	**9.35E−05**	
F7	avg.	**1.03E+02**	**1.03E+02**	1.00
	std. dev.	**9.16E+00**	1.07E+01	
F8	avg.	**4.80E+01**	4.87E+01	0.99
	std. dev.	7.02E+00	**4.82E+00**	
F9	avg.	5.08E−03	7.16E−03	0.50
	std. dev.	**1.79E−02**	2.48E−02	
F10	avg.	5.75E+03	**5.64E+03**	1.02
	std. dev.	**4.14E+02**	4.39E+02	
F11	avg.	5.69E+01	**5.27E+01**	1.08
	std. dev.	1.48E+01	**1.07E+01**	
F12	avg.	**1.29E+04**	1.34E+04	0.96
	std. dev.	**5.02E+03**	6.11E+03	
F13	avg.	2.41E+02	**2.08E+02**	1.16
	std. dev.	7.52E+01	**6.02E+01**	
F14	avg.	4.14E+01	**3.85E+01**	1.08
	std. dev.	**5.75E+00**	6.38E+00	
F15	avg.	7.81E+01	**6.38E+01**	1.22
	std. dev.	2.85E+01	**1.80E+01**	
F16	avg.	9.26E+02	**8.82E+02**	1.05
	std. dev.	2.11E+02	**2.10E+02**	
F17	avg.	6.31E+02	**6.20E+02**	1.02
	std. dev.	1.48E+02	**1.10E+02**	
F18	avg.	9.37E+01	**7.08E+01**	1.32
	std. dev.	4.90E+01	**2.71E+01**	
F19	avg.	5.24E+01	**3.98E+01**	1.32
	std. dev.	1.49E+01	**1.14E+01**	
F20	avg.	6.06E+02	**5.98E+02**	1.01
	std. dev.	**1.25E+02**	1.67E+02	
F21	avg.	**2.50E+02**	2.53E+02	0.99
	std. dev.	**6.63E+00**	7.61E+00	
F22	avg.	**5.15E+03**	5.54E+03	0.93
	std. dev.	2.09E+03	**1.61E+03**	
F23	avg.	**4.68E+02**	4.70E+02	1.00
	std. dev.	1.09E+01	**7.35E+00**	
F24	avg.	5.37E+02	**5.34E+02**	1.01
	std. dev.	**1.15E+01**	1.49E+01	
F25	avg.	**4.82E+02**	**4.82E+02**	1.00
	std. dev.	4.32E+00	**3.84E+00**	
F26	avg.	1.31E+03	**1.22E+03**	1.07
	std. dev.	1.44E+02	**1.16E+02**	
F27	avg.	5.32E+02	**5.27E+02**	1.01
	std. dev.	1.20E+01	**8.44E+00**	
F28	avg.	4.76E+02	**4.69E+02**	1.01
	std. dev.	2.35E+01	**1.99E+01**	
F29	avg.	4.61E+02	**4.55E+02**	1.01
	std. dev.	3.09E+01	**2.57E+01**	
F30	avg.	7.24E+05	**6.76E+05**	1.07
	std. dev.	9.36E+04	**8.44E+04**	
wins/ties/losses for RWS-L-SHADE				19/3/8

Table 4. Results for $D = 100$, laid out in the same fashion as Table 2.

function		L-SHADE	RWS-L-SHADE	IAR
F1	avg.	1.38E+03	**3.86E+02**	**3.58**
	std. dev.	9.48E+02	**2.36E+02**	
F2	avg.	3.74E+36	**9.33E+35**	4.01
	std. dev.	1.20E+37	**2.40E+36**	
F3	avg.	6.62E+02	**4.64E+02**	1.43
	std. dev.	3.42E+02	**1.95E+02**	
F4	avg.	2.15E+02	**2.14E+02**	1.01
	std. dev.	2.04E+01	**1.76E+01**	
F5	avg.	1.96E+02	**1.84E+02**	1.07
	std. dev.	**1.93E+01**	3.89E+01	
F6	avg.	2.44E−02	**1.51E−02**	1.62
	std. dev.	1.47E−02	**9.63E−03**	
F7	avg.	**3.20E+02**	3.21E+02	1.00
	std. dev.	2.03E+01	**1.53E+01**	
F8	avg.	1.86E+02	**1.83E+02**	1.02
	std. dev.	**3.50E+01**	3.71E+01	
F9	avg.	1.26E+00	**1.10E+00**	1.15
	std. dev.	8.65E−01	**6.56E−01**	
F10	avg.	**1.75E+04**	1.75E+04	1.00
	std. dev.	8.27E+02	**4.59E+02**	
F11	avg.	6.72E+02	**5.83E+02**	1.15
	std. dev.	1.01E+02	**9.75E+01**	
F12	avg.	**2.06E+05**	2.05E+05	1.00
	std. dev.	8.05E+04	**7.97E+04**	
F13	avg.	3.47E+03	**2.58E+03**	1.34
	std. dev.	9.86E+02	**7.43E+02**	
F14	avg.	2.76E+02	**2.66E+02**	1.04
	std. dev.	4.01E+01	**3.76E+01**	
F15	avg.	3.68E+02	**3.65E+02**	1.01
	std. dev.	7.42E+01	**6.91E+01**	
F16	avg.	3.51E+03	**3.37E+03**	1.04
	std. dev.	**3.01E+02**	3.33E+02	
F17	avg.	2.22E+03	**2.19E+03**	1.01
	std. dev.	2.88E+02	**2.61E+02**	
F18	avg.	3.65E+02	**3.28E+02**	1.11
	std. dev.	7.24E+01	**5.84E+01**	
F19	avg.	**1.83E+02**	1.94E+02	0.94
	std. dev.	3.39E+01	**2.46E+01**	
F20	avg.	**2.87E+03**	2.94E+03	0.98
	std. dev.	2.71E+02	**2.30E+02**	
F21	avg.	**3.82E+02**	3.82E+02	1.00
	std. dev.	6.67E+01	**6.06E+01**	
F22	avg.	**1.82E+04**	1.85E+04	0.98
	std. dev.	**5.43E+02**	6.83E+02	
F23	avg.	5.77E+02	**5.74E+02**	1.01
	std. dev.	2.48E+01	**2.12E+01**	
F24	avg.	**9.12E+02**	9.12E+02	1.00
	std. dev.	9.93E+00	**8.27E+00**	
F25	avg.	7.60E+02	**7.53E+02**	1.01
	std. dev.	**1.91E+01**	2.46E+01	
F26	avg.	3.36E+03	**3.30E+03**	1.02
	std. dev.	1.08E+02	**8.58E+01**	
F27	avg.	**6.34E+02**	6.39E+02	0.99
	std. dev.	2.18E+01	**1.95E+01**	
F28	avg.	5.59E+02	**5.45E+02**	1.03
	std. dev.	3.76E+01	**2.26E+01**	
F29	avg.	2.22E+03	**2.15E+03**	1.03
	std. dev.	2.92E+02	**2.27E+02**	
F30	avg.	3.31E+03	**3.01E+03**	1.10
	std. dev.	3.28E+02	**1.86E+02**	
wins/ties/losses for RWS-L-SHADE				21/5/4

competition on Real-Parameter Single Objective Optimisation Problems [22], is used. It contains 30 functions categorised into 4 groups including unimodal functions, multi-modal functions, hybrid multi-modal functions, and composite functions, summarised in Table 1.

For both L-SHADE and RWS-L-SHADE, the initial population size is set to $18 \times D$, where D is the dimensionality of the search space. The minimum population size is set to 4, the maximum number of function evaluations to $3000 \times D$, and the memory size to 5 for all experiments.

Each algorithm is run 25 times, and we report the average and standard deviation for objective function over these 25 runs. For each function, we also calculate the improved accuracy rate (IAR), defined as

$$IAR = \frac{\text{Error of L-SHADE}}{\text{Error of RWS-L-SHADE}},$$ (13)

with the errors calculated as the difference of the obtained objective function value and the known optimum. An IAR greater than 1 thus signifies that RWS-L-SHADE outperforms L-SHADE, whilst an IAR less than 1 indicates that L-SHADE performs better than RWS-L-SHADE (Table 4).

Table 2 gives the results of RWS-L-SHADE and L-SHADE for all functions and $D = 30$. From the table, we can see that RWS-L-SHADE outperforms L-SHADE in 18 cases and obtains similar results in 6 cases. In particular, RSW-L-SHADE performs better or similar compared to L-SHADE for 2 of the 3 unimodal functions, 14 of the 17 multimodal functions, and 8 of the 10 composite functions.

For $D = 50$, the results are given in Table 3. RWS-L-SHADE obtains and IAR greater or equal to 1 for 22 of the 30 functions. Also, RWS-L-SHADE gives the lowest standard deviation (and thus, best robustness) for 18 problems.

For $D = 100$, RWS-L-SHADE obtains even better results than for lower dimensionalities, giving better or equal performance compared to L-SHADE for 26 of the 30 functions, thus indicating that RWS-L-SHADE has better efficacy in higher dimensional search spaces. Also, RWS-L-SHADE yields a lowest standard deviation for 26 functions, indicating better robustness.

We also carry out a Wilcoxon signed-rank test between the two algorithms. The obtained p-values are 0.0432 for $D = 30$, 0.0427 for $D = 50$, and 0.0060 for $D = 100$, thus indicating that RWS-L-SHADE statistically outperforms L-SHADE for all dimensionalities.

Last but not least, Fig. 1 shows convergence curves of both algorithms, for three representative functions, F1, F15, and F30 for $D = 100$. As we can observe, RWS-L-SHADE converges faster than L-SHADE.

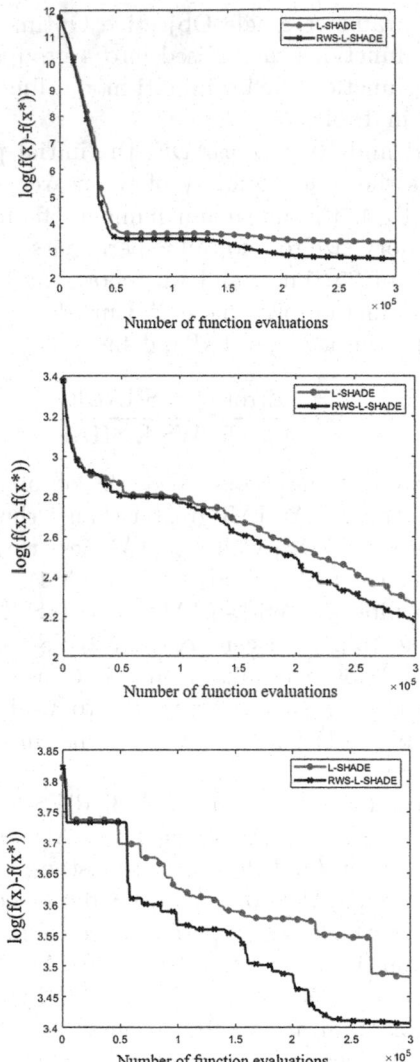

Fig. 1. Convergence plots for F1 (top), F5 (middle), and F20 (bottom).

5   Conclusions

In this paper, we have proposed RWS-L-SHADE, which extends L-SHADE with a roulette wheel selection (RWS) strategy for mutation. While the mutation operator in L-SHADE selects two individuals randomly, the RWS strategy of RWS-L-SHADE leads to better individuals being selected with a higher likelihood as an individual in the mutation operator. Experimental results on the CEC2017 benchmark functions and with dimensionalities of 30, 50, and 100 convincingly confirm RWS-L-SHADE to outperform L-SHADE. In future, we intend to extend the algorithm for multi-objective optimisation problems.

References

1. Awad, N.H., Ali, M.Z., Suganthan, P.N., Reynolds, R.G.: Differential evolution-based neural network training incorporating a centroid-based strategy and dynamic opposition-based learning. In: IEEE Congress on Evolutionary Computation, pp. 2958–2965. IEEE (2016)
2. Cai, Z., Gong, W., Ling, C.X., Zhang, H.: A clustering-based differential evolution for global optimization. Appl. Soft Comput. **11**(1), 1363–1379 (2011)
3. Fister, I., Fister, D., Deb, S., Mlakar, U., Brest, J., Fister, I.: Post hoc analysis of sport performance with differential evolution. Neural Comput. Appl. **32**(15), 10799–10808 (2018). https://doi.org/10.1007/s00521 018 3305 3
4. Liang, J., Qu, B., Suganthan, P.: Problem definitions and evaluation criteria for the CEC 2014 special session and competition on single objective real-parameter numerical optimization. Computational Intelligence Laboratory, Zhengzhou University, Zhengzhou China and Technical Report, Nanyang Technological University, Singapore 635 (2013)
5. Mohamed, A.W., Almazyad, A.S.: Differential evolution with novel mutation and adaptive crossover strategies for solving large scale global optimization problems. Appl. Comput. Intell. Soft Comput. **2017** (2017)
6. Mousavirad, S.J., Ebrahimpour-Komleh, H.: Human mental search: a new population-based metaheuristic optimization algorithm. Appl. Intell. **47**(3), 850–887 (2017). https://doi.org/10.1007/s10489-017-0903-6
7. Mousavirad, S.J., Rahnamayan, S.: Differential evolution algorithm based on a competition scheme. In: 14th International Conference on Computer Science and Education (2019)
8. Mousavirad, S.J., Rahnamayan, S.: Enhancing SHADE and L-SHADE algorithms using ordered mutation. In: 2020 IEEE Symposium Series on Computational Intelligence (SSCI), pp. 337–344. IEEE (2020)
9. Mousavirad, S.J., Rahnamayan, S.: Evolving feedforward neural networks using a quasi-opposition-based differential evolution for data classification. In: IEEE Symposium Series on Computational Intelligence (2020)
10. Mousavirad, S.J., Rahnamayan, S.: A novel center-based differential evolution algorithm. In: Congress on Evolutionary Computation. IEEE (2020)
11. Mousavirad, S.J., Rahnamayan, S.: One-array differential evolution algorithm with a novel replacement strategy for numerical optimization. In: International Conference on Systems, Man, and Cybernetics (2020)
12. Mousavirad, S.J., Rahnamayan, S., Schaefer, G.: Many-level image thresholding using a center-based differential evolution algorithm. In: Congress on Evolutionary Computation (2020)
13. Mousavirad, S.J., Schaefer, G., Korovin, I.: A global-best guided human mental search algorithm with random clustering strategy. In: International Conference on Systems, Man and Cybernetics, pp. 3174–3179 (2019)
14. Mousavirad, S.J., Schaefer, G., Korovin, I., Oliva, D.: RDE-OP: a region-based differential evolution algorithm incorporation opposition-based learning for optimising the learning process of multi-layer neural networks. In: 24th International Conference on the Applications of Evolutionary Computation (2021)

15. Mousavirad, S.J., Zabihzadeh, D., Oliva, D., Perez-Cisneros, M., Schaefer, G.: A grouping differential evolution algorithm boosted by attraction and repulsion strategies for masi entropy-based multi-level image segmentation. Entropy **24**(1), 8 (2022)
16. Rahnamayan, S., Tizhoosh, H.R., Salama, M.M.: Opposition-based differential evolution. IEEE Trans. Evol. Comput. **12**(1), 64–79 (2008)
17. Shi, Y., Eberhart, R.: A modified particle swarm optimizer. In: IEEE International Conference on Evolutionary Computation, pp. 69–73 (1998)
18. Storn, R., Price, K.: Differential evolution-a simple and efficient heuristic for global optimization over continuous spaces. J. Global Optim. **11**(4), 341–359 (1997)
19. Tanabe, R., Fukunaga, A.: Success-history based parameter adaptation for differential evolution. In: IEEE Congress on Evolutionary Computation, pp. 71–78. IEEE (2013)
20. Tanabe, R., Fukunaga, A.S.: Improving the search performance of shade using linear population size reduction. In: IEEE Congress on Evolutionary Computation, pp. 1658–1665. IEEE (2014)
21. Wang, X., et al.: Massive expansion and differential evolution of small heat shock proteins with wheat (triticum aestivum l.) polyploidization. Sci. Rep. **7**(1), 1–12 (2017)
22. Wu, G., Mallipeddi, R., Suganthan, P.: Problem definitions and evaluation criteria for the CEC 2017 competition on constrained real-parameter optimization. Technical report, Nanyang Technological University, Singapore (2016)
23. Zhang, J., Sanderson, A.C.: JADE: adaptive differential evolution with optional external archive. IEEE Trans. Evol. Comput. **13**(5), 945–958 (2009)

WebGE: An Open-Source Tool for Symbolic Regression Using Grammatical Evolution

J. Manuel Colmenar[1]([✉]) [iD], Raúl Martín-Santamaría[1] [iD],
and J. Ignacio Hidalgo[2] [iD]

[1] Rey Juan Carlos University, 28933 Móstoles, Spain
{josemanuel.colmenar,raul.martin}@urjc.es
[2] Complutense University, 28040 Madrid, Spain
hidalgo@ucm.es

Abstract. Many frameworks and libraries are available for researchers working on optimization. However, the majority of them require programming knowledge, lack of a friendly user interface and cannot be run on different operating systems. *WebGE* is a new optimization tool which provides a web-based graphical user interface allowing any researcher to use Grammatical Evolution and Differential Evolution on symbolic regression problems. In addition, the fact that it can be deployed on any server as a web service also incorporating user authentication, makes it a versatile and portable tool that can be shared by multiple researchers. Finally, the modular software architecture allows to easily extend *WebGE* to other algorithms and types of problems.

Keywords: Grammatical Evolution · Differential Evolution · Symbolic regression · Open-source software

1 Introduction

In the field of optimization and, more precisely, in the metaheuristics area, many software frameworks and libraries are available. Researchers may select among different alternatives implemented with different programming languages. For instance, jMetal [12], ECJ [26] or JCLEC-MO [24] are programmed in Java, LEAP [7] or EvoCluster [23] are coded in Python, PlatEMO [28] is coded in MATLAB and predtoolsTS is coded in R [6]. These are several of the most complete and recently available frameworks and libraries with different aims and scopes.

This work has been partially supported by the Spanish Ministerio de Ciencia, Innovación y Universidades (MCIU/AEI/FEDER, UE) under grants ref. PGC2018-095322-B-C22 and RTI2018-095180-B-I00; and Comunidad de Madrid y Fondos Estructurales de la Unión Europea with grants ref. P2018/TCS-4566, B2017/BMD3773 (GenObIA-CM) and Y2018/NMT-4668 (Micro-Stress - MAP-CM).

© Springer Nature Switzerland AG 2022
J. L. Jiménez Laredo et al. (Eds.): EvoApplications 2022, LNCS 13224, pp. 269–282, 2022.
https://doi.org/10.1007/978-3-031-02462-7_18

The common features among them are the availability of many algorithms and the possibility of adapting the framework to any target problem. However, there is one mandatory requirement: the researcher must have learnt how to program using the language of the selected framework. Besides, the majority of these libraries and frameworks lack of a friendly and modern graphical user interface (GUI) for an easier use.

Optaplanner is a constraint solver mainly specialized in scheduling and routing problems [10]. It provides different algorithms and metaheuristics like Tabu Search, Genetic Algorithms or Simulated Annealing, among others. The use of Optaplanner requires coding abilities, since any new problem has to be programmed either from scratch or using one of the provided examples as a template. Although Optaplanner provides GUI support for solution visualization for some examples, the experiments configuration and modelling has to be made on the source code.

Perhaps one of the most interesting optimization tools providing a GUI is HeuristicLab [13], which is an integrated environment for heuristic optimization. It includes many different types of algorithms and problem families, allowing the user also to create new problems using a template-based system. However, the main drawback of HeuristicLab is that it cannot be shared among researchers since it lacks of a user authentication system. Besides, it is developed for Windows and its portability to other operating systems like Linux or MacOS depends on third-party elements like Mono.

Therefore, despite that many optimization tools and frameworks are available for a researcher, none of them meet the following requirements: friendly GUI, no need of coding abilities, portability and shareable as a web service.

In this way, a new tool named WebGE, which stands for Web Grammatical Evolution, is proposed in this paper. WebGE provides a web-based GUI for experiments management, where the algorithm tuning can be performed using friendly web forms. Moreover, WebGE is packed using the Docker container technology, which allows WebGE to be run in any operating system. In addition, WebGE can be run on a shared server by multiple researchers, since it implements granular access controls, and it stores all the data and results in a relational database. Finally, WebGE is proposed as an open-source software, already available in https://github.com/GRAFO-URJC/WebGE, and specifically designed to be extended to other algorithms and problems.

In its current state of development, WebGE allows the user to work on symbolic regression problems. That is, problems devoted to finding models for a target variable from a given dataset. To this aim, not only does WebGE provide GE as an optimization algorithm, but it also provides the combination of GE with Differential Evolution (GE+DE). Therefore, WebGE can be used in any symbolic regression problem by any researcher with no programming background.

The rest of the paper is organized as follows. First, the integration of GE and DE into WebGE is explained in Sect. 2. Then, a general software description and the main features of WebGE are detailed in Sect. 3 and Sect. 4, respectively. In order to test the proposed application, a well-known benchmark is studied in Sect. 5. Finally, conclusions are drawn in Sect. 6.

2   Grammatical Evolution and Differential Evolution

Grammatical Evolution (GE) is a metaheuristic method belonging to the Genetic Programming family [22]. Its main advantage lies in being able to include particular knowledge of a problem into the grammar in order to guide the optimization process. Different GE implementations have been successfully applied to diverse problems like machine learning pipeline optimization [3,14], feature extraction on accelerometer data [19], glucose forecasting in diabetic patients [16] or energy demand estimation [18].

In particular, many of the works whose aim is to produce models tackle the problem as a symbolic regression process. This process begins with the compilation of a dataset composed by a set of input variables and a target variable to be modelled. This dataset is usually split in two: one part used in the process to obtain the models (training phase) and a second part devoted to assessing the quality of the obtained models (testing phase). The models are generated as mathematical expressions which produce series of data that are compared to the target variable one. The difference between the predicted values and the reference is called error, and there exists multiple metrics to measure it, such as the root mean squared error, the absolute error, the average error or R^2. Since WebGE is focused on this kind of problems, it is designed to ease the process of dataset division into training and test as well as the quality measure process, implementing all the previously mentioned metrics.

The GE implementation included in WebGE comes from the JECO library [1], which has been also used in several works like [16] and [18]. However, due to the modular design of WebGE, any other GE implementation could be integrated in the application.

In addition to GE, WebGE also provides the combination of GE with Differential Evolution (DE) proposed in [8] and called GE+DE. DE [27] is a metaheuristic algorithm very well suited for continuous optimization, which makes it interesting to ensemble with GE. In particular, the approach followed by the GE+DE implementation lets DE take care of the generation of constant values, which is a delicate task in GE [4,11], and allows GE to focus on the generation of parameterized models.

Figure 1 shows the optimization cycle performed by GE+DE in WebGE. As it can be seen, all the information related to an experiment is stored in the database. This way, the configuration of the experiment is loaded from the database and the algorithm begins its execution. Here, each generation of individuals in GE consists of a set of parameterized models generated under the guidance of the grammar. This population is sent to DE, which searches for the best parameter values of the models according to the selected objective function calculated on the training data. Once DE finishes executing, the models with parameter values are returned to GE and the evolutionary process is repeated until the maximum number of generations of GE is reached. At the end of the execution, the best model and its corresponding optimized parameters are stored in the database.

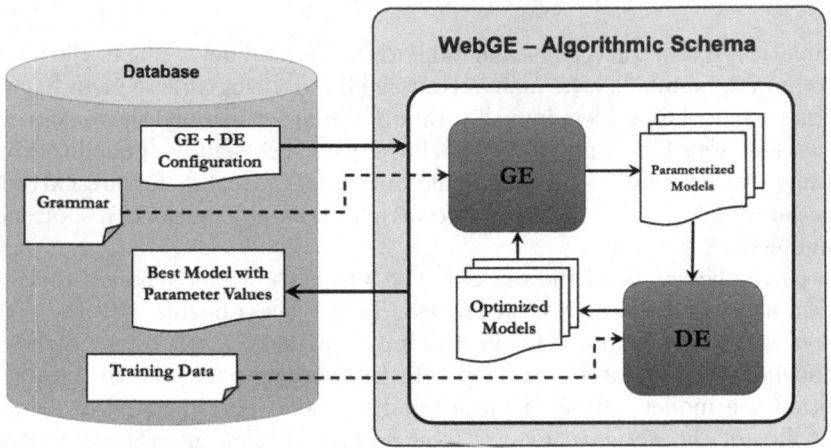

Fig. 1. GE+DE algorithm schema implemented in WebGE.

3   Software Description

WebGE has been developed as an open-source project under the GNU General Public License v3.0, and its source code is available at GitHub https://github.com/GRAFO-URJC/WebGE. Several students have contributed to this project from the very beginning, some of them working on WebGE for their final degree projects. We would like to acknowledge here their work and note that their contribution is credited at https://grafo-urjc.github.io/WebGE/.

Since the project follows the SOLID principles, in particular, *the Single responsibility principle*, the number of classes developed in this project is high, so no class diagram is provided. However, key design aspects and principal components are next described.

3.1   Modular Design

WebGE has been developed following a modular design whose aim is two-fold. On the one hand, in terms of code development, it allows the future extension of the functionalities of the application. On the other hand, the division into components allows a more efficient execution, identifying performance bottlenecks.

Figure 2 shows the modular design of WebGE in terms of component features. In particular, four main components, identified with white background in the figure, were developed. The first one, the *GUI + Endpoints* element, is the set of web-based interfaces which encompasses the communication with the user. These elements required the development of a whole set of endpoints which decouple the implementation of the GUI with the rest of the application, allowing a future extension to different interfaces. The second one is the *Persistence Wrapper*, which includes all the operations related to database storage and retrieval of information. The third one is the *Launcher*, which is the component that takes the information of the experiments (retrieved from the database by the *Persistence Wrapper*) and creates the experiment tasks to be run upon resources availability. As seen in the figure, the *Launcher* receives the command from the user through the corresponding endpoint. Then, it stores the generated experiment into a waiting queue, where the experiments are kept while no processors are available. Finally, the fourth component is the *Execution Engine*, which is in charge of monitoring the resource availability and to actually run the pending experiments.

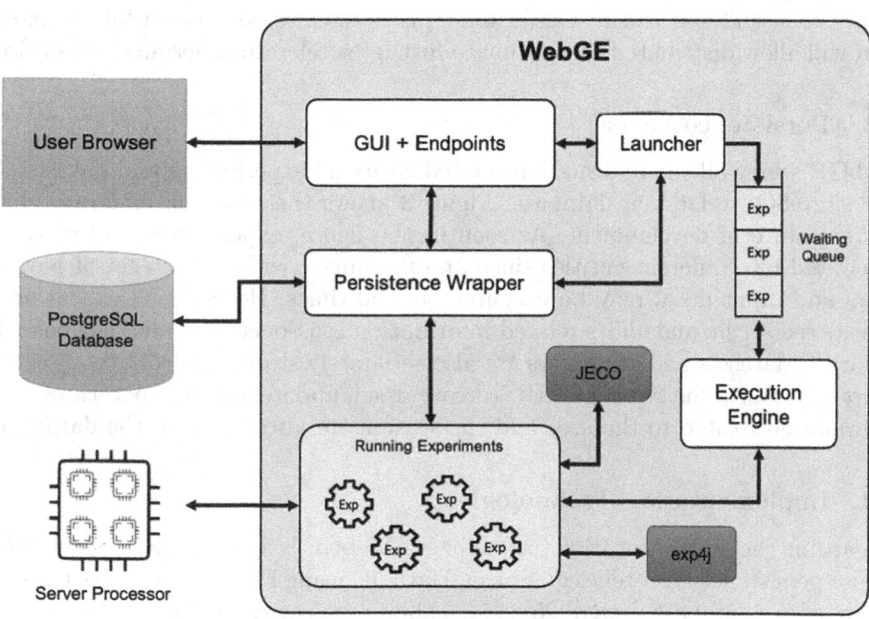

Fig. 2. WebGE components schema.

Notice also that the running experiments make use of two external libraries, identified with blue background in the figure: JECO and exp4j. JECO is the library which provides the evolutionary algorithms, as previously explained. The evaluation of model expressions is performed by exp4j [2]. This library provides a standard set of built-in functions and operators, allowing the developer to create new functions.

3.2   Parallel Execution

One of the most important features of WebGE is how the experiment execution is speed up by an automatic parallel executor. In particular, each experiment run is sliced into a set of execution tasks, which are actually run in parallel taking into account the resources available in the server where it is installed.

As seen in Fig. 2, this process is handled by the *Execution Engine* component, which takes the pending experiment runs from the waiting queue and generates the corresponding executable task.

The queue is implemented using RabbitMQ [17], which provides an asynchronous interface to reliable process all execution tasks. The queue is persistent, which means that in case of application failure or restart, the experiment can continue executing with a minimal loss of work (execution tasks dispatched but not committed are restarted). Moreover, using a neutral message broker such as RabbitMQ allows the application to completely decouple the experiment launch from the actual experiment execution. This is intentional, since a future extension will allow distributed computing to further accelerate experiment execution.

3.3   Persistence Layer

WebGE stores all the information related to users, experiments and datasets in a PostgreSQL relational database. Figure 3 shows the relational diagram of its current state of development. As seen in the figure, experiments and runs are separated into different entities since an experiment with a given set of parameters and input data, may be executed several times. Hence, each execution is considered a *run*, and all its related information is associated in the database. In addition, datasets and grammars are also separately stored, in order to allow the users to perform the typical CRUD (create, read, update, delete) operations. The information related to the user and the session are also stored in the database.

3.4   Implementation Technologies

Regarding the implementation technologies, WebGE is a Spring Boot application whose persistence layer relies on Spring Data [9], using Flyway to control the evolution of the database design [25] (Fig. 3 shows the corresponding history table). Besides, WebGE is packed as a Docker container, providing a *docker-compose* template to ease the deployment process. Notice that the use of Docker is currently considered one of the best practices for reproducible experimentation [5]. The latest Docker images of WebGE are also available at Docker Hub (https://hub.docker.com/r/jmcolmenar/webge) where releases are automatically generated by the continuous integration server, which allows a more efficient and less error-prone development process [20].

4   WebGE Most Relevant Features

Probably the most important feature of WebGE is that it provides a friendly web-based GUI supported by database storage. In addition, some other important features such as the integrated cross-fold validation and the detailed statistics are described in this section.

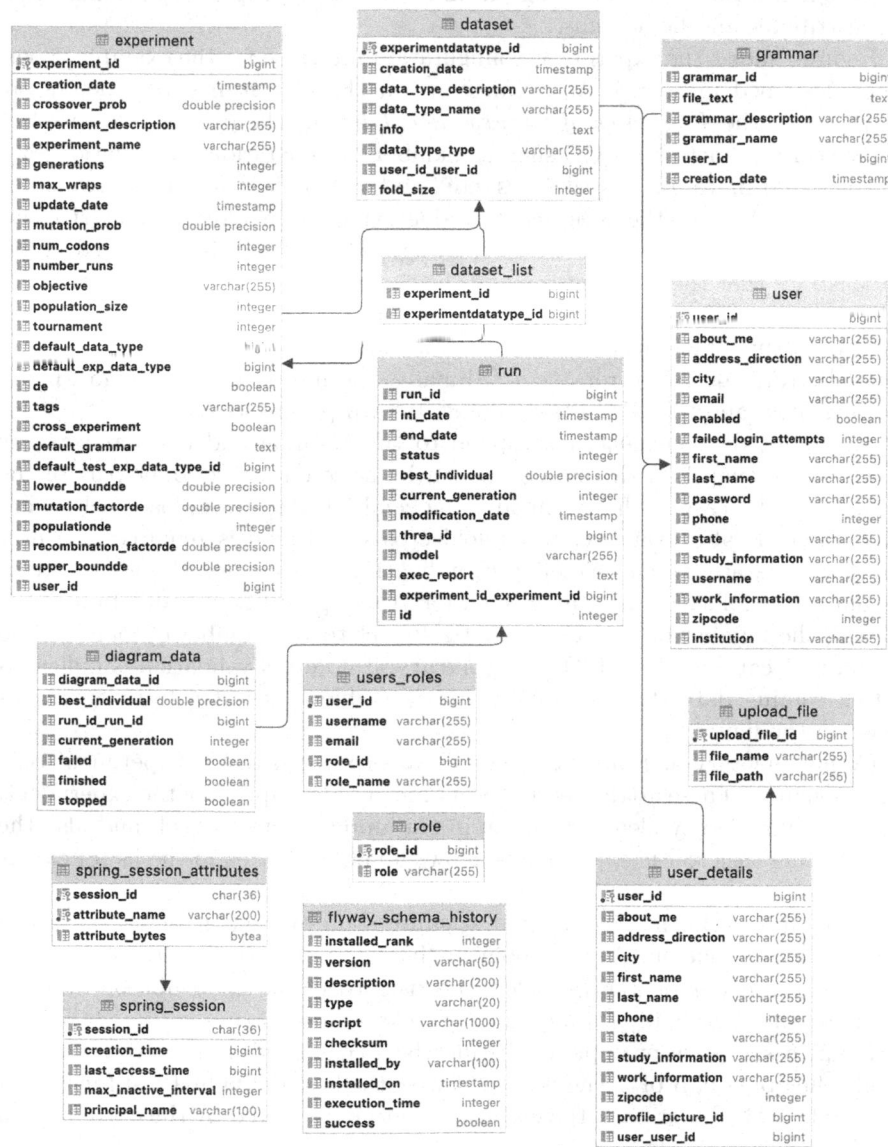

Fig. 3. WebGE relational database design.

4.1 GUI for Experiments Management

The design of WebGE pivots around the concept of experiment. An experiment includes four types of elements: algorithmic parameters, input data, complementary attributes and list of runs.

Figure 4 shows the experiment configuration interface for the example experiment described in Sect. 5. As it can be seen, the algorithmic parameters are shown under the *Properties of the experiment* label. The user is able to select the typical parameters for GE such as number of generations, population size, crossover and mutation probabilities, etc., and a specific field for the grammar, which is stored within the experiment. Besides, there is an additional interface for grammar management which allows creating, deleting and listing stored grammars. Notice that the *Copy* button and the combo box on its left allows the user to select and copy a stored grammar. In addition to these parameters, four different objective functions are available: root mean squared error (RMSE), mean squared error, absolute error and R^2. Finally, the hybrid GE+DE algorithm is also available, and the GUI allows tuning its own parameters.

The input data elements correspond to the training and test datasets. As shown in the snapshot, these datasets can be selected in the lower part of the form. As in the case of the grammars, a special interface to upload, delete and list datasets is also available. If a k-fold cross-validation is required, the user may *fold* a dataset when uploading it, indicating the number of folds to divide the data. If a folded dataset is selected for training and cross-validation run is marked, the number of runs is automatically set to the number of folds and no test dataset can be selected. On the contrary, if no cross-validation is indicated or the training dataset is not folded, it is possible to select a test dataset, as shown in the figure.

Complementary attributes can also be incorporated to the experiment configuration area. These attributes are the name and description of the experiment, which are mandatory elements shown in the upper part of Fig. 4, and also the tags. Adding tags to the experiments allows an easier search in the list of experiments.

Once an experiment is configured, it can be run by clicking on the *Save and run experiment* button. At that point, a list of runs is displayed as shown in Fig. 5. The buttons on the list allow the user to check the evolution and the results of a run, stop any (or all) runs, or delete finished runs.

Figure 6 shows an example of the interface that displays the evolution of a run. In the upper part of the figure, a plot with the best individual cost function is displayed. In the lower part, the execution report generated by the optimization algorithm is also shown.

4.2 Cross-fold Validation

In symbolic regression problems it is usual to perform cross-validation [15]. Therefore, WebGE incorporates the leave-one-out cross-validation. This procedure is automatically implemented in two ways: providing the ability to fold the

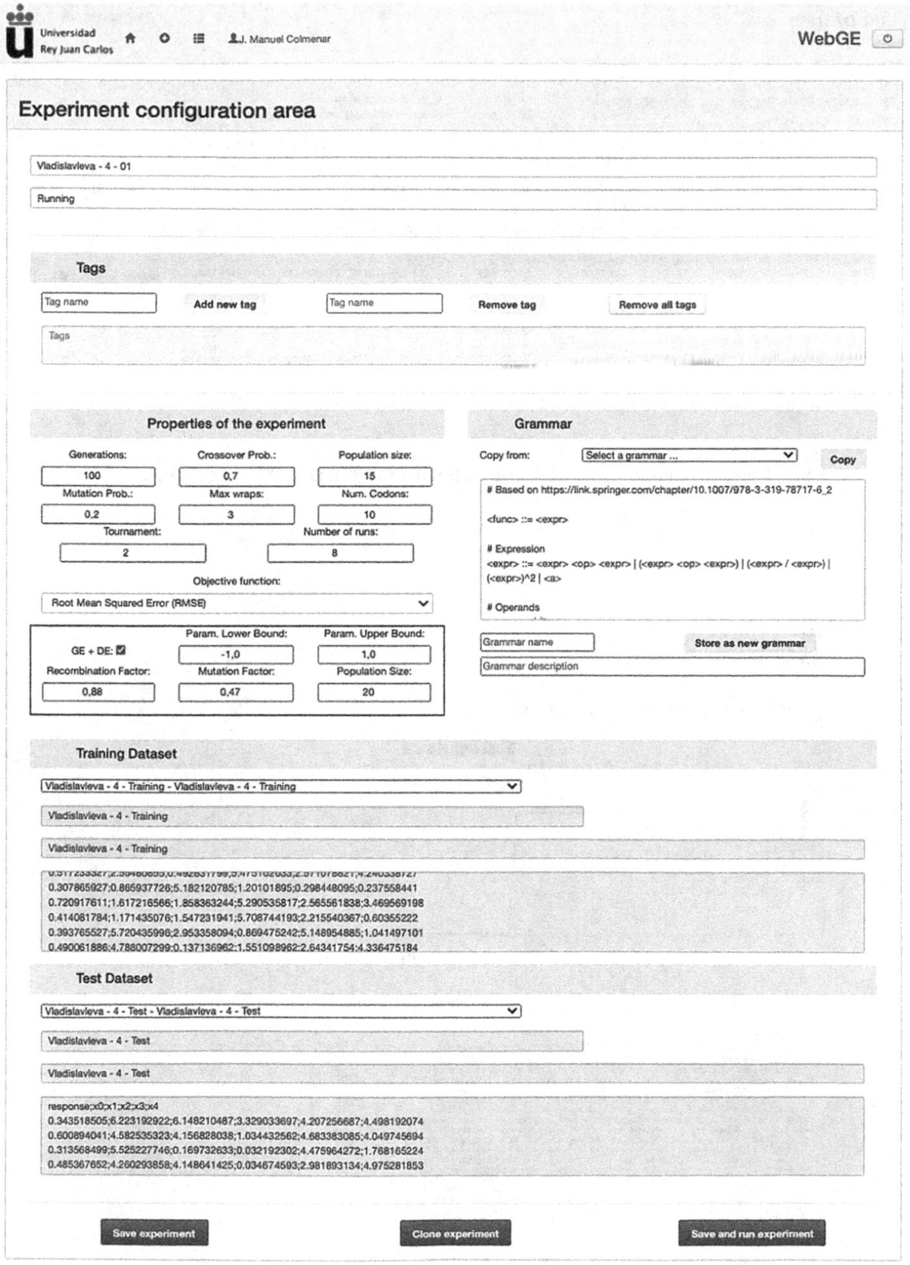

Fig. 4. Experiment configuration interface.

List of runs

Stop all runs

# Run	Best solution	Curr. Gen.	Status	Creation	Modified	Model
1	0.19607081347339866	100	FINISHED	30/03/2021 07:40:53	30/03/2021 08:55:30	(0.5937213093879579-(((X1*0.10757044732971653))^2)^2)
2	0.18984428576277848	100	FINISHED	30/03/2021 07:40:53	30/03/2021 08:57:41	(X5*-0.4441850075352228)+(X5^0.625321670383811)
3	0.20049722508807113	32	RUNNING	30/03/2021 07:40:53	30/03/2021 08:03:41	Waiting for finish
5	0.20037998263887036	100	FINISHED	30/03/2021 07:40:53	30/03/2021 08:57:03	-0.4759807259234329+((X2^0.017028733965999143))^2
6	0.18525510003001475	100	FINISHED	30/03/2021 07:40:53	30/03/2021 09:03:37	(X3^0.6241273571815942)+(X3*-0.44458303075789607)
7	0.19439413467958722	81	RUNNING	30/03/2021 07:40:53	30/03/2021 08:38:38	Waiting for finish
8	0.18525510003001483	100	FINISHED	30/03/2021 07:40:53	30/03/2021 08:58:17	(X3^0.6241273633210519)-(X3*0.444583035961619)

Avg.: 0.19136105638701545 **Std. Dev:** 0.0060064771995384505 **Min:** 0.18525510003001475 **Max:** 0.20037998263887036 Download stats Download predictions

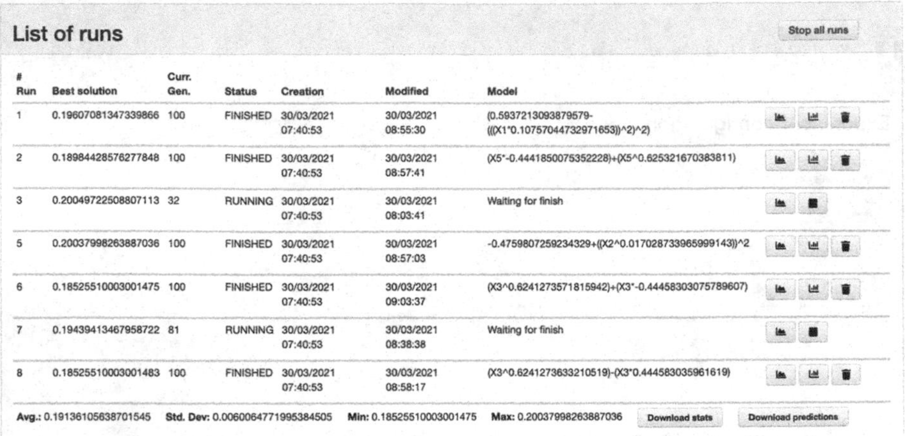

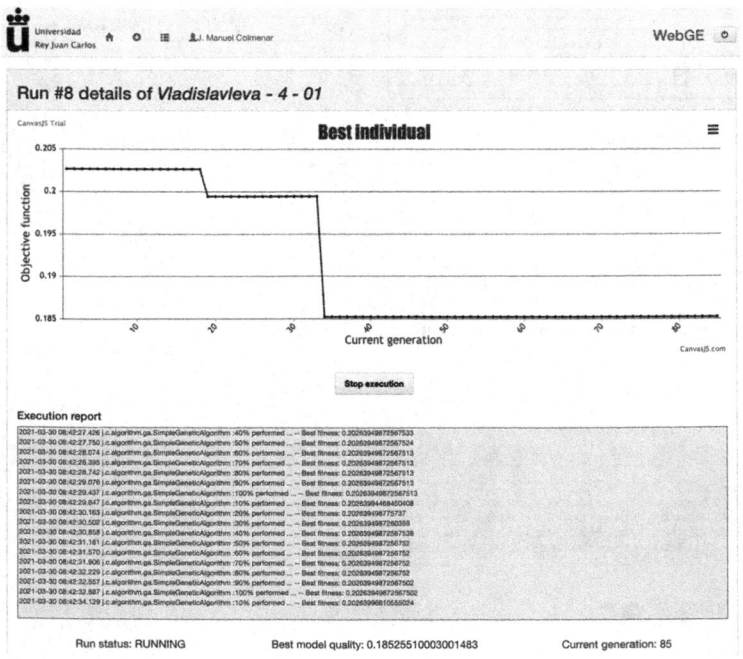

Fig. 5. List of runs for an experiment currently executing.

Fig. 6. Run progression interface.

dataset and automatically configuring the runs of the experiment to use the folds accordingly.

Regarding the folding of the dataset, this operation can be performed either when uploading the data file or once the data are stored. The user may select the number of folds (k) and, as recommended in the literature, WebGE randomly distributes the data in k folds of similar size.

If a folded training set is selected, a researcher may choose the cross-validation run. In this case, WebGE automatically configures the algorithm to execute k runs where, in run i, fold i is used for test while the rest of folds are used for training.

4.3   Detailed Statistics

Once a run is finished, the researcher may access to the detailed statistics of the model obtained in the run. Figure 7 shows a snapshot of this feature. As it can be seen, the average error, root mean squared error, absolute error, relative error and R^2 metrics are calculated for both the training and test datasets. A plot of the data generated by the model is also available for training and test, which is zoomed in the figure.

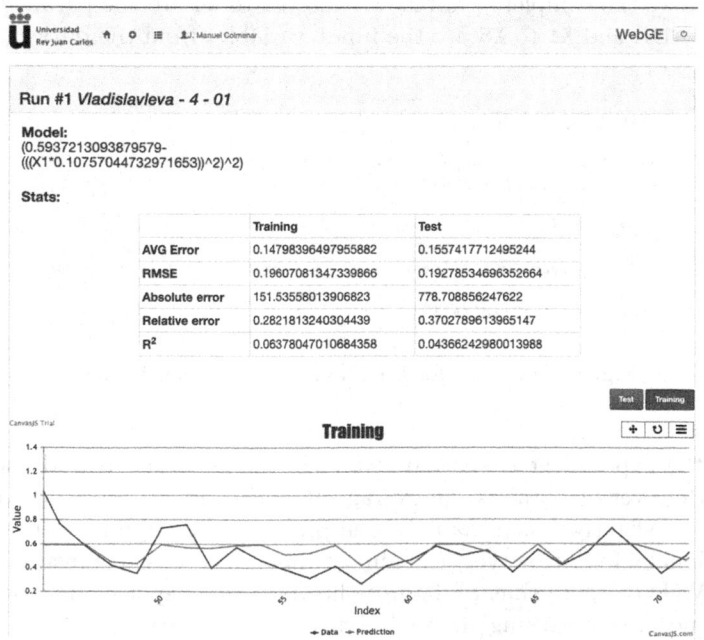

Fig. 7. Run statistics interface.

Moreover, in order to ease the work with the results, the user may download a CSV file with the detailed statistics obtained in all runs. This file is retrieved when clicking on the *Download stats* button shown in the lower part of Fig. 5. In addition, a CSV file with the predictions given by the models can be downloaded with the *Download predictions* button.

5   Use Case: Vladislavleva-4

In order to illustrate the use of WebGE, the well-known symbolic regression benchmark Vladislavleva-4 [29], also known as UBall5D, will be used. It is a synthetic benchmark with five input variables, where the target value is obtained with Eq. (1). The training dataset has 1024 entries where the five input variables values belong to the range $(0.05, 6.05)$. The validation dataset has 5000 entries and their input variables values belong to the range $(-0.25, 6.35)$.

$$\frac{10}{5 + \sum_i^5 (x_i - 3)^2} \tag{1}$$

In this example, the GE+DE algorithm is run using the parameter values shown in Fig. 4. The grammar used in this experiment is an adaption of the grammar for Vladislavleva-4 proposed in [21] where the constant value generation of GE has been replaced with parameters whose values will be explored by DE. Figure 8 shows the complete grammar where w1 to w4 are the parameters to be explored by DE and X1 to X5 are the input variables from the dataset.

```
<func> ::= <expr>

<expr> ::= <expr> <op> <expr> | (<expr> <op> <expr>) |
           (<expr> / <expr>) | (<expr>)^2          | <a>

<op> ::= +|-|*

<a> ::= (<var>^w1) | (<var> + w2) | (<var> * w3) | <var> | w4

<var> ::= X1|X2|X3|X4|X5
```

Fig. 8. Grammar for the Vladislavleva-4 benchmark.

After the experiment execution, WebGE was not able to find the optimal solution. However, it obtained an average RMSE value of 1.4643 for validation. This result is slightly worse than the one presented in [29], but using much less computational effort. However, the aim of this use case is to illustrate the ease of use of WebGE (note that all figures shown in this paper come from this use case) and not benchmarking the underlying JECO library.

6   Conclusions

In this paper, WebGE, an open-source tool for symbolic regression problem optimization based on Grammatical Evolution (GE), is presented. WebGE provides a friendly web-based user interface which allows researchers with no programming background to deal with this kind of problems. WebGE also includes as optimization algorithm the combination of GE with Differential Evolution, which allows a greater intensification of the search process.

WebGE can be easily deployed on any operating system since it has been packaged using the Docker container technology. In addition, all the information about experiments is synchronized to the persistence layer in real time, which allows WebGE to be used on a shared server by several concurrent users. The design of WebGE is modular and extensible, which allows the future integration of different algorithms to tackle new families of problems.

References

1. Adaptive and Bioinspired Systems Group: ABSys JECO (Java Evolutionary COmputation) library. https://github.com/ABSysGroup/jeco. Accessed 2021
2. Asseg, F., Chatterjee, S.: exp4j: a library for expression evaluation in Java. https://www.objecthunter.net/exp4j/index.html. Accessed 2021
3. Assunção, F., Lourenço, N., Ribeiro, B., Machado, P.: Evolution of scikit-learn pipelines with dynamic structured grammatical evolution. In: Castillo, P.A., Jiménez Laredo, J.L., Fernández de Vega, F. (eds.) EvoApplications 2020. LNCS, vol. 12104, pp. 530–545. Springer, Cham (2020). https://doi.org/10.1007/978-3-030-43722-0_34
4. Augusto, D.A., Barbosa, H.J., Barreto, A.M., Bernardino, H.S.: A new approach for generating numerical constants in grammatical evolution. In: Proceedings of the 13th Annual Conference Companion on Genetic and Evolutionary Computation, pp. 193–194 (2011)
5. Boettiger, C.: An introduction to Docker for reproducible research. ACM SIGOPS Oper. Syst. Rev. **49**(1), 71–79 (2015)
6. Charte, F., Vico, A., Pérez-Godoy, M.D., Rivera, A.J.: predtoolsTS: R package for streamlining time series forecasting. Prog. Artif. Intell. **8**(4), 505–510 (2019). https://doi.org/10.1007/s13748-019-00193-z
7. Coletti, M.A., Scott, E.O., Bassett, J.K.: Library for evolutionary algorithms in Python (LEAP). In: Proceedings of the 2020 Genetic and Evolutionary Computation Conference Companion, pp. 1571–1579 (2020)
8. Colmenar, J., Hidalgo, J., Salcedo-Sanz, S.: Automatic generation of models for energy demand estimation using grammatical evolution. Energy **164**, 183–193 (2018)
9. Davis, A.L.: Spring data. In: Spring Quick Reference Guide, pp. 43–59. Apress, Berkeley, CA (2020). https://doi.org/10.1007/978-1-4842-6144-6_6
10. De Smet, G., open source contributors: OptaPlanner User Guide. Red Hat, Inc. or third-party contributors (2006). https://www.optaplanner.org. Accessed 2021
11. Dempsey, I., O'Neill, M., Brabazon, A.: Constant creation and adaptation in grammatical evolution. In: Foundations in Grammatical Evolution for Dynamic Environments, vol. 194, pp. 69–104. Springer, Heidelberg (2009). https://doi.org/10.1007/978-3-642-00314-1_5
12. Durillo, J.J., Nebro, A.J.: jMetal: a Java framework for multi-objective optimization. Adv. Eng. Softw. **42**(10), 760–771 (2011)

13. Elyasaf, A., Sipper, M.: Software review: the HeuristicLab framework. Genet. Program. Evolvable Mach. **15**(2), 215–218 (2014). https://doi.org/10.1007/s10710-014-9214-4
14. Estévez-Velarde, S., Gutiérrez, Y., Almeida-Cruz, Y., Montoyo, A.: General-purpose hierarchical optimisation of machine learning pipelines with grammatical evolution. Inf. Sci. **543**, 58–71 (2021)
15. Fushiki, T.: Estimation of prediction error by using k-fold cross-validation. Stat. Comput. **21**(2), 137–146 (2011)
16. Hidalgo, J.I., et al.: Glucose forecasting combining Markov chain based enrichment of data, random grammatical evolution and bagging. Appl. Soft Comput. **88**, 105923 (2020)
17. Johansson, L., Dossot, D.: RabbitMQ Essentials: Build Distributed and Scalable Applications with Message Queuing Using RabbitMQ. Packt Publishing Ltd., Hawthorne, USA (2020)
18. Martínez-Rodríguez, D., Colmenar, J.M., Hidalgo, J.I., Villanueva Micó, R.J., Salcedo-Sanz, S.: Particle swarm grammatical evolution for energy demand estimation. Energy Sci. Eng. **8**(4), 1068–1079 (2020)
19. Mauceri, S., Sweeney, J., McDermott, J.: One-class subject authentication using feature extraction by grammatical evolution on accelerometer data. In: Yalaoui, F., Amodeo, L., Talbi, E.-G. (eds.) Heuristics for Optimization and Learning. SCI, vol. 906, pp. 393–407. Springer, Cham (2021). https://doi.org/10.1007/978-3-030-58930-1_26
20. Meyer, M.: Continuous integration and its tools. IEEE Softw. **31**(3), 14–16 (2014)
21. Nicolau, M., Agapitos, A.: Understanding grammatical evolution: grammar design. In: Ryan, C., O'Neill, M., Collins, J.J. (eds.) Handbook of Grammatical Evolution, pp. 23–53. Springer, Cham (2018). https://doi.org/10.1007/978-3-319-78717-6_2
22. O'Neill, M., Ryan, C.: Grammatical Evolution: Evolutionary Automatic Programming in an Arbitrary Language. Kluwer Academic Publishers, Norwell (2003)
23. Qaddoura, R., Faris, H., Aljarah, I., Castillo, P.A.: EvoCluster: an open-source nature-inspired optimization clustering framework. SN Comput. Sci. **2**(3), 1–12 (2021)
24. Ramírez, A., Romero, J.R., García-Martínez, C., Ventura, S.: JCLEC-MO: a Java suite for solving many-objective optimization engineering problems. Eng. Appl. Artif. Intell. **81**, 14–28 (2019)
25. Red Gate Software Ltd: Flyway open-source database migration tool. https://flywaydb.org/. Accessed 2021
26. Scott, E.O., Luke, S.: ECJ at 20: toward a general metaheuristics toolkit. In: Proceedings of the Genetic and Evolutionary Computation Conference Companion, pp. 1391–1398 (2019)
27. Storn, R., Price, K.: Differential evolution-a simple and efficient heuristic for global optimization over continuous spaces. J. Global Optim. **11**(4), 341–359 (1997)
28. Tian, Y., Cheng, R., Zhang, X., Jin, Y.: PlatEMO: a MATLAB platform for evolutionary multi-objective optimization. IEEE Comput. Intell. Mag. **12**(4), 73–87 (2017)
29. Vladislavleva, E.J., Smits, G.F., Den Hertog, D.: Order of nonlinearity as a complexity measure for models generated by symbolic regression via pareto genetic programming. IEEE Trans. Evol. Comput. **13**(2), 333–349 (2008)

A New Genetic Algorithm for Automated Spectral Pre-processing in Nutrient Assessment

Demelza Robinson[1] , Qi Chen[1($\boxtimes$)] , Bing Xue[1] , Daniel Killeen[2] ,
Keith C. Gordon[3] , and Mengjie Zhang[1]

[1] School of Engineering and Computer Science, Victoria University of Wellington,
PO Box 600, Wellington, New Zealand
{Robinsdeme,Qi.Chen,Bing.Xue,Mengjie.Zhang}@ecs.vuw.ac.nz
[2] The New Zealand Institute for Plant and Food Research Limited,
Nelson, New Zealand
daniel.killeen@plantandfood.co.nz
[3] Department of Food Science, University of Otago, Dunedin, New Zealand
keith.gordon@otago.ac.nz

Abstract. Vibrational spectroscopy can be used for rapid nutrient assessment of horticultural produce as a means of quality control. Most commonly, spectral data are calibrated against chemical reference data, which are acquired through resource-intensive analytical methods, using partial least squares regression (PLSR). Recently, genetic algorithms (GAs) have been applied to assist PLSR to construct high-performing models through feature selection and latent variable selection. The current approach relies on manually pre-processed data, which requires human expertise and produces inherent biases. To address this limitation, this paper aims to develop a new GA method for automatically selecting the most appropriate pre-processing techniques for specific tasks to bypass manual pre-processing. The results for infrared spectroscopy show the potential of this approach in out-performing manual pre-processing, while the Raman spectroscopy results are competitive, which demonstrates the utility of the approach in terms of saving time and resources.

Keywords: Genetic algorithm · Vibrational spectroscopy · Automated pre-processing

1 Introduction

The quality of many horticultural products depends on concentrations of key flavour, nutritional or bioactive molecules. If these components can be measured rapidly e.g. on a production line or at the time of harvesting, product quality can be improved and profits maximised [1]. However, accurate quantitation usually requires slow and expensive analyses using traditional analytical chemistry techniques such as high-performance liquid chromatography and gas chromatography. To overcome this, research over the past few decades has focused on

© Springer Nature Switzerland AG 2022
J. L. Jiménez Laredo et al. (Eds.): EvoApplications 2022, LNCS 13224, pp. 283–298, 2022.
https://doi.org/10.1007/978-3-031-02462-7_19

replacing traditional methods with more rapid analytical techniques, such as vibrational spectroscopy i.e. infrared, near infrared and Raman spectroscopy [2]. These methods are cheaper, more environmentally friendly, and the instruments are robust (no moving parts). They can also be coupled to fibre-optic probes (allowing in-line and on-line analysis), and are rapidly decreasing in both cost and size, making them well-suited to production environments [3,4]. On the other hand, a major disadvantage of vibrational spectroscopy is that spectral data cannot predict chemical compositions directly. Chemometrics solves this issue, by constructing regression models to relate structured variance in spectral data sets to accurate reference data acquired using traditional, accurate analytical methods [5]. Regression models produced in this way are then capable of predicting important chemistry in new samples using only rapidly acquired spectral data.

Partial least squares regression models (PLSR) have been used previously to calibrate the spectral data against the chemical reference data [6,7]. PLSR is a form of regression that undertakes dimensionality reduction. It finds new combinations of the features, that maximise the variance such that in theory, the most relevant information is extracted from the features. This forms what are commonly referred to as latent variables, as they are constructions of the original features in the latent space. These PLSR models can still be limited in performance due to the noisy, redundant and collinear nature of spectral data [8]. Accordingly, in [9] we developed a genetic algorithm (GA) enhanced PLSR method which extends GA for feature selection and latent variable selection to improve the performance of the PLSR model. This approach relies on the spectral data having been manually pre-processed by experts with domain knowledge. However, this can introduce biases into the results as the choice of which pre-processing methods to apply and in which order is quite subjective. Researchers have different preferences rather than being an agreed-upon consensus for how to apply these pre-processing techniques. In addition to requiring considerable domain knowledge, manual pre-processing is also very time-consuming. By enabling the algorithm itself to automate this pre-processing phase, such biases and resource constraints can be alleviated.

1.1 Goals

To address the limitations of manual spectral pre-processing, this paper will further develop the GA enhanced PLSR method by allowing it to automatically select the spectral pre-processing methods. To this aim, three objectives are set up to guide this work.

- investigate and determine which spectral pre-processing techniques should be made available for the algorithm to choose from,
- develop a new representation for pre-processing method selection, and
- develop a new search method to perform feature and latent variable selection, and pre-processing selection simultaneously and effectively.

1.2   Organisation

The remainder of this paper is arranged as follows. Section 2 is a literature review detailing the background and related work for this study. Section 3 presents a new GA method for pre-processing selection. In Sect. 4, the experimental design of this approach is described. Section 5 contains the results and discussions. Finally, Sect. 6 provides conclusions and highlights some future directions.

2   Background and Related Work

2.1   Vibrational Spectroscopy

Vibrational spectroscopic techniques have been previously applied in efforts to assess and quantify the phytochemical composition of New Zealand horticultural products [6,7]. These include Fourier transform (FT) Raman spectroscopy [10], FT-infrared (IR) spectroscopy [11] and near infrared (NIR) spectroscopy [12]. All spectra correspond to the infrared range of the spectrum and provide structural information. This includes characteristic vibrations known as the fingerprint region as well as information on the functional groups [13]. There exist fundamental differences between these spectroscopy techniques such that all three can offer complementary information [14]. The fingerprinting capacity of these techniques permits rapid, high-throughput and non-destructive nutrient quantification [15]. These factors allow vibrational spectroscopy to be implemented on the production line.

2.2   Partial Least Squares Regression

Chemometrics, i.e., the field of extracting information from chemical systems, relies on mathematical and statistical methods to extract relevant information from the data. To carry out the nutrient assessment task, useful information must be extracted from the spectral data and related to the wet chemistry reference data. This is a form of multivariate regression wherein each spectral intensity value is related to one bioactive component per model. Table 1 shows an example of the data used for multivariate regression relating Raman spectral data to the bioactive component, lupulone. Multivariate regression models relate the spectroscopic data to the reference data. These models take a given spectra across a range of wavenumbers as the input variables (or features), and the bioactive component as the output/target variable.

Partial least squares regression (PLSR) is a multivariate regression technique which is commonly used in Chemometrics [6,7]. PLSR relates the spectral and reference data to each other through a linear multivariate model. It constructs latent variables through searching for the multidimensional direction of the input variables/feature space that maximises the multidimensional variation in the target variables. The resultant latent variables model the covariance of the spectral features and are regressed against the bioactive component target variable [16].

Table 1. An example of Raman spectral data and the bioactive component of lupulone

	Bioactive component	Raman intensity by wavenumber					
	Lupulone	...	3498.8	3497.4	3496.0	3494.6	3493.1	...
Instances	1.858	...	0.8457	0.8442	0.8427	0.8411	0.8393	...
	1.67	...	0.8186	0.8170	0.8153	0.8135	0.8117	...
	...	...	...	...	...	...	...	...
	1.968	...	0.8522	0.8507	0.8494	0.8480	0.8467	...
	1.824		0.8311	0.8296	0.8282	0.8266	0.8247	

Given the spectral features X, and a given bioactive component Y, the PLSR models are typically represented as:

$$X = TP^T + E_1$$

$$Y = UQ^T + E_2$$

where T and U are projections of X and projections of Y, respectively. P and Q are orthogonal loading matrices; and E_1 and E_2 are the error terms, which are assumed to be random normal variables. The constructed PLSR model can then be used to predict the amount of each given bioactive component when provided with new spectral data samples.

2.3   Spectral Pre-processing

Pre-processing techniques applied to spectra, with the aim of removing unwanted noise or physical phenomena from the spectra, can be divided into three categories: baseline correction, normalisation and derivitisation. Due to the page limit, this subsection presents a brief outline of these three categories.

Baseline Correction: Baseline drift is an inherent issue in spectral data that arises due to interference or background effects during the collection process. It reduces the quality of the spectral signal and thus impairs the results in a multivariate analysis. Baseline drift must be corrected such that the true signal is available for regression [17]. Three baseline correction methods, which are commonly considered in the manual spectral data pre-processing procedure, are included in this work, i.e., adaptively iteratively re-weighted penalised least squares (AirPLS) [18], rubber band baseline correction [19] and asymmetric least squares (ALS) [20].

Normalisation: Normalisation is used to normalise the spectra along the wavenumber axis as absorbance is not always constant across this axis [21]. Three normalisation methods are included: min-max normalisation, standard normal variate (SNV) [22] and multiplicative scatter correction (MSC) [23].

Derivatisation: Derivatisation is used to enhance the features of overlapping bands [24]. This group of methods is often applied when lower-frequency (broad) features are noise, and higher-frequency (narrow) features contain the signal of interest. It enables useful information to be extracted from the higher-frequency features. In this work, Savitsky-Golay [25] with seven different window sizes, 5, 9, 13, 17, 21, 25 and 29, are available for the algorithm to select from.

2.4 PLSR for Nutrient Assessment

Killeen et al. [6] aimed to determine which of the three vibrational spectroscopy techniques, IR, NIR and Raman, is best suited to the rapid screening task by assessing 139 hops samples. PLSR models were generated to relate the data gathered by each spectroscopy technique to each bioactive component. Suitably high performing PLSR models were only obtained for a few of the bioactive components. Their work found that different vibrational techniques are more appropriate for different bioactive components when constructing PLSR models, e.g., both IR and Raman spectra could be related to Xanthohumol to achieve suitably high performance while NIR spectra could not be. A GA-enhanced PLSR approach which undertakes feature selection and latent variable selection has been developed for quantifying bioactive components in New Zealand fruit and plant from spectral data [9]. In [9], a feature selection process which intensities from the spectral data, is incorporated into the construction of latent variables, and the number of latent variables is also optimised for each PLSR model. Utilising a combination of feature selection and latent variable selection has shown to enable better generalisation performance in the vast majority of models.

3 The Proposed Approach

To give the GA-enhanced PLSR (GA-PLSR) method [9] an automatic pre-processing capability, this work extends the GA method by co-evolving two populations of individuals for different tasks. The new method in this work is named *GA-PLSR with pre-processing selection* (GA-PLSR-PPS). In GA-PLSR-PPS, the first population consists of individuals for feature and latent variable selection (FLVS). To address the task of pre-processing selection, another small population of individuals, which is referred to as the *PPS* population, will be *co-evolved* with the original population.

3.1 Representations for the Two Populations for Co-evolution

The representations of the individuals corresponding to the two populations are shown in Fig. 1. The top subplot illustrates the structure of the individuals in the FLVS population. The bottom subplot shows the structure of an individual belonging to the PPS population.

The FLVS population contains individuals used for feature selection and latent variable selection. The first n bits correspond to the number of features

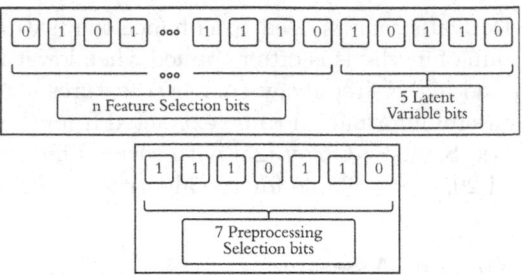

Fig. 1. Chromosomes of the proposed method.

Table 2. Decimilising the bits gives the corresponding pre-processing method for each category. The # methods includes applying no pre-processing technique in the category.

	# Methods available	# Bits
Baseline correction	4	2
Normalisation	4	2
Derivisation	8	3

while the last five bits correspond to the five latent variables. The feature selection bits simply determine whether or not a given spectral intensity value is included or excluded in the model, as reflected by 1 or 0, respectively. The five latent variable bits can be decimalised to produce an integer value ranging from 1 to 32. This denotes how many of the constructed latent variables should be included in the final PLSR model. This is less than or equal to the total number of features. The rationale behind latent variable selection is that PLS maximises the variance in the original feature space in latent variables. There are as many latent variables as features, and the variance they capture decreases with the first containing the most, and the last containing the least. Consequently, it is not necessary to include all latent variables in the final model. Moreover, latent variable selection also aids with the complexity of PLSR model.

Each pre-processing selection individual in the PPS population, which is also represented as bit strings, is decimalised to inform the way in which the spectral pre-processing techniques described in Subsect. 2.3 will be applied to the spectral data. These chromosomes consist of seven bits with the first two corresponding to baseline correction, the next two corresponding to normalisation and the last three corresponding to derivatisation. Each pre-processing category has a variety of methods available in it, including the option of not applying any method in this category. For a category with four methods available (i.e., baseline correction), this requires two bits as converting these to decimal yields four options. Similarly, for a category with eight options in it (i.e., derivatisation), this requires three bits as converting these to decimal gives eight options as shown in Table 2.

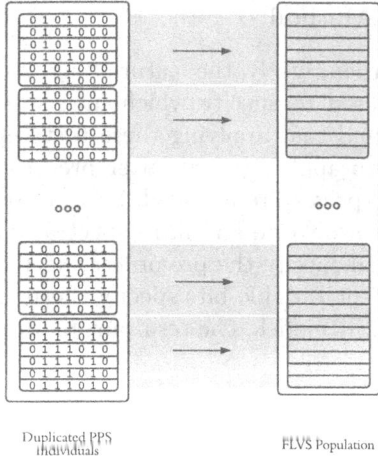

Duplicated PPS
Individuals

FLVS Population

Fig. 2. The population containing the pre-processing selection individuals (left) is created through duplication of a single smaller population to map each chromosome to many individuals in the feature selection & latent variable selection population (right).

As we can see, the search space for the optimal pre-processing regime is significantly smaller than that for the feature selection and latent variable selection task, with individuals that can be thousands of bits in size. To avoid inadequate selection pressure being applied to the PPS bits relative to the FLVS bits, and also to encourage the algorithm to consider both the PPS and FLVS tasks as equally important, the decision was made to split the two into separate populations.

3.2   Mapping of the Two Populations for Pairwise Evaluations

In the evolutionary process, the size of the PPS population needs to be smaller than that of the FLVS population as fewer unique combinations are possible, which means the additional computational effort of co-evolving a second population of the same size is unnecessary. However, the two populations of individuals need to be evaluated in a single sequential process, which refers to a pairwise evaluation that contains one PPS individual and one FLVS individual. To permit this pair-wise evaluation, the smaller PPS population is replicated to be the same size as the FLVS population. The replication, which bypasses having to create a large population of PPS individuals that are not necessarily unique from one another, enables mapping of one PPS individual to many individuals in the FLVS population as shown in Fig. 2. Note that the replication for the purpose of mapping is only used for the evaluation process, but not the evolutionary process of applying genetic operators. The latter would have a very high associated computational cost at no benefit to the performance, as the search space for the PPS task is significantly smaller.

3.3   The Evaluation Method

For each pairwise evaluation, firstly, the feature selection bits of the FLVS individual in the pair are used to specify which intensities or features are used in the model. It was found that applying these first resulted in better performance when compared to applying them after pre-processing. When noisy or irrelevant regions of the spectra are included, this can impair the ability of pre-processing methods which operate on entire spectra. Next, the pre-processing bits of the PPS individual specify the pre-processing regime. Thirdly, from the FLVS individual, the latent variable bits specify the number of latent variables to be included in the PLSR model. The resultant model can then be evaluated.

Kennard-Stone Sampling: A common trend in vibrational spectroscopic data is that it contains a small number of instances, due to the difficulty of collecting it. As a result, it is probable that any model constructed from the training data may struggle to elucidate all the information that is present in the unseen data. Evaluating the fitness of the GA individuals directly from the training performance of the PLSR models can achieve a low bias but with high variation, which can correspond to poor generalisation on unseen data. To avoid this limitation, in this work, the training data is split into training and fitness evaluation sets.

Kennard-Stone sampling [26], which allows the selection of samples with a uniform distribution over the feature space, is used to divide the original training data into a training set and a fitness evaluation set. Kennard-Stone operates by iteratively finding the most separated points in the training data. The metric for separation is Euclidean distance. This approach aims to allocate representative and diverse samples into the training set to learn from, with the remaining samples used in the fitness evaluation set. This works well in conjunction with PLSR [27], as this regression method constructs latent variables by maximising the variance. The chance of missing valuable information, which is needed for the model to generalise to unseen data, should therefore theoretically be reduced.

Both of the aforementioned training and fitness evaluation are used for training. The training set is used to construct the PLSR model. This model is fitted to the spectral data of the fitness evaluation set and the prediction performance is evaluated. This is more complex than a traditional training process in supervised learning. In addition to this, our approach uses a validation set for model selection, which will be described further in Sect. 4.

Fitness Function: The fitness function of the proposed GA method minimises three components. The first component is included as a metric for regression performance and the latter two in order to restrict model complexity. The number of spectral pre-processing techniques was not included in the fitness function since it is neither a direct measure of the performance nor a measure of the model complexity. The details of the fitness function are shown in Eq. (1)

$$fitness = w_1 \times MSE + w_2 \times \#Features + (1 - w_1 - w_2) \times \#LVs \quad (1)$$

Table 3. Hop data sets

Spectroscopy technique	#Features	#Instances	Bioactive components (target variables)
IR	679	139	Alpha Acids (HPLC), Alpha Acids (UV), Beta Acids (HPLC), Beta Acids (UV), Total Acids (HPLC), Total Acids (UV),
NIR	429		Cohumulone, Colupulone,
Raman	1524		Humulone, Lupulone, Xanthohumo

where MSE is the mean square error of the regression model on the fitness evaluation set, $#Features$ is the number of features, which is to be minimised in such a way features that do not offer much useful information should hypothetically be excluded from the model. $#LVs$ refers to the number of latent variables, which is minimised so that the PLSR model will use a small number of latent variables, and will further reduce the PLSR model complexity.

The two parameters w_1 and w_2 in the fitness function are set to 0.99 and 0.005 respectively. These values are set via empirical search with a bias to a larger value of w_1. The weights affect the influence of each component on the overall fitness of the individual. w_1 is weighted more highly as MSE is considered the most significant to the overall fitness. Specifically, there is no sense in minimising the number of features or latent variables at the expense of constructing an extremely low accuracy model. The fitness function is therefore a means of balancing the accuracy and complexity of the model.

4 Experiment Design

4.1 Datasets

Table 3 shows the data sets used in experiments. The spectroscopic data sets were acquired at Otago University, New Zealand. Reference data were generated by New Zealand Institute for Plant and Food Research and described in [6].

In the experiments, each data set are randomly divided into two sets with 80% of the data forming the training data and the remaining 20% forming the test set. 30 random splits are used for the 30 runs of experiments in each method. The training data is divided into three sets with Kennard-Stone sampling: 60% will be used as the training set, 20% will form a validation set and the remaining 20% will be the fitness evaluation set for GA. The training data is used to fit the PLSR model. The validation set is used at the end of training to select the best model, so as to prevent over-fitting to the training data.

4.2 Parameter Settings

Through a process of empirical search a population size of 50 and 50 generations was found to be suitable. Early stopping is included such that when the validation performance is observed to no longer be improving, the evolutionary process is stopped prematurely. In terms of the selection operators, individuals are chosen using tournament selection with a tournament size of three. In terms of genetic operators, crossover and mutation are applied to promote diversity. Two point crossover is used, with a rate of 0.5. Flip bit mutations are used with a rate of 0.2. These parameter settings were found to be the best combination in several trials of preliminary experiments.

The original PLSR technique and the GA-PLSR method [9], which is the baseline method of this work, are used as two benchmark techniques in the experiments for comparisons.

5   Results and Discussions

The results presented in this section include the training and the test performance in terms of the mean squared error (MSE) of the PLSR models. The mean and standard deviation of the training and the test MSEs obtained in the three methods are shown in Table 4. The smallest MSEs are highlighted in bold. Mann-Whitney U testing is used to test the statistical significance of any improvements. Additionally, the distribution of pre-processing techniques selected across the models is included. Finally, some plots illustrating the feature selection patterns are shown.

5.1   Comparisons on the Training and Test Performance

As shown in Table 4, the proposed method, GA-PLSR-PPS, generally achieves the best performance on most of training data where it outperforms PLSR on *all* the cases and achieves lower training MSEs than GA-PLSR-PPS on 30 out of the 33 comparisons. According to the Mann-Whitney U test, all the differences between these methods are statistically significant. Another finding is that, in the vast majority of cases, the standard deviation in GA-PLSR-PPS is also much smaller, which indicates that the pre-processing selection approach helps the PLSR model to predict the target variables with less variability. However, there exist examples, e.g., the Raman-specta and α-acids model, where the standard deviation of the MSE values is larger than the MSE value itself. This suggests that, for some models, the solution the model converges on is highly variable. This could be due to a limited number of instances being available relative to a large number of features, such that it is very challenging for the model to consistently obtain the same high performance over the 30 different training and test splits.

Considering the test performance, the pattern is different from that in the training data. The comparisons on the three PLSR models are very different across the three types of spectral data. GA-PLSR-PPS has worse test performance than the two baseline methods on most of the examined test sets in the NIR data. It is better on the Raman-based models where GA-PLSR-PPS has a competitive test MSE as that in GA-PLSR in most cases, but usually much

Table 4. Training and test MSEs in PLSR, GA-PLSR model and GA-PLSR-PPS.

Model	Bioactive Component	Training			Test		
		IR	NIR	Raman	IR	NIR	Raman
PLSR	α-acids (UV)	5.51 ± 2.54	4.81 ± 1.15	4.66 ± 2.12	2.57 ± 0.79	2.76 ± 0.82	3.29 ± 1.32
GA-PLSR		1.02 ± 0.72	1.12 ± 1.07	1.11 ± 0.87	1.54 ± 0.69	**2.23 ± 0.95**	**2.63 ± 1.01**
GA-PLSR-PPS		**0.11 ± 0.22**	**0.71 ± 1.86**	**0.31 ± 0.51**	**1.29 ± 0.55**	18.30 ± 8.32	2.84 ± 1.12
PLSR	β-acids (UV)	1.25 ± 0.25	1.83 ± 0.57	1.21 ± 0.30	0.75 ± 0.21	1.50 ± 1.18	0.75 ± 0.23
GA-PLSR		0.41 ± 0.22	0.58 ± 0.39	0.32 ± 0.20	0.58 ± 0.22	**0.86 ± 0.28**	**0.62 ± 0.26**
GA-PLSR-PPS		**0.07 ± 0.12**	**0.20 ± 0.33**	**0.05 ± 0.09**	**0.46 ± 0.19**	2.17 ± 1.14	0.66 ± 0.29
PLSR	Total-acids (UV)	11.53 ± 19.53	9.81 ± 13.35	7.65 ± 3.38	4.37 ± 1.38	3.73 ± 2.78	4.83 ± 1.51
GA-PLSR		1.44 ± 0.99	**1.34 ± 1.27**	0.77 ± 1.02	2.87 ± 1.38	**2.57 ± 0.94**	**4.34 ± 2.08**
GA-PLSR-PPS		**0.34 ± 0.53**	2.86 ± 4.70	**0.34 ± 0.69**	**2.27 ± 1.01**	26.60 ± 12.71	4.84 ± 2.16
PLSR	Xanthohumol	0.02 ± 0.01	0.02 ± 0.00	0.02 ± 0.00	0.01 ± 0.00	3.73 ± 2.78	1.00 ± 1.01
GA-PLSR		0.01 ± 0.00	0.01 ± 0.00	0.01 ± 0.00	0.01 ± 0.00	**0.01 ± 0.00**	**0.01 ± 0.00**
GA-PLSR-PPS		**0.00 ± 0.00**	0.02 ± 0.01	**0.00 ± 0.00**	0.01 ± 0.00	0.04 ± 0.02	**0.01 ± 0.00**
PLSR	Cohumulone	0.50 ± 0.17	0.70 ± 0.94	0.46 ± 0.12	0.18 ± 0.04	0.29 ± 0.10	0.28 ± 0.06
GA-PLSR		0.13 ± 0.06	0.23 ± 0.09	0.16 ± 0.08	0.16 ± 0.06	**0.27 ± 0.09**	**0.24 ± 0.12**
GA-PLSR-PPS		**0.02 ± 0.03**	**0.17 ± 0.21**	**0.04 ± 0.06**	**0.10 ± 0.04**	1.37 ± 0.86	0.25 ± 0.11
PLSR	Humulone	2.70 ± 0.78	3.58 ± 1.70	2.62 ± 0.66	1.45 ± 0.49	1.94 ± 1.12	1.81 ± 0.66
GA-PLSR		0.85 ± 0.43	0.74 ± 0.65	0.70 ± 0.58	1.09 ± 0.56	**1.67 ± 0.59**	**1.60 ± 0.56**
GA-PLSR-PPS		**0.12 ± 0.21**	**0.68 ± 1.32**	**0.17 ± 0.29**	**0.71 ± 0.35**	7.83 ± 4.79	**1.60 ± 0.73**
PLSR	Colupulone	0.29 ± 0.10	0.69 ± 1.04	4.26 ± 17.89	0.16 ± 0.04	0.44 ± 0.23	0.30 ± 0.10
GA-PLSR		0.12 ± 0.04	0.22 ± 0.10	0.14 ± 0.06	0.16 ± 0.05	**0.35 ± 0.22**	0.26 ± 0.09
GA-PLSR-PPS		**0.02 ± 0.02**	**0.11 ± 0.13**	**0.02 ± 0.03**	**0.11 ± 0.06**	0.70 ± 0.41	**0.18 ± 0.09**
PLSR	Lupulone	0.36 ± 0.07	0.43 ± 0.08	0.32 ± 0.07	0.17 ± 0.06	0.32 ± 0.19	0.18 ± 0.05
GA-PLSR		0.15 ± 0.04	0.25 ± 0.06	0.15 ± 0.03	0.17 ± 0.06	**0.23 ± 0.06**	**0.15 ± 0.05**
GA-PLSR-PPS		**0.01 ± 0.01**	**0.09 ± 0.11**	**0.02 ± 0.03**	**0.09 ± 0.05**	0.42 ± 0.28	0.16 ± 0.08
PLSR	α-acids (HPLC)	2.11 ± 0.21	1.89 ± 0.18	2.03 ± 0.21	2.45 ± 0.68	2.39 ± 0.68	2.96 ± 1.26
GA-PLSR		1.00 ± 0.62	**0.77 ± 0.76**	1.02 ± 0.80	1.39 ± 0.55	**2.17 ± 1.19**	**2.35 ± 0.97**
GA-PLSR-PPS		**0.16 ± 0.31**	1.09 ± 1.84	**0.23 ± 0.44**	**0.99 ± 0.43**	14.13 ± 7.55	2.52 ± 1.10
PLSR	β-acids (HPLC)	0.40 ± 0.03	0.58 ± 0.05	0.41 ± 0.05	0.55 ± 0.15	1.16 ± 0.93	0.63 ± 0.20
GA-PLSR		0.28 ± 0.14	0.38 ± 0.31	0.28 ± 0.19	0.45 ± 0.19	**0.73 ± 0.43**	0.54 ± 0.21
GA-PLSR-PPS		**0.06 ± 0.08**	**0.24 ± 0.32**	**0.06 ± 0.09**	**0.32 ± 0.14**	1.72 ± 0.90	**0.48 ± 0.20**
PLSR	Total-acids (HPLC)	2.77 ± 0.33	2.43 ± 0.26	3.25 ± 0.38	3.83 ± 1.27	3.39 ± 2.29	4.40 ± 1.68
GA-PLSR		1.34 ± 1.05	**1.01 ± 1.02**	1.15 ± 1.25	2.39 ± 1.01	**2.43 ± 0.89**	**3.86 ± 2.06**
GA-PLSR-PPS		**0.23 ± 0.44**	1.90 ± 3.44	**0.36 ± 0.63**	**1.88 ± 0.86**	23.84 ± 12.13	4.54 ± 1.87

better than PLSR. In these two types of spectral data, i.e., NIR and Raman, GA-PLSR significantly outperforms the proposed method in many cases. However, GA-PLSR-PPS has statistically significant improvements over PLSR and GA-PLSR in the IR spectral data.

A possible explanation for the very different outcomes reached by the proposed method across different spectroscopy types is that different spectra require different pre-processing techniques. It is possible that some of the necessary techniques are not included in this work. IR spectra are reported to require a less intensive pre-processing regime than Raman spectra [28]. This means less pre-processing is needed to obtain a suitably high performing model in IR. It also has been reported that the aspects of NIR spectra addressed by pre-processing can differ substantially from those in IR and Raman spectra [29]. Since there are a limited number of pre-processing methods available to apply in this work, it is possible that some methods that would aid in better Raman and NIR performance have not been included.

Another related finding is that, when comparing the training and test errors in the three type of spectroscopic data, the best performance is typically obtained on the spectral data of IR, which is much smaller than the other two types of spectroscopic data. This is particular the case in the models of the new method where the pre-processing methods for the three spectral data are different. This interesting pattern somehow provides the hint that the GA algorithm can be extended in future work to include additional pre-processing methods that could allow better pre-processing for Raman and NIR.

5.2   Analyses on the Pre-processing Selection

To investigate whether different pre-processing techniques are more suitable to optimising regression models on specific spectroscopic data, the pre-processing techniques selected by the proposed method are recorded and analysed in this section. Two examples of the distribution of pre-processing selection methods chosen by GA-PLSR-PPS for α-acids in the Raman and the IR spectroscopy models across the 30 runs are shown in Fig. 3a and Fig. 3b, respectively. Frequency (%) denotes the proportion of runs where the corresponding pre-processing technique was chosen.

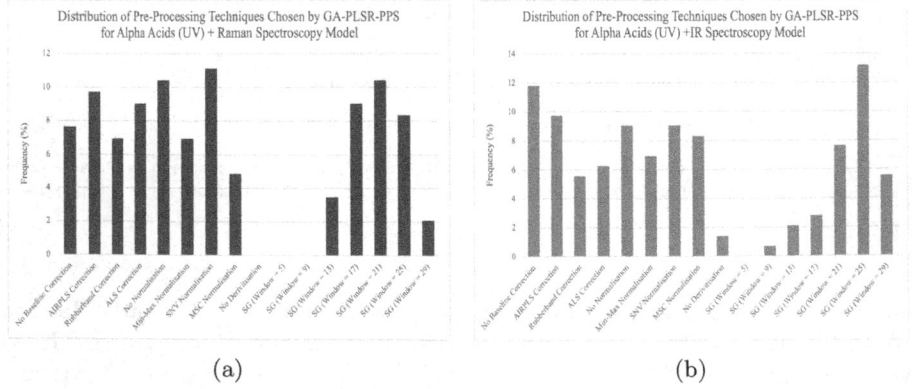

(a)   (b)

Fig. 3. The distribution of pre-processing methods chosen by GA-PLSR-PPS for the α-acids in Raman spectroscopy models (a) and IR spectroscopy models (b)

As shown in Fig. 3a, in the Raman spectroscopy models, GA-PLSR-PPS chooses to apply a baseline correction method approximately 75% of the time, as all four bars corresponding to baseline correction are approximately the same height. Among the baseline correction methods, AirPLS is the most popular one, while Rubberband correction is the least popular. This is interesting, as the GA-PLSR approach uses Rubberband baseline correction for all Raman models. In terms of normalisation, SNV normalisation is the most commonly chosen approach. A relatively large window size is favoured for Savitsky-Golay (SG) derivitisation, and derivitisation is always applied. This illustrates the benefit of enabling the algorithm to optimise the parameter settings for SG derivitisation.

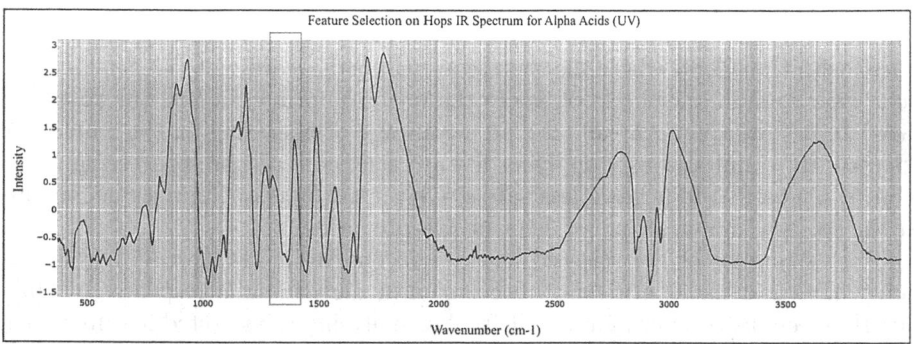

Fig. 4. The features (shown as grey bands) used in the IR model predicting α-acids (UV). (Color figure online)

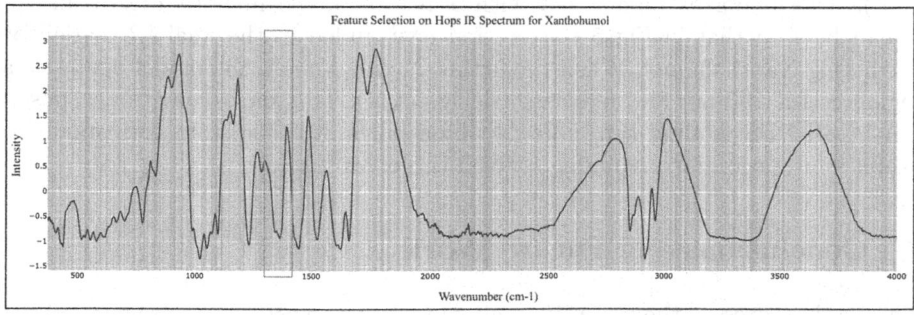

Fig. 5. The features (grey) used in the IR and Xanthohumol model. (Color figure online)

Considering the distribution of pre-processing methods chosen for the α-acids in the IR spectroscopy models as shown in Fig. 3b, different pre-processing methods are chosen compared to the Raman-based models. On average, the models choose to apply no baseline correction more often. It can also be seen that the normalisation approaches chosen in the IR-based models are more variable, rather than there being a most commonly chosen technique. Additionally, a longer SG derivitisation window is favoured on average compared to the Raman-based model. In some cases, GA-PLSR-PPS chose to apply no derivitisation in the IR-based models, which is not the case in the Raman-based models. This demonstrates how a targeted pre-processing approach can permit better predictive performance, compared to using the same manual pre-processing regime across all models.

5.3   Analyses on Feature Selection Results

In this section, an analysis of the feature selected in GA-PLSR-PPS is presented to give an insight of the distribution of these features on the raw spectra. Compared with GA-PLSR where the spectra are clipped according to a visual inspection with domain knowledge of where the useful information lies, the proposed GA-PLSR-PPS approach can offer an advantage of operating on the whole raw spectra. By simultaneously applying feature selection and pre-processing selection, it is possible to achieve less subjective means of selecting relevant regions of the spectra and applying subsequent pre-processing.

Two examples of the selected features in the IR models predicting the α-acids (UV) and the Xanthohumol are shown in Figs. 4 and 5, respectively. Only one IR spectrum is shown for simplicity. By analysing the selected feature which are shown in grey in the two figures, it is easy to observed that the chosen spectral features are distributed across the raw spectra as opposed to there being large sections included or excluded. There are isolated examples where narrow regions are excluded. However, in Fig. 4 a narrow section below 1500cm-1 is seen to have no features included in the final IR model for predicting α-acids (UV). This region is highlighted with a red rectangle in both Figs. 4 and 5. It can be seen that in Fig. 5 this section is included in the final IR model for predicting Xanthohumol. When predicting Xanthohumol, the model deems this region useful, but not for predicting α-acids (UV). This shows how the proposed GA-PLSR-PPS method can indicate which parts of the spectra correspond to useful information for predicting a given target variable.

6   Conclusions and Future Work

This work addressed the potential limitations of the existing GA-enhance PLSR method which relies on the manual pre-processing of the spectral data. The proposed automated spectral data pre-processing approach introduces a new population for pre-processing selection with a new co-evolutionary strategy into the existing GA-PLSR method. The utility of the proposed method can be seen in terms of significantly better training performance than the two baseline PLSR methods, and the competitive or better prediction performance for two of the three spectroscopy techniques examined. The rationale behind the improvements include that the new method is able to select a specific sequence of pre-processing approaches for each spectral model and can also operate on the whole spectra. When better performance is achieved, this illustrates how the manual pre-processing approach can be subjective. When comparable results are achieved, this is still useful in that it eliminates the need for domain knowledge and time spent manually pre-processing spectra.

One limitation of the proposed method is that it does not achieve good generalisation ability on the NIR data. Compared with the two other spectroscopy techniques, i.e., IR and Raman, the NIR models had a trend of overfitting which is apparent from the high training performance and comparably lower test performance. A further investigation into why this occurred would be a plausible next step. Further investigation is necessary to obtain competitive results for the NIR spectra. Moreover, some other potential approaches would be discouraging overfitting through restricting how many pre-processing methods can be applied and limiting model complexity

References

1. Noh, M.F.M., et al.: Recent techniques in nutrient analysis for food composition database. Molecules **25**(19), 4567 (2020)
2. Crocombe, R.A.: The future of portable spectroscopy. In: Portable Spectroscopy and Spectrometry 1: Applications, vol. 2, pp. 545–571 (2021)
3. Cortés, V., Blasco, J., Aleixos, N., Cubero, S., Talens, P.: Monitoring strategies for quality control of agricultural products using visible and near-infrared spectroscopy: a review. Trends Food Sci. Technol. **85**, 138–148 (2019)
4. Yang, D., Ying, Y.: Applications of Raman spectroscopy in agricultural products and food analysis: a review. Appl. Spectrosc. Rev. **46**(7), 539–560 (2011)
5. Tahir, H.E., et al.: Recent progress in rapid analyses of vitamins, phenolic, and volatile compounds in foods using vibrational spectroscopy combined with chemometrics: a review. Food Anal. Methods **12**(10), 2361–2382 (2019)
6. Killeen, D.P., Andersen, D.H., Beatson, R.A., Gordon, K.C., Perry, N.B.: Vibrational spectroscopy and chemometrics for rapid, quantitative analysis of bitter acids in hops (humulus lupulus). J. Agric. Food Chem. **62**(52), 12521–12528 (2014)
7. McIntyre, S.M., Ma, Q., Burritt, D.J., Oey, I., Gordon, K.C., Fraser-Miller, S.J.: Vibrational spectroscopy and chemometrics for quantifying key bioactive components of various plum cultivars grown in New Zealand. J. Raman Spectrosc. **51**(7), 1138–1152 (2020)
8. Amuah, C.L.Y., Teye, E., Lamptey, F.P., Nyandey, K., Opoku-Ansah, J., Adueming, P.O.-W.: Feasibility study of the use of handheld NIR spectrometer for simultaneous authentication and quantification of quality parameters in intact pineapple fruits. J. Spectroscopy **2019**, 1–9 (2019)
9. Robinson, D., et al.: Genetic algorithm for feature and latent variable selection for nutrient assessment in horticultural products. In: 2021 IEEE Congress on Evolutionary Computation (CEC), pp. 272–279 (2021)
10. Zhu, X., Tao, X., Lin, Q., Duan, Y.: Technical development of Raman spectroscopy: from instrumental to advanced combined technologies. Appl. Spectrosc. Rev. **49**(1), 64–82 (2014)
11. Paterova, A., Lung, S., Kalashnikov, D.A., Krivitsky, L.A.: Nonlinear infrared spectroscopy free from spectral selection. Sci. Rep. **7**(1), 42608 (2017)

12. Manley, M., Baeten, V.: Chapter 3 - spectroscopic technique: near infrared (NIR) spectroscopy. In: Sun, D.-W. (ed.) Modern Techniques for Food Authentication, 2nd edn., pp. 51–102. Academic Press (2018)
13. Brian Dyer, R., Woodruff, W.H.: Vibrational Spectroscopy. American Cancer Society (2011)
14. Hashimoto, K., Badarla, V.R., Kawai, A., Ideguchi, T.: Complementary vibrational spectroscopy. Nat. Commun. 10(1), 4411 (2019)
15. Hackshaw, K.V., Miller, J.S., Aykas, D.P., Rodriguez-Saona, L.: Vibrational spectroscopy for identification of metabolites in biologic samples. Molecules 25(20), 4725 (2020)
16. Rosipal, R., Krämer, N.: Overview and recent advances in partial least squares. In: Saunders, C., Grobelnik, M., Gunn, S., Shawe-Taylor, J. (eds.) SLSFS 2005. LNCS, vol. 3940, pp. 34–51. Springer, Heidelberg (2006). https://doi.org/10.1007/11752790_2
17. Liland, K.H., Rukke, E.-O., Olsen, E.F., Isaksson, T.: Customized baseline correction. Chemometr. Intell. Lab. Syst. 109(1), 51–56 (2011)
18. Zhang, Z.-M., Chen, S., Liang, Y.-Z.: Baseline correction using adaptive iteratively reweighted penalized least squares. Analyst 135(5), 1138–1146 (2010)
19. Shen, X., et al.: Study on baseline correction methods for the Fourier transform infrared spectra with different signal-to-noise ratios. Appl. Opt. 57(20), 5794–5799 (2018)
20. Peng, J., Peng, S., Jiang, A., Wei, J., Li, C., Tan, J.: Asymmetric least squares for multiple spectra baseline correction. Anal. Chim. Acta 683(1), 63–68 (2010)
21. Lasch, P.: Spectral pre-processing for biomedical vibrational spectroscopy and microspectroscopic imaging. Chemom. Intell. Lab. Syst. 117, 100–114 (2012)
22. Barnes, R.J., Dhanoa, M.S., Lister, S.J.: Standard normal variate transformation and de-trending of near-infrared diffuse reflectance spectra. Appl. Spectrosc. 43(5), 772–777 (1989)
23. Isaksson, T., Næs, T.: The effect of multiplicative scatter correction (MSC) and linearity improvement in NIR spectroscopy. Appl. Spectrosc. 42(7), 1273–1284 (1988)
24. Kolodziej, M., et al.: Classification of aggressive and classic mantle cell lymphomas using synchrotron Fourier transform infrared microspectroscopy. Sci. Rep. 9(1) (2019)
25. Press, W.H., Teukolsky, S.A.: Savitzky-Golay smoothing filters. Comput. Phys. 4(6), 669 (1990)
26. Pétillot, L., et al.: Calibration transfer for bioprocess Raman monitoring using Kennard stone piecewise direct standardization and multivariate algorithms. Eng. Rep. 2(11), e12230 (2020)
27. Li, H., Xu, Q., Liang, Y.: libPLS: an integrated library for partial least squares regression and discriminant analysis. PeerJ PrePrints 2, e190v1 (2014)
28. Byrne, H.J., Knief, P., Keating, M.E., Bonnier, F.: Spectral pre and post processing for infrared and Raman spectroscopy of biological tissues and cells. Chem. Soc. Rev. 45, 1865–1878 (2016)
29. Huang, J., Romero-Torres, S., Moshgbar, M.: Practical considerations in data pretreatment for NIR and Raman spectroscopy. Am. Pharm. Rev. 13, 116–127 (2010)

Evolutionary Computation in Edge, Fog, and Cloud Computing

Dynamic Hierarchical Structure Optimisation for Cloud Computing Job Scheduling

Peter Lane(✉)(iD), Na Helian(iD), Muhammad Haad Bodla(iD), Minghua Zheng(iD), and Paul Moggridge(iD)

Department of Computer Science, University of Hertfordshire, Hatfield, UK
{p.c.lane,n.helian,m.zheng2,p.m.moggridge}@herts.ac.uk,
haadbodla@gmail.com

Abstract. The performance of cloud computing depends in part on job-scheduling algorithms, but also on the connection structure. Previous work on this structure has mostly looked at fixed and static connections. However, we argue that such static structures cannot be optimal in all situations. We introduce a dynamic hierarchical connection system of sub-schedulers between the scheduler and servers, and use artificial intelligence search algorithms to optimise this structure. Due to its dynamic and flexible nature, this design enables the system to adaptively accommodate heterogeneous jobs and resources to make the most use of resources. Experimental results compare genetic algorithms and simulating annealing for optimising the structure, and demonstrate that a dynamic hierarchical structure can significantly reduce the total makespan (max processing time for given jobs) of the heterogeneous tasks allocated to heterogeneous resources, compared with a one-layer structure. This reduction is particularly pronounced when resources are scarce.

Keywords: Cloud computing · Dynamic hierarchical job scheduling structure · Genetic algorithms · Optimisation

1 Introduction

Cloud computing has enabled users to quickly access information technology infrastructure, platform, and software from a pool of heterogeneous resources via the Internet. The requirements of cloud computing, from a user's perspective, include minimising response time to user's requests and reducing cost. As a relatively new technology, cloud computing has several open challenges, including service performance variation, quality of service, and energy consumption. Service performance variation is subject to divergence in task load and diversity of resources [18]. To mitigate the effect of variations in load on servers, effective load balancing requires efficient job scheduling and load migration techniques. This paper focuses on reducing makespan, the time to complete given jobs.

© Springer Nature Switzerland AG 2022
J. L. Jiménez Laredo et al. (Eds.): EvoApplications 2022, LNCS 13224, pp. 301–316, 2022.
https://doi.org/10.1007/978-3-031-02462-7_20

Scheduling jobs to minimise makespan and/or response time is an NP-complete problem [15]. Recent work considers a hierarchical structure in which jobs are scheduled through sub-schedulers to servers. However, a fixed structure will not be optimal over time for all potential tasks. In this paper we introduce a dynamic solution to optimise the connections within such a job-scheduling structure in order to reduce the overall makespan, relying on an heuristic-search algorithm to find the optimal connection pattern.

A substantial amount of research has studied load balancing using two techniques [1,14]. First, a load migrator (load balancer) is used to migrate workloads across different computing resources to improve the performance and reliability of cloud services by balancing the workload on different servers. In contrast, a job scheduler is used to assign tasks to servers proactively in order to balance workloads. Detailed explanations of load migration and job scheduling are given in the following two paragraphs.

Load migration algorithms define mechanisms for distributing incoming tasks across a cluster of servers. For instance, if a server is overloaded, then a part of the load on that server can be migrated to another server which has spare capacity. To ensure the most efficient and shortest execution time of tasks, workload should be balanced among all servers. Through load migration, the cloud system can optimise resource usage, increase throughput, reduce response time, and avoid being overloaded on some servers. Many load migration algorithms have been proposed. The literature [14] assigns various categories to those algorithms, including static or dynamic, centralised or decentralised, cooperative or non-cooperative, and intelligent or non-intelligent. Nevertheless, load migration might be very expensive in time cost due to an intrinsic problem. It has a three-stage process of information collection, decision making, and data migration. These often lead to an increased response time of task requests.

Job scheduling has the same aim as load migration but uses a different approach to reduce response time. A job scheduler takes a job which can consist of various tasks and assigns the tasks to servers. Job scheduling is done in a forward-looking way to reduce the time cost and implementation complexity of load balancing, and based on the estimated time of task execution. The main approach to job scheduling is to algorithmically dictate the order of task execution. There are a plethora of algorithms for job scheduling, including global level or local level, static or dynamic, meta-task (batch) or individual (online) [1]. Reducing service response time depends not only on the job scheduling algorithm, but also on the job scheduling architecture. In most research, one-layer scheduling is a commonly used structure for job scheduling [1], comprising job generation, job scheduling by scheduler, and task execution by servers. Some literature [4,16,17] indicates that hierarchical job scheduling structures, which do not dispatch tasks to servers directly, but via middle layers, could improve service performance. All these hierarchical job-scheduling structures are fixed/static, even though it is logical to think that better performance might be achieved using a more dynamic system. Some work has considered dynamic migration of tasks, but not dynamically scheduling tasks. No prior research has been conducted on dynamically altering the connections within a hierarchical job-scheduling structure based on the current status of jobs and servers.

The objective of this research is to design a dynamic hierarchical structure for job scheduling. The proposed design constructs a hierarchical structure, by adding a sub-scheduler layer between scheduler and servers to form a two-layer scheduling structure, and dynamically sets the connections to optimise efficiency. For a data center with a big cluster of servers, managing those servers in a hierarchical structure could improve resource use. In order to adjust the multi-layer structure to task changes, dynamic connections are implemented between the sub-scheduling layer and servers. To optimise the connections dynamically, artificial intelligent (AI) algorithms are used to address this NP-complete problem with the purpose of reducing makespan: here, we compare genetic algorithms and simulated annealing. Our design aims to mitigate the effects of performance variation for heterogeneous tasks and resources. Moreover, this design potentially accommodates other service requirements in a cloud computing environment, due to its dynamic and flexible nature.

2 Related Work

Earlier work on hierarchical structure design includes job scheduling and load migration for load balancing under a cloud or grid environment [4,8,11,16,17, 19]. Two pieces of work have been conducted on how dynamic design could balance workload and adaptively make use of available resources to achieve a better response time for user requests [3,11]. In this section, we review some literature addressing two aspects of load balancing, through scheduling or load migration, with either a hierarchical structure or a dynamic design. First, we look at static hierarchical structures for job scheduling. Then we consider various dynamic systems which have been used in related applications, load migration, and similar areas of server capacity. As will be seen, none of these dynamic approaches covers the communication structure between the scheduler and servers.

A few studies have examined static hierarchical structure design for job scheduling [4,9,16,17]. The three-level hierarchical structure used in [17] is designed to improve scheduling performance. Their results show that the proposed method can achieve better execution efficiency and maintain load balancing in the system. However, because the structure of the whole system is static, it may not be flexible enough to achieve the best task assignment when the tasks have a range of sizes. In [9], a similar static hierarchical structure design to [17] for job scheduling is presented. In this work, Jobs are assigned to Schedulers which then allocate the tasks to Workers. Although certain task allocation rules are applied between Schedulers and Workers in order to achieve low latency (similar to makespan), they cannot guarantee (approximate) optimal task assignment due to the nature of a static structure design. A three-layer structure to manage resources is also proposed in [4]. In this design, a heuristic algorithm is used to schedule tasks in the hierarchical structure. In the same way as [17], this structure is static. The authors in [16] designed a hierarchical structure consisting of a global placement manager and local placement managers. The global placement manager identifies local clusters of servers to be used based on workload variability, and a local placement manager identifies the servers in the local pool based on the resource profile of the new workload and existing workload. Again, as with [17] and [4], the hierarchical structure is static with no optimisation.

A hierarchical management system designed for servers in a cloud is shown in [8]. Two types of servers are in this system: management and execution. A management server can dynamically change the structure. It solves the overutilisation problem by splitting a server node into two server nodes (the original server becomes one child and another child server is added). In contrast, it solves the underutilisation problem by removing a server node and distributing its children among its peers. The experimental results indicate that this hierarchical management system can improve response time and increase scalability. But this structure is on server side management, and the job scheduling structure is not explored. A dynamic hierarchical load balancing method that overcomes scalability challenges by creating multiple levels of load balancing domains is designed by [19]. This method considerably reduces the running time of the load balancing algorithm. This hierarchical structure is on servers and static. Similar to [8], the job scheduling structure is not investigated. Swarm optimisation algorithm is applied in [2] to migrate tasks across servers to achieve load balancing dynamically. Again, this work is focused on load migration, rather than job scheduling.

Some work has tried a dynamic approach to make a system adaptive to load changes. The authors in [13] proposed a set of heuristic scheduling algorithms, which could adapt to dynamical network states and changing traffic requirements, to maximise the network throughput by dynamically balancing data flow. Dynamic scheduling on the networking aspects of a sensor-cloud was proposed in [5] for optimising quality of service. Although their research proved that dynamic approaches could improve system performance, they did not consider load balancing on servers, which is the focus of our research in this paper. A dynamic heuristic algorithm for scheduling scientific workflow exhibits promising results in [12] but addresses a workflow scheduling problem. A new algorithm (Sigmoid) was proposed in [10] in order to dynamically load balance heterogeneous devices (servers), but the focus of this research is about how to split jobs to tasks which is different from ours.

The authors in [11] present a three-layer structure with the first layer being static, and the second and third dynamic. Their computational resources were dynamically available, and their algorithm could make best use of the resources. A decentralised approach which could migrate virtual machines by agent to achieve load balancing was devised in [3]. This paper developed a mathematical model for dynamic allocation of distributed resources in a data center. However, both of the aforementioned papers were focused on either solving the problem of load change or modelling the management of resource allocation. Neither of them investigated the way to schedule jobs.

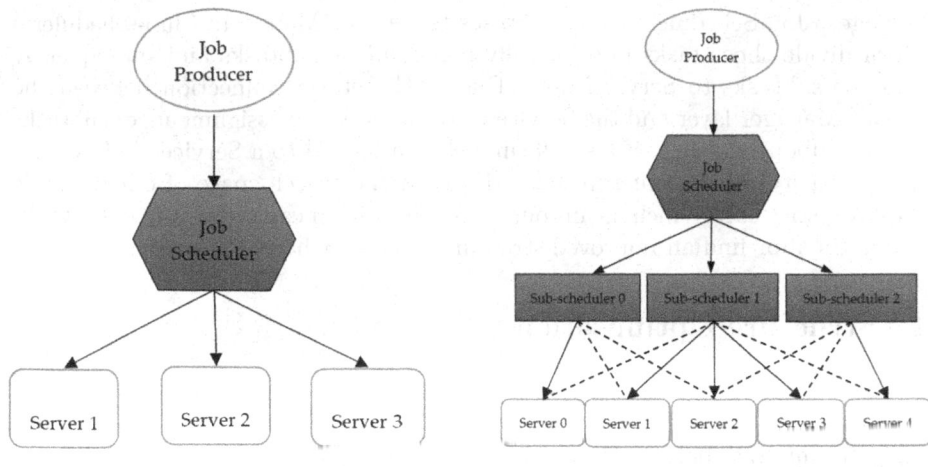

Fig. 1. One-layer structure   **Fig. 2.** Dynamic two-layer structure

All prior research has aimed at using static hierarchical structure or dynamic mechanisms for load migration, and to use static hierarchical structures or dynamic mechanisms for job scheduling. Those methods improve system performance in one way or another, but none of them is on designing dynamic hierarchical structure for job scheduling, which is the focus of the research presented in this paper.

3 Job Scheduling Structures

The scheduler accepts *jobs* from a Job Producer and allocates them (or their *tasks*) to individual *Servers*. The scheduler can take many forms. A one-layer scheduler structure has a single scheduler and multiple servers. The scheduler takes jobs and then assigns tasks to servers for processing. As illustrated in Fig. 1, once jobs have been created by the Job Producer, they are passed on to the Job Scheduler. It is at this point where a scheduler algorithm, such as round robin [6], is deployed. Here, we take this well-used structure as a baseline against which to compare our dynamic two-layer structure.

In contrast to the one-layer scheduler structure, our two-layer hierarchical structure introduces a new sub-scheduler layer between the Job Scheduler and Servers. As illustrated in Fig. 2, the Job Scheduler receives jobs from the Job Producer and assigns each job's tasks onto the Sub-schedulers. Each sub-scheduler then further assigns tasks to Servers according to a scheduler algorithm. The connection pattern between the Sub-schedulers and Servers can be optimised to improve performance, such as measured by makespan. In addition, the connections are varied dynamically to adapt to any change of jobs. Figure 2 uses solid lines to represent one possible connection pattern and dotted lines for another.

Our dynamic two-layer structure includes, as a special case, static hierarchical scheduling structures. For example, that introduced in [17] uses a Request

Manager (Job Scheduler) to assign tasks to Service Managers (Sub-schedulers) which divide those tasks into logically independent subtasks and subsequently load the subtasks to Server Nodes. Due to the static connection between the Service Manager layer and the Service Node layer, some assignments of subtasks may be suboptimal (e.g. if a small subtask is assigned to a Service Node with a large capacity). This structure from [17] is in the search space of our dynamic two-layer structure, which means our search algorithm can select it in cases when it is optimal or find an improved structure in cases where it is not.

4   Structure Optimisation

For our dynamic two-layer structure, achieving the best system performance requires the connection pattern between sub-schedulers and servers to be dynamically optimised to balance the loads across servers. Given a fixed number of sub-schedulers and a fixed number of servers, calculating all possible connections is an NP-complete problem. The optimal connection can be found by using a brute-force (BF) search algorithm, or through heuristic search algorithms, such as a genetic algorithm (GA) and simulated annealing (SA) algorithm. Makespan is the optimisation objective because, in general, makespan reflects the response time of user requests. The aim here is to demonstrate the effectiveness of dynamic optimisation of the connection patterns and so we do not, at this stage, consider network latency: future work will include network latency timings in the simulator and in the optimisation function.

In this optimisation context, the search space is not all the possible sets of connection patterns between sub-schedulers and servers because there are three constraints. First, each server can only be connected to one sub-scheduler. Second, each sub-scheduler must have at least one server. Third, the number of servers must be greater or equal to the number of sub-schedulers. For example, consider a system with three sub-schedulers, referred to as [0, 1, 2], and five servers, referred to as [0, 1, 2, 3, 4]. A connection pattern [2, 0, 1, 0, 1] indicates that Server 0 is attached to Sub-scheduler 2, Server 1 to Sub-scheduler 0, Server 2 to Sub-scheduler 1, Server 3 to Sub-scheduler 0, and Server 4 to Sub-scheduler 1. This is a valid connection pattern. But connection pattern [0, 0, 0, 0, 1] is not because Sub-scheduler 2 is not linked, and so it does not satisfy the connection constraints of our application.

The number of possible connections of the dynamic two-layer structure is:

$$S = P_n^m \times m^{(n-m)} \tag{1}$$

where S is number of possible connection structures, m is number of sub-schedulers and n is number of servers. P_n^m expresses randomly selecting m servers and then allocating one sub-scheduler to one server in a one to one relationship. The allocation order matters, so it is a permutation. For the remaining $(n - m)$ servers, each of them can be randomly allocated to any one of the sub-schedulers. Therefore the number of possible connection patterns for the remaining $(n - m)$ servers is $m^{(n-m)}$. Each allocation for the first m servers can be combined with every single allocation for the remaining $(n - m)$ servers, hence we need to multiply P_n^m with $m^{(n-m)}$ to form the total number of connection patterns.

We consider three search algorithms for finding a 'good' connection structure.

4.1 Brute Force Search Algorithm

A brute-force, or exhaustive, search algorithm, explores a search space by testing every potential candidate solution. As this algorithm systematically enumerates all candidates to find the optimal candidate, it may take too long if the number of candidates is large. As per Eq. (1), if m is 4 and n is 100, there are approximately 5.91×10^{65} candidate structures. Even with the most powerful computer in the world, it is impractical to examine all of them. Nevertheless, using a brute-force algorithm for smaller problems is a good way to compare the efficiency and accuracy of heuristic search algorithms of the same space.

Our brute-force search algorithm first sets the numbers of sub-schedulers, m and servers, n, before generating all possible structures connecting sub-schedulers and servers. Invalid structures are removed, based on the connection constraints of this application. After that, the algorithm checks each connection pattern to find the best candidate, the one with the smallest makespan.

4.2 Genetic Algorithm

Genetic algorithm (GA) uses a heuristic search, based on natural selection, to solve optimisation problems. A genetic algorithm iteratively modifies a population of individual solutions using operators such as selection, crossover, and mutation. For each iteration, GA selects from the better individual solutions in the current population (to be parents) and uses them to produce new solutions (children) in order to update the population. Over successive generations, the population evolves towards one containing even better solutions. The best solution is selected from the final population as the search result. GA is faster than the brute-force algorithm, especially when a search space is large, because it does not iterate through each candidate solution. However, GA may not precisely find the optimum, but an approximation.

In the experiments, a GA is used to optimise the connection pattern between sub-schedulers and servers. The GA works in a standard way, as follows:

1. **Initialisation**
   Four parameters are used. The population size, P, and the number of evolution cycles, e, control the GA process. The problem definition requires the number of sub-schedulers, m, and number of servers n. When producing each candidate, $m$schedulers are randomly allocated to the servers in a one to one relationship, then each of the remaining $(n - m)$ servers is randomly connected to one of the schedulers. This guarantees any generated pattern satisfies the connection constraints of this application.
2. **Fitness Function** The fitness of each individual connection pattern in a population is assessed using *makespan* as a fitness score.

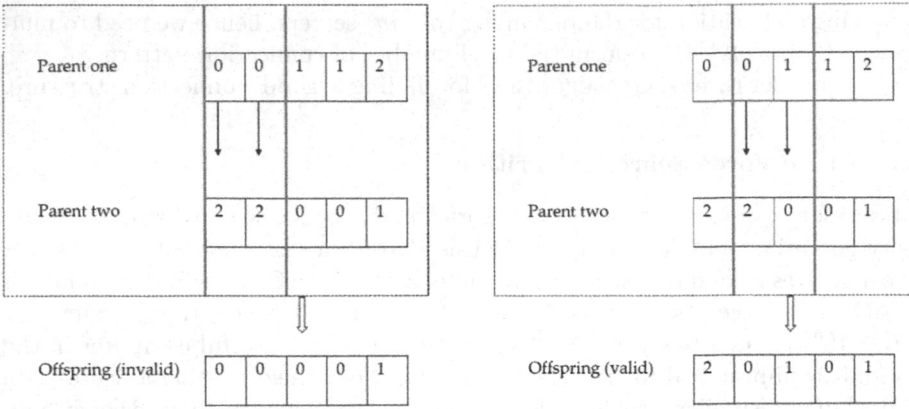

Fig. 3. Crossover (invalid)   **Fig. 4.** Crossover (valid)

3. **Selection** The fittest individuals from the population are selected, so as to pass their best traits to the next generation, ultimately to improve the overall population fitness. The process of selecting an individual as a parent is to randomly select a fixed number (tournament size t) of individuals from the population, and then choose the best candidate on makespan from the selected pool. A pair of individuals (parents) is selected based on their fitness scores for generating an offspring.

4. **Crossover** Two selected parents are used to produce one offspring, respecting the connection constraints of the hierarchical scheduling structure. The crossover method first randomly selects a start and end element point from the array representing the parents' connection patterns. The end point must be greater than the start point. Then, the array elements from the start point to the end point of parent one are replaced by parent two's elements at the corresponding positions to produce an offspring. Due to the constraints of the connection patterns, an offspring might be invalid. If this is the case, a new pair of parents will be generated and the whole crossover process will start over until a valid offspring is produced.

An example of this method is illustrated in Figs. 3 and 4. [0, 0, 1, 1, 2] and [2, 2, 0, 0, 1] are two single candidate connection patterns (parents). This structure consists of 3 sub-schedulers and 5 servers. In Fig. 3, the random start point is 0 and end point is 1. The elements between position 0 and position 1 of Parent one are then copied to Parent two at the corresponding elements to form an offspring [0, 0, 0, 0, 1]. This is an invalid offspring because it does not satisfy the connection constraints – Sub-scheduler 2 is not connected to any server. Hence, the crossover process is repeated until a valid offspring is generated. For example, with a start point 1 and an end point 2, the generated offspring [2, 0, 1, 0, 0] is valid, as illustrated in Fig. 4.

5. **Mutation** Some of the offspring are subject to a mutation on some of their genes (array element values): this process introduces some additional random variation into the population. A mutation probability parameter, p, determines the likelihood of mutating an offspring. As randomly changing the value of an element risks the new offspring becoming invalid, we mutate an offspring by randomly changing the position of each element.

Many factors can affect the performance and accuracy of a GA, including initialisation of the population of candidates, tournament size, crossover method, mutation method, mutation rate, termination condition, etc. The parameters of the GA used in this paper are given in Sect. 5.

4.3 Simulated Annealing Algorithm

Simulated annealing (SA) [7] is a widely applied optimisation algorithm, inspired by the process of heating and cooling material to reduce its defects. A temperature variable, T, is used to simulate the cooling process. When a simulated algorithm runs, T is initially set to a very high value. Later, the temperature is gradually reduced, like the cooling process of annealing, and candidate solutions are created and selected based in part on the current temperature. When the cooling process is finished, an approximately optimal candidate will have been found. SA is very simple to implement for complex problems compared with other evolutionary algorithms. Additionally, SA can avoid being stuck in local minima by (temporarily) accepting a worse variation as a new solution with a specified probability. In addition, SA can be very efficient because it does not iterate all candidates. However, in a similar way to GA, SA may not find the precise optimum, but a candidate which is very close to the optimum.

Here, we use a standard SA implementation:

1. The best pattern variable S_0 is assigned a random connection pattern.
2. C_0 is computed as the makespan of S_0, the current best solution makespan.
3. The new candidate S_1 is assigned a random connection pattern.
4. C_1 is computed as the makespan of S_1, the candidate makespan.
5. If C_1 is better than C_0, then accept the new candidate, setting $S_0 = S_1$.
6. Otherwise:
   (a) Compute acceptance probability p as $e^{\Delta E/T}$, where $\Delta E = C_1 - C_0$.
   (b) Generate a random number r
   (c) If $p > r$ then accept the new candidate, setting $S_0 = S_1$.
7. Reduce T by its cooling rate and repeat from step 2 as required.

When temperature is high, the chance of accepting a bad candidate is high; this means SA can explore broad regions of the search space and avoid becoming stuck in local minima. When the temperature is lower, the probability of accepting bad solutions reduces. The efficiency and accuracy of the simulated annealing algorithm depends on several factors, including candidate generator procedure, acceptance probability function, annealing schedule and initial temperature. For example, when the initial temperature value is set very high and reduced slowly, the optimum has a greater chance to be found, but at the cost of speed. On the other hand, if the initial temperature is very low, the cooling process might become stuck at a local, sub-optimal minimum.

5 Simulation Experiments and Results

We empirically investigate under which scenarios a dynamic hierarchical structure outperforms a one-layer structure for job scheduling in a cloud computing environment, and demonstrate that heuristic-based search algorithms, such as GA or SA, can be effective in finding optimal structures.

We present results from a series of simulations designed to investigate whether or not our dynamic hierarchical structure could improve makespan over one-layer structure for cloud computing job scheduling, particularly in a heterogeneous environment. First, we compare three search algorithms for their ability to dynamically find the optimal connection pattern between sub-schedulers and servers. Second, we investigate the impacts of heterogeneous servers, non-uniformed tasks and scrambled job patterns on makespan: because the connection pattern is formed by optimising on these different continuations, the resulting job-scheduling structure will be optimised for these heterogeneous situations. This is done by looking at the impacts of server processing power (UPS) dispersion, task size (U) dispersion, and incoming waves of jobs, respectively.

5.1 Setup

A simulator is used to evaluate our job schedulers and compute their *makespan* values. Here, we use ScheduleSim[1], a lightweight, open-source simulator specifically designed for job scheduling in a cloud environment. We use the following definitions throughout this section:

Consumer(server) A consumer represents a virtual machine (VM) server executing tasks in cloud environment. The server simply reduces the units of a task according to its unit per step (UPS) rating. A "fast" server will have a high UPS and therefore be able to quickly deplete a task of its units.

Producer A producer represents the point at which jobs enter the Cloud; it also produces tasks and dispatches them to a scheduler.

Task A task is the simulated load on a system. Task is measured in unit U.

Scheduler A scheduler assigns tasks to sub-schedulers or servers.

Step ScheduleSim works using discrete time steps. Each entity (producer/scheduler/server) should step once for a time step to be complete.

Unit(U) Unit is a measure metric for the size of a task.

Each experiment creates tasks and servers using a random distribution. In these experiments, we use a normal rather than constant distribution in order

---

[1] P. Moggridge, ScheduleSim, 2019. https://bitbucket.org/paulmogs398/schedulesim.

Table 1. Parameter values used for GA and SA algorithms

Genetic algorithm		Simulated annealing	
Initial population	50	Initial temperature	10,000
Mutation rate	0.015	Cooling rate	0.003
Tournament size	5		
Number of cycles	100		

Table 2. Parameter values for the job scheduler

Number of sub-schedulers	4	Task minimum size	20U
Number of servers	$\{4, 5, 6, 7, 8, 9\}$	Task maximum size	240U
Server pattern	Gaussian	Task size (μ, σ)	(200, 60) U
Server processing power (μ, σ)	(30, 0) UPS	Total Task Size	50,000U
Job pattern	Gaussian		

to obtain a heterogeneous set of tasks and servers. The continuous Gaussian distribution formula is used to create a distribution:

$$f(x) = \frac{1}{\sqrt{2\pi\sigma^2}} e^{-\frac{(x-\mu)^2}{2\sigma^2}} \tag{2}$$

where μ (mean) determines the position of the normal distribution, and σ^2 (variance) controls its width.

All experiments are measured using the metric of overall makespan, which is the latest finishing time of the simulation for given tasks.

$$\text{makespan} = \max_{j \in T} t_j \tag{3}$$

where, T is the set of all submitted tasks, and t_j is the time of task j to finish.

Each scheduler and sub-scheduler uses the round robin algorithm, which assigns tasks in circular order one by one. It is one of the most popular and simplest job scheduling algorithms, processing one task at a time, rather than many tasks at a time, so it is simple to implement and starvation-free. This makes the results easy to interpret.

The parameter values used for the GA and SA are presented in Table 1. (These values are given for the sake of reproducibility of results. In previous experiments a range of values was tried, and results were found to be robust to minor variations.) All results presented are the average of 20 runs to mitigate stochastic effects of the Gaussian distribution used to generate tasks and servers.

5.2 Experiment 1: Search Algorithm Comparison

This experiment compares three search algorithms (BF, GA and SA) on their ability to find good connection patterns. First, we want to know if the heuristic

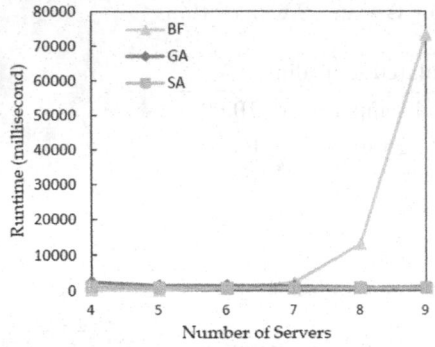

Fig. 5. Runtime for BF, GA and SA   **Fig. 6.** Makespan for BF, GA and SA

search algorithms can find solutions close to the optimum, and second get an idea of their efficiency. This experiment is on the hierarchical structure in Fig. 2.

Parameters defining the job scheduler are shown in Table 2. For this experiment, we used a Gaussian distribution for the job pattern, i.c. task size follows a Gaussian distribution and server processing power is uniform for all servers. Various server numbers are tested with the sub-scheduler number being fixed (as 4) for computational simplicity.

Results comparing the runtime and makespan, respectively, for the three search algorithms are shown in Figs. 5 and 6. In both graphs, the x-axis represents the number of servers and the y-axis the measured value. Runtime is the time taken for each algorithm to find the best connection pattern, and makespan is computed using the simulator.

As expected, the runtime for BF increases exponentially as the number of servers increases beyond 7, highlighting that it is not practical to use BF in real world applications. In contrast, the runtime for the GA or SA remains more or less the same, because these algorithms search a pre-set amount of the available search space, based on their parameters. For example, the GA always runs for 100 generations with a population size of 50.

Figure 6 shows that both GA and SA can find the approximate minimal makespan, because both of them have nearly the same makespan values as the BF, which has performed an exhaustive search. As the runtime efficiency of the two AI algorithms is much better than the BF, particularly when the number of servers is large, it is reasonable to use one or other of these AI algorithms to find a dynamic connection pattern for our application.

In the real world, the number of servers is normally high, depending on the size of a data center, so we now perform some experiments on larger numbers of servers to investigate the relative efficiency of GA against SA. All the parameter values are the same as in Table 2, except the number of servers now takes values of $\{100, 200, 300, 400, 500\}$ and total task size is now 900,000U.

The results in Fig. 7 show that runtime increases slightly with server number; we attribute this to the increased need to delete invalid solutions during the

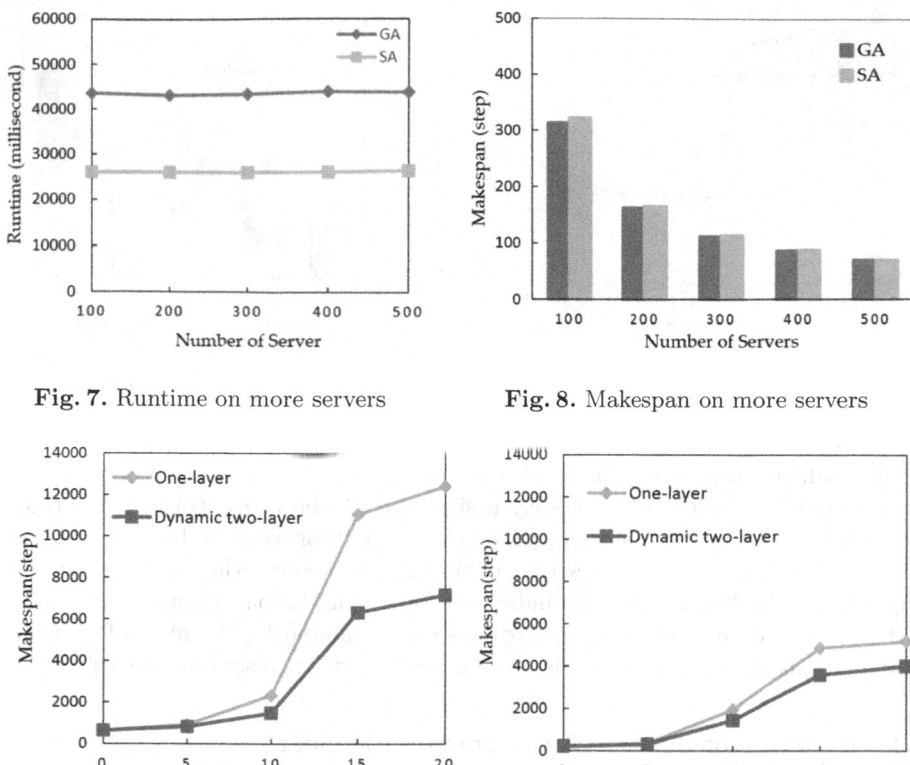

Fig. 7. Runtime on more servers   **Fig. 8.** Makespan on more servers

Fig. 9. Server dispersion impacts on makespan with 50 (left) and 150 (right) servers

search process. Moreover, it can be seen that the SA runtime is much shorter (around 40%) than that of the GA. Furthermore, as shown in Fig. 8, makespan difference between the GA and the SA is small: the GA is slightly better initially, but the two are the same after the number of servers increases to 300.

5.3 Experiment 2: Server Processing Power Dispersion Impact

For exploring the effect of server processing power dispersion, we alter the following parameters from Table 2: Number of servers in $\{50, 150\}$, Server processing power μ is fixed at 30, $\sigma \in \{0, 5, 10, 15, 20\}$, Total task size 900,000U. The results (obtained from the GA) are shown in Fig. 9.

First, we note that the dynamic two-layer structure always outperforms the one-layer structure, although the makespan difference between these two structures is smaller with a larger number of servers; not shown here, but this holds over a larger range of server numbers, and is because optimisation becomes less important when there are more resources available.

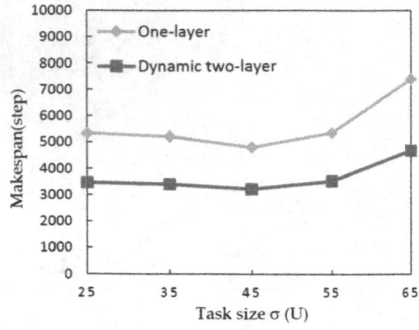

Fig. 10. Task size dispersion impact

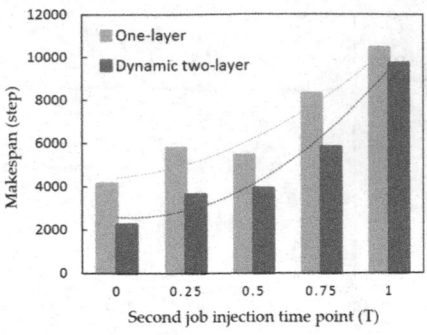

Fig. 11. Task complexity impact

Second, makespan lengthens with widening server processing power (larger σ), because it is harder to effectively assign tasks to heterogeneous servers than to uniform servers. Furthermore, the makespan improvement becomes bigger with σ widening, which implies that optimising the connection pattern is more important. The two results also indicate that the makespan improvement from a dynamic two-layer structure narrows with an increasing number of servers, which entails that optimisation is more necessary when resources are scarcer.

5.4 Experiment 3: Task Size Dispersion Impact

For exploring the effect of task size dispersion, we alter the following parameters from Table 2: Number of servers is set to 100, Server processing power μ is fixed at 30, Task size $\sigma \in \{25, 35, 45, 55, 65\}$, Total task size 1,800,000U.

Figure 10 demonstrates, again, that the makespan of the dynamic two-layer structure is smaller than that of the one-layer structure. In particular, makespan is relatively stable when σ is less than 45U, whilst the rate of makespan climbs after that. We could use an analogy to explain this. When there is a big size difference between a container and items, item occupation rate can be very high because small items fill gaps. On the other hand, when the item sizes are comparable to the container size, there would be more gaps left. Therefore, when the task size dispersion is small in a certain range, it is more likely that the servers are effectively occupied. However, when the dispersion is wider, there would be more tasks with the sizes comparable to server processing power, which might leave some servers underutilised. For this experiment, as compared to that in Table 2, we doubled total task size to generate more of the big tasks in order to show their impact (where the task size (U) is comparable to the servers processing speed (UPS)) on makespan.

5.5 Experiment 4: Job Complexity Impact

This experiment explores the impact of job complexity on makespan. Varying job complexity is achieved by injecting one job followed by another at various time points in order to get different degrees of overlay on execution times between the two jobs. Both jobs have the same Gaussian distribution, hence individually they will execute in approximately the same amount of time T. Here, T is used as the unit of measuring injection delay of the second job. The first job is always injected at time point 0 and the second job follows with various delays. Therefore, the second job injection time of 0T implies that the two jobs are fully overlapped, 1T means no overlap and 0.5T entails 50% overlap.

For exploring the effect of job complexity, we alter the following parameters from Table 2: Number of servers is set to 50 and Server processing power μ is fixed at 30. Second job injection time point is in $\{0, 0.25, 0.5, 0.75, 1.0\}T$.

In Fig. 11, the makespan improvement decreases with the increase of delay on the second job injection. Smaller delay entails more job complexity, because the two jobs have a bigger overlap. In contrast, a bigger delay implies less overall job complexity because the two jobs have less overlap. Therefore, we argue that the more complex the jobs are, the bigger the improvement in makespan from using the dynamic two-layer structure. It is worth mentioning that makespan rises with the delay degree of second job injection. This is because execution completion time of the second job increases with the degree of its injection delay.

6 Conclusion

This paper presents a dynamic hierarchical structure, which introduces sub-schedulers between scheduler and servers, to improve job scheduling makespan in a cloud environment. This novel design dynamically changes the connection pattern between sub-schedulers and servers, using an heuristic search algorithm, such as a GA, to find the appropriate connection pattern for a given set of jobs.

In our experiments, we first show that both GA and SA achieve near optimal connection patterns. Second, simulation experiments demonstrate that a dynamic two-layer structure outperforms a one-layer structure on overall job makespan, particularly when assigning heterogeneous tasks to heterogeneous resources. This effect is particularly pronounced when resources are scarce. Due to its dynamic and flexible nature, this hierarchical design could potentially meet other system design requirements, including geographical resource distribution, as well as clustering resources and tasks for various purposes.

Future work will aim to improve the GA/SA design and parameter values, optimise sub-scheduler number and layer number, deploy performant scheduling algorithms (such as Max-min) on all schedulers and sub-schedulers, as well as taking into account additional objectives, such as quality of service, energy consumption and execution cost.

References

1. Arunarani, A., Manjula, D., Sugumaran, V.: Task scheduling techniques in cloud computing: a literature survey. Future Gener. Comput. Syst. **91**, 407–415 (2019)

2. Balaji, K., Kiran, P.S., Kumar, M.S.: An energy efficient load balancing on cloud computing using adaptive cat swarm optimization. Mater. Today Proc. (2021, in press). https://doi.org/10.1016/j.matpr.2020.11.106

3. Benbrahim, S.E., Quintero, A., Bellaiche, M.: New distributed approach for an autonomous dynamic management of interdependent virtual machines. In: 2014 8th Asia Modelling Symposium, pp. 193–196. IEEE (2014)

4. Cao, J., Spooner, D.P., Jarvis, S.A., Nudd, G.R.: Grid load balancing using intelligent agents. Future Gener. Comput. Syst. 21(1), 135–149 (2005)

5. Chatterjee, S., Misra, S., Khan, S.U.: Optimal data center scheduling for quality of service management in sensor-cloud. IEEE Trans. Cloud Comput. 7(1), 89–101 (2019)

6. Dave, S., Maheta, P.: Utilizing round robin concept for load balancing algorithm at virtual machine level in cloud environment. Int. J. Comput. Appl. 94(4), 23–29 (2014)

7. Kirkpatrick, S., Gelatt, C.D., Vecchi, M.P.: Optimization by simulated annealing. Science 220(4598), 671–680 (1983)

8. Moens, H., Famaey, J., Latré, S., Dhoedt, B., De Turck, F.: Design and evaluation of a hierarchical application placement algorithm in large scale clouds. In: 12th IFIP/IEEE International Symposium on Integrated Network Management (IM 2011) and Workshops, pp. 137–144. IEEE (2011)

9. Ousterhout, K., Wendell, P., Zaharia, M., Stoica, I.: Sparrow: distributed, low latency scheduling. In: Proceedings of the Twenty-Fourth ACM Symposium on Operating Systems Principles, pp. 69–84 (2013)

10. Pérez, B., Stafford, E., Bosque, J., Beivide, R.: Sigmoid: an auto-tuned load balancing algorithm for heterogeneous systems. J. Parallel Distrib. Comput. 157, 30–42 (2021)

11. Reddy, K.H.K., Roy, D.S.: A hierarchical load balancing algorithm for efficient job scheduling in a computational grid testbed. In: 2012 1st International Conference on Recent Advances in Information Technology (RAIT), pp. 363–368. IEEE (2012)

12. Sahni, J., Vidyarthi, D.P.: A cost-effective deadline-constrained dynamic scheduling algorithm for scientific workflows in a cloud environment. IEEE Trans. Cloud Comput. 6(1), 2–18 (2018)

13. Tang, F., Yang, L.T., Tang, C., Li, J., Guo, M.: A dynamical and load-balanced flow scheduling approach for big data centers in clouds. IEEE Trans. Cloud Comput. 6(4), 915–928 (2016)

14. Thakur, A., Goraya, M.S.: A taxonomic survey on load balancing in cloud. J. Netw. Comput. Appl. 98, 43–57 (2017)

15. Ullman, J.D.: NP-complete scheduling problems. J. Comput. Syst. Sci. 10(3), 384–393 (1975)

16. Viswanathan, B., Verma, A., Dutta, S.: CloudMap: workload-aware placement in private heterogeneous clouds. In: Proceedings IEEE Network Operations and Management Symposium, pp. 9–16 (2012)

17. Wang, S.C., Yan, K.Q., Wang, S.S., Chen, C.W.: A three-phases scheduling in a hierarchical cloud computing network. In: 2011 Third International Conference on Communications and Mobile Computing, pp. 114–117. IEEE (2011)

18. Zhang, Q., Cheng, L., Boutaba, R.: Cloud computing: state-of-the-art and research challenges. J. Internet Serv. Appl. 1(1), 7–18 (2010). https://doi.org/10.1007/s13174-010-0007-6

19. Zheng, G., Bhatele, A., Meneses, E., Kale, L.V.: Periodic hierarchical load balancing for large supercomputers. Int. J. High Perform. Comput. Appl. 25(4), 371–385 (2011)

Optimising Communication Overhead in Federated Learning Using NSGA-II

José Ángel Morell[1(✉)], Zakaria Abdelmoiz Dahi[1,2], Francisco Chicano[1], Gabriel Luque[1], and Enrique Alba[1]

[1] ITIS Software, University of Malaga, Malaga, Spain
{jamorell,chicano,gabriel,eat}@lcc.uma.es, zakaria.dahi@uma.es
[2] Dep. Fundamental Computer Science and Its Applications, Fac. NTIC, University of Constantine 2, Constantine, Algeria
zakaria.dahi@univ-constantine2.dz

Abstract. Federated learning is a training paradigm according to which a server-based model is cooperatively trained using local models running on edge devices and ensuring data privacy. These devices exchange information that induces a substantial communication's load, which jeopardises the functioning efficiency. The difficulty of reducing this overhead stands in achieving this without decreasing the model's efficiency (contradictory relation). To do so, many works investigated the compression of the pre/mid/post-trained models and the communication rounds, separately, although they jointly contribute to the communication overload. Our work aims at optimising communication overhead in federated learning by (I) modelling it as a multi-objective problem and (II) applying a multi-objective optimization algorithm (NSGA-II) to solve it. To the best of the author's knowledge, this is the first work that (I) explores the add-in that evolutionary computation could bring for solving such a problem, and (II) considers both the neuron and devices features together. We perform the experimentation by simulating a server/client architecture with 4 slaves. We investigate both convolutional and fully-connected neural networks with 12 and 3 layers, 887,530 and 33,400 weights, respectively. We conducted the validation on the MNIST dataset containing 70,000 images. The experiments have shown that our proposal could reduce the communication by 99% and maintain an accuracy equal to the one obtained by the FedAvg Algorithm that uses 100% of communications.

Keywords: Federated learning · Evolutionary computation · Multi-objective optimisation

1 Introduction

Today's advances in Artificial Intelligence (AI) allow training machine learning models by exploiting the daily-generated data that was previously considered

© Springer Nature Switzerland AG 2022
J. L. Jiménez Laredo et al. (Eds.): EvoApplications 2022, LNCS 13224, pp. 317–333, 2022.
https://doi.org/10.1007/978-3-031-02462-7_21

useless [3]. Statista[1] has stated that there are 23.8 billion interconnected computing devices that are active in the world and will produce 149 zettabytes of data by 2024. Cisco[2] also estimated that at least 85 (10%) of the 850 zettabytes created in 2021 will be useful, while only 7 zettabytes of it will be stored. Indeed, most of this data cannot be stored/processed on the cloud despite the exponential increase of data demand and generation speed. In addition, data privacy prevents sharing it with third-parties (e.g. medical images). As promising solution to these issues, Federated Learning (FL) appeared. It is a learning paradigm that trains a shared model in a distributed manner while keeping private the data locally on edge devices. Federated Learning is being actively investigated and widely applied (e.g. medicine [6]). Its working mechanism induces a substantial communication overload that limits its applicability. It has been proven that this overhead is generated by several factors such as the number of devices participating in the learning process, the complexity of the model (e.g. number of layers, neurons, etc.), number of communication rounds, etc. [4]. Previous works have already investigated some of these factors in isolation to decrease the communication excess [1,8,9], although, one should note that the factors are jointly contributing to the communication overhead.

Achieving high-quality results requires performing efficient network training using substantial information and communication [4]. Thus, the main difficulty when reducing the communication cost stands in maintaining the same efficacy, due to the conflictual relation between both. Using classical exhaustive tools turns out to be computationally costly and time-consuming, due to the complexity of the problem and its multi-criterion nature. For such problem's class, stochastic algorithms such as metaheuristics and, in particular, the Non-Dominated Sorting Genetic Algorithm II (NSGA-II) are a promising alternative that provides a good trade-off between the solving efficiency and time consumption [2]. Bearing in mind the above-stated facts, our contributions stand in (I) modelling and formulating the Federated Learning Communication Overhead as a multi-objective Problem (FL-COP), (II) applying NSGA-II to solve it and (III) investigating within the same work the main parameters triggering communication overhead that the literature usually tackle separately. Our proposal has been assessed by simulating a server/client architecture of 4 devices, each one being tested with both convolutional and fully connected neural networks with 12 and 3 layers, 887,530 and 33,400 weights, respectively. The validation has been done on MNIST dataset containing 70,000 images of handwritten digits.

The rest of the paper is structured as follows. In Sect. 2, we present basic concepts of FL and communication-overhead reduction strategies. Section 3 introduces our proposal as well as our FL-COP formulation. Section 4 presents the experimental results and analysis. Finally, Sect. 5 concludes our work.

---

[1] www.statista.com/statistics/1101442/iot-number-of-connected-devices-worldwide.
[2] blogs.cisco.com/sp/five-things-that-are-bigger-than-the-internet-findings-from-this-years-global-cloud-index.

2   Fundamental Concepts

This section presents the basic concepts of federated learning and the communication reduction strategies.

2.1   Federated Learning

Originated from distributed deep learning [4], FL allows training a common model without compromising the users' data privacy. The latter are kept on local devices during the learning process. Instead of sharing the training data, the clients exchange their local models to help improve a global one (see Fig. 1).

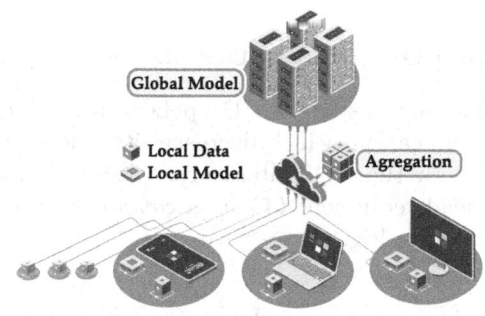

Fig. 1. Federated learning architecture.

Algorithm 1 sketches how the FederatedAveraging algorithm (FedAVG) [5] proceeds using a cluster of N clients, each with a learning rate of η. The variable S designates the set containing all clients, while C is a fraction representing a subset of selected clients from S, where $|S'| = (C \cdot N)$. The FedAVG acts in two synchronous steps, starting by generating a global model, say w_0, on the server. After that, it randomly chooses m participating clients where m is the maximum between $(C \cdot N)$ and 1. Each of the selected clients trains a local model similar to the global one during several local iterations $e = 1, \ldots, E$, where E is the communication interval. Once done, all local models are sent to the server in order to update the global one, where P_k is the weight of the kth client. The whole process is executed repeatedly during T iterations.

Algorithm 1. The federated averaging algorithm.

1: Initialise(w_0);
2: **for** $t = 1, ..., T$ **do**
3:     $m \leftarrow \max(C \cdot N, 1)$;
4:     $S' \leftarrow \text{random\_Pick}(S, m)$;
5:     **for all** clients $k \in S'$ in parallel **do**
6:         **for** $e \in 1, ..., E$ **do**
7:             $w_e \leftarrow w_{e-1} - \eta \nabla F(w_{e-1})$;
8:         **end for**
9:         $w_{t+1}^k \leftarrow w_e$;
10:    **end for**
11:    $w_{t+1} \leftarrow \sum_{k=0}^{m} P_k \cdot w_{t+1}^k / m$;
12: **end for**

2.2 Communication Overhead in Distributed Deep Learning

The FL workflow, like any Distributed Deep Learning (DDL), induces a substantial load of communication, which decreases its efficiency and applicability [7]. When going through the DDL literature, two main approaches exist for communication overhead reduction: (I) *data compression* and (II) *decreasing communication rounds* (see Fig. 2).

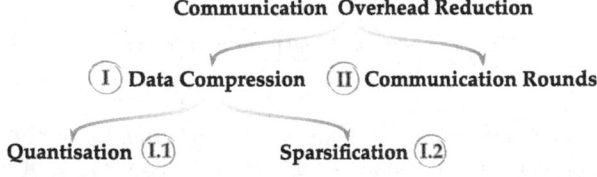

Fig. 2. Communication overhead reduction: taxonomy of the techniques.

Considering the first approach, the literature identifies two effective ways for compressing data: (I.1) quantisation [1] and (I.2) sparsification [8]. The first consists of representing the data using a low-precision/small-sized data type (e.g. bool). In contrast, the second approach transmits only essential values of each communication (about 1% of the overall values). Nonetheless, when using quantisation, the compression rate is low considering that the maximum compression ratio is limited to $(1/32)$ (32-bit-encoded data is frequently used in DDL). Also, when having fewer bits to carry the information, the models that use quantisation tend to have a slower convergence. Unlike quantisation, sparsification achieves a compression rate of $(1/100)$ without a significant modification of the model's convergence speed and final accuracy. Sparsification also comes backhanded since it introduces supplementary phases during the training process (e.g. sampling, de/compression, de/coding, etc.). This can affect the overall training efficiency, especially in battery-sensitive (e.g. smartphone) and low-performance (e.g. net-book) devices.

Moving now to the second communication-reduction technique, in vanilla FL [10] (i.e. standard FL), the communication happens at the end of each iteration ($E = 1$). A typical FL training of deep neural network takes hundreds of thousands of iterations. Enlarging the communication intervals would allow reducing the communication overhead. Therefore, FedAvg algorithm and its variants allow clients to perform multiple iterations of local training before updating the global model [10]. It has been proven that reducing the communication rounds increases the convergence speed. The communication interval in FedAvg is controlled by the hyperparameter E, which influences the model's trade-off between the accuracy and the training efficiency. Generally, a smaller E induces a better final accuracy, while a large E value accelerates the model's convergence. Therefore, experts would be needed to fine tune the communication interval E to allow the model to reach the best possible efficacy.

Most of the literature studies the communication-reduction approaches separately. Although, we believe that they are all equally important and jointly impact the communication rate. Therefore, as far as the authors' knowledge, we are the first to investigate all these approaches together within the same work.

3 Proposed Approach

This section presents our FL-COP formulation and the used NSGA-II solver.

3.1 The Proposed FL-COP Modelling and Formulation

Our formulation of the FL-COP is a bi-objective optimisation problem, where the two conflictual objectives consist of (I) minimising the communication overhead while (II) maximising the model's accuracy. It also assumed that each client in the architecture has a similar model (i.e. nodes, connections, layers, activation functions, etc.) as the one on the server. When mentioning the *local* and *global* models, we refer to the client's and server's models, respectively. Let us assume an architecture of one server connected to N clients. The model being trained has l layers L_i having n_i weights, where $i = 1, \ldots, l$. The FL-COP modelling is thought as a 4-levels communication-reduction scheme, where each layer represents when a given communication-reduction approach is applied. At the highest level, we identify the number of clients that will participate in training the global model, while the three remaining lower levels reflect the three communication-reduction approaches explained in Sect. 2.2: quantisation, sparsification and reducing the communication rounds (see Fig. 3).

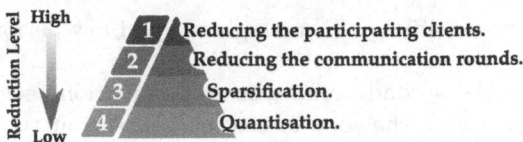

Fig. 3. The FL-COP modelling levels

The overall amount of communications happening during the FL learning process is proportional to the number of clients $m \in [1, N]$ that participate in training the global model. So, a first part of the FL-COP modelling stands in finding the number m of clients, selected randomly among the complete set S, and which will be the only ones sending their local models to the server. A second part of the FL-COP modelling consists in finding the number of training iterations $E \in [1, 1000]$ after which all the clients send their local models to the server. This variable determines the number of training steps that the clients perform before sending their local models (e.g. weights, gradients, etc.). It is important to note that for each client, the maximum number of training iterations allowed on overall is $(E \cdot T)$, where T is the maximum number of times the clients can send their local models to the server. The third part of the problem modelling consists in selecting, for each layer L_i having n_i weights, a percentage $\mu \in [0\%, 50\%]$ of the weights that will not be sent to the server.

Using the classical `FedAvg`, the weights are encoded with full precision (i.e. all their decimals) using 32 bits. Thus, the fourth, and final part of our FL-COP modelling consists in finding the optimal number of bits b_i allocated to encode the weights of each layer L_i in the model, where $i = 1, \ldots, l$. We also assume that ϖ_i and ϱ_i represent, respectively, the maximum and minimum values of the weights in the ith layer. Having b_i bits means that 2^{b_i} binary combinations can be created. We assign the all-ones and all-zeros combinations to encode the ϖ_i and ϱ_i values, respectively. The $(2^{b_i} - 2)$ remaining combinations will encode $(2^{b_i} - 2)$ values that are equally drawn from the interval $[\varpi_i, \varrho_i]$. Technically, the data that will be sent to the server will be the series of combinations that encodes each weight, as well as ϖ_i and ϱ_i. The server will perform the reverse mechanism to retrieve the full-precision weights. Each client will send to the server $\sum_{i=1}^{l}(n_i \cdot b_i) + 64$ bits instead of $\Theta = \sum_{i=1}^{l}(n_i \cdot 32)$ original bits.

Our formulation of the FL-COP is described using Eqs. (1)–(4). The first objective function $f_1(\overrightarrow{X})$ defined by Eq. (1) calculates the percentage of data reduction that the solution $\overrightarrow{X}$ achieves. Concretely, it is the sum of the percentage α and $\beta \in [0, 1]$ of data sent and received, respectively, by all the clients together from and to the server. These percentages are expressed with regard to the original data that would have been sent or received when no communication reduction is applied $(T \cdot N \cdot \Theta)$. The second objective function $f_2(\overrightarrow{X})$ defined by Eq. (2) evaluates the accuracy of the global model w_T^* at communication T (i.e. the last iteration) achieved via the solution $\overrightarrow{X}$. The server's model $w_T^* = \sum_{k=0}^{m} w_T^k / m$ is computed as the mean of the m local models obtained after T communications, while the accuracy is computed as the division of λ by ν, where λ and ν are the number of correct and total predictions made using the model w_T^*, respectively.

$$\underset{\overrightarrow{X}=\{x_1,\dots,x_d\}}{\text{Min}} \quad f_1(\overrightarrow{X}) = \frac{\alpha + \beta}{2} \tag{1}$$

$$\underset{\overrightarrow{X}=\{x_1,\dots,x_d\}}{\text{Max}} \quad f_2(\overrightarrow{X}) = \frac{\lambda}{\nu} \tag{2}$$

Where:

$$\alpha = \frac{1}{E} \cdot \frac{m}{N} \tag{3}$$

$$\beta - \frac{m}{N} \cdot \frac{1}{E} \cdot \sum_{i=1}^{l} \frac{b_i}{32} \cdot \frac{100 - \mu_i}{100} \cdot \frac{n_i}{\sum_{j=1}^{l} n_j} \tag{4}$$

Subject to:

$$m, E, \mu_i, b_i \in \mathcal{N}, 1 \le m \le N, 1 \le E \le 1000, 0 \le \mu_i \le 50, 1 \le b_i \le 32 \tag{5}$$

The Fig. 4(a) sketches a typical solution $\overrightarrow{X}$ of an FL-COP that trains a $l = 3$ layers model. On the other hand, Fig. 4(b) represents a concrete solution $\overrightarrow{X}$ for the same configuration using 20 training iterations, 2 rounds of client-server communications. During each round, only 90% of the weights of the 1st layer, 55% of the 2nd, and 98% from the 3rd are sent to the server. The weights sent from the 1st, 2nd and 3rd layers are encoded using 2, 20 and 15 bits, respectively.

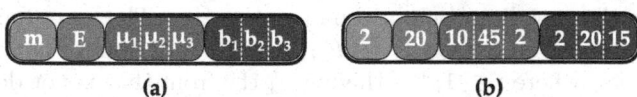

(a)   (b)

Fig. 4. A 3-layers model: (a) abstract and (b) concrete FL-COP solutions.

3.2 The Communication-Overhead Reduction Routine

To solve the FL-COP presented in Sect. 3.1, our approach consists in applying NSGA-II, a well-known evolutionary algorithm proposed by Deb et al. [2] and initially designed to tackle multi-objective problems. NSGA-II main contributions are the non-dominated sorting and the diversity-preservation heuristics with a computational complexity of $\mathcal{O}(MN^2)$ and $\mathcal{O}(MN\log N)$. Having a problem with M objectives, NSGA-II starts by randomly initialising a population of U individuals, let us say $\overrightarrow{X}=\{x_i, \dots, x_d\}$, where $i \in [1, d]$ and d is the size of the problem to be solved. Once this is done, NSGA-II enters in a loop until some stopping criterion is fulfilled. In the loop, it applies binary tournament selection, crossover and mutation to generate a population Q of U offspring. The union of both the parent and offspring populations, $R = P \cup Q$, will be used as input of a replacement operator in order to decide the solutions of the new population P' that will survive to the next iteration (see Algorithm 2).

Algorithm 2. The non-dominated sorting genetic algorithm II.

1: Set M objective functions $O_i/\ i \in \{1, \ldots, M\}$.
2: Set $\vec{X}$ a typical solution/ $\vec{X} = \{x_1, \ldots, x_d\}$, d number of variables to optimise.
3: Set $F = \{F_1, \ldots, F_K\}$, K the number of non-dominated fronts in the population.
4: $P \leftarrow$ Random_Generation(U);
5: **while** stopping criterion is not reached yet **do**
6:    $A \leftarrow$ Binary_Tournament_Selection(P, Crowded_Comparison);
7:    $B \leftarrow$ Crossover(A);
8:    $Q \leftarrow$ Mutation(B);
9:    $F \leftarrow$ Non_Dominated_Sorting($P \cup Q$);
10:   $P' \leftarrow \emptyset$;
11:   $i \longleftarrow 1$;
12:   **while** $(|P' \cup F_i| \leq U$ and $i \leq K)$ **do**
13:       $P' \leftarrow P' \cup F_i$;
14:       $i \leftarrow i + 1$;
15:   **end while**
16:   $F_i \leftarrow$ Descending_Sort_Crowding_Comparison(F_i);
17:   $P \leftarrow P' \cup F_i[1 : (U - |P'|)]$;
18: **end while**

The binary tournament selection is performed using the crowding-comparison heuristic, while the replacement step is based on the non-dominated sorting heuristic (also the crowding-comparison in some cases). The non-dominated sorting results is partitioned in a set $F = \{F_1, \ldots, F_K\}$ of K non-dominated fronts of increasing rank i, where $i \in [1, K]$. Having F_1 the front that is not dominated by any other one, while the remaining fronts are dominated by all the ones that have a lower rank. On the basis of the crowding-comparison operator, the NSGA-II favours solutions of lower rank if the solutions being compared belong to different fronts. On the other hand, if the solutions come from the same front, it advantages the solution having a higher crowding distance. For more details about the non-dominated-sorting and crowding-comparison operators, one should refer to the NSGA-II original work [2].

To solve the FL-COP, the NSGA-II is executed during a preliminary step in order to extract the optimal parameters of the FedAvg influencing the communication overhead. These parameters are: (I) the number E of training steps performed before sending the local model to the server, (II) the number m of clients participating in the training of the global model, (III) the number b of bits used to encode the weights of each layer of the local model, and (IV) the percentage μ of weights that will not be transmitted. In the following, we provide more details about each of the NSGA-II steps when solving the FL-COP.

Initialisation: The NSGA-II starts by initialising a population P of U solutions $\vec{X} = \{m, E, \mu_1, \ldots, \mu_l, b_1, \ldots, b_l\}$ of size $d = (2 \cdot l + 2)$ knowing that l is the number of layers in the trained model (see Fig. 5).

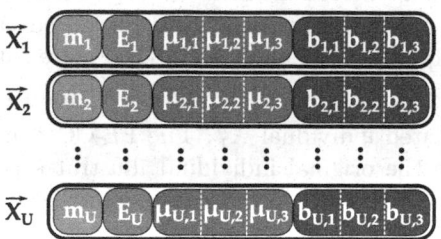

Fig. 5. NSGA-II population for solving the FL-COP

Selection: As a second step, NSGA-II performs a binary tournament selection on the parent population P to select the individuals that will undergo the breeding phases which will produce a population Q of U new offspring. The selection step creates a set A of $(N/2)$ pairs of parents, where the selection criterion is the crowding distance.

Crossover: Afterwards, according to a probability p_c, each pair of parents from the set A of selected ones will (or not) undergo the single-point crossover. Our aim is to prove that even using relatively-simple operators, the NSGA-II can still solve the FL-COP adequately. The crossover step will result in a new set B of U crossed offspring. It randomly chooses a switching point Ω from the interval $[1, d]$ and exchanges the solutions' substrings delimited by the variables at the position Ω and d. Figure 6 illustrates a single-point crossover applied on two individuals $\vec{X_1}$ and $\vec{X_2}$, representing a solution for FL-COP with a 3-layers model. Once applied, the crossover results in two new offspring $\vec{X_1'}$ and $\vec{X_2'}$.

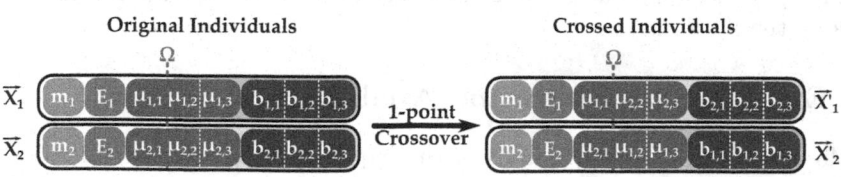

Fig. 6. Single-point crossover applied on FL-COP solutions

Mutation: During this step, each offspring in the set B obtained during the crossover phase will undergo (or not) a uniform mutation that is ruled by a probability p_m. The mutation phase results in a new population Q of U offspring. Like the crossover, our goal is to prove that even using operators of low complexity, the NSGA-II can still provide a meaningful efficiency.

The uniform mutation generates, for each variable of the solution being mutated, a random number from the interval $[\tau, \varrho]$, where τ and ϱ are the upper and lower bounds in which the variable being mutated can take valid values: $E \in [1, 100]$, $m \in [1, 4]$, $\mu \in [0, 50]$ and $b \in [1, 32]$. Figure 7 illustrates an example of the uniform mutation applied to an individual $\overrightarrow{X_1'}$ resulting from the crossover and producing a mutated individual $\overrightarrow{X_1''}$. The FL-COP in this case concerns a 3-layers mode, where the original individual illustrates a 200-iterations training, transfers 70% of the 2nd layer's weights and encodes the weights of 1st and the 2nd layers using 1 and 32 bits. Once mutated, the individual represents a 908-iterations training, will send 89% of the 2nd layer's weights and finally the weights of 1st and the 2nd layers will be encoded using 20 and 2 bits, respectively.

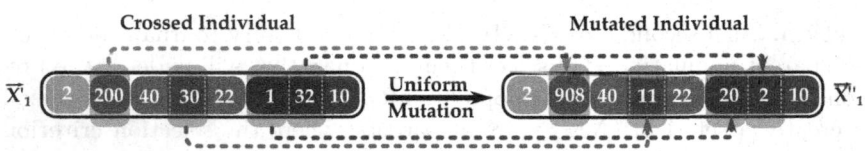

Fig. 7. Uniform mutation applied on FL-COP solutions

Replacement: Having the original population P as well as Q obtained after applying selection, crossover and mutation, a replacement step is applied in order to decide the composition of the population P' during the next iteration. Having $F = \{F_1, \ldots, F_K\}$ the set of non-dominated fronts obtained after applying the non-dominated sorting heuristic, P' will be filled by including the non-dominated fronts in an increasing rank until the $|P'| = N$. Let us admit that at some moment, one wants to include the ith front, but the union of both P' and F_i is greater than N. In this case, the solutions of the ith front are sorted in a descending order based on the crowding-comparison operator. Then, this P' will be filled by including the missing solutions from the best ones obtained after ranking the ith front.

4 Experimental Study and Analysis

In this section, we provide details of the experiments conducted to assess our proposal, as well as the obtained results and their discussion.

4.1 Problem Benchmarks and Experimental Settings

The implementation[3] has been made in **Python** version 3.8.8, while the execution has been done in the Picasso supercomputing center at the University of Malaga.

---

[3] https://github.com/NEO-Research-Group/flcop.

In particular, we used two types of hardware from a computation cluster: (I) a 24 × Bull R282-Z90 nodes: 128 cores (AMD EPYC 7H12 @ 2.6 GHz), 2 TB of RAM, Infini-Band HDR200 network, 3.5 TB of local-scratch disks. (II) a 4 × DGX-A100 nodes: 8 GPUs (A100 Tensor Core), 1 TB of RAM, Infini-Band FDR40 network and a 14 TB of local-scratch. We use a process with 128 cores and 400 GB of RAM to evaluate the solutions in parallel.

Our experiments have been thought of to assess our proposal's solving efficiency and scalability when dealing with different sizes of FL-COP benchmarks and its adaptability when dealing with different types of neural network models. Thus, we consider both convolutional and fully-connected Neural Network (NN) topologies with 12 and 3 layers and 887,530 and 33,400 weights, respectively. Our experiments have been done using the well-established MNIST dataset containing 70,000 images of handwritten digits. At the beginning of each execution, the initial weights of the models are drawn using the same seed. NSGA-II has been run using a population of 100 individuals, $p_c = 0.9$ and $p_m = (1/d)$. The size d of the solutions in the case of the convolutional model is 26, while in the fully-connected, it is 8. NSGA-II has been executed 30 times for each NN type, where it is executed during 300 iterations for the fully-connected model and 120 on the convolutional one.

The experimentation has been done by randomly distributing 60,000 images of the MNIST training set among the m clients. For simplicity, all clients have the same number of data. Each partition of the data remains private in each client throughout the learning process. We evaluate the final models obtained with the 10,000 images of the MNIST dataset. We conduct two different types of experiments. First, we apply NSGA-II for solving the FL-COP that uses a fully-connected neural network with 33,400 trainable parameters and 3 layers: one input (784), one intermediate (42), and one output layer (10). The middle layer and the output layer have a bias. We consider each bias of each layer as an independent array to optimise. Therefore, we have an array of weights of length 4 (i.e. [32928, 42, 420, 10]). In our experiments, we fix $l = 4$ where the two additional layers are of the bias. In the second experiment, we do the same on a convolutional neural network with 887,530 trainable parameters. In this case, we have a multidimensional array of length 12 (i.e. [800, 32, 25600, 32, 18432, 64, 36864, 64, 802816, 256, 2560, 10]). The termination criterion in both experiments is achieving one epoch (i.e. all clients trained with all their local data one time). In this experiment, we fix $l = 12$. In both cases, we simulate a server/client architecture of 4 clients (i.e. $N = 4$). All our results have been confirmed using a Wilcoxon test with Bonferroni correction and a significance level of 0.025.

4.2 Experimental Results and Discussion

Considering the fully-connected NN topology, Fig. 8 illustrates the Pareto fronts obtained by NSGA-II in 30 executions, while Fig. 9 represents the pseudo-optimal Pareto front that dominates all those obtained in 30 runs. Saying that for the convolutional we executed 25 executions. Similarly, Figs. 10 and 11 present the same information for the convolutional model. It can be seen in Figs. 8 and 9 that our proposal could reduce the communication to 35% in the worst solution, and to nearly 0% of communication, while maintaining accuracy above 0.94. Considering, Figs. 10 and 11, one can note that our approach could reduce the communication to 6% in the worst case, while it could achieve nearly 0% communication with an accuracy above 0.95.

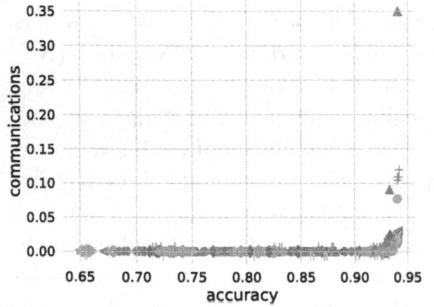

Fig. 8. Fully-connected NN: 30 executions' Pareto fronts (1 color/Pareto).

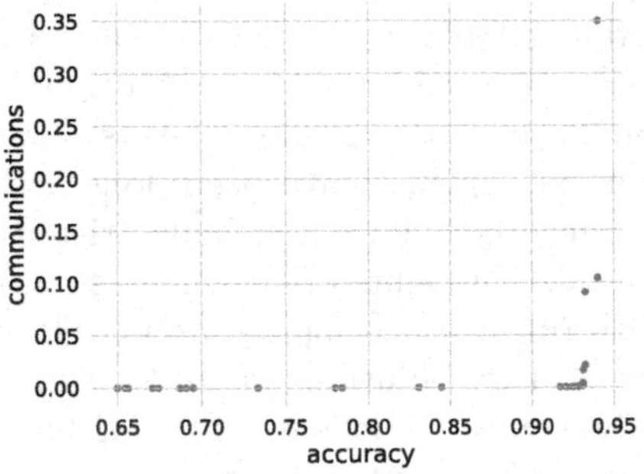

Fig. 9. Fully-connected NN: pseudo-optimal Pareto front.

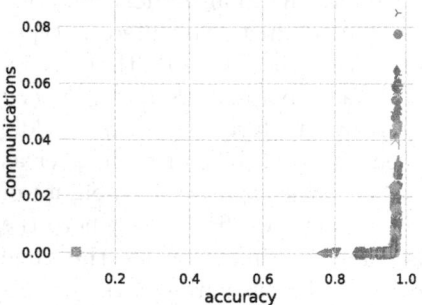
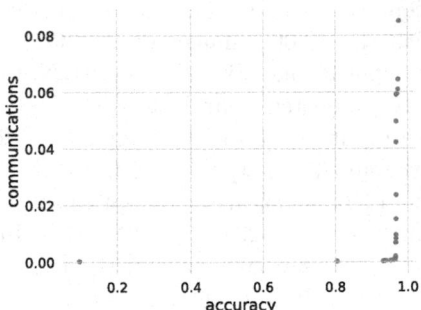

Fig. 10. Convolutional NN: 30 executions' Pareto fronts (1 color/Pareto).

Fig. 11. Convolutional NN: pseudo-optimal Pareto front.

Figures 12 and 13 illustrate the evolution of the average hypervolume of the Pareto fronts obtained by NSGA-II throughout one randomly selected, but yet representative execution when tackling the FL-COP using a fully-connected and convolutional neural network, respectively. The smooth evolution of hypervolume in the first iterations can be explained by the fact that NSGA-II starts with random low-quality individuals that can be quickly enhanced. Nonetheless, as the iterations go, the attained Paretos are of higher quality and difficult to enhance beyond iteration 40. Of course, more advanced hypotheses could be made to explain such behaviour, nonetheless it will be hard to confirm them without further in-depth analysis.

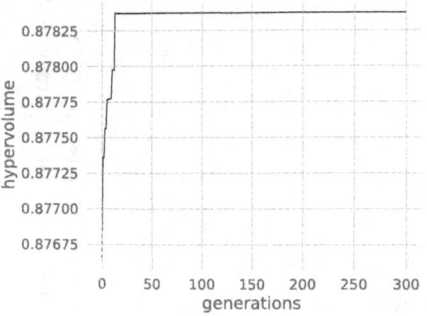

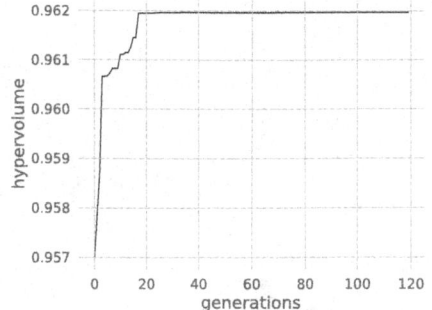

Fig. 12. Fully-connected NN: Hypervolume evolution through iterations.

Fig. 13. Convolutional NN: Hypervolume evolution through iterations.

Figure 14(a) illustrates the median, first, and third quartiles of the distribution of the solution's parameters distribution for the individuals obtaining the best accuracy in each final Pareto of the 30 executions. Figure 14(b) and (c) do the same for the slope and lowest-communication solutions. Figure 14(a)–(c) concerns the fully-connected NN, while Fig. 15(a)–(c) treats the convolutional NN. In Fig. 14(a) and (b), the solutions with the best accuracy and slope solutions seem to share a similar distribution pattern of the parameters. This could be explained by the fact that both are generally located in the elbow of the Pareto, as it can be seen in Fig. 9. In both types of solutions a large part of the clients are frequently involved in the training, but to cope with this, a clear reduction of the communication rounds is noticed. Moreover, it seems that NSGA-II allows a homogeneous compression of weights in terms of number of bits used to code them as well as the percentage of weights being transferred. An explanation of such observation is an attempt of the algorithm of compensating the lack of communication (low E) by sending more precise (b_i) and complete weights (μ_i) to help recover the model's accuracy. The results of the low-communication solution in Fig. 14(c) are quite self-explanatory since the algorithm involve an almost null number of clients in the training, as well as a clear high rate of weight compression. This will probably induce a low-accuracy solution as it can be seen in Fig. 14(i). Moving now to the convolutional NN in Fig. 15(a)–(c), NSGA-II produces solutions of different type where the number of participating clients is higher, while the weights' compression is non-homogeneous, but still meaningful. All the above stated-explanations can be mostly supported by Fig. 14(c)–(i) and Fig. 15(c)–(i). Another important conclusion that one can draw from Figs. 8, 9, 10 and 11, is that the parameters that have more impact on the final communication are m and E.

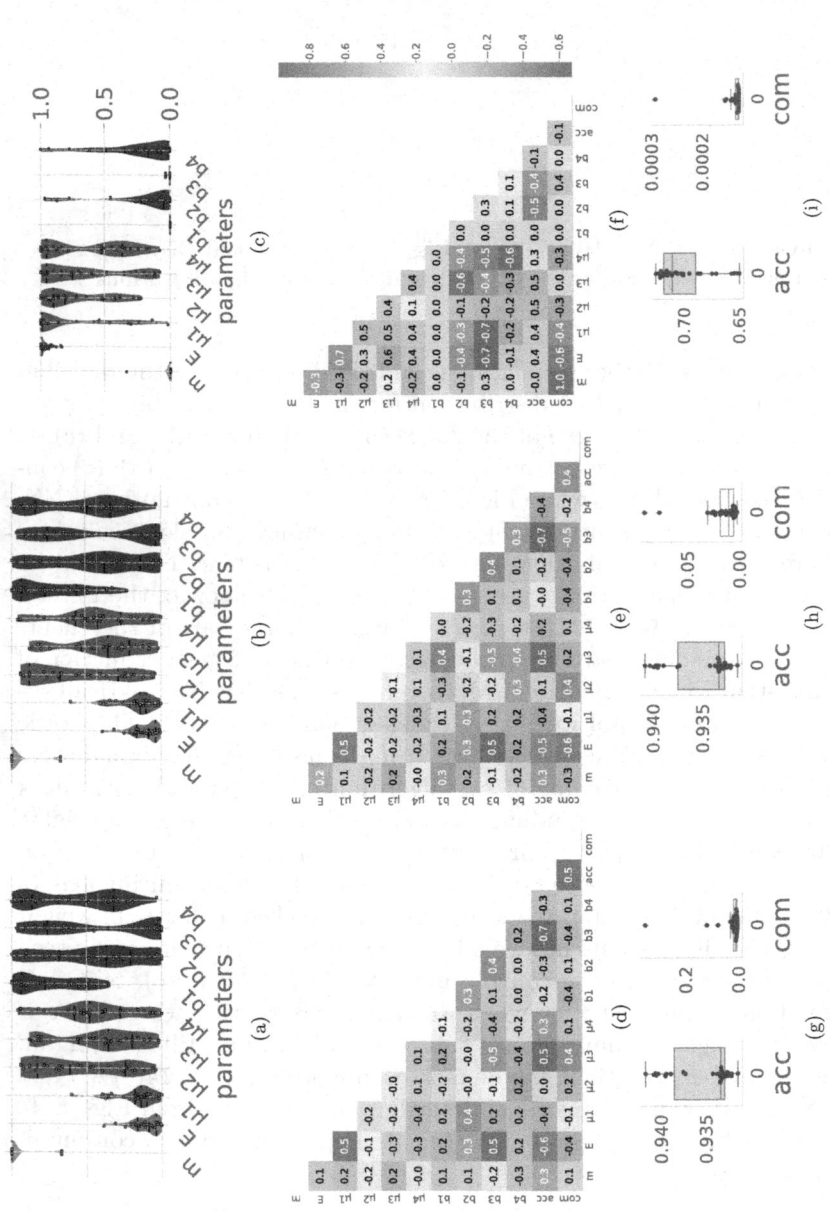

Fig. 14. Fully-connect NN: (a)–(c) Normalised parameters of the individuals with higher accuracy in final Pareto fronts. (d)–(f) Normalised parameters of the individuals with best correlation accuracy vs. communications in final Pareto fronts. (g)–(i) Accuracy and communications mean, std, max, min and quartiles.

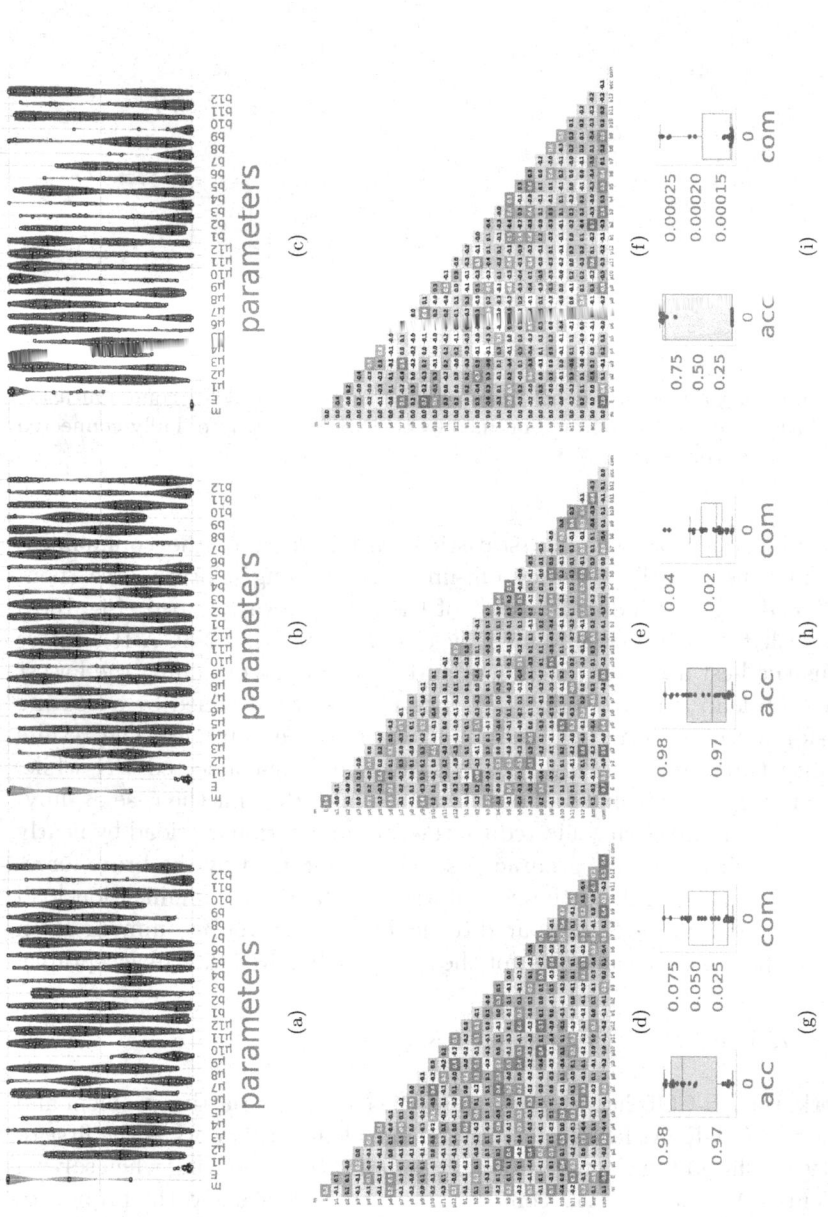

Fig. 15. Convolutional NN: (a)–(c) Normalised parameters of the individuals with higher accuracy in final Pareto fronts. (d)–(f) Normalised parameters of the individuals with best correlation accuracy vs. communications in final Pareto fronts. (g)–(i) Accuracy and communications mean, std, max, min and quartiles.

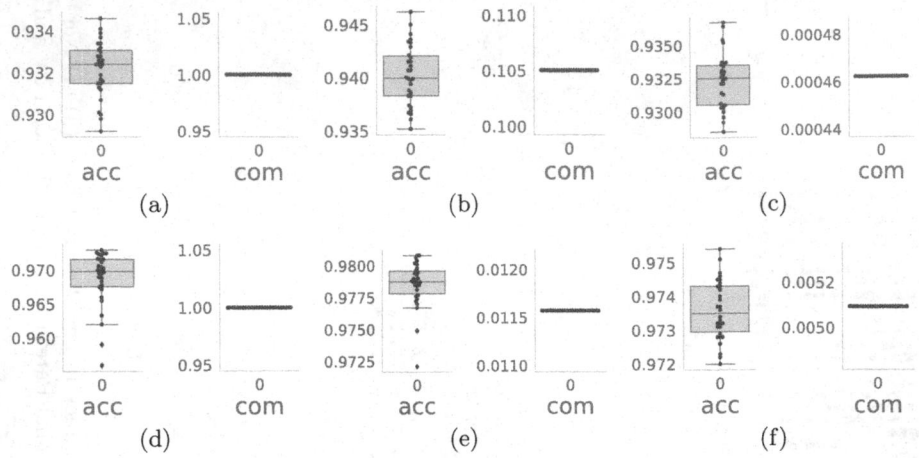

Fig. 16. Comparing the solution obtained using a brute-force (100%) communication with two solutions of the Pareto front obtained by the NSGA-II: (a)–(c) Fully-connected NN and (d)–(f) convolutional NN.

Figure 16(a)–(f) represents the statistical distribution of the communication load and the model's accuracy obtained when using a brute-force (non-optimised) FedAvg that induces a 100% of the communication load (Fig. 16(a) and (d)), with two solutions obtained by our proposal (i.e. NSGA-II), where one obtains the best accuracy (Fig. 16(b) and (e)) and one obtaining the lowest communication load (Fig. 16(c) and (f)). The results are obtained by executing the brute-force approach and the NSGA-II during 30 executions. The results are split into two categories: Fig. 16(a)–(c) for the fully-connected NN while, Fig. 16(d)–(f) for the convolutional one. One can note that for the case of fully-connected NN, our approach could reduce the communication overload by nearly 99% while achieving the same accuracy as the one obtained by the brute-force-100% communications. Taking the convolutional NN, the communication load has also been reduced by 99% compared to the brute-force-100% communication technique. Although the accuracy is not the same, we believe it is still acceptable.

5   Conclusions and Perspectives

In this work, the FL-COP has been formulated as a multi-objective problem and solved using NSGA-II. As far as the authors' knowledge, this work is the first to (I) investigate the add-in that evolutionary computation can bring when solving this problem and, to do so, (II) considers both the model's and the properties of used devices. The experiments have been made by simulating a server/client architecture using 4 devices. Both convolutional and fully-connected neural networks of 12 and 3 layers with 887,530 and 33,400 weights, respectively, have been researched. The validation has been done on the MNIST dataset containing 70,000 images. The Experiments have shown that our approach could outperform the

FedAvg algorithm using 100% of communications. We could reduce the communication by 99%, while maintaining an accuracy equal to the one obtained when using 100% of communications (i.e. brute-force). As for future work, we aim at testing our proposal using physically distributed devices and larger benchmarks.

Acknowledgments. This research is partially funded by the Universidad de Málaga, Consejería de Economía y Conocimiento de la Junta de Andalucía and FEDER under grant number UMA18-FEDERJA-003 (PRECOG); under grant PID 2020-116727RB-I00 (HUmove) funded by MCIN/AEI/ 10.13039/501100011033; and TAILOR ICT-48 Network (No 952215) funded by EU Horizon 2020 research and innovation programme. José Ángel Morell is supported by an FPU grant from the Ministerio de Educación, Cultura y Deporte, Gobierno de España (FPU16/02595). The authors thank the Supercomputing and Bioinnovation Center (SCBI) for their provision of computational resources and technical support. The views expressed are purely those of the writer and may not in any circumstances be regarded as stating an official position of the European Commission.

References

1. Alistarh, D., Grubic, D., Li, J.Z., Tomioka, R., Vojnovic, M.: QSGD: communication-efficient SGD via gradient quantization and encoding. In: Proceedings of the 31st International Conference on Neural Information Processing Systems, NIPS 2017, pp. 1707–1718 (2017)
2. Deb, K., Pratap, A., Agarwal, S., Meyarivan, T.: A fast and elitist multiobjective genetic algorithm: NSGA-II. IEEE Trans. Ev. Comp. **6**(2), 182–197 (2002)
3. LeCun, Y., Bengio, Y., Hinton, G.: Deep learning. Nature **521**, 436–444 (2015)
4. Mayer, R., Jacobsen, H.A.: Scalable deep learning on distributed infrastructures: challenges, techniques, and tools. ACM Comput. Surv. **53**(1), 1–37 (2020)
5. McMahan, B., Moore, E., Ramage, D., Hampson, S., y Arcas, B.A.: Communication-efficient learning of deep networks from decentralized data. In: Artificial Intelligence and Statistics, PMLR, pp. 1273–1282 (2017)
6. Sheller, M.J., et al.: Federated learning in medicine: facilitating multi-institutional collaborations without sharing patient data. Sci. Rep. **10**(1), 1–12 (2020)
7. Tak, A., Cherkaoui, S.: Federated edge learning: design issues and challenges. IEEE Network (2020)
8. Wangni, J., Wang, J., Liu, J., Zhang, T.: Gradient sparsification for communication-efficient distributed optimization. In: Proceedings of 32nd International Conference on Neural Information Processing Systems, pp. 1306–1316 (2018)
9. Xu, J., Du, W., Jin, Y., He, W., Cheng, R.: Ternary compression for communication-efficient federated learning. IEEE Trans. Neural Networks Learn. Syst. (2020)
10. Zhou, Y., Ye, Q., Lv, J.C.: Communication-efficient federated learning with compensated overlap-fedavg. IEEE Trans. Parallel Distr. Syst. (2021)

Evolutionary Machine Learning

Evolving Data Augmentation Strategies

Sofia Pereira, João Correia$^{(\boxtimes)}$, and Penousal Machado

CISUC, Department of Informatics Engineering, University of Coimbra,
3030 Coimbra, Portugal
{jncor,machado}@dei.uc.pt

Abstract. Deep Learning Algorithms are widely implemented and have reached state-of-the-art results in several scientific investigations. In medical images domain and Computer-Assisted Detection (CAD) systems, Convolution Neural Networks (CNNs) are the preferred deep network architecture. Despite getting good results, there are still some obstacles to overcome, namely the problem of overfitting. Lately, the Data Augmentation (DA) has been integrating the training pipeline of Deep Neural Networks to mitigate those issues. The effectiveness of classical image transformations in increasing the performance of classification tasks in medical imaging domain has been reported. However, the search for a suitable augmentation strategy is performed manually. This approach, mainly made of trial-and-error tasks can be very time-consuming and complex. Thereupon, a novel data augmentation approach is proposed. The approach is an Evolutionary Machine Learning approach that is able to automatically define an optimised DA strategy for each medical image classification task. Thus, the approach combines two algorithms, an evolutionary algorithm and combined with a deep learning algorithm to find suitable DA strategies. The results obtained demonstrate that the same or better level of performance is achieved when the Transformation Functions (TFs) and their parameters are defined automatically instead of manually.

Keywords: Data augmentation · Medical imagery · Evolutionary algorithm · Evolutionary machine learning · Convolutional neural network

1 Introduction

Nowadays, several medical exams used to evaluate the health status of the human body and diagnose pathologies are delivered in image formats such us pathological images, CT and MRI images, etc. [1] Computer-Assisted Detection (CAD) systems are specialised clinical decision support systems in medical imaging [2]. The rapid development of deep learning and computer vision has allowed a great advance for CAD systems. Deep neural networks are being widely applied to computer vision to perform tasks such as image segmentation, object detection and image classification. It is a expanding field thanks to the ease of access to data, increased computational power, and continued research into deep network architectures.

© Springer Nature Switzerland AG 2022
J. L. Jiménez Laredo et al. (Eds.): EvoApplications 2022, LNCS 13224, pp. 337–351, 2022.
https://doi.org/10.1007/978-3-031-02462-7_22

One of the most challenging improvements around deep learning models is the generalisation ability. When a model has low generalisation ability, overfitting is very likely to occur [3]. It is known that, typically, larger data sets help to obtain deep learning models with better generalisation. However, gathering big data sets can be a very challenging task due to all the effort required to manually collect and label the data. Data Augmentation (DA) is a technique used by the scientific community that helps to mitigate the overfitting issue [4]. DA allows to increase diversity and size of training data sets. In the image domain, this can be done by applying Transformations Functions (TFs) over the available images [5]. Classical image transformations are the most popular techniques among image data augmentation. These strategies are very straightforward and simple to implement, in addition to improving the performance of classification models by decreasing overfitting [6]. However, there is still no delineated procedure capable of finding the DA strategy that best fits each problem. At the moment, the existing procedures depend on the experience and time of the researchers and are done mainly by trial-and-error tasks [5].

With the above-mentioned aspects in mind, a novel approach combining algorithms of Machine Learning (ML), Evolutionary Algorithms (EAs) and classical Transformation Functions (TFs) is proposed. This Evolutionary Machine Learning (EML) approach is delivered as a framework and optimises the strategies with respect to the current task performance. Thus, this work contributions are as follows: i) presentation of an EML framework to perform an automatic search to define DA strategies using evolutionary strategies combined with the feedback of a Machine Learning algorithm; ii) evaluation and definition of suitable performance metrics for fitness assignment for the evolutionary strategy iii) instantiation of the framework on set of experiments on medical image classification tasks to yield the best DA strategy for a pre-determined dataset and model.

The remaining document is organised as follows: in Sect. 2 we present a brief overview of the literature on data augmentation applications for deep learning algorithms. In Sect. 3 we introduce and define the proposed approach. In Sect. 4 we describe the experimental procedure carried out to test the new approach and in Sect. 5 section we present and discuss the results obtained. Finally, conclusions and future work are available in the last section.

2   Related Work

DA, unlike other techniques developed to mitigate the overfitting problem, approaches the problem by its root, the training data set. DA strategy consists in the application of a set Transformation Functions (TFs) that can be applied to the original samples to inflate the data set size and to increase the diversity of samples by simulating more alterations that may occur in the real world [3]. DA strategies can be divided into two groups. The ones associated with basic image transformations such as geometric and colour space transformations and the others based on neural networks such as Generative Adversarial Network

(GAN) and neural style transfer [7]. Some of the most applied data augmentation methods are vertical and horizontal flips, vertical and horizontal translations, rotations, shearing, cropping, zooming, and scaling [8]. Basic transformations on input images require less computational time, are simpler to implement than methods based on neural networks and have regularly demonstrated effectiveness in improving image classification performance. [9,10]. In addition to being applied to mitigate the problem of overfitting, data augmentation can also be employed to solve unbalanced class distribution issues [3] and the susceptibility of models to suffer from adversarial attacks.

Kwasigroch et al. [11] applies data augmentation to perform up-sampling due to a highly imbalanced data set. The strategy used was a combination of random classical transformations: rotation, translations, zoom and flips. Zhang et al. [12] proposed a data augmentation strategy called mixup. This approach generates data by linearly interpolating images and corresponding labels from the available training data set. This produces unrealistic images, however, it has been shown to improve generalisation of state-of-the-art classification models. Chaitanya et al. [6] proposed an augmentation method based on GANs approaches that optimises the parameters of the generator by integrating unlabelled images and task loss. This process also results in augmented images somewhat unrealistic, despite improving the model's performance. These examples show that reproducing realistic images is not the only way of achieving performance improvements. Using an Evolutionary Machine Learning approach, Fernandes et al. [13] developed a framework that employs a supervisor module that uses an Evolutionary Computation approach to evolve sets of images drawn from Generative Adversarial Networks' latent space to augment the data set. The fitness function is based on the dissimilarity of the subsets generated by the Generative Adversarial Networks. This module handles the generated samples and chooses which set should be added to the training data set. The framework was tested in a bio-medicine multi-class problem with a small number of samples that provide a challenge to the different supervised classification approaches, achieving improvements in performance when the models were trained with the generated samples.

Usually, a tuning of transformation functions parameters is performed to better capture the more significant features when training a deep neural network. However, there is still no approach that implements an automatic search to find the most suitable parameters. At this point, they are either pre-defined empirically, done manually through trial-and-error tasks or are problem specific approaches.

3 Approach

The proposed approach consists in the automation of the search process for the most adequate DA strategy for each problem. To this end, a framework was developed using an Evolutionary Machine Learning approach. The evolutionary algorithm has the function of automatically generate data augmentation strategies. The evolutionary process is guided by an evaluation function. The role of the machine learning algorithm is to evaluate each of the proposed strategies. In this work, the evolutionary algorithm chosen was the evolutionary strategy and the machine learning algorithm is a convolutional neural network.

In this implementation of the evolutionary strategy, the representation of individuals is made using real numbers. The only variation operator present is mutation. During the selection process only one parent is selected in a deterministic and elitist way. Both parent and offspring are passed on to the next generation. To define the set of possible combinations of DA strategies, a collection of Transformation Functions (TFs) frequently mentioned in medical imaging literature was selected. Table 1 shows the number of TFs available in the framework at this moment. For more details on each TF in use refer to the documentation of imgaug[1] for more details. For the free parameters experiment, the TFs used are the 39 represented in Table 1. These functions can have up to 3 free parameters. A set of 136 TFs was extracted from these 39 functions by fixing parameters. For example, we can assign a parameter to the Add function, getting Add(10) or Add(20), etc. This last set is used in the fixed parameters experiment.

The approach is described by the following steps:

1. The evolutionary algorithm is fed by a set of transformation functions;
2. Through the evolutionary process, solutions formed by the supplied TF are generated;
3. Each solution is used to augment the training data set that will serve as input to the deep learning algorithm;
4. The DL algorithm will return the metrics (fitness function) obtained after training the classifier with the data set augmented by each of the solutions provided;
5. The evolutionary algorithm, based on the returned metrics, will follow up the evolution of the solutions;
6. New solutions are created via variation operators based on the previous ones and their respective fitness function;
7. The process is repeated from step 3 to 7 until the number of generations reaches the defined value.

Thus, during the evolutionary process, the genotype is translated to TF sequences that are generated from the previously selected set, which define a DA strategy. The evolutionary process is guided by an evaluation function that analyses each DA strategy to select the most suitable one. As an evaluation function, several metrics related to the validation data obtained after training a deep learning algorithm, can be used. The input image data set is previously submitted to the DA process through the application of the DA strategy under analysis. The next sections will detail, representation, variation operators and evaluation.

---

[1] https://imgaug.readthedocs.io/en/latest/.

Table 1. Transformation Functions (TF) selected.

Index	TF	Index	TF
0	Add	20	ScaleY
1	AddElementwise	21	TranslateX
2	AdditiveGaussianNoise	22	TranslateY
3	AdditiveLaplaceNoise	23	Rotate
4	AdditivePoissonNoise	24	ShearX
5	Multiply	25	ShearY
6	MultiplyElementwise	26	MultiplyBrightness
7	Cutout	27	AddToBrightness
8	Dropout	28	MultiplySaturation
9	CoarseDropout	29	AddToSaturation
10	ReplaceElementwise	30	GammaContrast
11	SaltAndPepper	31	LinearContrast
12	CoarseSaltAndPepper	32	HistogramEqualization
13	Salt	33	GaussianBlur
14	CoarseSalt	34	AverageBlur
15	Pepper	35	EnhanceContrast
16	CoarsePepper	36	EnhanceBrightness
17	Fliplr	37	EnhanceSharpness
18	Flipud	38	FilterSharpen
19	ScaleX		

3.1 Representation

In the approach with fixed parameters, TFs with predefined parameters were used. In this case, the genotype of each individual consists of positive integers ranging from 0 to $N - 1$, where N is the total number of previously selected TF. Each integer is seen as an index that matches a given FT. Following this logic, the phenotype is the TFs themselves. Figure 1a exemplifies an individual in these conditions.

In the free parameters, TFs with free parameters were used, i.e., the function parameters were also subject to modifications during the evolution process. The genotype of each individual is constituted by a tuple of two arrays. The first array, as in the previous experiment, is made up of positive integers ranging from 0 to $N - 1$. Each integer is seen as an index that matches a given TF with no associated parameters. The second array is composed by arrays of floats between 0 and 1. These values represent the parameters used by the functions represented in the first array. In other words, a gene is an integer that encodes a TF plus an array of floats corresponding to the parameters used by the TF. The phenotype of each individual is constituted by the TF with the respective parameters. Figure 1b exemplifies an individual in these conditions.

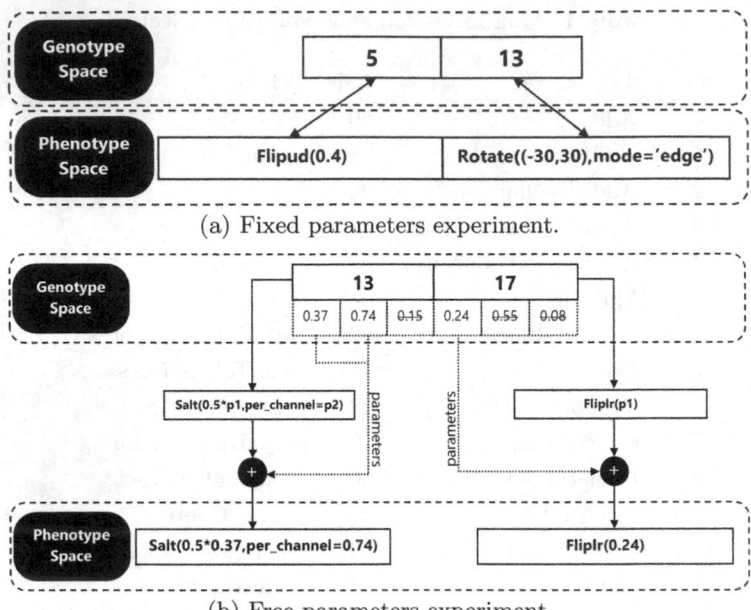

(a) Fixed parameters experiment.

(b) Free parameters experiment.

Fig. 1. Illustration of the genotype and the phenotype of an individual with 2 genes for both experiments.

3.2 Mutation

Using fixed parameters, mutation can occur through 3 different operations: addition, removal, alteration. Addition consists of inserting a gene with a random value in a random position of the individual. Removal consists of deleting a gene at a random position in the individual. The alteration consists in exchanging the value of a random gene of the individual for a random value. All operations have the same probability of occurring, however, removal is not possible if the individual has only one gene, nor addition if the individual already has 5 genes. Figure 2a displays an illustration of this mechanism.

For the free parameters experiments, mutation can occur at two distinct levels: at the level of the gene that encodes a TF or at the level of parameters. There is a 50% chance of the mutation occurring at the TF gene level and 50% at the parameter level. At the level of the gene encoding a TF, the mutation proceeds in an analogous manner to the fixed parameters experiment. Three operations are possible: addition, removal, alteration with equal probability. When addition occurs, along with the new gene, a new array of parameters corresponding to the new gene is also inserted. When removal occurs, the array of parameters related to the deleted gene is also discarded. In case of alteration, there are no changes in the parameters array. If the mutation occurs at the parameter level, the value of a random parameter is exchanged for a random number, keeping the first array of the tuple intact. Figure 2b displays an illustration of this mechanism.

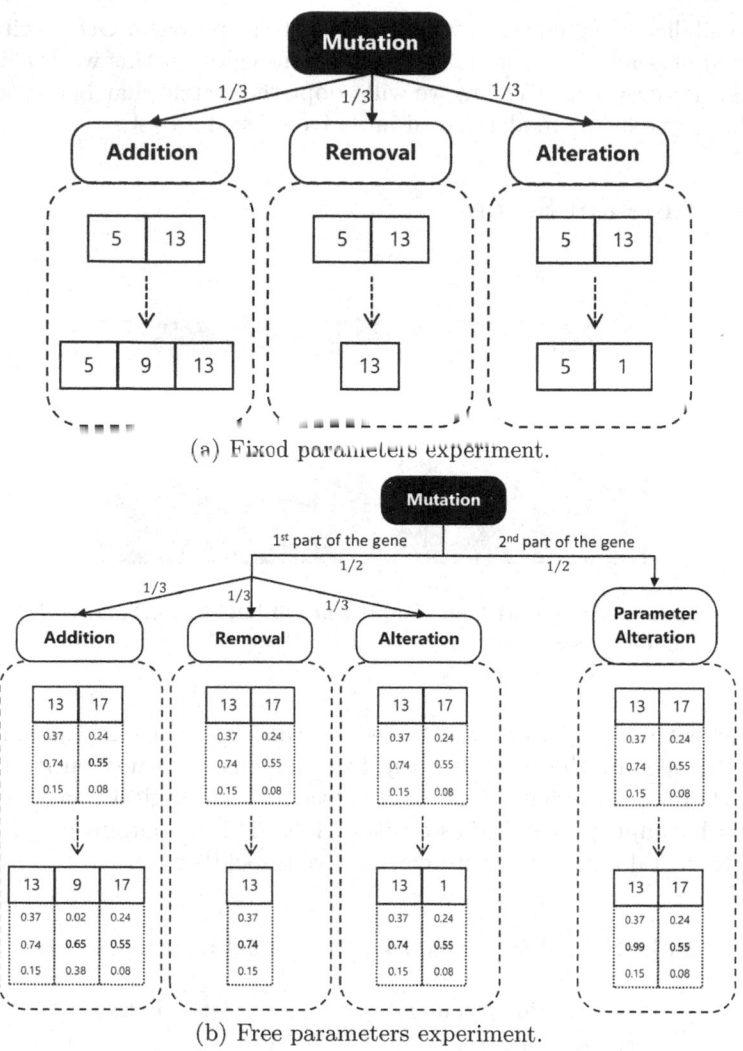

(a) Fixed parameters experiment.

(b) Free parameters experiment.

Fig. 2. Schematic representation of the mutation mechanism in both experiments.

3.3   Evaluation

The goal of the evolutionary process is to optimise performance on the test data set according to pre-defined performance metrics. However, during evolution there is no access to the test data set. Our approach will attempt to extrapolate the performance under test from the performance under validation. In order to achieve it, several metrics were collected for each of the TF used individually. Then, we studied the correlation between the test metrics and the validation metrics. The metrics tested are accuracy, loss, sensitivity, specificity, ROC AUC score, precision, F1 score and mean average precision. Note that this is one of many possibilities to guide the evolution using this approach. Other criteria can be inserted of combined to find suitable DA strategies. In this work it is out of scope testing several metrics but we will adopt the metric that better enhances the performance of the model overall in a given test data set.

4   Experimental Setup

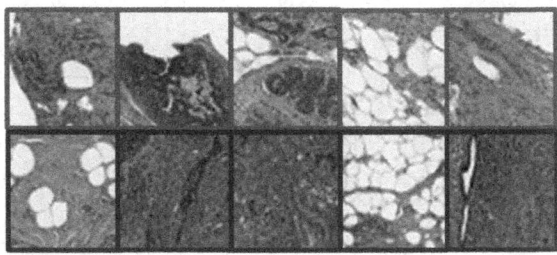

Fig. 3. Examples of samples without the presence of IDC (green) and with IDC (red). (Color figure online)

To implement our approach, we started by choosing a medical imaging data set and reproducing the training setup from an existing study on it. The data set consists of small patches that were extracted from digital images of breast tissue. Each sample is classified as positive (label 1) if it contains characteristics of invasive ductal carcinoma cells, or negative (label 0) otherwise.

Table 2. Number of train, test, IDC(−) and IDC(+) samples.

	No. of train samples	No. of test samples	
No. of IDC(−)	20 371	12 954	33 325
No. of IDC(+)	20 371	5 046	25 417
	40 742	18 000	58 742

The classification algorithm used is a convolutional neural network. The data set is available on the Kaggle platform[2], which is a repository of community published code and data. Figure 3 displays examples of samples with the presence of IDC (red) and without the presence of IDC (green). Table 2 describes the data set used. The classification algorithm used is a convolutional neural network and it's defined in Tables 3 and 4. A baseline was defined training the network without using any DA strategy.

---

[2] https://www.kaggle.com/paultimothymooney/breast-histopathology-images.

Table 3. Classification model parameters.

Parameter	Setting
Epochs	20
Batch size	32
Cross Validation Folds	5
Repetitions	10
Optimizer	adadelta
Learning rate	0.001

Table 4. Classification model layout.

Layer	Output shape	Parameters
Convolutional 2D	(,24,24,32)	896
Convolutional 2D	(,22,22,64)	18496
Max Pooling 2D	(,11,11,64)	0
Dropout	(,11,1,64)	0
Flatten	(,7744)	0
Dense	(,128)	991360
Dropout	(,128)	0
Dense	(,2)	256

Thereafter, we gather a set of TFs, having as reference their mention in the medical imaging literature and others that we consider pertinent in the domain of medical images. Each selected TF was applied individually over the data set to train the CNN during 20 epochs using 5 folds. Ten repetitions of the training were performed for each TF and several classification evaluation metrics were recorded. Examples of the resulting images after the application of some TFs are depicted in Fig. 4.

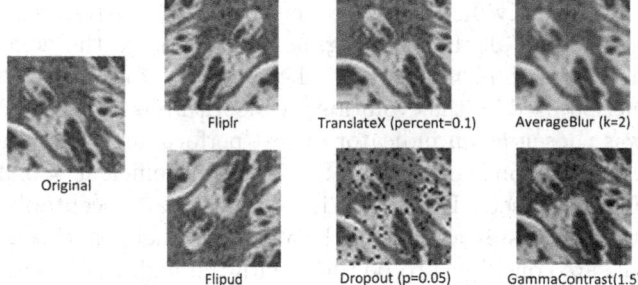

Fig. 4. Examples of the resulting images after applying some transformation functions.

To define the evaluation function, the study mention in Sect. 3.3 was performed and the results are displayed in Table 5. All metrics have a high Spearman coefficient (>0.7) showing a strong correlation between test and validation performance. Hence, validation metrics can be used as good indicators of the state of the evolutionary process.

Table 5. Spearman and Pearson correlations between validation and testing metrics for several classification metrics.

	Spearman correlation	Pearson correlation
Accuracy	0,714	0,823
Loss	0,887	0,934
Sensitivity	0,996	0,996
Specificity	0,997	0,998
ROC AUC Score	0,991	0,996
Precision	0,995	0,997
F1 Score	0,747	0,853
Mean Average Precision	0,981	0,990

Two experiments were carried out using different approaches to apply data augmentation strategies. In the first, the DA strategies are built with transformation functions that have predefined parameters. In the other, the transformation functions parameters are free, being subject to changes during the evolutionary process.

Table 6 resumes the evolutionary algorithm settings. For all experiments, the populations of the evolutionary strategy have a fix size of 5 individuals. In each generation, only one individual is selected as a parent, the fittest one. The parent, in addition to breeding all the offspring, also passes on to the next generation. In the designed algorithm, mutation is the only way to create offspring. The evaluation function used by the evolutionary algorithm is the ROC AUC metric. This metric was chosen as an indicator of test performance from the validation as it presents a very strong correlation (Spearman's coefficient = 0.991) between the results of the two sets. Besides, it is considered an acceptable and widely used metric that expresses a trade-off between specificity and sensitivity. The best individuals are considered to be those with the highest evaluation function value.

As a result of the trade-off between time and performance, during the evolution process the training data set is reduced to 1/4 the size of the original data set. Initialisation is done randomly by defining an individual with random gene size and values. For all the experiments, the framework was applied 30 times using 30 different random initialization seeds. Different performance metrics are gathered from the evolutionary process, while maintaining an comparison with the baseline model.

Table 6. Evolutionary Strategy Parameters.

Parameter	Setting
Nº of generations	50
Population size	5
Individual size	1 to 5 genes
Mutation operators	Addition, removal, alteration, parameter alteration (free)
Fitness function	Validation ROC AUC
Train data set size	1/4 original (40 742)

5 Experimental Results

A baseline was established by training the model with no augmentation. The training of the classification model was executed 10 times to obtain an average value with a standard deviation. Each execution was composed by 5 folds and the model was trained during 20 epochs. The ROC AUC test score of the baseline is 82,63 ± 1,09.

As already mentioned, two experiments were carried out. Figure 5 shows the average evolutionary optimisation curves for each experiment. Both curves stabilise at practically the same performance level.

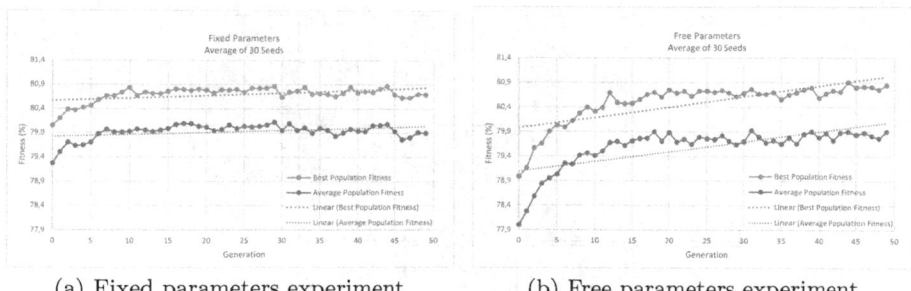

(a) Fixed parameters experiment   (b) Free parameters experiment

Fig. 5. Evolutionary optimisation curves obtained during 50 generations. Plots were obtained by averaging the results for each generation from 30 different seeds.

In Fig. 5b the stabilisation of the evolutionary optimisation curve takes longer when compared to Fig. 5a. As the search space for the free parameters experiment is so much higher than the fixed experiment, it is expected that the curve takes a little longer to stabilise. Also, for Fig. 5b, the performance increase across the generations is more notorious. The interpretation of this result lies on the fact that, for the fixed parameters experiment, the parameters were predefined considering the values and ranges used in the literature. This means that the improvement margin becomes smaller, and the evolution is less accentuated. On the other hand, as in free parameter experiment the parameters are free to take larger ranges of values, the initial solutions are not the most suitable. Thus, evolution starts with lower performances, which allows to a sharper rise in fitness.

For each seed in each experiment the best solution was found, i.e., the one with the highest fitness function value. To be able to compare both solutions with the baseline and between themselves, the solutions were subjected to the same procedure applied to obtain the baseline value. The training data set was augmented with the DA strategy defined by the solution being processed. Ten executions of the training of the classification model were executed. The best solution, i.e., the most suitable DA strategy for each experiment is the one with the highest value of Best ROC AUC Test Score. These strategies are shown below, in Figs. 6a and 7a, along with examples of images resulting from the application of those strategies (Figs. 6b, 7b):

Genotype	35	33	20	19
Phenotype	Flipud(0.1)	TranslateX(percent=(0.2), mode='edge')	Rotate((-50,50),mode='edge')	TranslateY(percent=(-0.1,0.1), mode='edge')

(a) Best DA strategy found.

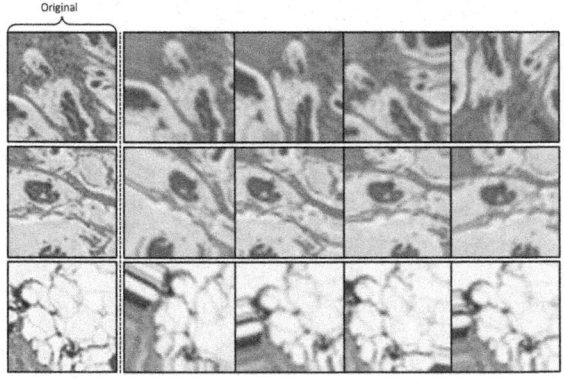

(b) Examples of the resulting images after applying the best DA strategy found. The original images are shown on the left.

Fig. 6. Results obtained with fixed parameters experiment.

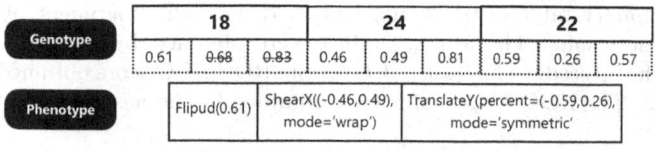

Genotype	18			24			22		
	0.61	0.68	0.83	0.46	0.49	0.81	0.59	0.26	0.57

Phenotype	Flipud(0.61)	ShearX((-0.46,0.49), mode='wrap')	TranslateY(percent=(-0.59,0.26), mode='symmetric'

(a) Best DA strategy found.

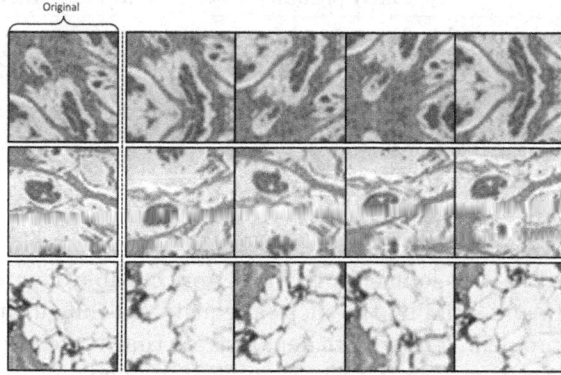

(b) Examples of the resulting images after applying the best DA strategy found. The original images are shown on the left.

Fig. 7. Results obtained with free parameters experiment.

Note that the data augmentation strategies found with the highest performances for each experiment are all made up only of transformation functions that belong to the geometric transformations group. According to the literature, this is the type of transformation that usually results in a greater increase in classification performance [14]. The results presented corroborate this statement. Table 7 shows the performance values for the two experiments. Both experiments outscored the baseline by 1.3%. The Average ROC AUC Test Score and the Best ROC AUC Test Score are very similar between the experiments. To assess this similarity the Mann-Whitney U test was applied [15]. The p-value between the fixed parameters experiment and the free parameters experiment is p = 0,976, much higher than the stipulated confidence level of 0.05. Therefore, the null hypothesis cannot be rejected so the results from the different experiments do not show statistically significant differences. However, Free Parameters Experiment is the one that utilises the most practical and automated architecture. Compared to the other experiment, it has the lowest average evolution time, in addition to being automated as no human intervention is required to define the parameters. The solution with the highest performance was also obtained by this experiment. As there are no significant differences in the results of the experiments, it can be concluded that the architecture used by the free parameters experiment is the most advantageous.

Table 7. Summary table of the results obtained for each experiment. All values are relative to test results. The average values were obtained by averaging the results obtained by the 30 seeds. The Best (AUC) is the ROC AUC score obtained by the best solution found. The ROC AUC scores were obtained by the mean of the 10 executions performed.

	Fixed parameters	Free parameters	Baseline
Avg (evolution time) (h)	35,4	29,8	–
Avg (AUC) (%)	83,34 ± 1,03	83,33 ± 1,05	82,63 ± 1,09
Best (AUC) (%)	83,69 ± 0,60	83,70 ± 0,64	

6   Conclusions and Future Work

We have presented the development of a framework capable of automatically augmenting data sets and defining the most adequate data augmentation strategy for each specific problem. The results demonstrate that there are no statistically significant differences between the performances obtained by the different experiments. In one of the experiments the functions used had predefined parameters, so there had to be a manual action to define them. In the other experiment, we tested the approach with free parameters, where there was human intervention on the selection of Data Augmentation Strategies. However, the results suggest that the performance of experiment is practically the same as that of the other. Nevertheless it can be concluded that the free parameters experiment presents the best conjugation between automation, computational time, effort and performance.

As the framework was designed with the aim of being flexible and being able to be applied to different situations, its configurations are easily modifiable. For future work, a logical next step would be studying the various configurations of the framework, such as population and individual's size and probabilities of the possible operations during the mutation mechanism, to look for relationships with performance. Another next step is testing the framework on multiple data sets and analyse its performance.

Acknowledgements. This research was partially funded by the project grant BEIS (Bridge Engineering Information System), supported by Operational Programme for Competitiveness and Internationalisation (COMPETE 2020), under the PORTUGAL 2020 Partnership Agreement, through the European Regional Development Fund (ERDF) and by national funds through the FCT - Foundation for Science and Technology, I.P., within the scope of the project CISUC - UID/CEC/00326/2020 and by European Social Fund, through the Regional Operational Program Centro 2020.

References

1. Skandarani, Y., Jodoin, P.M., Lalande, A.: GANs for medical image synthesis: an empirical study. vol. abs/2105.05318, May 2021

2. Eadie, L.H., Taylor, P., Gibson, A.P.: A systematic review of computer-assisted diagnosis in diagnostic cancer imaging. **81**, e70–e76 (2012)
3. Shorten, C., Khoshgoftaar, T.M.: A survey on image data augmentation for deep learning. **6**, 1–48 (2019)
4. Correia, J.a., Martins, T., Machado, P.: Evolutionary data augmentation in deep face detection. In: Proceedings of the Genetic and Evolutionary Computation Conference Companion, GECCO 2019, pp. 163–164. Association for Computing Machinery, New York (2019). https://doi.org/10.1145/3319619.3322053
5. Sánchez-Peralta, L.F., Picón, A., Sánchez-Margallo, F.M., Pagador, J.B.: Unravelling the effect of data augmentation transformations in polyp segmentation. **15**, 1975–1988 (2020)
6. Chaitanya, K., et al.: Semi-supervised task-driven data augmentation for medical image segmentation, vol. 68, p. 101934 (2021). https://www.sciencedirect.com/science/article/pii/S1361841520302048Y
7. Mikołajczyk, A., Grochowski, M.: Data augmentation for improving deep learning in image classification problem. In: 2018 International Interdisciplinary PhD Workshop (IIPhDW), pp. 117–122 (2018)
8. Kim, E.K., Lee, H., Kim, J.Y., Kim, S.: Data augmentation method by applying color perturbation of inverse psnr and geometric transformations for object recognition based on deep learning, vol. 10 (2020). https://www.mdpi.com/2076-3417/10/11/3755
9. Perez, L., Wang, J.: The effectiveness of data augmentation in image classification using deep learning (2017)
10. Wong, S.C., Gatt, A., Stamatescu, V., McDonnell, M.D.: Understanding data augmentation for classification: When to warp? In: 2016 International Conference on Digital Image Computing: Techniques and Applications (DICTA), pp. 1–6 (2016)
11. Kwasigroch, A., Mikołajczyk, A., Grochowski, M.: Deep neural networks approach to skin lesions classification - a comparative analysis. In: 2017 22nd International Conference on Methods and Models in Automation and Robotics (MMAR), pp. 1069–1074 (2017)
12. Zhang, H., Cisse, M., Dauphin, Y.N., Lopez-Paz, D.: mixup: beyond empirical risk minimization (2018)
13. Fernandes, P., Correia, J., Penousal, M.: Towards latent space exploration for classifier improvement. In: 24th European Conference on Artificial Intelligence (ECAI 2020) - ADGN20: First Workshop on Applied Deep Generative Networks (2020)
14. Hussain, Z., Gimenez, F., Yi, D., Rubin, D.: Differential data augmentation techniques for medical imaging classification tasks, vol. 2017, pp. 979–984, April 2018
15. Harris, J.E., Boushey, C., Bruemmer, B., Archer, S.L.: Publishing nutrition research: a review of nonparametric methods, part 3, vol. 108, pp. 1488–1496 (2008). https://www.sciencedirect.com/science/article/pii/S000282230801256X

Inheritance vs. Expansion: Generalization Degree of Nearest Neighbor Rule in Continuous Space as Covering Operator of XCS

Hiroki Shiraishi[1]([⊠])(iD), Yohei Hayamizu[2], Iko Nakari[1], Hiroyuki Sato[1], and Keiki Takadama[1]

[1] The University of Electro-Communications, Tokyo 182-8585, Japan
{hirowhite,iko0528}@cas.lab.uec.ac.jp, h.sato@uec.ac.jp,
keiki@inf.uec.ac.jp
[2] The State University of New York at Binghamton, New York 13902, USA
yhayami1@binghamton.edu

Abstract. This paper focuses on the covering mechanism which generates a new if-then rule when the input data does not match the rules in the *XCS Classifier System* (XCS), a rule-based machine learning system, and discusses how the new rule should be generated from the viewpoint of "inheritance" and "expansion" of the generalization degree of the nearest neighbor rule in the continuous space. For this purpose, this paper proposes the two covering mechanisms based on the "inheritance" and "expansion" of the generalization degree of the nearest neighbor rule and compares their results by applying them to *XCS for real-valued input spaces* (XCSR). Through the intensive experiments on three types of problems with the different characteristics, the following implications have been revealed: (1) the new rules should be generated by inheriting the generalization degree of the nearest neighbor rule in comparison with expanding it in the continuous space; and (2) XCSR with the "inheritance" based covering mechanism achieves higher classification accuracy with fewer rules than the conventional XCSR, which achieves higher classification accuracy than XCSR with the "expansion" based covering mechanism.

Keywords: Learning classifier systems · XCS classifier system · Covering mechanism · Generalization degree · Inheritance

1 Introduction

The *XCS Classifier System* (XCS) [22] is an evolutionary classifier system that obtains accurate and generalized rules by genetic algorithms (GA) [8] and reinforcement learning techniques [15], and is one of the most widely studied types of *Learning Classifier System* (LCS) [9]. Since XCS needs to learn a set of rules that cover the entire state-action space, the generalization degree (called *generality*) of

© Springer Nature Switzerland AG 2022
J. L. Jiménez Laredo et al. (Eds.): EvoApplications 2022, LNCS 13224, pp. 352–368, 2022.
https://doi.org/10.1007/978-3-031-02462-7_23

rules (*i.e.,* the coverage of the state-action space of rules) has a significant impact on the learning performance of XCS [7]. However, when generating new rules by the covering mechanism, the generalization degree is normally determined at random based on a uniform random number, which often deviates from the optimal generalization degree of the rules. As a result, such mechanism may generate the rules with the significantly low generalization (*i.e.,* the over-specific rule) that can only be applied to a specific input [17,19] and/or the rules with the significantly high generalization (*i.e.,* the over-general rule) that causes significant degradation of system performance [1,10,11,20]. To address this issue, Tadokoro *et al.* [16] proposed the *Local Covering* mechanism, which starts to find the rule that can match the input with the minimum modification of the attributes in the rule (hereafter, we call this rule as the *nearest neighbor rule*) and generates new rules by generalizing the mismatch attributes between the input and the nearest neighbor rule instead of generating new rules with the uniform random. In [16], XCS with the Local Covering mechanism showed achieves higher classification accuracy than the conventional XCS on binary input problems.

What should be noted here is that the appropriate generalization of the rules in the continuous space is more difficult than the binary (or discrete) space because of a lot of variation of the generalization degree in the rule (*e.g.,* $\alpha < x < \beta$, where x is the attribute and α and β are arbitrary values in the continuous space). This means that the Local Covering mechanism still generates the over-general rule in the continuous space because the change to the don't care symbol in the mismatch attributes between the input and the nearest neighbor rule maximumly expands the generalization degree in the rule (*e.g.,* $-\infty < x < \infty$). To overcome this problem, this paper proposes the concept of the "inheritance" of generalization degree in the rule in addition to the "expansion" of generalization degree in the rule. For clear understanding, the former generalization changes from $\alpha < x < \beta$ to $\gamma < x < \delta$ under $|\alpha - \beta| = |\gamma - \delta|$, while the latter generalization changes from $\alpha < x < \beta$ to $\gamma < x < \delta$ under $|\alpha - \beta| < |\gamma - \delta|$). To implement these concepts, this paper proposes the two types of covering mechanisms, *LCGE* (Local Covering based on Generality Expansion) mechanism and *LCGI* (Local Covering based on Generality Inheritance) mechanism, where the former mechanism expands the generalization degree of the input-nearest rule while the latter mechanism inherits the generalization degree of the input-nearest rule. This paper applies them to *XCS for real-valued input spaces* (XCSR) [23] to compare their results.

The rest of this paper is organized as follows. Section 2 describes an overview of XCS, and Sect. 3 explains Local Covering for the binary space. Section 4 proposes the two types of covering mechanisms for the continuous space, and Sect. 5 conducts the experiments with three different types of problems. The experimental results are reported in Sect. 6 and discussed in Sect. 7. Finally, Sect. 8 gives conclusions of this paper.

2   XCS Classifier System

The *XCS Classifier System* (XCS) [22] is a rule-based machine learning system based on the concept of accuracy, which aims to build a ruleset [P] (population) with maximum accuracy and generalization performance. XCS employs a Q-learning-inspired reinforcement learning algorithm (RL) and a genetic algorithm (GA) to update and evolve a set of if-then rules called *classifiers*, which represent the relationships between inputs and outputs, to obtain readable representations of knowledge.

2.1   Classifier

The classifier of XCS consists of the condition part C (IF part), the action part A (THEN part), the prediction p, the prediction error ϵ, the fitness F, the numerosity num, and some other parameters[1], as shown in Fig. 1. XCS evolves the condition part C and the action part A using GA's crossover/mutation operator. The prediction p is updated to get closer to the received actual reward ρ by the system, and the prediction error ϵ represents the absolute error between p and ρ. The fitness F indicates the accuracy computed from ϵ. The numerosity num indicates the subsumed number of classifiers that can be subsumed (described later).

The expression form of the condition part C differs depending on the assumption of the input to be handled. For example, in the case of binary inputs, the condition part C is represented as a fixed-length bit string with a *ternary alphabet representation* of $\{0, 1, \#\}$ as in Fig. 1. The "$\#$" is a wildcard symbol indicating that it can take any value (0 or 1), called *Don't Care*. The ratio of "$\#$" to the dimensionality of C indicates the generalization degree, *i.e.*, the *generality* of the classifier.

On the other hand, when dealing with real-valued inputs, the condition part C is represented by an interval-based representation that allows continuous values. The XCS based on this representation is often called the *XCS for real-valued input spaces* (XCSR) [23]. One of the typical interval representations, *Unordered Bound Representation* (UBR) [14], introduces two attributes of boundary variables p_i and q_i, and represents the matching interval of dimension i in the condition part of the classifier as $[\min(p_i, q_i), \max(p_i, q_i))$. That is, the condition part of the XCSR classifier is represented by an n-dimensional hyperrectangle, and its hypervolume corresponds to the generalization degree. Note that either p_i or q_i corresponds to the upper or lower bound of the matching interval of dimension i, and in general, each value of p_i and q_i is set in the range $[0, 1]$ [14]. In this paper, we employ XCSR, which adopts the UBR representation.

2.2   Learning Flow

XCS(R) consists of three parts: the Performance Component, the Reinforcement Component, and the Discovery Component. While the Performance Component

---

[1] The experience exp, the time stamp ts, and the action set size as.

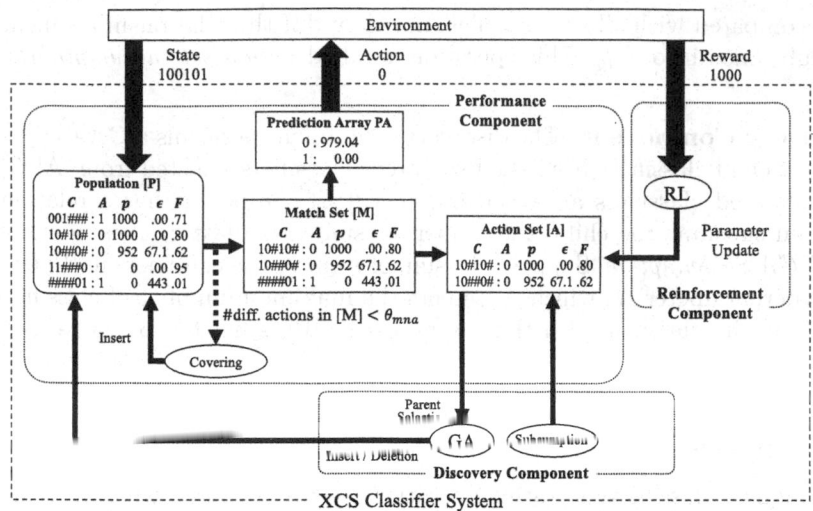

Fig. 1. The architecture of the XCS classifier system.

and the Reinforcement Component are repeated, the Discovery Component is executed when certain conditions are satisfied. Note that the explanation in this section is limited to the minimum due to the limitation of space. For further details, please refer to [6,22] and [23].

Performance Component. In the performance component, the system observes the state input of the environment, checks it against all classifiers in the ruleset (*population*: [P]), and forms a *match set* [M] of classifiers whose condition part matches the input. If there are no more than θ_{mna} different actions in [M], generate a new classifier with a condition part that matches the state and an action part that is not present in [M]. This operation is called *covering*, which means that there are at least θ_{mna} classifiers with different action parts in [M]. Then, the prediction p for each action is weighted by the fitness F of the classifier in [M] and stored in the *prediction array* PA, and the action (output) is determined by ε-greedy selection based on the prediction in PA.

Reinforcement Component. In the reinforcement component, XCS receives a positive reward r_{max}/negative reward r_{min} from the environment as an evaluation value for the execution of the action, if the selected action and the correct label for the actual input match/mismatch. In order to update the classifiers that contributed to the action selection, classifiers with the same action part as the selected action are extracted from [M] to form the *action set* [A]. Once [A] is formed, the prediction p, the prediction error ϵ, the fitness F, and the action set size as of the classifiers in [A] are updated by the *Widrow-Hoff delta rule* [21] with reference to the received rewards. Finally, the most general classifier cl_{mg} in

[A] is compared with all the classifiers in [A], and if the relationship is inclusive, it is subsumed into cl_{mg}. This operation is called *action set subsumption*.

Discovery Component. The discovery component performs a GA to generate two new child classifiers from the two parent classifiers selected from [A]. Then, the generated classifiers are inserted into [P]. If the parent has a relationship that can subsume the child, the former subsumes the latter. This operation is called *GA Subsumption*. Suppose the sum of *num* of the classifiers in [P] exceeds the hyperparameter N, which represents the maximum number of classifiers. In that case, the classifier with the low fitness in [P] is deleted by roulette-wheel selection.

2.3 Covering

Covering is a classifier-generating operation that is triggered when there are no more than θ_{mna} different actions in [M]. The condition part C of the classifier generated by covering, called *covering classifier* is set to the same value as the state input. After that, the attribute of each dimension is replaced by # with probability $P_{\#}$ when dealing with a ternary alphabet representation, and the lower/upper bounds of the matching interval are set by subtracting/adding $U_{[0,r_0)}$ to the state input of each dimension when dealing with an interval representation. Here, $P_{\#}$ and r_0 are hyperparameters that significantly affect the performance of XCS(R), determining the generalization degree of the covering classifier, and should be changed according to the applied problem [2,19]. Also, $U_{[0,r_0)}$ means a real-valued uniform random number between 0 and r_0 (r_0 is excluded). The action part A is set randomly from the values not yet found in [M]. The prediction p, the prediction error ϵ, and the fitness F are set to hyperparameters p_I, ϵ_I, and F_I, respectively.

3   Local Covering

3.1   Overview

Local Covering [16] is a covering method for binary inputs without using the hyperparameter $P_{\#}$. This method refers to the condition part of the nearest neighbor classifier to the input and generates a condition part that generalizes (replaces #) the mismatch with the input. The condition part of the covering classifier generated by Local Covering is set to a higher generalization degree than that of the classifiers in at least [P], so that the generation of over-specific classifiers by covering does not occur. Local Covering also applies minimal generalization (generating #) based on the Hamming distance to existing rules, preventing the generation of over-general classifiers.

3.2 XCS-LCPCI

XCS-LCPCI (*XCS with Local Covering for Previously Covered Inputs*) [16] is the XCS that adopts Local Covering only when generating the second or subsequent covering classifier for a given input. Since Local Covering refers to existing classifiers and selects the one with the closest condition part, there must be a sufficient number of classifiers in the ruleset [P] before Local Covering is performed. Thus, conventional covering based on probability $P_\#$ is activated for inputs that have never been observed in the past, and for other inputs, Local Covering is activated. In other words, XCS-LCPCI performs a global search in the covered region of the classifier by conventional covering with $P_\#$ in the early stage of training, and performs local search by Local Covering without $P_\#$ in the latter stage of training to achieve efficient search space coverage.

4 Proposed Method: Local Covering for Real Values

In this paper, we propose an extension of Local Covering for real-valued inputs (hereafter referred to as the proposed covering). The proposed covering refers to the condition part of the classifier in [P] that is closest to the input, and sets it as the condition part of the covering classifier by expanding (increasing the generalization degree) or moving (inheriting the generalization degree) it to match the input. The algorithm of the proposed covering is shown in Algorithm 1.

4.1 Procedure of Covering Classifier Generation

The generation of covering classifiers in the proposed covering is performed in the following flow (Algorithm 1).

1. For all classifiers in [P], compute the *Nearest Neighbor Distance* (NND, described later in Sect. 4.2) between the input state σ and the condition part C (line 5).
2. Enumerate the classifiers with the smallest NND among [P] and store their condition parts in the array *selectedConds* (lines 6–12).
3. Set the condition part taken randomly from *selectedConds* to the condition part C of the covering classifier (line 18). However, if *selectedConds* is empty, the covering classifier is generated by the Standard Covering based on the hyperparameter r_0 (lines 14–16).
4. Modify the condition part C of the covering classifier to match the input state σ based on the condition generation operation (lines 19–25, described later in Sects. 4.3 and 4.4).

4.2 Nearest Neighbor Distance (NND)

We introduce the *Nearest Neighbor Distance* (NND) as a measure of the distance between a state to a condition part that does not match it. NND denotes the

Algorithm 1. LOCALCOVERINGFORREALVALUES([P], [M], σ)

1: *selectedConds* ← empty array
2: D ← +inf   ▷ Maximum NND
3: a ← random action not present in [M]
4: **for each** classifier *cl* in [P] **do**
5:     d ← NEARESTNEIGHBORDISTANCE($cl.C, \sigma$)
6:     **if** $cl.A = a$ and $cl.exp \geq 1$ and $d \leq D$ **then**
7:         **if** $d < D$ **then**
8:             $D \leftarrow d$
9:             *selectedConds* ← empty array
10:         **end if**
11:         *selectedConds*.append($cl.C$)
12:     **end if**
13: **end for**
14: **if** *selectedConds* is empty **then**
15:     **return** STANDARDCOVERING([M], σ)
16: **end if**
17: cl_c ← new classifier
18: $cl_c.A \leftarrow a$; $cl_c.p \leftarrow p_I$; $cl_c.\epsilon \leftarrow \epsilon_I$; $cl_c.F \leftarrow F_I$;
    $cl_c.exp \leftarrow 0$; $cl_c.as \leftarrow 1$; $cl_c.num \leftarrow 1$;
    $cl_c.ts$ ← current iteration count;
    $cl_c.C$ ← random in *selectedConds*
19: **if** *doGeneralityExpansion* **then**
20:     DOGENERALITYEXPANSION($cl_c.C, \sigma$)
21: **end if**
22: **if** *doGeneralityInheritance* **then**
23:     DOGENERALITYINHERITANCE($cl_c.C, \sigma$)
24: **end if**
25: **return** cl_c

---

minimum Euclidean distance between the hyperrectangular condition part C and the input σ in state space. The algorithm to calculate the NND is shown in Algorithm 2.

4.3 Condition Generation Operation Based on Generalization Degree Expansion of Nearest Neighbor Rules

The condition part of the covering classifier by the condition part generation operation based on the generalization degree expansion of the nearest neighbor rule is determined in the following flow (Algorithm 3).

1. In each dimension i, when the condition part $C[i]$ of the covering classifier that does not include the input $\sigma[i]$ satisfies $C[i].u \leq \sigma[i]$, then $C[i].u = \min(1, \sigma[i] + d_{gen})$. Here, since each dimension of the classifier condition part in XCSR is represented by a half-open interval of $[l, u)$, the hyperparameter d_{gen} is added to the upper bound of the interval $C[i].u$ in order to generate a condition part with the minimum generalization degree that includes $\sigma[i]$. (lines 2–4).

Algorithm 2. NEARESTNEIGHBORDISTANCE(C, σ)

1: $d^2 \leftarrow 0$
2: **for** $i = 0...(\#\sigma - 1)$ **do**
3:     **if not** $C[i].l \leq \sigma[i] < C[i].u$ **then**
4:         $d^2 \leftarrow d^2 + \min((\sigma[i] - C[i].l)^2, (\sigma[i] - C[i].u)^2)$
5:     **end if**
6: **end for**
7: **return** $|d|$

Algorithm 3. DOGENERALITYEXPANSION(C, σ)

1: **for** $i = 0...(\#\sigma - 1)$ **do**
2:     **if** $C[i].u \leq \sigma[i]$ **then**
3:         $C[i].u \leftarrow \min(1, \sigma[i] + d_{gen})$
4:     **end if**
5:     **if** $C[i].l > \sigma[i]$ **then**
6:         $C[i].l \leftarrow \max(0, \sigma[i] - d_{gen})$
7:     **end if**
8: **end for**

---

2. In each dimension i, if the condition part $C[i]$ of the covering classifier that does not include the input $\sigma[i]$ satisfies $C[i].l > \sigma[i]$, then $C[i].l = \max(0, \sigma[i] - d_{gen})$. As in step 1, the hyperparameter d_{gen} is subtracted from the lower bound of the interval $C[i].l$ to generate a condition part with a minimal generalization degree that includes $\sigma[i]$ (lines 5–7).

Hereafter, the proposed covering that introduces this operation is called LCGE-Covering (*Local Covering for real-valued inputs based on Generality Expansion*). With LCGE-Covering, the condition part of the covering classifier is determined by the green color of the generalization degree of the nearest neighbor rule (Classifier 3) for an input such as Fig. 2a, which is minimally expanded until the input is covered. This operation is the very concept of Local Covering for binary inputs. The XCSR that introduces LCGE-Covering is denoted as XCSR-LCGE (*i.e., doGeneralityExpansion = yes* in Algorithm 1). Note that LCGE-Covering in XCSR-LCGE, like XCS-LCPCI, is triggered only when the second or later covering classifier matching a certain input is generated.

4.4 Condition Generation Operation Based on Generalization Degree Inheritance of Nearest Neighbor Rules

The condition part of the covering classifier by the condition part generation operation based on the generalization degree inheritance of the nearest neighbor rule is determined in the following flow (Algorithm 4).

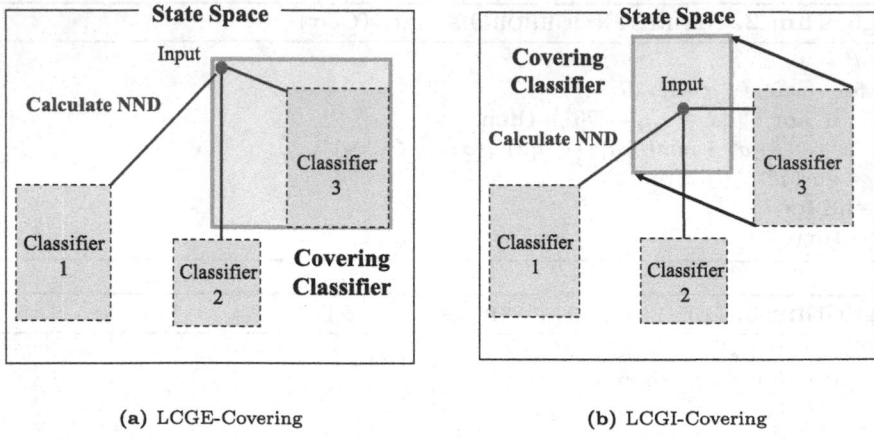

(a) LCGE-Covering   (b) LCGI-Covering

Fig. 2. Covering classifier generation of proposed covering. (Color figure online)

1. Calculate the center point $c = C[i].l + (C[i].u - C[i].l)/2$ of the condition part $C[i]$ of the covering classifier in each dimension i (line 2).
2. In each dimension i, when the condition part $C[i]$ of the covering classifier that does not include the input $\sigma[i]$ satisfies $C[i].u \leq \sigma[i]$, find the spread in the covering $r = \sigma[i] - c$. Then, add r to each of $C[i].l, C[i].u$ (lines 3–7).
3. In each dimension i, when the condition part $C[i]$ of the covering classifier that does not include the input $\sigma[i]$ satisfies $C[i].l > \sigma[i]$, find the spread in the covering $r = c - \sigma[i]$. Then, subtract r from each of $C[i].l, C[i].u$ (lines 8–12).

Hereafter, the proposed covering that introduces this operation is called LCGI-Covering (*Local Covering for real-valued inputs based on Generality Inheritance*). With LCGI-Covering, the green condition part of the nearest neighbor rule (Classifier 3) for an input such as Fig. 2b is determined as the condition part of the covering classifier by moving it (inheriting the generalization degree) until the input is covered. The XCSR that introduces LCGI-Covering is denoted as XCSR-LCGI (*i.e., doGeneralityInheritance = yes* in Algorithm 1). Note that LCGI-Covering in XCSR-LCGI, like XCS-LCPCI and XCSR-LCGE, is triggered only when the second or later covering classifier matching a certain input is generated[2].

5   Experiment

In order to verify the performance of the proposed methods, XCSR-LCGE and XCSR-LCGI, experiments are conducted on three types of problems: the *Real-Valued Multiplexer* (RMUX) problem [23], the *Class-Imbalanced Real-Valued*

---

[2] To implement XCSR-LCGE and XCSR-LCGI, replace the function LocalCovering of XCS-LCPCI in [16] with Algorithm 1.

Algorithm 4. DoGeneralityInheritance(C, σ)

1: **for** $i = 0...(\#\sigma - 1)$ **do**
2:     $c \leftarrow C[i].l + (C[i].u - C[i].l)/2$
3:     **if** $C[i].u \le \sigma[i]$ **then**
4:         $r \leftarrow \sigma[i] - c$
5:         $C[i].l \leftarrow \max(0, C[i].l + r)$
6:         $C[i].u \leftarrow \min(1, C[i].u + r)$
7:     **end if**
8:     **if** $C[i].l > \sigma[i]$ **then**
9:         $r \leftarrow c - \sigma[i]$
10:         $C[i].l \leftarrow \max(0, C[i].l - r)$
11:         $C[i].u \leftarrow \min(1, C[i].u - r)$
12:     **end if**
13: **end for**

Multiplexer (IRMUX) problem [13], and the *Paddy-Leaf* classification problem [20]. The RMUX problem is one in which the distribution of inputs is uniform, the IRMUX problem is one in which there is a strong bias in the dimension-wise distribution of inputs to determine the correct class, and the Paddy Leaf problem is a multi-class classification problem with real-world data.

5.1 Experiment 1: Real-Valued Multiplexer (RMUX) Problem

The n-RMUX problem is a kind of benchmark problem used to evaluate the performance of real-valued LCSs. It is a randomly generated n-dimensional real-valued input $\sigma \in [0, 1)^n$ for which one of $\{0, 1\}$ is the correct answer. Here, n is defined as $n = k + 2^k$ with any integer k. The correct answer class is determined by taking the following steps: (1) Convert the input $\sigma = (\sigma_0, \sigma_1, ..., \sigma_{n-1})$ to XCSR into a bit sequence $b = b_0 b_1 ... b_{n-1}$ based on the threshold value of 0.5. Specifically, if the input element σ_i is less than 0.5, then $b_i = 0$, and if it is greater than 0.5, then $b_i = 1$. (2) Let d be the value obtained by converting the address bits $(b_0...b_{k-1})$ of the first half k dimensions into decimal numbers. (3) The bit value of b_{k+d} is the correct answer class.

In this paper, we deal with the 11-RMUX problem with $k = 3$ [3]. The number of alternating explore (train)/exploit (test) steps is set to 100,000.

5.2 Experiment 2: Class-Imbalanced RMUX (IRMUX) Problem

The n-IRMUX problem is a n-RMUX problem with an imbalance in the ratio of the correct answer class $\{0,1\}$. Here, the system first generates the input σ as in the n-RMUX problem, and then computes its correct answer class. If σ belongs to the majority class, σ will always be accepted as input to the XCSR. On the other hand, if σ belongs to the minority class, σ will be sent to XCSR

---

[3] For example, in the case of the bit sequence $b = 11101010101$, $d = (b_0 b_1 b_2)_2 = (111)_2 = 7$, so the correct answer class is determined to be $b_{k+d} = b_{3+7} = b_{10} = 1$.

with a probability of $1/2^i$. If it is not sent out, generate σ again and perform the same operation. Here, i is the *imbalance level*, a parameter that indicates the imbalance ratio of the correct answer class.

In this paper, we deal with the 6-IRMUX problem with $i = 5$. The number of alternating explore/exploit steps is set to 200,000. The expected number of inputs for each class during training in this setup is approximately 193,960 for the majority class and 6040 for the minority class. However, in the exploit steps, the same inputs are given as in the standard RMUX problem without imbalance for each class.

5.3   Experiment 3: Paddy Leaf Classification Problem

The Paddy Leaf dataset[4] consists of three real-valued attributes representing the average values of the red, green, and blue color channels in 6000 photographs of paddy leaf and four different class labels based on nitrogen fertilizer recommendations. The paddy leaf classification is the task of training and classifying the dataset [20]. The number of photos for each class in this dataset is 1500, and the classes are well balanced. Each attribute value is normalized to the range $[0, 1]$ for input to XCSR. The number of alternating explore/exploit steps is set to 100,000.

5.4   Experimental Setup

For the XCSR hyperparameters in the 11-RMUX[5] and 6-IRMUX[6] problems, we employ the standard settings [13,23], but for the three hyperparameters θ_{sub}, β, and ϵ_0, we adopt approximate theoretical values based on *XCS Learning Theory* [12]. In order to satisfy the assumptions of XCS Learning Theory, MAM (*Moyenne Adaptive Modifée*) [18] in the update equation of the prediction p and the prediction error ϵ is turned off. For the Paddy Leaf classification problem[7], we employ the same hyperparameter values as in the previous study [20]. Commonly in all problems, uniform crossover is assigned as the crossover method in GA, and tournament selection ($\tau = 0.4$ [5]) is used for parent selection. The reward value for each XCSR is $r_{max} = 1000$ for a correct answer and $r_{min} = 0$ for an incorrect answer. The hyperparameter d_{gen} employed in XCSR-LCGE is set to 0.01. We run 30 trials of independent experiments with different random seeds and record the mean and standard deviation of the evaluation values (described below) overall exploit steps. In all experiments, we evaluate the performance

---

[4] https://www.kaggle.com/torikul140129/paddy-leaf-images-aman(02.08.2022).

[5] 11-RMUX: $N = 20,000$, $\alpha = 0.1$, $\beta = 0.229242$, $\delta = 0.1$, $\nu = 5$, $\theta_{mna} = 2$, $\theta_{GA} = 12$, $\theta_{del} = 20$, $\theta_{sub} = 15$, $\epsilon_0 = 109.918427$, $\chi = 0.8$, $\mu = 0.04$, $p_I = 0.01$, $\epsilon_I = 0.01$, $F_I = 0.01$, $FitnessReduction = 0.1$, $m_0 = 0.1$, $r_0 = 1.0$, $doASSubsumption = yes$, $doGASubsumption = yes$.

[6] 6-IRMUX: Analogous to 11-RMUX, except: $N = 2000$, $\beta = 0.030691$, $\theta_{GA} = 192$, $\theta_{sub} = 217$, $\epsilon_0 = 9.085638$, $doASSubsumption = no$.

[7] Paddy Leaf: Analogous to 11-RMUX, except: $N = 6400$, $\beta = 0.2$, $\theta_{mna} = 4$, $\theta_{GA} = 48$, $\theta_{del} = 50$, $\theta_{sub} = 50$, $\epsilon_0 = 1.0$, $m_0 = 0.5$.

Table 1. Overall results for classification tasks, *i.e.,* Reward, System Error (SysErr) and Population Size ($\|[P]\|$). All entries show the mean ± standard deviation of 30 trials, and the letters in parentheses indicate groups with statistically significant differences, *i.e.,* groups with p-values $< \alpha = 0.01$ (the best value is denoted as "A" and the worst value as "C"). The arrows indicate the increase or decrease in the index relative to the conventional method, XCSR. The green shadings and the bold entries indicate significant improvements relative to XCSR.

11-RMUX	Reward	SysErr	$\|[P]\|$
XCSR-LCGI	**866.22** ↑ ±15.93 (A)	**208.29** ↓ ±21.02 (A)	**10051.83** ↓ ±313.66 (B)
XCSR-LCGE	653.34 ↓ ±51.43 (C)	409.62 ↑ ±44.96 (C)	**8447.95** ↓ ±732.22 (A)
XCSR	832.61 ± 23.93 (B)	251.75 ± 29.84 (B)	10648.63 ± 272.36 (C)
6-IRMUX	Reward	SysErr	$\|[P]\|$
XCSR-LCGI	**806.11** ↑ ±11.07 (A)	**202.28** ↓ ±20.77 (A)	**902.81** ↓ ±27.20 (B)
XCSR-LCGE	641.56 ↓ ±63.79 (C)	357.67 ↑ ±62.11 (C)	**929.31** ↓ ±42.79 (A)
XCSR	761.72 ± 30.47 (B)	240.07 ± 28.55 (B)	1003.79 ± 27.62 (C)
Paddy Leaf	Reward	SysErr	$\|[P]\|$
XCSR-LCGI	**821.02** ↑ ±23.37 (A)	**264.19** ↓ ±20.48 (A)	2423.74 ↑ ±171.61 (B)
XCSR-LCGE	753.09 ↓ ±31.72 (C)	308.14 ↑ ±19.96 (C)	**2092.19** ↓ ±167.80 (A)
XCSR	797.60 ± 11.47 (B)	282.81 ± 8.94 (B)	2383.27 ± 107.77 (B)

of the conventional method, XCSR, and the proposed methods, XCSR-LCGE and XCSR-LCGI, using the following three metrics. (1) Received reward value (Reward), *i.e.,* the classification accuracy scaled to the range [0,1000]. (2) System Error (SysErr), *i.e.,* the absolute error between the received actual reward value and the reward value predicted by XCSR. (3) The number of classifiers, *i.e.,* the size of the ruleset [P] ($\|[P]\|$). These evaluation values are analyzed by *Levene's test*, and if *homoscedasticity* is positive, *One-Way ANOVA* and *Tukey-HSD* post-hoc test are conducted in pairs. If *homoscedasticity* is negative, *Welch-ANOVA* and *Games-Howell* post-hoc test are conducted in combination.

6   Result

Table 1 shows the evaluation values for all problems. Figure 3 shows the moving average of the evaluation values (received reward value, system error, and the number of classifiers) for XCSR, XCSR-LCGE, and XCSR-LCGI. The horizontal axis shows the number of exploit steps and the vertical axis shows the average evaluation value for over 30 trials. The error bars in each graph represent the 95% confidence interval.

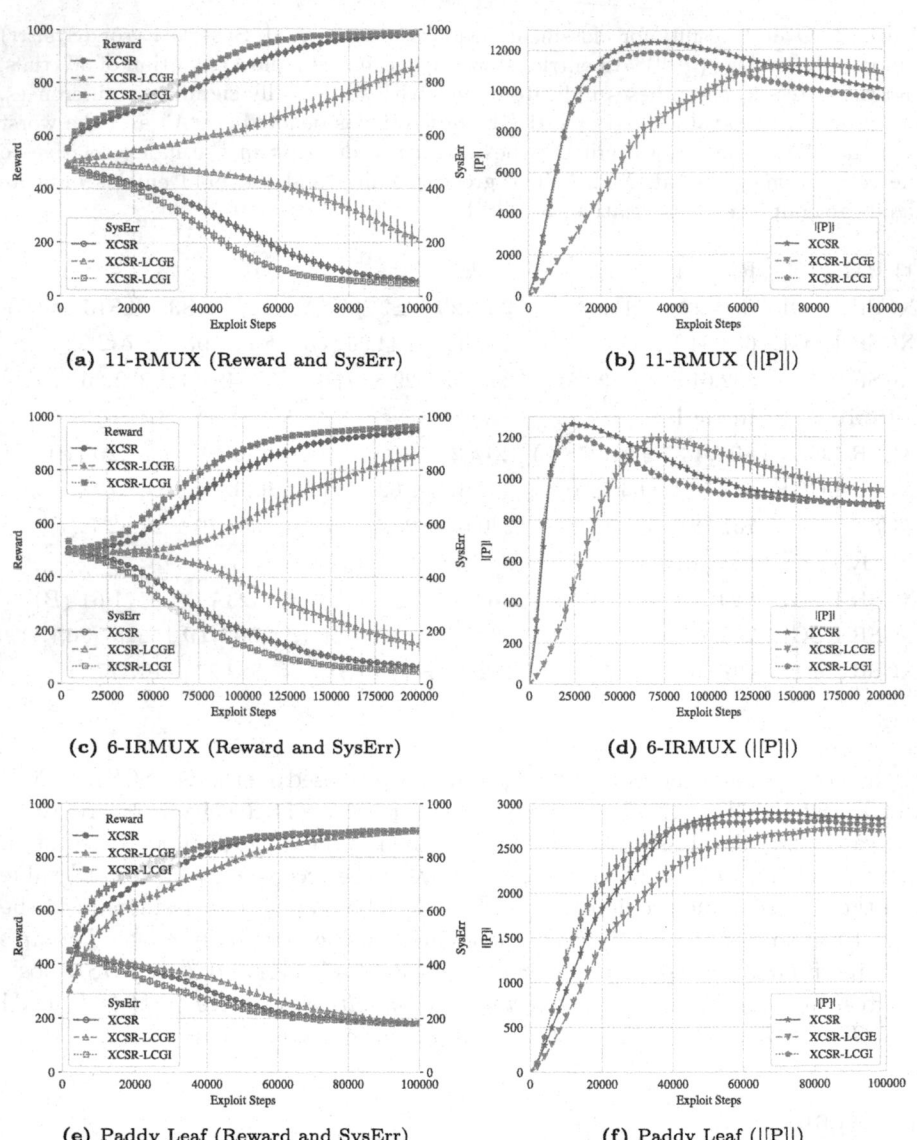

Fig. 3. Learning curve plots of the conducted experiments.

7   Discussion

As Table 1 shows, the performance of the proposed method, XCSR-LCGI, is statistically over than the conventional method, XCSR, in all evaluated values for all problems. Moreover, as Fig. 3 shows, XCSR-LCGI outperforms XCSR in all exploit steps except for the Paddy Leaf problem, $|[P]|$. On the other hand, the other proposed method, XCSR-LCGE, significantly underperformed the other two methods on all problems.

Figures 3b, 3d and 3f show that the size of the ruleset [P] generated by XCSR-LCGE has a slower increase than the other two methods for all problems. This is considered to be because by setting the generalized condition part of the nearest neighbor rule as the condition part of the covering classifier, the classifier with a higher generalization degree than necessary covers most of the state space, and the number of times the covering classifier is generated for unknown inputs is suppressed. To test this hypothesis, we investigated the average generality of the ruleset for the 11-RMUX and the 6-IRMUX problems as shown in Fig. 4. The generality of the ruleset of XCSR-LCGE in the early stages of training is significantly higher than that of the other two methods, and it is consistently highly general throughout training. In addition, since there is no explicit mechanism to handle over-general classifiers in the current XCSR [20], we conclude that the inaccurate over-general classifiers generated by the excessive generalization pressure of LCGE-Covering contribute to the action decisions of XCSR-LCGE more than necessary and cause significant degradation in the system performance of it.

On the other hand, Table 1 shows that XCSR-LCGI is the best in all evaluation indices for the 11-RMUX and 6-IRMUX problems. Unlike the Paddy Leaf problem, in these binary classification problems, the appropriate generalization degree of the required classifier is uniformly determined regardless of the class[8], so the LCGI-Covering property, which inherits the generalization degree of the nearest neighbor rule without any change, contributes to the improvement of the system performance. Figure 4 also shows that the generality of XCSR-LCGI is higher than that of conventional XCSR and lower than that of XCSR-LCGE, whose ruleset is flooded with over-general classifiers. This shows that XCSR-LCGI is capable of efficient and well-balanced state space exploration by using both conventional covering (global search) with hyperparameter r_0 and LCGI-Covering (local search) without r_0. However, in terms of the number of rules for the Paddy-Leaf problem (rule generalization performance), the performance of XCSR-LCGI is comparable to that of conventional XCSR, which is due to the characteristic of the multi-class classification problem, where the optimal generality differs depending on the class. Because LCGI-Covering does not consider the prediction p of the rule when determining the nearest neighbor rule for the input, it is suggested to refer to a rule whose prediction is close to $r_{min} = 0$ in some cases. We expect to solve this problem by introducing the prediction threshold parameter to refer to rules with a prediction of $r_{max} = 1000$ or close to it.

---

[8] For example, in the case of the 11-RMUX problem, the generality of each rule in the optimal ruleset [O] [4] is uniformly $0.5^{k+1} = 0.5^4 = 0.0625$, regardless of the class. Similarly, in the case of 6-IRMUX, it is uniformly $0.5^{k+1} = 0.5^3 = 0.125$.

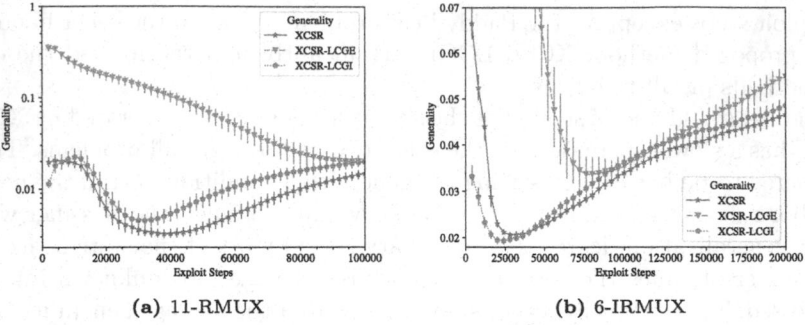

(a) 11-RMUX   (b) 6-IRMUX

Fig. 4. Learning curve plots of Generality in [P].

8   Concluding Remarks

This paper aims to achieve a suitable generalization of the rules for XCSR in continuous space by improving the covering mechanism. To achieve this purpose, we proposed two covering operators. One is LCGE-Covering, which generates new rules by "expanding" the generalization degree of nearest neighbor rules, and the other is LCGI-Covering, which generates new rules by "inheriting" the generalization degree of nearest neighbor rules. The experimental results show that only LCGI-Covering is effective in three problems of different types, and is more significant than conventional XCSR in terms of evaluation metrics of received reward value and system error in all problems, and records fewer classifiers in 11-RMUX and 6-IRMUX problems.

Future work includes applying LCGI-Covering to UCS (*sUpervised Classifier System*) [3]. UCS forms the *Best Action Map* that learns only the optimal action, instead of XCSR, which forms the *Complete Action Map* that learns the entire action space in each state. It is expected that UCS with LCGI-Covering performs better, especially in multi-class classification problems such as the Paddy Leaf problem, because the covering mechanism is not triggered based on references to classifiers whose reward predictions are zero (*i.e.*, the condition-action pairs are entirely incorrect). In addition, we plan to evaluate the performance of XCSR-LCGI on more classification problems, especially on more real-world classification datasets.

References

1. Barry, A.M.: The stability of long action chains in XCS. Soft. Comput. **6**(3–4), 183–199 (2002)
2. Behdad, M., French, T., Barone, L., Bennamoun, M.: On principal component analysis for high-dimensional XCSR. Evol. Intel. **5**(2), 129–138 (2012)
3. Bernadó-Mansilla, E., Garrell-Guiu, J.M.: Accuracy-based learning classifier systems: models, analysis and applications to classification tasks. Evol. Comput. **11**(3), 209–238 (2003)

4. Butz, M.V., Kovacs, T., Lanzi, P.L., Wilson, S.W.: How XCS evolves accurate classifiers. In: Proceedings of the Third Genetic and Evolutionary Computation Conference (GECCO-2001), pp. 927–934. Citeseer (2001)
5. Butz, M.V., Sastry, K., Goldberg, D.E.: Tournament selection: stable fitness pressure in XCS. In: Cantú-Paz, E., et al. (eds.) GECCO 2003. LNCS, vol. 2724, pp. 1857–1869. Springer, Heidelberg (2003). https://doi.org/10.1007/3-540-45110-2_83
6. Butz, M.V., Wilson, S.W.: An algorithmic description of XCS. Soft. Comput. **6**(3–4), 144–153 (2002)
7. Fredivianus, N., Prothmann, H., Schmeck, H.: XCS revisited: a novel discovery component for the eXtended classifier system. In: Deb, K., et al. (eds.) SEAL 2010. LNCS, vol. 6457, pp. 289–298. Springer, Heidelberg (2010). https://doi.org/10.1007/978-3-642-17298-4_30
8. Goldberg, D.E.: Genetic Algorithms in Search, 1st edn. Optimization and Machine Learning. Addison-Wesley Longman Publishing Co., Inc., USA (1989)
9. Holland, J.H.: Escaping brittleness: The possibilities of general-purpose learning algorithms applied to parallel rule-based systems. Machine learning, an artificial intelligence approach **2**, 593–623 (1986)
10. Kovacs, T.: Towards a theory of strong overgeneral classifiers. In: Foundations of Genetic Algorithms 6, pp. 165–184. Elsevier (2001)
11. Lanzi, P.L.: An analysis of generalization in the XCS classifier system. Evol. Comput. **7**(2), 125–149 (1999)
12. Nakata, M., Browne, W.N.: Learning optimality theory for accuracy-based learning classifier systems. IEEE Trans. Evol. Comput. **25**(1), 61–74 (2020)
13. Orriols-Puig, A., Bernadó-Mansilla, E.: Bounding XCS's parameters for unbalanced datasets. In: Proceedings of the 8th Annual Conference on Genetic and Evolutionary Computation, pp. 1561–1568 (2006)
14. Stone, C., Bull, L.: For real! XCS with continuous-valued inputs. Evol. Comput. **11**(3), 299–336 (2003)
15. Sutton, R.S.: Learning to predict by the methods of temporal differences. Mach. Learn. **3**(1), 9–44 (1988)
16. Tadokoro, M., Hasegawa, S., Tatsumi, T., Sato, H., Takadama, K.: Local covering: adaptive rule generation method using existing rules for XCS. In: 2020 IEEE Congress on Evolutionary Computation (CEC), pp. 1–8. IEEE (2020)
17. Tadokoro, M., Sato, H., Takadama, K.: XCS with weight-based matching in VAE latent space and additional learning of high-dimensional data. In: 2021 IEEE Congress on Evolutionary Computation (CEC), pp. 304–310. IEEE (2021)
18. Venturini, G.: Adaptation in dynamic environments through a minimal probability of exploration. In: Proceedings of the Third International Conference on Simulation of Adaptive Behavior: from Animals to Animats 3, pp. 371–379 (1994)
19. Wada, A., Takadama, K., Shimohara, K., Katai, O.: Analyzing parameter sensitivity and classifier representations for real-valued XCS. In: Kovacs, T., Llorà, X., Takadama, K., Lanzi, P.L., Stolzmann, W., Wilson, S.W. (eds.) IWLCS 2003-2005. LNCS (LNAI), vol. 4399, pp. 1–16. Springer, Heidelberg (2007). https://doi.org/10.1007/978-3-540-71231-2_1
20. Wagner, A.R.M., Stein, A.: On the effects of absumption for XCS with continuous-valued inputs. In: Castillo, P.A., Jiménez Laredo, J.L. (eds.) EvoApplications 2021. LNCS, vol. 12694, pp. 697–713. Springer, Cham (2021). https://doi.org/10.1007/978-3-030-72699-7_44
21. Widrow, B., Hoff, M.E.: Adaptive switching circuits. Stanford Univ Ca Stanford Electronics Labs, Technical report (1960)

22. Wilson, S.W.: Classifier fitness based on accuracy. Evol. Comput. **3**(2), 149–175 (1995)
23. Wilson, S.W.: Get real! XCS with continuous-valued inputs. In: Lanzi, P.L., Stolzmann, W., Wilson, S.W. (eds.) IWLCS 1999. LNCS (LNAI), vol. 1813, pp. 209–219. Springer, Heidelberg (2000). https://doi.org/10.1007/3-540-45027-0_11

Detecting Nested Structures Through Evolutionary Multi-objective Clustering

Cristina Y. Morimoto[1]([✉])(iD), Aurora Pozo[1](iD), and Marcílio C. P. de Souto[2](iD)

[1] Federal University of Paraná, Curitiba, PR, Brazil
`cristina.morimoto@ufpr.br`, `aurora@inf.ufpr.br`
[2] LIFO/University of Orléans, Orléans, France
`marcilio.desouto@univ-orleans.fr`

Abstract. The evolutionary multi-objective algorithms have been widely applied for clustering. However, in general, the detection of heterogeneous nested clusters remains challenging for clustering algorithms. This paper proposes an adaptation of the connectedness criterion used as an objective function in established Evolutionary Multi-Objective Clustering approaches (EMOCs). This adaptation can improve the conflict between the objective functions, and then it promotes the detection of nested clusters. We performed experiments with four EMOCs (MOCK, MOCLE, Δ-MOCK, and EMO-KC) that provide different features. These different EMOCs have different initialization methods and representation schemes, allowing us to analyze how the proposed objective function can contribute to detecting nested clusters. Our results show that our adapted objective function promotes a general gain in the performance of all these algorithms.

Keywords: Multi-objective clustering · Nested data clustering · Evolutionary multi-objective optimization · Clustering methods · Data mining

1 Introduction

Complex data allow multiple data interpretations in which multiple clustering approaches can describe alternative aspects that characterize the data in different views [14]. The Evolutionary Multi-Objective Clustering approaches (EMOCs) have been widely applied to extract patterns and provide these multiple views, allowing to analyze alternative aspects that characterize the data [6,8,9,13]. However, the use of EMOCs to detect nested structures is still under-explored in the literature, especially to detect heterogeneous data structures.

Some EMOCs were applied to detect heterogeneous data structures with the generation of solutions with multiple partitions [6,8,9]. They used multiple criteria (e.g., compactness and connectedness) as objective functions to deal

This work was partially supported by the National Council for Scientific and Technological Development (CNPq).

J. L. Jiménez Laredo et al. (Eds.): EvoApplications 2022, LNCS 13224, pp. 369–385, 2022.
https://doi.org/10.1007/978-3-031-02462-7_24

with datasets with different types of clusters. However, no studies have widely evaluated them to detect nested data structures and analyze how their objective functions impact this task.

In this study, we propose a modification of the connectedness criterion adopted for established EMOCs to improve the detection of a different number of clusters, especially in nested clusters. The connectivity index used by these approaches has limitations to detect some multi-level solutions, such as nested clusters, in a single run. This modified objective function was evaluated in four EMOCs: MOCK [9], MOCLE [6], Δ-MOCK [8], and EMO-KC [16], in which we analyze how the different strategies adopted in these algorithms can contribute (or hamper) to detect nested clusters. We performed experiments on fifteen datasets, which yielded promising results using the modified connectivity in all these algorithms.

The remainder of this paper is organized as follows. In Sect. 2, we present the main concepts concerning MOCK, Δ-MOCK, MOCLE, and EMO-KC, considering their representation, initialization strategy, optimization strategy, objective functions, crossover, and mutation operators. In Sect. 3, we describe some general issues around the connectedness criterion used as an objective function in MOCK, Δ-MOCK, MOCLE and introduce the proposed modification in this index. Section 4 presents the datasets used in the experiments, the specific configuration and settings of the compared methods, and the performance assessment adopted. Then, in Sect. 5, we present and discuss the results of our experimental evaluation of the use of this modified connectivity index. Finally, Sect. 6 highlights our main findings and discusses future works.

2   Background

A nested cluster refers to a cluster that is composed of sub-clusters or multi-level data structures. Formally, given a partition $\pi = \{c_1, \ldots, c_k\}$, for any $c_a, c_b \in \pi$ either they are non-overlapping ($c_a \cap c_b = \emptyset$) or one of them includes the other ($c_a \subseteq c_b$ or $c_b \subseteq c_a$), which is equivalent to assert that $c_a \cap c_b \in \{\emptyset, c_a, c_b\}$ [1]. For example, Fig. 1 depicts the Venn diagram of the nested data structures presented in the set $X = \{x_1, x_2, x_3, x_4, x_5\}$, where $\pi_1 = \{\{x_1, x_2\}, \{x_3, x_4, x_5\}\}$ and $\pi_2 = \{\{x_1, x_2\}, \{x_3\}, \{x_4, x_5\}\}$ represent the solutions at different levels in the hierarchy, in which π_1 has two clusters and π_2 has three clusters.

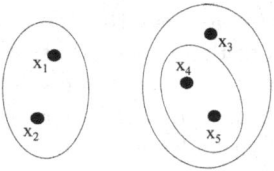

Fig. 1. Venn diagram of the nested data structures

Hierarchical clustering is the most traditional nested clustering strategy applied to produce a sequence of clusterings in which each cluster is nested into the next cluster in the sequence. This kind of approach presents a hierarchical grouping of the objects, which can be viewed as finding multiple partitions. However, the different clustering solutions obtained at different hierarchical levels differ only in their granularity [11,17].

Two well-known hierarchical clustering algorithms used in our analysis are the Single-Linkage (SL) and Group Average-Linkage (AL). In both the SL and AL algorithms, each object starts out standing as an individual cluster, and a series of merge operations is followed until it reaches the top with a single cluster. The main difference between these algorithms is the distance measure used to compute proximity between the pairs of clusters used to define the closest pair of sub-sets that are merged. SL considers the minimal distance between two objects of a cluster pair to define the closest sub-sets, and AL considers the average distance of all observations of pairs of clusters [17].

In contrast to hierarchical clustering, the EMOCs provide a diverse set of solutions considering different aspects of the data structures. This study analyzes the capabilities of four EMOCs to provide a diverse set of solutions that include nested clusters by using a modified connectedness criterion. The EMOCs analyzed are: MOCK [9], MOCLE [6], Δ-MOCK [8], and EMO-KC [16].

2.1   MOCK, Δ-MOCK, MOCLE and EMO-KC

MOCK (Multi-Objective Clustering with automatic K-determination) [9] and MOCLE (Multi-Objective Clustering Ensemble) [6] are well-known EMOCs. Δ-MOCK was introduced by [8] to improve the scalability of MOCK [9]. EMO-KC (Evolutionary Multi-objective Optimization-k-clustering) was described in [16], introducing an adapted sum of squared distances (SSD) to improve the generation of multiple solutions with a different number of clusters. These approaches present different representation encodings, initialization strategies, and/or evolutionary operators to optimize clustering criteria. In Sect. 5, we analyze how some of these different features can contribute to detecting different data structures, including nested clusters, based on a modified connectedness criterion (Sect. 3). In the following, we present more details of these EMOCs, considering the initialization strategy, representation encoding, optimization strategy, objective function, crossover, and mutation operators. Our analysis focuses on the ability of these algorithms to generate a set of solutions containing high-quality partitions. Thus, we will not be concerned with the selection of a final solution to be presented to the data expert.

Initialization Strategy. The generation of the initial population in MOCK consists of two methods: (i) Minimum Spanning Tree (MST) derived partitions, based on a measure called *degree of interestingness* (DI), and (ii) k-means (KM) [12] derived partitions. Δ-MOCK only uses one method to generate the initial population, the MST-derived partitions. In [6], MOCLE considered partitions generated by Single-Linkage (SL), Average-Linkage (AL), KM, and Shared

Nearest Neighbor-based clustering (SNN) [5]. In contrast, EMO-KC considers a random choice of the points in the dataset to define the initial centroids.

Representation. MOCK introduced the locus-based adjacency graph representation, in which a solution is described as a vector of genes, and each gene g_i can take an integer value between 1 and n; if a value j is assigned to the ith gene, it can be interpreted as a link between the data points i and j, i.e., i and j belong to the same cluster. Δ-MOCK introduced two reduced locus-based adjacency graph representations, Δ-locus and Δ-binary; these schemes can significantly reduce the length of the genotype by using the concepts of MST and DI according to the length of the encoding defined by a user-defined parameter (δ). MOCLE uses a label-based encoding that considers labels for each object in the partition. At last, EMO-KC uses a centroid-based encoding, in which the genes represent the coordinates of the cluster centroids.

Optimization Strategy. In terms of the optimization strategy, MOCK [9], MOCLE [6], Δ-MOCK [8], and EMO-KC [16] use traditional multi-objective evolutionary algorithms (MOEA) to optimize clustering criteria as objective functions. MOCK relies on the Pareto envelope-based selection algorithm version II (PESA-II) [3] in the optimization; in contrast, MOCLE, Δ-MOCK [8] and EMO-KC [16] use the Non-dominated Sorting Genetic Algorithm II (NSGA-II) [4]. Both these MOEA, PESA-II and NSGA-II, use the Pareto dominance relation to rank and select the solutions in evolutionary optimization. The Pareto dominance is an important concept used in our analysis, that can be defined as follows: Let $\mathbf{x}_1$ and $\mathbf{x}_2$ be two feasible solutions; $\mathbf{x}_1$ is said to dominate $\mathbf{x}_2$ (denoted as $\mathbf{x}_1 \prec \mathbf{x}_2$), if the following two conditions are satisfied [2]: (i) $\forall i \in \{1, 2, \ldots, z\}: f_i(\mathbf{x}_1) \leq f_i(\mathbf{x}_2)$, (ii) $\exists j \in \{1, 2, \ldots, z\}: f_j(\mathbf{x}_1) < f_j(\mathbf{x}_2)$.

Objective Functions. The analyzed EMOCs use two objective functions. MOCK and MOCLE optimize the overall deviation (dev) and connectivity index (con) as objective functions [6,9]. The dev is computed according to (1), where π represents a partition, $\mathbf{x}_i$ denotes an object in the cluster $\mathbf{c}_k$, $\boldsymbol{\mu}_k$ is the centroid of cluster $\mathbf{c}_k$, and $d(.,.)$ refers to the selected distance function.

$$dev(\pi) = \sum_{\mathbf{c}_k \in \pi} \sum_{\mathbf{x}_i \in \mathbf{c}_k} d(\mathbf{x}_i, \boldsymbol{\mu}_k), \tag{1}$$

The con is defined according to (2), where n is the number of objects in the dataset, L is the parameter that denotes the number of nearest neighbors that contributes to the connectivity, a_{ij} is the jth nearest neighbor of the object $\mathbf{x}_i$, and $\mathbf{c}_k$ is a cluster in the partition π.

$$con(\pi) = \sum_{i=1}^{n} \sum_{j=1}^{L} f(\mathbf{x}_i, a_{ij}), \text{where } f(\mathbf{x}_i, a_{ij}) = \begin{cases} \frac{1}{j}, \text{if } \nexists \mathbf{c}_k : \mathbf{x}_i, a_{ij} \in \mathbf{c}_k \\ 0, \text{otherwise} \end{cases} \tag{2}$$

Δ-MOCK also optimizes the *con*, but employs the intra-cluster variance (*var*) instead of the *dev* as an objective function. The *var* is defined according to (3), where π denotes a partition, n is the number of objects in the dataset, $\mathbf{x}_i$ is an object in the cluster $\mathbf{c}_k$, $\boldsymbol{\mu}_k$ is the centroid of the cluster $\mathbf{c}_k$, and $d(.,.)$ is the selected distance function [8].

$$var(\pi) = \frac{1}{n} \sum_{\mathbf{c}_k \in \pi} \sum_{\mathbf{x}_i \in \mathbf{c}_k} d(\mathbf{x}_i, \boldsymbol{\mu}_k)^2 \tag{3}$$

At last, EMO-KC optimizes an adapted sum of squared distances (SSD) and the number of clusters (k) as objective functions. None of the other EMOCs use the number of clusters as an objective function. In terms of the *SSD*, it is computed in the same way as (3) multiplied by n (the number of objects in the dataset). The adapted SSD, here denoted as *var'*, is computed according to (4) [10].

$$var' = (1 - exp^{-1 \times (SSD)}) - k \tag{4}$$

All these objective functions should be minimized in the optimization.

Crossover and Mutation Operators. MOCK and Δ-MOCK use the standard uniform crossover and a neighborhood-biased mutation scheme [9]. MOCLE uses the Hybrid Bipartite Graph Formulation (HBGF) [7] as crossover operator, and no mutation is employed [6]. EMO-KC relies on the standard operators of the NSGA-II: simulated binary crossover and polynomial mutation [16].

3   An Improved Connectivity Index

In our studies, we verified that, according to the setting of the neighborhood size parameter (L), the connectivity index formulation could limit the detection of some data structures. For example, a dataset with well-separated nested data structures, as ds2c2sc13 (Fig. 2), can produce several solutions with optimal *con*. Consequently, when EMOCs select which solutions to keep, the decision will be taken based on the other criteria. For instance, if we consider optimizing two objective functions, where the *var* or *dev* is applied along with *con*, the partitions with a lower number of clusters would be discarded in the selection. The reason is that this solution is dominated by other solutions with lower *var* or *dev* because those solutions have a higher number of clusters. In term of the dataset ds2c2sc13, the algorithms that use these pairs of objective functions, such as [6,8,9], may not find the true partition[1] of the S1 (when they use L = 10), because it is dominated by S2—the true partition of the S1, Fig. 2a, has

---

[1] The True Partition or ground truth is the labeled data that forms the real partition, the underlying structure of the data; and S denotes the hierarchy level of the partitions, in which S1 represents the partition with the lowest number of clusters (a high-level partition), and a higher S refers to a partition with a low level of hierarchy.

(dev = 63.038, con = 0) or (var = 0.013, con = 0); S2, Fig. 2b, has (dev = 33.457, con = 0) or (var = 0.004, con = 0); and S3, Fig. 2c, has (dev = 24.075, con = 7.519) or (var = 0.002, con = 7.519). A behavior that could also occur in other datasets with well-separated but no compact clusters.

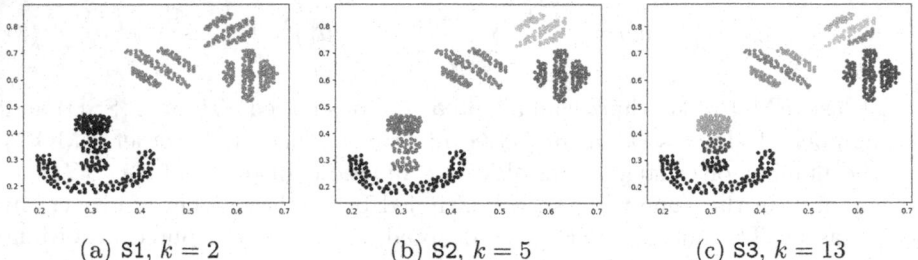

(a) S1, $k = 2$ (b) S2, $k = 5$ (c) S3, $k = 13$

Fig. 2. True partitions of the artificial dataset ds2c2sc13

It is important to note that the use of another setting for L can lead to the detection of the S1. However, it can generate the dominance of the other clustering levels. Thus, in order to deal with this problem, we propose in (5) a slight but effective modification of the definition of the con (2):

$$con'(\pi) = con(\pi) + \left(\frac{k}{n \times L} \right) \tag{5}$$

where k is the number of clusters in partition π, n is the number of objects in the dataset, and L is the number of nearest neighbors that contribute to connectivity. The term $(n \times L)$ ensures that the number of clusters k will be mapped to a value lower or equal to $\frac{1}{L}$, taking values in the interval $]0, \frac{1}{L}]$. That it is required to maintain the ordinal relationship between the best and the worst connectivity results. So, this modification will only affect the solutions that have the same outcome for the sum of the penalties of connectedness (2).

Intuitively, the new term added to con to yield con' will produce a new dominance relation that distinguishes the partitions with the same value of con but with a different number of clusters—a scenario that can occur in nested clusters, as in ds2c2sc13 dataset.

In other words, con' contains the information regarding the sum of the penalties of connectedness (the primary criterion), in the same way as the original con, added to the information about the number of clusters (the secondary criterion). The added information will affect the dominance evaluation of the solutions with the same outcome for the sum of the penalties of connectedness. Since the order relation regarding connectivity will only be modified in this group of solutions.

4   Experimental Design

This section presents the methodology employed to evaluate the adapted objective function, the used datasets, the experimental setup of the EMOCs, and the indicator applied for the performance assessment.

4.1   Datasets

Regarding the datasets, Table 1 summarizes the main characteristics of the fifteen datasets used in our experiments. In this table, n is the number of objects, d is the dimension of the dataset (number of attributes), S is the number of true partitions, i.e., the number of different levels of the (nested) data structures, and k* is the number of clusters of each data structure. S is also applied as an identifier for each true partition, where the associated number refers to the hierarchy level of the partitions, in which S1 represents the partition with the lowest number of clusters, and S4 the partition with the highest number of clusters. These datasets were divided into five groups (G1, G2, G3, G4, and G5), considering the general features evaluated in our analysis.

G1 and G2 contain datasets with several different properties, such as different data structures, number of observations, and distribution. G1 contains artificial datasets (20d-60c, Aggregation, D31), and G2 contains real datasets (Iris, Libras, UKC1). These groups are used to verify the general impact of the *con'* in comparison with *con* when applied in different datasets with a single true partition.

G3 contains artificial datasets with nested data structures and well-separated clusters (ds2c2sc13 and Spiralsquare). In this group, besides comparing the use of *con* and *con'*, we will analyze the capabilities of the *con'* in the EMOCs in relation to the hierarchical clustering algorithms SL and AL.

G4 also contains artificial datasets with nested clusters, but they have several different properties, such as cluster shapes, distributions, and proximity between the clusters. In this group, we have the Monkey dataset and three new datasets, Bear, Glassesman and Stomata, Fig. 3. These three datasets were used for the first time here[2]. In particular, Bear and Glassesman contain different types of sub-sets, in which the lowest (hierarchy) level of structures (S3) contains nested clusters along with other sub-sets. For example, the overlapping clusters that represent the nose and mouth in the S3 of the dataset Glassesman are sub-sets of one general cluster, but such clusters are not nested at this level of data. The same occurs in the clusters that represent the eyes in the dataset Bear, they are sub-sets in the S3, but these clusters are not nested structures.

At last, G5 contains real datasets, (Golub, Glass, and Leukemia), that present more than one specified true partition, and may present nested data structures. The analysis of these groups will provide a general view of how the *con'* impacts the clustering performance of the EMOCs in complex datasets. As well, we will analyze the results of the EMOCs with regard to the hierarchical clustering algorithms SL and AL.

---

[2] Available at https://github.com/cymorimoto/newdatasets.

Table 1. Dataset characteristics

#	Dataset	n	d	|S|	k	Description
G1	20d-60c	4,395	20	1	60	20d-60c has 60 ellipsoidal clusters with arbitrary elongation and orientation distributed in a 20-dimensional space.
	Aggregation	788	2	1	7	Aggregation consists of heterogeneous structures with clusters of varied sizes and shapes.
	D31	3,100	2	1	31	D31 contains 31 equal sizes and spread clusters that are slightly overlapping and distributed randomly in a 2-dimensional space.
G2	Iris	150	4	1	3	Iris contains 3 clusters (types of iris plant) that contain an equal number of observations.
	Libras	360	90	1	15	Libras is composed of representations of different hand movements in the Brazilian Sign Language (LIBRAS).
	UKC1	29,463	2	1	11	UKC1 is a dataset with a very large number of objects related to street-level crime in the U.K.
G3	ds2c2sc13	588	2	3	2, 5, 13	ds2c2sc13 contains three different structures: S1 represents two well-separated clusters; S2 and S3 combine different types of clusters.
	Spiralsquare	1,500	2	2	2, 6	Spiralsquare contains two true partitions: S1 represents two well-separated clusters, and S2 contains 2 spirals and 4 Gaussian-like clusters.
G4	Monkey	4,000	2	4	2, 3, 5, 8	Monkey has a set of clusters with different sizes and shapes that represent a monkey head. S1 contains two major clusters. S2 and S3 present clusters with different granularities of the S1.
	Bear	1,480	2	3	2, 5, 11	Bear contains clusters with different dispersion and distributions, considering clusters obtained from the datasets Pathbase and ds3c3sc6.
	Glassesman	5,878	2	3	3, 4, 5	Glassesman contains heterogeneous structures with clusters of varied sizes and shapes, including clusters presented in the datasets Engytime and twoDiamonds.
	Stomata	2,376	2	3	2, 8, 16	Stomata was designed and inspired by the cells found in the epidermis of leaves, named stomata. It contains three data structures: S1 considers two internal cells surrounded by the other cells, S2 represents each cell as a cluster, S3 distinguishes the cells and their nucleus.
G5	Glass	214	9	3	2, 5, 6	Glass is a benchmark dataset, that contains glass attributes used to identify the type of glass.
	Golub	72	3,571	2	2, 3	Golub refers to gene expression data from the leukemia micro-array study.
	Leukemia	327	271	2	3, 7	Leukemia also refers to gene expression. Both Golub and Leukemia have a small number of objects (distributed in clusters of very different sizes), but a large number of attributes, typical of bio-informatics data.

The datasets 20d-60c and UKC1 were introduced in [8]. ds2c2sc13, Glass, Golub, Iris, Leukemia, Libras, Monkey, and Spiralsquare were obtained from the Clusters evaluation benchmark repository[3]. D31 and Aggregation were obtained from the Clustering basic benchmark repository[4]. Besides that, the datasets that compose the Bear and Glassesman were obtained from these two repositories.

---

[3] Available at http://lasid.sor.ufscar.br/clustersEvaluationBenchmark.
[4] Available at http://cs.uef.fi/sipu/datasets.

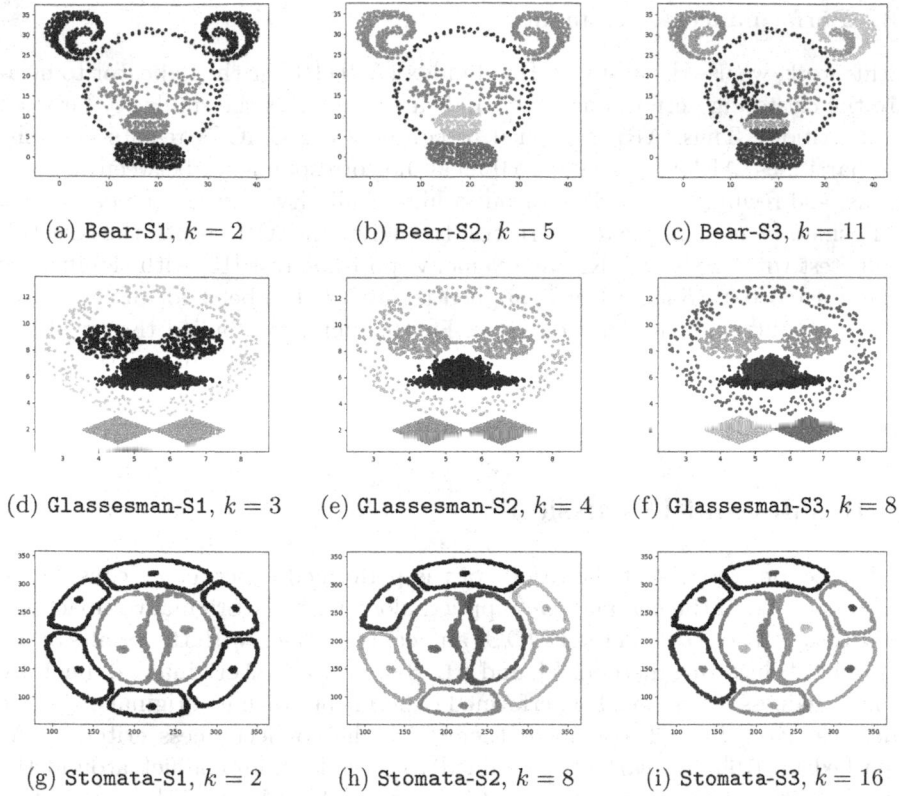

(a) Bear-S1, $k = 2$ (b) Bear-S2, $k = 5$ (c) Bear-S3, $k = 11$

(d) Glassesman-S1, $k = 3$ (e) Glassesman-S2, $k = 4$ (f) Glassesman-S3, $k = 8$

(g) Stomata-S1, $k = 2$ (h) Stomata-S2, $k = 8$ (i) Stomata-S3, $k = 16$

Fig. 3. New artificial datasets with nested data structures

4.2 Experimental Setup

We employed the same general settings as reported in [8,9] to execute MOCK and Δ-MOCK. Regarding the Δ-MOCK representation, in this paper, we used the Δ-locus scheme with δ defined as a function of $\sim 5/\sqrt{n}$, where n is the number of objects in the dataset—this function is one of the heuristics employed in [8]. Concerning the MOCLE, we used the general setting as in [6], considering the NSGA-II as MOEA and HBGF as a crossover operator. At last, for EMO-KC, we applied the same general setting presented in [16]. Furthermore, for every approach, including the hierarchical algorithms SL and AL, we applied the Euclidean distance as a distance function, and we adjusted the other parameters required to produce partitions containing clusters in the range $\{2, 2k^*\}$, in the same way as MOCK and Δ-MOCK.

For the EMOCs, the L parameter applied in the *con* and *con'* was set $L = 10$ for all the experiments. Finally, as such algorithms are non-deterministic, we executed the experiments 30 times.

4.3   Performance Assessment

In this work, we use the adjusted Rand index (ARI) [10] as the indicator to measure the clustering performance. This indicator measures the similarity between two partitions. Thus, ARI is applied to compare the EMOCs results with the true partitions. ARI results close to 0 mean no correspondence between the partitions, and results close to 1 point out a high similarity between the partitions.

Besides, we use a non-parametric test to analyze the ARI results, the Kruskal-Wallis test with the Tukey-Kramer-Nemenyi post-hoc test [15] with significance level alpha = 0.01. Such a test is applied to analyze the behavior of each algorithm on a different problem (dataset). Furthermore, we applied the Friedman and Bergmann-Hommel Post Hoc hypothesis test [15] with alpha = 0.05. This last combination of tests is applied to compare the overall performance of the algorithms in the datasets with nested clusters.

5   Results and Discussion

In this section, we present the results of the performed experiments considering the comparison of the *con* and *con'* applied along with the original compactness index (*dev*, *var*, or *var'*) in the EMOCs and compare them with the results of the hierarchical clustering methods SL and AL. Since EMO-KC originally did not use a connectedness index, we also performed experiments with its original objective functions (*var'*, *k*) and associated them with the connectedness criterion. As described by [16], the *var'* was designed to provide more conflict around the number of the clusters; we consider that a general analysis of these objective functions can provide insights about the performance of the purpose modification of the *con* to produce conflict around the compactness criterion.

Table 2 presents the ARI of the best partition found by SL and AL, and the average ARI of the best partitions of MOCLE, MOCK, Δ-MOCK, and EMO-KC found in experiments using 2 objective functions, considering their original compactness criterion (*dev*, *var*, or *var'*) associated with *con* and *con'*, as objective functions. For EMO-KC, it also presents the results regarding the original objective functions presented in [16], (*var'*, *k*). The ARI highlighted in boldface represents the best values found for each evolutionary multi-objective approach, considering the comparison of the different objective functions. Furthermore, underlined ARI points out the results with a significant difference according to the Kruskal-Wallis test. In the case of the SL and AL results, the ARI highlighted in boldface represents the result where these algorithms found the best ARI compared to the EMOCs.

Table 2. The ARI of the best partition found by SL and AL, and the average ARI of the best partition found by MOCLE, MOCK, Δ-MOCK and EMO-KC. Average of 30 executions for the EMOCs

#	Datasets	S	SL	AL	MOCLE		MOCK		Δ-MOCK		EMO-KC		
			-	-	dev,con	dev,con'	dev,con	dev,con'	var,con	var,con'	var,k	var',con	var',con'
G1	20d-60c	-	0.0007	0.2601	**0.8989**	**0.8989**	0.7819	0.7810	**0.9003**	0.8998	0.5350	**0.5554**	0.5525
	Aggregation	-	0.8089	**1.0000**	**1.0000**	**1.0000**	0.9935	0.9908	**0.9671**	0.9656	0.7898	0.9346	**0.9535**
	D31	-	0.2124	0.9307	**0.9523**	**0.9523**	0.9030	0.9044	0.7456	0.7291	0.8136	0.8274	**0.8313**
G2	Iris	-	0.5681	0.7592	**0.8284**	**0.8284**	0.7707	0.7891	0.7709	0.7869	0.7370	0.7543	0.7437
	Libras	-	0.0224	0.3346	**0.3346**	**0.3346**	0.3942	0.3973	0.3865	0.3886	0.2762	0.2949	0.2933
	UKC1	-	1.0000	0.9415	**1.0000**	**1.0000**	0.9985	1.0000	0.9962	0.9995	0.9574	0.9577	0.9498
G3	ds2c2sc13	S1	1.0000	1.0000	0.6840	1.0000	0.3810	1.0000	0.3520	1.0000	1.0000	0.6828	1.0000
		S2	1.0000	1.0000	1.0000	1.0000	1.0000	1.0000	0.9520	1.0000	0.8775	0.8777	0.8841
		S3	0.8724	0.6648	1.0000	0.9690	0.8710	0.8380	0.8720	0.8720	0.5889	0.6613	0.6622
	Spiralsquare	S1	1.0000	1.0000	0.8888	1.0000	0.5711	1.0000	0.5711	1.0000	0.9663	1.0000	1.0000
		S2	0.9283	0.5410	0.9971	0.9971	0.9980	0.9973	0.9986	0.9987	0.4742	0.5940	0.5011
G4	Monkey	S1	0.5122	0.4479	0.5131	0.4566	0.3377	0.5653	0.4544	0.8654	0.3737	0.6124	0.6076
		S2	0.8551	0.2279	0.8267	0.8551	0.7776	0.8351	0.7776	0.9292	0.2881	0.4468	0.4629
		S3	0.8341	0.5305	0.8341	0.8341	0.7610	0.7640	0.7960	0.7960	0.4272	0.5197	0.5037
		S4	0.8708	0.6713	0.8707	0.8707	0.8628	0.8404	0.8737	0.8719	0.6078	0.6997	0.6622
	Bear	S1	0.0042	0.1266	0.2858	0.2858	0.4252	0.4181	0.4142	0.4135	0.2179	0.2565	0.2431
		S2	0.0675	0.7194	0.7194	0.7194	0.7138	0.7195	0.7220	0.7219	0.5883	0.6752	0.7048
		S3	0.3895	0.6842	0.6842	0.6842	0.8061	0.8097	0.7798	0.7793	0.6349	0.6760	0.6781
	Glassesman	S1	0.8775	0.8269	0.8775	0.8775	0.7920	0.7857	0.7889	0.7890	0.7909	0.8077	0.8097
		S2	0.5944	0.5048	0.9691	0.9691	0.9271	0.9291	0.9549	0.9529	0.9034	0.9273	0.9228
		S3	0.2403	0.5048	0.8428	0.8428	0.8505	0.8467	0.8791	0.8798	0.7954	0.8155	0.8214
	Stomata	S1	0.0214	0.0382	0.0382	0.0382	0.5986	0.5946	0.8635	0.8311	0.0005	0.0124	0.0108
		S2	0.7233	0.2966	0.7233	0.7233	0.6987	0.7025	0.7970	0.8368	0.3269	0.3708	0.3574
		S3	0.7783	0.2620	0.9190	0.9190	0.7356	0.6805	0.8952	0.8573	0.2992	0.3368	0.3455
G5	Glass	S1	0.0536	0.0536	0.6468	0.6468	0.5418	0.5424	0.5620	0.5663	0.6099	0.6086	0.6077
		S2	0.1057	0.4918	0.5043	0.5043	0.4338	0.4359	0.4605	0.4552	0.4608	0.4951	0.4999
		S3	0.0403	0.2488	0.2980	0.2980	0.2060	0.2030	0.2050	0.2020	0.2205	0.2295	0.2307
	Golub	S1	-0.0026	-0.0139	0.4193	0.4193	0.7884	0.8054	0.5410	0.5469	0.4630	0.7203	0.7406
		S2	-0.0108	0.6473	0.6473	0.6473	0.8816	0.8714	0.5569	0.5676	0.5615	0.7055	0.7126
	Leukemia	S1	-0.0037	0.3346	0.3295	0.3295	0.3049	0.4133	0.3040	0.4097	0.2352	0.2945	0.3004
		S2	0.0224	0.3346	0.7589	0.7589	0.7767	0.7767	0.7706	0.7708	0.5922	0.7201	0.7180
Significant Wins			-	-	2	3	3	6	2	8	-	0	4

The results point out that the con' improves the general performance of all the EMOCs, as shown in Table 2. The row **Significant Wins** presents the number of the datasets in which one pair of objective functions win over the other pair (as indicated by the statistical test), considering the objective functions with con' or con. For instance, for MOCLE, the objective functions (dev, con) has 2 significant wins while the pair (dev, con') has 3 significant wins. In general, Δ-MOCK and MOCK are algorithms that have the major significant wins with the con', where MOCK has 6 significant wins and Δ-MOCK 8 significant wins.

By analyzing each group of datasets, we obtained more details about how the *con'* impacted the results of the studied EMOCs. For example, in general, the use of the *con'* does not impact the results in the datasets with a single true partition, as the datasets present in G1 and G2, in which the results of all EMOCs were very close to that present with *con*. Only for the dataset UKC1 (present in G2), the use of the *con'* provided a significant gain of ARI in MOCK and Δ-MOCK.

On the other hand, for the datasets with nested data structures and well-separated clusters, as presented in G3, we have the greatest improvement of the clustering results by using the *con'*. For example, the well-separated structures S1 in the datasets ds2c2sc13 and SpiralSquare were detected in all studied approaches, and Δ-MOCK was able to detect S2 in the datasets ds2c2sc13.

A particular case occurred in the S3 of the datasets ds2c2sc13, where the use of the *con'* caused a significant loss in the ARI in MOCLE and MOCK, when compared with the results of *con*. In this case, the parameter L is still impacting the dominance around the true partition. However, in MOCLE, our general results for this dataset are higher than others reported in the related work, as in [6] or results provided by hierarchical methods SL and AL.

In contrast, the results in the G4 and G5 were diverse. For example, we obtain an ARI gain in the S1 of the dataset Leukemia in MOCK and Δ-MOCK. However, the dominance of the true partition and the influence of the L parameter are still impacting the results in the datasets Monkey and Stomata, in which we obtained gain of the ARI in some partitions and an ARI decrease in other ones. An analysis of the Pareto front (PF) of these datasets is presented in Sect. 5.1, to detail how the *con'* impacts the optimization and to explain these results. At last, for the other datasets in these two groups (G4 and G5) there are not any significant differences by using *con* or *con'*. Regarding the results of the SL and AL, in general, the best results found by them for G4 and G5 were worse or equal to the results found by MOCLE.

In general, this minor loss in MOCLE, MOCK, and Δ-MOCK by using *con'* is not so significant when compared to the general ARI gain in the datasets as presented in Table 2 (**Significant Wins** row). Furthermore, it promotes a significant increase in the performance of the EMO-KC without any loss.

Furthermore, it is important to observe that the Δ-MOCK provides the highest ARI for datasets with multiple true partitions. That also is pointed in the Critical Difference Diagram, Fig. 4, which shows the performance comparison of the strategies according to the Friedman and Bergmann-Hommel Post Hoc hypothesis test, in which Δ-MOCK has the best rank with *con'*.

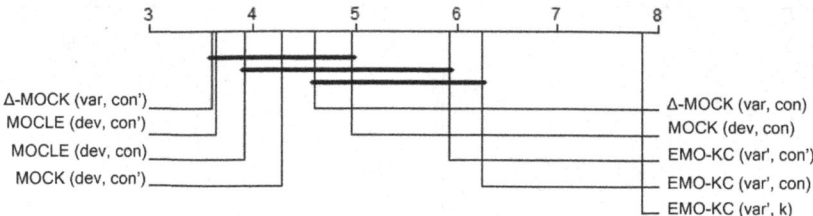

Fig. 4. Critical Difference Diagram of the EMOCs considering the different pairs of objective functions. The bold horizontal lines link the strategies that had statistically equivalent performance among them at a confidence level of 95%, and the lower the rank the better performance of an approach.

Additionally, we also performed experiments with EMO-KC using three objectives, (var', con, k) and (var', con', k), that produces equivalent ARI results to the pair (var', con'). Since there is not a significant difference between the overall performance of the EMO-KC using two or three objective functions, we do not display these last results. Nevertheless, it is important to note that, based on these results, in EMO-KC, the use of the con' provides evidence that the conflict around the number of clusters is improved, even though the general ARI gain is not so robust.

5.1 The Impact of the con' in the Optimization

As above-mentioned, the use of the con' promoted a general gain in the ARI; however, it also caused a loss in the ARI in some datasets. In this context, to analyze how the con' impact the optimization, we look over the Pareto front of the datasets Monkey and Stomata.

Figure 5 presents the Pareto front of the datasets Monkey generated by MOCLE, in which the red points represent each true partition at different levels, as a reference for comparison. In Fig. 5b we can observe an increase of the solutions in the region of the true partition of the S1, S2, and S3 when compared with Fig. 5a. For the S1, we observed that the solution that ARI = 0.5131 was dominated by other solutions when con' is applied, but it generates new solutions near the true partition, making it possible to apply other methods (local search) to improve the results. In particular, to detect the partition S4 in Monkey requires further exploration of the region with smaller var in MOCLE. It is important to note that MOCLE has an inner property that reduces the number of solutions while producing new solutions generated by the ensemble-based crossover. It is the main reason for the small number of solutions in the Pareto front when compared to the other EMOCs.

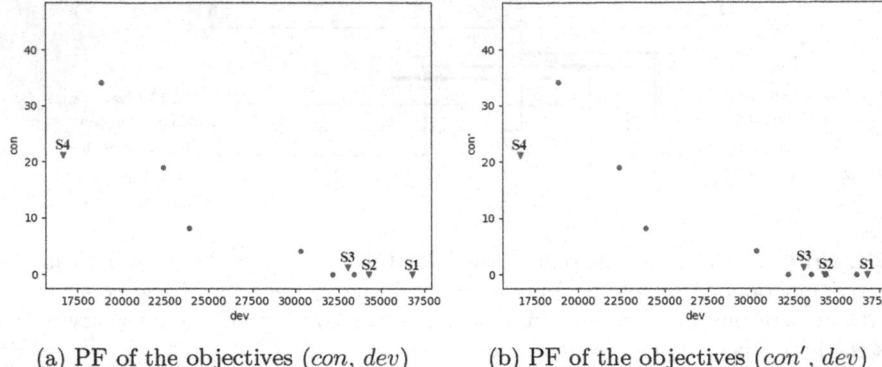

(a) PF of the objectives (*con*, *dev*)   (b) PF of the objectives (*con′*, *dev*)

Fig. 5. Front of the final population of the dataset Monkey generated by MOCLE (Color figure online)

Figure 6 presents the Pareto front of this same dataset generated by MOCK. In the sub-figures, we also observe similar behavior to the MOCLE, in which using *con′* promoted the increase of solutions in the region of the high-level structures that are close to the true partition. However, for the S4 in Monkey, the *L* parameter is still affecting the general performance of the MOCK, in which the use of the *con′* improved the convergence of the solutions.

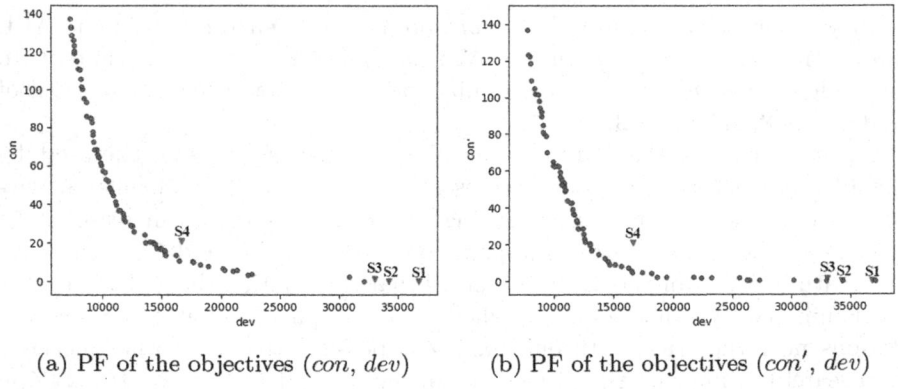

(a) PF of the objectives (*con*, *dev*)   (b) PF of the objectives (*con′*, *dev*)

Fig. 6. Front of the final population of the dataset Monkey generated by MOCK

For the dataset Stomata, in Δ-MOCK *con′* promote a better distinction of the solutions around the true partition of the S2, S3. Since the *con* for these structures is the same (*con* = 9.8115), the *con′* generates more diversity of solutions in which the convergence is better than in *con*. In this context, this distinguishing of solutions improve of the ARI in the S2. For S1 it promoted the increase of solutions in the region of the high-level structures that are close to the true

partition, as illustrated in Fig. 7. The general loss in the ARI of the S1 and
S3 using *con'* occurred because we have new solutions in the front with better
convergence but still need some local exploitation to get a better ARI.

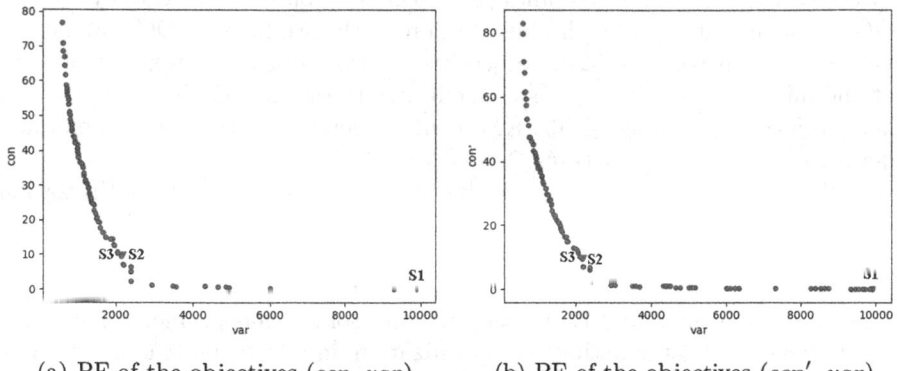

(a) PF of the objectives (*con*, *var*)   (b) PF of the objectives (*con'*, *var*)

Fig. 7. Front of the final population of the Stomata generated by Δ-MOCK

It is important to note that in the Δ-MOCK, besides using the *con'*, the initialization strategy also had an important role in its general results. We observe
that initialization strategies that include KM, as in MOCK and MOCLE, could
generate solutions that dominate other promising solutions with nested data
structures. Besides that, the reduced encoding used in Δ-MOCK did not affect
the clustering performance, in which Δ-MOCK is the more scalable approach
with good ARI results.

On the other hand, in EMO-KC, the use of the random initialization and
centroid-based representation had difficulties in detecting concentric clusters,
such as the two spirals in the Spiralsquare, or clusters with close centroid and
elongated data structures.

In summary, by using *con'* the optimization of the solutions was improved,
which promoted more diversity of the solutions, including the regions of the high
level of the nested data structure, and increased the convergence of the solution;
however, some aspects of the EMOCs, such as initialization and representation,
can impact the detection of the nested clusters.

6   Conclusion

In this study, we provide an analysis regarding the use of EMOCs for nested
data structures. Furthermore, we deal with a problem in the definition of the
connectivity index, in which several different partitions could present the same
optimal value (*con* = 0) depending on the considered neighborhood size (L).
In this scenario, the decision would be essentially taken based on the other
objective function in evolutionary multi-objective optimization. To tackle this

problem, we presented a modified version of the connectivity index called con'. The results obtained with con', in terms of ARI and the ability to find nested cluster structures, are promising. In particular, there is a significant increase of the ARI in artificial datasets that present well-separate nested structures.

Besides the meaningful advantages in the scalability described by [8], Δ-MOCK demonstrated to be the best option of the studied EMOCs for nested clustering by using the con' as an objective function. In this context, we observe that the initialization strategy also contributes to the Δ-MOCK results, where other initialization strategies, like KM, could generate partitions that dominate other ones with nested structures.

Furthermore, we demonstrate how this modification impacts the optimization process by presenting the plot of the Pareto Front of the EMOCs, evidence that the con' improves the generation of a more diverse and convergent set of solutions.

Our results also showed that there are still some open problems regarding the L parameter still impacting the optimization, in which the true partition is dominated, deserving more studies. For future work, we consider that an analysis of different values of L can provide the extent of the results that depend on L.

We also introduce three new datasets (`Bear`, `Glassesman`, `Stomata`) that present a great challenge for the studied EMOCs, that could be explored in future works.

References

1. Bertrand, P., Diatta, J.: Multilevel clustering models and interval convexities. Discret. Appl. Math. **222**, 54–66 (2017)
2. Coello, C.A.C., Lamont, G.B., Veldhuizen, D.A.V.: Evolutionary Algorithms for Solving Multi-Objective Problems (Genetic and Evolutionary Computation). Springer, Heidelberg (2006)
3. Corne, D.W., Jerram, N.R., Knowles, J.D., Oates, M.J.: PESA-II: region-based selection in evolutionary multiobjective optimization. In: Proceedings of the 3rd Annual Conference on Genetic and Evolutionary Computation, pp. 283–290. Morgan Kaufmann Publishers Inc., San Francisco (2001)
4. Deb, K., Pratap, A., Agarwal, S., Meyarivan, T.: A fast and elitist multiobjective genetic algorithm: NSGA-II. IEEE Trans. Evol. Comput. **6**(2), 182–197 (2002)
5. Ertoz, L., Steinbach, M., Kumar, V.: A new shared nearest neighbor clustering algorithm and its applications. In: Workshop on Clustering High Dimensional Data and its Applications at 2nd SIAM International Conference on Data Mining, pp. 105–115 (2002)
6. Faceli, K., de Leon Ferreira de Carvalho, A.C.P., de Souto, M.C.P.: Multi-objective clustering ensemble. Int. J. Hybrid Intell. Syst. **4**(3), 145–156 (2007). http://content.iospress.com/articles/international-journal-of-hybrid-intelligent-systems/his00047
7. Fern, X.Z., Brodley, C.E.: Solving cluster ensemble problems by bipartite graph partitioning. In: Proceedings of the Twenty-First International Conference on Machine Learning, p. 36. ACM (2004)

8. Garza-Fabre, M., Handl, J., Knowles, J.: An improved and more scalable evolutionary approach to multiobjective clustering. IEEE Trans. Evol. Comput. **22**(4), 515–535 (2017)
9. Handl, J., Knowles, J.: An evolutionary approach to multiobjective clustering. IEEE Trans. Evol. Comput. **11**(1), 56–76 (2007)
10. Hubert, L., Arabie, P.: Comparing partitions. J. Classification **2**(1), 193–218 (1985)
11. Jain, A.K., Murty, M.N., Flynn, P.J.: Data clustering: a review. ACM Comput. Surv. **31**(3), 264–323 (1999). https://doi.org/10.1145/331499.331504, http://doi.acm.org/10.1145/331499.331504
12. MacQueen, J., et al.: Some methods for classification and analysis of multivariate observations. In: Proceedings of the Fifth Berkeley Symposium on Mathematical Statistics and Probability, vol. 1, pp. 281–297. Oakland, CA, USA (1967)
13. Mukhopadhyay, A., Maulik, U., Bandyopadhyay, S.: A survey of multiobjective evolutionary clustering. ACM Comput. Surv. (CSUR) **47**(4), 61 (2015)
14. Muller, E., Gunnemann, S., Farber, I., Seidl, T.: Discovering multiple clustering solutions: grouping objects in different views of the data. In: 2012 IEEE 28th International Conference on Data Engineering, pp. 1207–1210. IEEE (2012)
15. Pohlert, T.: PMCMR: Calculate Pairwise Multiple Comparisons of Mean Rank Sums, May 2018. https://CRAN.R-project.org/package=PMCMR
16. Wang, R., Lai, S., Wu, G., Xing, L., Wang, L., Ishibuchi, H.: Multi-clustering via evolutionary multi-objective optimization. Inf. Sci. **450**, 128–140 (2018). https://doi.org/10.1016/j.ins.2018.03.047
17. Xu, R., Wunsch, D.: Survey of clustering algorithms. IEEE Trans. Neural Networks **16**(3), 645–678 (2005). https://doi.org/10.1109/TNN.2005.845141

Integrating Safety Guarantees into the Learning Classifier System XCS

Tim Hansmeier[✉][ID] and Marco Platzner[ID]

Paderborn University, Paderborn, Germany
{tim.hansmeier,platzner}@upb.de

Abstract. On-line learning mechanisms are frequently employed to implement self-adaptivity in modern systems. With more widespread use in technical systems that interact with their physical environment, e.g. cyber-physical systems, the fulfillment of safety requirements is increasingly gaining attention. We focus on the learning classifier system XCS with its human-interpretable rules and propose an approach to integrate safety guarantees into its rule base. We leverage the interpretability of XCS' rules to internalize the safety-critical knowledge, as opposed to related work, which relies on an external safety monitor. The experimental evaluation shows that such manually injected knowledge not only gives safety guarantees but aids the learning mechanism of XCS. Especially in complex environments where XCS is struggling to find the optimal solution, the use of hand-crafted forbidden classifiers leads to a performance that is up to 41.7 % better than with an external safety monitor.

Keywords: Safety · Safe reinforcement learning · LCS · XCS

1 Introduction

Autonomous systems need to continuously adapt to environmental changes that could not be foreseen by the system designer, which is only achievable through the introduction of learning capabilities. Consequently, the employment of on-line learning algorithms is explicitly considered in systematic design paradigms for autonomous systems like Organic [10] or Self-aware Computing [9]. These design paradigms are increasingly being applied when designing self-adaptive Cyber-Physical Systems (CPS), leading to several new challenges which have rarely been addressed so far [3]. One of the most important ones is to design CPS which have a high degree of autonomy, but still exhibit safety guarantees to prevent catastrophes during operation. Since a CPS has a real-world impact, it is able to permanently harm itself, its environment or living beings in its

This work was partially supported by the German Research Foundation (DFG) within the Collaborative Research Centre On-The-Fly Computing (GZ: SFB 901/3) under the project number 160364472.

© Springer Nature Switzerland AG 2022
J. L. Jiménez Laredo et al. (Eds.): EvoApplications 2022, LNCS 13224, pp. 386–401, 2022.
https://doi.org/10.1007/978-3-031-02462-7_25

surrounding. However, as on-line learning mechanisms are inherently explorative, a self-aware CPS might execute such harmful actions purely out of curiosity. Proper means to prevent such safety-critical events are part of current research in the field of machine learning and artificial intelligence [2].

Since continuous runtime monitoring is often infeasible and detailed a priori knowledge of the operational environment is lacking, autonomous systems are typically designed for unsupervised operation, for which learning techniques from the domain of Reinforcement Learning (RL) are commonly used. One class of RL techniques that is often proposed for the use in self-aware systems are Learning Classifier Systems (LCS) [12], most notably the classifier system XCS [15], which, for instance, has already been applied on adaptive traffic control [11] or the autonomous management of multi-core system on chips [17]. XCS evolves a set of rules, termed classifiers, and in contrast to pure Q-learning, XCS has the ability to generalize over different environmental states and summarize them under one rule if they are sufficiently similar. This enables XCS to propose suitable actions even for situations that it has never seen before, as long as they depict resemblance with known circumstances. As a side effect, such generalized "if-then" rules are well interpretable by humans, which gives the rule base that is evolved by XCS a high degree of explainability.

Existing work in the field of safe reinforcement learning is often relying on an additional subsystem external to the learning algorithm, that is equipped with pre-defined knowledge and constantly monitors the situation and the behavior of the learner to intervene whenever harmful actions are about to be executed [5]. In this work, we focus specifically on learning classifier systems and propose an approach in which the safety-critical knowledge is directly embedded into the knowledge base of the learning classifier system XCS, making an external monitor superfluous. To achieve this, we make use of the interpretability of the rules inside XCS and introduce the concept of *forbidden classifiers*, which are rules that do not propose an action but instead prevent it from being executed in safety-critical situations. So far, the introduction of safety guarantees has rarely been discussed in the context of LCS or XCS in specific, and to the best of our knowledge, this work is the first to leverage the human-interpretability of XCS' rule base by systematically introducing hand-crafted classifiers.

The paper continues with Sect. 2 by outlining related work in the field of safe RL. Section 3 provides an overview of XCS, giving the details that are necessary to understand the concept of forbidden classifiers and the necessary algorithmic modifications, which are presented in Sect. 4. In Sect. 5, the approach is experimentally compared to an XCS with a straight-forward external safety monitor on three reinforcement learning problems that are common in XCS research. Finally, Sect. 6 concludes the paper and outlines future work.

2   Related Work

According to a survey by García and Fernández [5], the approaches to introduce safety constraints into RL techniques can be categorized either as a modification of the optimization criterion or of the exploration process. Techniques from the first category do not solely focus on maximizing the (discounted) payoff received from the environment, as most standard RL algorithms do, but also take into account some measure of risk, e.g. the variance of the payoff that is received. However, these approaches are often overly pessimistic, leading to non-optimal problem solutions and only improve the level of safety if reaching a catastrophic state results in an increase of the risk measure, which is not necessarily the case for arbitrary problem domains. Further, such approaches only reduce the probability of the occurrence of risky situations but do not give any strict guarantees.

Completely avoiding catastrophic states from the beginning of operation is only possible if external knowledge is used, as otherwise the only way for a learning system to identify harmful actions is to try them out at least once. Techniques that belong to the second category of safe RL approaches and modify the exploration process make use of this. The external knowledge can either be inserted into the system via a pre-deployment learning phase with manually created samples and policies or through teacher advice at runtime. While the first approach does not require an additional subsystem to be deployed in the final system, it is not able to give strict guarantees, either, because the exploratory nature of on-line learning can still lead to catastrophic states during operation. On the other hand, a teacher that is constantly active can give such strict guarantees, at least if the teacher-learner relationship is designed such that safety-critical advice is mandatory for the learner.

Closely related to the concept of teacher-advised RL is the approach of shielded learning, as proposed by Alshiekh et al. [1]. The shield constantly monitors the state of the environment and the action that the learner is proposing. Whenever an action is proposed that is considered harmful in the current state of the environment, the shield is overriding the action to prevent catastrophic consequences. In [1], a formal method with proven correctness and minimal interference is presented to synthesize a shield for Markovian Decision Problems (MDP) based on safety requirements given as temporal logic specification. Experimental results indicate that a shield is not only providing safety guarantees but also improves the speed of learning.

For XCS in specific, Tomforde et al. [14] proposed a teacher-like approach, in which newly created classifiers must first pass a safety check in a simulation environment before being inserted into the population. In contrast to this, our approach differs in the way the safety-critical knowledge is integrated into the system. While both teacher and shielding approaches require an additional subsystem that is external to the learning algorithm and constantly monitors it, we insert the knowledge directly into the learning base of XCS with manually created rules and ensure the safety guarantees through minor algorithmic modifications without the need to learn from interactions with an external subsystem.

3   XCS

Since its initial development, several different variants of XCS with various (subtle) implementational differences have been proposed over the past decades,

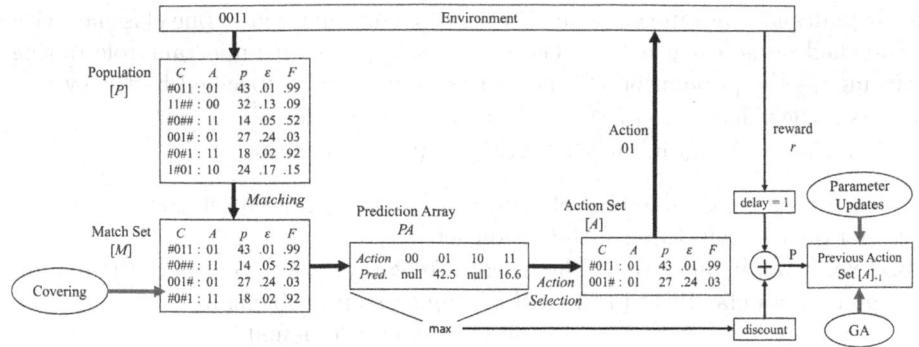

Fig. 1. Illustration of XCS taken from [16], with a 4-bit input, available actions $A \in \{00, 01, 10, 11\}$ and the minimum number of actions θ_{mna} been set to 2. The environment provides the sensory input 0011 and XCS selects action 01 for execution. Within single-step environments, the discounting mechanism and the previous action set $[A]_{-1}$ are not necessary, as the GA and the parameter update take place on the current action set $[A]$ using the immediate reward r.

which is why *the* standard XCS does not exist. Hence, we restrict our discussion to the most common version of XCS as described in [4], which we also have used for our experimental evaluation in Sect. 5. XCS is able to tackle both single- and multi-step problems, with the latter depicting environmental inputs that are not independent of each other and requiring a series of actions to solve a problem instance. Even though our description of XCS given in this section covers the most important aspects, it is still not exhaustive since XCS contains a plethora of parameters and corresponding algorithmic mechanisms. Hence, the focus lies on the aspects which are relevant for understanding the concept of forbidden classifiers and their implications on XCS' learning behavior.

The main components and mechanisms of XCS are shown in Fig. 1. XCS interacts with the environment according to a reinforcement learning paradigm, i.e. it receives a sensory input $\sigma(t) \in \{0,1\}^L$ encoded as a bitstring of length L, decides for an action $\alpha(t) \in \{\alpha_1, \ldots, \alpha_n\}$ and receives a scalar reward r from the environment after $\alpha(t)$ has been executed. The reward is then used to guide the learning process inside XCS. The core of XCS is a population $[P]$ of classifiers, which can be interpreted as if-then rules. Each classifier cl consists of a condition $C \in \{0, 1, \#\}^L$, which defines for which inputs the classifier applies. The don't cares '#' allow XCS to generalize and create classifiers that match multiple inputs. Each classifier proposes exactly one action $\alpha \in \{\alpha_1, \ldots, \alpha_n\}$ for all situations that match to its condition. In addition, the classifier keeps an estimate of the payoff p that is expected to be received when the classifier matches a situation and α is executed. The parameter ϵ represents an estimate of the prediction error that is associated with the payoff prediction and is used to calculate the fitness F, which guides the GA to evolve accurate and general classifiers. For reasons of computational efficiency, identical classifiers do not occur multiple times in the population but are summarized to one classifier with a so-called numerosity $n > 1$. The numerosity plays an important role during learning, as the population $[P]$ has a maximum size N, specified by the system designer, such that $\sum_{cl \in [P]} cl.n \leq N$ is always fulfilled.

Upon receiving an input $\sigma(t)$, XCS executes four steps:

1. Identification of matching classifiers to form the match set $[M]$.
2. Action selection to form the action set $[A]$.
3. Execution of the action. The received reward is used to update the parameters of all classifiers in $[A]$ ($[A]_{-1}$ in multi-step problems).
4. Applying the GA to $[A]$ ($[A]_{-1}$ in multi-step problems).

Overall, the goal of XCS is to evolve a population of classifiers that are accurate and maximally general. If successful, this leads to a population of minimal size which is able to predict for every situation the best-suited action. The goal of evolving such a population is inherent to each of the four main steps.

1) Creation of Match Set. The whole population is searched for classifiers whose condition C matches the current input $\sigma(t)$. Matching classifiers are added to the match set $[M]$. By default, it is required that all available actions are present in $[M]$, but this can be changed by setting the configurable parameter θ_{mna}, which specifies the minimum number of different actions that must be present in $[M]$. In case not enough actions are present, a new matching classifier is randomly created via a covering procedure.

2) Action Selection. Once all matching classifiers have been identified, one of their actions must be selected for execution. In order to choose actions that maximize the payoff, XCS builds the prediction array PA, which holds for every action the fitness-weighted average of the payoff predictions of all classifiers in $[M]$ that propose this action. Since the fitness of a classifier is based on its prediction accuracy, more accurate payoff predictions are weighted higher. Several action selection strategies are possible, but most commonly used is an ϵ-greedy strategy, in which either pure exploitation, i.e. selecting the action with the highest prediction, or pure exploration, i.e. random selection, is conducted. After an action has been selected, the action set $[A]$ is formed with all classifiers of $[M]$ that propose the selected action.

3) Parameter Update. After the action has been executed in the environment, the parameters of the classifiers in $[A]$ are updated. In the case of a single-step problem, XCS uses the immediate reward r as payoff, i.e. $P = r$, to update the classifiers in $[A]$. In multi-step problems, the update takes place on the previous action set $[A]_{-1}$ and calculates the Payoff P as sum of the previous immediate reward and the discounted maximum of the current prediction array, i.e. $P = r_{-1} + \gamma \cdot max(PA)$. This ensures that immediate rewards received at later points in time are propagated to preceding classifiers, which have paved the way for receiving this reward. The prediction p, its error estimate ϵ, and the fitness F of each classifier are updated according to the Widrow-Hoff delta rule, as described in [4]. In XCS, the fitness F is not based on the magnitude of the predicted payoff, but on the accuracy of the prediction, i.e. the error estimate ϵ.

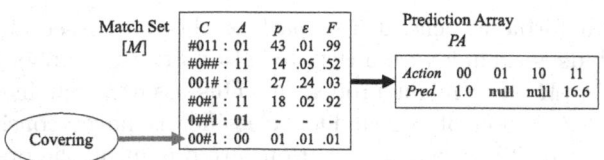

Fig. 2. The same situation as already shown in Fig. 1, but with a forbidden classifier (marked red) that is preventing the selection of the action 01. θ_{mna} is still set to 2, so a new classifier is created via covering for one of the actions that have not yet been part of PA. (Color figure online)

4) Genetic Algorithm. During exploration iterations, the GA is executed on $[A]$, or on $[A]_{-1}$ in multi-step problems, if the average time period since the classifiers in the action set participated in a GA is greater than the configurable threshold θ_{GA}. If so, two classifiers are selected from $[A]$ via a roulette-wheel selection based on the fitness F and two offspring classifiers are created with two-point crossover and mutation. Before inserting both offspring into the population, *GA subsumption* takes place. If the parent classifiers are both experienced and accurate, and the offspring covers only a subset of one of its parents' conditions and proposes the same action, it is not inserted into the population. Instead, the numerosity n of the respective parent is incremented by one.

Whenever a new classifier is added to the population, and the maximum population size N is exceeded, a classifier is selected for deletion from the population $[P]$ with a roulette-wheel selection. The deletion vote of each classifier is high if it is often occurring in large action sets and has a low fitness. In case a classifier with a numerosity $n > 1$ is selected for deletion, the classifier is not removed from the population, but its numerosity is decremented by one.

4 Forbidden Classifiers

To prevent XCS from taking catastrophic actions, we leverage the interpretability of the classifiers given by their condition→action rule structure and propose the use of rules that *forbid* an action in certain situations instead of proposing it. Consequently, we term this kind of XCS rules *forbidden classifiers*. Such classifiers model situations in which a certain action is harmful or has catastrophic consequences and thus should not be executed. To extend XCS with the ability to exhibit safety guarantees via the integration of forbidden classifiers, the following algorithmic modifications have been employed:

1. Forbidden classifiers are inserted into the classifier population $[P]$ upon initialization of XCS. The numerosity n of a forbidden classifier is set to the constant value of 1, representing the minimal value that the numerosity can take. The remaining parameters of a forbidden classifier, e.g. the payoff prediction p, are of no further relevance to XCS due to the prohibitive nature of the classifier.

2. Whenever a forbidden classifier is part of the match set $[M]$, the action that it forbids is excluded from the prediction array PA, regardless of other classifiers in $[M]$ that might propose it. This assures that in safety-critical situations the action of a forbidden classifier is never considered during action selection. If less than θ_{mna} actions are present in the prediction array PA, covering is applied until either θ_{mna} actions are part of PA, as shown in the example given in Fig. 2, or no additional actions are available.

3. Forbidden classifiers do not participate in the GA. Considering that their sole purpose is to prevent the execution of harmful actions, there exists no apparent need to evolve new classifiers from them. In the most common variant of XCS, participation in the GA is prevented by design, as the GA works on the action set $[A]$ that forbidden classifiers are never part of.

4. XCS keeps a list of all forbidden classifiers, and whenever a new offspring is generated by the GA, it is checked if the offspring is subsumed by one of the forbidden classifiers. If this is the case, the offspring is not inserted into $[P]$. This additional subsumption mechanism is necessary because the existing subsumption mechanisms work solely on the action set $[A]$, in which a classifier that is subsumed by a forbidden classifier will never be present.

5. Forbidden classifiers are made transparent to the deletion mechanism, i.e. they do not give a deletion vote during the roulette-wheel selection. Due to their non-participation in the selection, forbidden classifiers are never deleted from the population, preserving the safety guarantees that they represent.

Overall, forbidden classifiers can be characterized as static and in a sense passive, as they are never deleted, do not participate in the GA and their action is never selected for execution. As such, the concept of forbidden classifiers can be incorporated not only into XCS but into all learning classifier systems that depict a condition→action structure, e.g. into XCSR with its interval-based real-valued conditions [13] or XCS with code-fragment based conditions [6]. The classifiers required to fulfill given safety guarantees can either be created manually by the system designer using domain knowledge or even automatically, e.g. using the approach proposed in [1] for shield synthesis. However, the latter approach requires an MDP model of the environment, which for some environments is not necessary, as the necessary forbidden classifiers can easily be handcrafted. In any case, it must be assured that no forbidden classifier is entered incorrectly and that in all situations at least one action remains for XCS to execute.

Even though the concept of using static rules based on domain knowledge to prevent harmful actions from being executed is rather straightforward and closely related to the concept of an external shield, we still argue that it is favorable to internalize such knowledge into the learning mechanism. With an external shield, the safety-critical knowledge embedded into it must still be internalized through learning, potentially wasting time and computing resources. By manually inserting correctly generalized forbidden classifiers, XCS is relieved from the burden of finding adequate generalizations, at least in the safety-critical niches. Thus, forbidden classifiers promise to incorporate the safety guarantees into the classifier population in a minimally intrusive way without "distracting" XCS from the actual problem-solving.

5 Experimental Results

To investigate if forbidden classifiers can hold this promise, our experimental evaluation is mainly guided by two questions:

1. How do forbidden classifiers impact the learning process of XCS, especially that of the GA?
2. Compared to the shielding approach, does the insertion of forbidden classifiers provide XCS with any advantage?

To investigate both questions, we evaluate our approach in three different learning environments that are common in the field of XCS research. Tackling question (1), we first consider the 6-Multiplexer problem, which is a trivial binary classification problem that does not require any safety guarantees. However, it allows for investigating the effects that forbidden classifiers have on the learning mechanisms inside XCS, as the classifier population can easily be interpreted and evaluated because of the small input and action space of the 6-Multiplexer.

Concerning question (2), the use of forbidden classifiers is evaluated in two maze environments, where XCS is navigating a robot in a maze to find a target field with a minimal number of steps. In such robot navigation tasks, it must under all circumstances be avoided that the robot crashes into an obstacle to avoid damage to itself and its environment. We guarantee this by inserting appropriate forbidden classifiers and compare our approach to XCS with an external shield. With Woods1 and Maze4, two maze environments of different complexity are investigated, as Woods1 is a very small environment requiring few steps to reach the target field, while Maze4 is larger and contains more obstacles.

The shielding approach presented in [1] assumes that the learner is capable of updating multiple policies in parallel. Since there exists no straightforward way to implement this for XCS, different approaches to realize an external shield are thinkable. While parameter updates on multiple action sets in parallel are possible, it is, among other aspects, unclear whether the GA that is executed on one action set should insert its offspring classifiers into the other action sets in case they match. If so, the order in which the GA is executed on the different action sets might have an effect on the learning progress. We opted for a shield that provides the smallest, or most negative, reward possible in the environment every time XCS chooses an unsafe action. Afterward, XCS is called again, until a valid action is chosen, i.e. all parameter updates are fully sequential and the shield is not able to force XCS to select a particular action. Since parameter updates take place both during exploration and exploitation, it is guaranteed that eventually a valid action is selected by XCS on its own. With the additional XCS invocations imposed by the shield, XCS is given additional learning opportunities to internalize the knowledge embedded in the shield.

For our experiments, we have employed the Python implementation scikit-XCS [18] and extended it with the ability to solve multi-step problems, which is required for the maze environments. To avoid the problem of overgeneralization, a phenomenon frequently occurring in multi-step environments [8], the specify operator of Lanzi [7] has been employed in the maze environments with $N_{Sp} = 20$ and $P_{Sp} = 0.5$. Apart from a discount factor γ of 0.71, which is only relevant for multi-step problems, and maximum population sizes N tailored to each learning environment, we kept the default parameterization of XCS as specified by scikit-XCS[1]. As exploration strategy, the common ϵ-greedy strategy with an exploration probability p_{explr} of 0.5 has been used. Since the operating performance during exploration trials is not allowing any conclusions about the state of learning, only the performance during exploitation trials is reported. To obtain statistically meaningful results, each experiment has been repeated 100 times with different random seeds, with the problem environments and XCS using separate random number generators.

5.1   6-Multiplexer

The 6-Multiplexer is a classification problem in which XCS is presented six randomly drawn bits as input and has to choose between the two actions $\{0, 1\}$. Of the input bits, the first two represent index bits that point to one of the remaining four bits, which XCS has to output correctly. In case XCS outputs the correct value, it receives a reward of 1,000 and otherwise 0. For the 6-Multiplexer problem, the optimal population consists of 16 classifiers, each having three bits specified – the two index bits and the bit that they point to. Overall, this results in 8 classifiers that propose the correct action, and 8 classifiers proposing the wrong action, which must also be present in the population since XCS learns to accurately predict the reward for all state/action pairs.

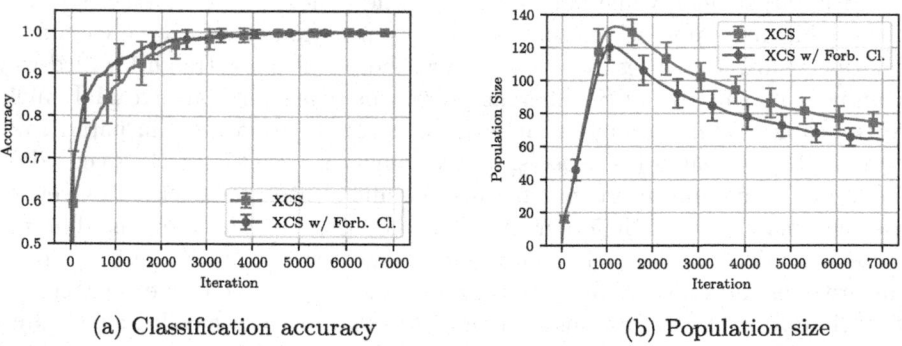

(a) Classification accuracy   (b) Population size

Fig. 3. Experimental results obtained on the 6-Multiplexer problem. Results are averages over 100 trials and shown as a moving average over 200 samples. The error bars visualize the observed standard deviation.

---

[1] That is $\beta = 0.2$, $\alpha = 0.1$, $\nu = 5$, $\mu = 0.04$, $\delta = 0.1$, $p_I = 10$, $\epsilon_I = 0$, $f_I = 0.01$, $P_\# = 0.5$, $\epsilon_0 = 10$, $\chi = 0.8$, $\theta_{GA} = 25$, $\theta_{sub} = 20$, $\theta_{del} = 20$, $\gamma = 0.71$, $DoGaSubsumption = True$, $DoActionSetSubsumption = False$.

To investigate the effect of forbidden classifiers on the learning mechanism of XCS, we initialized the population with the two forbidden classifiers $001\#\#\#\to 0$ and $10\#\#1\#\to 0$, which represent the case that either the first or the third bit is indexed and has the value 1, but action 0 is selected. Since forbidding the wrong action in a binary classification problem inevitably leads to the selection of the correct action, these two forbidden classifiers already provide the classifier population with 25% of the problem solution. With random guessing as the baseline, it can be expected that XCS with forbidden classifiers achieves a classification accuracy that is about 12.5% points better than XCS alone.

Figure 3a shows the development of the classification accuracy both for the standard XCS and XCS with added forbidden classifiers. The maximum population size N has been set to 200 and overall 7,000 iterations have been performed. At the beginning of the experiment, the advantage of XCS with the two forbidden classifiers is indeed around 12.5% points but is reducing quickly in the following iterations as the standard XCS is learning the knowledge represented by the forbidden classifiers as well. At the end of the experiment, both variants of XCS reliably achieve perfect classification accuracy. The size of the macro-classifier population, i.e. the number of distinct classifiers in the population not considering their numerosities n, develops as shown in Fig. 3b. With forbidden classifiers, the population stays consistently smaller, indicating better generalization since the problem can be solved with fewer classifiers.

C	α	n	p	ϵ	F
#0#### :	0	13	1000	0	0.673
11###1 :	0	12	0	0	0.881
11###1 :	1	11	1000	0	0.657
01#1## :	1	11	1000	0	0.773
0##0## :	0	9	1000	0	0.465
$\cdots$					
001### :	0	1	-	-	-
10##1# :	0	1	-	-	-

Fig. 4. Excerpt of an exemplary classifier population that has been evolved by the end of a trial. The classifiers are sorted according to their numerosity n in descending order. Manually added forbidden classifiers are marked red. (Color figure online)

The reason for the improved generalization capabilities becomes apparent when manually inspecting the evolved classifier populations. Figure 4 shows an excerpt of a population that has been evolved through one of the trials. Most distinctly, the classifier $\#0\#\#\#\#\to 0$ dominates the classifier population in terms of numerosity and predicts the correct payoff of 1,000 with a prediction error ϵ of 0, i.e. with perfect accuracy. Normally, it would be considered as overgeneralized, since classifiers that predict the correct payoff in the 6-Multiplexer problem can at maximum be generalized to have 3 bits specified. The only specified bit is the second index bit, meaning that in matching situations the bit to predict is either the first or third bit of the remaining bits. However, in case one of these bits is indexed and has the value 1, one of the forbidden classifier matches and choosing the action 0 is prevented, i.e. whenever the wrong action is proposed, the forbidden classifiers become active and prevent its action from being selected. The mechanism of forbidding actions in certain niches of the environment thus enables XCS to generalize classifiers into these niches, potentially reducing the population size. Hence, the introduction of forbidden classifiers not only acts as a filter for the action selection but can aid XCS' learning process by opening generalization possibilities that normally do not exist.

5.2   Woods1

The Woods1 environment represents a simple robot navigation task and is shown in Fig. 5. The environment contains three different types of fields: Empty fields, obstacles (denoted as rocks), and a single target field, termed food. At the beginning of each run, the robot is placed on a random empty field. The goal of XCS is to learn the shortest path to the food from each field of the map. At each time step, XCS is provided with the types of the eight surrounding fields, each encoded with two bits. As action, XCS has to decide towards which of the eight surrounding fields a step should be made. Upon reaching the food, a reward of 1,000 is provided, and for all other steps a reward of 0. If the map is left, the robot re-enters on the opposite side of the map. Since only the last step that reaches the food is receiving a positive reward, XCS must make use of its multi-step learning capabilities and evolve a chain of classifiers through the discounting mechanism. Since this type of learning is more complex than the single-step learning applied on the 6-Multiplexer, a larger maximum population size N of 800 has been employed. As it is common practice for such learning problems, a maximum number of steps per run has been defined to escape unsuccessful runs, e.g. if the robot is stuck in a loop. If the food has not been reached after 30 steps, the run is prematurely stopped.

To avoid crashing into obstacles, which in a real-world environment could damage the robot or its surroundings, we have introduced eight forbidden classifiers into the population. Due to the generalized conditions of the classifiers, eight forbidden classifiers are fully sufficient, as each classifier is responsible for preventing a crash in one direction, e.g. one forbidden classifier has a condition which matches whenever a rock is northern of the robot and then prevents the

```
. . . . . .
. R R F .
. R R R .
. R R R .
. . . . . .
```

Fig. 5. The Woods1 environment. Empty fields are denoted by dots, while obstacles are represented by rocks ('R'). The target field is the food ('F').

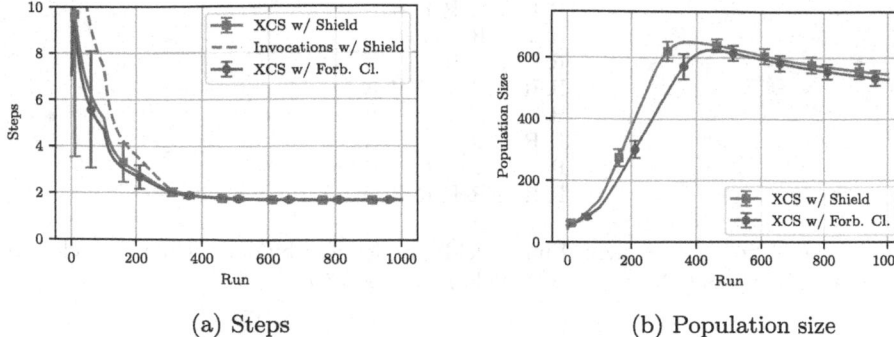

(a) Steps   (b) Population size

Fig. 6. Experimental results obtained on the Woods1 problem. Results are averages over 100 trials and shown as a moving average over 100 runs. The error bars visualize the observed standard deviation.

action "move north". As a comparison, we have employed our shielded version of XCS, where the additional learning iterations that take place when the shield rejects an action do not number among the maximum step limit. This gives the shielded XCS additional learning opportunities to internalize the safety-critical knowledge embedded in the shield.

Figure 6a shows the number of steps needed to reach the food in the Woods1 environment throughout our experiment with 1,000 runs. Both employed variants of XCS show similar behavior and reach close to optimal performance, which lies at around 1.7 steps. That the shortest path is not always found could be related to the parameter configuration of XCS, which has not been specifically optimized for the problem domain. During the earlier runs, the shielded XCS requires slightly more steps than its counterpart, showing that it must first internalize the external knowledge embedded in the shield. To achieve this, it requires considerably more invocations of XCS. Normally, each step is associated with one invocation of XCS, but with the external shield, a step towards an obstacle leads to an invocation as well, even though the step is not actually carried out. The process of internalizing the shield knowledge can also be observed in the development of the population size shown in Fig. 6b, where the population with integrated forbidden classifiers is consistently smaller, with the gap closing toward the end of the experiment. Overall, this indicates that XCS with forbidden classifiers requires a smaller computational budget, as for XCS this mainly depends on the number of invocations and the population size.

5.3 Maze4

To evaluate the approach of introducing forbidden classifiers in a more complex navigation task, the Maze4 environment shown in Fig. 7 has been employed. In contrast to the Woods1 environment, it is larger and thus requires longer paths to reach the food. This makes the problem harder to solve for XCS, as it has to discount the reward for reaching the food over a longer chain of classifiers.

```
R R R R R R R
R . . R . . F R
R R . . R . . R
R R . R . . R R
R . . . . . . R
R R . R . . . R
R . . . . R . R
R R R R R R R R
```

Fig. 7. The Maze4 environment. Empty fields are denoted by dots, while obstacles are represented by rocks ('R'). The target field is the food ('F').

Hence, classifiers that match positions distant to the target field will receive weaker reward signals even if they propose a suitable action. Therefore, the maximum population size N has been set to 1,200.

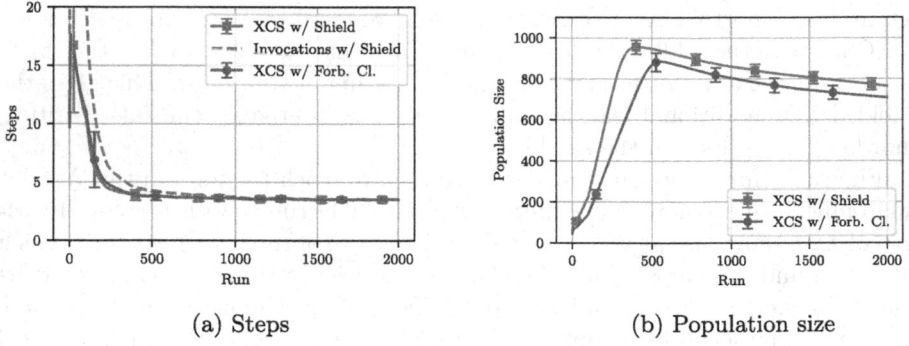

(a) Steps   (b) Population size

Fig. 8. Experimental results obtained on the Maze4 problem. Results are averages over 100 trials and shown as a moving average over 100 runs. The error bars visualize the observed standard deviation.

Figure 8a shows the average number of steps required to reach the food, with the optimum being at 3.5 steps. Both compared variants of XCS depict quick progress at the beginning of the experiment, followed by gradual improvements until the optimal solution is reached at the end of the experiment after 2,000 runs. An inspection of the development of the population size, shown in Fig. 8b, leads to the already seen observation that the use of forbidden classifiers is beneficial for creating smaller classifier populations.

So far, XCS with forbidden classifiers achieved similar results than with an external shield. Common to both evaluated maze environments is that XCS was able to solve the problem in the majority of trials, and since the shielded XCS underwent additional invocations to learn the knowledge embedded in the shield, the use of forbidden classifiers did not seem to offer any advantage in terms of learning speed. However, we noticed during our experimental evaluation that this is no longer the case if XCS is struggling to solve the problem, especially in the

Maze4 environment. Therefore, we conducted two additional experiments in the Maze4 environment, where in one case the specify operator has been disabled, making XCS susceptible to overgeneralization, and in the other case a smaller population has been employed.

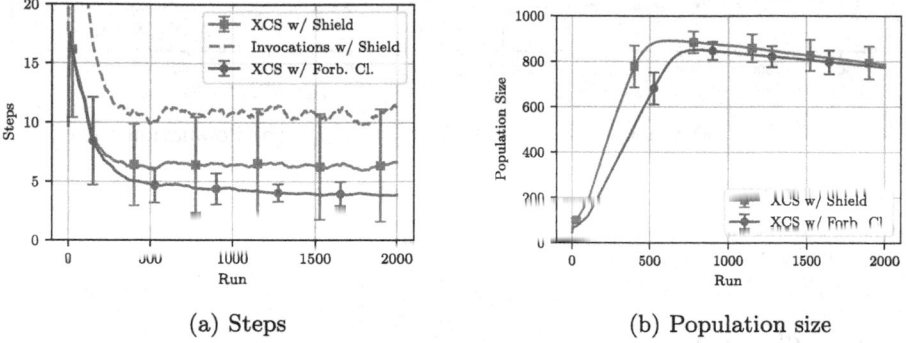

(a) Steps   (b) Population size

Fig. 9. Experimental results obtained on the Maze4 problem with the deactivated specify operator. Results are averages over 100 trials and shown as a moving average over 100 runs. The error bars visualize the observed standard deviation.

Figure 9a shows the steps required to reach the food when the specify operator is disabled and XCS is affected by overgeneralization. With the external shield, XCS is not reaching the optimal solution, at least not reliably, as depicted by the high standard deviation. Further, the gap between taken steps and XCS invocations is not closing. With forbidden classifiers, on the other hand, around 41.7 % fewer steps are needed at the end of the experiment with a standard deviation that is considerably lower. One explanation could be that overgeneralization into the environmental niches covered by the forbidden classifiers is sanitized. This effect was already discussed and evaluated in Sect. 5.1 for the 6-Multiplexer, where new generalization opportunities were created. In the case of an overgeneralized classifier, this could mean that it is no longer overgeneralized, but in the presence of the forbidden classifiers actually represents an accurate and adequately generalized classifier.

When a considerably smaller maximum size of the population is employed, i.e. $N = 400$, the shielded XCS is not able to derive the optimal solution, either, as depicted in Fig. 10a. Again, the insertion of forbidden classifiers enables XCS to reliably evolve a close to optimal solution that requires 27.6 % fewer steps at the end of the experiment. However, both compared variants show similar behavior in terms of actual population size, as shown by Fig. 10b. Apparently, the manually created forbidden classifiers are a more efficient representation of the safety-critical niches than the classifiers that XCS evolves on its own.

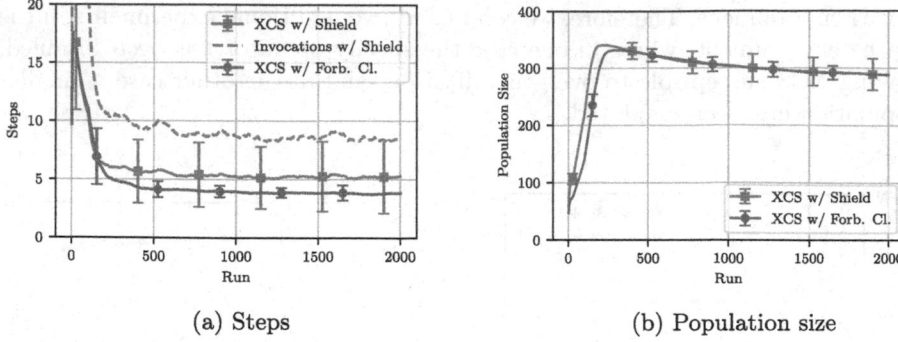

(a) Steps

(b) Population size

Fig. 10. Experimental results obtained on the Maze4 problem with $N = 400$. Results are averages over 100 trials and shown as a moving average over 100 runs. The error bars visualize the observed standard deviation.

6 Conclusion

We have proposed a concept to integrate safety guarantees into the learning classifier system XCS through the injection of domain knowledge in the form of *forbidden classifiers*, thereby leveraging the unique interpretability of XCS' rule base. In contrast to related work, which implements safety guarantees by deploying an external shield or teacher that constantly monitors the behavior of the RL algorithm, our implementation embeds the safety guarantees directly into the classifier population of XCS. Apart from showing that their introduction requires only minor algorithmic modifications to XCS, our experimental evaluation revealed that forbidden classifiers

- open generalization opportunities not present with an external shield, overall leading to smaller classifier populations,
- relieve XCS from the computational burden of internalizing the safety-critical knowledge embedded in an external shield, and
- can considerably improve the operating performance in situations in which XCS is struggling to find the optimal solution.

The last observation might be relevant especially for real-world environments, where it can often be the case that XCS is not able to derive a perfectly accurate payoff estimation for each state/action pair. Therefore, the deployment of forbidden classifiers in application settings is relevant future work. Further, it could be promising to systematically investigate how knowledge in general could be manually added into XCS, e.g. by bootstrapping the population with classifiers that represent a suitable, but not necessarily optimal heuristic. In contrast to the injection of forbidden classifiers, the algorithmic modifications necessary for such an approach are not intuitive, as such added classifiers should provide strong guidance to XCS at the beginning, but not restrict it from finding a better solution in the long term.

References

1. Alshiekh, M., Bloem, R., Ehlers, R., Könighofer, B., Niekum, S., Topcu, U.: Safe reinforcement learning via shielding. In: Proceedings of the AAAI Conference on Artificial Intelligence, vol. 32 (2018)
2. Amodei, D., Olah, C., Steinhardt, J., Christiano, P.F., Schulman, J., Mané, D.: Concrete problems in AI safety. CoRR abs/1606.06565 (2016)
3. Bellman, K., et al.: Self-aware cyber-physical systems. ACM Trans. Cyber-Phys. Syst. 4(4), 1–26 (2020)
4. Butz, M.V., Wilson, S.W.: An algorithmic description of XCS. Soft Comput. 6(3–4), 144–153 (2002)
5. García, J., Fernández, F.: A comprehensive survey on safe reinforcement learning. J. Mach. Learn. Res. 16(42), 1437–1480 (2015)
6. Iqbal, M., Browne, W.N., Zhang, M.: Reusing building blocks of extracted knowl edge to solve complex, large-scale boolean problems. IEEE Trans. Evol. Comput. 18(4), 465–480 (2014)
7. Lanzi, P.L.: A Study of the Generalization Capabilities of XCS. In: Bäck, T. (ed.) Proceedings of the 7th International Conference on Genetic Algorithms, East Lansing, MI, USA, July 19–23, 1997, pp. 418–425. Morgan Kaufmann (1997)
8. Lanzi, P.L.: An analysis of generalization in the XCS classifier system. Evol. Comput. 7(2), 125–149 (1999)
9. Lewis, P.R., Platzner, M., Rinner, B., Tørresen, J., Yao, X. (eds.): Self-aware Computing Systems. Springer International Publishing (2016)
10. Müller-Schloer, C., Schmeck, H., Ungerer, T. (eds.): Organic Computing — A Paradigm Shift for Complex Systems. Springer Basel (2011)
11. Prothmann, H., Tomforde, S., Branke, J., Hähner, J., Müller-Schloer, C., Schmeck, H.: Organic traffic control. In: Organic Computing - A Paradigm Shift for Complex Systems, pp. 431–446. Springer Basel (2011)
12. Stein, A., Tomforde, S.: Reflective learning classifier systems for self-adaptive and self-organising agents. In: 2021 IEEE International Conference on Autonomic Computing and Self-Organizing Systems Companion (ACSOS-C), pp. 139–145 (2021)
13. Stone, C., Bull, L.: For real! xcs with continuous-valued inputs. Evol. Comput. 11(3), 299–336 (2003)
14. Tomforde, S., Brameshuber, A., Hahner, J., Müller-Schloer, C.: Restricted on-line learning in real-world systems. In: 2011 IEEE Congress of Evolutionary Computation, CEC 2011, pp. 1628–1635 (2011)
15. Wilson, S.W.: Classifier fitness based on accuracy. Evol. Comput. 3(2), 149–175 (1995)
16. Wilson, S.W.: Generalization in the XCS classifier system. In: Koza, J., et al. (ed.) Genetic Programming 1998: Proceedings of the Third Annual Conference, pp. 665–674. Morgan Kaufmann, San Franciso (1998)
17. Zeppenfeld, J., Herkersdorf, A.: Applying autonomic principles for workload management in multi-core systems on chip. In: Proceedings of the 8th International Conference on Autonomic Computing, pp. 3–10. ACM, New York (2011)
18. Zhang, R.F., Urbanowicz, R.J.: A scikit-learn compatible learning classifier system. In: Proceedings of the 2020 Genetic and Evolutionary Computation Conference Companion. ACM, New York (2020)

ANN-EMOA: Evolving Neural Networks Efficiently

Steven Künzel$^{(\boxtimes)}$ (iD) and Silja Meyer-Nieberg (iD)

Bundeswehr University Munich, Werner-Heisenberg-Weg 39,
85577 Neubiberg, Germany
{steven.kuenzel,silja.meyer-nieberg}@unibw.de

Abstract. Multi-objective neuroevolution is a research field of growing importance within reinforcement learning. This paper introduces
ANN-EMOA, a novel multi-objective neuroevolutionary algorithm that is
inspired by nNEAT and aims at high efficiency, usability, and comprehensibility. To that end it applies a simple encoding and efficient variation
operators. Diversity plays a key role in evolutionary computation. For
this reason, we apply the Riesz s-energy to foster diversity explicitly.
This paper also develops a new efficient approach to determine the individual Riesz s-energy contribution of each solution within a set. To assess
the performance of the new ANN-EMOA it is compared to nNEAT and NEAT-
MODS, two multi-objective variants of NEAT, in the multi-objective Double Pole Balancing problem. While other domains and more complex test
cases need to be investigated, these promising first results show that ANN-
EMOA does not only converge faster and to higher quality-levels than its
competitors, but it also maintains more compact network-genomes and
shows convincing performance even with comparably small populations.

Keywords: Neuroevolution · Genetic algorithms · Neural networks

1 Introduction

Machine Learning – especially neural networks, see e.g. [1] for an introduction,
represents an important subfield of artificial intelligence. Finding the appropriate network topology for a given task still remains a challenging problem. Owing
to its importance, the problem has a long research tradition. Today the general
area of neural architecture search (NAS) comprises techniques as reinforcement
leaning, gradient-based methods, Bayesian optimization and neuroevolution. Following [2], methods and research in NAS comprise the dimensions: search space
– how are the architectures presented or encoded, search strategy – how is the
space explored, and performance estimation. The contributions of this paper
focus on the first two aspects. The general challenge in finding the appropriate

Supplementary Information The online version contains supplementary material
available at https://doi.org/10.1007/978-3-031-02462-7_26.

J. L. Jiménez Laredo et al. (Eds.): EvoApplications 2022, LNCS 13224, pp. 402–417, 2022.
https://doi.org/10.1007/978-3-031-02462-7_26

network structure becomes even more pronounced when more than one objective must be considered. When these are at least partly conflictive, the task needs to be addressed as a multi-objective problem – a continually active research field in its own right. Here as well in the general area of neural architecture search, maintaining and fostering diversity is of immense importance.

Multi-objective tasks are often approached with evolutionary algorithms or similar techniques, see e.g. [3]. While they often lead to relieable results, the additional problem of finding and maintaining appropriate control parameter settings of these algorithms themselves arises, see for example [4].

This paper focuses on neuroevolution [5] a hybrid technique where neural networks are encoded in genomes which are mutated and recombined throughout the evolutionary process. It develops a novel multi-objective neuroevolutionary algorithm, ANN-EMOA which introduces new, compact genetic encoding for neural networks. Additionally, ANN-EMOA is highly adaptable and usable by following the concept of q-procedures, introduced in [6]. To foster the maintenance of diversity on population level explicitly, a new efficient algorithm to determine the individual Riesz s-energy contribution, recently proposed by [7], of each solution of a set is introduced. To tackle the problem of finding appropriate control parameters ANN-EMOA employs the EARPC [8] algorithm which adapts these dynamically. The main research question that the paper addresses is whether the novel ANN-EMOA can bring advantages over existing algorithms in terms of performance.

This paper is structured as follows: In the following, we present a short literature review concerning the current state of the art – focusing on the encoding and multi-objective neuroevolution. Section 3 introduces ANN-EMOA, a novel multi-objective neuroevolutionary algorithm similar to nNEAT but with a simpler encoding. Furthermore, it has a stronger focus on SMS-EMOA than nNEAT does. This results in an overall simplification and better performance concerning multi-objective optimization problems. In Sect. 4 the multi-objective Double Pole Balancing (moDPB) problem is introduced. Furthermore, an experimental comparison of ANN-EMOA against two reference algorithms is carried out. Note that the experimental analysis is intended as a proof-of-concept. Based on our results ANN-EMOA will be tested on a greater variety of more complex problems in future work. These and further research directions along with a summary of our findings are provided in Sect. 5.

2 Neuroevolution

Neuroevolution comprises several distinct techniques [5]. One of the most prominent representatives of *topology and weight evolving artificial neural networks* (TWEANNs) is the *NeuroEvolution of Augmenting Topologies* (NEAT) algorithm by Stanley and Miikkulainen [9]. It employs a direct encoding of networks where each link is assigned a unique `innovation id`, depending on the time of creation of the link within the evolutionary process [9]. A neural network thus can be described by a list of links, where one gene contains the following information: `innovation id`, `weight`, and `status` (enabled or disabled) of the link.

By innovation ids NEAT allows to identify common and distinct parts of two genomes without prior analysis and provides a solution to the variable genome length [10, p. 20] and competing conventions problems [9]. In a recent paper Stanley et al. [5] provide an overview over neuroevolution, different encodings, trends, and challenges.

There have been different approaches on direct and indirect encoding of neural networks since the appearance of NEAT. Dürr et al. [11] applied an analog genetic encoding which describes a neural network by a string of terminals and non-terminals. While the method is robust and was shown to be superior to NEAT in the Double Pole Balancing task, its implementation is complex and of high computational cost [11]. For example, the recombination of two networks first requires determining their respective common elements by searching similar parts within the genome strings [12]. With HyperNEAT Stanley et al. [13] introduced an indirect encoding variant of NEAT allowing to develop comparably large neural networks with small effort. Subject of evolution are connective compositional pattern-producing networks (CPPNs) which take two coordinates (the neuron ids) as input and return the weight of the corresponding link between those. Based on the weights of all neuron pairs, a neural network phenotype can be decoded [13]. Another indirect encoding has been proposed by van Steenkiste et al. [14]. Their wavelet-based encoding aims at reducing the dimension of the decision space by exploiting spatial regularities. In an experimental comparison on the *Arcade Learning Environment* [15] the novel encoding was able to outperform the CPPN-based HyperNEAT [14].

In practice many problems entail multiple conflicting objectives that cannot be optimized concurrently. A solution then represents a trade-off among these [3, p. 195f]. The concept arising naturally in the context of multi-objective optimization is the Pareto dominance[1] [3, p. 196f]. While this allows to determine whether one solution is better than another, it does not yield an answer for non-dominating solutions. Many evolutionary multi-objective algorithms (EMOA) apply a secondary measure to find a ranking among such solutions [7,16,17]. Effectively, such a measure can be any procedure that assigns a quality-value to a solution, based on its fitness, diversity, or other properties. While NEAT in its default variant [9] is not able to cope with multiple objectives simultaneously, several multi-objective derivatives have emerged in the last years. We briefly introduce three recent algorithms:

NEAT-PS [18] is based on the default NEAT procedure, employing the Pareto Strength approach to translate fitness vectors into fitness scalars. These can be operated by NEAT without any modification.

NEAT-MODS [19] combines NEAT with concepts of NSGA-II [16] for multi-objective optimization and furthermore introduces a new survivor selection procedure, focused on maintaining a high diversity level in the population.

---

[1] An objective vector a dominates another vector b if a is equal or better than b in all objectives and better than b in at least one objective [3, p. 196].

nNEAT [6,20] has been proposed by the authors of this paper. It merges NEAT with the more recent SMS-EMOA [17]. nNEAT aims at high performance and usability, for example achieved by a default procedure for parameter control.

Please note that while today deep neural networks increasingly move into the focus of researchers [21], this work is dedicated to traditional neuroevolution with the aim for developing the smallest possible network architectures. Thereby the dimension of the search space is kept manageable and hence the neural networks can be trained with comparably small computational effort. However, the proof-of-concept in this work can be extended to the more complex case of deep neuroevolution, as will be outlined in Sect. 5.

3   ANN-EMOA

In recent works the combination of neuroevolutionary algorithms like NEAT with evolutionary multi-objective algorithms as SMS-EMOA has been found to lead to promising results [6,19,20,22]. SMS-EMOA, or more precisely the concept behind, can be seen as a general framework for multi-objective optimization [17], independent of the genotype at hand. It is based on a steady-state population of fixed size where one offspring solution is added to the population per epoch. After evaluation, the worst member of the population is identified by non-dominated ranking. If the worst front contains more than one solution, a gradation among those is established by the Hypervolume[2] each solution dominates exclusively [23].

ANN-EMOA is based on nNEAT [6], it evolves recurrent neural networks with the focus on small architectures. As an important aspect ANN-EMOA also relies on the concept of q-procedures for translating multi-dimensional objective vectors into quality scalars. An instance of a q-procedure only requires to define a procedure for translation: It can be non-dominated ranking, a quality indicator [17,24], a diversity measure [7,16] or even a solution-architecture based measure. To resolve potential conflicts of equal q-values, multiple q-procedures can be applied hierarchically as well as concurrently. q-procedures are highly usable as they allow the user to select and combine different measures that determine what *solution quality* means in a uniform way. Thus, different combinations of measures can be applied with small implementation effort. We introduced the concept of q-procedures in [6] and extended it in [20, Section 3.2] to where we refer the reader for further details.

ANN-EMOA extends nNEAT in the following components to foster the usability and efficiency of the algorithm:

- novel variation operators (Sect. 3.1),
- a time-invariant representation (Sect. 3.2),
- promoting diversity explicitly with Riesz s-energy (Sects. 3.3 and 3.4), and
- dynamic parameter control (Sect. 3.5).

---

[2] One of the standard measures in multi-objective optimization, see [17].

3.1   Reduced Complexity in Variation

The variation operators in `ANN-EMOA` are inspired by NEAT [9]. However, the goal is to reduce their complexity and increase the comprehensibility. There are four mutation operators:

Add Link. Adds a link between two unconnected neurons. Links must not end in neurons of the input layer.

Split Link. Selects a link l between neurons s and e. Link l is *removed* and a new neuron n is created. Furthermore, two new links l_1 from s to n and l_2 from n to e are created.

Remove Link. Selects a link to be removed. The following conditions must be fulfilled: Either the link is self-recurrent, i.e., starts and ends in the same neuron. Or there is at least one other non-self-recurrent link that starts at the link's starting neuron. Same for the end neuron. These conditions ensure that the function of the network remains, and no neurons are isolated.

Perturb Link Weights. Randomly modifies the weight(s) of one or more links in a network.

One idea of NEAT [9] was to keep disabled links as these may be beneficial in later stages of the evolutionary process. However, during our research we found that the benefit of disabling links over removing links is negligible. Removing replaced links can avoid unnecessary growth of the genomes without proportional increment of the network's performance, also referred to as bloat [25].

During recombination of two networks `ANN-EMOA` determines all links that occur in both networks. These are called *common* genes. All genes that do only occur in one of the networks are referred to as *non-common* genes. To create an offspring genome, the link weights of common genes are copied randomly from either parent or averaged. Non-common genes are copied from the more fit parent; hence the less fit parent does not contribute its non-common genes to the offspring at all. The only exception is when both parents are found to be equally fit, then all genes are copied into the offspring. That procedure is like the one in NEAT [9].

3.2   A Time-Invariant Encoding/Representation

The initial node-based encoding of neural networks in NEAT [9] is highly efficient and comprehensible. However, it requires a so-called *innovation manager* that is responsible for tracking all topological mutations occurring in the networks. Its main task is to provide any link with its corresponding `id`. It depends on the time of creation of the link. Thereby two networks from different runs of NEAT cannot be compared to each other without accessing information of the innovation managers. While the concept of the innovation manager represents an adequate solution, it also reduces NEAT's comprehensibility. Additionally, in multi-threaded environments its implementation must be thread safe.

For `ANN-EMOA` we do also employ a node-based encoding where the genome of a network only contains the `ids` and `weights` of the links it consists of. A genome neither requires the corresponding starting and ending neuron ids nor a reference to an authority like the innovation manager. This is due to a link id itself allows to be decoded into its corresponding neuron ids[3]. Furthermore, links may not be disabled in ANN-EMOA and thereby genes do not have to store this information. Links are either existing or not present in a genome.

Just as NEAT [9], ANN-EMOA also starts with minimal networks with i input and o output neurons. These are numbered subsequently starting at 0 for the first input neuron and ending at $i + o - 1$ for the last output neuron. A link's id depends on the ids of the neurons it is connecting, take for example a network with $i = 4$ and $o = 2$: The id of the link between neurons 0 (input) and 4 (output) equals $(0, 4)$. Note that the neuron created through the split-operation gets assigned the id of the link it "replaces". If for example the link between neurons 0 and 4 was split, the resulting neuron would take the id $(0, 4)$. Furthermore, two new links: $(0, (0, 4))$ and $((0, 4), 4)$ would be created, connecting the three affected neurons. This naming strategy brings the advantage that every neuron- and link-id is unique and can be reproduced independent of its time of creation. Thereby no innovation manager is necessary. Concerning the implementation of ANN-EMOA the drawback is that ids must be stored as Strings instead of Ints. However, this disadvantage is negligible.

3.3 Fostering Convergence and Diversity

In [20] we suggested a combination of quality and diversity measures: On the one hand non-dominated ranking (NDR) [16] as first and the $R2$ indicator [24] as secondary measure. And on the other hand, the crowding distance (CD) [16] as diversity measure. The combination is denoted as (NDR + $R2$, CD). Although it has led to promising results in [20], the crowding distance is not suitable to determine the diversity in three- or higher dimensional spaces [26]. Recently Falcón-Cardona et al. [7,27] suggested the Riesz s-energy E_s as a considerable and efficient diversity measure. For a set of solutions A and an exponential $s > 0$ it is defined as:

$$E_s(A) = \sum_{a \neq b \in A} d(a, b)^{-s},$$

where $d(a, b)$ represents the Euclidean distance between a and b in objective space [7], and s defines the "contribution pressure"[4]. To determine the individual E_s-contribution of each solution the set-subset-approach is applied. The contribution of a solution $a \in A$ equals:

$$E_s(a) = E_s(A) - E_s(A \setminus \{a\}) \ [7].$$

This results in a total runtime of $O\left(|A|^3\right)$. In the next section we propose an efficient procedure to determine the individual E_s-contribution.

---

[3] However, to save the decoding-effort one could also store this redundant information in the genes.

[4] Higher values of s lead to a more pronounced penalization of smaller distances.

3.4   Determining the Exact E_s-Contribution Efficiently

The idea behind this algorithm is simple: Instead of determining the E_s-value for the full set and its subsets in separate loops, a single iteration maintains an array c of size $|A|$ where the individual contributions are initialized with zero each. An outer loop iterates over the indices $i = 0 \longrightarrow |A| - 2$ and an inner loop over the indices $j = i + 1 \longrightarrow |A| - 1$. Then the squared Euclidean distance[5] d^2 between the objective vectors $A[i]$ and $A[j]$ is determined. If d^2 is smaller than a predefined distance threshold d_{min}, it is set $d^2 = d_{min}$. This is necessary to establish an upper bound for the individual E_s-contribution as will be explained below.

The energy e between $A[i]$ and $A[j]$ equals $e = \left(d^2\right)^{-s}$. The individual contributions $c[i]$ and $c[j]$ are each incremented by e. The larger the resulting value e, the smaller is the Euclidean distance between the two solutions. Hence, a solution with small total value in c is located further from its neighbors on average than a solution with a large total value in c. After the outer loop terminated the worst possible value c_{max} is determined:

$$c_{max} = \left(d_{min}\right)^{-s} \cdot \left(|A| - 1\right).$$

It is applied as upper bound for the c-values of the solutions and thereby enables scaling the resulting contributions between zero and one. The final contribution c of a solution at index i equals:

$$c = 1 - \frac{c[i]}{c_{max}}.$$

Thereby the best solutions, i.e., the ones with larger distance to the remaining solutions will have a contribution close to one while the contribution of solutions in more densely populated regions will approach zero. The total runtime complexity of the algorithm equals $O\left(|A|^2\right)$. Our implementation of the Riesz s-energy will be applied throughout the experiments in this paper, see Sect. 4.

3.5   Parameter Control

The behavior of ANN-EMOA is determined by twelve numeric control parameters, for example the Mutation Probability. Due to space restrictions, we will not provide details here but refer the reader to the supplementary material provided. Aside from nNEAT [6] ANN-EMOA is the first neuroevolutionary algorithm that explicitly controls its parameters automatically employing the EARPC algorithm [8]. This increases the performance of ANN-EMOA and furthermore strongly contributes to its usability as it relieves the user from setting the parameter values. Note that the population size μ is not controlled automatically but determined by the user initially.

---

[5] We employ the squared Euclidean distance as it avoids the computationally expensive sqrt-operation which has no influence on the individual E_s-contribution.

In general, control parameters strongly influence the performance of any genetic algorithm [28, p. 271]. The optimal values do not only depend on the problem instance but also the state of the evolutionary process. For example, a high Mutation Probability may be beneficial in preliminary stages, but later, smaller values may contribute best. Parameter control addresses these issues by dynamically selecting the "best" parameter values. EARPC, a popular parameter control approach has been introduced by Aleti and Moser [8]. It operates on numeric parameters exclusively and divides the range of values into two subranges with individual selection probabilities. Every epoch a new parameter value is drawn randomly from either range. Previous experiments underlined that it contributes positively to the performance of neuroevolutionary algorithms [20, Section 6.3].

3.6 Procedure

The procedure of ANN-EMOA starts by randomly initializing the population with minimal networks. These are evaluated and sorted with the q-procedure provided. Then the main loop is entered: First, the control parameters are updated, followed by λ pairs of solutions being selected via SUS [3, p. 84] to become parents. Then λ offspring is created through the described variation operators[6] and is added to the population. The offspring (or the entire population – depending on the problem at hand) is evaluated and the population is sorted again. Then the worst λ solutions are discarded. If the termination condition is fulfilled, the main loop is left and ANN-EMOA terminates. The remaining components of ANN-EMOA are described in Table 1.

Table 1. Components of ANN-EMOA equal to nNEAT.

Population	Steady-state, fixed size, user-defined
Initialization	Minimal networks, random weights
Fitness functions	User-defined
Parent selection	Rank-based Stochastic Universal Sampling (SUS) [3, p. 84]
Survivor selection	Deterministically, discard worst
Termination	Fixed number of evaluations, user-defined

4 Experimental Analysis

To investigate the performance of the novel ANN-EMOA we apply it on the multi-objective Double Pole Balancing problem (moDPB) [22]: A cart is placed at the center of a track of limited length; two poles are mounted on the cart with a hinge. The neural network controller must decide whether to move the cart to left or

---

[6] Every variation operator is applied with a certain probability, also controlled by EARPC.

right to keep the poles balanced without leaving the track. There are three fitness functions: 1. time in balance, 2. number of directional changes of the cart, 3. mean deviation from the track center. Consult the supplementary material for more information on the fitness functions. While the first fitness function represents the main goal, the two latter objectives force the controller to behave energy efficient [22]. The three fitness functions are conflicting. As reference, we compare ANN-EMOA to NEAT-MODS [19] and nNEAT [6] since these have been identified as the most promising variants in our earlier research. The parameters of the moDPB experiment take the default values defined by Stanley [10, Table A.3 on p. 190]. Initially, the poles are inclined by random angles within $[-9°, 9°]$ to add noise to the experiment. The noise is controlled with multi-objective Standard Error Dynamic Resampling (moSEDR) [29] which evaluates solutions more often when the information about their fitness is less reliable. Every solution is evaluated between three and 15 times. The experiment is carried out in two difficulties: 1. with six inputs to the neural networks, containing the velocities of the poles and the cart, and 2. with three inputs and no velocities being provided. The latter is seen as the more challenging task as it requires the neural network to store the velocities internally. Three different population sizes $\mu \in \{20, 100, 500\}$ are investigated. Every experiment is repeated for 100 times. An experiment terminates after 10,000 (with velocities, easier task) or 50,000 (without velocities, harder task) evaluations, respectively. To compare and assess the performance of ANN-EMOA and the reference algorithms, we consider the following measures:

Average Time to a Solution. How many solutions had to be evaluated on average until a solution of minimum fitness F^* has been found. We define $F^* = \left(f_1 = 0, f_2 = \frac{1}{20}, f_3 = \frac{1}{12}\right)$, i.e., the controller must not fail in balancing the poles at any time (f_1), it is allowed to change the cart movement direction at most every 20th time step (f_2) and it must ensure the cart not to stay away more than 20 cm from the track center on average (f_3).

Success Rate. Partition of the experiment repetitions where a solution of minimum fitness F^* has been found before termination.

Mean Best Quality. Average quality of the final Pareto front. As "quality" we define the covered Hypervolume in this experiment.

Runtime Behavior. Development of the average quality of the Pareto front (here: covered Hypervolume) throughout the evolutionary process.

Mean Genome Size. Average size (neurons and links) of the final Pareto front. Due to space restrictions we forego a detailed architectural analysis. Furthermore, it would be of limited value as the considered algorithms are focused on finding the smallest possible architectures with only a few hidden neurons. Hence we consider the genome size as sufficient.

A statistical analysis will be carried out for all measures except the success rate and the runtime behavior. More information about the considered measures can be found in [20, p. 94 – 97]. We conduct a two-stage test [20, p. 107f]; in both stages the Skillings-Mack test [30] – a general variant of the Friedman test – will be applied. The first stage determines whether there is a statistically significant difference among any pair of the three algorithms. If such a difference was found, the second stage is a pairwise comparison of the algorithms. The null hypothesis says in both stages that there are no significant differences among any algorithms. Finally, the obtained p-values must be corrected to reduce the risk of a false rejection; we employ Shaffer's algorithm [31] to that end. As significance level we define $\alpha = 0.05$. Details about the statistical comparison can be found here [20, p. 107f]. As q-procedures for ANN-EMOA we employ the combination (NDR + R2, RE) where RE represents the Riesz s-energy according to Sects. 3.3 and 3.4. Additionally, the control parameters of ANN-EMOA, nNEAT and NEAT-MODS are dynamically set with EARPC to allow a fair comparison.

Note that the algorithm names are abbreviated in Tables 2 to 5 as follows: nN = nNEAT, MODS = NEAT-MODS and ANN = ANN-EMOA. Additionally, in Tables 4 and 5, small italic numbers in the last line of a cell denote statistically significant differences to the value(s) in the corresponding column(s). The < and > signs denote the relation of the corresponding mean ranks.

Table 2. (Success Rate) Partition of successful repetitions of the experiment in percent. Recall that μ describes the population size. Left: moDPB with velocities (after 10,000 eval.). Right: moDPB without velocities (after 50,000 eval.).

μ	nN	MODS	ANN		μ	nN	MODS	ANN
20	99	88	99		**20**	30	4	65
100	100	68	100		**100**	54	7	83
500	97	31	100		**500**	56	4	57

Table 3. (Average Time to a Solution) Average number of evaluations until a solution of adequate quality had been found. Numbers in brackets result from experiments with success rate < 50%. No statistically significant differences occurred. Left: moDPB with velocities (after 10,000 eval.). Right: moDPB without velocities (after 50,000 eval.).

μ	nN	MODS	ANN		μ	nN	MODS	ANN
20	1307 ± 1031	3369 ± 2238	1443 ± 1080		**20**	(28523) ± 11417	(32459) ± 13150	21984 ± 11562
100	2380 ± 944	4768 ± 2584	2179 ± 922		**100**	21197 ± 11309	(35711) ± 5987	24326 ± 12824
500	6122 ± 1642	(5993) ± 2411	6262 ± 1619		**500**	34433 ± 9195	(38548) ± 6854	32982 ± 10558

The success rates of the algorithms are printed in Table 2. It shows the proportion of experiment runs where a solution of minimum fitness F^* could be found within 10,000 and 50,000 evaluations, respectively. Concerning moDPB with velocities nNEAT and ANN-EMOA reached success rates close to 100 %. NEAT-

Table 4. (Mean Best Quality) Average Hypervolume covered by the final Pareto front. Left: moDPB with velocities (after 10,000 eval.). Right: moDPB without velocities (after 50,000 eval.).

μ	nN	MODS	ANN
20	0.994 ± 0.048 $2^>$	0.973 ± 0.093 $1^<, 3^<$	0.998 ± 0.008 $2^>$
100	0.999 ± 0.001 $2^>$	0.924 ± 0.18 $1^<, 3^<$	0.999 ± 0.001 $2^>$
500	0.996 ± 0.005 $2^>, 3^<$	0.73 ± 0.29 $1^<, 3^<$	0.997 ± 0.005 $1^>, 2^>$

μ	nN	MODS	ANN
20	0.763 ± 0.312 $2^>, 3^<$	0.248 ± 0.293 $1^<, 3^<$	0.893 ± 0.227 $1^>, 2^>$
100	0.862 ± 0.269 $2^>, 3^<$	0.451 ± 0.377 $1^<, 3^<$	0.977 ± 0.017 $1^>, 2^>$
500	0.929 ± 0.155 $2^>$	0.637 ± 0.282 $1^<, 3^<$	0.955 ± 0.085 $2^>$

Table 5. Average topology (quantity) of solutions belonging to the final Pareto front (N. = Neurons, L. = Links.). Left: moDPB with velocities (after 10,000 eval.). Right: moDPB without velocities (after 50,000 eval.).

μ		nN	MODS	ANN
20	N.	7.3 $2^<, 3^>$	8.0 $1^>, 3^>$	7.1 $1^<, 2^<$
	L.	6.9 $2^<, 3^>$	9.0 $1^>, 3^>$	6.3 $1^<, 2^<$
100	N.	7.1 $2^<, 3^>$	8.9 $1^>, 3^>$	7.0 $1^<, 2^<$
	L.	6.4 $2^<, 3^>$	10.9 $1^>, 3^>$	6.1 $1^<, 2^<$
500	N.	7.1 $2^<, 3^>$	7.7 $1^>, 3^>$	7.0 $1^<, 2^<$
	L.	6.2 $2^<, 3^>$	7.7 $1^>, 3^>$	6.0 $1^<, 2^<$

μ		nN	MODS	ANN
20	N.	5.2 $2^<$	6.7 $1^>, 3^>$	5.1 $2^<$
	L.	6.8 $2^<$	10.4 $1^>, 3^>$	6.4 $2^<$
100	N.	5.3 $2^<, 3^>$	8.2 $1^>, 3^>$	4.7 $1^<, 2^<$
	L.	7.3 $2^<, 3^>$	14.6 $1^>, 3^>$	5.5 $1^<, 2^<$
500	N.	5.0 $2^<, 3^>$	7.3 $1^>, 3^>$	4.7 $1^<, 2^<$
	L.	6.3 $2^<, 3^>$	11.8 $1^>, 3^>$	5.0 $1^<, 2^<$

MODS was successful more often in case of smaller populations. The more difficult moDPB without velocities shows a deterioration of NEAT-MODS success rates to circa five percent. ANN-EMOA always succeeds in most experiment runs for $\mu \leq 100$. In case of $\mu = 500$ nNEAT and ANN-EMOA were successful similar often. Table 3 provides an overview how many solutions had to be evaluated until a solution of fitness F^* has been found. Only successful experiment runs are considered. In moDPB with velocities the difference between nNEAT and ANN-EMOA is negligible, both do also show a small standard deviation. However,

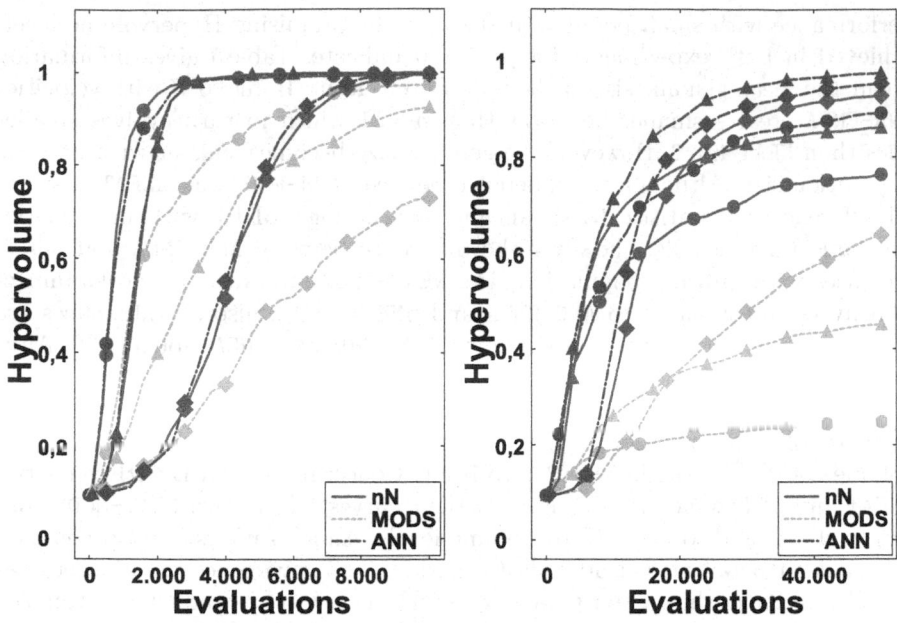

Fig. 1. (Runtime Behavior) Average Hypervolume covered by the Pareto front after the corresponding number of evaluations. Left: moDPB with velocities. Right: moDPB without velocities. Markers: Circle ($\mu = 20$), Triangle ($\mu = 100$), Diamond ($\mu = 500$).

NEAT-MODS requires the most evaluations in all cases and has a larger standard deviation. Concerning the moDPB without velocities only ANN-EMOA has a success rate of more than 50 % for $\mu = 20$. In case of $\mu = 100$ nNEAT requires slightly fewer evaluations than ANN-EMOA, however nNEAT's success rate is considerably lower. When the success rates of both are similar, as it is the case for $\mu = 500$, ANN-EMOA requires fewer evaluations than nNEAT, but with a larger standard deviation. Altogether the differences between nNEAT and ANN-EMOA are negligible, NEAT-MODS always requires most iterations and has the lowest success rate. A general trend observable is that the number of evaluations grows with increasing population size. This can be explained: In a large population more offspring is created per epoch. Hence, fewer epochs are carried out within the same number of evaluations. Table 4 depicts the mean Hypervolume covered by the final Pareto front. While in the moDPB with velocities all algorithms, except NEAT-MODS for $\mu = 500$, reached levels > 0.9, the situation is different in moDPB without velocities: Hypervolume levels around 0.9 are only achieved by ANN-EMOA and nNEAT (only $\mu \geq 100$). It is discernible that all algorithms, except ANN-EMOA for $\mu = 500$, achieve better Hypervolume levels with increasing population size. Additionally, ANN-EMOA and nNEAT reach statistically significant better Hypervolume levels than NEAT-MODS in all cases. Furthermore, the difference between ANN-EMOA and nNEAT is significant for $\mu = 500$ in moDPB with velocities and $\mu < 500$ in moDPB without velocities. ANN-EMOA exhibits sufficient performance with small population sizes, as the promising Hypervolume levels achieved in both experiments for $\mu \leq 100$ indicate. Table 5 gives information about the mean genome size of the final Pareto front. In moDPB with velocities ANN-EMOA creates smaller networks than nNEAT which in turn evolves smaller ones than NEAT-MODS. However, the genome size becomes smaller with growing population size. Although the difference between ANN-EMOA and nNEAT is small, all differences are statistically significant. Concerning moDPB without velocities the same ranking holds, however the difference between ANN-EMOA and nNEAT increases with growing population size. NEAT-MODS grows the networks unnecessarily large compared to ANN-EMOA and nNEAT. Here, also all differences are significant with the only exception being ANN-EMOA and nNEAT for $\mu = 20$. Altogether ANN-EMOA can keep the average genome size comparably small. Especially in moDPB with velocities often only the link weights were modified without any other topological mutations. Fig. 1 shows the development of the mean Hypervolume covered by the known Pareto front. Concerning moDPB with velocities nNEAT and ANN-EMOA converge fastest, their curves are similar. NEAT-MODS converges slower and to lower Hypervolume levels. All algorithms have in common that smaller populations lead to faster and higher convergence than larger ones do. This is due to the greater number of epochs that can be conducted within the same number of evaluations. In moDPB without velocities all algorithms reach lower Hypervolume levels, NEAT-MODS shows the most pronounced differences to the first experiment. All algorithms profit from larger populations which have led to the best Hypervolume levels here. However, ANN-EMOA achieves the best level with $\mu = 100$, followed by $\mu = 500$. Its worst final Hypervolume level is still better than the one approached by nNEAT for $\mu \leq 100$. These findings underline that ANN-EMOA is promising with comparably small populations.

The findings of the two experiments are that although the difference between ANN-EMOA and nNEAT often is small, ANN-EMOA performed best. It does not only achieve the best mean qualities and success rates but also finds networks of comparably small size. We assume that this is due to our decision to allow links being removed instead of disabled. Furthermore, ANN-EMOA seems to be very efficient with small population sizes. NEAT-MODS has resulted in the poorest performance in both experiments. Especially in moDPB without velocities its performance has dropped significantly. We assume that this is due to the speciation procedure of NEAT-MODS which depends on the weights and topologies of the networks as well as certain control parameters. These are crucial for the number of resulting species and thereby the diversity of the population. The overall procedure is prone to errors by that. However, maintaining high diversity levels is crucial in moDPB without velocities.

5   Summary and Outlook

Neural architecture search represents an important task in artificial intelligence. This paper focused on neuroevolution and introduced ANN-EMOA, a novel multi-objective neuroevolutionary algorithm. Its main advantages, compared to the reference algorithms nNEAT and NEAT-MODS, are that its procedure is simpler and achieves better results in shorter time in the multi-objective Double Pole Balancing experiments carried out in this paper. Due to the new variation operators, it creates much smaller network genomes on average which allows a faster evaluation and thereby increases the speed of the evolutionary process in practice. ANN-EMOA also has been found to be very efficient with small populations. Therefore, it represents a crucial step to make neuroevolution more practicable and applied in science and the industry. However, the limited experimental analysis does not allow to draw general conclusions. To that end, ANN-EMOA will be tested on different and more complex tasks in future. Another important direction for future work will be the application of ANN-EMOA as a deep neuroevolutionary algorithm. The effectiveness of deep neuroevolution has already been shown before [32,33]. Further future work entails e.g., improvements concerning the usability, for example cetting user-defined parameters like the population size μ automatically or introducing dynamic termination conditions, see for example [34]. Also, the proposed algorithm to determine the individual Riesz s-energy contribution of solutions will be further improved: A first step is to adjust the minimum considered distance d_{min} automatically.

References

1. Haykin, S.: Neural Networks and Learning Machines, 3rd. vol. 3, p. 906 (2008). ISBN: 9780131471399
2. Elsken, T., Metzen, J.H., Hutter, F.: Neural architecture search: a survey. J. Mach. Learn. Res. **20** (2019). ISSN: 15337928. arXiv: 1808.05377
3. Introduction to Evolutionary Computing. NCS, Springer, Heidelberg (2015). https://doi.org/10.1007/978-3-662-44874-8
4. Katoch, S., Chauhan, S.S., Kumar, V.: A review on genetic algorithm: past, present, and future. Multimed. Tools Appl. **80**(5), 8091–8126 (2020). https://doi.org/10.1007/s11042-020-10139-6
5. Stanley, K.O., et al.: Designing neural networks through neuroevolution. Nature Mach. Intell. **1**(1), 24–35 (2019). https://doi.org/10.1038/s42256-018-0006-z. ISSN: 25225839
6. Künzel, S., Meyer-Nieberg, S.: Coping with opponents: multi-objective evolutionary neural networks for fighting games. Neural Comput. Appl. **32**(17), 13885–13916 (2020). https://doi.org/10.1007/s00521-020-04794-x
7. Falcón-Cardona, J.G., Coello, C.A.C., Emmerich, M.: CRI-EMOA: a pareto-front shape invariant evolutionary multi-objective algorithm. In: Lecture Notes in Computer Science, vol. 11, 411. 2019, pp. 307–318 (2019). 12598–1. https://doi.org/10.1007/978-3-030-_25. ISBN: 9783030125974
8. Aleti, A., Moser, I.: Entropy-based adaptive range parameter control for evolutionary algorithms. In: GECCO 2013 - Proceedings of the 2013 Genetic and Evolutionary Computation Conference (2013). https://doi.org/10.1145/2463372.2463560
9. Stanley, K.O., Miikkulainen, R.: Evolving neural networks through augmenting topologies. Evol. Comput. **10**(2), 99–127 (2002). https://doi.org/10.1162/106365602320169811. ISSN: 10636560

10. Stanley, K.O.: Efficient evolution of neural networks through complexification. Ph.D. thesis. The University of Texas at Austin, p. 227 (2004). http://nn.cs.utexas.edu/keyword?stanley:phd04

11. Dürr, P., Mattiussi, C., Floreano, D.: Neuroevolution with analog genetic encoding. In: Runarsson, T.P., Beyer, H.-G., Burke, E., Merelo-Guervós, J.J., Whitley, L.D., Yao, X. (eds.) PPSN 2006. LNCS, vol. 4193, pp. 671–680. Springer, Heidelberg (2006). https://doi.org/10.1007/11844297_68

12. Mattiussi, C.: Evolutionary synthesis of analog networks. Ph.D. thesis (2005). https://doi.org/10.5075/epfl-thesis-3199

13. Stanley, K.O., D'Ambrosio, D.B., Gauci, J.: A hypercube- based encoding for evolving large-scale neural networks. Artif. Life 15(2), 185–212 (2009). https://doi.org/10.1162/artl.2009.15.2.15202. ISSN: 10645462

14. Van Steenkiste, S., et al.: A wavelet-based encoding for neuroevolution. In: GECCO 2016 - Proceedings of the 2016 Genetic and Evolutionary Computation Conference. 2016, pp. 517–524. https://doi.org/10.1145/2908812.2908905. ISBN: 9781450342063

15. Bellemare, M.G., et al.: The arcade learning environment: an evaluation platform for general agents. J. Artif. Intell. Res. 47, 253–279 (2013). https://doi.org/10.1613/jair.3912. ISSN: 10769757

16. Deb, K., et al.: A fast and elitist multiobjective genetic algorithm: NSGA-II. IEEE Trans. Evol. Comput. 6(2), 182–197 (2002). https://doi.org/10.1109/4235.996017. ISSN: 1089778X

17. Emmerich, M., Beume, N., Naujoks, B.: An EMO algorithm using the hypervolume measure as selection criterion. In: Coello Coello, C.A., Hernández Aguirre, A., Zitzler, E. (eds.) EMO 2005. LNCS, vol. 3410, pp. 62–76. Springer, Heidelberg (2005). https://doi.org/10.1007/978-3-540-31880-4_5

18. Van Willigen, W., Haasdijk, E., Kester, L.: A multi-objective approach to evolving platooning strategies in intelligent transportation systems. In: GECCO 2013 - Proceedings of the 2013 Genetic and Evolutionary Computation Conference, pp. 1397–1404 (2013). https://doi.org/10.1145/2463372.2463534. ISBN: 978-1-45031-963-8

19. Abramovich, O., Moshaiov, A.: Multi-objective topology and weight evolution of neuro-controllers. In: 2016 IEEE Congress on Evolutionary Computation, CEC 2016. 2016, pp. 670–677 (2016). https://doi.org/10.1109/CEC.2016.7743857. ISBN: 9781509006229

20. Künzel, S.: Evolving artificial neural networks for multi-objecitve tasks. Ph.D. thesis. Bundeswehr University Munich (2021). https://doi.org/10.13140/RG.2.2.19743.07843, https://athene-forschung.unibw.de/138617

21. Dargan, S., Kumar, M., Ayyagari, M.R., Kumar, G.: A survey of deep learning and its applications: a new paradigm to machine learning. Archives Comput. Methods Eng. 27(4), 1071–1092 (2019). https://doi.org/10.1007/s11831-019-09344-w

22. Künzel, S., Meyer-Nieberg, S.: Evolving artificial neural networks for multi-objective tasks. In: Sim, K., Kaufmann, P. (eds.) EvoApplications 2018. LNCS, vol. 10784, pp. 671–686. Springer, Cham (2018). https://doi.org/10.1007/978-3-319-77538-8_45

23. Beume, N., Naujoks, B., Emmerich, M.: SMS-EMOA: multiobjective selection based on dominated hypervolume. Europ. J. Oper. Res. 181(3), 1653–1669 (2007). https://doi.org/10.1016/j.ejor.2006.08.008. ISSN: 0377-2217

24. Hansen, M.P., Jaszkiewicz, A.: Evaluating the quality of approximations to the non-dominated set. In: IMM Technical Report IMM-REP-1998-7 (1998)

25. Trujillo, L., et al.: Neat genetic programming: controlling bloat naturally. Inf. Sci. **333**, 21–43 (2016). https://doi.org/10.1016/j.ins.2015.11.010. ISSN: 00200255
26. Kukkonen, S., Deb, K.: Improved pruning of non-dominated solutions based on crowding distance for bi-objective optimization problems. In: 2006 IEEE Congress on Evolutionary Computation, CEC 2006. 2006, pp. 1179–1186 (2006). https://doi.org/10.1109/cec.2006.1688443. ISBN: 0780394879
27. Falcon-Cardona, J.G., Ishibuchi, H., Coello Coello, C.A.: Exploiting the trade-off between convergence and diversity indicators. In: 2020 IEEE Symposium Series on Computational Intelligence, SSCI 2020. 2020, pp. 141–148 (2012). https://doi.org/10.1109/SSCI47803.2020.9308469. ISBN: 9781728125473
28. Doerr, B., Neumann, F. (eds.): Theory of Evolutionary Computation. Natural Computing Series. Springer, Cham (2020) 978–3-030-29413-7. https://doi.org/10.1007/978-3-030-29414-4
29. Siegmund, F., Ng, A.H.C., Deb, K.: Standard error dynamic resampling for preference-based evolutionary multi-objective optimization. Technical report COIN Laboratory, Michigan State University (2016)
30. Skillings, J.H., Mack, G.A.: On the use of a friedman-type statistic in balanced and unbalanced block designs. Technometrics **23**(2), 171–177 (1981). https://doi.org/10.1080/00401706.1981.10486261. ISSN: 15372723
31. Shaffer, J.P.: Modified sequentially rejective multiple test procedures. J. Am. Stat. Assoc. **81**(395) (1986). https://doi.org/10.1080/01621459.1986.10478341. ISSN: 0162–1459
32. Miikkulainen, R., et al.: Evolving deep neural networks. In: Artificial Intelligence in the Age of Neural Networks and Brain Computing (2018). https://doi.org/10.1016/B978-0-12-815480-9.00015-3
33. Such, F.P., et al.: Deep neuroevolution: genetic algorithms are a competitive alternative for training deep neural networks for reinforcement learning. In: December 2018. arXiv: 1712.06567
34. Liu, Y., Zhou, A., Zhang, H.: Termination detection strategies in evolutionary algorithms: a survey. In: GECCO 2018 - Proceedings of the 2018 Genetic and Evolutionary Computation Conference, pp. 1063–1070 (2018). https://doi.org/10.1145/3205455.3205466. ISBN: 9781450356183

Augmenting Novelty Search
with a Surrogate Model to Engineer
Meta-diversity in Ensembles of Classifiers

Rui P. Cardoso[1(✉)], Emma Hart[2], David Burth Kurka[1], and Jeremy Pitt[1]

[1] Imperial College London, London, UK
{rui.cardoso,d.kurka,j.pitt}@imperial.ac.uk
[2] Edinburgh Napier University, Edinburgh, Scotland
e.hart@napier.ac.uk

Abstract. Using Neuroevolution combined with Novelty Search to promote behavioural diversity is capable of constructing high-performing ensembles for classification. However, using gradient descent to train evolved architectures during the search can be computationally prohibitive. Here we propose a method to overcome this limitation by using a *surrogate* model which estimates the behavioural distance between two neural network architectures required to calculate the sparseness term in Novelty Search. We demonstrate a speedup of 10 times over previous work and significantly improve on previous reported results on three benchmark datasets from Computer Vision—CIFAR-10, CIFAR-100, and SVHN. This results from the expanded architecture search space facilitated by using a surrogate. Our method represents an improved paradigm for implementing horizontal scaling of learning algorithms by making an explicit search for diversity considerably more tractable *for the same bounded resources.*

Keywords: Diversity · Ensemble · Novelty search · Surrogate

1 Introduction

Ensemble performance is fundamentally dependent on both the accuracy of individual base learners and the diversity between them [4]. However, techniques to promote diversity are typically only implicit, such as training the models on different subsets of the data or starting from different random initialisations. In previous work [2], we proposed a method that *explicitly* searched for diversity amongst a set of base learners by making use of metrics for measuring *behavioural* diversity. However, a fundamental limitation of this approach was its computational complexity, with a costly step of training all the neural network models in the population at each step of the search. Such time and computational demands compromise the goal of our approach, which is to develop learning algorithms which scale horizontally, namely with models which can be distributed across

© Springer Nature Switzerland AG 2022
J. L. Jiménez Laredo et al. (Eds.): EvoApplications 2022, LNCS 13224, pp. 418–434, 2022.
https://doi.org/10.1007/978-3-031-02462-7_27

many low-cost machines. The claims to tackling the unwieldiness of Deep Learning (DL) algorithms with more scalable solutions were undermined by the fact that our approach was so computationally intensive.

In order to overcome the costly step of training each model, we have introduced a surrogate model [15] into our method. We use this surrogate model, pretrained on a sample drawn from the search space of neural network architectures, to get an estimate of the *error* distance between two neural networks given architectural descriptors, *without* training these networks. Whereas this calculation had previously been a very costly step, this technique renders it essentially instantaneous. This produces a speedup of 10 times compared to the previous approach when the same parameters are used, *without loss of performance*. By changing the parameters to expand the search space of neural network architectures we have considerably improved on previous results reported on three benchmark datasets from Computer Vision—CIFAR-10, CIFAR-100, and SVHN. Using a surrogate model has enabled us to search a wider space of neural network architectures and run the Novelty Search procedure for longer.

The major contribution of this paper is that it proposes an improved paradigm for implementing horizontal scaling of learning algorithms. Explicitly creating diversity amongst the members of an ensemble establishes a sound criterion for distributing these models. By improving the method with a surrogate model in the way described above, our approach makes an explicit search for diversity considerably more tractable *for the same bounded resources*.

2   Background

In previous work [2], we proposed a method that *explicitly* searched for behavioural diversity amongst a set of base learners. This used a Novelty Search (NS) algorithm in conjunction with Neuroevolution, in which novelty was determined by novel metrics that explicitly measured behavioural diversity. The evolved behaviourally diverse ensembles outperformed both their individual learners and ensembles created with techniques that only implicitly promote diversity. This work also enabled us to study and compare different definitions of diversity. However, a fundamental limitation was its computational complexity. In order to calculate the behavioural distance between two models, we need to compare the classification errors that they make on a validation data set. This requires first training each member of the current population of neural network models on a training data set with gradient descent at each iteration of the NS. If computational resources are limited, this very time-consuming step can be prohibitive. This poses significant challenges because it restricts the search to only a few iterations at best and renders the problem intractable at worst. Here we overcome this difficulty by augmenting the NS with a *surrogate model*.

Combining an evolutionary algorithm (EA) with a surrogate modelling function has been common in the literature for many years, e.g. in single-objective optimisation [19], multi-objective optimisation [14], and particularly in expensive optimisation [23]. A first surrogate model for neural network optimisation was introduced by Gaier *et al.* [5] and used in conjunction with the NEAT [16] algorithm for evolving the weights and topology of a neural network. This paper used a surrogate distance-based model, employing a genotypic compatibility distance metric that is part of NEAT. The approach has been quickly adopted in the literature using a range of surrogates and a variety of methods to evolve networks. There are several examples of approaches that use surrogates to estimate the performance of an architecture. For example, in 2017 Deng *et al.* proposed the Peephole algorithm [3], which predicted the performance of a convolutional neural network based on its architecture information: a long-short term memory (LSTM) neural network was used to train the model. Stork *et al.* [17] extended a Cartesian Genetic Programming method called CPGANN to evolve neural networks using surrogate-based optimisation to reduce the number of fitness evaluations required. They used a Kriging model [9] as the surrogate. In [18], a Random Forest algorithm (RF) was used as a surrogate to predict the performance of a CNN architecture—the authors proposed a method for describing a CNN as a set of features which were used as input to the RF. In [15], the authors use a surrogate benchmark for neural architecture search (NAS).

In contrast, Hagg *et al.* [7] introduce a more flexible method for building a surrogate model that is independent of network topology: rather than describing the neural network architecture, they introduce a *phenotypic* metric which measures the difference in output between two neural networks given the same input sequence. The difference is used in a Kriging surrogate model. Our proposed approach is conceptually closest to that of Hagg. For a given neural network, we calculate a behavioural vector that describes its behaviour on a dataset (see Sect. 3.2). We then propose a RF surrogate model that is used to estimate the distance between the behavioural vectors produced by any two neural networks, as this value is required to drive a NS algorithm.

3   Methods and Materials

We use NS to evolve an ensemble of behaviourally diverse neural network models. The NS operates over a space of architectures defined by a set of hyperparameters. Unlike our previous work [2], where the neural networks in a generation had to be trained with gradient descent at each iteration of the NS in order to calculate the behavioural distances between each pair of architectures, these distances are now *estimated* by a surrogate model which is pretrained on a sample drawn from the space of neural network architectures. The most diverse models are added to the final ensemble, which is then trained on the input data. We evaluate the method against our previous method and compare the performance obtained with different diversity metrics. The following subsections go into detail about each of these steps.

3.1   Neural Network Architectures

The architectures evolved by our procedure are *residual neural networks* [8] based on the wide architectures proposed by [22]. They are of the same kind as those we used in our previous work. Figure 1a shows a generic neural network and Fig. 1b illustrates a generic residual block. Please refer to our previous paper [2] for a more detailed description of these architectures.

The *hyperparameters* of each network are evolved by NS. Each individual in the population is defined by a variable-length vector, depending on the number of blocks r: $[J, C, O^1, ..., O^r, D^1, ..., D^r]$, where J is a Boolean value indicating whether the network should be trained jointly or separately if it is in the final ensemble, C is the output size of the first convolution, O^i is the output size of block i, and D^i its dropout probability. Each individual is mapped to a Pytorch module [13] for implementation purposes. The *parameters* of each network are randomly initialised and then optimised by a standard gradient descent procedure.

In order to preprocess the input to the surrogate model, we *normalise* the representation described above in the following way: we first *rescale* the elements in all positions so that they lie between 0 and 1. The first element is the Boolean value indicating whether the neural network should be trained jointly or separately, so it need not be normalised. Then, given that the representations have variable length depending on the number of residual blocks in each neural network, we *pad* the vector so that it has fixed length, corresponding to the maximum possible number of residual blocks, by adding an appropriate number of elements equal to 0 before the sequence of block output sizes and before the sequence of dropout probabilities. Therefore, if the number of residual blocks in the network is r and the maximum number of blocks is R; if the maximum and minimum sizes of the first convolution in the network are C_{max} and C_{min}, respectively; if the maximum and minimum sizes of each residual block are O_{max} and O_{min}, respectively; and if the maximum and minimum dropout probability of each block are D_{max} and D_{min}; then the normalised representation of neural network m_i is:

$$\mathbf{norm\_rep}_i = \left[J_i, \frac{C_i - C_{min}}{C_{max} - C_{min}}, \right.$$
$$0, ..., 0, \frac{O_i^{R-r} - O_{min}}{O_{max} - O_{min}}, ..., \frac{O_i^R - O_{min}}{O_{max} - O_{min}},$$
$$\left. 0, ..., 0, \frac{D_i^{R-r} - D_{min}}{D_{max} - D_{min}}, ..., \frac{D_i^R - D_{min}}{D_{max} - D_{min}} \right] \quad (1)$$

Where there are $R-r$ elements equal to 0 before the sequence of block output sizes and before the sequence of dropout probabilities.

3.2   Diversity Metrics

In order to calculate novelty scores, which are used as the objective function by the NS, we have considered six different diversity metrics, five of which we have defined ourselves. These metrics are calculated between each pair of individual neural network architectures. We have used three of these metrics in the previous version of our procedure [2].

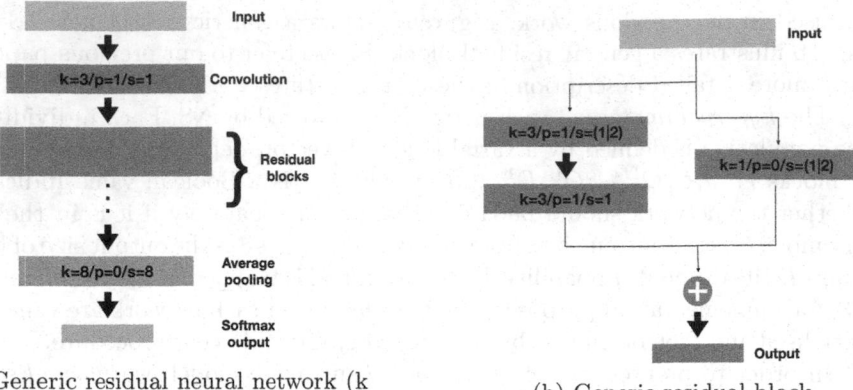

(a) Generic residual neural network (k = kernel size; p = padding; s = stride)

(b) Generic residual block

Fig. 1. Generic topology of individual neural networks

Let $\boldsymbol{y}_i$ be the vector of predictions for model m_i with each prediction y_i^n for data point $\boldsymbol{x}^n$ being a class label in $\{1..C\}$. Let $\boldsymbol{p}_i$ be a binary vector where $p_i^n = 1$ if the prediction y_i^n is correct and $p_i^n = 0$ otherwise. Let N^{11}, N^{00}, N^{01}, and N^{10}, respectively, be the total number of test instances where two models are both correct, both incorrect, and when one is correct and the other is not. The first diversity metric we consider is the *proportion of different errors* between two models when at least one of them is *correct*. We expect it to provide insight into the divergence between the errors made by two models. It is defined as:

$$\text{prop}_{i,j}^1 = \frac{N^{01} + N^{10}}{N^{11} + N^{01} + N^{10}} \qquad (2)$$

The second diversity metric we consider is very similar and is the *proportion of different errors* between two models when at least one of them is *incorrect*. We have first proposed this metric in our previous work [2], defining it as:

$$\text{prop}_{i,j}^2 = \frac{N^{01} + N^{10}}{N^{00} + N^{01} + N^{10}} \qquad (3)$$

The third metric we propose is the *harmonic mean* between these two proportion metrics. This is a sound way of averaging the two proportion metrics into a single metric so that they are both taken into account. It is defined as:

$$\text{prop}_{i,j}^{\text{harm}} = \frac{2 \cdot \text{prop}_{i,j}^1 \cdot \text{prop}_{i,j}^2}{\text{prop}_{i,j}^1 + \text{prop}_{i,j}^2} \qquad (4)$$

We also consider a widely used metric (e.g. [10,12,20]) defined as the *disagreement* between two models, i.e. the proportion of test instances where one is correct and the other is not. We take this metric into account since it expresses how commonly two models disagree on any test instance. It is defined as:

$$\text{dis}_{i,j} = \frac{N^{01} + N^{10}}{N^{00} + N^{01} + N^{10} + N^{11}} \qquad (5)$$

Consider now the *two's complement* of the binary vector of correct predictions p_i, w_i, i.e. the binary vector of *wrong* predictions. The next metric we propose is the *cosine distance* between the binary vectors of wrong predictions made by two models m_i and m_j. Like $\text{prop}^1_{i,j}$ and $\text{prop}^2_{i,j}$, we consider this metric because it is a measure of the distance between the errors made by two models. We have defined it as:

$$\cos\text{-}dist_{i,j} = 1 - \frac{w_i \cdot w_j}{\|w_i\|\|w_j\|} \tag{6}$$

At last we consider a metric of *architectural* diversity. Take the *normalised* vector which represents each individual neural network, as described in Sect. 3.1. Let its size be L. To obtain an architectural representation, we simply remove the first element from the normalised representation, i.e. the Boolean value indicating whether or not the neural network should be trained separately or jointly. Thus, referring to Eq. 1, the architectural representation of model m_i is:

$$\text{arch\_rep}_i = \text{norm\_rep}_i^{\{1..L-1\}} \tag{7}$$

We then define *architectural distance* between neural networks m_i and m_j as the cosine distance between their normalised architectural representations:

$$\text{arch\_dist}_{i,j} = 1 - \frac{\text{arch\_rep}_i \cdot \text{arch\_rep}_j}{\|\text{arch\_rep}_i\|\|\text{arch\_rep}_j\|} \tag{8}$$

These metrics determine the *behavioural distance* between two neural network models, which is used to calculate the novelty scores that guide the NS procedure, as explained in Sect. 3.5. Note that the metrics $\text{prop}^2_{i,j}$ and $\cos\text{-}dist_{i,j}$ focus more closely on the instances where the models made a *prediction error*. In our previous work [2], these two metrics have led to better performance than the others. Here we are interested in learning whether the same pattern can be observed with our new improved version of the NS procedure.

3.3 Surrogate Model to Estimate Distances

The NS requires novelty scores to be determined, which in turn require the distances between pairs of neural networks in the current population to be calculated. However, calculating the exact distance values between two neural network models entails first training the models on the input data with gradient descent and then evaluating them on a validation dataset, as we did in previous work [2]. This can be a very costly step if computational resources are limited, which constrains the NS to only a few iterations and the population to a small size—as the neural networks have to be trained in parallel for efficiency. Here we overcome this limitation by pretraining a Random Forest [1] surrogate model which estimates the behavioural distances between a pair of neural network models.

Note that the estimates of behavioural distances produced by the surrogate model do not need to be very accurate. This is because, when calculating the novelty score of a particular individual neural network, we only need to know relative distances in order to determine nearest neighbours. This means that the surrogate model need only capture the general trends of growth of the distance values, even if the actual values are not very precise. This makes the use of a surrogate model very appropriate with no need for a very complex model. Figure 2 shows the differences between the previous method for calculating exact distance values, shown in Fig. 2a, and the current method using a surrogate model, shown in Fig. 2b. Calculating exact distance values is a very costly step, potentially requiring several GPU hours depending on the length of training. In contrast, estimating these distances by means of the surrogate model is an instantaneous process, once the surrogate model has been trained on sample data beforehand.

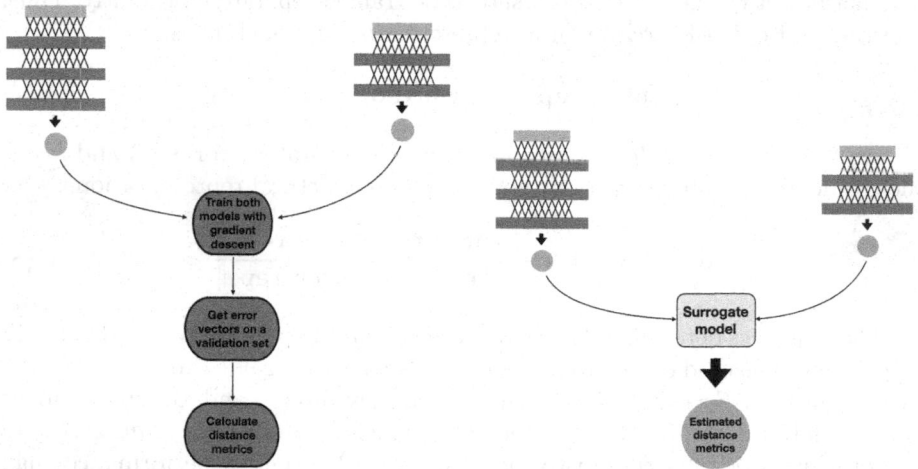

(a) Steps required to calculate exact distance values between two neural networks

(b) Estimating the distance values between two neural networks with a surrogate model

Fig. 2. Difference between calculating and estimating distance values

3.4   Pretraining the Surrogate Model

The surrogate model must be trained beforehand so that it can be used effectively during the NS to estimate the distance values between two neural network models. To do this, we draw a sample of neural networks from the search space of architectures defined by the set of hyperparameters used with the NS method. We first train each of these neural networks with gradient descent and calculate their error vectors on a validation data set. We then build random pairs of neural networks and calculate the exact distance values, for all six metrics considered, between them as a function of either their error vectors or their architectural descriptors, as explained in Sect. 3.2. Finally, we construct a data set on which we fit a Random Forest regressor [1] which takes as input the normalised representations of two neural network architectures, as per Sect. 3.1, and has six

---

Algorithm 1. Pretraining the surrogate model on sample architectures

    draw a sample S of neural networks from the search space defined by J, $[C_{min}, C_{max}]$, $[O_{min}, O_{max}]$, and $[D_{min}, D_{max}]$;

    train(S);   ▷ Models trained jointly or separately according to the value of J_i

    for neural network model $m_i \in S$ **do**

        get error vector e_i on validation set $\mathcal{D}_{val}$;

    end for

    build $\frac{\|S\|^2 - \|S\|}{2}$ unique pairs of neural networks;

    initialise dataset $\mathcal{D}_{dists} \leftarrow \emptyset$;

    for each pair m_i, m_j **do**

        $\boldsymbol{d} \leftarrow$ all 6 distance values;   ▷ calculated as per Section 3.2

        norm_rep$_i$, norm_rep$_j$ are the normalised representations of m_i and m_j;

        add data point $x \leftarrow \{\text{norm\_rep}_i, \text{norm\_rep}_j, \boldsymbol{d}\}$ to $\mathcal{D}_{dists}$;

    end for

    train random forest model rf on $\mathcal{D}_{dists}$;

    return random forest model rf;

---

outputs: the estimates of the distance values for all six metrics considered. We have selected a Random Forest model due to its low complexity and because we expect it to generalise well on new data, given that it is an ensemble model. Algorithm 1 describes the process of training this surrogate model in pseudocode.

3.5 Novelty Search Algorithm

Our algorithm for building an ensemble implements NS as described by [11], applying it to our problem domain. Algorithm 2 presents the pseudocode for this procedure. The original training data is split into two sets, one for training and one for validation. The training set is used to train the final ensemble; it is also used to train the sample of neural network architectures drawn from the search space that is in turn used to pretrain the surrogate model. Whereas training each of these sample neural networks makes use of the entire training set, pretraining the surrogate model only requires the validation set, which is used to calculate exact distance values between pairs of neural networks.

Selection in NS is driven by the novelty score, which computes the sparseness at any point in the behavioural space. This sparseness is defined by one of the distance metrics of Sect. 3.2. Areas with denser clusters of visited points are considered less novel and therefore rewarded less. This is defined as the average distance to the K-nearest neighbours of a point, calculated with respect to the other individuals in the current generation and to a stored *archive* of previously sampled solutions. Hence, the novelty score is calculated as:

$$NS_i = \frac{1}{k} \sum_{k=0}^{K} \text{div\_metric}(m_i, \mu_k) \tag{9}$$

Where μ_k is the kth-nearest neighbour of m_i with respect to the diversity metric div_metric$_{i,j}$, selected from the metrics defined in Sect. 3.2.

Algorithm 2. Ensemble evolution through NS

---

randomly initialise population *pop*;
$archive \leftarrow \emptyset$;
$elite\_archive \leftarrow \emptyset$;
draw $\mathcal{D}_{train}$ and $\mathcal{D}_{val}$ from training set $\mathcal{D}$;
set evolution iterations *epochs*;
set archive sample size n_A;
set final ensemble size *ensemble_size*;
surrogate model s_{div} pretrained as per Section 3.3;
select diversity div_metric$_{i,j}$ from Section 3.2;
for *epochs* **do**
    for $m_i, m_j \in pop \times pop \cup archive : m_i \neq m_j$ **do**
        div_metric$_{i,j} \approx s_{div}(m_i, m_j)$
    end for
    for $m_i, m_j \in pop \times elite\_archive$ **do**
        div_metric$'_{i,j} \approx s_{div}(m_i, m_j)$
    end for
    for $m_i \in pop$ **do**
        $NS_i \leftarrow \frac{1}{k} \sum_{k=0}^{K}$ div_metric(m_i, μ_k) ▷ Equation 9
        $NS'_i \leftarrow \sum_{m_j \in elite\_archive}$ div_metric$'(m_i, m_j)$
    end for
    $sample \leftarrow$ random_sample(pop, n_A)
    $archive \leftarrow archive \cup sample$
    $el\_best \leftarrow \max(pop, NS'_i)$
    $elite\_archive \leftarrow elite\_archive \cup \{el\_best\}$
    $s \leftarrow$ tournament_select(pop, NS_i)
    $pop \leftarrow$ mutate(s)
end for
for $m_i, m_j \in elite\_archive \times elite\_archive : m_i \neq m_j$ **do**
    div_metric$^*_{i,j} \approx s_{div}(m_i, m_j)$
end for
for $m_i \in elite\_archive$ **do**
    $NS^*_i \leftarrow \sum_{m_j \in elite\_archive : m_i \neq m_j}$ div_metric$^*(m_i, m_j)$
end for
$ensemble \leftarrow \max(elite\_archive, NS^*_i, ensemble\_size)$
train$(ensemble)$;   ▷ Models trained jointly or separately according to the value of J_i

---

Individuals are selected for reproduction on the basis of their novelty scores using a tournament selection procedure. In the interests of promoting divergence and avoiding convergence, reproduction only uses mutation. Mutation either adds or removes a randomly chosen residual block from an individual, modifying input/output sizes at the mutation point as necessary; changes the output size and dropout probability of a random block; or swaps two consecutive blocks chosen at random.

After evaluating the entire population, n_A randomly chosen individuals are added to the *archive*, following the method suggested in [6]. In addition, the individual from the population with the highest *elite score*, calculated in a similar fashion to the novelty score, is added to an *elite archive*. After running the NS for the specified number of iterations, a subset of this elite archive is selected as the final ensemble. This subset is chosen so as to maximise the average distance amongst its members. The final ensemble is then trained by gradient descent, the *only time* when this parameter optimisation takes place.

3.6   Evaluation of Evolved Ensembles

In order to evaluate the performance of the evolved ensemble, we use the stacking technique [21], which trains a linear model to weight the predictions of each individual learner. This linear model is trained for a configurable number of iterations on the validation set mentioned in Sect. 3.5. This is to avoid overfitting the test set.

3.7 Baseline: Previous Method

The approach we present here is an improvement of the method that we first proposed in [2]. We use this as a baseline against which we compare the new method. In its main aspects, the previous method is similar to the new method, with the notable difference that it does not make use of a surrogate model to estimate the distance between two neural network models. As discussed previously, this original method calculates exact distances between each pair of neural networks by first training all the models in the current generation with gradient descent and then getting their error vectors by evaluating them on a validation set. As an additional difference, the previous method would calculate at each iteration an *ensemble selection metric* for each member of the population and then add to the final ensemble the single best-scoring neural network in each generation. The new method maintains an *elite archive*, to which a sample of neural networks from each generation with highest novelty score with respect to this archive, which we call *elite score*, is added at each iteration; novelty scores with respect to the final elite archive are calculated at the end for each of its members and this *ensemble score* is used to select a subset of neural networks which will make up the final ensemble.

We compare the new method proposed in this paper with our previous approach along two main lines. Firstly, we seek to understand whether there is a speedup with the new method as a result of increased efficiency when looking for solutions of similar complexity to those found with the previous method. Secondly, we investigate whether the new method can be used to produce better solutions, i.e. solutions of higher complexity and leading to better final performance. This means that we are interested in investigating whether there is both a quantitative improvement, i.e. being able to do *more of* what could be done with the previous method thanks to a more efficient use of computational resources, and a qualitative improvement, i.e. being able to do *more than* what could be done with the previous method by tackling solutions that were previously unfeasible or intractable.

4 Experiments

This section describes the two sets of experiments carried out for comparing the new NS method, which makes use of a surrogate model to estimate the distance between models as described previously, with the previous method, which instead calculates exact distance values by first training all the models in the population with gradient descent and then determining their error vectors on a validation set. We compare the methods based on resource usage, namely runtime, for similar parameter settings and model complexity, as well as on their ability to scale to larger search spaces and search for more complex models.

4.1   Test Set 1: Resource Usage for Similar Complexity

In this set of tests, we investigate the total time required to run each of the two methods when they are looking for *solutions of the same complexity* and running for the same number of iterations. We wish to determine the speedup that can be gained with the new method, which makes use of a surrogate model to overcome the need for training all the models in the current generation with gradient descent in order to calculate novelty scores. We run both the new and the previous methods on CIFAR-10 and fix the parameters, as shown in the second column of Table 1. We conjecture that in these conditions our new method not only results in a speedup due to the use of a Random Forest surrogate model, but also outputs ensembles of similar performance. This is expressed by Hypotheses 1 and 2.

Hypothesis 1 (Runtime of previous NS method vs new method enhanced with a surrogate model). *Enhancing the NS procedure with a Random Forest surrogate model pretrained to estimate the distance between models, and thereby their novelty scores, results in a speedup compared with our previous method, which calculates exact distance values and novelty scores, when constructing ensembles of the same complexity.*

Hypothesis 2 (Performance achieved with the previous NS method vs the new method with a surrogate model). *When looking for solutions of the same complexity, the new NS procedure, using a surrogate model, outputs ensembles which do not perform worse than those constructed by our previous method, even though the new method only* estimates *distance values and novelty scores.*

4.2   Test Set 2: Expanding the Search Space

Using a surrogate model to speed up the procedure has enabled us to both search for solutions of higher complexity and run the NS for longer. In this set of experiments, we apply the new method to three benchmark datasets from the Computer Vision (CV) literature—CIFAR-10, CIFAR-100, and SVHN—and test it with all diversity metrics previously defined in Sect. 3.2. We also compare the results achieved with the new method to the best results observed with the previous method. The parameters that we use with the new method are shown in the third column of Table 1; they correspond to the *expanded* search space made possible by the use of a surrogate model. We expect to see further evidence of what we observed in previous work [2] regarding *error diversity* metrics, namely that those diversity metrics which focus more closely on the instances where the models make prediction errors lead to higher-performing ensembles. This is expressed by Hypothesis 3. We also expect the new method to lead to higher-performing ensembles than those constructed with the previous method, since the use of a surrogate model makes it feasible to expand the search space and run the NS for longer. This is expressed by Hypothesis 4.

Table 1. Novelty search parameters for both test sets

Parameter	Test set 1: runtime comparison (both methods)	Test set 2: expanded search space (new method only)
Iterations	10	100
Final ensemble size	11	40
Population size	30	100
Diversity metric	$\cos\_dist_{i,j}$	All from Sect. 3.2
Number of blocks	2:6	2:6
Number of channels in the first convolution	4:16	4:16
Number of channels in residual blocks	24:32	16·64
Dropout probability in residual blocks	0.1:0.4	0.1:0.9
Number of neighbours K	3	15
Size n_A of archive sample	5	10
Size of tournament for selection	10	50

Hypothesis 3 (Better performance with metrics that focus on error instances). *In a similar fashion to what we have observed with our previous method, running the NS procedure with the distance metrics that focus more closely on the instances where the models make prediction errors leads to higher-performing ensembles than when more generic diversity metrics are employed.*

Hypothesis 4 (Performance achieved with the previous NS method vs the new method with a surrogate model). *The new NS method enhanced with a surrogate model makes it possible to search a larger space of more complex neural network architectures and, therefore, outputs higher-performing ensembles than the best ones constructed by our previous method.*

5   Results and Discussion

In this section, we present the results of the two sets of experiments described in Sect. 4. We then discuss these results and whether the hypotheses formulated above can be rejected.

5.1   Hypothesis 1

Table 2 shows the median value, calculated after 10 independent runs, of the time required to run both the previous NS method and the new method, which makes use of a surrogate model, with the same parameters. These results show that the new method is about 10 times faster than the original NS method.

Table 2. Median results over 10 runs of the previous NS method and the new NS method with a surrogate model on CIFAR-10 (test set 1)

Runtime of NS	48760.5 s
Runtime of NS with surrogate model	4871 s
Training a sample of architectures	28970.5 s
Building a dataset and training the Random Forest surrogate model	18113.5 s
Accuracy achieved by NS	82.245%
Accuracy achieved by NS with surrogate model	83.885%

A Mann-Whitney significance test shows that this difference is significant at the 1% level. This supports the claim of Hypothesis 1 that enhancing the NS method with a Random Forest surrogate model to estimate the distances between models speeds up the search for diverse models and the construction of a diverse ensemble. For reference, we also report in Table 2 the median time, over 10 runs, required to train a sample of 40 neural network architectures on CIFAR-10, as well as to build a dataset and train the Random Forest surrogate model as per Algorithm 1. Note that these two runtimes are a *one-off cost* and that, in order to pretrain the surrogate model for our experiments, we have trained a total of 3200 sample architectures by running several processes in parallel on a cluster, each training 40 architectures.

5.2   Hypothesis 2

Table 2 also shows the median accuracy, calculated after 10 independent runs, achieved by ensembles constructed by both the previous NS method and the new method, when these are executed with the same parameters. The results show that the ensembles constructed by the new method do not perform worse than those constructed by the original method, which calculates exact values for the distance metrics and novelty scores. In fact, we observe that the new method leads to slightly better performance. A Mann-Whitney significance test shows that this difference is significant at the 1% level. This corroborates Hypothesis 2, which claims that there is no loss in performance when using the new method and its surrogate estimates. Besides the use of surrogate models, the major difference between the previous and the new method is the way a subset of all the models is selected to be in the final ensemble. As explained before, the previous method applies an *ensemble selection metric* at each iteration of the NS, whereas the new method keeps an *elite archive*, from which the final ensemble is selected in an additional step at the end of the procedure. It seems that the ensemble selection procedure of the new method is the cause behind the better performance achieved by its ensembles.

Table 3. Median accuracy over 10 runs of ensembles constructed by the new method (test set 2). Best results highlighted. Best results with the original NS shown for comparison

Dataset	Diversity metric	Final ensemble accuracy	Best accuracy with original NS (from [2])
CIFAR-10	$prop^1_{i,j}$	67.295%	83.51%
	$prop^2_{i,j}$	90.605%	
	$prop^{harm}_{i,j}$	83.975%	
	$dis_{i,j}$	86.28%	
	$cos\_dist_{i,j}$	90.11%	
	$arch\_dist_{i,j}$	80.4%	
CIFAR-100	$prop^1_{i,j}$	28.725%	45.42%
	$prop^2_{i,j}$	63.05%	
	$prop^{harm}_{i,j}$	63.41%	
	$dis_{i,j}$	63.18%	
	$cos\_dist_{i,j}$	63.035%	
	$arch\_dist_{i,j}$	49.83%	
SVHN	$prop^1_{i,j}$	78.825%	91.435%
	$prop^2_{i,j}$	94.8%	
	$prop^{harm}_{i,j}$	89.775%	
	$dis_{i,j}$	90.675%	
	$cos\_dist_{i,j}$	94.79%	
	$arch\_dist_{i,j}$	90.68%	

5.3   Hypothesis 3

Table 3 shows the median accuracy, after 10 runs, of ensembles evolved by the new NS procedure extended with a surrogate model, for all six diversity metrics of Sect. 3.2 and all three datasets considered. We observe that on CIFAR-10 and SVHN, the metrics $prop^2_{i,j}$ and $cos\_dist_{i,j}$ lead to the highest-performing ensembles. Mann-Whitney tests show that the difference to the other metrics is statistically significant. On CIFAR-100, this is observed additionally with the metrics $prop^{harm}_{i,j}$ and $dis_{i,j}$.

The metrics $prop^2_{i,j}$ and $cos\_dist_{i,j}$ are the two that focus more closely on the instances where the two models being compared make prediction errors. Additionally, the metric $prop^{harm}_{i,j}$ depends on the value of $prop^2_{i,j}$. These observations back the claim of Hypothesis 3 that error diversity metrics lead to better-performing ensembles compared to more generic diversity metrics. This confirms what we observed in our previous work [2].

5.4 Hypothesis 4

The last column of Table 3 shows the best performance achieved by ensembles evolved with our previous NS method. These results show very clearly that the new method constructs higher-performing ensembles than our previous procedure, with the most considerable difference being observed on CIFAR-100 and CIFAR-10. Mann-Whitney tests reveal that, for each dataset, the difference between the best results achieved by the new method and the best achieved by the previous method is indeed statistically significant. This difference results from the fact that the new method, thanks to its use of a surrogate model, is able to *search a wider space of neural network architectures*, even though it runs on *the same bounded resources*. We conclude that this supports Hypothesis 4.

6 Conclusions and Future Work

This paper has extended on previous work [2], which proposed an innovative NS method to build behaviourally diverse ensembles of classifiers. The previous method had signposted an innovative way to construct high-performing ensembles by explicitly searching for diversity. However, its application in practice had been hampered by limitations in the amount of available computational resources, since it involved a time-consuming step of training all networks in each generation of the NS with gradient descent. Our new method overcomes this limitation by using a pretrained surrogate model to estimate the distance between neural network architectures, necessary to calculate novelty scores, without the need to train them. In this way, we can obtain an approximate speedup of 10 times w.r.t. the previous method when running them both with the same parameters, *without loss of classification accuracy*. We can also construct better-performing ensembles thanks to the expanded architecture search space facilitated by using a surrogate. We have confirmed previous observations that error diversity metrics lead to better-performing ensembles than more generic metrics.

Our method thus represents an improved paradigm for implementing horizontal scaling of learning algorithms. It makes an explicit search for diversity considerably more tractable than our original approach *for the same bounded resources*. In future work, we will extend the current method by implementing a local competition (LC) variant, so that it is possible to include objectives of accuracy into the NS. This will enable us to further study the relationship between diversity and classification accuracy and to investigate trade-offs between the two. We will also propose more definitions of diversity with new and improved metrics, which will provide more insight into what makes a good diversity metric that fits the task of constructing a diverse high-performing ensemble.

References

1. Breiman, L.: Random forests. Mach. Learn. **45**(1), 5–32 (2001)
2. Cardoso, R.P., Hart, E., Kurka, D.B., Pitt, J.V.: Using novelty search to explicitly create diversity in ensembles of classifiers. In: Proceedings of the Genetic and Evolutionary Computation Conference, GECCO 2021, pp. 849–857. Association for Computing Machinery, New York (2021)
3. Deng, B., Yan, J., Lin, D.: Peephole: predicting network performance before training. arXiv preprint arXiv:1712.03351 (2017)
4. Dietterich, T.G.: Ensemble methods in machine learning. In: Kittler, J., Roli, F. (eds.) MCS 2000. LNCS, vol. 1857, pp. 1–15. Springer, Heidelberg (2000). https://doi.org/10.1007/3-540-45014-9_1
5. Gaier, A., Asteroth, A., Mouret, J.B.: Data-efficient neuroevolution with kernel-based surrogate models. In: Proceedings of the Genetic and Evolutionary Computation Conference, pp. 85–92 (2018)
6. Gomes, J., Mariano, P., Christensen, A.L.: Devising effective novelty search algorithms: a comprehensive empirical study. In: GECCO 2015 - Proceedings of the 2015 Genetic and Evolutionary Computation Conference (2015)
7. Hagg, A., Zaefferer, M., Stork, J., Gaier, A.: Prediction of neural network performance by phenotypic modeling. In: Proceedings of the Genetic and Evolutionary Computation Conference Companion, GECCO 2019, pp. 1576–1582. Association for Computing Machinery, New York (2019)
8. He, K., Zhang, X., Ren, S., Sun, J.: Deep residual learning for image recognition. In: Proceedings of the IEEE Conference on Computer Vision and Pattern Recognition, pp. 770–778 (2016)
9. Chilès, J.-P., Desassis, N.: Fifty years of kriging. In: Daya Sagar, B.S., Cheng, Q., Agterberg, F. (eds.) Handbook of Mathematical Geosciences, pp. 589–612. Springer, Cham (2018). https://doi.org/10.1007/978-3-319-78999-6_29
10. Kuncheva, L.I., Whitaker, C.J.: Measures of diversity in classifier ensembles and their relationship with the ensemble accuracy. Mach. Learn. **51**, 181–207 (2003). https://doi.org/10.1023/A:1022859003006
11. Lehman, J., Stanley, K.O.: Abandoning objectives: evolution through the search for novelty alone. Evol. Comput. **19**, 189–223 (2011)
12. Pasti, R., De Castro, L.N., Coelho, G.P., Von Zuben, F.J.: Neural network ensembles: immune-inspired approaches to the diversity of components. Nat. Comput. **9**(3), 625–653 (2010)
13. Paszke, A., Gross, S., Chintala, S., et al.: Automatic differentiation in PyTorch. In: Advances in Neural Information Processing Systems 32 (2019)
14. Ruan, X., Li, K., Derbel, B., Liefooghe, A.: Surrogate assisted evolutionary algorithm for medium scale multi-objective optimisation problems. In: Proceedings of the 2020 Genetic and Evolutionary Computation Conference, pp. 560–568 (2020)
15. Siems, J., Zimmer, L., Zela, A., et al.: NAS-Bench-301 and the case for surrogate benchmarks for neural architecture search (2020)
16. Stanley, K.O., Miikkulainen, R.: Evolving neural networks through augmenting topologies. Evol. Comput. **10**(2), 99–127 (2002)
17. Stork, J., Zaefferer, M., Bartz-Beielstein, T.: Improving NeuroEvolution efficiency by surrogate model-based optimization with phenotypic distance kernels. In: Kaufmann, P., Castillo, P.A. (eds.) EvoApplications 2019. LNCS, vol. 11454, pp. 504–519. Springer, Cham (2019). https://doi.org/10.1007/978-3-030-16692-2_34

18. Sun, Y., Wang, H., Xue, B., et al.: Surrogate-assisted evolutionary deep learning using an end-to-end random forest-based performance predictor. IEEE Trans. Evol. Comput. **24**, 350–364 (2019)
19. Tong, H., Huang, C., Minku, L.L., Yao, X.: Surrogate models in evolutionary single-objective optimization: a new taxonomy and experimental study. Inf. Sci. **562**, 414–437 (2021)
20. Van Krevelen, R.: Error diversity in classification ensembles. Ph.D. thesis (2005)
21. Wolpert, D.H.: Stacked generalization. Neural Netw. **5**, 241–259 (1992)
22. Zagoruyko, S., Komodakis, N.: Wide residual networks (2016)
23. Zhou, Z., Ong, Y.S., Nair, P.B., et al.: Combining global and local surrogate models to accelerate evolutionary optimization. IEEE Trans. Syst. Man Cybern. Part C (Appl. Rev.) **37**(1), 66–76 (2006)

Neuroevolution of Spiking Neural P Systems

Leonardo Lucio Custode🆔, Hyunho Mo🆔, and Giovanni Iacca$^{(\boxtimes)}$🆔

Department of Information Engineering and Computer Science, University of Trento, Trento, Italy
giovanni.iacca@unitn.it

Abstract. Membrane computing is a discipline that aims to perform computation by mimicking nature at the cellular level. Spiking Neural P (in short, SN P) systems are a subset of membrane computing methodologies that combine spiking neurons with membrane computing techniques, where "P" means that the system is intrinsically parallel. While these methodologies are very powerful, being able to simulate a Turing machine with only few neurons, their design is time-consuming and it can only be handled by experts in the field, that have an in-depth knowledge of such systems. In this work, we use the Neuroevolution of Augmenting Topologies (NEAT) algorithm, usually employed to evolve multi-layer perceptrons and recurrent neural networks, to evolve SN P systems. Unlike existing approaches for the automatic design of SN P systems, NEAT provides high flexibility in the type of SN P systems, removing the need to specify a great part of the system. To test the proposed method, we evolve Spiking Neural P systems as policies for two classic control tasks from OpenAI Gym. The experimental results show that our method is able to generate efficient (yet extremely simple) Spiking Neural P systems that can solve the two tasks. A further analysis shows that the evolved systems act on the environment by performing a kind of "if-then-else" reasoning.

Keywords: Neuroevolution · NEAT · Membrane computing · Spiking P systems · OpenAI Gym

1 Introduction

Membrane computing is a branch of natural computing initiated by Păun in 1998 [1]. The goal of membrane computing is to perform computations by emulating nature at the *cellular* level. In the area of membrane computing, membrane systems (also called P systems) indicate models that have parallel and distributed computation capability. Spiking Neural P (in short, SN P) systems [2,3] incorporate the idea of spiking neurons (and spike trains) into P systems. SN P systems, differently from other combinations of neural models and P systems [4], use *time* as a source of information in the computation, similarly to what happens in

© Springer Nature Switzerland AG 2022
J. L. Jiménez Laredo et al. (Eds.): EvoApplications 2022, LNCS 13224, pp. 435–451, 2022.
https://doi.org/10.1007/978-3-031-02462-7_28

biological brains. Moreover, it has been proved that a SN P system can simulate a Turing machine, given a sufficient number of neurons [5–7].

While SN P systems have proven to be applicable to a wide variety of problems, their design is (almost always) currently done manually by an expert. Very few attempts have been made at automating this step [8,9]. The lack of automatic design methodologies, of course, represents a *bottleneck* in the development of the field, as a lot of work (and time) is needed to design such systems.

Here, we employ a well-known neuroevolutionary algorithm, namely the Neuroevolution of Augmenting Topologies (NEAT) [10], to automatically design SN P systems for a given task. More specifically, we modify the original NEAT algorithm to handle the parameters of a specific type of SN P systems by increasing the number of parameters contained in the genotype and adapting them to the parameters of this type of neurons.

To the best of our knowledge, our work represents the first attempt to use neuroevolution to *fully* design SN P systems. In fact, while other approaches for the automatic design of SN P systems do exist, they limit the parameters that can be optimized, e.g., by fixing the topology [8] or, by fixing the rules [9]. Our approach, instead, allows to optimize all the SN P system's parameters (except for the number of rules) simultaneously. In this sense, it reduces the need of experts for designing such systems, which in turn may foster a broader applicability of the SN P systems.

In a nutshell, the goal of this paper is to address two main research questions:

- Is it possible to evolve SN P systems by using (a modified version of) the NEAT algorithm?
- How do SN P systems evolved with NEAT compare to other methodologies for classic control tasks?

Our results on two classic control tasks, namely MountainCar-v0 and CartPole-v1 from OpenAI Gym [11], show that our approach is able to produce SN P systems of good quality that are competitive with the state of the art.

The rest of the paper is structured as follows. The next section introduces the background concepts on SN P systems. Section 2 summarizes the related work, followed by Sect. 4, which describes the proposed method to evolve SN P systems. Section 5 presents the numerical results and their analysis. Finally, Sect. 6 draws the conclusions of this work and suggests future works.

2   Background

In the neurophysiological behavior of biological neurons, a neuron transmits an electric pulse, a spike, via its synapses. In particular, the spiking neurons considered in our study carry information by means of the number and the timing of the spikes rather than the size and the shape of each spike, assuming that all the spikes of a spiking neuron are identical.

A P system is a computing device based on the progression of objects in a membrane structure initialized with a specific number of objects in each membrane. The system then operates using the rules present in the membrane until its computation is finished. After finishing the computation, the result is provided as the number of objects in each membrane.

A neural-like P system is a P system that has its compartments structured as in a neural net. The behavior of a neural-like P system is based on the state of its neurons and their interactions. Finally, SN P systems are a class of neural-like P systems that apply the idea of spiking neurons.

In the following, we describe the formal definition of a SN P system with the same notions of regular languages used in [2]. A standard SN P system of degree $m \geq 1$, is formally defined as follows:

$$\Pi - (O, \upsilon_1, \ldots, \upsilon_m, syn, l_{in}, l_{out}) \tag{1}$$

where:

- $O = a$ is a singleton alphabet, where a represents a spike;
- $\sigma_1, \ldots, \sigma_m$ are neurons, each one defined as: $\sigma_i = (n_i, R_i)$, $1 \leq i \leq m$, being n_i the number of spikes initially present in σ_i and R_i a finite set of rules in σ_i, respectively;
- syn is the set of synapses, where each synapse is defined in $\{1, \ldots, m\} \times \{1, \ldots, m\}$;
- I_{in} and I_{out} indicate, respectively, the sets of input and output neurons, with each of them being a mutually exclusive subset of $\{\sigma_1, \ldots, \sigma_m\}$.

At each timestep, the state of the system is updated based on the number of spikes and the set of rules in each neuron.

The first type of rule $(E/a^c \rightarrow a^p; d)$ is called *spiking rule* (or *firing rule*). In the formula, E is a regular language over O; c and $p < c$ denote the number of spikes consumed and the number of spikes generated, respectively, when the spiking rule is applicable; d is the "refractory" period that forces the neuron to wait d timesteps between two consecutive spikes. Denoting with g the number of spikes contained in a neuron σ_i, the spiking rule above can be interpreted as follows: the rule is applicable only if the number of spikes g is greater than or equal to the number of spikes to be consumed c (i.e., $g \geq c$). At a certain timestep, if the number of spikes g contained in σ_i is above the threshold c, then σ_i consumes c spikes to fire the neuron and $g - c$ spikes remain in the neuron. After immediately emitting p spikes, the neuron cannot fire for the following d timesteps.

The second type of rule $(E/a^f \rightarrow \lambda)$ is referred to as *forgetting rule*. In the formula, f denotes the exact number of spikes needed to apply the forgetting rule, and λ represents an empty string. At a certain timestep, if a neuron σ_i contains *exactly* f spikes and the spiking rule is not applicable, then the neuron consumes f spikes without producing any spikes.

Concerning the set of synapses, syn, its elements have the form of $(j, i, w_{j,i})$ where $1 \leq j, i \leq m$, $j \neq i$ denote the neuron indexes and the weight on synapse

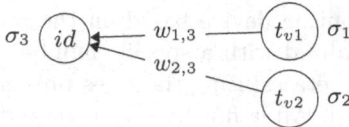

Fig. 1. Graphical representation of an example SN P system.

(j, i), denoted by $w_{j,i}$, is an integer. Thus, *syn* describes the topology of the connections among neurons, and their weights. Figure 1 shows the graphical representation of a SN P system used in the rest of the paper. Each node in the graph represents a neuron σ_i. Each neuron is either an input or an output neuron: $I_{in} = \{\sigma_1, \sigma_2\}$ and $I_{out} = \{\sigma_3\}$. The text in the input nodes, t_{v1} or t_{v2}, indicates the number of spikes for the corresponding input variables obtained from the environment of the given task. Each output neuron, which has a given *id*, produces a response to the environment by emitting spikes. Lastly, the edges between neurons show the synapses *syn*, and the weight of each synapse is specified on each edge.

One additional note is that a neuron can use only one rule at each timestep. If there are more than two applicable rules for a neuron σ_i at any timestep, one of them is chosen non-deterministically with the same probability.

3   Related Work

Spiking Neural P Systems. One of the earliest applications of SN P systems was to use the spike trains as language generators: in [3], a binary language generating device was introduced, based on the distances between spikes. Then, several extensions of this idea were investigated in [12], to get a language over an alphabet with as many symbols as the number of concurrently generated spikes.

SN P systems have also been used to solve several computationally hard problems. The baseline idea, i.e., to activate the exponentially large number of inactive neurons in polynomial time, was proposed by [13], to solve a SAT problem. This idea was further extended in [14,15], by starting with an exponentially large precomputed workspace instead of producing an exponential workspace in polynomial time. After that, a variant of SN P systems called Optimization Spiking Neural P systems (OSNP systems) was introduced in [16], to obtain an analytic solution for the knapsack problem, which is known to be NP-complete. This variant introduces a "guider" that adaptively adjusts rule probabilities to solve combinatorial optimization problems. OSNP systems were further improved in [17] to address the Travelling Salesman Problem (TSP), which is known to be NP-hard. The major difference w.r.t. [16] is that they employ a genetic algorithm (GA) to adjust the rule probabilities instead of following the guide algorithm specified in [16].

In [18], Ionescu et al. applied SN P systems for simulating logical gates. They encoded the Boolean values, 0 and 1 respectively, into one and two spikes, as inputs given to one neuron. This study inspired the application of SN P systems for designing the arithmetic logic unit used in CPUs. For instance, a variant of SN P systems using anti-spikes was proposed in [19], to perform arithmetic operations such as addition and subtraction, as well as logic operations such as AND, OR and NOT. Further improvements to this approach were made in [20], with the introduction of asynchronous parallelism. XOR and NAND gate operations were later simulated by Song et al. in [21], using SN P systems with astrocyte-like control.

Another important area of application of SN P systems is fuzzy reasoning, particularly in the area of fault diagnosis in power systems. In [22], Peng et al. proposed a fuzzy reasoning Spiking Neural P system (FRSNP system) to handle fuzzy diagnosis knowledge and reasoning, which are indispensable for some fault diagnosis applications. Their proposed method includes several mechanisms ouch as fuzzy logic and a new firing mechanism, and was tested on the fault diagnosis of a transformer. Such FRSNP system was further improved in [23] and [24]. In particular, the method proposed in [23], called adaptive FRSNP system (AFRSNP system), is able to adjust the weights in the fault diagnosis model automatically. In [24], the efficiency of the AFRSNP system was further improved by optimizing the learning algorithm by means of particle swarm optimization (PSO).

Other works have used SN P systems and their variants to perform pattern recognition tasks. In this area, [25,26] used SN P systems to implement a parallel method for image skeletonizing. The proposed method, based on SN P systems with weights that are associated with the synapses, was used as a thinning algorithm for skeletonizing binary images. SN P systems have been used also for fingerprint recognition. For this task, [27] proposed a double-layer self-organized SN P system that can adaptively create and delete neurons present in the different layers. Finally, Song et al. [28] introduced a variant called SN P system with Hebbian learning function to recognize English letters. The use of the learning function, which enables a dynamic update of the neuron connections during the computation, allowed the proposed method to obtain promising results compared to traditional neural networks based optimized by back-propagation.

While previous attempts to the automatic design of SN P systems do exist [8, 9], they still require an expert to specify the hyperparameters of these algorithms. Our approach, instead, removes almost all the hyperparameters, leaving only the number of rules (inside each neuron) as parameter that the user has to specify (besides the hyperparameters for NEAT).

Neuroevolution. Neuroevolution, that is the application of evolutionary algorithms to optimize neural networks, is a growing field in Computational Intelligence. In this area, Neuroevolution of Augmenting Topologies (NEAT) [10] is a well-established technique that is capable to optimize both the parameters and topology of neural networks. The earliest applications of NEAT concerned evolutionary learning in control problems, such as pole balancing [10] and pole chasing [29].

Later, NEAT has been widely used in games, in particular to obtain optimal strategies to decide which action to take given as input a description of the current state of the game. In [30], NEAT was used to play Go by evaluating where the next stone should be placed. Taylor et al. [31] employed NEAT for the same task, but their proposed method accelerates learning by using transfer learning. A real-time version of NEAT (rtNEAT) was introduced in [32] to evolve neural networks in real time, so that the proposed method makes agents improve their behavior while the game is being played. This method was tested on a game called neuroevolving robotic operatives (NERO), to improve the competitiveness of a virtual robot team in real time.

More recently, NEAT has been applied to the automatic design of deep neural networks (DNNs). In 2019, Miikkulainen et al. proposed a further extension of NEAT, called CoDeepNEAT [33], that can be applied to DNNs to perform coevolutionary optimization of topology, network components, and hyperparameters. They evaluated their method on various tasks: object recognition, language modeling and automated image captioning. CoDeepNEAT achieved promising results comparable to human-designed networks.

NEAT and its variants discussed above are restricted to traditional (connectivist) neural networks, including DNNs. In contrast, [34,35] applied NEAT to spiking neural networks (SNNs). In particular, in [34] a powerful neuromorphic hardware called SpiNNaker was used for evolving neural controllers based on SNNs through NEAT. In [35], NEAT was used to develop a recurrent spiking controller that can solve nonlinear control problems in continuous domains. The proposed method was evaluated on a pole balancing task, demonstrating that the learning speed of the evolved spiking controller is significantly faster than that of a traditional neural network that makes use of a sigmoidal activation function.

Finally, we should note that while NEAT is one of the most popular neuroevolutionary algorithms, it is not the only one. In fact, several other methods have been recently proposed for evolutionary neural architecture search [36–38]. However, most of these methods evolve neural networks by composing high-level blocks, while the aim of our work is to evolve neural networks by optimizing them at the level of single neurons.

4   Method

In order to validate our approach, we consider here the automatic design of basic SN P systems (i.e., not the advanced variants discussed in Sect. 3). This choice was made to ensure that NEAT is able to deal at least with the simplest SN P systems. Moreover, for simplicity we consider neurons that have exclusively one spiking rule and one forgetting rule, and we set the initial number of spikes to $n_i = 0$. These assumptions make these neurons a specialization of the more general type of neurons that can be used in SN P systems. While these neurons may be significantly less expressive than the general ones, they allow us to adapt, with minimal effort, the NEAT algorithm. However, the reduced expressiveness of SN P systems with only two rules per neuron may reduce their performance in some tasks where higher expressiveness (for each neuron) is required.

Another assumption we make regards the weights: in fact, we evolve SN P systems whose connections may have a (either positive or negative) weight, as in [6]. The weight acts as a multiplier for the number of tokens that are produced by the neurons, i.e., given t_o tokens in output from neuron o and a connection from o to p with weight $w_{o,p}$, the total number of spikes given in input to the neuron p is $t_p = t_o \cdot w_{o,p}$.

While, ideally, given an input we want to wait that the SN P system has finished the computation (i.e., no more spikes are produced in the system), this may significantly slow down the evolution. To speed up the evolutionary process, we constrain the computation in the SN P system to 100 timesteps.

4.1 NEAT

The NEAT algorithm starts the evolutionary process from minimal networks that, as the evolutionary process goes on, are complexified by means of mutation and crossover. Each network is encoded through two lists: the list of nodes and the list of connections. Moreover, the algorithm allows to perform efficient neuroevolution by allowing the detection of *similar* individuals, and preserves diversity by using niching. More details on these aspects of the algorithm can be found in [10].

4.2 Genotype

We adapt the original NEAT algorithm[1] to our purpose by including into the genotype the following parameters of every neuron σ_i:

- $c_i \in \mathbb{N} \setminus \{0\}$: no. of spikes needed to fire the neuron;
- $p_i \in \mathbb{N} \setminus \{0\}$: no. of output spikes;
- $d_i \in \mathbb{N}$: minimum delay between two subsequent spikes for the neuron;
- $f_i \in \mathbb{N}$: no. of spikes needed to activate the forgetting rule;
- $w_{j,i} \in \mathbb{Z}$: weight of synapse (j, i) for each synapse in input to the neuron.

Moreover, we also optimize the set of connections *syn*.

4.3 Phenotype

When translating a genotype into a phenotype (i.e., an instance of SN P systems), the following constraints are handled:

- $p_i \leq c_i$: if $p_i > c_i$ then it is set to $p_i = c_i$;
- $f_i < c_i$: if $f_i \geq c_i$ then it is set to $f_i = c_i - 1$.

---

[1] Our code, which is based on the **neat-python** package [39], is publicly available online at https://github.com/leocus/snps.

Moreover, since during the initialization of the genotypes the values for the parameters are sampled from Gaussian distributions, to create the SN P systems the parameters are cast into integers by taking the floor of the number.

Each resulting phenotype consists then of a SN P system containing k neurons, where k is the number of non-hidden nodes from the genotype. Subsequently, a connection is created for each enabled connection gene, connecting the created previously neurons.

4.4   Input Features

The tasks considered in our experimentation use real-valued input data, which are incompatible with the data type employed in SN P systems. To mitigate this problem, we perform a very simple conversion from floating-point encoding to integer encoding. The conversion consists in normalizing each input in its range of variation:

$$\bar{x}_k = \frac{x_k - min(x_k)}{max(x_k) - min(x_k)} \tag{2}$$

Then, we convert the normalized input into a number of input spikes as follows:

$$t_k = \lfloor k \cdot \bar{x}_k \rfloor \tag{3}$$

In our experiments, we empirically set k to 20. Finally, the input spikes $\{t_0, \ldots, t_n\}$, where n is the number of inputs, are fed into the SN P system.

4.5   Fitness Evaluation

After the genotype-phenotype mapping is applied, each SN P system is evaluated in n_{ep} episodes, where an episode consists in our case in a simulation of the control task at hand. At each timestep of a simulation, we transform the raw inputs coming from the simulator into integer features, using the procedure described in the previous subsection; then, we feed the features into the SN P system, which produces an output vector containing the number of spikes for each output neuron (with size equal to the number of actions). The action performed by the SN P system is then the argmax of the output vector.

When the n_{ep} episodes have been completed, the phenotype is assigned a fitness equal to the mean score across the n_{ep} episodes.

Note that, when evaluating the SN P system, each neuron cannot contain a negative number of spikes. So, when $s_i < 0$, the number of spikes inside that neuron is reset to $s_i = 0$.

5   Results

To test the capabilities of our approach, we evolve SN P systems on two classic control tasks, namely MountainCar-v0 and CartPole-v1, taken from the OpenAI Gym library [11]. While these two tasks may seem trivial for testing modern neuro-evolutionary approaches, it must be noted that, to our knowledge, this is the first application of SN P systems as controller for control tasks, thus their ability to work in these scenarios is yet to be determined. The MountainCar-v0 task is a "driving" task, where there is a car, initially in a valley, that has to reach the rightmost hill. To do so, the agent must learn how to build momentum by swinging between the two hills. In this case, the agent takes in input both the horizontal position and velocity of the car and must produce a decision $a \in \{0, 1, 2\}$ that corresponds to: accelerate to the left, do not accelerate, and accelerate to the right, respectively. In this task, each timestep gives a reward of -1 points, so the quicker the agent solves the task, the higher the score. The task is considered solved if the agent reaches a score $s > -110$. On the other hand, the CartPole-v1 environment consists in a classic pole-balancing task. In this task, the agent takes the following inputs: horizontal position (x), horizontal velocity (y), angle of the pole (θ), angular velocity of the pole (ω), and must produce a binary decision $a \in \{0, 1\}$ that consists in moving the cart to the left or to the right, respectively. The reward given to the agent at each timestep consists in 1 point. If the pole falls, the simulation is terminated, If the maximum score is reached (500), the simulation is terminated. The task is considered as solved if the agent reaches a score $s > 475$. Table 1 shows the parameters used to evolve SN P systems with the NEAT algorithm. The same parameters were used on both the MountainCar-v0 and the CartPole-v1 tasks.

Table 1. Parameters used for the NEAT algorithm.

Parameter	Value	Parameter	Value
Population size	300	Generations	300
Init c_i	$\sim \mathcal{N}(30, 10)$	Init p_i	$\sim \mathcal{N}(30, 10)$
Init d_i	$\sim \mathcal{N}(30, 10)$	Init f_i	$\sim \mathcal{N}(30, 10)$
Init weight	$\sim \mathcal{N}(0, 3)$	c_i range	[1, 100]
p_i range	[1, 100]	d_i range	[0, 100]
f_i range	[1, 100]	Weight range	[−10, 10]
Mutation power	$\sim \mathcal{N}(0, 3)$	Mutation rate	0.2
Replacement rate	0.1	Add connection rate	0.5
Remove connection rate	0.5	Add node rate	0.5
Remove node rate	0.5	Toggle "enable" rate	0.1
Max stagnation period	20		

Table 2 shows the results obtained in 10 independent runs on the two tasks. Note that, to speed up the computation, we set a stopping criterion based on the score: if the agent achieves a mean score that is greater or equal than a threshold, the evolution is stopped. For the CartPole-v1 task, the maximum score is 500, while the minimum score required for solving the task is 475. In this case, we set the threshold to 499. On the other hand, for the MountainCar-v0 task there is no maximum score, and the minimum score required is -110. Here, we set the threshold to -105. While this stopping criterion may hinder reaching the global maximum, it significantly speeds up the evolution process. As shown in the table, our approach is able to solve the task in 100% of the cases (computed on 10 independent runs) for both tasks.

Figures 2a and 2b show the fitness trends averaged over the 10 independent runs for each task, where it can be observed that the proposed method is fairly robust across multiple runs, and converges quite quickly (after about 40 generations in the case of MountainCar-v0, 20 in the case of CartPole-v1).

Table 2. Descriptive statistics of the results obtained by the proposed method in 10 independent runs on the two tested tasks.

Task	Min	Mean	Median	σ	Max	Solved
MountainCar-v0	-105.18	-104.56	-104.37	0.34	-104.14	10/10
CartPole-v1	491.52	497.49	498.05	2.86	500.00	10/10

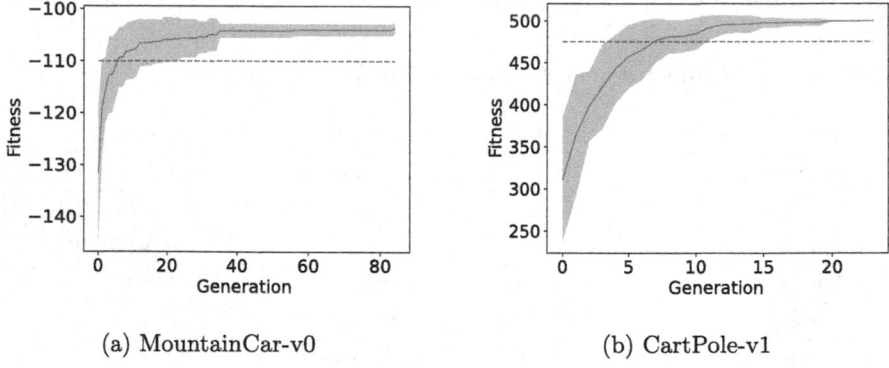

(a) MountainCar-v0   (b) CartPole-v1

Fig. 2. Fitness trend (mean $\pm$ std. dev. across 10 independent runs) for the best individuals found during the evolutionary process on the two tested tasks. The dashed line represents the "solved" threshold.

5.1   Analysis of the Solutions

In the following, we analyze the best SN P system evolved for each task, trying to gain insights on the evolved policies.

MountainCar-v0. Figure 3 shows the diagram of the best SN P system obtained for this task, where t_x and t_v represent the number of tokens (i.e., spikes) obtained from the real-valued inputs coming from the simulator (x and v

stand for position and velocity of the car, respectively) by means of the "translation" process described in Sect. 4.4. Each circle with a number inside represents a spiking P neuron, where the number represents its id. Each circle with a symbol inside represents an input neuron, where the symbol identifies the spikes generated from the corresponding input. Ignoring Neuron 0 and 1 (that do not contribute to the control of the agent), the evolved parameters are:

- Neuron 2 (Accelerate to the right):
  $c_i = 43$, $p_i = 31$, $d_i = 31$, $f_i = 14$

The policy shown in Fig. 3 acts as an "if-then-else" policy. In fact, since we choose the action based on the argmax, two scenarios may happen:

- Neuron 2 does not produce any spikes in this case, the action taken is 0 (Accelerate to the left). This happens because the argmax function returns the first index that contains the max.
- Neuron 2 produces one or more spikes: in this case, the argmax would be 2 and the agent will accelerate to the right.

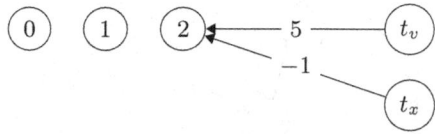

Fig. 3. Best SN P system evolved for the MountainCar-v0 task.

CartPole-v1. Figure 4 shows the best SN P system obtained for this task, which follows the same representation of Fig. 3. In this case, the evolved parameters are:

- Neuron 0 (Accelerate to the left):
  $c_0 = 31$, $p_0 = 2$, $d_0 = 22$, $f_0 = 26$
- Neuron 1 (Accelerate to the right):
  $c_1 = 47$, $p_1 = 36$, $d_1 = 25$, $f_1 = 8$

This policy seems more complex than the one produced for the MountainCar-v0 task. In fact, here we observe that all the input variables (t_x, t_v, t_w and t_θ) contribute (positively) to the output of Neuron 1, while only two variables, namely t_x and t_θ, contribute to the output of Neuron 0, and t_θ contributes negatively to it.

However, by inspecting how Neuron 0 works, we can conclude that also this policy can be reduced to an "if-then-else". In fact, for each spike in t_x 2 spikes will be added to this neuron, and, for each spike in t_θ, 5 spikes will be removed from it. Since a neuron cannot contain a negative number of spikes, the number of spikes that Neuron 0 can contain is:

$$s_{0,in} = max(0, 2(t_x - t_\theta)) \tag{4}$$

This means that, after all the spikes in t_θ have been consumed (assuming $t_x > t_\theta$) 2 spikes will be added to $s_{0,in}$ at each timestep. Then, three scenarios can happen:

- $t_x \leq t_\theta \Rightarrow s_{0,in} = 0$;
- $t_x - t_\theta \in [0, 13) \Rightarrow s_{0,in}$ will be too small to trigger any rule;
- $t_x - t_\theta \in [13, 20] \Rightarrow s_{0,in}$ will be equal to 26 at a certain point, so the *forgetting* rule will be applied. The number of spikes that can be added in the following steps will be insufficient to trigger any rule.

This means that Neuron 0 will never fire any spike and, for this reason, also this SN P system represents an "if-then-else" policy, similarly to what we obtained for the MountainCar-v0 task.

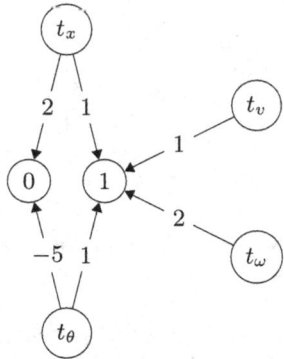

Fig. 4. Best SN P system evolved for the CartPole-v1 task.

5.2   Comparison with State of the Art

Tables 3 and 4 compare our approach to the state of the art on the two tasks (from either the OpenAI Gym Leaderboard[2] or the literature). We observe that our approach is competitive with the state of the art. Moreover, the fact that we used a stopping criterion based on the fitness may have hindered the discovery of a better performing method on the MountainCar-v0 task.

---

[2] github.com/openai/gym/wiki/Leaderboard.

Table 3. Comparison of our approach to the state of the art on the MountainCar-v0 task. The boldface indicates the best values known so far.

Source	Method	Score
Zhiqing Xiao[a]	Closed-form policy	−102.61
Keavnn[b]	Soft Q Networks [40]	−104.58
Harshit Singh[c]	Deep Q Network	−108.85
Colin M[d]	Double Deep Q Network	−107.83
Amit[e]	SARSA	−105.99
Anas Mohamed[f]	SARSA	−109.41
Custode & Iacca [41]	Decision Tree	**−101.72**
Ours	SN P system	−104.14

[a] github.com/ZhiqingXiao/OpenAIGymSolution
[b] github.com/StepNeverStop/RLs, accessed: 3 aug 2021.
[c] github.com/harshitandro/Deep-Q-Network
[d] github.com/CM-Data/Noisy-Dueling-Double-DQN-MountainCar
[e] github.com/amitkvikram/rl-agent
[f] github.com/amohamed11/OpenAIGym-Solutions

Table 4. Comparison of our approach to the state of the art on the CartPole-v1 task. The boldface indicates the best values known so far.

Source	Method	Score
Meng et al. [42]	Deep Q Network	327.30
Meng et al. [42]	Tree-Backup (λ)	494.70
Meng et al. [42]	Importance-sampling	498.70
Meng et al. [42]	Qπ	489.90
Meng et al. [42]	Retrace (λ)	461.10
Meng et al. [42]	Policy discrepancy w/β	499.90
Meng et al. [42]	Policy discrepancy w/η	493.20
Meng et al. [42]	Watkins's Q (λ)	484.30
Meng et al. [42]	Policy discrepancy w/β	494.90
Meng et al. [42]	Policy discrepancy w/η	493.30
Meng et al. [42]	Peng & Williams's Q (λ)	496.70
Meng et al. [42]	Policy discrepancy w/β	**500.00**
Meng et al. [42]	Policy discrepancy w/η	499.40
Meng et al. [42]	General Q (λ)	499.90
Meng et al. [42]	Policy discrepancy w/β	**500.00**
Meng et al. [42]	Policy discrepancy w/η	**500.00**
Xuan et al. [43]	Deep Q Network	98.33
Xuan et al. [43]	Bayesian Deep RL	113.52
Xuan et al. [43]	Bayesian Deep RL weighted	136.75
Beltiukov [44]	K-FAC	321.00
Custode & Iacca [41]	Decision Tree	**500.00**
Ours	SN P system	**500.00**

6 Conclusions

Spiking Neural P (in short, SN P) systems are a computational tool from the field of membrane computing that gained a lot of attention in recent years. Several variants of these systems have been proposed for different applications. However, until now, these systems have almost always been manually derived from experts. The approach presented here aims to automate the design step, so that SN P systems can be automatically produced for a given task with no (or little) supervision from an expert. We tested our method on two classic control tasks, and found that our approach is able to solve both tasks in all the runs. Moreover, it produces controllers whose performances are competitive with the state of the art.

The most important limitation of this work is that we assumed rather simplified SN P systems. In particular, we limited the number of rules that each neuron can employ. In future work, we expect to remove this limitation. Moreover, we considered only a basic version of SN P systems, while it would be interesting to combine our approach with more advanced variants such as OSNP systems [16], in order to perform both "architectural search" and real-time parameter optimization.

Other relevant future directions include, for instance: 1) testing our approach on different (and more challenging) tasks; 2) applying techniques to allow SN P systems to work with non-integer inputs; 3) evolving SN P systems that, instead of using the number of spikes for each neuron, use the difference between two spikes as the output, as proposed in [2]; 4) extend our approach to evolve also the number of timesteps allowed in the computation; and 5) test novel approaches for the neuro-evolution of SN P systems, such as Cartesian genetic programming [45].

References

1. Păun, G.: Computing with Membranes. J. Comput. Syst. Sci. **61**(1), 108–143 (2000)
2. Ionescu, M., Păun, G., Yokomori, T.: Spiking neural P systems. Fundamenta informaticae **71**(2, 3), 279–308 (2006)
3. Păun, G., Pérez-Jiménez, M.J., Rozenberg, G.: Spike trains in spiking neural P systems. Int. J. Found. Comput. Sci. **17**(04), 975–1002 (2006)
4. Martín-Vide, C., Pazos, J., Păun, G., Rodríguez-Patón, A.: A new class of symbolic abstract neural nets: tissue P systems. In: Ibarra, O.H., Zhang, L. (eds.) COCOON 2002. LNCS, vol. 2387, pp. 290–299. Springer, Heidelberg (2002). https://doi.org/10.1007/3-540-45655-4_32
5. Pan, L., Zeng, X.: A note on small universal spiking neural P systems. In: Păun, G., Pérez-Jiménez, M.J., Riscos-Núñez, A., Rozenberg, G., Salomaa, A. (eds.) WMC 2009. LNCS, vol. 5957, pp. 436–447. Springer, Heidelberg (2010). https://doi.org/10.1007/978-3-642-11467-0_29
6. Wang, J., Hoogeboom, H.J., Pan, L., Păun, G., Pérez-Jiménez, M.J.: Spiking neural P systems with weights. Neural Comput. **22**(10), 2615–2646 (2010)

7. Wang, X., Song, T., Gong, F., Zheng, P.: On the computational power of spiking neural P systems with self-organization. Sci. Rep. **6**(1), 1–16 (2016)
8. Dong, J., Stachowicz, M., Zhang, G., Cavaliere, M., Rong, H., Paul, P.: Automatic design of spiking neural P systems based on genetic algorithms. Int. J. Unconventional Comput. **16**(2/3), 201–216 (2021)
9. Casauay, L.J.P., et al.: A framework for evolving spiking neural P systems. Int. J. Unconventional Comput. **16**, 121–139 (2021)
10. Stanley, K.O., Miikkulainen, R.: Evolving neural networks through augmenting topologies. Evol. comput. **10**(2), 99–127 (2002)
11. Brockman, G., et al.: OpenAI Gym (2016)
12. Chen, H., Ishdorj, T.-O., Paun, G., Pérez Jiménez, M.deJ.: Spiking neural P systems with extended rules. In: 4h Brainstorming Week on Membrane Computing (BWMC), vol. I, pp. 241–265. ETS de Ingeniería Informática, 30 de Enero-3 de Febrero, Fénix Editora, Sevilla (2006)
13. Chen, H., Ionescu, M., Ishdorj, T.-O.: On the efficiency of spiking neural P systems. In: 4h Brainstorming Week on Membrane Computing (BWMC), vol. I, pp. 195–206. ETS de Ingeniería Informática, 30 de Enero-3 de Febrero, Sevilla (2006)
14. Ishdorj, T.-O., Leporati, A.: Uniform solutions to SAT and 3-SAT by spiking neural P systems with pre-computed resources. Nat. Comput. **7**(4), 519–534 (2008)
15. Leporati, A., Gutiérrez-Naranjo, M.A.: Solving subset sum by spiking neural P systems with pre-computed resources. Fundamenta Informaticae **87**(1), 61–77 (2008)
16. Zhang, G., Rong, H., Neri, F., Pérez-Jiménez, M.J.: An optimization spiking neural P system for approximately solving combinatorial optimization problems. Int. J. Neural Syst. **24**(05), 1440006 (2014)
17. Qi, F., Liu, M.: Optimization spiking neural P system for solving TSP. In: Gu, X., Liu, G., Li, B. (eds.) MLICOM 2017. LNICST, vol. 227, pp. 668–676. Springer, Cham (2018). https://doi.org/10.1007/978-3-319-73447-7_71
18. Ionescu, M., et al.: Some applications of spiking neural P systems. Comput. Inform. **27**(3+), 515–528 (2008)
19. Peng, X.W., Fan, X.P., Liu, J.X.: Performing balanced ternary logic and arithmetic operations with spiking neural P systems with anti-spikes. Adv. Mater. Res. **505**, 378–385 (2012)
20. Hamabe, R., Fujiwara, A.: Asynchronous SN P systems for logical and arithmetic operations. In: International Conference on Foundations of Computer Science (FCS), The Steering Committee of the World Congress in Computer Science (2012). 1
21. Song, T., Zheng, P., Dennis Wong, M.L., Wang, X.: Design of logic gates using spiking neural P systems with homogeneous neurons and astrocytes-like control. Inf. Sci. **372**, 380–391 (2016)
22. Peng, H., Wang, J., Pérez-Jiménez, M.J., Wang, H., Shao, J., Wang, T.: Fuzzy reasoning spiking neural P system for fault diagnosis. Inf. Sci. **235**, 106–116 (2013)
23. Tu, M., Wang, J., Peng, H., Shi, P.: Application of adaptive fuzzy spiking neural P systems in fault diagnosis of power systems. Chin. J. Electron. **23**, 87–92 (2014)
24. Wang, J., Peng, H., Tu, M., Pérez-Jiménez, J.M., Shi, P.: A fault diagnosis method of power systems based on an improved adaptive fuzzy spiking neural P systems and PSO algorithms. Chin. J. Electron. **25**(2), 320–327 (2016)
25. Díaz-Pernil, D., Peña-Cantillana, F., Gutiérrez-Naranjo, M.A.: A parallel algorithm for skeletonizing images by using spiking neural P systems. Neurocomputing **115**, 81–91 (2013)

26. Song, T., Pang, S., Hao, S., Rodríguez-Patón, A., Zheng, P.: A parallel image skeletonizing method using spiking neural P systems with weights. Neural Process. Lett. **50**(2), 1485–1502 (2019)
27. Ma, T., Hao, S., Wang, X., Rodríguez-Patón, A.A., Wang, S., Song, T.: Double layers self-organized spiking neural P systems with anti-spikes for fingerprint recognition. IEEE Access **7**, 177562–177570 (2019)
28. Song, T., Pan, L., Wu, T., Zheng, P., Wong, M.L.D., Rodríguez-Patón, A.: Spiking neural P systems with learning functions. IEEE Trans. Nanobiosci. **18**(2), 176–190 (2019)
29. Pardoe, D., Ryoo, M., Miikkulainen, R.: Evolving neural network ensembles for control problems. In: Genetic and Evolutionary Computation Conference, pp. 1379–1384 (2005)
30. Stanley, K.O., Miikkulainen, R.: Evolving a roving eye for Go. In: Deb, K. (ed.) GECCO 2004. LNCS, vol. 3103, pp. 1226–1238. Springer, Heidelberg (2004). https://doi.org/10.1007/978-3-540-24855-2_130
31. Taylor, M.E., Whiteson, S., Stone, P.: Transfer via inter-task mappings in policy search reinforcement learning. In: International Joint Conference on Autonomous Agents and Multiagent Systems (AAMAS), pp. 1–8 (2007)
32. Stanley, K.O., Bryant, B.D., Miikkulainen, R.: Real-time neuroevolution in the NERO video game. IEEE Trans. Evol. Comput. **9**(6), 653–668 (2005)
33. Miikkulainen, R., et al.: Evolving deep neural networks. In: Artificial Intelligence in the Age of Neural Networks and Brain Computing, pp. 293–312. Elsevier (2019)
34. Vandesompele, A., Walter, F., Röhrbein, F.: Neuro-evolution of spiking neural networks on SpiNNaker neuromorphic hardware. In: Symposium Series on Computational Intelligence (SSCI), pp. 1–6. IEEE (2016)
35. Qiu, H., Garratt, M., Howard, D., Anavatti, S.: Evolving spiking neural networks for nonlinear control problems. In: Symposium Series on Computational Intelligence (SSCI), pp. 1367–1373. IEEE (2018)
36. Suganuma, M., Shirakawa, S., Nagao, T.: A genetic programming approach to designing convolutional neural network architectures. In: Genetic and Evolutionary Computation Conference, pp. 497–504 (2017)
37. Assunção, F., Lourenço, N., Machado, P., Ribeiro, B.: DENSER: deep evolutionary network structured representation. Genet. Program Evolvable Mach. **20**(1), 5–35 (2019)
38. Lu, Z., et al.: NSGA-net: neural architecture search using multi-objective genetic algorithm. In: Genetic and Evolutionary Computation Conference, pp. 419–427 (2019)
39. McIntyre, A., Kallada, M., Miguel, C.G., da Silva, C.F.: neat-python. https://github.com/CodeReclaimers/neat-python
40. Liu, J., Gu, X., Liu, S., Zhang, D.: Soft Q-network. arXiv:1912.10891 [cs] (2019)
41. Custode, L.L., Iacca, G.: Evolutionary learning of interpretable decision trees (2021)
42. Meng, W., Zheng, Q., Yang, L., Li, P., Pan, G.: Qualitative measurements of policy discrepancy for return-based deep Q-network. IEEE Trans. Neural Netw. Learn. Syst. **31**, 4374–4380 (2019)
43. Xuan, J., Lu, J., Yan, Z., Zhang, G.: Bayesian deep reinforcement learning via deep kernel learning. Int. J. Comput. Intell. Syst. **12**(1), 164–171 (2018)
44. Beltiukov, R.: Optimizing Q-learning with K-FAC algorithm. In: van der Aalst, W.M.P., et al. (eds.) AIST 2019. CCIS, vol. 1086, pp. 3–8. Springer, Cham (2020). https://doi.org/10.1007/978-3-030-39575-9_1

45. Miller, J.F., Harding, S.L.: Cartesian genetic programming. In: Genetic and Evolutionary Computation Conference - Companion, pp. 2701–2726 (2008)

Self-adaptation of Neuroevolution Algorithms Using Reinforcement Learning

Michael Kogan[ID], Joshua Karns[ID], and Travis Desell[(✉)][ID]

Rochester Institute of Technology, Rochester, NY 14623, USA
tjdvse@rit.edu

Abstract. Selecting an appropriate neural architecture for a given dataset is an open problem in machine learning. Neuroevolution algorithms, such as NEAT, have shown great promise in automating this process. An extension of NEAT called EXAMM has demonstrated the capability to generate recurrent neural architectures for various time series datasets. At each iteration the evolutionary process is furthered using randomly selected mutation or crossover operations, which are chosen in accordance with pre-assigned probabilities. In this paper we present a self-adapting version of EXAMM that incorporates finite action-set learning automata (FALA), a reinforcement learning technique. FALA is used to dynamically adjust the aforementioned probabilities, thereby guiding the evolutionary process, while also significantly reducing the number of required hyperparameters. It is also demonstrated that this approach improves the performance of the generated networks with statistical significance. Furthermore, the evolution of the adapted probabilities is analyzed to gain further insight into the inner workings of EXAMM.

Keywords: Neuroevolution · Reinforcement learning · FALA

1   Introduction

Inspired by the architecture of the cerebral cortex, neural networks are arguably the most powerful machine learning model in use today [16]. The success of neural networks can be largely attributed to their flexible architecture, which can be readily adapted to suit the data at hand. This has enabled neural networks to succeed in problem spaces that other machine learning approaches find intractable, such as image recognition [11] and natural language processing [4]. The process of designing an appropriate neural architecture for most problem sets is rather complex, and so it is commonplace for machine learning practitioners adapt existing architectures instead of developing them from scratch. Often times networks pre-trained on generic data are tweaked on the target data in a process known as transfer learning [20].

Determining an appropriate neural architecture for a given dataset is an open problem, but one technique that has had considerable success is known as

© Springer Nature Switzerland AG 2022
J. L. Jiménez Laredo et al. (Eds.): EvoApplications 2022, LNCS 13224, pp. 452–467, 2022.
https://doi.org/10.1007/978-3-031-02462-7_29

neuroevolution. This approach takes inspiration from natural evolutionary processes to automate the design of artificial neural networks [3]. Neuroevolution algorithms can generate architectures that are significantly different from hand-crafted neural networks and have demonstrated the potential to outperform even the best human-designed architectures [3]. One of the most famous neuroevolution algorithms is known as NeuroEvolution of Augmenting Topologies (NEAT) [17]. NEAT is a genetic approach where complex neural architectures are progressively evolved from a simple baseline construct using pre-defined genetic operations. While NEAT was initially designed for regular feed forward neural networks, it has recently been extended to evolve both convolutional [2] and recurrent [13] architectures. The latter of these is known as Evolutionary eXploration of Augmented Memory Models (EXAMM), and the improvement of this algorithm is the topic of this paper.

Work by Alba and Tomassini has shown that evolutionary algorithms can be greatly sped up through the incorporation of an island speciation strategy [1], and this has been implemented in EXAMM [13]. Lyu et al. have extended this concept further and incorporated island extinction and repopulation [10], which has been shown to outperform both the baseline island speciation strategy as well as NEAT's original speciation strategy. Recently, work by Radaideh and Shirvan has demonstrated that rule-based reinforcement learning can be used to guide evolutionary algorithms in constrained optimization [14]. In this paper, we propose an extension of EXAMM that incorporates a reinforcement learning technique known as Finite Action-set Learning Automate (FALA) to guide the evolutionary process. To our knowledge reinforcement learning has not yet been applied to this class of algorithms.

2 EXAMM

At the core of EXAMM is a master process that maintains a population of Recurrent Neural Network (RNN) genomes. This population is spread out across several islands, with each island containing an identical number of genomes. At each iteration, EXAMM selects an island in a round-robin manner, and then randomly selects a genome from that island to evolve. It chooses a genetic operation in accordance with pre-defined operational probabilities, and evolves the selected genome using the chosen operation. The genetic operations used in EXAMM and associated probabilities are listed in Table 1 (for further details please consult [13]). The evolved genome is trained for a given number of back propagation iterations using one of the worker nodes and its fitness is evaluated based on the validation data set performance. EXAMM then attempts to re-insert the trained genome into the island from which it was initially picked. If the fitness of the newly trained genome is superior to that of the worst genome in the selected island, the worst genome is dropped and the newly generated genome is added to the population of that island. This process continues until the maximum number of allowed genomes is generated. Certain operations will result in the insertion of new recurrent edges into the architecture of a given genome. The depths of these recurrent edges are drawn from a uniform probability distribution, which returns values between the minimum and maximum recurrent depths as set by the programmer prior to running the algorithm.

Table 1. Genetic Operations in EXAMM and their Default Probabilities

Operation	Prob.	Operation	Prob.	Operation	Prob.
Clone	0.07	Add edge	0.07	Split node	0.07
Add recurrent edge	0.07	Enable edge	0.07	Merge node	0.07
Disable edge	0.07	Add node	0.07	Crossover	0.20
Enable node	0.07	Disable node	0.07	Island crossover	0.10

2.1 Opportunities for Improvement

It is not unreasonable to assume that certain genetic operations are more likely to result in well-performing genomes, and that the preferred operations will change over time as the population evolves. As such, guiding the evolutionary process by adjusting the probabilities associated with these genetic operations has the potential to improve the quality of generated networks. Operations that generate well-performing genomes should have their probabilities increased, whereas operations that produce undesirable genomes should have their probabilities decreased.

Additionally, treating all islands equally by cycling through them in a round-robin manner may not be the optimal approach. The populations of different islands tend to be heterogeneous with some islands containing better-performing genomes than others. Selecting the better islands more frequently could speed up the learning process by targeting desirable populations. Furthermore, it is possible that not all recurrent edge depths are pertinent to a given dataset, and so dynamically adjusting the probabilities associated with the selection of different depths could also speed up the learning process.

3 Finite Action-Set Learning Automata

Finite Action-set Learning Automata is a reinforcement learning algorithm which evolves a vector of discrete probabilities associated with a set of actions. This approach has shown promise in the field of robotics engineering, where it has been used for behavioral design [15] and controller parameter tuning [6,8]. A FALA process can be described by the following quadruple:

$$\Gamma = (\alpha, p(t), r, \tau) \tag{1}$$

In the formulation above, the four components represent the following [9]:

1. A finite set of possible actions referred to as the action set. For the purpose of this paper it will be defined as $\alpha = \{\alpha_1, \alpha_2, \dots, \alpha_n\}$. The action selected at time point t is defined as $\alpha(t)$.

2. The probabilities associated with each action at time point t, which are defined as a probability vector $p(t) = (p_t(\alpha_1), p_t(\alpha_2), \ldots, p_t(\alpha_n))$. This vector is initialized at time step 0 to the starting probability values.
3. A reinforcement function r which will observe the results of the action taken at time t and generate a reinforcement signal r_t.
4. A learning function τ which will update the action probability vector based on the selected action, the current action probabilities, and the generated reinforcement signal, such that $p(t+1) = \tau(\alpha(t), p(t), r_t)$.

FALA is an iterative algorithm with a relatively simple flow. At each time step t, the parent algorithm selects an action $\alpha(t)$ in accordance with the probability vector $p(t)$ and carries out that action against the environment. The result of this action is interpreted by the reinforcement function r and an appropriate reinforcement signal r_t is generated. The probability vector is then updated in accordance with the learning function τ. The process repeats until the maximum number of iterations is reached or the action probabilities converge to a single value. In some cases the latter outcome is undesirable, and there exist a number of techniques that can be used to prevent this from happening [7]. The most obvious approach is to enforce a minimum value for each action probability, which is guaranteed to prevent convergence. Other techniques include negative reinforcement, proportional reinforcement, and staged learning.

3.1 Action Probability Minimums

Placing a floor on the values of the action probabilities ensures that no action will be completely removed as a possibility. The slack available to the learning algorithm is proportional to the sum of the differences between the starting and minimum probabilities of all actions. The decision of where to set the probability minimums comes down to trial and error as these values are effectively a set of hyperparameters tuned by the programmer.

3.2 Negative Reinforcement

Negative reinforcement is the punishment applied to actions that have undesirable outcomes and results in their probabilities being adjusted downwards. While it may be used at any point during the learning process, it is often applied only once an action probability exceeds a certain degradation threshold, which is generally set to be equal to or greater than the initial probability for a given action. The degradation thresholds for each action are determined by the programmer, and are another set of hyperparameters that need to be tuned. Negative reinforcement, when applied selectively to probabilities exceeding their degradation thresholds, makes it very difficult for any given action probability to dominate the others.

Algorithm 1. FALA Psuedocode

$iterations\_in\_stage$ = 0
for t **in** 0..$MAX\_ITERATIONS$ **do**
   for each Γ^i **do**
      $\alpha^i(t) = select\_action(\alpha^i, p^i(t))$
      $execute(\alpha^i(t))$
   end for
   for each Γ^i **do**
      $r_t^i = r^i(\alpha^1(t), \ldots, \alpha^n(t))$ # potential for negative reinforcement
      $r_t^i = scale(r_t^i, p^i(t))$ # proportional reinforcement, [OPTIONAL]
      $p^i(t+1) = \tau(\alpha^i(t), p^i(t), r_t^i)$ # must enforce probability floors
   end for
   if $iterations\_in\_stage$ == $ITERATIONS\_PER\_STAGE$ **then**
      $iterations\_in\_stage$ = 0
      for each Γ^i **do**
         $p^i(t+1) = p^i(0)$ # may wrap with logic block, [OPTIONAL]
         $adjust(prob\_floors^i, degradation\_thresholds^i)$ # [OPTIONAL]
      end for
   end if
end for

3.3 Proportional Reinforcement

Proportional reinforcement means scaling the reinforcement signal based on the current probability of the action being reinforced. The positive reinforcement for high action probabilities is scaled downwards and the positive reinforcement for low probabilities is scaled upwards. The opposite is true for negative reinforcement. This helps actions with lower probabilities learn rapidly and slows the learning of actions with higher probabilities. The reinforcement may be scaled either to the starting probability of an action or to the largest value in the probability vector. This makes it progressively more difficult to learn as the probabilities for certain actions increase, and if used alongside negative reinforcement can significantly impede the learning of high probability actions.

3.4 Staged Learning

In staged learning, the FALA process is broken down into a number of stages. Each stage consists of a certain number of iterations, and a stage transition occurs at the end of each stage. During a stage transition the action probabilities may be reset to their initial values, thereby forgetting all the learning that took place in the previous stage. This allows the algorithm to re-learn from scratch, enabling it to continuously adapt to a changing environment. Optionally, the minimum values and the degradation thresholds for each action probability may be altered during a stage transition. This increases the slack afforded to the learning algorithm as it advances through the stages, and makes learning easier by applying negative reinforcement at increasingly higher thresholds.

3.5   A Game of FALA

A game of FALA refers to situations where at every timestep, the parent algorithm takes not one but several interdependent actions, meaning it is impossible to attribute the outcome to any single action [12]. The actions are chosen independently but due to the inherent interdependence, the choice of action from one process may affect the desirability of the others. Reinforcement is applied based on the cumulative result, where each FALA process has it's own reinforcement function, but the outcome being interpreted is identical across all learning processes. The algorithm reinforces desirable combinations of actions from all processes and not the most desirable actions in each process. Mathematically, a game of FALA is nothing more than a list of quadruples: $\mathbf{\Gamma} = (\Gamma^1, \Gamma^2, \ldots, \Gamma^n)$, where n is the total number of concurrent FALA processes [18]. A high level pseudocode implementation of the complete FALA algorithm as described in this section is presented below in Algorithm 1.

4   Applying FALA to EXAMM

A single iteration of the combined FALA-EXAMM algorithm consists of a single EXAMM iteration augmented with a FALA component to adjust the action probabilities. It is composed of the following steps:

1. Choosing an island according to the island probabilities
2. Selecting a random genome from that island
3. Evolving it using an operation selected according to the genetic operation probabilities and training it for 10 back propagation iterations
4. Evaluating it to determine whether it should be inserted
5. Updating the probabilities of all actions accordingly

As EXAMM progresses through the evolutionary process it maintains a population of candidate genomes spread across a certain number of islands, with each island maintaining a given number of genomes. These two parameters are used to calculate an appropriate number of iterations per stage, which is a multiple of the product of the number of islands and the number of genomes per island. The number of iterations per stage is a constant calculated once at the beginning of the evolutionary process. It is used to determine when a stage transition occurs, and by extension the total number of stages in a given run. The performance of a newly generated genome is evaluated when EXAMM attempts to insert it back into the island it was picked from, and a reinforcement signal is calculated based on two factors: the fitness of the genome and its ranking within the island. Positive reinforcement is applied if the ranking is non-negative, which indicates that the genome was successfully inserted. Negative reinforcement is applied if the genome was not inserted and if the probability associated with the operation that generated that genome exceeds its degradation threshold. The probabilities of all actions are then updated using their respective learning functions, which ensure that no probability drops below its minimum threshold.

Stage transitions occur as necessary and the associated operations take place. The evolutionary process needs to warm up for approximately 100 iterations before probability adjustment is viable, so no learning takes place during the first stage. The starting action probabilities, minimum probabilities, degradation thresholds, and reinforcement functions are determined independently for each FALA process. For all learning processes, the positive reinforcement signal is calculated as a weighted average of a genome's ranking and its test set performance, whereas the negative reinforcement signal is a fixed constant. EXAMM's varying probabilities were adapted by FALA as follows:

Genetic Operation Probabilities: The action set and initial probability vector $p(0)$ are detailed in Table 1. There are no additional considerations as this process maps naturally to the FALA framework.

Island Selection Probabilities: At each iteration, EXAMM initially selects an island in a round-robin manner, choosing a genome from the island at random and evolving it. This is modified so that after the first stage is complete, EXAMM selects an island according to an evolving distribution. The action set is the collection of all islands, and the probability vector stores the associated probabilities. The probability vector is initialized uniformly with each island receiving a starting probability equal to $1/N_{islands}$. The reinforcement signal has an additional component, granting a small reward to the island with the best genome after each iteration.

Recurrent Depth Probabilities: The depth of a newly inserted recurrent edge is initially drawn from a uniform distribution, with values ranging between the minimum and maximum recurrent depths set by the programmer. Newly created recurrent edges are marked so that the learning process can identify which recurrent edges belong exclusively to the new genome, ensuring that only those depths that contributed to the improvement are reinforced. The action set is the list of all possible recurrent depths, and the probabilities are initialized uniformly to $1/N_{depths}$ for each depth. Some operations may result in the addition of multiple recurrent edges in a single iteration. The reinforcement function accounts for this by dividing the total reinforcement by the number of new recurrent edges and reinforcing each depth accordingly.

4.1   Exploration vs Exploitation

The exploration vs exploitation dilemma is common in machine learning, and is certainly present in this context. When it comes to FALA, exploration generally implies slower learning to allow the algorithm to explore various combinations without getting too entrenched in initial positive results. On the other hand, exploitation suggests faster learning where actions that initially lead to good results are strongly reinforced. It is possible to play these concepts off of one another: one option is to have fast learning with generous slack but to frequently reset the probabilities by having short stages. This allows the algorithm to rapidly increase the probabilities of the actions that are doing well while the frequent resets during stage transitions force the algorithm to re-discover desirable actions. Another option is to have slower learning and longer stages. This approach takes longer to learn but it is more robust to outlier situations where an action which is not generally desirable is reinforced due to some low-probability event. This tradeoff is controlled by adjusting the various FALA hyperparameters. It is interesting to note that a slower learning approach with no stage transitions may end up being more exploitative in the long run, as the probabilities get fixed over time. An approach that includes staging to ensure that probabilities are reset is going to be more exploratory due to the ability to reconfigure parameters to the current state of the evolutionary process more quickly.

5   Experimental Approach

In order to evaluate the efficacy of FALA in improving the performance of EXAMM, three different configurations were considered: Base EXAMM to provide a point of comparison, a slower learning approach, and a faster learning approach. Each configuration was executed 10 times, and each run was configured to use five islands each with a population of five genomes. Initial probabilities for genetic operations were set in accordance with the values shown in Table 1. The minimum recurrent depth was set to 1 and the maximum recurrent depth was set to 10. The runs were programmed to terminate when 2000 genomes were generated. The FALA hyperparameters for each configuration are summarized in Table 2. It is noteworthy that the number of hyperparameters for a given learning process is lower than that of base EXAMM, and may be reduced even further through additional refinement of the algorithm. This fact highlights an additional benefit of the proposed approach: there are less hyperparameters for the programmer to tune. Furthermore, the number of FALA hyerparameters is relatively constant and does not increase as new actions are introduced.

The best genome from each run was recorded and the results were compared using the Mann-Whitney U test to determine whether the difference between the run configurations was significant. All runs were carried out using the coal-fired powerplant dataset, which is a collection of minute- by-minute readings taken from 12 sensors in a coal powerplant burner over 120 days. Readings from sensors 1 to 11 are used as predictor variables and the reading from sensor 12 (main flame intensity) is the response.

6   Results

The experimental results are presented in Table 3 and visualized in Fig. 1. The comparisons of different populations using the Mann-Whitney U test are presented in Table 4. These results demonstrate that the extension of EXAMM with FALA was successful in improving the overall performance. It is clear that the

Table 2. Run configurations

Approach	Learning process	Stages	Learning rates	Min Probs	Degr Thres	Neg Reinf
Slower learning	Islands	4	0.005	0.2, mult by 0.965 every stage	0.2, add 0.02 every stage	0.3, add 0.05 every stage, scaled to degr thres
	Genetic	Unstaged	0.0005	Mutation: 0.055 CO: 0.17 Island CO: 0.09	Mutation: 0.075 CO: 0.21 Island CO: 0.105	0.5, scaled to degradation threshold
	Depths	Unstaged	0.0005	0.08	0.12	0.5
Faster learning	Islands	10	0.007	0.2, sub 0.00625 every stage	0.2, add 0.00625 every stage	0.5, scaled to degr thres
	Genetic	10	0.002	Mutation: 0.04 CO: 0.14 Island CO: 0.06	Mutation: 0.10 CO: 0.26 Island CO: 0.14	1.0, scaled to degr thres
	Depths	10	0.002	0.06	0.12	1.0

means and medians of the slower and faster learning configurations are better than those of the base algorithm. The result of the Mann-Whitney U test between the base configuration and slower learning configuration is significant at p = 0.05. The p-value of the Mann-Whitney U test between the base configuration and the faster learning configuration is relatively low but not considered significant at any commonly-accepted threshold.

Table 3. Experimental results - lowest test set MSE from each run

	Base EXAMM	Slower learning	Faster learning
Run 1	0.001418	0.0013	0.001778
Run 2	0.001584	0.001291	0.00119
Run 3	0.00178	0.001629	0.001228
Run 4	0.001472	0.001221	0.00127
Run 5	0.001492	0.001168	0.001436
Run 6	0.001342	0.00138	0.001271
Run 7	0.001254	0.000952	0.001356
Run 8	0.001456	0.001374	0.001526
Run 9	0.001966	0.001399	0.001218
Run 10	0.0012	0.001134	0.001536
Mean	0.0014964	**0.0012848**	0.0013809
Median	0.001464	**0.0012955**	0.0013135
Min	0.0012	**0.000952**	0.00119
Max	0.001966	**0.001629**	0.001778
Variance	$5.4163 * 10^{-8}$	$\mathbf{3.3335 * 10^{-8}}$	$3.5243 * 10^{-8}$

Assuming that the slower learning approach is indeed superior to the faster learning approach, it would appear that in the context of EXAMM, exploration

Table 4. Comparison of results from Table 3 using the Mann-Whitney U test

Population 1	Population 2	P-Value
Base EXAMM	Slower learning	**0.03546**
Base EXAMM	Faster learning	0.2799
Slower learning	Faster learning	0.4359

is preferred to exploitation. This could be because exploitation is difficult in such a dynamic environment, where the population of genomes is constantly being renewed. Another explanation could be that favoring any one selection too heavily is detrimental for the evolution of new genomes, implying that a successful genome is evolved when all selections are used at least somewhat evenly. To better understand which selections were more successful, we can analyze the plots of the different probabilities over time from different runs. Figures 2 and 3 show the evolution of genetic operation, island selection, and recurrent depth probabilities for the best runs in both the slower and faster learning approaches respectively. Figures 4 and 5 show the same graphs but for the worst run from each configuration.

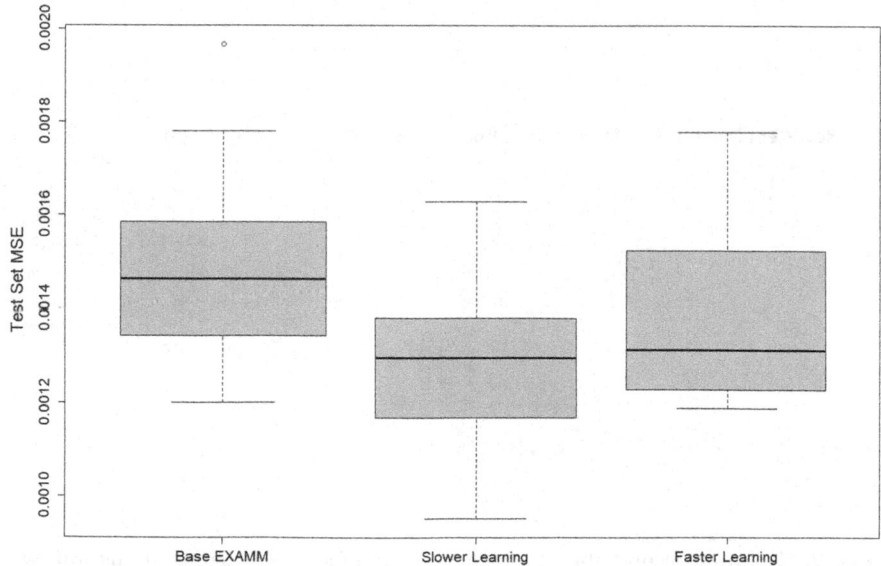

Fig. 1. Distribution of results from Table 3 by configuration

The difference between the slower and faster approaches is evident through the evolution of genetic operation and recurrent depth probabilities. This provides a good visualization of the exploration vs exploitation tradeoff discussed

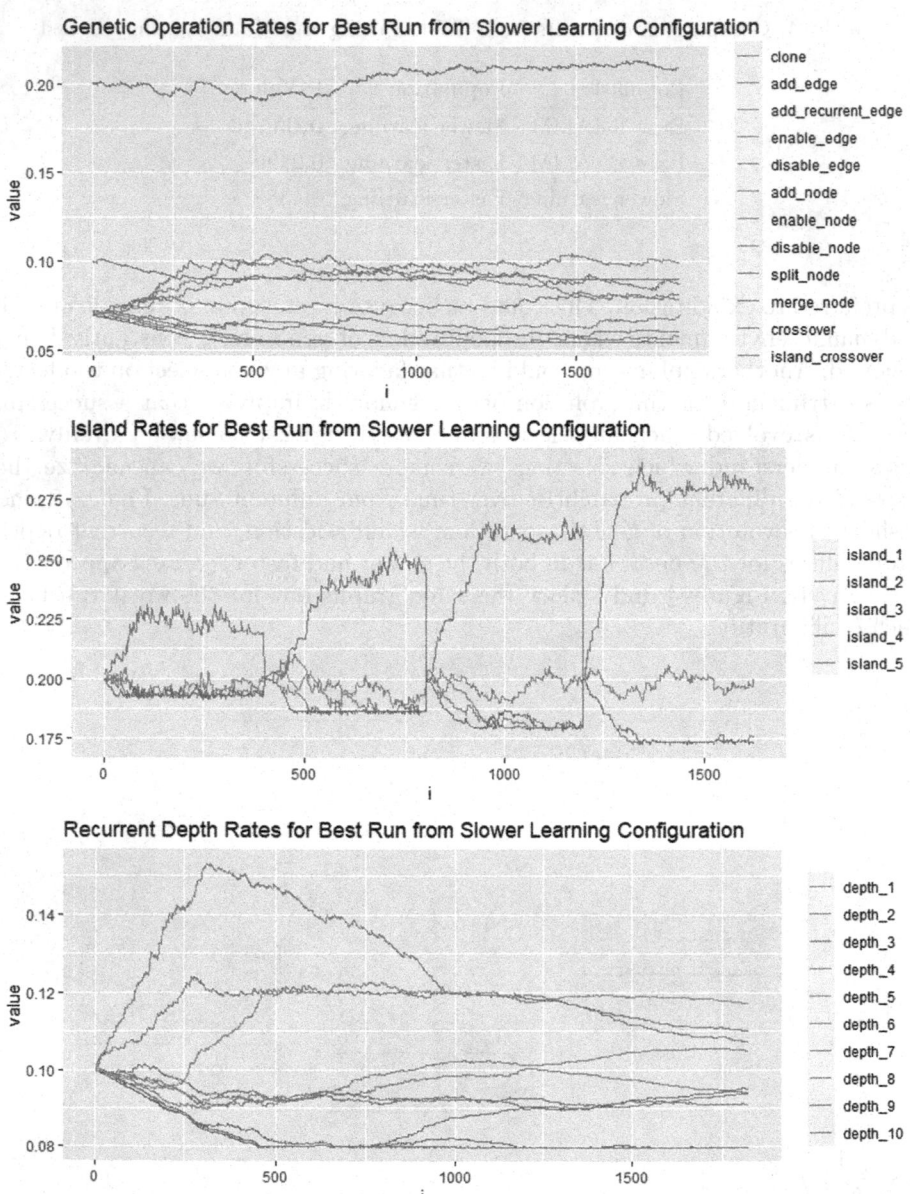

Fig. 2. Evolution of probabilities from best run of slower learning configuration

in Sect. 4.1. The evolution of the island selection probabilities is actually some-what comparable, apart from the fact that one has four stages and the other has ten. This makes sense because their hyperparameter configuration as out-lined in Table 2 is almost identical. It is interesting that for both of the worst runs, the island selection learning process has trouble settling on a best island,

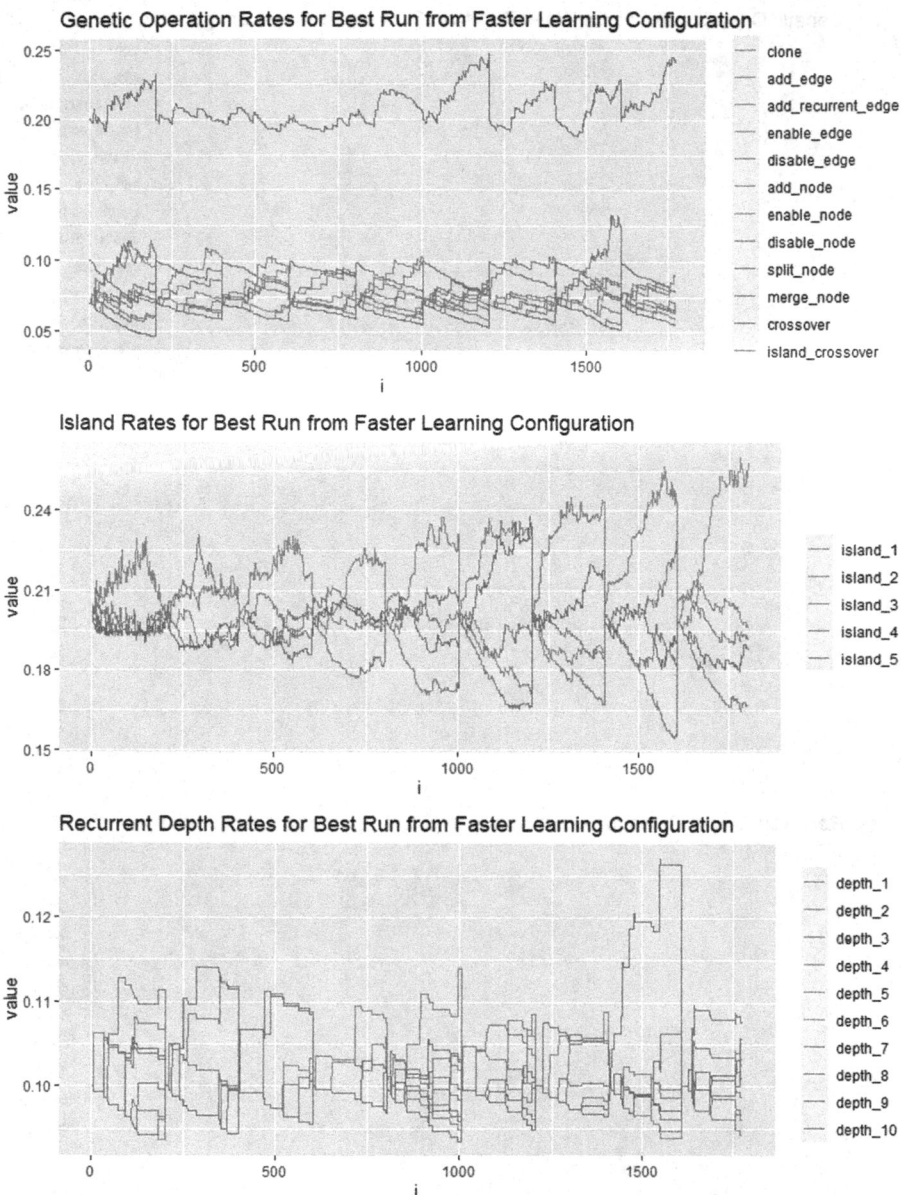

Fig. 3. Evolution of probabilities from best run of faster learning configuration

whereas for the best runs the learning process settles on an island (or two) early on. A potential explanation for this is that the performance of a run is largely dependent on how well it starts. If the algorithm has a poor start, we tend to observe multiple islands with comparable genomes that will compete with each other throughout the learning process. However, if the algorithm has a good

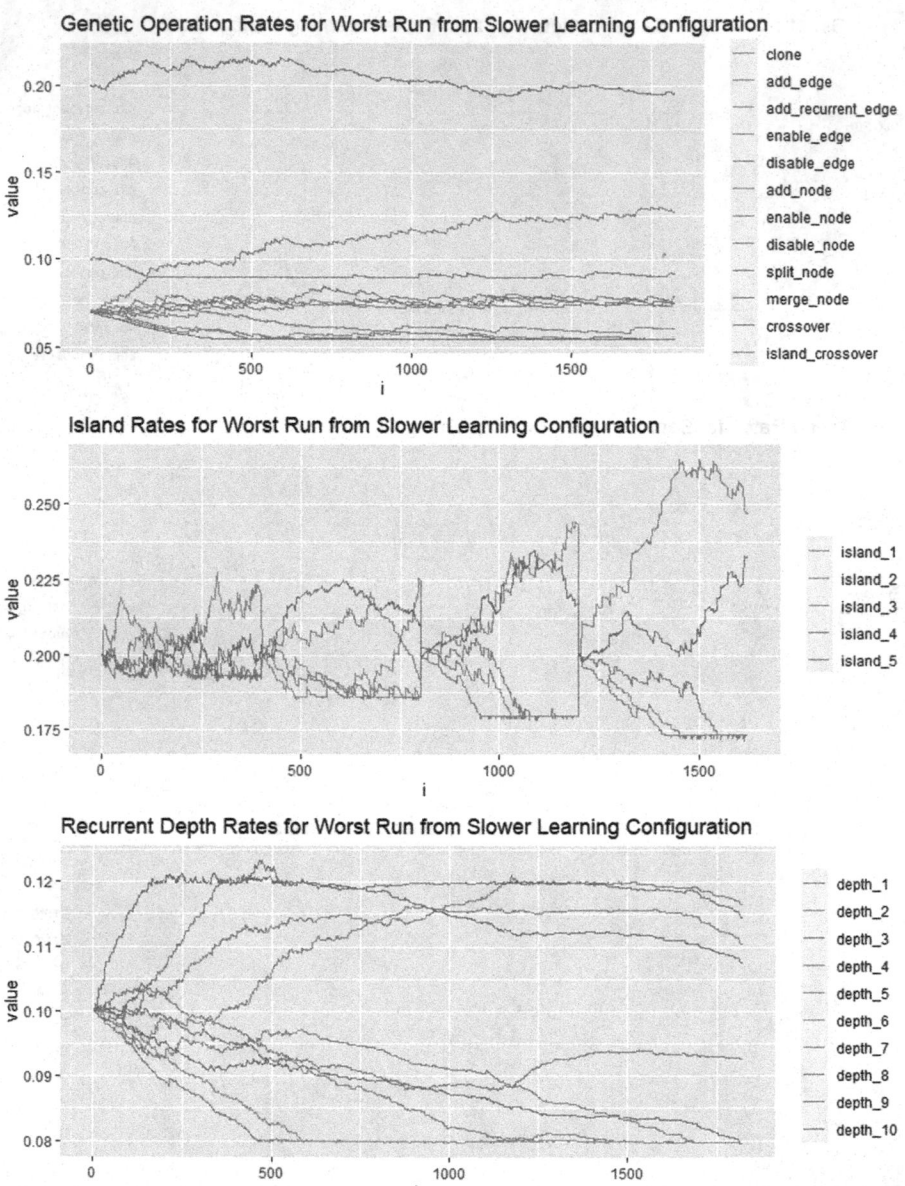

Fig. 4. Evolution of probabilities from worst run of slower learning configuration

start, there will likely be one or two islands with genomes that are superior to the populations of the other islands, and this is reflected in the probabilities of those islands being heavily reinforced.

Probability dynamics for genetic operations are better visualized in the faster configuration, and it is evident that no single operation dominates the others,

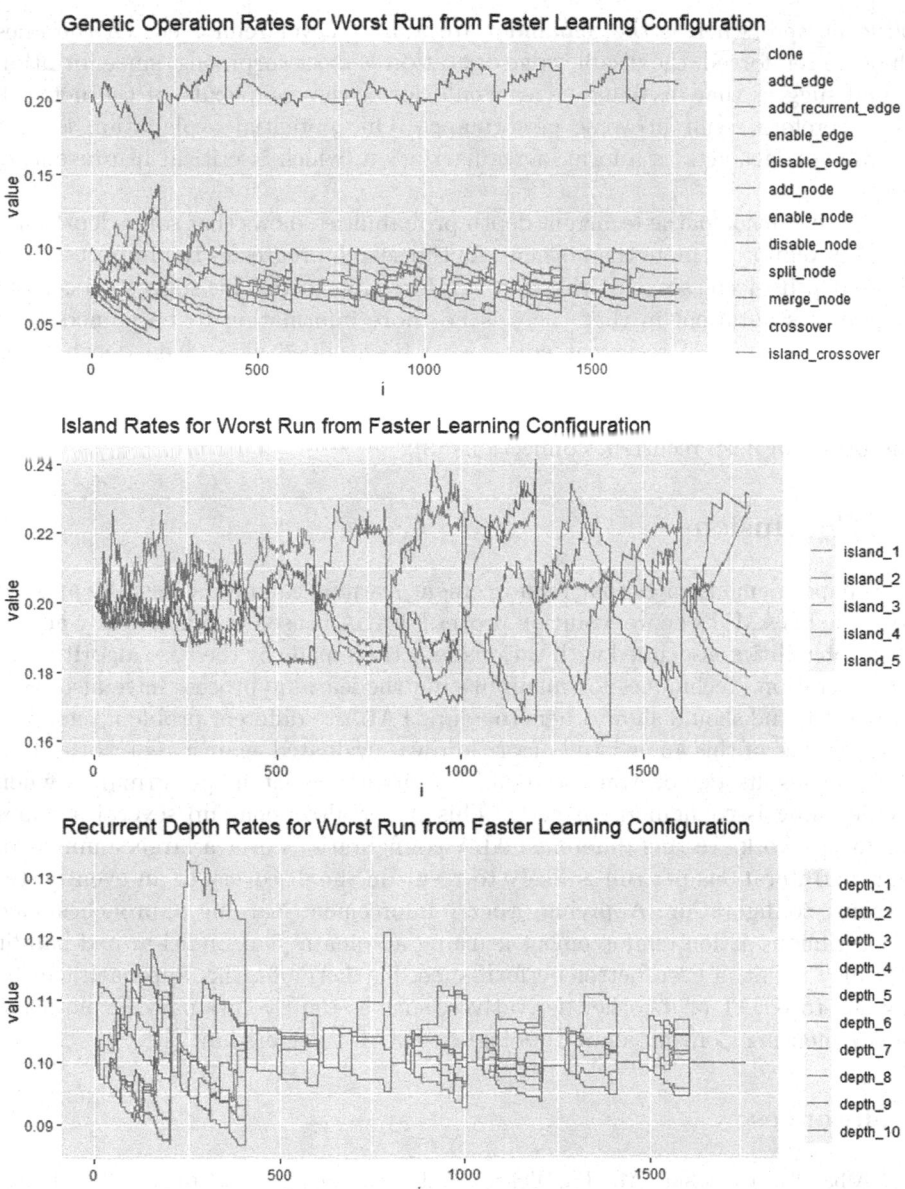

Fig. 5. Evolution of probabilities from worst run of faster learning configuration

despite the high learning rates and frequent resets. That being said, there are operations that almost never see their probabilities increased, implying that they are less likely than the others to generate good genomes. The crossover operation seems to always perform well, as do the clone, add recurrent edge, and add node operations. The add edge, disable edge, and island crossover operations also perform well, whereas the remainder are almost never reinforced. Of the ones that are reinforced, the disable edge operation is most surprising, since intuition would suggest that disabling edges would lower the complexity of the network and therefore result in worse performance. One potential explanation is that disabling edges acts as a form of regularization, which is critical in preventing overfitting.

The evolution of the recurrent depth probabilities shows that some depths are initially reinforced more than others. As the search progresses the importance of those depths decreases and others are reinforced instead. This suggests a certain degree of saturation, in that there is no more information to be gained after a certain number of recurrent edges have been added at a given depth. It is noteworthy that larger depths appear to have been consistently reinforced in both learning configurations, corroborating findings by Desell *et al.* regarding the utility of deep recurrent connections [3].

7   Conclusion

This paper demonstrates that reinforcement learning can be successfully applied to guide EXAMM's neuroevolution process. FALA is used to dynamically adjust the probabilities associated with various selections made by the base algorithm at each iteration. Techniques for manipulating the learning process have also been presented, and should allow others to adapt FALA to different problem sets. The performance of this augmented algorithm was evaluated against real-world data and the results demonstrate a significant improvement in performance when the learning is configured correctly. This paper also opens up several avenues for future work. Testing different FALA configurations over a larger number of runs and iterations per run is likely to result in the discovery of an even better learning configuration. Applying other reinforcement learning approaches such as continuous action reinforcement learning automata [5] or multi-armed bandit [19] may result in even better performance. Furthermore, this approach can be applied to NEAT or its other derivatives, such as the Evolutionary Exploration of Augmenting Convolutional Topologies (EXACT) algorithm [2].

References

1. Alba, E., Tomassini, M.: Parallelism and evolutionary algorithms. IEEE Trans. Evol. Comput. **6**, 443–462 (2002)
2. Desell, T.: Large scale evolution of convolutional neural networks using volunteer computing, pp. 127–128 (2017)
3. Desell, T., ElSaid, A., Ororbia, A.: An Empirical Exploration of Deep Recurrent Connections Using Neuro-Evolution, pp. 546–561 (2020)
4. Floridi, L., Chiriatti, M.: GPT-3: its nature, scope, limits, and consequences. Minds Mach. **30**, 1–14 (2020)
5. Howell, M., Best, M.: On-line PID tuning for engine idle-speed control using continuous action reinforcement learning automata. Control Eng. Pract. **8**, 147–154 (2000)

6. Jardine, P.T., Kogan, M., Givigi, S.N., Yousefi, S.: Adaptive predictive control of a differential drive robot tuned with reinforcement learning **33**(2), 410–423 (2018)
7. Jardine, P.: A Reinforcement Learning Approach to Predictive Control Design: Autonomous Vehicle Applications. Ph.D. thesis, May 2018
8. Jardine, P.T., Givigi, S.N., Yousefi, S.: Experimental results for autonomous model-predictive trajectory planning tuned with machine learning. In: 2017 Annual IEEE International Systems Conference (SysCon), pp. 1–7 (2017)
9. Kogan, M., Jardine, P.T., Givigi, S.N.: Architecture for testing learning-based autonomous vehicle control design. In: 2018 Annual IEEE International Systems Conference (SysCon), pp. 1–7 (2018)
10. Lyu, Z., Karns, J., ElSaid, A., Desell, T.: Improving neuroevolution using island extinction and repopulation, May 2020
11. Matuszewski, J., Rajkowski, A.: The use of machine learning algorithms for image recognition. In: Radioelectronic Systems Conference 2019, vol. 11442, pp. 412–422 (2020)
12. Narendra, K.S., Thathachar, M.A.L.: Learning automata - a survey. IEEE Trans. Syst. Man Cybern. SMC-**4**(4), 323–334 (1974)
13. Ororbia, A., ElSaid, A., Desell, T.: Investigating recurrent neural network memory structures using neuro-evolution. In: Proceedings of the Genetic and Evolutionary Computation Conference, pp. 446–455 (2019)
14. Radaideh, M.I., Shirvan, K.: Rule-based reinforcement learning methodology to inform evolutionary algorithms for constrained optimization of engineering applications. Knowl. Based Syst. **217**, 106836 (2021)
15. Barros dos Santos, S.R., Givigi, S.N., Nascimento, C.L.: Autonomous construction of multiple structures using learning automata: description and experimental validation. IEEE Syst. J. **9**(4), 1376–1387 (2015)
16. Sejnowski, T.J.: The unreasonable effectiveness of deep learning in artificial intelligence. In: Proceedings of the National Academy of Sciences, vol. 117(48), pp. 30033–30038 (2020)
17. Stanley, K.O., Miikkulainen, R.: Evolving neural networks through augmenting topologies. Evol. Computat. **10**(2), 99–127 (2002)
18. Thathachar, M.A.L., Sastry, P.S.: Networks of Learning Automata: Techniques for Online Stochastic Optimization. Springer-Verlag, Berlin, Heidelberg (2003). https://doi.org/10.1007/978-1-4419-9052-5
19. Vermorel, J., Mohri, M.: Multi-armed bandit algorithms and empirical evaluation. In: Gama, J., Camacho, R., Brazdil, P.B., Jorge, A.M., Torgo, L. (eds.) ECML 2005. LNCS (LNAI), vol. 3720, pp. 437–448. Springer, Heidelberg (2005). https://doi.org/10.1007/11564096_42
20. Weiss, K., Khoshgoftaar, T., Wang, D.: A survey of transfer learning. J. Big Data **3**, May 2016

Soft Computing Applied to Games

Automating Speedrun Routing: Overview and Vision

Matthias Groß[1]([✉]) [ID], Dietlind Zühlke[2] [ID], and Boris Naujoks[2] [ID]

[1] Advanced Media Institute, TH Köln, Köln, Germany
matthias.gross2@th-koeln.de
[2] Institute for Data Science, Engineering, and Analytics, TH Köln, Köln, Germany
{dietlind.zuehlke,boris.naujoks}@th-koeln.de

Abstract. Speedrunning in general means to play a video game fast, i.e. using all means at one's disposal to achieve a given goal in the least amount of time possible. To do so, a speedrun must be planned in advance, or routed, as referred to by the community. This paper focuses on discovering challenges and defining models needed when trying to approach the problem of routing algorithmically. To do so, this paper is split in two parts. The first part provides an overview of relevant speedrunning literature, extracting vital information and formulating criticism. Important categorizations are pointed out and a nomenclature is built to support professional discussion. The second part of this paper then refers to the actual speedrun routing optimization problem. Different concepts of graph representations are presented and their potential is discussed. Visions for problem solving are presented and assessed regarding suitability and expected challenges. Finally, a first assessment of the applicability of existing optimization methods to the defined problem is made, including metaheuristics/EA and Deep Learning methods.

Keywords: Speedrun · Routing · Graph models · Metaheuristics

1 Introduction

What remains after you finished your favorite game for the first time? There are different options like changing to another game, play it once again in a different way or getting deeper into the game and try to solve it faster. And then once again, even faster. This approach ultimately leads to speedrunning, i.e. trying to solve the game as fast as possible. That has become a rapidly growing sub-community around video games. One might consider special techniques, referred to as glitches, to speed up the run or strictly object to these. Either way, for a lot of players, speedrunning presents the ultimate challenge.

The charity driven speedrunning live event *Awesome Games Done Quick 2020* has attracted a maximum of 237,523 concurrent viewers [2] and from 2011 to 2022 the event's donation total has been raised from 52,519.83 to 3,419,397.85 USD [1]. Despite its popularity, scientific research on this specific field of e-sports is still only found sporadically. In one of the first works that formally

© Springer Nature Switzerland AG 2022
J. L. Jiménez Laredo et al. (Eds.): EvoApplications 2022, LNCS 13224, pp. 471–486, 2022.
https://doi.org/10.1007/978-3-031-02462-7_30

covered speedrunning, Newman describes it as *"[...] concerned with completing videogames in as speedy a time as possible"* [27].

To do so, speedrunners traverse the progression mechanics of a given game as quickly as possible. This can mean very different things depending on the game, from going around a racing track optimally to solving puzzles in the most efficient order.

Being a discipline building heavily on optimization, it's only natural that speedrunning has become subject of various approaches to enhance this optimization algorithmically. As diverse as the game mechanics are, so are the means by which they have been tried to be supported. Finding successful policies for finishing a game is often approached using (Deep) Reinforcement Learning (see e.g. [36]). For speedrunners, however, policies from Reinforcement Learning are a too detailed level of strategy. They prefer to fix only the way points between helpful events and leave the rest to the skill of the player. Thus we will focus on a different approach that allows optimizing the order of certain events and way points in the game's progression.

During the course of this work, Nintendo's famous Action-Adventure title *The Legend of Zelda: Ocarina of Time (OoT)* will be used as a working example.

The first part of this paper introduces speedrunning in general and builds understanding of the associated *routing* problem, the main focus of this paper. In *Speedruns and Categories* (2), this work elaborates on the general description of speedrunning as well as the terminology as used in the community. The *Related Work* section (3) focuses on existing scholarly works regarding speedrunning, including nomenclature to support professional discussions. Connections between these works are made and criticism is formulated at the end of the section.

Equipped with this knowledge, the second part of this paper approaches the actual speedrun routing optimization problem. The *Envisioned Models and Challenges* (4) section presents approaches and ideas to formalize the problem. Different models are envisioned and emerging challenges are critically discussed. After these approaches to define the underlying problem, the *Prospective Solutions* (5) section elaborates on possible approaches to design algorithms for this problem, also taking metaheuristics/EA and Deep Learning methods into account. Finally, the paper closes with *Conclusions and Outlook* (6). All game titles used are listed in the *Ludography* at the end of this paper.

2   Speedruns and Categories

Speedrunning in general means to play a game fast, i.e. using all means at one's disposal to achieve a given goal in the least amount of time possible. However, what is at one's disposal and what constitutes the given goal can vary widely. To be comparable to each other and to give consistency for the runners, there are rules imposed on speedruns. These rules are decided upon by the game's community, often by means of polls. For a run to be listed on its leaderboard, it must comply to these rules, proven by uploaded footage of the corresponding runs. As the largest accumulation of speedrun footage and leaderboards, the website speedrun.com will be used as point of reference here [4]. Some of the rules concern the form of proof each runner must submit with a claimed completion time, but this paper will only take the rules imposed on actual gameplay into account. There are some site-wide rules that apply to all runs on speedrun.com, e.g. the ban of any hardware manipulation. There can also be game-wide rules all runs of this particular game must comply to. Most of the site's leaderboards additionally list multiple *categories* for the same game. These categories are defined by different rulesets for a game's speedruns, each with their own leaderboard. Although multiple games might share similar *categories*, this does not necessarily mean any similarity beyond the names. The underlying rulesets can differ largely from game to game. Popular categories for many games are *any%*, *100%* and *Glitchless*.

Any%. Runs under the category *any%* usually don't have any additional rules imposed on them, other than getting from a defined starting state to a defined ending state by any means possible, while respecting community-wide rules. The name is derived from the fact, that any percentage of the game might be completed before ending the run. This also usually means that these speedruns involve a lot of unconventional gameplay using programming mistakes and inconsistencies in the game to lower the completion time as much as possible. This unrestricted use of such *glitches* and *exploits* – both of which will be discussed in more detail later on in *Related Work* (3) – often change the gameplay to a degree that makes the run look very different from a regular playthrough of the game.

100%. Runs need to accomplish every objective in the game before completing it. The definition of "every objective in the game" has to be agreed upon by the community and defined in the category's ruleset. If the game provides a completion percentage display, e.g. in a saving, loading or status menu, it is used as the authoritative reference most of the time. The nature of these runs vary heavily depending on the game being run. As there often are a lot of objectives to achieve, the usage of glitches is prevalent.

Glitchless. Sometimes, newfound glitches completely trivialize a game's speedrun or parts of it. Other times, the game's community finds a speedrun without glitches more appealing to do or watch [33]. In such cases, the glitches in question can be banned in the categories' rulesets, or *Glitchless* leaderboards can be created. As the name suggests, *Glitchless* categories prohibit the use of *glitches*, leading to runs more similar to casual playthroughs. Again, what constitutes a *glitch* is agreed upon by the community, as it is not a simple distinction [30]. Alternatively, the ruleset can list all allowed techniques, all banned techniques, or both.

Categories can be combined as applicable, e.g. *OoT* has, among others, an *any%*, a *100%*, a *Glitchless any%* and a *Glitchless 100%* leaderboard on speedrun.com. The resulting difference in permitted techniques and completion requirements have a heavy impact on the run.

Another important categorization of speedruns is the differentiation between a *Tool Assisted Speedrun* (TAS) and a *Real-Time Attack* (RTA) run. RTA runs are performed by a human in real-time. Although the term "speedrunning" can be seen as more of an umbrella term, RTA runs are the most prevalent form of speedrunning and are generally tied to the term unless specified. A TAS, in contrast, is prerecorded and assisted by tools, such as being able to create exact game state save points at arbitrary times. Every input for every frame or simulation step in the game is handcrafted and scripted beforehand and then played back to create a perfectly optimized speedrun. TASs are often created to demonstrate what could be theoretically achieved in a perfect run, or just for entertainment. A variation of TASs are *Low Optimization Tool Assisted Demonstrations* (LOTADs as referred to in the speedrunning community), which usually demonstrate more exotic ways of playing a game or are used as reference for other runs.

One challenge that every speedrun, regardless of category, has to accommodate is *routing*. With regard to speedrunning, routing is the act of planning out a playthrough of a game in a fashion that needs the least amount of time possible. Often, the game is broken down to individual events, depending on the game (locations or characters to visit, items to collect, checkpoints to drive by etc.). Then, a route trough the game world has to be constructed, that covers all events needed for game completion. In many cases, specific vital events have preconditions in form of other events, that have to be incorporated into the route. A good route is crucial for optimizing any given speedrun. That said, this statement of course holds only true as long as the player has any choice on the ordering of the events. Perfectly linear games may not have a strong focus on routing, if any at all.

3   Related Work

Aside from the categories and nomenclature presented earlier, there are also some scholarly works that categorize different aspects of speedrunning. Besides the gain in public interest in speedrunning, only a small amount of such works on this field has emerged approximately in the last decade. While a lot of these works focus on the interesting narrative and sociological consequences of the emergence of speedrunning as a mode of play [15,18,19,27,28,30,32–34], this body also includes terminology and covers practical problems of speedrunning.

Newman and Scully-Blaker both introduce nomenclature regarding the nature of speedruns and the involved activities employed by speedrunners.

3.1   Newman's Activity Categories

In a recent work, Newman [28] presents a thorough insight into the narrative consequences of speedrunning on the example of *OoT*. They also introduce categories of speedrunning activity: *hidden affordances, exploiting inconsistencies* and *manipulation and reconstruction.*

Hidden Affordances. In Newman's definition *hidden affordances* can be broken down to optimally utilizing the designerly intended game mechanics.

For example in *OoT* going backwards is faster than going forwards, so speedruns almost always consist of Link (the protagonist of most of the games from the *Legend of Zelda* franchise) walking backwards a lot of the time, which is referred to as *backwalking*[1]. While Newman also states that *"'designerly intent' is, obviously, difficult to assert"*, they use as much official documentation about the game as possible to do so.

Exploiting Inconsistencies. Is the most expressive designation. Newman details it as *"exploit[ing] programming bugs and systemic errors in the code's design and execution"* [28]. An example of this is *clipping*, a technique used in speedruns of many games. *Clipping* refers to the penetration of walls or other geometry in the game world the player is not supposed to go through.

Manipulation and Reconstruction. The third category of activity is denoted *manipulation and reconstruction*. Newman describes it as follows:

> *"The outcome of the technique is the creation of a connection in code that is subsequently rendered in polygonal space between regions in the game's spatial and narrative architecture that are as palpably unintended by OoT's developers as they are disruptive and apparently injurious to the integrity of the game's myth."* [28]

Activities of this category include techniques that consciously make use of the game's inner mechanics a player would normally be unaware of. Creating techniques like these in many cases requires extensive knowledge about the game's logic and/or code. The *Reverse Bottle Adventure* (*RBA*) technique for example makes use of knowledge about the IDs that each item in the game internally is identified by. With this, executing very specific actions in the game can add or remove specific items from the player's inventory, based on the ID or amount of other items already in the player's inventory [3].

These categories can help to define the earlier introduced term *glitch*. Although there is no universal consensus in the community, activities that fall into Newman's categories of *exploiting inconsistencies* and *manipulation and reconstruction* are often referred to as *glitches*. This work will follow this specification.

3.2   Scully-Blaker's Speedrun Categories

Scully-Blaker [32] distinguishes between *finesse runs* and *deconstructive runs*, regarding the degree of glitch exploitation and sequence breaking.

Finesse runs are detailed as *"runs in which the player interacts with the game as an extreme extension of what a game designer may consider an 'ideal' player [...] largely respecting the game's 'narrative boundaries' while navigating them with an extreme level of efficiency"* [32]. Put differently, one could say a *finesse run* consists of playing a game as intended – as far as ascertainable – but very fast.

---

[1] To put it in Newman's words: *"That which is in front of Link is space already consumed."* [28].

Deconstructive runs on the other hand do not maintain any boundaries. These runs are *"runs in which the player exploits glitches within the game to break scripted sequences"* [32], skipping and reordering game content at will, as long as it serves the primary directive: speed.

3.3   Routing

As outlined at the end of the Introduction (1), speedruns have to be planned in advance, a procedure referred to as *routing*. Newman [27] writes:

> *FPSs [First-Person-Shooters] are favoured because of their apparent non-linearity and the scope they seem to afford gamers to invent and create their own routes and develop their own styles that move them through the gameworld."*

Newman goes on to also include Role-Playing-Games (RPGs) into these thoughts.

Speedrun routes consist of a number of events, connected by dependencies, starting with the defined opening event and ending with one or multiple possible ending events. The problem of finding the ordering that takes the least amount of time possible is combinatorial in nature. Some approaches to formalize and algorithmically address the problem of speedrun routing have been documented [21–23] but it is still underrepresented in research. In these approaches, routing is usually defined as a graph shortest path problem.

Lafond [23] takes the action-platform title *Mega Man* as well as its successors *Mega Man 2* and *Mega Man 3* as examples. One key feature of these 2D Jump'n'Run variants is the aquisition of a new power after clearing each stage of the game. These powers can be used to speed up subsequent stages. All stages have to be cleared, while the order in which the stages are cleared is mostly left to the player. Lafond defines sets of stages $\mathbb{S} = \{S_1, ..., S_n\}$, with each stage $S_i = \{s_1, ..., s_k\}$ being a set of events on which time can be saved during a run[2]. This save depends on previously completed stages, giving $s_j : \mathbb{S} \rightarrow \mathbb{N}$ as a function mapping each previously completed stage to a specific save for each event(See footnote 2). They form a dependency graph for a given game to be optimized, providing a route with maximum time save through the game. Even under favorable assumptions, Lafond [23] proves this problem to be W[2]-hard [12].

A more informal approach has been used by Iškovs [21] with the working example of the popular game *The Elder Scrolls III: Morrowind* As a typical open-world RPG title, there is a lot of traveling, character skill development and Non-Player-Character (NPC) relationships involved in the game mechanics. The category Iškovs takes on is *all factions*, consisting of becoming the leader of all possible factions of NPCs in the game. Taking into account the mentioned aspects among others, they formed an extensive *quest dependency graph*. Iškovs then continues to use an evolutionary algorithm to generate and improve on possible routes through the game progression. A lot of handcrafted customizations

---

[2] Symbols altered from original to prevent ambiguities.

are then applied to minimize the run's time even further. This yields a very specific routing tool for a single game. At the time of writing, there is only one entry in the corresponding leaderboard on speedrun.com, the creator of which learned the route from Iškovs' efforts [39].

Another informal example of algorithmic routing has been conducted by speedrunner JaV [22], who breaks down a track in the car racing game *TrackMania Nations Forever* to a variation of the travelling salesperson problem. Track completion in this game is done by passing a number of checkpoints. In JaV's approach, these checkpoints are represented as a complete weighted digraph. They then manually remove specific implausible edges from the graph and use a genetic algorithm [5] to generate near-optimal solutions. The used algorithm had to be adapted to avoid certain implausible combinations and the solutions had to be curated, i.e. manually test-driven in the game to check for plausibility. Though this adaptions resulted in a specialized tool, it is one of the rare examples that led to an improvement on the leaderboard times of a game.

Taking it to the extreme, the process of routing can further be broken down to single simulation steps of a game's engine. In 2009, a competition has been conducted with *"The focus [...] on developing controllers that could play a version of Super Mario Bros as well as possible"* [38]. The competition brought forth a number of submitted tools and algorithms to automate playing a variant of the popular game *Super Mario World*. The winner of the competition modeled the game state of each simulation step as a node of a graph, and each possible next state (depending on different inputs) as adjacent nodes. Then, the shortest path to the goal – or, as an approximation within available knowledge, to the right screen edge – was determined through a variation of the A* algorithm [17]. The second and third best contributors of the competition used similar approaches.

3.4   Criticism

Glitches are often what differentiate Scully-Blaker's categories of runs from each other. Scully-Blaker's definitions of run categories can be supported by Newmans terminology of activities. While *finesse runs* drive *hidden affordances* to the extreme but leave the scripted game sequence and narrative structures mostly intact, *deconstructive runs* employ glitches and make heavy use of them, often distorting and/or completely reassembling the event sequence. However, these two categories are not to be seen as distinct. The given descriptions are more likely two extremes of the space in which speedruns and their categories range.

Given this relation between Newman's and Scully-Blaker's terminology, connections to the leaderboard categories can be drawn as well. As *any%* speedruns often seek to use whatever means possible to get to the defined end state, usage of glitches is prevalent. More so, the used glitches can often be classified as Newman's *exploiting inconsistencies* or *manipulation and reconstruction* activities, therefore these runs often range more in the realm of *deconstructive runs*. On the other side of the spectrum there are the *Glitchless* leaderboard categories, which are governed by strict rules, mostly oriented on narratively and designerly intentions. As such, these runs range more in the realm of *finesse runs*.

In a recent work, Ricksand [30] expresses disagreement with Scully-Blaker's categorizations, listing lack of a definition of the term *glitch*, lack of definition of completion requirements and inconsistencies in definition of an ideal player as reasons. However, considering distinctions already made in this paper, Scully-Blakers categories are considered suitable here. As a slight addition, Scully-Blaker's definition of an "ideal player" can be extended by means of Newman's *hidden affordances*, which then would include more techniques challenged by Ricksand like *backwalking*.

Ricksand also questions the current procedure of rule acquisition in the speedrunning community. In their work, they suggest that the current process of voting on rulesets is flawed because it results in rules being arbitrary. However, the argument can be made that speedrunning in its very essence is arbitrary. Ultimately, Ricksand proposes the following model:

"Mechanic m in game g is allowed in a glitchless speedrun if and only if use of m does not contradict the fictional truth regarding the world in which the story of g takes place." [30]

Not only does this not account for any technical glitch that doesn't get reflected by anything in the game world, it also introduces the concept of fictional truth in order to legitimize glitchless rulesets. Fictional truth however is partly up to the audience of a narrative [7], rendering it arbitrary.

If nothing else, this disagreement shows the difficulty in deciding what to categorize as a glitch, in scholarly work as well as in the speedrunning community.

The 2009 Mario AI Competition [38] did yield a number of good performing agents. This competition can be seen as a TAS creation competition: Although there were no input prerecordings and the competition scoring incentivised total progression amount rather than speed, the submitted agents often played the game extremely fast. Those efforts resemble TASs rather than RTA runs and have mostly been focusing on what will be called *operational level routing* or *operational routing* in this paper, i.e. optimally planning or deciding on specific inputs at specific frames or simulation steps to reach a given short term goal or state. This focus and the success of the developed tools is partly due to a property of the 2D-platformer nature of the *Super Mario* titles – going right on the screen is almost always a save indication of progress. In fact, this metric has been used by the 2009 competition's winner as an heuristic for the A* algorithm [38]. Optimizing the traversal of the game's progression mechanic meant going right as quickly as possible. However, this holds true only for the specific variation of the *Super Mario* title used (consisting only of horizontally oriented levels, progressing to the right) and only if isolated levels are taken into account. In contrast, to minimize the total time of an entire game's completion as the previously presented works did, *strategic level routing* or *strategic routing* is employed. As a result, speedruns of games which rely heavily on *strategic routing* could not yet benefit from these advances. The strategic layer of planning the traversal between levels – or, more generally speaking, between different events, from a speedrun's opening event to an ending event – is generally what is referred to as *routing* in the speedrunning community as it is in this work.

In many cases, routing is not a one-time process. Especially with *deconstructive runs* [32] with a high amount of *inconsistency exploiting* and *manipulation and reconstruction* activities, routes are significantly reworked with new glitches and exploits being found, due to their often disruptive nature regarding the game's intended event sequence. This act of altering the game's intended order of progression by means of glitches is referred to as *sequence breaking* in the speedrunning community. This leads to a challenge when defining the underlying routing problem, as these often unpredictable changes in routing possibilities have to be accounted for. *Glitchless* speedruns can mostly be characterized as *finesse runs* [32], as the use of glitches is prohibited in these rulesets. Therefore, routing underlies more consistent structures and can easier be defined formally.

Some challenging aspects of speedrunning have been engaged by the presented works, others are yet left unaccounted for.

4   Envisioned Models and Challenges

With the knowledge compiled to this point, the actual speedrun routing optimization problem can be approached. To do so, this problem has to be formalized and defined as a model. Given the outlined nature of the routing problem, graph representations will be favored. The presented models are considered visions that can be elaborated and improved on rather than final solutions. Again, *OoT* will be used as the working example and results may be less applicable to other games or genres.

4.1   Weighted Game Event Digraph

As a first approach, a weighted game digraph is assumed

$$G = (V, E, w), \tag{1}$$

with nodes $V = \{v_1, ..., v_n\}$ as the set of all events relevant to the game's progression. Each possible traversal between these events make up the set of directed edges $E = \{e_1, ..., e_m\}$ between the nodes representing the given events. Edges are weighted with a function $w : E \rightarrow \mathbb{R}$, assigning each edge the time it takes to traverse between the in-game events in the given direction. Edge directions are considered as this traversal time can differ by direction. Note that the term traversal is used as opposed to travel, as node traversal does not necessarily involve in-game movement. The routing process would then mean to find the shortest path between node v_α resembling the starting event and v_ω representing the ending event – or a set V_ω of ending events –, traversing all events $V_r \subseteq V$ deemed required by the category's ruleset.

This rudimentary model already introduces several challenges:

1. *Defining the Nodes*: "Events relevant to the game's progression" is not defined and it is nontrivial to do so.

2. *Defining the Edges*: The extent of "each possible traversal between events" has underlying restrictions that have to be defined as well.
3. *Dynamic Weights*: Edge weights are not consistent but rather change with graph traversal.
4. *Repeatable Events*: A subset of the events can be repeated once triggered, while others can not, further increasing the complexity.
5. *Multiobjective Optimization*: The model does not account for any dimension other than time in a possible route.

These challenges will be elaborated on in succession, followed by possible ways to accommodate them within the problem model.

Defining the Nodes. As for the definition of "events relevant to the game's progression", in the spirit of optimization it could be argued that only the items needed to qualify for the category should be involved, i.e. reducing (1) to

$$G_r = (V_r, E_r, w_r) \,, \tag{2}$$

with $E_r \subseteq E$ being the reduced set of edges remaining between V_r – weighted by time of traversal w_r –, as everything else would constitute a detour and thus cost additional overall time. Routing could be accomplished by finding the shortest topological ordering of G_r. However, an optimization frequently employed by speedrunners renders this reduction unsuitable.

OoT speedruns make heavy use of the resources at the player's disposal, including but not limited to Link's health, explosives (bombs and bombchus) and (in-game) time of day. A detour early on in a route to get more explosives can significantly speed up the rest of the route, as many travelling times can be drastically reduced due to glitches using bombs as a key component. For example, the bombchu item – a moving, explosive device – is not required by many OoT categories, yet many speedruns start out their route by collecting the nearest supply of bombchus in order to speed up the rest of the route enough to make it worthwhile. Thus, in fact, the opposite argument could be made: Expanding the set of events by all possible item pickups and refill locations as well as taking resource management into account can minimize the time even further. This obviously would increase the problem's complexity.

Defining the Edges. The definition of "each possible traversal between events" depends on the focus of the route. Considering only RTA viable routes this can be put as "humanly doable and complying to the category's ruleset". Finding all possible edges and assigning weights to them would require manually checking for category viability and timing the traversal between any two events in the game in both directions, which would be very time consuming and error prone. To counteract this complexity increase, nodes can be clustered by spatial proximity. Events and items that are very close to each other and/or obtainable by similar means can be clustered to single nodes. Another measure can be to approximate traversal times or using duration categories as integer numbers instead of actual timings, trading reduced complexity for uncertainties.

Dynamic Weights. As suggested above, edge weights are not static, but can and frequently do change with the already traversed nodes and in some cases (e.g. when in-game time of day is involved) with time. Even when considering only required events V_r, the items collected can also enhance the player's ability to traverse the game world in a speedy fashion in different ways they may or may not be intended for. This effectively changes the edge weights w while traversing the game graph. This *hidden gain* can lead to faster overall routes by taking detours early on that may have been undiscovered when only taking the starting weights into account.

Lafond [23] partly accommodated this by weighting the graph edges differently. For weighting an edge from stage S_i to S_j they use the time saved in S_j as the result of doing S_i beforehand, rather than the time needed to traverse between the stages. Considering multiple previously cleared stages and thus possibly multiple ways of clearing a stage, it is assumed that for every event the immediate best alternative is used.

The *OoT* equivalent for this would be to weight an edge from event v_i to v_j by the time saved by traversing v_i before v_j. This could constitute a way to significantly simplify the game graph, as many items and events do not provide any time save and thus have no impact on traversal speed.

Another way to handle dynamic edge weights would be to model game states rather than events as the graph's nodes. Depending on the information considered for defining the states, this can lead to a very large amount of nodes. This is elaborated on in section *Weighted Game State Graph* (4.2).

Repeatable Events. Some of the events mentioned above are repeatable, especially when considering item pickups and refills. For example, grass patches in *OoT* can drop bomb pickups when destroyed, and the patch regrows every time the location is reloaded. Thus, to repeat and benefit from a desired event, it might be required to traverse to another node and back to the desired event's node. Other events are one-time only. This holds true for all pickups and items from chests as well as for most story progression events. These events' nodes can be removed from consideration once traversed. Section *Weighted Game State Graph* (4.2) discusses the possible implications of this on a graph representation.

Multiobjective Optimization. The objective of speedrunning this far has only been considered a one-dimensional problem, solely dictated by time. While speed is in fact the primary directive of speedrunning, there is another dimension that can be taken into account: difficulty. Many speedrunning techniques require very precise inputs, sometimes precise to the frame. Some techniques aren't even possible to be performed by humans but are exclusive to TASs. Similarly, other aspects can be taken into consideration when constructing a route, like the aforementioned *hidden gain* or resource management. A consideration of possible changes to a graph model is given in section *Vector Valued Game Graph* (4.3).

4.2 Weighted Game State Graph

One way to engage the challenge of *Dynamic Weights* (4.1) is to consider the nodes V_s of a state graph G_s, reflecting game states, rather than events:

$$G_s = (V_s, E_s, w_s) \tag{3}$$

Similar to a *Weighted Game Event Digraph* (4.1) from (1), the edges E_s represent traversals between different states of the game, weighted by the time taken to do so w_s. If these states account for all necessary elements described above (location, items, item counts, time of day, etc.) this would eliminate edge weight dynamics. However, this would lead to an immense increase of nodes and edges. In fact, the presence of repeatable events would likely render this graph infinite.

4.3 Vector Valued Game Graph

The challenge of *Multiobjective Optimization* (4.1) can be addressed by a graph

$$G_m = (V, E, \mathbf{w}_m), \tag{4}$$

with a number n of weighting dimensions, which would yield vector valued edge weights and thus a vector valued weighting function $\mathbf{w}_m : E_m \to \mathbb{R}^n$. For example, a second vector element can be used to assign difficulty ratings to graph traversals. This can be used to parameterize possible routing algorithms by the runner's skill level. When including TASs, very high difficulty ratings can be assigned to TAS-only traversals.

Another possible application for vector valued edge weights is to represent the *hidden gain* introduced in *Dynamic Weights* (4.1) as a vector element. This would constitute an approximation, but could be a useful reduction of complexity. Weight dynamics could be replaced by this. In case of repeatable events, this *hidden gain* value might decrease when the corresponding node or edge is traversed, to prevent indefinite looping. Also, difficulty and *hidden gain* ratings can be combined to yield three-dimensional vectors.

Vector valued edge weights open up the problem to multiobjective optimization [13,25] and pathfinding algorithms [29,35], resulting in non-dominated sets of possible routes, possibly further increasing creativity and idea sparking in speedrun routing.

The events V and their traversals E can also be replaced by states V_s and their respective traversals E_s from (3), resulting in a combination with a *Weighted Game State Graph* (4.2), introducing all advantages and disadvantages pointed out there.

5 Prospective Solutions

After a model has been chosen to represent the routing problem, an optimization method has to be found. All concepts outlined are again specific to *OoT*, applicability to other games and genres has to be assessed. Good solutions should consider the challenges identified in *Envisioned Models and Challenges* (4) and look out for new, unidentified challenges. Of course, solutions should also respect the complexity of a given run. For example, a *glitchless* route for a mostly linear game with *finesse run* characteristics resembles different challenges than an *any%* route for a highly non-linear game.

When modeling an *Event Graph* with traversal times as edge weights, conventional pathfinding algorithms like Dijkstra's [10] or A* [17] are unsuitable given the edge dynamics. Same applies to algorithms like Multiobjective A* [35] in the case of vector valued edge weights. For edge weights to become static, a *State Graph* with extensively detailed states is needed, vastly increasing the amount of nodes.

Considering time save as edge weight similar to Lafond's approach [23] can simplify the graph. However, some edge dynamics will still prevail, as some techniques need multiple items to be collected or events to be triggered beforehand.

Two of the presented works on speedrun routing [21,22] employed evolutionary algorithms (EAs) [5,8,14] to handle the complexity involved. An approach using EAs for (multiobjective) optimization [6,9] seems promising, including assertions to exclude implausible combinations. Next to the implementation of different EA approaches, also the use of other metaheuristics is envisioned [16,37]. For example, ant colony optimization algorithms [11] have already proved their suitability to solve pathfinding problems [26,31].

Another possible approach is to conduct the traversal by an agent, deciding further progression after every edge traversal. Decision making can be conducted by above mentioned approaches, combinations of them, or by other means, such as a neuronal network taking information about the graph and previous traversal as input. Applications of machine learning for decision making in game agents – especially in form of deep reinforcement learning – suggest potential in this approach [20,24,40]. As deep reinforcement learning in a huge action space is still challenging, one can think of a combined approach, where the EA population is used to narrow done the potential action space for the agent in a certain status. However, at least reinforcement learning as a de facto standard in optimizing game play will be the benchmark for our EA solution.

With regard to edge dynamics, an agent must take Link's inventory, the previously traversed nodes and edges as well as other relevant information into account to maintain the Markov property.

6   Conclusions and Outlook

Routing is an integral part of most speedrun categories, regardless of the game. As such, having tools at disposal to automate this process can greatly help speedrunners and support speedrunning very complex games and genres. In addition, such tools are expected to be transferable to complex problems from logistics, huge vehicle routing instances from globally operating companies for example. However, as was shown in this work as well as the given references, speedrun routing is hard, whether done manually or algorithmically.

Given the complexity of *OoT*'s routing setting and the fact that despite its age new, heavily exploitable glitches and techniques are being found to this day, it is unreasonable to assume that a route will ever be found that cannot be improved on in the future. In fact this can be said for most games with enough combinatorial complexity to its progression mechanics. In these cases, routing algorithms can be seen as a tool for speedrunners to support their routing efforts by new influences. Moreover, there are speedrunners who enjoy manual routing as an important part of speedrunning. Solving this problem computationally would, if possible at all, render this process obsolete.

This work provides a first overview of works important for the subject area of speedrunning. These are critically discussed and relevant terminology is pointed out to support professional discussion. Moreover, different approaches are presented to model the identified speedrunning routing optimization problem and corresponding challenges are depicted. Finally, prospective solutions for the challenges are outlined.

Future works can discuss further on one or more of the outlined approaches for modeling and solving the problem of speedrun routing. Applying the findings to other games and genres can uncover further possibilities not apparent when only focusing on a single game.

As a final note, the topic suggests itself for competitions. Participants can compete in different ways, such as finding the most accurate route representation model, solving given representations, or of course creating a full route through a given game. In the latter case, support of actual speedrunners would be needed to assess plausibility.

Ludography

The Legend of Zelda: Ocarina of Time (OoT) [Nintendo 64] Developer/ Publisher: Nintendo, Japan (1998)
Super Mario World [SNES] Developer/Publisher: Nintendo, Japan (1990)
Mega Man/2/3 [NES] Developer/Publisher: Capcom, Japan (1987/1988/1990 resp.)
The Elder Scrolls III: Morrowind [PC (Windows), Xbox] Developer: Bethesda Softworks (2002), Publisher: Ubisoft, France
TrackMania Nations Forever [PC (Windows)] Developer: Nadeo (2008), Publisher: Focus, France; Enlight, USA; Deep Silver, Germany; Digital Jesters, UK

References

1. GDQ Tracker - Event List. https://gamesdonequick.com/tracker/events/. Accessed 6 Feb 2022
2. GDQStat.us. https://gdqstat.us/previous-events/agdq-2020/?series=0. Accessed 6 Feb 2022
3. Reverse Bottle Adventure - ZeldaSpeedRuns. https://www.zeldaspeedruns.com/ oot/ba/reverse-bottle-adventure. Accessed 6 Feb 2022
4. speedrun.com. https://www.speedrun.com/oot. Accessed 6 Feb 2022

5. Bäck, T., Hammel, U., Schwefel, H.P.: Evolutionary computation: comments on the history and current state. IEEE Trans. Evol. Comput. **1**(1), 3–17 (1997). https://doi.org/10.1109/4235.585888

6. Coello, C.A.C., Lamont, G.B., van Veldhuizen, D.A.: Applications Of Multi-Objective Evolutionary Algorithms. World Scientific Press, Singapore. 2. edn. (2007)

7. Currie, G.: Fictional truth. Philos. Stud. **50**(2), 195–212 (1986). https://doi.org/10.1007/BF00354588

8. De Jong, K.A.: Evolutionary Computation: A Unified Approach. MIT Press, Cambridge (2016)

9. Deb, K.: Multi-Objective Optimization Using Evolutionary Algorithms. Wiley, Hoboken (2001)

10. Dijkstra, E.W.: A note on two problems in connexion with graphs. Numer. Math. **1**, 269–271 (1959)

11. Dorigo, M., Di Caro, G.: Ant colony optimization: a new meta-heuristic. In: Congress on Evolutionary Computation (CEC99), vol. 2, pp. 1470–1477 (1999). https://doi.org/10.1109/CEC.1999.782657

12. Downey, R.G., Fellows, M.R.: Parameterized Complexity. Springer (1999). https://doi.org/10.1007/978-1-4612-0515-9

13. Ehrgott, M.: Multicriteria Optimization. Springer, 2nd edn. (2005). https://doi.org/10.1007/3-540-27659-9

14. Eiben, A.E., Smith, J.E.: Introduction to Evolutionary Computing. Natural Computing Series. Springer, 2. edn. (2015). https://doi.org/10.1007/978-3-662-44874-8

15. Ford, D.: Speedrunning: transgressive play in digital space. In: Nordic DiGRA 2018 (2018). https://doi.org/10.13140/RG.2.2.12357.91369

16. Gendreau, M., Potvin, J.Y., et al.: Handbook of Metaheuristics, vol. 3. Springer (2019). https://doi.org/10.1007/978-3-319-91086-4

17. Hart, P.E., Nilsson, N.J., Raphael, B.: A formal basis for the heuristic determination of minimum cost paths. IEEE Trans. Syst. Sci. Cybern. **4**(2), 100–107 (1968). https://doi.org/10.1109/TSSC.1968.300136

18. Hay, J.: Fully optimized: the (Post)human art of Speedrunning. J. Posthuman Stud. **4**(1), 5–24 (2020). https://doi.org/10.5325/jpoststud.4.1.0005

19. Hemmingsen, M.: Code is law: subversion and collective knowledge in the ethos of video game speedrunning. Sport, Ethics Philos. 1–26 (2020). https://doi.org/10.1080/17511321.2020.1796773

20. Huang, S., Bamford, C., Ontanon, S., Grela, L.: Gym-μRTS: toward affordable full game real-time strategy games research with deep reinforcement learning. In: IEEE Conference on Games (CIG) (2021). https://doi.org/10.13140/RG.2.2.18639.82081

21. Iškovs, A.: Travelling murderer problem: planning a morrowind all-faction speedrun with simulated annealing (2018). https://www.kimonote.com/@mildbyte/travelling-murderer-problem-planning-a-morrowind-all-faction-speedrun-with-simulated-annealing-part-1-41079/. Accessed 6 Feb 2022

22. JstAnothrVirtuoso: Finding the Optimum Nadeo Cut... With Science!! (2019). https://www.youtube.com/watch?v=1ZsAjvO9E1g. Accessed 6 Feb 2022

23. Lafond, M.: The complexity of speedrunning video games. In: Ito, H., Leonardi, S., Pagli, L., Prencipe, G. (eds.) Fun with Algorithms (FUN), vol. 100, pp. 27:1–27:19. Schloss Dagstuhl–Leibniz-Zentrum fuer Informatik (2018). https://doi.org/10.4230/LIPIcs.FUN.2018.27

24. Lample, G., Chaplot, D.S.: Playing FPS games with deep reinforcement learning. CoRR (2016). http://arxiv.org/abs/1609.05521

25. Miettinen, K.: Nonlinear Multiobjective Optimization. Kluwer (1999)
26. Mocholi, J.A., Jaen, J., Catala, A., Navarro, E.: An emotionally biased ant colony algorithm for pathfinding in games. Expert Syst. Appl. **37**(7), 4921–4927 (2010). https://doi.org/10.1016/j.eswa.2009.12.023
27. Newman, J.: Playing with Videogames. Routledge, London (2008)
28. Newman, J.: Wrong warping, sequence breaking, and running through code. J. Jpn Assoc. Digital Humanit. **4**(1), 7–36 (2019). https://doi.org/10.17928/jjadh.4.1_7
29. Rajabi-Bahaabadi, M., Shariat-Mohaymany, A., Babaei, M., Ahn, C.W.: Multiobjective path finding in stochastic time-dependent road networks using nondominated sorting genetic algorithm. Expert Syst. Appl. **42**(12), 5056–5064 (2015). https://doi.org/10.1016/j.eswa.2015.02.046
30. Ricksand, M.: "Twere well it were done quickly": what belongs in a glitchless speedrun? Game Stud. **21**(1) (2021). http://gamestudies.org/2101/articles/ricksand. Accessed 6 Feb 2022
31. Rishiwal, V., Yadav, M., Arya, K.V.: Finding optimal paths on terrain maps using ant colony algorithm. Int. J. Comput. Theory Eng. **2**(3), 416–419 (2010). https://doi.org/10.7763/IJCTE.2010.V2.178
32. Scully-Blaker, R.: A practiced practice: speedrunning through space with de Certeau and Virilio. Game Stud. **14**(1) (2014). http://gamestudies.org/1401/articles/scullyblaker. Accessed 6 Feb 2022
33. Scully-Blaker, R.: Re-Curating the Accident: Speedrunning as Community and Practice. Masters thesis, Concordia University (2016)
34. Scully-Blaker, R.: The Speedrunning museum of accidents. Kinephanos (Preserving Play, Special Issue), 71–88 (2018). https://www.kinephanos.ca/2018/the-speedrunning-museum-of-accidents/. Accessed 6 Feb 2022
35. Stewart, B.S., White, C.C.: Multiobjective A*. J. ACM **38**(4), 775–814 (1991). https://doi.org/10.1145/115234.115368
36. Szita, I.: Reinforcement learning in gamecurrs. In: Wiering, M., van Otterlo, M. (eds.) Reinforcement Learning: State-of-the-Art, pp. 539–577. Springer (2012). https://doi.org/10.1007/978-3-642-27645-3_17
37. Sörensen, K.: Metaheuristics-the metaphor exposed. Int. Trans. Oper. Res. **22**(1), 3–18 (2015). https://doi.org/10.1111/itor.12001
38. Togelius, J., Karakovskiy, S., Baumgarten, R.: The 2009 Mario AI competition. In: Congress on Evolutionary Computation (CEC), pp. 1–8. IEEE Press (2010). https://doi.org/10.1109/CEC.2010.5586133
39. Volvy: Reddit post about the Morrowind all factions speedrun route (2018). www.reddit.com/r/speedrun/comments/9u1r9o/using_ai_to_grind_out_routes/e91dg6w/. Accessed 6 Feb 2022
40. Ye, D., et al.: Mastering complex control in MOBA games with deep reinforcement learning. In: AAAI Conference on Artificial Intelligence **34**(04), 6672–6679 (2020). https://doi.org/10.1609/aaai.v34i04.6144

Co-evolution of Spies and Resistance Fighters

Johanna Lange[iD], Mario Stanke[iD], and Marc Ebner[(✉)][iD]

Institut für Mathematik und Informatik, Universität Greifswald,
Walther-Rathenau-Str. 47, 17487 Greifswald, Germany
{mario.stanke,marc.ebner}@uni-greifswald.de

Abstract. We use an evolution strategy to evolve game strategies for resistance fighters as well as spies for the popular card game "The Resistance". In our experiment, players only communicate via observable actions. Players are judged by how they behave and not by what they say. Resistance fighters observe the behavior of all game players and try to deduce who is a spy by maintaining a score that represents who is likely to be a spy. Players likely to be spies are not taken on a mission. Spies use probabilities for their behavior. We use co-evolution to evolve resistance fighters and spies. Fitness plots seem to indicate that no progress is being made, i.e. we clearly see the Red Queen Effect in our experiments. However, the master tournament and current individual vs ancestral opponents method show that evolutionary progress is being made.

Keywords: Co-evolution · Red Queen effect · Gameplay strategies

1 Introduction

We use co-evolution to evolve game strategies for the popular game "The Resistance". This is interesting at several different levels. It allows creating artificial agents that are capable of playing the game. Aside from this aspect it will also help to understand what structure successful strategies need to have in order to win the game. Our research may also help to understand how human players are playing the game and aid in creating artificial human-like behavior.

The game "The Resistance" can be played by a group of 5 to 10 people. It is a card based game that relies on secret identities. Each player is either a spy or a member of the resistance. All spies work together to win the game. The same holds for the resistance fighters. Each spy knows the identity of the other spies. However, the resistance fighters do not know the identity of the spies. During gameplay, the players interact via unrestricted oral communication. Members of the resistance need to draw conclusions on the identity of the spies while spies try to deceive the members of the resistance. Researchers (started by a workgroup at the Dagstuhl Seminar "Artificial and Computational Intelligence in Games") previously developed AI agents to play this game [29]. These agents are analyzed

© Springer Nature Switzerland AG 2022
J. L. Jiménez Laredo et al. (Eds.): EvoApplications 2022, LNCS 13224, pp. 487–502, 2022.
https://doi.org/10.1007/978-3-031-02462-7_31

in detail by Taylor [31]. Some of the agents use opponent modeling or include expert rules. The players that are evolved here also include expert rules and score keeping to model other player's behavior. For this study, communication only occurs via observed behavior of the players. What they say is irrelevant. What the players say might be a lie to deceive others.

Our experiments show the Red Queen Effect. Even though fitness no longer improves, we show that evolution still continues and players are improving their abilities to play the game.

2   Background

Co-evolution has been used by many researchers to improve solutions. Experiments include evolution of sorting networks [18], pursuit-evasion tactics [3,21], 3D Tic-Tac-Toe players [27], Tron players [16], virtual plants [6–8], Quidditch playing teams [4], car racing controllers [32], combat algorithms to control space ships [14], predator/pray behavior [15], or Pac Man/ghost controllers [1]. Experiments have also been carried out in the field of evolutionary robotics [12,22]. Most co-evolutionary research uses two separate populations. However, some researchers have worked with a single population of individuals. For instance, in the experiments with virtual plants [7], plants of a single population compete for sunlight.

When evolutionary experiments use co-evolution, the fitness of one population depends on the performance of the other population. After several generations, it may appear as if evolution has come to a halt because fitness no longer increases. However in reality, the populations might be engaged in an arms race where both populations improve a particular trait even though fitness remains constant [5]. A classic example is the evolution of predator and prey, e.g. cheetahs and gazelles. Both populations may improve their respective ability (to hunt prey/to evade a predator) while the number of offspring, i.e. their fitness stays constant. This is called the Red Queen effect [5,23,25,26,33].

The Red Queen effect also appeared in our experiments. Fitness seems to level off after several generations. Different methods have been developed to analyze evolutionary progress. Luke and Wiegand [20] have analyzed single population co-evolution objective measures. Stanley and Miikkulainen [30] have introduced the dominance tournament. Ficici and Pollack [10] proposed a memory mechanism to maintain useful traits when using co-evolution. De Jong [19] suggested the incremental pareto-coevolutionary archive to evaluate evolutionary progress. We use current individual vs ancestral opponent [2] as well as the master tournament [11] to show that evolution has not come to a halt. Players are still improving their ability to play the game. Next, we briefly review the rules of the game.

3   Rules of the Game

A group of resistance fighters (between 5 and 10 people) is trying to fight an evil government. The resistance fighters perform up to 5 secret missions. A subset of the group is chosen to go on a mission. Resistance fighters try to exclude the spies from the mission. If a spy is allowed to go on a mission, he may sabotage the mission. If a mission has been sabotaged by a spy then it has failed to reach its objective from the resistance fighters' point of view. Thus, spies try to become a member of the mission while resistance fighters try to discern the spies' identities. Resistance fighters want to make sure that players likely to be spies are not taken along on a mission. The game is won by the resistance fighters if three of their missions have succeeded. It is won by the spies if three of the missions have been sabotaged or if the group cannot agree on who to send on the next mission.

For the research conducted here, we assume 5 players. If the group consists of 5 players then 3 of them are actual resistance fighters while 2 of them are spies. The identities are determined at random. At the very beginning of the game the spies get to know the identity of the other spies. The actual resistance fighters do not know the identity of the spies.

Table 1. Number of persons on the team.

Mission no	1	2	3	4	5
No. of team members	2	3	2	3	3

One person of the 5 players starts of as team leader. The team leader needs to assemble a team that will then go on a secret mission. Resistance fighters want the mission to succeed while spies want the mission to fail by sabotaging it. The size of the team is shown in Table 1 for each of the 5 missions. The team leader is free to choose any players for the team. He may or may not select himself to be a member of the team. Obviously, if the team leader is an actual resistance fighter, he wants to avoid selecting spies to be on the team. If the team leader is a spy he may or may not include the other spy to be on the team. If two spies were selected into the team for the very first mission, then both could go ahead with the mission. By deliberately not sabotaging the mission, they might gain the trust of the other game players and convince them that they are not spies while in reality they are spies that will then sabotage future missions.

Once the team leader has selected the players that will be on the team, all players vote simultaneously (yes or no) on whether this team is OK to go on a mission. This allows other players to hinder this team from going on a mission. Resistance fighters may vote against the team if they believe that spies are on the team and that the mission is likely to fail because of this. Spies may vote against the team if they know that no spy is on the team. Obviously if a spy votes against the team then the spies need to come up with a cover story on why they think that the mission is going to fail with the current team members. If the majority vote is "yes" then the team will go on a mission. If the majority vote is "no" then the team leader is assumed to have lost trust of the group. The next player in turn will become a new team leader. This team leader will then try to assemble another team to go on a mission. Again the team will be voted on by all of the players. This process repeats until either the team will go on a mission or until 5 consecutive teams have been rejected. If this happens, the spies have won because the resistance group has lost trust in each other.

Then the assembled team will go on a mission. Actual resistance fighters have no interest in sabotaging the mission. Only spies need to decide whether or not to sabotage a mission. A spy may decide not to sabotage in order to gain the trust of the actual resistance fighters. A mission will fail as soon as one person sabotages the mission. The entire gameplay is summarized in Algorithm 1. Interestingly, "The Resistance" is not completely balanced if all players make each decision independently and uniformly distributed on the set of choices (team choice, voting, sabotaging). Our computations have shown that the resistance will win the game under these conditions with a probability of approximately 56.6%.

4   Strategies for Spies and Resistance Fighters

Communication between all of the players is very important for this game. It allows actual resistance fighters to possibly identify spies because spies need to lie when playing the game. Spies need to deceive resistance fighters into believing that they are resistance fighters too. Obviously, this aspect of the game is incredibly difficult to model. That's why we do not include any communication between players. Players, i.e. actual resistance fighters, need to deduce who might be a spy by observing the behavior of other players. If a group goes on a mission and the mission fails, then at least one member of the group has to be a spy. Thus, every time a mission is carried out, the resistance fighters gain some knowledge about the identities of the other players (spy or not a spy).

Because of this, our actual resistance fighters maintain a score for each player. This score is taken as the likelihood of this particular member being a spy. It is updated after each mission. As a team leader, the resistance fighter chooses the players with the lowest score values to go on a mission. The team leader is always part of the team. If it's the first mission, the team leader selects himself and randomly chooses the second player for the team (because no knowledge is available yet). If the team contains game players with a high score value then actual resistance fighters would vote against this team. The main question is how to modify this score based on the decisions made by the players during the game. One obvious choice would be to increase this score for all members of the team by a certain amount if the mission fails. But there are additional options on how to modify this score.

We have used an evolution strategy [9, 24, 28] to optimize a set of real values r_i with $i \in \{1, ..., 14\}$ for each actual resistance fighter. These values are used to modify the score of each player depending on the choices made by the other players during the game. At the beginning of a game each resistance fighter sets the score of all other players to 0. The resistance fighter also keeps a score for himself. His own score value is set to -1000. During the game, the behavior of the other players is monitored. Depending on their behavior during the assembly of

---

Algorithm 1: The Resistance gameplay algorithm

Distribute roles randomly over $\{Player_1,...,Player_5\}$
$MissionsLeader \leftarrow Player_1$
$nMissionsSabotaged \leftarrow 0$
$nMissionsCompleted \leftarrow 0$
$nMission \leftarrow 1$
while *! Game over* **do**   // next mission
   $nVotingRound \leftarrow 1$
   $nYesVotes \leftarrow 0$
   while $nYesVotes < 3$ *&* $nVotingRound \leq 5$ **do**
      if $nMission = 1$ **then**
         | $Team \leftarrow \{MissionLeader\} \cup \{$Random selection of players$\}$
      else
         | $Team \leftarrow$ Select players based on score
      end
      $nYesVotes \leftarrow \forall_i : Player_i$ votes (acc/reject)
      if $nYesVotes < 3$ **then**
         $nVotingRound++$
         $MissionLeader \leftarrow$ select next player
      end
   end
   if $nVotingRound = 6$ **then**
      Game over: Spies have won the game
   else   // do mission
      $\forall_j Player_j \in Team$: If a spy, sabotage (yes/no)
      if *mission has been sabotaged* **then**
         | $nMissionSabotaged++$
      else
         | $nMissionCompleted++$
      end
      if $nMissionSabotaged = 3$ **then**
         | Game over: Spies have won the game
      end
      if $nMissionCompleted = 3$ **then**
         | Game over: Resistance has won the game
      end
      $\forall_j Player_j \in Team$: If not a spy, update score
      $MissionLeader \leftarrow$ select next player
      $nMission++$
   end
end

---

the team (rating of team leader), the round of voting (rating of voting behavior) and during the mission (rating of the members of the team) the score is modified after the mission has been completed. The values r_i with $i \in \{1,...,7\}$ are added to the score of a team member whenever the member shows the respective behavior. All behavior options are shown in Table 2. The values r_i with

Table 2. Rating of the team leader and of the members who went on a mission.

	Parameter	Team size	Own player is a member of the team	# of sabotage acts	Note
Team member rating	r_1	2	no	1	at least one spy is a member of the team
		3	yes	1	at least one of the other members is a spy
	r_2	3	no	1	
	r_3	3	no	2	
	r_4	2	yes	0	
	r_5	2	no	0	
	r_6	3	yes	0	
	r_7	3	no	0	at least one spy is a member of the team
Leader rating	r_8	3	no	0	at least one spy is a member of the team
	r_9	3	yes	0	
		2	N/A	0	
	r_{10}	N/A	N/A	1/2	

$i \in \{8, ..., 10\}$ are added to the score of a team leader whenever applicable. The voting behavior also has an influence on the score. The parameters used to judge this behavior is shown in Table 3. In certain game situations (see Table 4) spies can be identified with 100% accuracy after a mission has failed. Whenever a spy is identified, its score is increased by 1000 points.

Table 3. Rating of voting behavior.

Parameter	Own player is a member of the team	Mission successful	5th voting round	Note
r_{11}	N/A	N/A	Yes	Voted against last team
r_{12}	N/A	Yes	No	Voted against successful team
r_{13}	N/A	No	No	Voted against a team that included a spy
r_{14}	N/A	N/A	No	Voted for team of size 3 & not part of team (spy)

The algorithm used by the actual resistance fighters to determine whether to vote for (yes) or against (no) a team is shown in Algorithm 2.

The genotype of a spy consists of the parameters s_i with $i \in \{1, ..., 10\}$. All 10 parameters represent probabilities that are used by the algorithm to control spy behavior. The description of all 10 probabilities is shown in Table 5.

Table 4. Identification of spies.

Team size	Own player is a member of the team	# of sabotage acts	Note
2	Yes	1	One spy identified
2	No	2	Two spies identified
3	Yes	2	Two spies identified

Algorithm 2: Actual resistance fighter voting.

Sort players according to scores
if (
Player is MissionLeader OR // own team
nVotingRound=5 OR // last chance
nMission=1 // no knowledge available
) **then**
| Vote Yes
else
| **if** *The two players with highest scores not on the team* **then**
| | Vote Yes
| **else**
| | Vote No   // presumably one or two spies
| **end**
end

Table 5. Rating of voting behavior.

Parameter	Note
s_1	Prob. to select random other player as 2nd team member (team size 2)
s_2	Prob. to select random other players as team members (team size 3)
s_3	Probability to accept team in 5th voting round
s_4	Probability to accept team if only one spy is a member of the team
s_5	Probability to accept team if two spies are members of the team
s_6	Probability to accept team (remaining option)
s_7	Probability to sabotage (two spies in team with team size = 2)
s_8	Probability to sabotage (two spies in team with team size = 3)
s_9	Probability to sabotage first mission (only spy in team)
s_{10}	Probability to sabotage mission (only spy in team)

Algorithms (3–5) show how these probabilities are used by the spy to make decisions. Algorithm 3 controls how a team is assembled. Algorithm 4 controls how a spy votes on the assembled team. Algorithm 5 controls whether or not a spy sabotages the mission (Fig. 1).

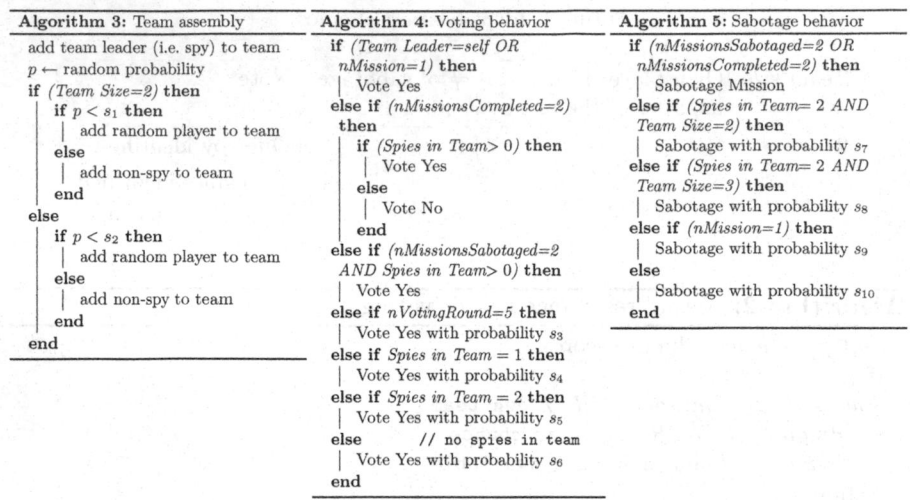

Fig. 1. Spy behavior algorithms.

5   Co-evolution of Strategies

We have used the DEAP (Distributed Evolutionary Algorithms in Python) framework [13] to evolve playing strategies for resistance fighters and spies. A player does not know which role he will be assigned to. Thus, successful players have to master both roles, the role of a resistance fighter and the role of a spy.

We work with two populations. One population codes for resistance fighters (14 parameters for each individual as shown in Table 2 and Table 3). The second population codes for spies (10 probabilities as listed in Table 5). Resistance fighters will have to compete against spies and vice versa. Therefore, the fitness of a resistance fighter depends on how well the spy is able to play the game. The fitness of a spy depends on how well the resistance fighter is able to play the game. We use hall of fame to evaluate players (spies as well as resistance fighters). The best individual of a generation enters the hall of fame [27]. In hall of fame evaluation, an individual of the current population is evaluated against 100 randomly selected opponents from the hall of fame. We simply count how many times a player has won the game and then compute the probability to win the game by dividing by the number of games played. Each individual plays 100 games to evaluate its fitness.

We use a (μ, λ) evolution strategy to evolve the player's strategies. For our experiments, we have used 21 parents producing 150 offspring resulting in a selection pressure of approx. 14% (a reasonable setting according to Gerdes et al. [17]). The parameters of this evolution strategy are shown in Table 6. For the first generation, the parameters r_i are randomly selected from the range $[-1, 1]$. A negative value for a given action enhances the belief that the player who carried out the action is an actual resistance fighter. A positive value enhances the belief that the player is a spy if he carried out the given action. Spy parameters s_i are

Table 6. Evolutionary strategy parameters.

Parameter	Value	Note
G	200	Number of generations
μ	21	Number of parents
λ	150	Number of offspring
p_{cross}	10%	Crossover (uniform) probability
σ_r	0.1	Mutation standard deviation (resistance fighter)
σ_s	0.1	Mutation standard deviation (spy)
n_{HoF}	100	Number of hall of fame opponents

taken from the range $[0, 1]$ since they represent probabilities. Crossover was used with probability p_{cross}.

Note that a simple scaling of the resistance fighter's parameters r_i will not result in a different belief of who might be a spy because the scores of all players are sorted before making a decision. That's the reason why we did not use step size adaptation for this experiment. Standard evolution strategy mutation was used (addition of normally distributed values). The standard deviation was σ_r for resistance fighters and σ_s for spies. The standard deviation for the mutation of spies was reduced by 0.5% after each generation.

We performed 10 evolutionary runs for this experiment. Figure 2 shows average fitness for resistance fighters (blue) as well as spies (red). The dark plot shows average fitness of the population while the lighter plot shows the maximum fitness per generation. In the beginning, resistance fighters perform worse than the spies. In other words, our randomly generated spies are more likely to

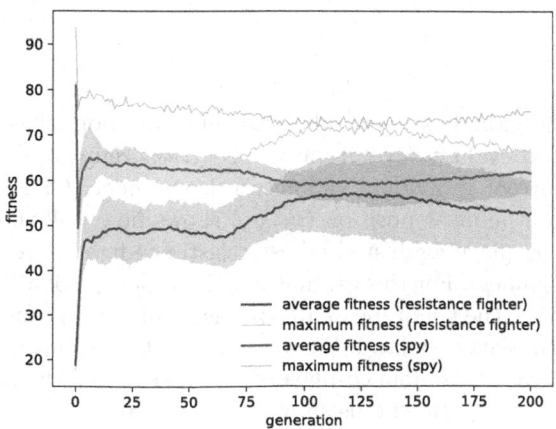

Fig. 2. Average and maximum fitness of spies and resistance fighters (10 runs). The shaded area shows the standard deviation. (Color figure online)

win than a randomly generated resistance fighter. Spies of the first generation win approximately 80% of the games.

A reason for this behavior might be that the search space we have used for the resistance fighters is much larger and that random adjustments of the scores is bound to perform poorly. In contrast, a randomly generated spy may still have good chances of winning the game. Within a small number of generations, the fitness of resistance fighters improves considerably. After 12 generations the fitness of both populations seems to stagnate. If we only look at the fitness plot, we might get the impression that evolution has come to a halt quite soon. However, this has not happened here. Constant fitness might be a result of the so called Red Queen effect that has appeared previously in other co-evolutionary experiments [7,11]. We show that even though fitness remains approximately constant, individuals are still improving.

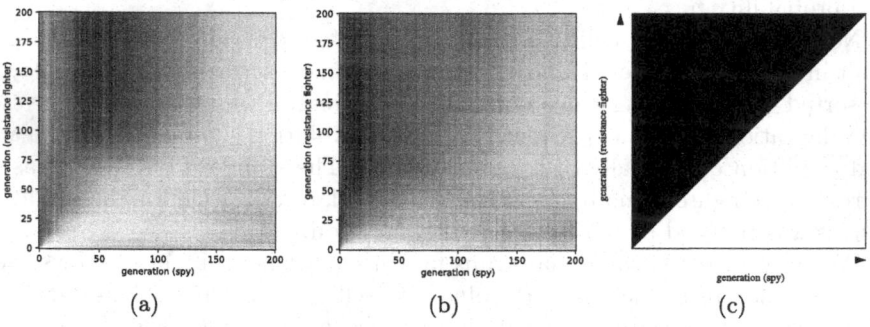

Fig. 3. Average percentage of winning (current individual vs ancestral opponent). (a) Co-players have identical behavior. (b) Results from additional experiments where co-players have different behavior. (c) Idealized perfect CIAO plot.

6   Evaluation

We use current individual vs ancestral opponent evaluation [2] as well as a master tournament [11] to evaluate evolutionary progress. Figure 3(a) shows the evaluation data of current individual vs ancestral opponent (CIAO) averaged over 10 runs. Each data point at position (g_x, g_y) shows how well the best spy taken from generation g_x plays against the best resistance fighter from generation g_y (averaged of 100 games). For this evaluation, the genotype of all resistance fighters is identical. Also, the genotype of all spies is identical. In other words, players know how their co-players behave. We also ran a set of experiments where players have to play against random co-players also taken from the hall of fame. The CIAO plot for these experiments is shown in Fig. 3(b).

The darker the pixel, the better the resistance fighter performed. A black pixel corresponds to a perfect performance of resistance fighters while a white pixel corresponds to a perfect performance for spies. The CIAO plot indicates that both spies and resistance fighters of later generations perform better if placed against opponents from earlier generations. For comparison, an idealized perfect CIAO plot for our experiments is shown in Fig. 3(c).

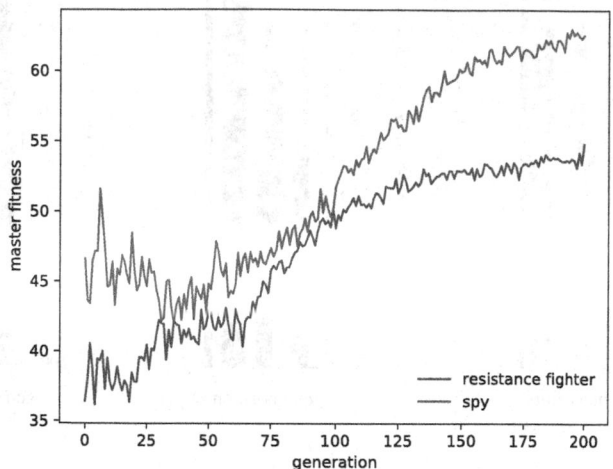

Fig. 4. Master tournament data averaged over 10 runs.

We also used the master tournament of Floreano and Nolfi [11] to evaluate evolutionary progress. For the master tournament, the best individual of a given generation is evaluated 100 times against all opponents who made it into the hall of fame. Given the full range of opponents (weak as well as fully adapted to the given domain), individuals can be ranked on an absolute scale. The master tournament plot is shown in Fig. 4. This plot clearly shows that evolutionary progress is being made up to generation 200. In particular, the spies improve considerably compared to their performance at the beginning of the experiment.

It is also helpful to look at elite bitmaps [2]. Elite bitmaps of the resistance fighters are shown in Fig. 5(a) and (b). Elite bitmaps of the spies are shown in Fig. 5(c). The elite bitmaps visualize the genotype of the individuals over all generations. For each generation, the best individual is shown from a single run. Generation 0 is shown at the top while generation 200 is shown at the bottom. We only show data from every fourth generation. For the spy elite bitmap, a black pixel corresponds to a 0 while a white pixel corresponds to a 1. The genotype of resistance fighters contains positive as well as negative values. That's why two plots are shown. The absolute values of the parameters r_i are shown on the left hand side while the sign is shown in the plot on the right hand side. A positive value is denoted by a back pixel. A negative value is denoted by a white pixel. Since only the relative values are important for the performance of the resistance fighter, the genotype of each elite resistance fighter is normalized by dividing all values by the absolute maximum parameter value.

The genotype of the best evolved individual (resistance fighter and spy) of generation 200 of run 1 is shown in Fig. 6. We also show the average values of all genes

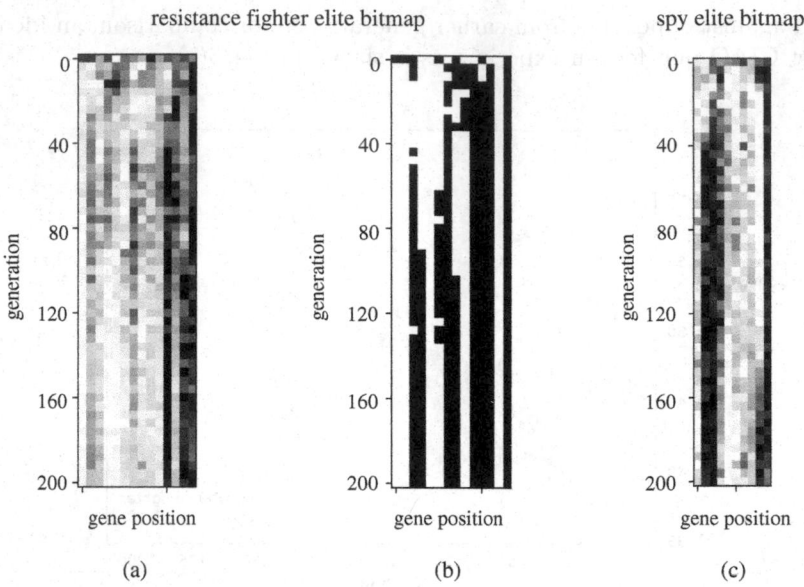

resistance fighter elite bitmap   spy elite bitmap

(a)   (b)   (c)

Fig. 5. Genotype of the best resistance fighter taken from every fourth generation from a single run. Grayscale indicates absolute value of the parameter (a). Sign of the parameter: Positive = black, negative = white (b). Genotype of the best spy (c). Black represents 0, white represents 1.

over the 10 runs. The standard deviation is also shown. In addition, we show how often the sign of a given parameter agrees with the sign of the average value for that parameter (resistance fighter only). Parameters r_1, r_4 and r_5 do not seem to be very relevant. These parameters have a very small value compared to the other parameters. In hindsight, this result can be explained. Parameter r_1 is used to score players if a sabotage happened but it is not clear who might be a spy (team size = 2 but own player was not a member of the team or team size = 3 and own player was a member of the team). Parameter r_4 and r_5 are used when a 2 player team goes on a mission and the mission succeeds. Apparently the strongest information is obtained when a player votes against the mission in the fifth voting round (parameter r_{11}) or if a player voted for a team of size 3 but is not a member of the team, i.e. must be a spy (parameter r_{14}).

The standard deviation is particularly small for parameters s_3 and s_6. The parameter s_3 is the probability for a spy to vote "Yes" during the fifth voting round. A vote of "No" is penalized by the resistance fighters. Therefore, it is of no surprise that the spies learn to vote mostly "Yes" during the fifth voting round. Parameter s_6 is the probability for a spy to vote for a team with no spies being members. It makes sense for a spy to oppose such teams. Spies also have learned not to vote yes (s_5 close to zero probability) on teams with two spies being members. Voting "Yes" on such a team would make it easy for resistance fighters to discern who is a spy. It is better to have only one spy on a team. Spies

Par.	Run 1	Avg.	Std. Dev	Sign
r_1	-0.06	0.06	0.18	4
r_2	-0.37	-0.32	0.32	7
r_3	0.32	0.09	0.20	7
r_4	0.02	-0.19	0.30	6
r_5	-0.07	-0.27	0.27	10
r_6	-0.15	-0.04	0.23	7
r_7	0.27	0.25	0.27	9
r_8	0.29	-0.04	0.35	6
r_9	-0.24	-0.15	0.25	6
r_{10}	0.11	-0.01	0.31	3
r_{11}	0.92	0.47	0.22	10
r_{12}	0.34	0.34	0.25	9
r_{13}	-0.88	-0.71	0.41	9
r_{14}	1.00	0.66	0.30	10
s_1	0.27	0.31	0.34	
s_2	0.92	0.45	0.37	
s_3	0.98	0.97	0.04	
s_4	0.16	0.29	0.32	
s_5	0.00	0.16	0.29	
s_6	0.15	0.04	0.05	
s_7	0.13	0.15	0.20	
s_8	0.28	0.27	0.12	
s_9	0.91	0.56	0.44	
s_{10}	0.66	0.71	0.33	

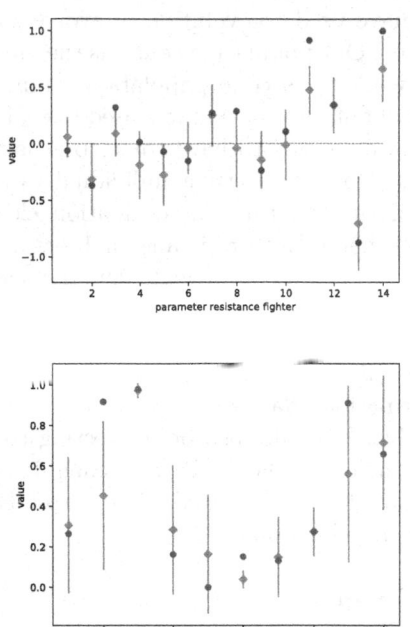

Fig. 6. Genotype of the best individual from generation 200 from run 1. Average value and standard deviation for the parameters (Par.) are computed over all 10 runs. The last column shows the number of times the sign agrees with the sign of the average value (only relevant for resistance fighters). The plots visualize the data shown on the left. The dots represent the values of the best individual from run 1.

also learn to mostly sabotage the very first mission (s_9) and all other missions (s_{10}). Note that the standard deviation for parameter s_9 is relatively high. This might be an indicator that there are other strategies such as not sabotaging the very first mission and thereby gaining the trust of the resistance fighters.

Players that are perfectly adapted to each other would win with a probability of 50%. We performed 10.000 games using the genotype shown in Fig. 6. This resulted in approximately 44% wins for the resistance fighters and 56% wins for the spies. For some of the best individuals taken from generation 200 of our experiments, spies ended up winning between 60% and 70% of the games. It could be that the search space that we have used here favors spies, i.e. it is easier for spies to find good solutions. After all, only probabilities are encoded in the spy genotype while the genotype coding for resistance fighters contain negative as well as positive values.

It could also be that no single optimal strategy exists. If this is the case, the strategy needs to be adapted to the type of opponent given [22].

7 Conclusion

We have used co-evolution to evolve game strategies for the game "The Resistance". One population contains the individuals representing the resistance fighters, while the second population contains the individuals representing the spies. Hall of fame evaluation was used to evaluate individuals. After a few generations, evolution seems to have come to a halt. This is a signature of the Red Queen effect. However, current individual vs ancestral opponent evaluation as well as the master tournament evaluation clearly show that evolutionary progress is being made. Both resistance fighters and spies continually improve their gameplay. In future experiments communication between players could be included. An interesting research question could be whether oral communication has any relevance to this game, i.e. whether it might provide an advantage to win the game. The difficulty in including communication is of course to separate reliably information that is true and helpful from information that is transmitted to deceive. It would also be interesting to evolve Genetic Programming strategies that include higher-order reasoning, i.e. strategies that maintain probabilities on who might be suspected to be a spy. However, this is outside the scope of the present submission.

References

1. Cardona, A.B., Togelius, J., Nelson, M.J.: Competitive coevolution in ms. pac-man. In: IEEE Congress on Evolutionary Computation, pp. 1403–1410. IEEE (2013)
2. Cliff, D., Miller, G.F.: Tracking the Red Queen: measurements of adaptive progress in co-evolutionary simulations. In: Morán, F., Moreno, A., Merelo, J.J., Chacón, P. (eds.) ECAL 1995. LNCS, vol. 929, pp. 200–218. Springer, Heidelberg (1995). https://doi.org/10.1007/3-540-59496-5_300
3. Cliff, D., Miller, G.F.: Co-evolution of pursuit and evasion II: simulation methods and results. In: Maes, P., Mataric, M.J., Meyer, J.A., Pollack, J., Wilson, S.W. (eds.) From Animals to Animats 4: Proceedings of the 4th International Conference on Simulation of Adaptive Behavior, pp. 506–515. The MIT Press, Cambridge (1996)
4. Crawford-Marks, R., Spector, L., Klein, J.: Virtual witches and warlocks: A quidditch simulator and quidditch-playing teams coevolved via genetic programming. In: Keijzer, M. (ed.) Late Breaking Papers at the 2004 Genetic and Evolutionary Computation Conference. Seattle, Washington (2004)
5. Dawkins, R., Krebs, J.R.: Arms races between and within species. Proc. R. Soc. Lond. B **205**, 489–511 (1979)
6. Ebner, M.: Evolution and growth of virtual plants. In: Banzhaf, W., Ziegler, J., Christaller, T., Dittrich, P., Kim, J.T. (eds.) ECAL 2003. LNCS (LNAI), vol. 2801, pp. 228–237. Springer, Heidelberg (2003). https://doi.org/10.1007/978-3-540-39432-7_25
7. Ebner, M.: Coevolution and the Red Queen effect shape virtual plants. Genet. Program Evolvable Mach. **7**(1), 103–123 (2006)
8. Ebner, M., Grigore, A., Heffner, A., Albert, J.: Coevolution produces an arms race among virtual plants. In: Foster, J.A., Lutton, E., Miller, J., Ryan, C., Tettamanzi, A. (eds.) EuroGP 2002. LNCS, vol. 2278, pp. 316–325. Springer, Heidelberg (2002). https://doi.org/10.1007/3-540-45984-7_31

9. Eiben, A.E., Smith, J.E.: Introduction to Evolutionary Computing. Springer, Berlin (2007). https://doi.org/10.1007/978-3-662-05094-1

10. Ficici, S.G., Pollack, J.B.: A game-theoretic memory mechanism for coevolution. In: Cantú-Paz, E., et al. (eds.) GECCO 2003. LNCS, vol. 2723, pp. 286–297. Springer, Heidelberg (2003). https://doi.org/10.1007/3-540-45105-6_35

11. Floreano, D., Nolfi, S.: God save the red queen! competition in co-evolutionary robotics. In: Koza, J.R., et al. (eds.) Genetic Programming 1997: Proceedings of the Second International Conference on Genetic Programming, pp. 398–406. Morgan Kaufmann Publishers, San Francisco (1997)

12. Floreano, D., Nolfi, S., Mondada, F.: Competitive co-evolutionary robotics: from theory to practice. In: Pfeifer, R., Blumberg, B., Meyer, J.A., Wilson, S.W. (eds.) From Animals to Animats 5: Proceedings of the Fifth International Conf. on Simulation of Adaptive Behavior, pp. 515–524. The MIT Press, Cambridge (1998)

13. Fortin, F.A., De Rainville, F.M., Gardner, M.A., Parizeau, M., Gagné, C.: DEAP: evolutionary algorithms made easy. J. Mach. Learn. Res. **13**, 2171–2175 (2012)

14. Francisco, T., dos Reis, G.M.J.: Evolving combat algorithms to control space ships in a 2D space simulation game with co-evolution using genetic programming and decision trees. In: GECCO Workshop Proceedings: Defense Applications of Computational Intelligence, Atlanta, GA, pp. 1887–1892. ACM, New York (2008)

15. Francisco, T., dos Reis, G.M.J.: Evolving predator and prey behaviours with co-evolution using genetic programming and decision trees. In: GECCO Workshop Proceedings: Defense Applications of Computational Intelligence, Atlanta, GA 12-16, pp. 1893–1900. ACM, New York (2008)

16. Funes, P., Sklar, E., Juillé, H., Pollack, J.: Animal-animat coevolution: using the animal population as fitness function. In: Pfeifer, R., Blumberg, B., Meyer, J.A., Wilson, S.W. (eds.) From Animals to Animats 5: Proceedings of the 5th International Conference on Simulation of Adaptive Behavior, pp. 525–533. MIT Press, Cambridge (1998)

17. Gerdes, I., Klawonn, F., Kruse, K.: Evolutionäre Algorithmen. Vieweg Verlag, Wiesbaden (2004)

18. Hillis, W.D.: Co-evolving parasites improve simulated evolution as an optimization procedure. In: Langton, C.G., Taylor, C., Farmer, J.D., Rasmussen, S. (eds.) Artificial Life II, SFI Studies in the Sciences of Complexity, pp. 313–324. Addison-Wesley, Reading (1991)

19. Jong, E.D.: The incremental pareto-coevolution archive. In: Deb, K. (ed.) GECCO 2004. LNCS, vol. 3102, pp. 525–536. Springer, Heidelberg (2004). https://doi.org/10.1007/978-3-540-24854-5_55

20. Luke, S., Wiegand, R.P.: Guaranteeing coevolutionary objective measures. In: Foundations of Genetic Algorithms VII, pp. 237–251. Morgan Kaufman, San Francisco (2002)

21. Miller, G.F., Cliff, D.: Protean behavior in dynamic games: arguments for the co-evolution of pursuit-evasion tactics. In: Cliff, D., Husbands, P., Meyer, J., Wilson, S.W. (eds.) From Animals to Animats III: Proceedings of the 3rd International Conference on Simulation of Adaptive Behavior, pp. 411–420. The MIT Press, Cambridge (1994)

22. Nolfi, S., Floreano, D.: Co-evolving predator and prey robots: Do 'arms races' arise in artificial evolution? Artif. Life **4**(4), 311–335 (1998)

23. Ochoa, G., Jaffé, K.: On sex, mate selection and the Red Queen. J. Theor. Biol. **199**, 1–9 (1999)

24. Rechenberg, I.: Evolutionsstrategie '94. Frommann-Holzboog, Stuttgart (1994)

25. Ridley, M.: The Red Queen: Sex and the Evolution of Human Nature. Penguin Books, New York (1994)
26. Rosenzweig, M.L., Brown, J.S., Vincent, T.L.: Red queens and ESS: the coevolution of evolutionary rates. Evol. Ecol. 1, 59–94 (1987)
27. Rosin, B.: New methods for competitive coevolution. Technical report CS96-491, Cognitive Computer Science Research Group, Department of Computer Science and Engineering, University of California, San Diego, La Jolla, CA (1996)
28. Schwefel, H.P.: Evolution and Optimum Seeking. Wiley, New York (1995)
29. Spronck, P., Cowling, P., Champandard, A., Lanzi, P.L., Paiva, A.: AI for modern board games. In: Lucas, S.M., Mateas, M., Preuss, M., Spronck, P., Togelius, J. (eds.) Artificial and Computational Intelligence in Games, pp. 58–59. Dagstuhl Publishing, Schloss Dagstuhl (2012)
30. Stanley, K.O., Miikkulainen, R.: The dominance tournament method of monitoring progress in coevolution. In: Barry, A. (ed.) Genetic and Evolutionary Computation Conference Workshop Program, pp. 242–248 (2002)
31. Taylor, D.P.: Trust-based, multi-agent board games with imperfect information with Don Eskridge's "the resistance". Technical report Bachelor of Science Project, University of Derby, School of Computing & Mathematics (2014)
32. Togelius, J., Burrow, P., Lucas, S.M.: Multi-population competitive co-evolution of car racing controllers. In: Srinivasan, D., Wang, L. (eds.) IEEE Congress on Evoutionary Comp., Singapore, pp. 4043–4050. IEEE Press, Piscataway (2007)
33. Valen, L.V.: A new evolutionary law. Evol. Theory 1, 1–30 (1973)

Deep Catan

Brahim Driss[(✉)] and Tristan Cazenave

LAMSADE, Université Paris-Dauphine, PSL, CNRS, Paris, France
brahimdriss.bd@gmail.com, Tristan.Cazenave@dauphine.psl.eu

Abstract. Catan is a popular multiplayer board game that involves multiple gameplay notions: stochastic elements related to the dice rolls as well as to the initial placement of resources on the map and the drawing of development cards, strategic notions for the placement of the cities and the roads which call upon topological and shape recognition notions and notions of expectation of gains linked to the probabilities of the rolls of the dice. In this paper, we develop a policy for this game using a convolutional neural network. The used deep reinforcement learning algorithm is Expert Iteration [2] which has already given excellent results for Alpha Zero and its descendants.

Keywords: Multiplayer board game · Convolutional neural network · Deep reinforcement learning algorithm · Expert Iteration · Alpha Zero

1 Introduction

Monte Carlo Tree Search (MCTS) [7, 11] has been used in two-player complete information games. Modern board games such as Catan have more complex rules and deal with incomplete information due to dice rolls or drawing cards. MCTS has already been applied to Catan with success [15]. In this paper we address the use of deep neural network in combination with MCTS to play Catan. The combination of Deep Reinforcement Learning with MCTS gave strong computer players for Go, Chess and Shogi with Alpha Zero [14] and was further applied to many games with the Polygames framework [6]. The underlying Deep Reinforcement Learning for these systems is Expert Iteration [2]. In this paper we advocate that the combination of deep neural networks trained from zero knowledge in combination with MCTS can outperform MCTS alone.

The paper is organized as follows. The second section recalls related work. The third section presents Deep Reinforcement Learning of Catan. The fourth section gives experimental results.

2 Background and Related Work

2.1 Monte Carlo Tree Search

Monte Carlo Tree Search (MCTS) is a general search algorithm that was initially designed for the game of Go [7]. The most popular MCTS algorithm is Upper Confidence bounds applied to Trees (UCT) [11]. UCT is the standard MCTS algorithm. It uses the mean of the previous random playouts to guide the beginning of the current

© Springer Nature Switzerland AG 2022
J. L. Jiménez Laredo et al. (Eds.): EvoApplications 2022, LNCS 13224, pp. 503–513, 2022.
https://doi.org/10.1007/978-3-031-02462-7_32

playouts. There is a balance between exploration and exploitation when choosing the next move to try at the beginning of a playout. Exploitation tends to choose the move with the best mean, while exploration tends to try alternative and less explored moves to see if they can become better. The principle of UCT is optimism in face of uncertainty. It chooses the action with the UCB formula:

$$argmax_a\left\{Q(s,a) + C\sqrt{\frac{log(N(s))}{N(s,a)}}\right\} \tag{1}$$

where $N(s,a)$ is the number of simulations of the node, $N(s)$ the number of simulations of the parent node (state s before taking action a), $Q(s,a)$ the winrate of the action a in the state s (number of wins/number of simulations) and C the UCB bandit exploration coefficient.

The All Moves As First heuristic (AMAF) [3] is a heuristic that was used in Gobble, the first Monte Carlo Go program [5]. It consists in updating the statistics of the moves of a position with the result of a playout, taking into account all the moves that were played in the playout and not only the first one.

In Catan, standard MCTS with 10,000 simulations could still be beaten easily, [15] used a heuristic action selection procedure inside the MCTS and virtual wins.

In order to improve the level of play, instead of randomly sampling moves, probabilities can be associated to moves according to their type, building a city or settlement having higher chances when available to be selected than building the road for example and skipping the turn being the lowest, since building is always rewarding for the player and skipping when building is available is generally a bad idea.

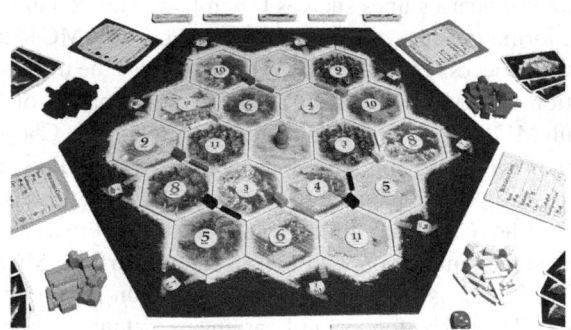

Fig. 1. A Catan board

2.2 Rules of Catan

In Catan, players compete to colonize an island represented by a board (Fig. 1) of hexagonal tiles. There are 5 resource types - Brick, Lumber, Ore, Grain, and Wool - which can be spent to make various actions. The first player to reach 10 Victory Points (VP) or more is considered the winner. VP can be acquired by various means: placing settlements (1VP) or cities (2VP) on the board, having the longest road or largest army

(2VP), or special development cards (1VP). The island of Catan is represented as a board of 19 land hexagonal tiles called hexes, randomly placed when setting up the game. Tiles can either represent a desert, or produce one of the 5 resources, in which case they will be assigned a number between 2 and 12. We will call the edge of a hex a path, and its corner an intersection.

At the beginning of the game, each player places 2 settlements, each with an adjacent road. Settlements must be placed on intersections and can not be next to one another. During each turn, a player can take a sequence of actions if they have the required resources for it:

- Build a Road: 1*Brick + 1*Lumber
- Build a Settlement: 1*Brick + 1*Lumber + 1*Grain + 1*Wool
- Build a City: 3*Ore + 1*Grain
- Buy a Development Card: 1*Ore + 1*Grain + 1*Wool
- Trade resources with the bank.

For trading the default ratio is four of the same resource for any one resource, but having a settlement or city on a harbor can reduce the rate to 3:1 or 2:1. There is no simple winning strategy. Basically, a stable and varied production of resources is beneficial to obtain VP. Thus, players should prioritize placing their settlements in intersections surrounded by balanced resources and high production chance (with numbers around 7), near promising unexploited areas or on interesting harbors. Buying development cards is a bit of a lucky dip as the player is buying blindly from a stack of cards but they are all beneficial to him. Development cards have different bonuses such as blocking one hex from producing resources and stealing a card from one of its neighbors, adding a victory point, building 2 free roads for example.

2.3   Related Work

Previous works on Catan used various methods. The earliest agent used Model Trees trained through self-play [13], multi-agent systems were also able to obtain strategic game play [4]. Szita, Chaslot, and Spronck [15] used Monte Carlo Tree Search in a perfect-information variation of the game. Other works explored other aspects of the game. Afantenos [1] focuses on strategic conversation concerning bargaining negotiation in the game of Catan. Guhe [10] focuses on persuasions using empirical data from game simulations from a game theory point of view. Other papers also used Deep Reinforcement Learning, two of them focused only on a subset of actions that is trading, using Deep Q-Learning [8] and Deep Q-Learning with LSTM [16]. A third paper [9] also used Deep Reinforcement Learning with an agent using a variation of Advantage Actor Critic in a 2 player version of the game.

In this paper, we use Monte Carlo Tree Search combined with deep neural networks within the original rules, i.e., imperfect information and 4-player game, without domain knowledge, refusing and never initiating trades with other players, but continuing to trade with the bank. The used methods and objectives are different from previous works, since there is no Advantage Actor Critic [12] and the game is not the 2-player version [9]. We think that the 2-player version is very different from the 4-player one. Having 3 players playing before the next turn is not the same as having only one. There will be a higher chance for instance to be blocked by enemies and lose good spots for settlements, there will also be less place to expand because of building constraints. We are also playing the entire game not only focusing on a single part of it, on the contrary of [8] and [16]. Using Monte Carlo Tree Search [15] did not give good results after 10,000 simulations, this is why we propose a Local Value Estimation network that improves on Monte Carlo Tree Search, learning AMAF statistics during self-play. We also show that Monte Carlo Tree Search can be further improved using deep reinforcement learning with Expert Iteration.

3 Deep Reinforcement Learning of Catan

3.1 Supervised Learning of the Value Network

We train a value network using games played by MCTS. The value network takes a random state of the game as an input and predicts the MCTS winrates of the root nood for the 4 players given by the average of the rollouts from that state. The output of the network is a softmax over the 4 outputs giving the winning probabilities for the 4 players (Fig. 2).

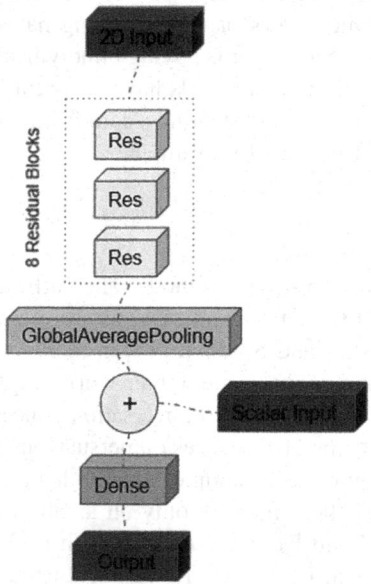

Fig. 2. Value network architecture

Since Catan has multiple information unrelated to the board (e.g. visible victory points, development cards or resources left), the network used two inputs: A 2-dimensional input for the board and a scalar input for game information. The 2 dimensional input is transformed using residual blocks to process the board. A regular board of Catan contains 19 hexes, 72 paths and 54 intersections. We employ a similar architecture to the Alpha Zero network. We pass in the board position as a 23×13 image and use convolutions to construct a representation of the position.

Table 1. Neural network 2D input channels

Board (2D channels)		29
Roads		1(x4)
Settlements		1(x4)
Cities		1(x4)
Ports		2
Resources types		1(x5)
Resources odds		1(x5)
Robber odds		1(x5)
Vector input		
Game phase		12
Visible VP		11(x4)
Resources cards		11(x4)
Development cards		11(x4)
Resources left		21(x5)
Largest army		1(x4)
Longest road		1(x4)
Buildings left		
	Roads left	16(x4)
	Settlements left	6(x4)
	Cities left	5(x4)

The Catan board is different from boards in games such as Chess and Go. In these games all cells are similar. Catan cells are hexagons and the board topology of Catan is unlike that of Hex or Havannah for instance. The hexes, path and intersections have different roles and features. We split features of different types in 29 channels to prevent the convolution from processing them in the same way. The description of the different channels is shown in Table 1.

We also use the brick coordinate [9]. We use a 5×3 kernel which makes the neighbors comparable to the actual neighbors on the hexagonal board. The mapping is explained in Fig. 3.

The kernel used in the convolutions of the residual blocks is the 5×3 brick coordinate kernel. The optimizer is Adam (Learning rate = 0.001), ReLU activation functions in the hidden layers with Dropout (0.3 and 0.5 rate), Softmax in the output layer. The loss is the Mean Absolute Error.

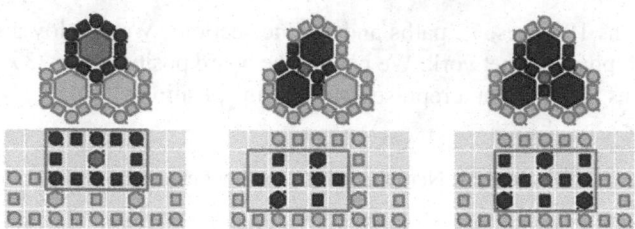

Fig. 3. 5 × 3 kernel on brick coordinate

3.2   Local Value Estimation

A second neural network with dense layers and ReLU activations, was trained on AMAF statistics obtained during rollouts. The purpose of this neural network is to evaluate the player moves to reduce the breadth of the search in the tree, without removing the simulations in the playout phase, by modifying the UCB bandit value, adding an evaluation term:

$$argmax_a \left\{ Q(s,a) + C\sqrt{\frac{log(N(s))}{N(s,a)}} + C_2 \times eval \right\} \qquad (2)$$

where $eval$ is this neural network prediction of the value of the move. Instead of focusing on the board as a whole state, the network evaluates the possible moves locally, giving more data to train on that is less complex (n pairs of moves and scores instead of a single state and its value). It is updated after each training iteration with data from the self-play games. The inputs of the network are the moves and their local corresponding features as shown in Fig. 4 and the output is the prediction of the AMAF score of MCTS playouts. For buildings, the neighbor hexes of the construction are the features. This network will be used in the UCTNet experiment and be compared to UCT without the network evaluation, provided with the same budget of rollouts.

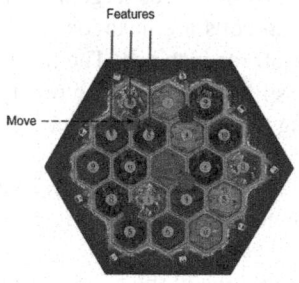

Fig. 4. Input example for the Local Value network

3.3 Expert Iteration of the Value Network

Compared to imitation learning techniques, Expert Iteration (ExIt) is enriched by an expert improvement step. Improving the expert player and then solving the imitation learning problem allows us to exploit the fast convergence properties of imitation learning, even in contexts where no strong player was originally known, such as when learning from scratch.

At each iteration i, the algorithm proceeds as follows: we create a set S_i of game states by playing the $\hat{\pi}_{i-1}$ learner. In each of these states, we use our expert to compute an imitation learning target at s (e.g., the expert's $\pi_{i-1}^*(a|s)$ action); the state-target pairs (e.g., $(s, \pi_{i-1}^*(a|s))$) form our dataset D_i. We train a new apprentice $\hat{\pi}_i$ on D_i (learning by imitation). Then, we use our new apprentice to update our expert $\pi_i^* = \pi^*(a|s; \hat{\pi}_i)$ (expert improvement).

4 Experimental Results

4.1 Importance of the Budget

The budget allocated to the different MCTS is important. An experiment was performed to verify the impact of the number of rollouts on the MCTS performance. Four matches of 200 games were performed, opposing an improved MCTS (number of rollouts multiplied by 2) against 3 normal MCTS. The results of this experiment are shown in Table 2:

Table 2. Matches between MCTS using twice as many rollouts as its 3 opponents.

Match	Games won	Winrate
100 vs 50	70	35%
200 vs 100	56	28%
400 vs 200	63	31.5%
800 vs 400	59	29.5%

We can see that the MCTS with the most rollouts always has a positive winrate (higher than 25%). A higher number of rollouts improves the level of the MCTS.

4.2 Training Performance

First Iteration. The neural network is trained on the scores of 800 games played in self-play (4 different players using the same method in order to find the best move) and evaluated on 200 games that do not exist in the training data at each ExIt iteration. To avoid having similar states, all games are mixed, and the network is trained on minibatches of 64 states. At the end of the iteration, the network will have learned about 50,000 game states with their Monte Carlo scores.

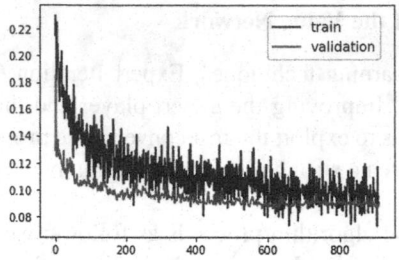

Fig. 5. Evolution of the network loss during training at the 1st iteration

Figure 5 shows that at the beginning of the training the network has an average loss of 0.2 in the predictions which eventually stabilizes at 0.1 at the end of the training over the 1,000 games played. The network is then able to make Monte Carlo score predictions for the 4 players with an average error of 0.1 without simulating the game.

Second Iteration. The second iteration is the step where the network trained in the first iteration is used in an MCTS and replaces the rollouts by a single neural network evaluation.

The new scores obtained are the labels of a new neural network. The new neural network is then trained on 1,000 games of self-play by MCTS using the first network.

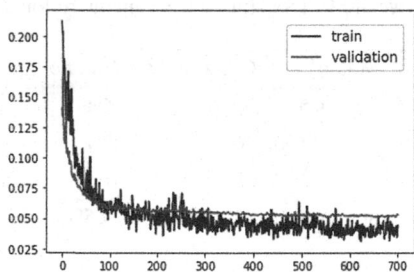

Fig. 6. Evolution of the network loss during training at the 2nd iteration

We notice as shown in the Fig. 6 that the network starts with the same error of 0.2 and stabilizes at a lower error than the first iteration (0.05 in the test compared to 0.1 in the first iteration).

We also notice that the training of the new network is less noisy, with smaller variance of the error during the training.

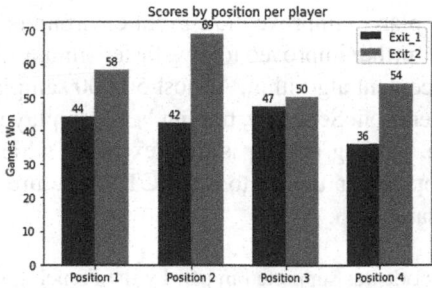

Fig. 7. Results of the games played between the two models

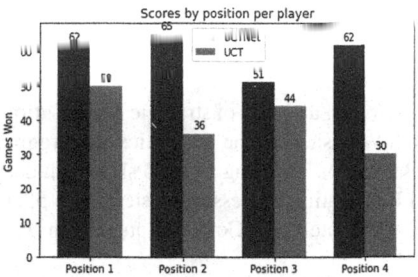

Fig. 8. Results of the games played between UCT and UCTNet

4.3 Evaluation of Neural Networks

Value Network. 400 games opposing the two value networks (ExIt_1 is the first iteration and ExIt_2 the second) in 2 vs 2 were played, with the same budget of rollouts given to the two networks, the positions (1, 2, 3, 4) of the players are drawn randomly at the beginning of the game. The network of the second iteration won 231 games out of 400 played (58% winrate) against the network of the first iteration. Results of the games are shown in Fig. 7.

Local Value Estimation Network. 400 games opposing UCT (Classic MCTS algorithm) and UCTNet (UCT using the Local Value Estimation), with the same budget of rollouts given to the two UCT, the positions (1, 2, 3, 4) of the players are drawn randomly at the beginning of the game. UCTNet won 240 games out of 400 played (60% winrate) against UCT. The results are shown in Fig. 8.

5 Conclusion and Future Work

The first experiment shows that Expert Iteration can improve the level of play in a game involving chance. Since the Monte Carlo evaluations of the network were less noisy, the training in the second iterations had less variance. The second experiment evaluates moves instead of entire boards and therefore was easier to train. Using such a network, UCTNet was able to defeat UCT in 60% of games.

For future work, we plan to improve the current environment. Firstly, we are using Python, the code can be further improved to give faster simulations in order to do more iterations of our reinforcement algorithm. Almost 50,000 samples were needed for the value network at each iteration. Secondly, trading between players is a social and interesting part of the game. Adding trading is the next step, which would require some changes in the game loop and an update to the MCTS structure, since trading involves multiple players in the same turn.

Acknowledgment. This work was supported in part by the French government under management of Agence Nationale de la Recherche as part of the "Investissements d'avenir" program, reference ANR19-P3IA-0001 (PRAIRIE 3IA Institute).

References

1. Afantenos, S., et al.: Developing a corpus of strategic conversation in the settlers of catan. In: SeineDial 2012-The 16th Workshop on the Semantics and Pragmatics of Dialogue (2012)
2. Anthony, T., Tian, Z., Barber, D.: Thinking fast and slow with deep learning and tree search. In: Advances in Neural Information Processing Systems, pp. 5360–5370 (2017)
3. Bouzy, B., Helmstetter, B.: Monte-Carlo Go developments. In: ACG. IFIP, vol. 263, pp. 159–174. Kluwer (2003)
4. Branca, L., Johansson, S.J.: Using multi-agent system technologies in settlers of catan bots. In: Agent-based Systems for Human Learning and Entertainment (ABSHLE) (2007)
5. Brügmann, B.: Monte Carlo Go. Technical report (1993)
6. Cazenave, T., et al.: Polygames: improved zero learning. ICGA J. **42**(4), 244–256 (2020)
7. Coulom, R.: Efficient selectivity and backup operators in Monte-Carlo tree search. In: van den Herik, H.J., Ciancarini, P., Donkers, H.H.L.M.J. (eds.) CG 2006. LNCS, vol. 4630, pp. 72–83. Springer, Heidelberg (2007). https://doi.org/10.1007/978-3-540-75538-8_7
8. Cuayáhuitl, H., Keizer, S., Lemon, O.: Strategic dialogue management via deep reinforcement learning, vol. abs/1511.08099. Springer (2015)
9. Gendre, Q., Kaneko, T.: Playing catan with cross-dimensional neural network. In: Yang, H., Pasupa, K., Leung, A.C.-S., Kwok, J.T., Chan, J.H., King, I. (eds.) ICONIP 2020. LNCS, vol. 12533, pp. 580–592. Springer, Cham (2020). https://doi.org/10.1007/978-3-030-63833-7_49
10. Guhe, M., Lascarides, A.: The effectiveness of persuasion in the settlers of catan. In: 2014 IEEE Conference on Computational Intelligence and Games, pp. 1–8. IEEE (2014)
11. Kocsis, L., Szepesvári, C.: Bandit based Monte-Carlo planning. In: Fürnkranz, J., Scheffer, T., Spiliopoulou, M. (eds.) ECML 2006. LNCS (LNAI), vol. 4212, pp. 282–293. Springer, Heidelberg (2006). https://doi.org/10.1007/11871842_29
12. Mnih, V., et al.: Asynchronous methods for deep reinforcement learning. In: Balcan, M.F., Weinberger, K.Q. (eds.) Proceedings of The 33rd International Conference on Machine Learning. Proceedings of Machine Learning Research, vol. 48, 20–22 June 2016, pp. 1928–1937. PMLR, New York. https://proceedings.mlr.press/v48/mniha16.html
13. Pfeiffer, M.: Reinforcement learning of strategies for settlers of catan. In: Proceedings of the International Conference on Computer Games: Artificial Intelligence, Design and Education. Citeseer (2004)
14. Silver, D., et al.: Mastering chess and shogi by self-play with a general reinforcement learning algorithm. CoRR abs/1712.01815 (2017)

15. Szita, I., Chaslot, G., Spronck, P.: Monte-Carlo tree search in settlers of catan. In: van den Herik, H.J., Spronck, P. (eds.) ACG 2009. LNCS, vol. 6048, pp. 21–32. Springer, Heidelberg (2010). https://doi.org/10.1007/978-3-642-12993-3_3

16. Xenou, K., Chalkiadakis, G., Afantenos, S.: Deep reinforcement learning in strategic board game environments. In: Slavkovik, M. (ed.) EUMAS 2018. LNCS (LNAI), vol. 11450, pp. 233–248. Springer, Cham (2019). https://doi.org/10.1007/978-3-030-14174-5_16

Machine Learning and AI in Digital Healthcare and Personalized Medicine

Vectorial GP for Alzheimer's Disease Prediction Through Handwriting Analysis

Irene Azzali[1], Nicole Dalia Cilia[2,4], Claudio De Stefano[2],
Francesco Fontanella[2(✉)], Mario Giacobini[1], and Leonardo Vanneschi[3]

[1] Department of Veterinary Sciences, University of Torino, Largo Paolo Braccini 2,
10095 Grugliasco, TO, Italy
[2] Department of Electrical and Information Engineering (DIEI), University
of Cassino and Southern Lazio, Via G. Di Biasio 43, 03043 Cassino, FR, Italy
fontanella@unicas.it
[3] NOVA Information Management School (NOVA IMS),
Universidade Nova de Lisboa, Campus de Campolide, 1070-312 Lisbon, Portugal
[4] Instituto for Computing and Information Sciences, Radboud University,
Toernooiveld 212, 6525 Nijmegen, EC, The Netherlands

Abstract. Alzheimer's Disease (AD) is a neurodegenerative disease
which causes a continuous cognitive decline. This decline has a strong
impact on daily life of the people affected and on that of their relatives. Unfortunately, to date there is no cure for this disease. However,
its early diagnosis helps to better manage the course of the disease with
the treatments currently available. In recent years, AI researchers have
become increasingly interested in developing tools for early diagnosis of
AD based on handwriting analysis. In most cases, they use a feature
engineering approach: domain knowledge by clinicians is used to define
the set of features to extract from the raw data. In this paper, we present
a novel approach based on vectorial genetic programming (VE_GP) to
recognize the handwriting of AD patients. VE_GP is a recently defined
method that enhances Genetic Programming (GP) and is able to directly
manage time series in such a way to automatically extract informative
features, without any need of human intervention. We applied VE_GP to
handwriting data in the form of time series consisting of spatial coordinates and pressure. These time series represent pen movements collected
from people while performing handwriting tasks. The presented experimental results indicate that the proposed approach is effective for this
type of application. Furthermore, VE_GP is also able to generate rather
small and simple models, that can be read and possibly interpreted.
These models are reported and discussed in the Last part of the paper.

Keywords: Alzheimer's disease · Artificial intelligence · Handwriting
analysis · Vectorial genetic programming

1 Introduction

Alzheimer's Disease (AD) is a progressive neurologic disorder that causes the
brain to shrink (atrophy) and brain cells to die. Currently accounting for 60–80%

© Springer Nature Switzerland AG 2022
J. L. Jiménez Laredo et al. (Eds.): EvoApplications 2022, LNCS 13224, pp. 517–530, 2022.
https://doi.org/10.1007/978-3-031-02462-7_33

of dementia cases worldwide, it causes a continuous decline in thinking, behavioral and social skills, and it strongly affects a person's ability to function independently. Unfortunately, there is no remedial cure, thus early diagnosis can be of paramount importance for delaying the progression of the disease. Recently, researchers have shown that patients affected by AD have altered spatial organization and poor movement control. In particular, handwriting, which is the result of a complex network of cognitive, kinesthetic and perceptive motor skills, can be significantly compromised. Therefore, the observation of motor activities can be crucial for AD's early diagnosis: in this perspective, analyzing the characteristics and anomalies of handwriting is a relevant step. In the Last few years, there has been a growing focus on the use of Artificial Intelligence to investigate the peculiarities and irregularities of handwriting [10, 13, 14].

These studies have highlighted the importance, and the difficulty, of extracting suitable features and/or understanding the complex interactions between multiple features in handwriting. In particular, one of the major issues is the identification of features that allow the system to distinguish the natural handwriting alterations due to age from those caused by neurodegenerative disorders. The typical approach is to analyse data coming from handwriting in the form of time series, and then extract the features from the recorded handwriting spatial coordinates, a process that nowadays is mainly manual. For instance, Cilia and colleagues extracted a set of informative features, such as velocity, acceleration and jerk, from temporal data of the bi-dimensional coordinates of the handwriting (indicated as X and Y coordinates in the following) and the pen pressure. They used those features to classify AD patients [8]. Feature extraction, however, can be an extremely hard task. In particular, the selection of a limited set of variables from the available ones can strongly limit the possibility to generate combinations, that allow us to obtain more informative features. This paves the way for the application of vectorial genetic programming, VE_GP [4], to handwriting analysis of AD patients.

VE_GP is a recently defined method that enhances Genetic Programming (GP) [19, 24] by introducing the possibility of directly using vectors (like for instance time series) as terminal symbols in the trees encoding the individuals, and by proving a rich and easily extensible set of primitive operations to manage vectors. In this way, time series do not need to be collapsed "a priori" and can be used directly by the algorithm. Furthermore, the inclusion of aggregating parametric functions in the primitive set allows VE_GP to automatically generate informative features from data, without any need of human intervention. Last but not least, VE_GP is often able to generate models that are simple enough to be read and interpreted, in some cases offering some important insights on the application [2, 3]. This is a crucial element in the medical field, where trust is at the basis of the employment of machine learning models in practice [1].

In this work, we explore the use of VE_GP to classify patients affected by AD, on the basis of their ability to accomplish nine well-defined task related to handwriting. Those tasks were introduced in [8], and are described in Table 1, later in this paper. The participants involved in the study performed the required actions on a graphic tablet, that acquires data in the form of X and Y coordinates and pressure. Differently from [8], VE_GP allows us to make predictions directly from X, Y and pressure. In order to demonstrate the advantages of VE_GP, we compare the results obtained by VE_GP against those of the features used in [8]. It is useful to remark that VE_GP and the other methods used for the comparison use exactly the same raw data, i.e. the sequences of points (represented in terms of X, Y coordinates and pressure) produced by the participants in carrying out the tasks: while VE_GP automatically generates the features from the raw data, the other methods require a preliminary phase of feature extraction to be carried out manually.

The paper is organized as follows. After a brief overview of the research activities related to our study (Sect. 2), Sect. 3 introduces materials and methods; Sect. 4 reports the experimental results; finally, Sect. 5 concludes the paper and proposes ideas for future research.

2 Related Work

Thanks to their search-ability, Evolutionary Algorithms [23] have been often employed in health applications, in particular Genetic Algorithms (GAs) and GP [11,12]. Concerning GP, it has been used in a wide range of applications. For example, [6] proposed a constrained-syntax GP-based algorithm for discovering classification rules in medical data. The authors tested their approach on five datasets and achieved better results than decision trees. In [7], the authors presented an novel approach based on Geometric Semantic GP (GSGP) to solve a problem related to the physio-chemical properties of proteins, involving the prediction on these properties in tertiary structure. GSGP uses new genetic operators that allows the definition of an unimodal error surface for any supervised learning problem, independently from the complexity and size of the underlying data set. The authors proved that the proposed approach was more effective than Artificial Neural Networks and Support Vector Machines. GP has also been used as a tool to support medical decisions for treating rare diseases. In particular, in [5] the authors developed and tested six predictive data models using GP. The models was integrated into a web-application to be used by therapists for supporting them in their patients care and treatments activities. More recently, Parziale et al. used a Cartesian GP to automatically identify people affected by the Parkinson's disease (PD) through the analysis of their handwriting and drawings [22]. The proposed approach allowed them to support the diagnosis of PD through explicit classification models which provided results that can be interpreted by physicians. The comparison results classification modelsa twofold goal: proved that shown that GP has also been used as a tool .

GAs have been widely used in medical applications, in most of the medical specialities, among which medical imaging, rehabilitation medicine, and health care management [15]. As concerns neurodegenerative diseases, in [26] the authors used a multi-objective GA to find the relevant volumes of the brain related to the Alzheimer's disease, whereas in [18] a GA was used to search the optimal set of neuropsychological tests, to be used to build a system for the prediction of the Alzheimer's disease. More recently, in [10] the authors presented a GA-based system to improve the performance of the system previously developed for congnitive impairments through handwriting analysis.

From the brief literature review outlined above, we can observe that to the best of our knowledge, there are no other studies in which a GP-based approach is used for the prediction of AD through handwriting analysis.

3   Materials and Methods

This section presents the computational methods and the data used in our study. In particular, in Subsect. 3.1 we introduce VE_GP and in Subsect. 3.2 we describe the characteristics of the employed dataset.

3.1   Vectorial Genetic Programming

Vectorial genetic programming (VE_GP) is a recently developed approach of GP to properly deal with time series as predictors or targets. VE_GP allows vectors as terminals, providing a suitable representation for time series, and supplies a set of primitive functions to operate on vectors. In the continuation of this section, we begin with a description of these parametrizable primitives, first presenting the functions of arity 1 and then the functions of arity 2. Next, we discuss other peculiar characteristics of VE_GP, that distinguish it from traditional GP, such as the population initialization and the parameter mutation, that allows to change the parametrization of the primitive functions. Finally, we discuss how we employed VE_GP for classification in this work. Further details on VE_GP can be found in [4].

Functions of Arity 1. VE_GP's primitive functions, when applied to vectors, typically return a scalar value that, informally speaking, "represents" the elements of the vector, or "aggregates" them into one single value, able to capture some characteristics of the vector. For this reason, in some cases, they will be referred to as aggregating functions. VE_GP provides several aggregating functions, along with their parameters. Parameters define the part of the vector (or "window") where the function will be applied. The window can slide all the vector. Let $[v_1, v_2, ..., v_p, ..., v_q, ..., v_n]$ be a vector and let p and q be two indexes, with $p < q$; then $f_{p,q}(v) = f([v_p, ..., v_q])$. In case the window extends to not existing elements of the vector, these elements are not included in the calculation. To provide a numerical example let us consider $v = [1, 2.5, 4.3, 0.7, 1.6]$ and $(p, q) = (2, 3)$. The function V_sum$_{2,3}$ applied to v returns V_sum$_{2,3}(v) = [1, 2.5, 4.3, 0.7, 1.6]$. For a more detailed explanation of these functions see Section 3, Table 3 of [4].

Functions of Arity 2. Regarding functions of arity 2, they are simply the extension to vectors of the arity 2 functions between scalars often used by traditional GP. In particular, when the inputs of a function are two vectors of length greater than 1, the shortest is completed with the null element of the function up to the length of the longest before applying the function itself. Differently, when a scalar and a vector of length greater than 1 are the inputs, the scalar is initially replicated up to the length of the other vector input.

To provide a numerical example let us consider $v = [1, 2.5, 4.3, 0.7, 1.6]$ and $w = [0.8, 3.6, 1.9]$. The function VSUMW applied to v and w returns $\text{VSUMW}(v, w) = [1.8, 6.1, 6.2, 0.7, 1.6]$. To clarify, see Section 3 of [4].

Initialization. VE_GP proposes a peculiar initialization method, with the objective of using in a correct way the aggregating functions included in the primitive set. Basically, a portion of n_1 individuals are forced to apply aggregating functions only to vectorial variables. This feature ensures a pool of individuals that properly use aggregating functions., The remaining n_2 individuals are generated with one of the classical initialization techniques [24] without any forcing. Both n_1 and n_2 are decided by the user.

Parameter Mutation. The genetic operator of parameter mutation (PM) is developed in VE_GP in order to let the evolution find the most informative windows of time. PM simply looks in an individual for parametric aggregating functions, randomly selects one of them and randomly changes one of its parameters. The parameter is mutated without violating the rule that $p < q$.

VE_GP for Classification and Fitness Evaluation. VE_GP was originally introduced to deal with time series forecasting. However, it can be easily extended to classification. Classification, in fact, can be performed using the individuals as discriminating functions at threshold 0, as for instance in [17]. In this way, evaluations ≥ 0 correspond to class 0, while the others correspond to class 1. Given that we need a scalar output to apply this threshold function, in this work if an individual returns a vector, a mean is applied to it, in order to obtain a scalar.

3.2   Dataset

The data used in this work were gathered by highly qualified AD specialists, following a rigorous procedure. The first step is devoted to acquisition and recording of subjects' handwriting. This is done by recording three values (X, Y and Z) for each point, at a constant sampling rate equal to 200 Hz. The first two coordinates (X and Y) are the point position in the two-dimensional space representing the surface where the writing is produced, while the third one (Z) is a measure of the pressure exerted by the subject at that point. Such a measure assumes a positive value when the pen is resting on the sheet and a null value when the pen is detached, up to a maximum distance of 3 cm from the sheet, beyond which the system is not able to receive information. A specialized application was used

Table 1. The tasks involved in the analysis.

Task#	Description
1	Copy the letters 'l', 'm' and 'p'
2	Copy the letters 'n', 'l', 'o' and 'g' on the adjacent rows
3	Write exactly four joined lowercase cursive letters 'l', in a single smooth movement
4	Write exactly four joined lowercase pairs of cursive letters 'le', in a single smooth movement
5	Copy the word "foglio" ("sheet" in Italian)
6	Copy the word "foglio" above a line
7	Copy the word "mamma" ("mum" in Italian)
8	Copy the word "mamma" above a line
9	Write a simple sentence under dictation

to acquire these data while the subjects were writing on a A4 sheet, fasten to the tablet surface. Furthermore, since handwriting skills could be influenced by sex, age, education level and type of work, this information is also stored. The participants to the study were recruited with the support of the geriatric ward, Alzheimer unit, of the "Federico II" hospital, in Naples (Italy). Standard clinical tests, such as the Mini-Mental State Examination (MMSE), were used to distribute the subjects into two groups of healthy controls and patients with a cut off of 27. All participants were right-handed and comfortably positioned at a distance of approximately 70 cm from the sheet.

The database used in this paper includes 130 subjects, each performing the nine tasks defined in the protocol presented in [9], and described in Table 1.

These tasks analyse the patients abilities in repeating complex graphic gestures, with a semantic meaning. As in [9], while performing Task #1, the subjects must copy three letters; the letters ('l', 'm', 'p') were chosen because they have rather diverse graphic shapes and they force the participant to draw in ascending and descending direction while writing [28]. Task #2 consists in copying four letters ('n', 'l', 'o' and 'g') on adjacent rows. Its objective is to test the spatial organization abilities of the participant [21]. Tasks #3 and #4 require participants to write continuously for four times, in cursive, the letter 'l' and the bigram 'le', respectively [16]. This task is used to test motion control alterations. Tasks #5, #6, #7 and #8 consist in copying words, which is the most explored activity in the analysis of handwriting for people affected by cognitive impairments [16,20,28]. Moreover, to observe the spatial organization ability, the copy of the same word with and without a cue is required. Finally, in Task #9, participants are asked to write a simple sentence under dictation, namely "Il bambino gioca a palla" (the child is playing ball). This is a short sentence with a complete meaning and easy to memorize. The rationale of this task is that the movements of AD patients can be altered because of the lack of visualization of the stimulus (as for the copy tasks).

The experiments of VE_GP were performed starting from the raw kinematic and pressure time series (X, Y and Z values). While in [8] a process of feature extraction was applied to those series, here they are given as input to VE_GP without any further intervention. For each task, we used 130 files (one per participant). Each file contained the points (coordinates X and Y, and pressure Z) recorded by the graphic tablet while the participant was performing that task. Finally, to distinguish between AD patients and the control group, we properly named each file.

4   Experimental Results

To evaluate the performance of VE_GP, we performed 30 independent runs for each of the nine tasks, involving a five-fold cross-validation technique to split the dataset into a training, a validation and a test set. For each task, at the beginning of each run, we partitioned the available data into 5 folds of identical size. Then, one random fold is selected as the test set, 3 out of the remaining folds are randomly selected as the training set, while the 4th is used as the validation set. The validation set in a population based technique is fundamental to select the best individual, which is the proposed solution to the problem. To evolve individuals, we selected the accuracy on the training set as the fitness measure. Accuracy is defined as the percentage of correctly classified subjects, therefore the problem is turned into a maximization one. In order to simulate some of the features extracted by the standard classification approach, we included in VE_GP innovative terminal functions. Among these, we have V_normpq that returns the 2-norm of the sub-vector defined by the range $[p, q]$; V_distpq, that returns the distance between the element of position p and the element of position q of the input vector; V_length, that returns the length of the vector (to gain knowledge about the duration of the task); and V_duration0, that returns the time of the on air stretch. The experimental parameters used for VE_GP are provided in Table 2. Due to the exploratory nature of the work parameters are not tuned, but based on default values.

To assess the effectiveness of our system we carried out three sets of experiments. In the first, we assessed the generalization ability of VE_GP, whereas in the second we compared our results with those obtained in [8]. Finally, in the third set, we analyzed the features extracted by VE_GP. The experiments are detailed in the following subsections.

4.1   Testing the Generalization Ability of VE_GP

To assess the generalization ability of VE_GP we have plotted its accuracy box-plots on the learning set (i.e. the union of the training and validation sets) and on the test set. These plots are shown in Fig. 1. These plots allow us to visually detect possible overfitting. Looking at the median accuracy, we can observe that there is no relevant discrepancy between the results obtained on the learning and on the test set, and this suggests that the models generated by VE_GP are not affected by overfitting.

Table 2. Parameters setting of the algorithm of VE_GP.

Parameter	Value
Maximum numb. of generations	50
Population size	200
Genetic operators	crossover, mutation, parameter mutation
Genetic operators probabilities	0.5, 0.1, 0.4
Initialization	$n_1 = 30\%$, $n_2 = 70\%$ Ramped Half-and-Half
Functions set	VSUMW, V_W, VprW, VdivW, V_maxpq, V_minpq, V_meanpq, V_sumpq, V_normpq, V_distpq, V_length, V_duration0
Terminals	input variables + randomly generated numbers (as constants)
Fitness	accuracy
Parents selection	Lexicographic Parsimony Pressure
Elitism	keep best
Maximum depth of the individuals	17

4.2 Comparative Study

To test the effectiveness of our approach, we compared its performance with the one of the methods presented in [8]. In that study, the authors used a standard feature engineering approach, extracting features suggested by clinicians, such as velocity, acceleration and pressure. Motivated by the fact that many previous studies report significant differences in patients' motor performance between in-air and on-paper traits, the authors extracted three groups of features:

- On-paper: the features extracted from the written traits (i.e. during pen-down and the sunsequent pen-up);
- In-air: the features extracted from the in-air traits (i.e. those acquired by the system when the pen is lifted from the sheet, within the maximum acquisition distance of 3 cm). These movements characterize the planning activity for positioning the pen tip between two successive written traits;
- All: these features are computed without distinguishing between in-air and on-paper traits. In practice, for each task, each feature was extracted averaging the values on both in-air and on-paper traits. Note that this kind of features allow the merge of the information contained in the on-paper and in-air traits.

As in [8], for each of these categories, we trained a Decision Tree (DT) classifier, using the C4.5 algorithm [25]. We used the five-fold cross-validation technique to split data into a training and a test set, and we performed 30 independent runs for each of the nine tasks. To statistically validate the results of the comparison between VE_GP and DT, given that data are not normally distributed, we used the Wilcoxon rank-sum test, which is a non-parametric statistical test. Table 3 shows the accuracies achieved on the test set (expressed in percentages) of VE_GP and DT on the three feature categories used to represent handwriting

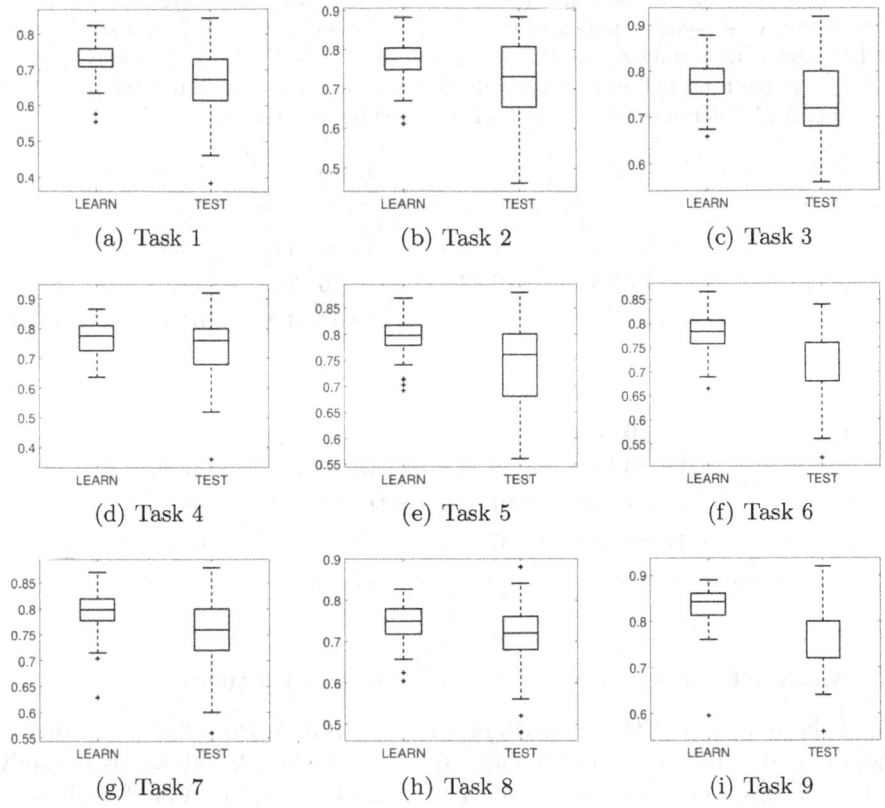

Fig. 1. Boxplots of learn and test accuracy of VE_GP classifiers.

movements. The values shown are averaged over the thirty independent runs. The symbols '+'/'−' denote a VE_GP's statistically significant win/loss, respectively, whereas the symbol '≈' denotes a tie (i.e. the differences between VE_GP and DT are not statistically significant), according to the Wilcoxon test (with $\alpha = 0.05$).

From the table, we can observe that in terms of win/loss comparison VE_GP outperformed the compared approach on the three categories of features, while for about half of tasks performance differences are not statistically significant. From the table, we can also observe that, for four out of the nine studied tasks, VE_GP achieved the best performance with respect to the three compared sets of features (bold values in column 2), whereas for four tasks VE_GP "tied" with at least one of the feature categories (starred values in column 2). Finally, it is worth observing that VE_GP achieved the best overall performance (tasks #7 and #9), confirming its effectiveness as a feature extraction and classification method.

Table 3. Results of the comparison between VE_GP, proposed here, and DT, proposed in [8]. A '+' symbol indicates that VE_GP is better than DT in a statistically significant way, according to the Wilcoxon test (with $\alpha = 0.05$); a '−' indicates that VE_GP is worse than DT in a statistically significant way; a '≈' indicates that there is no statistical difference between the VE_GP results and the ones of DT.

Task#	VE_GP		All			In-air			On-paper		
	Avg	Dev	Avg	Dev		Avg	Dev		Avg	Dev	
1	66.15	8.3	72.98	3.4	−	72.56	3.5	−	72.87	3.7	−
2	72.13*	8.8	73.01	3.1	≈	63.76	3.7	+	74.96	2.6	≈
3	**73.60**	8.8	65.67	4.5	+	63.44	3.2	+	64.89	3.2	+
4	70.80*	12.1	72.95	3.8	≈	68.49	2.9	≈	74.42	3.1	≈
5	73.33*	8.4	69.34	4.7	+	64.19	3.3	+	72.15	3.7	≈
6	**72.40**	6.9	67.46	3.3	+	63.64	3.4	+	68.67	3.9	+
7	**76.40**	7.2	65.58	3.8	+	66.49	3.4	+	67.86	3.6	+
8	69.60*	9.0	69.01	2.9	≈	66.87	3.3	≈	69.41	2.6	≈
9	**76.80**	8.2	66.87	3.3	+	66.18	2.8	+	68.36	3.4	+
Win/Tie/Loss	−		5/3/1			6/2/1			4/4/1		

4.3 Models Evolved by VE_GP and Selected Features

One of the most interesting characteristics of VE_GP is that it is often able to generate models that are simple enough to be readable. This allows us to study the features used in those models, to understand the most meaningful windows of time chosen by the model and, in some cases, depending on the complexity of the model, even to interpret the interaction between variables. The best models evolved by VE_GP for each task are:

– Task #1:

$$\text{V\_mean}_{1020,3644}\left(2Z + \text{V\_sum}_{1481,3255}(Y) - \text{V\_max}_{1651,3440}(Z)\right) - \text{V\_max}_{913,4594}(Z)$$

– Task #2:

$$\text{V\_dist}_{1295,2200}(Y + 0.028) - \text{V\_mean}_{1500,6199}(Y) + \text{V\_mean}_{865,1091}(Y)$$

– Task #3:

$$\text{Mean}\left(0.407Z + 0.451Z - 0.204\text{V\_sum}_{695,6003}(Z)\right)$$

– Task #4:

$$\text{Mean}\left(X - \text{V\_max}_{1765,3358}(X)\right)$$

– Task #5:

$$\text{V\_sum}_{1929,2388}\left(XZ - \text{V\_sum}_{317,1325}(Y)\right) - \text{V\_sum}_{317,2935}(Y)\text{V\_dist}_{4177,5745}(Z)$$

– Task #6:

$$\text{V\_mean}_{1019,6212}\Big(Z - \text{V\_mean}_{753,3858}(Y)\Big) - \text{V\_mean}_{753,5387}(Z)$$

– Task #7:

$$\text{Mean}\Big(Z - \text{V\_dist}_{5376,5464}(Z) - \text{V\_dist}_{1235,4276}(Z)\Big)$$

– Task #8:

$$\text{Mean}\Big(X - \text{V\_norm}_{782,3432}(Y)\Big)$$

– Task #9:

$$\text{Mean}\Bigg(\Bigg[Y - \Bigg(\Bigg(\frac{0.074}{\text{V\_dist}_{6180,6297}(Y)\text{V\_length}(Z) + 1} + \text{V\_duration0}(Z)\Bigg)\cdot$$

$$\frac{1}{\text{V\_max}_{6267,6449}(Y) - 0.137 \cdot Y - Y}\Bigg) - 0.137\text{V\_max}_{370,6331}(Y) -$$

$$\text{V\_max}_{6227,6449}(Y - 1)\Bigg] \Bigg/ \Bigg[2Y - \text{V\_max}_{6227,6449}(Y) +$$

$$0.137\text{V\_max}_{2551,2898}(Y) + \text{V\_length}(Z)\Bigg]\Bigg)$$

Except for the model evolved for Task #5, all the other models do not use all the features, but a subset of them. This fact suggests that each task is characterized by peculiar movements of the hand. Most of the models do not involve the new terminal functions, which is a surprising fact, since they were built for the purpose. The interesting properties of the time series are thus caught only by the classical aggregating functions. Furthermore, it is possible to observe a significant difference between the models evolved for Task #5 and Task #7: remembering that both these tasks consist in writing a single word (i.e. the Italian words "foglio" and "mamma", respectively), we observe that, while the model evolved for Task #5 considers all the existing variables, the model evolved for Task #7 uses only pressure (i.e. variable Z). This fact suggests a bigger difficulty for the subjects in writing the word "foglio", compared to the word "mamma". Besides the fact that "foglio" contains more diverse letters than "mamma", we could also speculate that "foglio" contains letters that expand in the vertical dimension, which suggests that the Y variable may be important in this case. The most complex model is undoubtedly the one evolved for Task #9, probably because the task requires the writing of a whole sentence.

5   Conclusions and Future Work

In this article, we have proposed the use of vectorial genetic programming (VE_GP) for predictive modeling in Alzheimer's Disease (AD) through the analysis of handwriting. Thanks to its ability of incorporating vectorial information, including time series, directly into the evolved model, VE_GP has revealed a very appropriate method for this type of application, paving the way for its practical use as a tool for early diagnosis. More specifically, we have used data extracted from the handwriting of AD patients and a control group, while performing nine tasks. VE_GP has generated models that have an accuracy that is comparable, and for some tasks (four out of the studied nine) even better, than those obtained using three set of features typically used in the literature to represent handwriting movements. Furthermore, VE_GP proved its effectiveness by achieving the best overall performance on the nine tasks. Finally, VE_GP has generated small and simple predictive models, that can be understood by domain experts, and that can provide useful insights. Given that trust has been recently identified as a crucial element for the large scale employment of Artificial Intelligence in the medical field [1], and given that trust is directly related to models' interpretability, these characteristics of VE_GP reveal of paramount importance, and pave the way for a larger scale employment of VE_GP for predictive modeling of AD through handwriting analysis. It was also worth pointing out that the results presented in this work are very promising since they come from an exploratory work. Therefore, we expect for them a large room for improvement.

In the future, we plan to extend our study to larger datasets, including cohorts of patients coming from different medical institutions and countries. In our vision, models validation on data that have significantly different characteristics compared to the ones that have been used during training is a crucial step to assess the generalizability of the approach, and thus its potential use as a decision support system in hospital and clinics. In order to stress as much as possible the possibility of producing models with a good generalization ability, we are currently working on an extension of VE_GP to incorporate regularization techniques such as soft target and functional complexity reduction. The approach has recently revealed very successful for traditional genetic programming [27]. Last but not least, we are planning to extend our investigation to other types of degenerative diseases, different from AD.

Acknowledgments. This work was partially supported by FCT, Portugal, through funding of projects BINDER (PTDC/CCI-INF/29168/2017) and AICE (DSAIPA/DS/0113/2019).

This work was also supported by MIUR (Minister for Education, University and Research, Law 232/216, Department of Excellence).

References

1. Asan, O., Bayrak, A.E., Choudhury, A.: Artificial intelligence and human trust in healthcare: focus on clinicians. J. Med. Internet Res. **22**(6) (2020). https://doi.org/10.2196/15154
2. Azzali, I., Vanneschi, L., Bakurov, I., Silva, S., Ivaldi, M., Giacobini, M.: Towards the use of vector based gp to predict physiological time series. Appl. Soft Comput., **89** (2020). https://doi.org/10.1016/j.asoc.2020.106097

3. Azzali, I., Vanneschi, L., Mosca, A., Bertolotti, L., Giacobini, M.: Towards the use of genetic programming in the ecological modelling of mosquito population dynamics. Genet. Program Evolvable Mach. **21**(4), 629–642 (2020). https://doi.org/10.1007/s10710-019-09374-0

4. Azzali, I., Vanneschi, L., Silva, S., Bakurov, I., Giacobini, M.: Review of classification using genetic programming. In: Genetic Programming, EuroGP 2019, Lecture Notes in Computer Science (2019)

5. Bakurov, I., Castelli, M., Vanneschi, L., Freitas, M.J.: Supporting medical decisions for treating rare diseases through genetic programming. In: Kaufmann, P., Castillo, P.A. (eds.) EvoApplications 2019. LNCS, vol. 11454, pp. 187–203. Springer, Cham (2019). https://doi.org/10.1007/978-3-030-16692-2_13

6. Bojarczuk, C.C., Lopes, H.S., Freitas, A.A., Michalkiewicz, E.L.: A constrained-syntax genetic programming system for discovering classification rules: application to medical data sets. Artif. Intell. Med. **30**(1), 27–48 (2004)

7. Castelli, M., Vanneschi, L., Manzoni, L., Popović, A.: Semantic genetic programming for fast and accurate data knowledge discovery. Swarm Evol. Comput. **26**, 1–7 (2016)

8. Cilia, N.D., De Stefano, C., Fontanella, F., Molinara, M., Scotto Di Freca, A.: Handwriting analysis to support alzheimer's disease diagnosis: a preliminary study. In: Vento, M., Percannella, G. (eds.) CAIP 2019. LNCS, vol. 11679, pp. 143–151. Springer, Cham (2019). https://doi.org/10.1007/978-3-030-29891-3_13

9. Cilia, N., De Stefano, C., Fontanella, F., Scotto Di Freca, A.: An experimental protocol to support cognitive impairment diagnosis by using handwriting analysis. In: Procedia Computer Science, Proceeding of The 8th International Conference on Current and Future Trends of Information and Communication Technologies in Healthcare (ICTH), pp. 1–9. Elsevier (2019)

10. Cilia, N.D., De Stefano, C., Fontanella, F., Di Freca, A.S.: Using genetic algorithms for the prediction of cognitive impairments. In: Castillo, P.A., Jiménez Laredo, J.L., Fernández de Vega, F. (eds.) EvoApplications 2020. LNCS, vol. 12104, pp. 479–493. Springer, Cham (2020). https://doi.org/10.1007/978-3-030-43722-0_31

11. De Falco, I., Tarantino, E., Cioppa, A., Fontanella, F.: An innovative approach to genetic programming-based clustering. Adv. Soft Comput. **34**, 55–64 (2006)

12. De Falco, I., Tarantino, E., Della Cioppa, A., Fontanella, F.: A novel grammer-based genetic programming approach to clustering. In: Proceedings of the ACM Symposium on Applied Computing, vol. 2, pp. 928–932 (2005)

13. Diaz, M., Ferrer, M.A., Impedovo, D., Pirlo, G., Vessio, G.: Dynamically enhanced static handwriting representation for parkinson's disease detection. Pattern Recogn. Lett. **128**(204–210) (2019)

14. Garre-Olmo, J., Faundez-Zanuy, M., de Ipiña, K.L., Calvo-Perxas, L., Turro-Garriga, O.: Kinematic and pressure features of handwriting and drawing: Preliminary results between patients with mild cognitive impairment, alzheimer disease and healthy controls. Curr. Alzheimer Res. **14**, 1–9 (2017)

15. Ghaheri, A., Shoar, S., Naderan, M., Hoseini, S.S.: The applications of genetic algorithms in medicine. Oman Med. J. **30**(6), 406–416 (2015)

16. Impedovo, D., Pirlo, G.: Dynamic handwriting analysis for the assessment of neurodegenerative diseases: a pattern recognition perspective. IEEE Reviews in Biomedical Engineering, pp. 1–13 (2018)

17. Jabeen, H., Baig, A.: Review of classification using genetic programming. In: International Journal of Engineering Science and Technology (2010)

18. Johnson, P., et al.: Genetic algorithm with logistic regression for prediction of progression to alzheimer's disease. BMC Bioinform. **15**(S11) (2014)

19. Koza, J.R.: Genetic programming: On the programming of computers by means of natural selection. In: MIT Press, Cambridge (1992)
20. Onofri, E., Mercuri, M., Archer, T., Ricciardi, M.R., F.Massoni, Ricci, S.: Effect of cognitive fluctuation on handwriting in alzheimer's patient: a case study. Acta Medica Mediterranea **3**, 751 (2015)
21. Onofri, E., Mercuri, M., Salesi, M., Ricciardi, M., Archer, T.: Dysgraphia in relation to cognitive performance in patients with Alzheimer's disease. J. Intellectual Disability-Diagnosis Treatment **1**, 113–124 (2013)
22. Parziale, A., Senatore, R., Della Cioppa, A., Marcelli, A.: Cartesian genetic programming for diagnosis of parkinson disease through handwriting analysis: performance vs. interpretability issues. Artif. Intell. Med. **111**, 101984 (2021)
23. Petrowski, A., Ben-Hamida, S.: Evolutionary algorithms. In: Wiley-ISTE (2020)
24. Poli, R., Langdon, W., McPhee, N.: A field guide to genetic programming. Published via http://lulu.com and freely available at http://www.gp-field-guide.org.uk (2008)
25. Quinlan, J.R.: C4.5: Programs for Machine Learning (Morgan Kaufmann Series in Machine Learning). Morgan Kaufmann, San Francisco (1993)
26. Valenzuela, O., Jiang, X., Carrillo, A., Rojas, I.: Multi-objective genetic algorithms to find most relevant volumes of the brain related to alzheimer's disease and mild cognitive impairment. Int. J. Neural Syst. **28**(09) (2018)
27. Vanneschi, L., Castelli, M.: Soft target and functional complexity reduction: A hybrid regularization method for genetic programming. Expert Syst. Appl. **177**, 114929 (2021)
28. Werner, P., Rosenblum, S., Bar-On, G., Heinik, J., Korczyn, A.: Handwriting process variables discriminating mild alzheimer's disease and mild cognitive impairment. J. Gerontology: PSYCHOLOGICAL SCIENCES **61**(4), 228–36 (2006)

Negative Selection Algorithm for Alzheimer's Diagnosis: Design and Performance Evaluation

Giuseppe De Gregorio[1(✉)], Antonio Della Cioppa[1,2,3], and Angelo Marcelli[1,2]

[1] DIEM, University of Salerno, Via Giovanni Paolo II 132, 84084 Fisciano, SA, Italy
{gdegregorio,adellacioppa,amarcelli}@unisa.it
[2] CINI, National Laboratory of Artificial Intelligence and Intelligent Agents,
University of Salerno Unit, Fisciano, SA, Italy
[3] ICAR-CNR, Via P. Castellino 111, 8131 Naples, Italy

Abstract. We present a method for discriminating between healthy subjects and Alzheimer's diseases patients from on-line handwriting. Departing from the current state of the art methods, that adopts machine learning methods and tools for building the classifier, we propose to apply the Negative Selection Algorithm. The major advantage of the proposed method in comparison with others machine learning techniques is that it requires only data by healthy subjects to build the classifier, thus avoiding to collect patient data, as requested by competing techniques. Experiments results involving data produced by 175 subjects show that the proposed method achieves state-of-the-art performance.

Keywords: Immunocomputing · Alzheimer's disease diagnosis ·
Handwriting analysis

1 Introduction

Neurodegenerative diseases (NDs) affect millions of people worldwide, and among them, Alzheimer's disease is the most frequent one. It is the most common cause of dementia in people over the age of 65, and since it is age-related, the increment of the lifespan due to the continuous improvement in the medical field will make its impact on both individuals and society even greater. According to the World Alzheimer Report released by the Alzheimer's Disease International, in 2015 it affected more than 40 million people worldwide, and it is estimated this number will increase to 131.5 million by 2050 [39].

In this context, there has been an increasing interest in developing CI methodologies for the early diagnosis of Alzheimer's disease [30,43]. Most of the approaches exploit biomarkers useful for characterising the specific neurodegenerative disorder. Clinical, imaging, biochemical, and genetic are four categories of biomarkers that have been suggested for the early diagnosis. However, the predicting values of biomarkers have been questioned [28,29], they are either expensive, invasive, may require a long time to be administered, and need tools

© Springer Nature Switzerland AG 2022
J. L. Jiménez Laredo et al. (Eds.): EvoApplications 2022, LNCS 13224, pp. 531–546, 2022.
https://doi.org/10.1007/978-3-031-02462-7_34

and materials mostly available in hospitals and specialized labs. Deep learning based approaches, moreover, require a large amount of data that are difficult to collect and whose scarcity is challenging to mitigate even by transfer learning or data augmentation [2]. All these issues make them impractical to be used for routine check-ups of aging people.

To overcome these problems, other approaches have investigated the neural processes involved in motor learning and execution, as well as the deterioration of motor processes involved in writing and drawing in subjects affected by NDs, highlighting the most distinctive features of patient's movements [5,23,33,41,42,44–46]. The findings provided by these methods is that handwriting analysis might provide a cheap, quick, and non-invasive method for both diagnosing the diseases and evaluating its progression, and therefore may represent a viable solution to develop a test that can be administered routinely by family doctor or even by trained caregivers, thus allowing a continuous monitoring along the years, avoiding overloading hospitals and specialized medical personnel and reducing the cost for both diagnosis and treatment.

Based on these findings, machine learning methods combining a variety of tasks, features, and classifiers to discriminate between healthy subjects and Alzheimer's disease (hereinafter AD) patients have been proposed in the literature [6,8,9,11,12,31,32,47]. All of them, however, need samples from both populations for telling apart the two classes. Collecting AD patient samples is a much more demanding task collecting data from healthy subjects, for various reasons. It has been reported that both societal and self-stigma are still a major barrier to the diagnosis [18], making patients reluctant to disclose their pathology to others. Patients very often need assistance to move around or perform other tasks and thus need to be reached at their living places. Administering the test, moreover, may need the assistance of the caregiver or even specialized medical personnel. Those mentioned above are possible reasons why there are a few datasets publicly available, and most of them include a few dozen subjects, are often unbalanced and the two populations may differ significantly in their characteristics [15,36,37].

To tackle the data acquisition problems mentioned above, We introduce the negative selection algorithms (NSA) [19], to discriminate between AD patients and healthy subjects. It belongs to the "one class classification" algorithms, i.e., classification algorithms trained by using only samples from one of them. The ability to learn a classifier capable of discriminating between AD patients and healthy subjects using only handwriting samples of the latter, thus overcoming the drawbacks mentioned above on collecting samples from both populations for training, represents the primary motivation for our study. To the best of our knowledge, there are only two previous works in the literature that exploit this class of algorithms to support the automatic diagnosis of diseases, particularly NSA, and none of them regards the diagnosis of neurodegenerative diseases. The performance of the proposed method is evaluated and compared on a large dataset recently collected within the DARWIN project[1] with Random Forest

---

[1] The dataset is publicly available on the following page: http://webuser.unicas.it/fontanella/darwin/.

and Support Vector Machine classifiers, as they are among the most performing classifiers. The paper is organised as follows. In Sect. 2, we review the previous work on the automatic diagnosis of AD, to highlight the variety of features, classifiers, and datasets used by different authors to develop and evaluate their proposed solutions, and then the literature on using NSA for supporting the automatic diagnosis of diseases. In Sect. 3, we summarize the main features of the biological immune system and how they are simulated in the NSA algorithm to implement the learning process. Then, we discuss the main variant of the NSA algorithm proposed in the literature, how they fit within the general context of one-class learning algorithms, and derive the rationale for our implementation. Section 4 illustrates the experiments we performed to evaluate the effectiveness of the proposed system and report the obtained results. Finally, conclusions and future research directions are discussed in Sect. 5.

2 Related Work

One of the early attempts to look at handwriting for discriminating AD and patients affected by Mild Cognitive Impairment (MCI) from healthy controls is reported in [48], which exploits both kinematic measures of the handwriting process across different tasks and cognitive functioning tests. The study involved 31 subjects with MCI, 22 with mild Alzheimer's disease, and 41 healthy controls, and each performed different copying tasks. The authors use discriminant analysis for the classification and report a 72% accuracy.

As handwritten signatures are well-learned movements, that are executed in a highly automatic mode, they might prove to be a good benchmark, as AD impairs planning, memory, and motor control, which are all recruited for movements execution, they have been exploited for the early diagnosis of AD in [38]. Signatures are represented by twelve features extracted from their Log-Normal decomposition, depicting the maximum speed and the number of Log-Normal divided by the total time for signing, and the number of peaks of the speed during the signing. Authors have used CART, bagging CART, and SVM with the linear kernel for classifier signatures of AD patients, and the best performance was obtained by the bagging CART, which achieved an accuracy of 96.7%, with sensitivity and specificity equal to 96.5% and 96.8% respectively.

The work reported in [17] aimed at discriminating among AD patients, MCI patients, and control subjects by comparing their handwriting kinematics extracted from both writing and drawing tasks. The former included copying a sentence, writing a dictated sentence, and spontaneous writing, while the latter consisted of coping two and-three dimensions drawings and executing the clock drawing test. By using discriminant analysis as a classification algorithm, the authors have analysed the most discriminating features as well as the capability of each task to discriminate between different groups and concluded that the discriminatory features depended on the type of groups to be discriminated.

In [25], the authors propose an approach for characterising early Alzheimer's. They have exploited the loop velocity trajectory (full dynamics) through a temporal clustering based on K medoids. Then, classification is performed by aggregating the clusters' contributions by probabilistically combining the discriminative power of each. Experiments were performed on a dataset including samples produced by 27 early-stage Alzheimer's patients and 27 healthy subjects. They report as best performance an overall accuracy of 74% with a specificity of 72.2% and sensitivity of 75.6%.

The work reported in [8] addresses specifically the protocol for data acquisition. The proposed protocol includes different writing tasks, namely copy of words, letters, and sentences, and the authors aimed at evaluating the kinematic properties of both on-paper and in-the movements performed by the subject, thus including most of the tasks used by other authors. Both types of data were used for training two different classifiers: the Random Forest and the Decision Tree. They achieved an overall score greater than 70% in classifying AD patients. In a later study [9], they extend the experimental work by considering an Artificial Neural Network and a Support Vector Machine and showed that on-air features have the greatest weight in the classification of patients by contributing to increase the accuracy of the classification phase.

Finally, in [22], five representative tasks used in neuropsychological tests, namely spontaneous writing, trail marking test with numbers and characters, crossed pentagons, and clock drawing, were investigated for extracting meaningful features to classify AD and mild cognitive impairment subjects (MCI). Data were collected from 71 senior, including MCI, AD, and control subjects. Using a three-class classification model based on a generalised linear model with a logit link function and combining the features of multiple tasks, they obtain an overall accuracy of 74.6%.

With [32], the problem of automatic diagnosis of Parkinson's disease through the analysis of handwriting and drawing samples is addressed. Different machine learning techniques are compared by evaluating the classification accuracy and the degree of interpretability of the different techniques. The work focuses on the interpretability of the classification process since a clear and entirely explicable process can be easily understood and adopted by physicians. The techniques analyzed are Decision Tree (DT), Random Forest (RF), Support Vector Machine (SVM), and it is shown how DT achieves good results by providing a high level of interpretability. In [31], the authors use Cartesian Genetic Programming (CGP) as a classification method and show that this method performs better in terms of both accuracy and interpretability. Very recently, however, features extracted by hidden layers of deep learning architectures have been adopted to discriminate with very good performance between patients and controls [1,3].

As with regards to the NSA applications in the medical field, and particularly for supporting automatic diagnosis, we found only two works using RNSA for diagnosing epilepsy, breast cancer, and liver pathologies. In [4], a hybrid method, called Adaptive Particle Swarm Negative Selection (APSNS), is used for the automatic diagnosis of epileptic seizures in EEG signals. APSNS uses RNSA to generate the detectors and the PSO to force them to explore the space of epileptic signals and maintain diversity and generality among them. EEG signals are represented by a feature vector extracted from the discrete wavelet transform of the signals. The performance has been evaluated on publicly available EEG datasets that have healthy and seizure signals, and the results reveal that the APSNS accuracy compares favourably to other state of the art methods proposed in the literature for EEG signals discrimination.

A variant of the RNSA called V-detector [24], which supports variable size of detector radius, is adopted in [27] to randomly generate the sets of detectors with the aim of maximizing the coverage area of each detector. Experimental results on Breast Cancer Wisconsin and BUPA Liver Disorder datasets and on biomedical data show that the V-Detector can achieve the highest detection rates and lowest false alarm rates in comparison with Artificial Neural Network and Sequential Minimal Optimization, achieving an accuracy of 98.95%, 74.44%, and 71.64%, respectively, on the three datasets.

3   The Proposed Method

The Negative Selection Algorithm (hereinafter NSA) was firstly introduced by Stephanie Forrest and her colleagues in 1994 [16] by drawing an analogy with the way the biological immune system learns to discriminate body cells from foreign cells and pathogens. In biological immune system, all newly produced T-cells undergo a process of negative selection in the thymus, where the T-cells binding with self antigens (body cells) are destroyed. When the mature T-cells are released to the blood circle, they can only bind with non-self antigens (foreign cells/pathogens). NSA simulates the T-cells censoring process in the thymus. It starts with a set of self antigens A, and proceeds by generating a set of detectors, D, that only recognize the complement of A. A detector d is said to recognize a self antigen if its affinity is below a given threshold. Once a detector that does not recognize any of the self antigen of A is found, it is stored in D, and the process is iterated until a given number of detectors is discovered. These detectors are eventually applied to new data in order to classify them as being self or non-self. Thus, when used for pattern recognition, it implements a prototype-based approach, where the prototypes of one class are learned by using only the samples of the others. In the original NSA implementation, detectors and antigens were bit strings, the former were generated randomly, and the Hamming distance measured their affinity. Although later implementations adopting r-contiguous bits, r-chunks, landscape-affinity matching to evaluate the affinity showed better performance, it was soon realized that the major drawback that limited the applications of NSA was the use of string matching to compute the affinity. To cope with it, Gonzales et al. introduced the real value negative selection algorithm (RNSA) [19]. It allowed representing detectors and antigens by real-valued vectors, so that the affinity could be evaluated by using Euclidean, Minkowski, or other distances between real values. Since then, many variants of the RNSA have been proposed for many different applications, showing state of the art performance. An exhaustive review of the literature on RNSA and its variants envisaging an empirical comparison with 22 different one class classifiers on an ad-hoc dataset, as well as on 17 datasets is reported in [20].

Algorithm 1: Negative Selection Algorithm

Input   : the set S of m self-samples in R^n space, the number of detectors N and the self-radius σ

Output: the detector set D

1 Inizialize D to the empty set;
2 **repeat**
3 | Create a random vector x drawn from $[0,1]^n$;
4 | **for** *every s_i in S, $i = 1, 2, \ldots, m$* **do**
5 | | $d_i = \texttt{Distance}(x, s_i)$;
6 | | **if** $d_i > \sigma$ *for all i* **then**
7 | | | Add x to D;
8 | | **end**
9 | **end**
10 **until** D *contains N valid detectors*;

The experimental results reported there show that RNSA performance is definitively state of the art, and in many cases, is better and faster than other methods.

In this study, we adopt canonical RNSA implementation, i.e. detectors are generated randomly in the representation space. Detectors and self strings are normalized into the real-valued feature space $[0,1]^n$, where n denotes the number of features, and the affinity is calculated by the euclidean distance.

The RNSA algorithm has 2 parameters, the number n of detectors and the affinity threshold δ, that need to be set depending on the problem at hand, as it will be discussed in the following Section. Algorithm 1 reports the procedure adopted by NSA to seek for the N detectors.

4   Experimental Results

In this section, we report the results of the experiments we conducted to evaluate the performance of the proposed methodology. We first describe the data set used in the experiments, provide details on the implementation of the proposed method as well as of the competing ones we have considered, and then report the performance achieved by using the complete feature set. Then, considering that one of the NSA's drawbacks is its difficulty dealing with very-high dimensional spaces, we apply a feature selection method and report and compare the performance achieved by the three classifiers on this reduced dataset.

4.1   Dataset

The dataset was collected by using the protocol proposed in [7]. It includes 25 different tasks, which are grouped into four categories, in increasing order of difficulty:

- *Graphic tasks*; to test the ability to write elementary moves, connect dots, and draw simple figures;
- *Copy and Reverse Copy tasks*; aimed at evaluating the ability to reproduce complex graphic gestures, such as letters, words, and numbers;
- *Memory tasks*; the purpose is to test the variation of the graphic section, keeping in memory a word, a letter, a graphic gesture, or a motor plan.
- *Dictation*; to test the patient's ability to use the working memory during the variation of writing tasks.

The data were recorded and digitized using a Bamboo Wacom tablet. This tablet can be used to write on a regular white sheet of paper and simultaneously digitize the handwriting result. It was chosen to reproduce the pen-and-paper setting the individuals are more familiar with, thus avoiding any adaptation of his motor skills to performance proficiently, as it would have been the case with a stylus, or the finger, on the tablet screen. This is of paramount importance because the deterioration of the motor performance of previously learned movements induced by ND diseases may be confused with the poor performance that is observed in the early stages of learning to execute a familiar task in a different setting. The data collection campaign involved 174 individuals, of which 89 are Alzheimer's patients, and 85 are healthy subjects. All subjects were recruited so as to match the patients and the control group for age, education level, type of occupation, and gender. The two populations show similar mean values and standard deviations for age, sex, and years of education. The features computed on each sample resulting from the execution of the tasks are summarised in Table 1. Thus, each sample is represented by a feature vector of 450 real values. A reduced dataset including 107 features for each sample was eventually obtained by applying the feature selection method described below.

4.2   Features Selection

Features play a key role in creating predictive models. Having a large number of features may lead to overfitting and, at the same time, will cause the curse of dimensionality, i.e., the dimensions of search space for the problem will increase. Feature Importance is a technique that can be used to decide which features are most important and which features are least important for predicting the target variable. In this study, we have used the Searching for Uncorrelated List of Variables (SULOV) method, which is very similar to the Minimum Redundancy feature selection algorithm [34], to find out highly correlated features and remove them from the dataset. The method works as follows: it finds out all the pairs whose correlation is larger than a given threshold and then computes their Mutual Information Score (MIS) with respect to the class. Then, it detects for each pair of correlated features and eliminates the feature with the lower MIS score. The features that have the least correlation and highest MIS scores with each other are then collected and passed to Recursive XGBoost (RXGBoost) with the aim to determine the best features. RXGBoost splits at random all features in the dataset into training and validation sets, then selects the top

Table 1. Features used for the experiment. Features are grouped in on-paper and on-air, depending on whether they are extracted when the pen touches the paper or not. Feature names ending in P denote the former, while those ending in A denote the latter.

Feature	Name	Description
Total Time	TT	Total time required to perform the entire task
Air Time	AT	Time spent near the sheet, with the tip of the pen not in contact with the sheet
Paper Time	PT	Time spent on the sheet, with the tip of the pen in contact with the sheet
Mean Speed on-paper	MSP	Average speed recorded on paper
Mean Speed in-air	MSA	Average of the speeds recorded near the sheet
Mean Acceleration on-paper	MAP	Average of accelerations recorded on paper
Mean Acceleration in-air	MAA	Average of the accelerations recorded near the sheet
Mean Jerk on-paper	MJP	Average of the jerk recorded on the sheet
Mean Jerk in-air	MJA	Average of the jerk recorded near the sheet
Pressure Mean	PM	Average of the pressure levels exerted on the sheet
Pressure Var	PV	Variance of pressure levels exerted on the sheet
GMRT on-paper	GMRTP	Generalization of the Mean Relative Tremor defined in [35] computed on in-air movements. It considers the top left corner of the sheet as the center for the computation
GMRT in-air	GMRTA	Generalization of the Mean Relative Tremor defined in [35] computed on on-paper movements. It considers the top left corner of the sheet as the center for the computation
Mean GMRT	GMRT	Average of GMRTP and GMRTA
Pendowns Number	PWN	The number of pendowns
Max X Extension	XE	Maximum extension recorded along the X axis
Max Y Extension	YE	Maximum extension recorded along the Y axis
Dispersion Index	DI	It measures the dispersion of the drawing on the sheet; a fully covered sheet will correspond to an index equal to one while a completely empty sheet will have an index equal to zero

features on the training set by using the validation set for preventing overfitting. This step is repeated a number of times (usually 5). Finally, the top features selected at the end of each step are combined and, eventually, duplicated features are removed. In this way, RXGBoost selects the least number of features from the dataset to build a high performing model.

The feature reduction technique, as already mentioned allowed us to significantly diminish the dataset dimensions, from 450 to 107 features with a reduction of 76.22%. Table 2 and Table 3 offer a synthetic view of the feature selection results.

Table 2 shows the number of features selected for each task. In the table, tasks are sorted according to the number of selected features, showing first the tasks with the highest number. The data reported in the table show that the distribution of the number of features selected per task is not uniform, with the most feature-rich task envisaging 7 features and the least one only 2.

Table 2. The number of features selected for each task after the feature selection. Before the feature selection, there were 18 features for each task.

Task 8	Task 2	Task 3	Task 4	Task 12	Task 16	Task 23	Task 5	Task 21
7	6	6	6	6	6	6	5	5
Task 24	Task 22	Task 9	Task 10	Task 15	Task 20	Task 25	Task 19	Task 14
4	4	4	4	4	4	4	4	4
Task 17	Task 6	Task 13	Task 18	Task 1	Task 7	Task 11		
3	3	3	3	2	2	2		

We can take another look at the results of the feature selection by considering that the original set of features can be divided into six classes: temporal, kinematic, pendown, spatial, pressure, tremor measurement. Additionally, a split can be made between in-air and on-paper features depending on whether they are extracted when the pen tip touches the paper or not. Table 3 shows the number of features selected for each class. The data in the table show that kinematic features are the most selected ones, followed by the temporal, spatial, tremor, and pressure ones, while the pendown is among the least selected feature. They also reveal a predominance of features extracted in-air compared to those extracted on-paper. Moreover, it is interesting to note that the most selected features are the air_time and the mean_jerk (both on-paper and in-air). This is in agreement with observations by other authors, according to which the time spent on air between successive movements tend to be longer for patients than for healthy subjects [40], and that the former show less fluency in the movements than the latter [26].

4.3   Implementation Details

To accomplish the feature selection task, we have used the Python library *Featurewiz* by setting the correlation threshold value equal to 0.5.

The experiments were conducted using RNSA and two of the most commonly used classification models that have shown exemplary performance in the prior art. The two classification models are the Random Forest (RF) [21] and the Support Vector Machine (SVM) [10]. For the experiments, we used the machine learning tool WEKA [49] and used the models available in the standard library of the tool. To use the SVM model, we preprocessed the data by applying a standardization technique with a mean of 0 and a standard deviation of 1.

Table 3. The number of features selected for each feature class.

Feature class	Feature	#Selected
Temporal	air_time	9
	paper_time	4
	total_time	6
Kinematic	mean_speed_on_paper	4
	mean_acc_on_paper	7
	mean_jerk_on_paper	9
	mean_speed_in_air	3
	mean_acc_in_air	6
	mean_jerk_in_air	9
Pendown	num_of_pendown	4
Spatial	disp_index	8
	max_x_extension	4
	max_y_extension	4
Pressure	pressure_mean	6
	pressure_var	8
Tremor	mean_gmrt	4
	gmrt_on_paper	4
	gmrt_in_air	8

As regards the hyper-parameters RF and SVM depend on, we have effected a greedy search around the default parameters given in the WEKA implementation. Anyway, the default values have exhibited the best performance.

For the RNSA, after a preliminary tuning phase, we have chosen the pairs $(n = 16000, \delta = 2.28)$ and $(n = 4000, \delta = 5.03)$ for the complete and the reduced datasets, respectively.

4.4   Performance Evaluation

As regards RF and SVM, the training was performed following a ShuffleSplit cross-validation technique, with a training set equal to 80% of the data and a number of folds equal to 15, while for RNSA, we used 80% of the healthy subjects for the training set and the remaining 20% along with 100% of the patients samples for the test set. Moreover, to perform a fair comparison, we have performed 15 RNSA runs by selecting at random the samples of the healthy subjects to be included in the training set.

Table 4 reports the mean, the standard deviation, the maximum, and the minimum of the accuracy achieved across the 15 folds in case of the RF and the SVM, over the 15 runs in case of the RNSA. The values in Table 4 show that all the classifiers achieved good performance on the complete dataset, with

Table 4. Average accuracy, standard deviation, maximum and minimum obtained on the Complete Dataset (CD) and the Reduced Dataset (RD).

	RF		SVM		RNSA	
	CD	RD	CD	RD	CD	RD
Average accuracy	88.00%	91.62%	86.69%	87.62%	89.58%	96.58%
Standard deviation	5.73%	5.15%	4.93%	4.66%	3.80%	1.82%
Max accuracy	94.29%	100%	91.75%	94.29%	97.48%	98.32%
Min accuracy	71.43%	82.86%	74.29%	77.14%	84.87%	92.44%

Table 5. Average sensitivity and specificity on the complete dataset (CD) and the reduced dataset (RD).

	RF		SVM		RNSA	
	CD	RD	CD	RD	CD	RD
Sensitivity	88.88%	88.88%	83.33%	88.88%	83.14%	100%
Specificity	88.23%	94.11	88.23%	88.23%	100%	86.67%

less than 3 points of difference between the mean accuracy of the best and the worst. They also show the positive impact on the performance of the feature selection: all the classifiers exhibited a performance enhancement, but the RNSA benefits most, improving its performance by 7.0%, almost twice the improvement exhibited by the RF classifier and more the nine times that of the SVM. These results are also in accordance with the observation, already mentioned, that RNSA suffers the curse of dimensionality more than other machine learning algorithms. Eventually, they show that on both datasets, the proposed method exhibits better performance than the competitors.

The most relevant results, however, follow from Table 5 that reports the average sensitivity and specificity of the classifiers on both datasets. They show that in every run, the RNSA achieved a 100% sensitivity (i.e., all the AD patients were classified as such), while few healthy subjects (3 or 4 depending on the run) were misclassified as patients. The RF was capable of correctly classifying all the test set only for one fold, while in all the other cases, there were 1 or 2 samples of each class that were misclassified. In the case of the SVM, there were 2 or 3 misclassified samples of each class on each fold.

5   Conclusions

The motivation for the study presented in this paper was to overcome the problem of data collection, which we believe represents one of the major obstacles to adopting machine learning methods for developing automatic tools for supporting the diagnosis of Alzheimer's disease. Departing with the prevalent literature approaches, which focus on generating *more data*, we turned our attention toward machine learning methods that require *less data*. Then, considering that, in our case as in many other diagnostic processes in medical application, the major difficulties with data acquisition are those concerning the collection of patent's samples to train the system, and that the classification problem is ultimately binary by nature (is there any evidence that the subject is affected from the diseases?), we consider the Negative Selection Algorithm as a viable approach to address the problem.

The main feature that makes NSA appealing is that it takes only samples of one class for learning the model of the other class, meaning that in our case, it can be trained by using only samples produced by healthy subjects. Those kinds of data, in the specific application as well as in many others based on behavioural data, are much easier to collect. Additionally, with respect to other one-class classifiers, it has exhibited comparable or better performance in terms of accuracy, being much faster than most of the competitors. Last but not least, it has only two parameters to set, namely the number of detectors and the affinity threshold.

The NSA, however, suffers from the curse of dimensionality: as the number of dimensions of the representation space increases, so increases the number of detectors to be used. Thus the problem of estimating that number arises and still stands today as an open challenge to the scientific community. In any case, reducing the dimensions of the representation space would be beneficial in terms of computational times, as a smaller number of detectors must be generated to achieve a given performance. We have then resorted to feature selection and adopted the SULOV algorithm for the implementation.

The performance of the method we have developed has been evaluated on one of the larger data sets publicly available we have contributed to collect, and compared with those achieved by two competing approaches, adopting an RF and an SVM classifier. The results show that the proposed method outperforms the competing ones on both the complete and the reduced datasets and also that it benefits more than its competitors from feature selection. Those results confirm that NSA may represent a viable solution to reduce the efforts for collecting training data and improving performance regarding the current state of the art classifiers.

In our opinion, the most remarkable result is that the proposed method achieves a 100% sensitivity on every run. This result has two major implications. From a medical point of view, this makes the test a viable solution for a routine check of aging individuals, for early diagnosis and continuous monitoring of patients, as suggested by the international health organizations. It is non-invasive, easy to perform, requires only a cheap device to collect the data, and the software can be installed on small computing devices, such as a tablet, thus making the whole system easy to carry over. Scientifically, as the healthy subjects and the patients were collected in such a way that the two populations have some degree of matching in their demographic features, this result confirms that the handwriting alterations induced by neurodegenerative diseases do not necessarily need to be characterized by their own features, but rather than a good model of the handwriting of a *proper* population of healthy subjects is all we need to discriminate between individuals of the two population. In other words, one class classifiers may achieve similar (or better) performance than traditional machine learning tools by using a much smaller dataset, including only healthy subjects samples. This is also very encouraging from the application point of view because the profile of the patient's population is already available in terms of age, level of education, sex, nutrition habits, anamnestic data, and so forth. Thus, selecting the *proper* population for collecting their samples would be much easier, thus providing the scientific community with the data they need for further signs of progress.

The relevance of the results, however, must be considered with caution, considering the experimental setting. Further investigations will be therefore focused on designing and implementing an extensive testing and eventually improving the robustness of the method, as well as considering a wider set of classifiers. A systematic investigation on the role of the hyper-parameters values on the performance is already on-going at the time of writing, and preliminary results seem very promising. The large reduction in the number of features and the corresponding performance improvements we have recorded suggests that there is a significant redundancy in the tasks included in the protocol. The detailed analysis of the feature selection has shown that there are some tasks that contribute more feature than others to the reduced feature set. This results deserve to be further investigated with the aim of defining if there exist a smaller number of tasks and features that allow to achieve the best performance, so as to simplify the test administration.

Eventually, the results have been obtained by using a canonical implementation of the RNSA, but there are variants proposed in the literature that automatically choose the affinity threshold. Investigating the role played on their performance by the number of detectors will be the next topic of our future work. In this regard, we plan to use the same methodological approach we have used in previous work [13, 14] to investigate the role of niche radius and population size on the performance of niching methods based on fitness sharing.

References

1. A V, A.S., Lones, M.A., Smith, S.L., Vallejo, M.: Evaluation of recurrent neural network models for parkinson's disease classification using drawing data. In: 2021 43rd Annual International Conference of the IEEE Engineering in Medicine Biology Society (EMBC), pp. 1702–1706 (2021). https://doi.org/10.1109/EMBC46164. 2021.9630106
2. Agarwal, D., Marques, G., de la Torre-Díez, I., Franco Martin, M.A., García Zapiraín, B., Martín Rodríguez, F.: Transfer learning for alzheimer's disease through neuroimaging biomarkers: a systematic review. Sensors **21**(21) (2021)
3. Alissa, M., et al.: Parkinson's disease diagnosis using convolutional neural networks and figure-copying tasks. Neural Comput. Appl. **34**(2), 1433–1453 (2022). https:// doi.org/10.1007/s00521-021-06469-7

4. Ba-Karait, N.O., Shamsuddin, S.M., Sudirman, R.: Eeg signals classification using a hybrid method based on negative selection and particle swarm optimization. In: Proceedings of the 8th International Conference on Machine Learning and Data Mining in Pattern Recognition, pp. 427–438 (2012)
5. Broderick, M.P., Van Gemmert, A.W., Shill, H.A., Stelmach, G.E.: Hypometria and bradykinesia during drawing movements in individuals with parkinson's disease. Exp. Brain Res. **197**(3), 223–233 (2009)
6. Cavaliere, F., Della Cioppa, A., Marcelli, A., Parziale, A., Senatore, R.: Parkinson's disease diagnosis: towards grammar-based explainable artificial intelligence. In: 2020 IEEE Symposium on Computers and Communications (ISCC), pp. 1–6 (2020). https://doi.org/10.1109/ISCC50000.2020.9219616
7. Cilia, N.D., De Stefano, C., Fontanella, F., Di Freca, A.S.: An experimental protocol to support cognitive impairment diagnosis by using handwriting analysis. Procedia Comput. Sci. **141**, 466–471 (2018)
8. Cilia, N.D., De Stefano, C., Fontanella, F., Molinara, M., Scotto Di Freca, A.: Handwriting analysis to support Alzheimer's disease diagnosis: a preliminary study. In: Vento, M., Percannella, G. (eds.) CAIP 2019. LNCS, vol. 11679, pp. 143–151. Springer, Cham (2019). https://doi.org/10.1007/978-3-030-29891-3_13
9. Cilia, N.D., De Stefano, C., Fontanella, F., Molinara, M., Scotto Di Freca, A.: Using handwriting features to characterize cognitive impairment. In: Ricci, E., Rota Bulò, S., Snoek, C., Lanz, O., Messelodi, S., Sebe, N. (eds.) ICIAP 2019. LNCS, vol. 11752, pp. 683–693. Springer, Cham (2019). https://doi.org/10.1007/978-3-030-30645-8_62
10. Cortes, C., Vapnik, V.: Support-vector networks. Mach. Learn. **20**(3), 273–297 (1995)
11. De Gregorio, G., Desiato, D., Marcelli, A., Polese, G.: A multi classifier approach for supporting Alzheimer's diagnosis based on handwriting analysis. In: Del Bimbo, A., Cucchiara, R., Sclaroff, S., Farinella, G.M., Mei, T., Bertini, M., Escalante, H.J., Vezzani, R. (eds.) ICPR 2021. LNCS, vol. 12661, pp. 559–574. Springer, Cham (2021). https://doi.org/10.1007/978-3-030-68763-2_43
12. De Stefano, C., Fontanella, F., Impedovo, D., Pirlo, G., di Freca, A.S.: Handwriting analysis to support neurodegenerative diseases diagnosis: a review. Pattern Recogn. Lett. **121**, 37–45 (2019)
13. Della Cioppa, A., De Stefano, C., Marcelli, A.: On the role of population size and niche radius in fitness sharing. IEEE Trans. Evol. Comput. **8**(6), 580–592 (2004)
14. Della Cioppa, A., De Stefano, C., Marcelli, A.: Where are the niches? dynamic fitness sharing. IEEE Trans. Evol. Comput. **11**(4), 453–465 (2007)
15. Drotár, P., Mekyska, J., Rektorová, I., Masarová, L., Smékal, Z., Faundez-Zanuy, M.: Evaluation of handwriting kinematics and pressure for differential diagnosis of parkinson's disease. Artif. Intell. Med. **67**, 39–46 (2016)
16. Forrest, S., Perelson, A.S., Allen, L., Cherukuri, R.: Self-nonself discrimination in a computer. In: Proceedings of 1994 IEEE Computer Society Symposium on Research in Security and Privacy, pp. 202–212 (1994)
17. Garre-Olmo, J., Faúndez-Zanuy, M., López-de Ipiña, K., Calvó-Perxas, L., Turró-Garriga, O.: Kinematic and pressure features of handwriting and drawing: preliminary results between patients with mild cognitive impairment, alzheimer disease and healthy controls. Curr. Alzheimer Res. **14**(9), 960–968 (2017)
18. Gautier, S., Rosa-Neto, P., Morais, J.a., Webster, C.: World Alzheimer Report 2021: Journey through the diagnosis of dementia. ADI, London, UK (2021)

19. Gonzalez, F., Dasgupta, D., Kozma, R.: Combining negative selection and classification techniques for anomaly detection. In: Proceedings of the 2002 Congress on Evolutionary Computation, CEC 2002, vol. 1, p. 705–710 (2002)
20. Gupta, K.D., Dasgupta, D.: Negative selection algorithm research and applications in the last decade: A review (2021)
21. Ho, T.K.: Random decision forests. In: Proceedings of 3rd International Conference on Document Analysis and Recognition, vol. 1, pp. 278–282 (1995). https://doi.org/10.1109/ICDAR.1995.598994
22. Ishikawa, T., et al.: Handwriting features of multiple drawing tests for early detection of Alzheimer's disease: a preliminary result. In: MedInfo, pp. 168–172 (2019)
23. Jankovic, J.: Parkinson's disease: clinical features and diagnosis. J. Neurol. Neurosurgery Psychiatry **79**(4), 368–376 (2008)
24. Ji, Z., Dasgupta, D.: V-detector: an efficient negative selection algorithm with "probably adequate" detector coverage. Inf. Sci. **179**(10), 1390–1406 (2009)
25. Kahindo, C., El-Yacoubi, M.A., Garcia-Salicetti, S., Rigaud, A.S., Cristancho-Lacroix, V.: Characterizing early-stage Alzheimer through spatiotemporal dynamics of handwriting. IEEE Signal Process. Lett. **25**(8), 1136–1140 (2018)
26. Kawa, J., Bednorz, A., Stepień, P., Derejczyk, J., Bugdol, M.: Spatial and dynamical handwriting analysis in mild cognitive impairment. Comput. Biol. Med. **82**, 21–28 (2017)
27. Lasisi, A., Ghazali, R., Herawan, T.: Chapter 11 - application of real-valued negative selection algorithm to improve medical diagnosis. In: Al-Jumeily, D., Hussain, A., Mallucci, C., Oliver, C. (eds.) Applied Computing in Medicine and Health, pp. 231–243. Emerging Topics in Computer Science and Applied Computing, Morgan Kaufmann, Boston (2016)
28. Le, W., Dong, J., Li, S., Korczyn, A.D.: Can biomarkers help the early diagnosis of parkinson's disease? Neurosci. Bull. **33**(5), 535–542 (2017)
29. Li, T., Le, W.: Biomarkers for parkinson's disease: How good are they? Neurosci. Bull. **36**(2), 183–194 (2020)
30. Myszczynska, M.A., et al.: Applications of machine learning to diagnosis and treatment of neurodegenerative diseases. Nat. Rev. Neurol. **16**, 440–456 (2020)
31. Parziale, A., Senatore, R., Della Cioppa, A., Marcelli, A.: Cartesian genetic programming for diagnosis of parkinson disease through handwriting analysis: performance vs. interpretability issues. Artif. Intell. Med. **111**, 101984 (2021)
32. Parziale, A., Della Cioppa, A., Senatore, R., Marcelli, A.: A decision tree for automatic diagnosis of parkinson's disease from offline drawing samples: Experiments and findings. In: Ricci, E., Rota Bulò, S., Snoek, C., Lanz, O., Messelodi, S., Sebe, N. (eds.) Image Analysis and Processing - ICIAP 2019, pp. 196–206 (2019)
33. Parziale, A., Senatore, R., Marcelli, A.: Exploring speed-accuracy tradeoff in reaching movements: a neurocomputational model. Neural Comput. Appl. **32**, 13377–13403 (2020)
34. Peng, H., Long, F., Ding, C.: Feature selection based on mutual information criteria of max-dependency, max-relevance, and min-redundancy. IEEE Trans. Pattern Anal. Mach. Intell. **27**(8), 1226–1238 (2005)
35. Pereira, C.R., et al.: A step towards the automated diagnosis of parkinson's disease: analyzing handwriting movements. In: 2015 IEEE 28th International Symposium on Computer-Based Medical Systems, pp. 171–176 (2015)
36. Pereira, C.R., Weber, S.A.T., Hook, C., Rosa, G.H., Papa, J.P.: Deep learning-aided parkinson's disease diagnosis from handwritten dynamics. In: 2016 29th SIBGRAPI Conference on Graphics, Patterns and Images (SIBGRAPI), pp. 340–346 (Oct 2016)

37. Pereira, C.R., et al.: A new computer vision-based approach to aid the diagnosis of Parkinson's disease. Comput. Methods Programs Biomed. **136**, 79–88 (2016)
38. Pirlo, G., Diaz, M., Ferrer, M.A., Impedovo, D., Occhionero, F., Zurlo, U.: Early diagnosis of neurodegenerative diseases by handwritten signature analysis. In: Murino, V., Puppo, E., Sona, D., Cristani, M., Sansone, C. (eds.) ICIAP 2015. LNCS, vol. 9281, pp. 290–297. Springer, Cham (2015). https://doi.org/10.1007/978-3-319-23222-5_36
39. Prince, M., Wimo, A., Guercet, M., Ali, G.C., Wu, Y.T., Prina, M.: World Alzheimer Report 2015: The Global Impact of Dementia. ADI, London, UK (2015)
40. Rosenblum, S., Engel-Yeger, B., Fogel, Y.: Age-related changes in executive control and their relationships with activity performance in handwriting. Hum. Mov. Sci. **32**(2), 363–376 (2013)
41. Senatore, R., Marcelli, A.: A neural scheme for procedural motor learning of handwriting. In: International Conference on Frontiers on Handwriting Recognition. pp. 659–664. Springer (2012)
42. Senatore, R., Marcelli, A.: A paradigm for emulating the early learning stage of handwriting: performance comparison between healthy controls and parkinson's disease patients in drawing loop shapes. Hum. Mov. Sci. **65**, 89–101 (2019)
43. Tanveer, M., et al.: Machine learning techniques for the diagnosis of Alzheimer's disease: a review. ACM Trans. Multimedia Comput. Commun. Appl. **16**(1s), 1–35 (2020)
44. Teulings, H.L., Contreras-Vidal, J.L., Stelmach, G.E., Adler, C.H.: Parkinsonism reduces coordination of fingers, wrist, and arm in fine motor control. Exp. Neurol. **146**(1), 159–170 (1997)
45. Teulings, H.L., Stelmach, G.E.: Control of stroke size, peak acceleration, and stroke duration in parkinsonian handwriting. Hum. Mov. Sci. **10**(2–3), 315–334 (1991)
46. Van Gemmert, A., Adler, C.H., Stelmach, G.: Parkinson's disease patients undershoot target size in handwriting and similar tasks. J. Neurol. Neurosurgery Psychiatry **74**(11), 1502–1508 (2003)
47. Vessio, G.: Dynamic handwriting analysis for neurodegenerative disease assessment: A literary review. Appl. Sci. **9**(21), 4666 (2019)
48. Werner, P., Rosenblum, S., Bar-On, G., Heinik, J., Korczyn, A.: Handwriting process variables discriminating mild Alzheimer's disease and mild cognitive impairment. J. Gerontol. B Psychol. Sci. Soc. Sci. **61**(4), P228–P236 (2006)
49. Witten, I.H., Frank, E., Hall, M.A., Pal, C.J.: Data Mining, Fourth Edition: Practical Machine Learning Tools and Techniques, 4th edn. Morgan Kaufmann Publishers Inc., San Francisco (2016)

Evolutionary Computation in Image Analysis, Signal Processing and Pattern Recognition

Ground-Truth Segmentation of the Spinal Cord from 3T MR Images Using Evolutionary Computation

Mohamed Mounir EL Mendili[1,2], Noémie Villard[2], Brice Tiret[2], Raphaël Chen[2], Damien Galanaud[3], Benoit Magnin[3], Stéphane Lehericy[3,4], Pierre-François Pradat[2,5], Evelyne Lutton[6,7], and Salma Mesmoudi[7,8,9(✉)]

[1] APHM, Hôpital Universitaire Timone, CEMEREM, Marseille, France
[2] Aix Marseille Univ, CRMBM, CNRS, Marseille, France
[3] APHP, Groupe Hospitalier Pitié-Salpêtrière, Service de Neuroradiologie, Paris, France
[4] Sorbonne Universités, UPMC Univ Paris 06, UMR-S975, Inserm U975, CNRS UMR7225, Centre de recherche de l'Institut du Cerveau et de la Moelle épinière – CRICM, Centre de Neuroimagerie de Recherche – CENIR, Paris, France
[5] APHP, Groupe Hospitalier Pitié-Salpêtrière, Département des Maladies du Système Nerveux, Paris, France
[6] INRAE, UMR MIA 518, EKINOCS Team, AgroParisTech, Paris, France
[7] Institut des Systèmes Complexes, Paris, France
salma.mesmoudi@univ-paris1.fr
[8] MATRICE Project, Sorbone Univ Paris 1, Paris, France
[9] Centre Européen de Sociologie et de Science Politique (CESSP UMR-8209), Paris, France

Abstract. Spinal cord atrophy is one of the neuroimaging features associated with neurodegenerative diseases, inflammatory diseases and trauma. MR images segmentation can be used to assess cord atrophy, with varying degrees of manual intervention. However the accuracy of segmentation results highly depends on the operator's experience: there is a clear need for methods that simplifies and facilitates expert intervention, while providing an accurate quantification of cord atrophy. We propose and test here a ground-truth segmentation based on a simple evolutionary algorithm. EAcord integrates a set of segmentation methods with varying accuracy and manual intervention (manual, semi-automated and automated methods), as well as knowledge about the spinal cord anatomy, its relative location, immediate surrounding environment and shape at C2 vertebral level. A lighter version, EAcord-light, is also proposed, using only segmentations from semi-automated and automated methods as inputs. An experimental analysis of both algorithms showed an improved reproducibility and similar or even better accuracies compared to manual outlining. More interestingly, in some cases, EAcord-light produced a ground-truth segmentation with minimal expert intervention.

Keywords: MRI · Spinal cord atrophy · Evolutionary algorithm · Ground-truth segmentation

---

N. Villard, B. Tiret, R. Chen—Equal contribution.

© Springer Nature Switzerland AG 2022
J. L. Jiménez Laredo et al. (Eds.): EvoApplications 2022, LNCS 13224, pp. 549–563, 2022.
https://doi.org/10.1007/978-3-031-02462-7_35

Abbreviations

ALS	Amyotrophic lateral sclerosis
ASM	Active surface method
STAPLE	Simultaneous truth and performance level estimation
CoV	Coefficient of variation
CSA	Cross-sectional area
CSF	Cerebrospinal fluid
DTbM	Double threshold-based method
EA	Evolutionary algorithm
GbM	Gradient-based method
SCI	Spinal cord injury
ROI	Region of interest
TbM	Threshold-based method

1 Introduction

Amyotrophic lateral sclerosis (ALS) shows pronounced cord atrophy from MR images at C2 vertebral level correlating with clinical disability [1]. Several segmentation methods with varying degrees of manual intervention, reproducibility and accuracy have been proposed to assess spinal cord atrophy from MR images [2–5]. However, characterizing the accuracy of these segmentation methods is an important challenge since their use could influence, in the pathological context, disease physiopathological mechanisms understanding, clinical and chirurgical intervention decision-making process [6–10].

Various algorithms to assess segmentation accuracy have been proposed [2–5]. However, to fulfill these algorithm's hypotheses, only sets of segmentations resulting from unbiased, independent and trained experts on segmentation and automated methods can be used, which is highly constraining. For this reason, an algorithm that characterizes segmentation methods accuracy with minimal expert intervention is desired.

Ground-truth segmentation refers to the closest segmentation to real features that are extracted from an image. In the context of biomedical imaging, a segmentation is considered as ground-truth when radiologists agreed that it covers most accurately features of interest when compared to existing methods [11, 12]. Ground-truth segmentations are based on digital phantoms: MR images of plastic phantoms of known dimensions with manual outlining from unbiased, independent and trained experts or with automated outlining based on inputs sets of manual segmentations by experts [2–4, 11–19]. But it is recognized that phantoms images are insufficiently realistic to represent the complexity of the spinal cord MR images, especially in the pathological context.

Additionally, manual outlining is time consuming, subject to intra- and inter-observer variability, and its accuracy highly depends on operator experience [20]. To circumvent the constraint of an excessive dependence on experts for the construction of a ground-truth segmentation, we propose a method based on evolutionary algorithms (EA).

EA have been used for various biomedical image segmentations [21–25]. They have the advantage to be able to simultaneously integrate, thanks to appropriate representations, a priori knowledge about the human anatomy, the relative location, the immediate surrounding and known shape of an organ, in the same way an expert uses it to perform manual segmentations of a medical image.

In this work, we explore the use of an EA as ground-truth generator. Our EA (EAcord) was optimized to generate a ground-truth segmentation using pre-existing segmentations. EAcord is initialized with a population of contours automatically computed using nine segmentation methods (methods are detailed in Sect. 2.4). The genetic engine then aims at optimizing a combination of two fitness functions: (1) Otsu's segmentation (V_1) that minimizes the intra-class variance between the foreground and the background of the MRI images [26]; (2) Sobel operator (V_2) that performs a 2D spatial gradient measurement on the image [27].

The paper is organized as follows: Sect. 2 details the experimental setup, the MRI protocol, the evolutionary algorithm EAcord and the description of the nine methods used to generate a ground-truth segmentation. Section 3 presents the results obtained with EAcord and its variant EAcord-light that uses only segmentation from semi-automated and automated methods. A discussion of these results and a conclusion are given in Sects. 4 and 5.

2 Materials and Methods

2.1 Patients

Thirty subjects were randomly selected from our database of spinal cord 3T-MRI scans performed at CENIR, Pitié-Salpétrière Hospital, Paris, France. Subjects included 10 healthy volunteers (mean age ± SD: 45.7 ± 15.8 years, 6 females), 10 patients with ALS (mean age ± SD: 54.5 ± 10.9 years, 3 females) and 10 patients with spinal cord injury (SCI) (mean age ± SD: 46.4 ± 18.2 years, 2 females). The local Ethics Committee of our institution approved all experimental procedures (Paris-Ile de France Ethical Committee under the 2009-A00291-56 registration number), and a written informed consent was obtained from each participant. Imaging examinations have been conducted according to the principles expressed in the latest revision of the Declaration of Helsinki.

2.2 MRI Acquisition

Scans were performed using a 3T MRI system (TIM Trio 32-channel, Siemens Health-care, Erlangen, Germany) using a body coil for signal excitation and a neck/spine coil for signal reception. The spinal cord was imaged using a sagittal T2-weighted three-dimensional (3D) turbo spin echo image with slab selective excitation (SPACE). Imaging parameters were: isotropic voxel size $0.9 \times 0.9 \times 0.9$ mm^3; FOV $= 280 \times 280$ mm^2; 52 sagittal slices; TR/TE $= 1500/120$ ms; flip angle $= 140°$; generalized auto-calibrating partially parallel acquisition (GRAPPA) with acceleration factor R $= 3$; turbo factor $= 69$; acquisition time $= 6$ min [28]. For a comprehensive description of the advantages of using SPACE sequence over other sequences to image the spinal cord, the reader is referred to [29–34].

2.3 Data Preprocessing

Data were corrected for non-uniformity intensity using Minc-Toolkit N3 [35]. The operator had then to choose a region of interest interactively using an in-house Matlab® script (The Mathworks Inc., Natick, MA, USA) as follows:

The operator drew in the sagittal slice, where the spinal cord was the most median at C2 vertebral level, a line perpendicular to the spinal cord at the largest spinal canal position of the C2 vertebral level [36]. This line corresponds to a plane perpendicular to the spinal cord. Slices parallel to this reference plane were resampled to a resolution of $0.3 \times 0.3 \times 0.3$ mm³ using 3D cubic interpolation in order to minimize partial volume effect between the spinal cord and the CSF [12, 18, 36]. Three slices were automatically selected on either side of the plane including the reference plane to be segmented, and then cropped, leading to a total of 7 slices with a FOV = 80×100 pixels. Figure 1 shows an example of data preprocessing for a SCI patient.

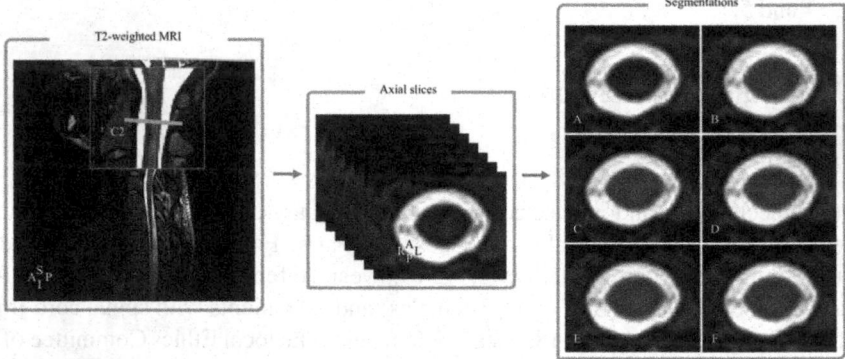

Fig. 1. T2-weighted MRI panel: T2-weighted MRI mid-sagittal slice for a spinal cord injury (SCI) patient after non-uniformity intensity correction (N3). A plane perpendicular to the spinal cord was drawn from the largest spinal canal position of the C2 vertebral level (Maya blue line). Axial slices panel: Seven axial slices were selected from resampled data. Segmentations panel: Resulting segmentations of the mid-axial slice (A) from the experienced operator (orange) using: (B) Manual outlining. (C) threshold-based method (TbM). (D) gradient-based method (GbM). (E) active surface method (ASM) (Color figure online)

2.4 Evolutionary Algorithm

Spinal cord MR images resulting from the preprocessing step were segmented by two operators blinded to the type of data. One operator was experienced with spinal cord segmentation techniques (i.e. 5 years experience) and the other operator had no previous experience (i.e. had been given brief instructions on five representative data sets different from those used in the present study). The following segmentation methods were used:

1. Semi-automated threshold-based method (TbM), from Losseff et al. (1996) [3].
2. Semi-automated gradient-based method (GbM), from Tench et al. (2005) [4].

3. Active surface method (ASM) from Horsfield et al. (2010) [2]. This method is based on image smoothing followed by an active surface model constrained by intensity gradient image information.
4. Automated double threshold-based method (DTbM) from El Mendili et al. [18].
5. Manual segmentation was also carried out by the two operators in the images resulting from the preprocessing step; MOp1 and MOp2 for segmentations made by the experienced and the non-experienced operators, respectively.

For a comprehensive description of the previous segmentation methods, the reader is referred to [2–4, 12, 18]. Figure 1 (right) shows an example of these segmentations. These methods have a variable accuracy and are considered as the initial population of the EAcord, with the aim of improving the accuracy of these segmentations.

Encoding Solutions. The individuals of the population are a set of candidate contours represented in polar coordinates centered on the center of mass of the segmented area (computed at the initialization and kept fixed during the EAcord evolution). They are encoded with a series of 90 radius (R) with angular resolution of 4° corresponding to the position of the contour on this radius. A solution is thus a chromosome of 90 real values R_1 to R_{90}.

Initial Population. Initial contours were computed from the results of the nine segmentation methods ($2\times$Manual, $2\times$TbM, $2\times$GbM, $2\times$ASM and $1\times$DTbM). 990 contours were produced this way. The initial population was completed to 1000 individuals by generating ten additional random perturbations on the previous ones considering known mean error in cross-sectional area (CSA) estimation (literature analysis): -3% for TbM [2, 3, 18], 3% for GbM [12], 9% for ASM [2, 12], and -1% for DTbM [12, 20].

The 990 contours are constructed by randomly choosing a radius per angle from the nine available radiuses (Fig. 2). Radiuses from manual segmentation have the biggest probability to be chosen (1/4) because compared to other methods, this segmentation is supposed to be closer to the ground-truth. Radiuses from the DTbM segmentation have the same probability to be chosen as TbM and GbM (3/34), even if they have a smaller mean error in CSA estimation than DTbM [2–4, 12, 20]. DTbM is thus not privileged in comparison to TbM and GbM. Due to the mean error in CSA estimation made by ASM compared to other methods, the probability to choose a radius from ASM segmentations is three times lower than for TbM, GbM and DTbM (1/34).

Genetic Operators. A basic random mutation and a two-points crossover are used. The EAcord parameters and operators are detailed in Table 1 and Fig. 2.

Fitness Function. The fitness function approximates with automatic procedures a series of known facts experts are able to express when manually separating spinal cord (hypointense tissue) from the corticospinal fluid (hyperintense tissue) (see Fig. 1). We thus use the two following components:

1. *Otsu's segmentation (V_1):* This threshold-based method seeks the optimal grey level threshold (T) that minimizes the intra-class variance between the foreground and the background of the MRI images [26]. The optimal threshold separates the spinal cord from the CSF.

2. *Sobel operator (V₂):* This classical edge detection method performs a 2D spatial gradient measurement on the image and emphasizes regions of high spatial frequency that mainly correspond to the edge between the spinal cord and the CSF [27].

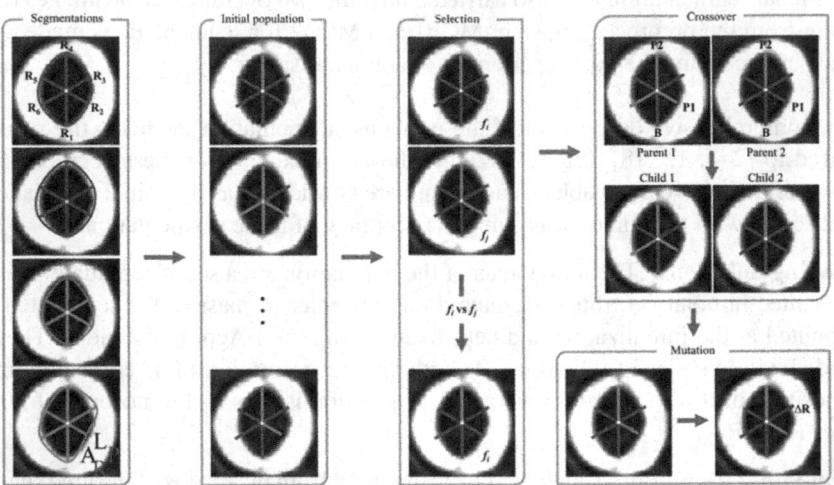

Fig. 2. Overview of the EA-cord operators. Segmentations panel: for simplification, only 4 segmentations and 6 radiuses are shown in the figure (R_1 to R_6). Initial population panel: new contours were constructed by choosing randomly a radius per angle from the four available segmentations. Selection panel: fi and fj correspond to fitness values of the contours i and j, respectively. Crossover panel: B represents the beginning of the contour and P1 and P2 represent the two crossover points. Mutation panel: ΔR segment in red represents the mutation that occurs in the randomly selected radii, i.e. blue radii. A, anterior; L, left; R, right; P, posterior. (Color figure online)

According to each couple of criteria values (V_1, V_2), a global fitness is computed for each individual (contour) $f = V_1 + V_2$. The combination of the two criteria could be seen as a cooperation between a threshold-based and a gradient-based approach.

Each mutation causes a change in the value of a radius R_i (with $1 \leq I \leq 90$). Each change of a radius leads to a new contour "chromosome" with new values of contrast and gradient of the image (v1 + v2).

To show that the segmentation improvement is not dependent on the high accuracy of the initial population contours (i.e. manual segmentations), we run the EAcord on a lighter initial population that includes only segmentations from semi-automated and automated methods (EAcord-light). The parameter setting is the same as EAcord except for the probability of a radius to be chosen from each segmentation method. Radiuses from DTbM, TbM and GbM are chosen with a probability of 3/16 and 1/16 for ASM.

Evaluation

Convergence. The algorithm converged when the last 100 best contour fitness values changed by less than 100 (without unit) compared to the previous generation, which represented less than 0.5% of the lowest fitness value of all generated contours. Otherwise, the algorithm stopped automatically at the 1000th generation.

Reproducibility. The used EAcord and EAcord-light were repeated 30 times for the whole data sets. Reproducibility was evaluated using coefficient of variation (CoV). CoV was defined as the ratio of the standard deviation to the average of CSA values across repeated measurements.

Independency from Manual Outlining. Student's paired t-test was performed to test the difference in CSA estimation between EAcord and EAcord-light.

Ground-Truth Value.
Visual Evaluation. Two neuroradiologists (senior: D. G; junior: B. M.), evaluated EAcord segmentations as well as the other manual outlining from the experienced and non-experienced operators (MOp1 and MOp2, respectively). They were blinded to the type of data and methods. They classified each segmentation method result from the best one (rank = 1) to the worst one (rank = 3). Two segmentations or more could have the same rank if the neuroradiologist considered their accuracies similar.

The ground-truth value of each method (EAcord, MOp1 and MOp2) was quantified by calculating the rank difference or contrast between couples of segmentation methods. These contrasts included (EAcord - MOp1) and (MOp1 - MOp2). The last contrast quantified the experience value difference between the experienced and non-experienced operators.

Quantitative Evaluation. Accuracies of EAcord and EAcord-light segmentations as well as manual segmentations from experienced and non-experienced operators were compared with ground truth segmentation generators STAPLE and SbA [11, 13]. STAPLE and SbA ground truth segmentations are computed using five manual segmentations from experienced operator (each made at least 2 weeks apart), and the automated method DTbM [11].

Accuracy was qualitatively evaluated taking STAPLE/SbA segmentations as ground-truths: (1) by using the Dice similarity coefficient (DSC), which evaluates the performance of segmentation methods by measuring their spatial overlaps [37]; (2) by using Hausdorff distance (HD), which quantify the dissemblance between two contours [38]; (3) by computing the relative CSA estimation error.

Student's paired t-test was used to evaluate the difference in CSA estimation between STAPLE/SbA, EAcord, EAcord-light, MOp2 and MOp1 (Bonferroni-corrected). Furthermore, Student's paired t-test was performed to test the difference in DSC values between EAcord, EAcord-light, MOp1 and MOp2 using STAPLE/SBA segmentations as ground-truths (Bonferroni-corrected).

2.5 Complexity

A contour was made of 90 radiuses. Each radius length could minimally vary within the interval [0–40] voxels (i.e. FOV = 80 × 100). There are thus 40^{90} possible contours. Several years of computation time are needed for a systematic scan of the search space.

Table 1. Parameters setting for EAcord. The parameter values were set experimentally. f_i and f_j are the fitness value of contours i and j, respectively.

Parameter	Value
Initial population	1000 contours
Selection	Selection rate: 0.95 Replacement rate: 0.9
Crossover	Two points crossover rate: 0.9
Mutation	Random with probability of 1/90 rate: 0.5
Stopping criterion	*If convergence, 100 generations without change if not, 1000 generations*
Repetitions	*30*

2.6   Computation Time

The computation time was 3.03 ± 0.10 h/subject for an EAcord segmentation and 3.01 ± 0.12 h/subject for an EAcord-light one. This was achieved using a 64-bit Quad-core (Intel® Xeon®, processor speed: 2.67 GHz) workstation. Computation times were not optimized as the whole procedure was coded in Matlab® language without parallelization.

3   Results

All statistical analyses were performed in Matlab®. Values are expressed as means $\pm$ standard deviation.

3.1   Independency from Manual Outlining

Student's paired t-test showed no significant difference in CSA estimation between EAcord and EAcord-ligth ($P = 0.75$). The difference in CSA estimation between EAcord and EAcord-ligth was 0.33 ± 0.25 mm^2.

3.2   Ground-Truth Value

Visual Evaluation. Classification results for the whole population are presented in Tables 2 and 3. The average rank of the classification made by the senior neuroradiologist was 1.66 ± 0.71 for EAcord, 1.66 ± 0.75 for MOp1 and 2.40 ± 0.86 for MOp2. The average rank of the classification made by the junior neuroradiologist was 1.13 ± 0.43 for EAcord, 2.23 ± 0.57 for MOp1 and 2.46 ± 0.68 for MOp2.

Ranks difference between methods classifications made by the senior neuroradiologist for the whole population were: 0.00 ± 1.20 for EAcord-MOp1 and -0.73 ± 1.39 for MOp1-MOp2. Ranks difference between methods classification made by the junior

Table 2. Results of the visual classification made by two neuroradiologists for EAcord, MOp1 and MOp2 segmentations.

Method	Mean ± SD rank	
	Neuroradiologist 1[+]	Neuroradiologist 2[++]
EAcord	1.66 ± 0.71	1.13 ± 0.43
MOp1*	1.66 ± 0.75	2.23 ± 0.57
Mop2	2.40 ± 0.86	2.46 ± 0.68

[+]Senior
[++]Junior
*Experienced operator

Table 3 Mean ± SD rank difference between two segmentations according to the classification made by the two neuroradiologists.

Methods	Rank difference	
	Neuroradiologist 1[+]	Neuroradiologist 2[++]
EAcord - MOp1*	0.00 ± 1.20	−1.10 ± 0.76
MOp1* - MOp2	−0.73 ± 1.39	−0.23 ± 1.14

[+]Senior
[++]Junior
*Experienced operator

neuroradiologist were: -1.10 ± 0.76 for EAcord-MOp1 and -0.23 ± 1.14 for MOp1-MOp2. To summarize, EAcord segmentations were considered as accurate as manual segmentations from the experienced operator. The experienced operator produced more accurate segmentation than the non-experienced one.

Quantitative Evaluation with STAPLE. STAPLE segmentation was considered as the ground-truth for accuracy quantification. Accuracy results are shown in Table 4. DSC value was 97.64 ± 0.76 % for EAcord, 97.45 ± 0.78 % for EAcord-light, 97.40 ± 0.56 % for MOp1 and 96.37 ± 1.21 % for MOp2. HD value was 0.77 ± 0.38 % for EAcord, 0.90 ± 0.59 for EAcord-light, 0.89 ± 0.50 for MOp1 and 1.04 ± 0.45 for MOp2, taken STAPLE segmentation as the ground-truth. EAcord, EAcord-light, MOp1 and MOp2 underestimated the CSA measured by STAPLE segmentation (considered as the ground-truth) by 1.32 ± 2.68 %, 1.34 ± 2.72 %, 1.49 ± 2.28 % and 3.85 ± 3.42 %, respectively.

There was no significant difference in CSA estimation between STAPLE $(71.88 \pm 8.19 \text{ mm}^2)$, EAcord and EAcord-light (Student's paired t-test; Bonferroni correction; $P_{STAPLE-EAcord/EAcord-light} > 0.0125$). There was a significant difference in CSA estimation between STAPLE, MOp1 and MOp2 (Student's paired t-test; Bonferroni correction; $P_{STAPLE-MOp1} = 10^{-3}$ and $P_{STAPLE-MOp2} < 10^{-5}$). Based on mean DSC

Table 4. Accuracy of mean ± SD cross-sectional area measurements. STAPLE and SbA segmentations were taken as ground-truths.

Measurements	Methods			
	EAcord	EAcord-light	MOp1*	MOp1*
CSA estimation (mm²)	71.03 ± 8.92	71.01 ± 8.91	70.81 ± 8.14	69.11 ± 8.17
STAPLE				
DSC (%)	97.64 ± 0.76	97.45 ± 0.78	97.4 ± 0.56	96.37 ± 1.21
HD (pixel)	0.77 ± 0.38	0.90 ± 0.59	0.89 ± 0.50	1.04 ± 0.45
Error in CSA estim (%)	−1.32 ± 2.68	−1.34 ± 2.72	−1.49 ± 2.28	−3.85 ± 3.42
SbA				
DSC (%)	97.89 ± 0.66	97.69 ± 0.70	97.92 ± 0.60	96.48 ± 1.14
HD (pixel)	0.75 ± 0.40	0.88 ± 0.40	0.58 ± 0.33	1.18 ± 0.45
Error in CSA estim (%)	−0.95 ± 2.55	−0.97 ± 2.59	−1.11 ± 2.18	−3.48 ± 3

*Experienced operator

values, EAcord and EAcord-light showed no significant difference in accuracy compared to MOp1 (Student's paired t-test; $P_{MOp1-EAcord} = 0.05$ and $P_{MOp1-EAcord-light} = 0.70$). MOp2 showed a significant lower accuracy than MOp1 (Student's paired t-test; Bonferroni correction; $P_{MOp1-MOp2} < 10^{-5}$).

Quantitative Evaluation with SbA. SbA segmentation was considered as the ground-truth for accuracy quantification. Accuracy results are shown in Table 4. DSC value was 97.69 ± 0.66 % for EAcord, 97.69 ± 0.70 % for EAcord-light, 97.92 ± 0.60 % for MOp1 and 96.48 ± 1.14 % for MOp2, taken SbA segmentation as the ground-truth. HD value was 0.75 ± 0.40 for EAcord, 0.88 ± 0.40 for EAcord-light, 0.58 ± 0.33 for MOp1 and 1.18 ± 0.45 for MOp2, taken SbA segmentation as the ground-truth. EAcord, EAcord-light, MOp1 and MOp2 underestimated the CSA measured by SbA segmentation (considered as the ground-truth) by 0.95 ± 2.55 %, 0.97 ± 2.59 %, 1.14 ± 2.22 % and 3.48 ± 3.53 %, respectively.

There was no significant difference in CSA estimation between SbA (71.61 ± 8.12 mm²), EAcord and EAcord-light (Student's paired t-test; $P_{SbA-EAcord} = 0.06$, $P_{SbA-EAcord-light} = 0.06$). There was a significant difference in CSA estimation between SbA and MOp1 and MOp2 (Student's paired t-test; Bonferroni correction; $P_{SbA-MOp1} < 10^{-2}$, $P_{SbA-MOp2} < 10^{-4}$). Based on mean DSC values, EAcord and EAcord-light showed no significant difference in accuracy than MOp1 (Student's paired t-test; $P_{MOp1-EAcord} = 0.77$, $P_{MOp1-EAcord-light} = 0.10$). MOp2 showed a significant lower accuracy than MOp1 (Student's paired t-test; Bonferroni correction; $P_{MOp1-MOp2} < 10^{-5}$).

In sum, there was a difference of accuracy evaluation for EAcord by the two neuroradiologists, mainly due to their difference in experience. The two neuroradiologists considered EAcord segmentations similar to better than the manual outlining made by the experienced operator, the current reference method for ground-truth [2, 3, 12, 17,

18]. As expected, manual segmentations from experienced operator showed higher accuracy than those from the non-experienced one, underlining the benefit of experience in segmentation [29]. These results were confirmed by quantitative evaluations based on STAPLE and SbA as ground-truths.

4 Discussion

A Ground-truth segmentation has been computed thanks to a simple implementation of an evolutionary algorithm initialized either with a set of manual, semi-automated and automated methods (EAcord) or with only semi-automated and automated methods (EAcord-light). Reproducibility of EAcord and EAcord-light as well as their differences in cross-sectional area (CSA) estimations were computed. EAcord ground-truth segmentations were visually and quantitatively compared to those provided by manual outlining.

EAcord showed high reproducibility over 30 simulations. Its reproducibility outperformed the other segmentation methods, including manual outlining made by experts, regardless of their experience difference [2–4, 18, 39]. More interestingly, when considering only automated and semi-automated methods as an initialization, EAcord-light showed similar reproducibility as EAcord. The difference in CSA estimation between EAcord and EAcord-light segmentations was not significant, suggesting that manual outlining made by experts may not be necessary to construct ground-truth segmentation.

The EAcord-light method may be a good alternative to manual outlining. It may be particularly useful when following an atrophy rate corresponding to small changes in short periods in ALS, SCI and possibly in MS (the usual median decline of cord area at C2 vertebral level is around 2.2% per year for the secondary progressive form, 1.5% per year for the primary progressive form and 1.3% per year for the relapsing remitting form) [40]. Recent technical developments in MRI allow accurate exploration of larger spinal cord regions: the corresponding volumetric segmentations may be used as initialization for EAcord-light [2, 12, 17, 41–44]. This is important as an accurate monitoring is needed in pathologies that affect large portions of the spinal cord, for instance neurodegenerative, trauma and neuroinflammatory diseases [8, 28, 45–50], as well as for constructing an accurate spinal cord template facilitating groups-atrophy quantification [12, 42].

5 Conclusion

EAcord actually integrates various important priors:

It is initialized with a combination of reference segmentations from the literature with variable accuracies (from −1% to 9% error in CSA estimation) [2–4, 12, 20], leading to the creation of a probabilistic prior on the search space similar to the STAPLE approach [11]. In that regard, other segmentation methods could be introduced in order to enhance the accuracy of the result.

EAcord then combines and modifies, based on image characteristics, the proposed segmentations along generations to reach an optimal solution, i.e. a ground-truth segmentation. The choice of combining gradient and threshold-based methods in the fitness computation was largely inspired by the observation of the way an experienced operator delineates the spinal cord in MR images.

The experimental analysis presented in this paper is a proof of concept: even with a very simple evolutionary engine, results are competitive enough for building accurate ground-truth segmentations of the spinal cord while minimizing manual intervention from experts.

In addition, EAcord is currently based on a mono-objective approach using the sum of a threshold-based (Otsu's method) and gradient-based (Sobel operator) segmentation evaluation [26, 27], the precise tuning of the balance between these two components has not been addressed very acutely here. In a future work, we will consider a multi-objective scheme for dealing with a variety of segmentation criteria [51, 52]. Regarding execution time, the code will be optimized and more efficient selection methods will be tested to shorten convergence time. This work is also a prior validation for a future interactive evolutionary scheme, where experts will be involved along the process in a more efficient and user-friendly way.

Acknowledgments. We are grateful to all subjects and their relatives. We thank Kevin Nigaud, Romain Valabrègue, Alexandre Vignaud, Frédéric Humbert, Christelle Macia, Mélanie Didier and Eric Bardinet from CENIR for helping with the acquisitions. We thank Drs Mark A. Horsfield for providing us the trial version of Jim 6.0. We thank Tania Mancheno for her manuscript's English language revision.

This study was supported by the Association Française contre les Myopathies (AFM) and the Institut pour la Recherche sur la Moelle épinière et l'Encéphale (IRME). The research leading to these results has also received funding from the program "Investissements d'Avenir" ANR-10-IAIHU-06.

References

1. Branco, L.M., De Albuquerque, M., De Andrade, H.M., Bergo, F.P., Nucci, A., et al.: Spinal cord atrophy correlates with disease duration and severity in amyotrophic lateral sclerosis. Amyotroph Lateral Scler Frontotemporal Degener. **15**(1–2), 93–97 (2014)
2. Horsfield, M.A., et al.: Rapid semi-automatic segmentation of the spinal cord from magnetic resonance images: application in multiple sclerosis. Neuroimage **50**(2), 446–455 (2010)
3. Losseff, N.A., et al.: Spinal cord atrophy and disability in multiple sclerosis. A new reproducible and sensitive MRI method with potential to monitor disease progression. Brain **119**(Pt 3), 701–708 (1996)
4. Tench, C.R., Morgan, P.S., Constantinescu, C.S.: Measurement of cervical spinal cord cross-sectional area by MRI using edge detection and partial volume correction. J. Magn. Reson. Imaging **21**(3), 197–203 (2005)
5. Zivadinov, R., et al.: Comparison of three different methods for measurement of cervical cord atrophy in multiple sclerosis. AJNR Am. J. Neuroradiol. **29**(2), 319–325 (2008)
6. Abdel-Aziz, K., et al.: Evidence for early neurodegeneration in the cervical cord of patients with primary progressive multiple sclerosis. Brain **138**(Pt 6), 1568–1582 (2015)
7. Healy, B.C., et al.: Approaches to normalization of spinal cord volume: application to multiple sclerosis. J. Neuroimaging **22**(3), e12–e19 (2012)
8. Klein, J.P., et al.: A 3T MR imaging investigation of the topography of whole spinal cord atrophy in multiple sclerosis. AJNR Am. J. Neuroradiol. **32**(6), 1138–1142 (2011)

9. Yang, J., et al.: Statistical modeling approach to quantitative analysis of interobserver variability in breast contouring. Int. J. Radiat. Oncol. Biol. Phys. **89**(1), 214–221 (2014)
10. Kosztyla, R., et al.: High-grade glioma radiation therapy target volumes and patterns of failure obtained from magnetic resonance imaging and 18F-FDOPA positron emission tomography delineations from multiple observers. Int. J. Radiat. Oncol. Biol. Phys. **87**(5), 1100–1106 (2013)
11. Warfield, S.K., Zou, K.H., Wells, W.M.: Simultaneous truth and performance level estimation (STAPLE): an algorithm for the validation of image segmentation. IEEE Trans. Med. Imaging. **23**(7), 903–921 (2004)
12. El Mendili, M.M., et al.: Fast and accurate semi-automated segmentation method of spinal cord MR images at 3T applied to the construction of a cervical spinal cord template. PLoS ONE **10**(3), e0122224 (2015a)
13. Rohlfing, T., Maurer, J.C.R.: Shape-based averaging. IEEE Trans. Image Process. **16**(1), 153–161 (2007)
14. Pohl, K.M., et al.: Logarithm odds maps for shape representation. Med. Image Comput. Comput. Assist. Interv. **9**(Pt 2), 955–963 (2006)
15. Heckemann, R.A., Hajnal, J.V., Aljabar, P., Rueckert, D., Hammers, A.: Automatic anatomical brain MRI segmentation combining label propagation and decision fusion. Neuroimage **33**(1), 115–126 (2006)
16. Bergo, F.P.G., França Jr., M.C., Chevis, C.F., Cendes, F.: SpineSeg: a segmentation and measurement tool for evaluation of spinal cord atrophy. In: CISTI 2012 (7ª Conferencia Ibérica de Sistemas y Tecnologia de Información), Madrid, Spaiwn, vol. 2, pp. 400–403, June 2012
17. De Leener, B., Kadoury, S., Cohen-Adad, J.: Robust, accurate and fast automatic segmentation of the spinal cord. Neuroimage **98**, 528–536 (2014)
18. El Mendili, M.M., et al.: Validation of a semiautomated spinal cord segmentation method. J. Magn. Reson. Imaging. **41**(2), 454–459 (2015)
19. Asman, A.J., Bryan, F.W., Smith, S.A., Reich, D.S., Landman, B.A.: Groupwise multi-atlas segmentation of the spinal cord's internal structure. Med. Image Anal. **18**(3), 460–471 (2014)
20. El Mendili, M.M., et al.: Spinal cord atrophy quantification: comparison of segmentation methods for 3T MRI T2-weighted images. In: Proceeding of the 20th Annual Meeting of the Organization of the Human Brain Mapping (OHBM), Hamburg, Germany. Abstract 2944 (2014)
21. Garg, G., Juneja, S.: Brain tumor segmentation using genetic algorithm and FCM clustering approach. Int. J. Comput. Appl. **49**(2), 24–27 (2012)
22. Sharma, M., Mukharjee, S.: Brain tumor segmentation using hybrid genetic algorithm and artificial neural network fuzzy inference system (ANFIS). Int. J. Fuzzy Logic Syst. **2**(4), 31–42 (2012)
23. Miranda Teixeira, G., Ramalho Pommeranzembaum, I., de Oliveira, B.L., Lobosco, M., Weber dos Santos, R.: Automatic segmentation of cardiac MRI using snakes and genetic algorithms. In: Bubak, M., van Albada, G.D., Dongarra, J., Sloot, P.M.A. (eds.) ICCS 2008. LNCS, vol. 5103, pp. 168–177. Springer, Heidelberg (2008). https://doi.org/10.1007/978-3-540-69389-5_20
24. Xie, F., Bovik, A.C.: Automatic segmentation of dermoscopy images using self-generating neural networks seeded by genetic algorithm. Pattern Recogn. **46**(3), 1012–1019 (2013)
25. Yuan, X., Situ, N., Zouridakis, G.: Automatic segmentation of skin lesion images using evolution strategies. Biomed. Sign. Process. Control **3**(3), 220–228 (2008)
26. Otsu, N.: A threshold selection method from gray-level histograms. Automatica **11**(285–296), 23–27 (1975)
27. Sobel, I., Feldman, G.: A 3x3 isotropic gradient operator for image processing. In: A Talk at the Stanford Artificial Project, pp. 271–272 (1968)

28. Cohen-Adad, J., et al.: Demyelination and degeneration in the injured human spinal cord detected with diffusion and magnetization transfer MRI. Neuroimage **55**(3), 1024–1033 (2011)
29. Del Grande, F., Chhabra, A., Carrino, J.A.: Getting the most out of 3 tesla MRI of the spine. Rheumatology Netw. (2012)
30. Meindl, T., et al.: Magnetic resonance imaging of the cervical spine: comparison of 2D T2-weighted turbo spin echo, 2D T2*weighted gradient-recalled echo and 3D T2-weighted variable flip-angle turbo spin echo sequences. Eur. Radiol. **19**(3), 713–721 (2009)
31. Lee, S., et al.: MRI of the lumbar spine: comparison of 3D isotropic turbo spin-echo SPACE sequence versus conventional 2D sequences at 3.0 T. Acta Radiol. (2014)
32. Lichy, M.P., et al.: Magnetic resonance imaging of the body trunk using a single-slab, 3-dimensional, T2-weighted turbo-spin-echo sequence with high sampling efficiency (SPACE) for high spatial resolution imaging: initial clinical experiences. Invest. Radiol. **40**(12), 754–760 (2005)
33. Rodegerdts, E.A., et al.: 3D imaging of the whole spine at 3T compared to 1.5T: initial experiences. Acta Radiol. **47**(5), 488–493 (2006)
34. Tins, B., Cassar-Pullicino, V., Haddaway, M., Nachtrab, U.: Three-dimensional sampling perfection with application-optimised contrasts using a different flip angle evolutions sequence for routine imaging of the spine: preliminary experience. Br. J. Radiol. **85**(1016), e480–e489 (2012)
35. Sled, J.G., Zijdenbos, A.P., Evans, A.C.: A nonparametric method for automatic correction of intensity nonuniformity in MRI data. IEEE Trans. Med. Imaging **17**(1), 87–97 (1998)
36. Lundell, H., et al.: Independent spinal cord atrophy measures correlate to motor and sensory deficits in individuals with spinal cord injury. Spinal Cord **49**(1), 70–75 (2011)
37. Dice, L.R.: Measures of the amount of ecologic association between species. Ecology **26**, 297–302 (1945)
38. Huttenlocher, D.P., Klanderman, G., Rucklidge, W.J.: Comparing images using the Hausdorff distance. IEEE Trans. Pattern Anal. Mach. Intell. **15**(9), 850–863 (1993)
39. Yiannakas, M.C., et al.: Feasibility of grey matter and white matter segmentation of the upper cervical cord in vivo: a pilot study with application to magnetization transfer measurements. Neuroimage **63**(3), 1054–1059 (2012)
40. Lukas, C., et al.: Cervical spinal cord volume loss is related to clinical disability progression in multiple sclerosis. J. Neurol. Neurosurg. Psychiatry (2014)
41. Chen, M., et al.: Automatic magnetic resonance spinal cord segmentation with topology constraints for variable fields of view. Neuroimage **83**, 1051–1062 (2013)
42. Fonov, V.S., et al.: Framework for integrated MRI average of the spinal cord white and gray matter: the MNI-Poly-AMU template. Neuroimage **15**(102 Pt 2), 817–827 (2014)
43. Kawahara, J., McIntosh, C., Tam, R., Hamarneh, G.: Globally optimal spinal cord segmentation using a minimal path in high dimensions. In: 2013 IEEE 10th International Symposium on Biomedical Imaging (ISBI), pp. 848–851. IEEE (2013)
44. McIntosh, C., Hamarneh, G., Toom, M., Tam, R.C.: Spinal cord segmentation for volume estimation in healthy and multiple sclerosis subjects using crawlers and minimal paths. In: 2011 First IEEE International Conference on Healthcare Informatics, Imaging and Systems Biology (HISB), pp. 25–31. IEEE (2011)
45. Cohen-Adad, J., et al.: Involvement of spinal sensory pathway in ALS and specificity of cord atrophy to lower motor neuron degeneration. Amyotroph Lateral Scler Frontotemporal Degener. **14**(1), 30–38 (2013)
46. El Mendili, M.M., et al.: Multi-parametric spinal cord MRI as potential progression marker in amyotrophic lateral sclerosis. PLoS ONE **9**(4), e95516 (2014)
47. Katsuno, M., et al.: Clinical features and molecular mechanisms of spinal and bulbar muscular atrophy (SBMA). Adv. Exp. Med. Biol. **685**, 64–74 (2010)

48. Liu, W., et al.: In vivo imaging of spinal cord atrophy in neuroinflammatory diseases. Ann. Neurol. **76**(3), 370–378 (2014)
49. Wijesekera, L.C., Leigh, P.N.: Amyotrophic lateral sclerosis. Orphanet. J. Rare Dis. **4**, 3 (2009)
50. Wyndaele, M., Wyndaele, J.J.: Incidence, prevalence and epidemiology of spinal cord injury: what learns a worldwide literature survey? Spinal Cord. **44**(9), 523–529 (2006)
51. Fourman, M.P.: Compaction of symbolic layout using genetic algorithms. In: Proceedings of the 1st International Conference on Genetic Algorithms, pp. 141–153. L. Erlbaum Associates Inc. (1985)
52. Schaffer, J.D.: Multiple objective optimization with vector evaluated genetic algorithms. In: Proceedings of the 1st International Conference on Genetic Algorithms, pp. 93–100. L. Erlbaum Associates Inc. (1985)

Applications of Nature-Inspired Computing for Sustainability and Development

A Machine Learning-Based Approach for Economics-Tailored Applications: The Spanish Case Study

Zakaria Abdelmoiz Dahi[1,2]([✉]), Gabriel Luque[1], and Enrique Alba[1]

[1] ITIS Software, Edificio Ada Byron, University of Malaga, Málaga, Spain
zakaria.dahi@uma.es, {gabriel,eat}@lcc.uma.es
[2] Department of Fundamental Computer Science and Its Applications, Fac. NTIC, University of Constantine 2, Constantine, Algeria
zakaria.dahi@univ-constantine2.dz

Abstract. The continuous evolution of economy hinders the decision-making process in this field. The former requires sophisticated techniques, and thus, manual and empirical methods are becoming increasingly obsolete. In this paper, we propose a computationally-supported approach that performs an economic profiling of cities based on their economic features and also a prediction of the future evolution of these economical metrics. Our contributions include (I) a data-ingestion module to extract, transform and load data, (II) a profiling module that achieves an unsupervised classification via a new distance-based cellular genetic algorithm and K-means and (III) a prediction unit based on long short-term memory artificial neural networks. Our proposal is tested on Spain, analysing all its 52 cities, where we use 33 types of real-world economic data that have been recorded monthly for fifteen years. All data has been obtained from the Spanish National Institute of Statistics. Our experiments show that the 52 cities could be clustered into only three economic profiles. This decrease in the complexity of entities to be considered allows managers at several levels and countries to take faster and more accurate decisions by dealing with few profiles rather than treating each city apart. Also, we found that each profile contains repetitive similarity patterns that are not only determined by economics but also indirectly ruled by the cities' geo and demographic situations. Results also showed our prototype's promising economic predictions.

Keywords: Economy · Machine learning · Metaheuristics

1 Introduction

Economy is a key domain that can be affected by numerous unpredictable factors requiring rapid and adequate measures, whereas the world's fast evolution increases the complexity of economical decisions (e.g. budget allowance). Thus, to take such decisions, the use of exhaustive and empirical methods is

© Springer Nature Switzerland AG 2022
J. L. Jiménez Laredo et al. (Eds.): EvoApplications 2022, LNCS 13224, pp. 567–583, 2022.
https://doi.org/10.1007/978-3-031-02462-7_36

becoming inefficient, while Artificial Intelligence (AI) appears as a promising alternative [1], especially that nowadays, valuable, raw and diverse information is continuously recorded. The question is: *how AI and raw data can empower computationally-affordable AI-assisted decision-making systems in economy?*

As a tentative answer to this question, we make a realistic assumption that the decision-taking complexity in economy comes principally from two problems: (I) the number of entities and constraints to be considered (e.g. companies, budget, etc.) and (II) the long-term planning. Therefore, as a solution to the first issue, we believe that reducing the complexity of one (or both) of entities/constraints will ease making faster appropriate decisions. To do so and considering that the "sparse, raw and non-annotated/unlabeled" nature of economic data hinders the use of supervised and semi-supervised learning, the unsupervised one through K-means clustering appears as a reputed computationally-cheap but still efficient solution. The K-means' initial centroids quality is one of the main factors that determines its efficiency [11]. Such task can be modeled as an optimisation problem, where Evolutionary Computation (EC) has already proven its solving efficiency for suchlike problems [3]. Moving to the second problem, we think that providing relevant predictions can help managers take more accurate decisions. Bearing in mind this, Long Short-Term Memory Artificial Neural Networks (LSTM ANN) have already proved their efficacy [6].

Our contribution consists in proposing a prototype of a nation-wise AI-assisted profiling and prediction approach based on both unsupervised and supervised Machine Learning (ML). The unsupervised part is done via K-means, where initial centroids are selected using a newly-proposed Distance-based cellular Genetic Algorithm (DcGA) that considers the clustering properties. Our proposal creates groups of one country's cities having each similar economic features (e.g. employment situation) [12]. On the other hand, the supervised ML unit consists of an LSTM ANN. This will help managers to make adequate long-run decisions by considering a few economic profiles rather than treating each city apart. We propose an entire profiling prototype, so our contributions range from an Extract-Transform-Load mechanism (ETL), including data acquisition, pre- and post-processing (e.g. feature selection) to the EC-ML process that results in an economic profiling and prediction via ML. Our approach has been tested on Spain, studying all its 52 cities, where each city has data recorded monthly for a period of 15 years (2003–2017). The economic profiling is made using 33 types of real-world economic information obtained from the Spanish National Institute of Statistics (Instituto Nacional de Estadística INE) (https://www.ine.es/). When going through the literature of ML in economics (e.g. [8]), we believe that our work is the first to combine these functionalities together in such a way.

The remainder of the paper is structured as follows. Section 2 introduce our approach, while Sect. 3 presents our experiments. Finally, Sect. 4 concludes the paper.

2   The Proposed Approach

Our work presents both *methodological* and *practical* contributions. First, we have performed several experiments to design each component of both our profiling and prediction units (e.g. feature selection, data series reduction, the hybrid clustering algorithms, etc.). Second, we present here the practical implementation of our proposal, including data acquisition, data formatting and pre-processing, data clustering pre-processing, profiling and prediction (see Fig. 1).

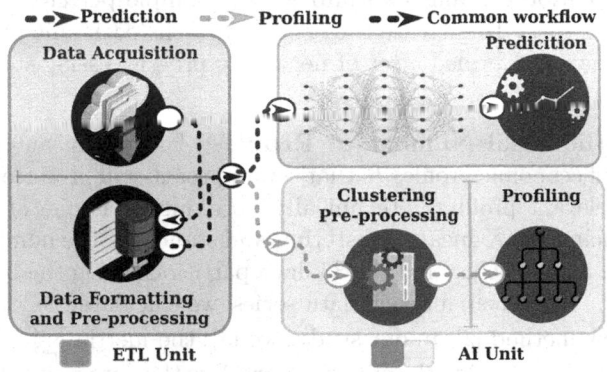

Fig. 1. The proposed economics-tailored approach

2.1   Data Acquisition, Formatting and Pre-processing

We take the ETL architecture given in [7] as a reference, although we neglect its in-depth technicality considering that our ETL purpose, architecture and data size are different. We focus on I) extracting raw data, II) making the data usable (i.e. cleaning and formatting), and III) loading the data to the database. To ensure our proposal's applicability to other datasets, we used real-world economic data obtained from INE, which is an official Spanish organism in charge of national and EU statistics-related duties. The manual downloading of INE data is not suitable for us since getting and updating data is time-consuming. This hinders automating the whole approach and prevents non-experts from using it. Therefore, we have implemented our own code that uses the open INE API (available at: https://www.ine.es/), to automatically download data.

Some INE data/attributes can mislead the profiling or the prediction. Thus, data formatting is a key process where we organised our data as a relational database with *schema* (id: int, location_name: varchar(x), serie_name: varchar(x), year: int, period: int, value: float), and *functional dependencies* {id $\longrightarrow$ location_name, id $\longrightarrow$ serie_name, id $\longrightarrow$ year, id $\longrightarrow$ period, id $\longrightarrow$ value}. Our approach obtained 30,000 data series from which we identified 50 relevant ones. We found that: I) some data was not available for some/all cities, II) the period's range of data recording, and III) the recording's starting date differ between cities. Thus, we updated our proposal to automatically remove these data inconsistencies. We obtained an homogeneous dataset of 52 Spanish cities, where each city has 33 types of economic data series concerning employment, goods, etc. that were recorded monthly from 2003 to 2017. All our intermediary and final data are available at [4].

2.2   Data-Clustering Pre-processing

The data we use is organised in 33 partitions (one partition per data series), where each partition contains 52 samples (one sample per city) with a size of 15 * 12 features each. To increase the accuracy of the economic profiling, in the following, we have performed a set of necessary pre-processing steps.

Identifying the Ideal Number of Profiles: Extracting automatically the ideal number of economic profiles K that our approach will produce is paramount to ensure a relevant profiling. Technically speaking, it is one of the principal factors influencing the K-means' clustering and stands for the number of clusters K used. Thus, first, we clustered the 33 data partitions using the DcGA-Kmeans for $K = \{1, \ldots, 10\}$. Then, for each data series, we calculate the ideal value of K using the elbow method [2]. In our study, we use the inertia $\sum_{i=1}^{K} \sum_{X \in C_i} \|X - m_i\|^2$ as a metric for selecting the elbow, where X is the data point that belongs to a given cluster C_i and m_i is the centroid of the ith cluster. Finally, we consider the median of the 33 elbows as the number of clusters K to be used in our proposal. It is worth stating that more complex strategies could be used for defining K (e.g. assigning a dynamic K value). But, we used the above-introduced technique to ensure that our proposal is computationally affordable and data-independent.

Features' Reduction: Each of the 33 data partitions contains 52 samples of 180 features size each (see Fig. 2). Using samples with such a size can mislead average-based techniques such as the K-means that is sensitive to outliers. Considering these facts and to open new perspectives in feature reduction techniques, we devised a heuristic to reduce the features that will be considered during the clustering. To do so, for each of the 33 data partitions, we perform the clustering using the value K found in the previous paragraph. Then, we perform the same clustering by replacing each Z consecutive features (i.e. Z consecutive months within the same year) by their mean value, where $Z = \{2, 3, 4, 6, 12\}$. So, we will perform clusterings on data samples of sizes 90, 60, 45, 30 and 15 features instead of 180. After, for each of the 33 partitions, we compute the distance between the original clustering done using 180 features and those achieved using 90, 60, 45, 30 and 15 features. Thus, for each of the 33 data series, we will obtain 5 distances.

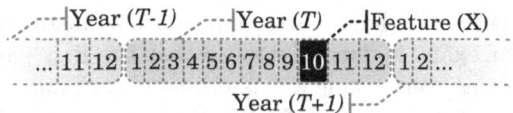

Fig. 2. Illustration of features sample

Since the distance calculation method is the same regardless to the number of features, let us consider, as an example, the distance between a clustering using 180 features and the one using 90 features. Assuming that, from previous step, both clusterings using 180 or 90 features will result in K clusters each, say the two sets $S_1 = \{A_1, \ldots, A_K\}$ and $S_2 = \{B_1, \ldots, B_K\}$, respectively. In this case, we will have $K!$ configurations of the following form· $\{\{A_1, B_u\}, \ldots, \{A_K, B_q\}\}$, where $u, \ldots, q$ are one of the $K!$ possible permutations without repetitions of K digits. For each one of the $K!$ configurations, we sum-up the distances between each pair of clusters $\{A, B\}$ (see Eq. (1)) included in that configuration. Then, the distance between the clustering using 180 features and the one using 90 features will be the minimum accumulative distances among all the $K!$ computed ones (one accumulative distance per configuration). The whole process is repeated for all the 33 data partitions. Once done, we select the number F of features (90, 60, 45, 30 or 15) that is statistically producing clusterings that are the closest (i.e. having the lowest distance) to the original clustering using 180 features and the 33 data series (see Algorithm 1).

$$\texttt{distance}(A, B) = \texttt{cardinality}((A \cup B) - (A \cap B)) \tag{1}$$

One should bear in mind that during the feature reduction step, we make the hypothesis that the original clustering produced using 180 features is *noise-free*. Thus, our feature reduction technique is designed, on purpose, to select the features that produce the closest clustering to the one with 180 features.

Algorithm 1. Feature reduction heuristic

---

for each of the 33 data series **do**
    for $Z \in \{2, 3, 4, 6, 12\}$ **do**
        Compute the distance between clusterings using 180 and $(180/Z)$ features.
    end for
    Extract the Z producing the smallest distance to the clustering using 180 features.
end for
Return Z that is statistically greater.

Data Series Reduction: Among the 33 economic data series, it is possible that more than two series result in the same useless clustering (i.e. economic profiles) which is computationally-penalising for our approach. Thus, we devised a heuristic to identify data series that provide different profiling from one another (see Algorithm 2). We start by forming $(\frac{33!}{(33-2)!}/2)$ pairs using the 33 data series, then, for each data series of those pairs, we perform a clustering using K clusters and F features. For example, let us suppose that the first pair contains the 1st and the 2nd data series. Performing a clustering on both data series will result in K clusters each, say $S_1 = \{A_1, \ldots, A_K\}$ and $S_2 = \{B_1, \ldots, B_K\}$, respectively. The same process is repeated for the rest of the pairs. After, for each of the pairs, we compute the distance between their two clusterings S_x and S_y, where x and $y \in \{1, \ldots, 33\}$ are the indices of the series that constitutes the processed pair. The distance calculation is made as explained in Eq. (1). Once this has been achieved, we extract the list of Series considered Similar (SS). Data series of a given pair are said to be similar if the distance between their respective clusterings S_x and $S_y \leqslant \epsilon$. Considering our offline experiments, we have set ϵ to 26 since we admit that two clusterings are considered similar if they classify differently (i.e. in different clusters), at most, 25% of the cities (i.e. 13 cities). Knowing that each city that is classified differently in two clusterings will result in a distance of 2, in our case, 13 cities will generate a distance of 26.

Algorithm 2. Data series reduction heuristic

Form ($\frac{33!}{(33-2)!}$ / 2) pairs of data series.
Compute the distance between the clusterings in each pair.
Extract, to SS, all the pairs that are considered similar.
Save, in MS, the data series that are not included in one of the extracted pairs.
repeat
    Rank the data series in the extracted pairs.
    Save, in MS, the data series with the highest occurrence and delete all the pairs in SS in which it is included.
until $\{$SS $= \emptyset\}$
Return MS.

Once this has been done, we save, in a list MS (Maintained Series), the data series not included in SS. Then, we rank the data series according to the number of times they are included in one of the pairs of SS. After, we move the data series with the highest occurrence to the MS list, and we delete all the pairs that it is part of in the SS. We repeat this process until SS becomes empty. At the end of this process, the list MS is the list of series that will be used.

2.3   Economic Profiling via Data Clustering

This section describes our proposal's AI module, where we perform a clustering using the K clusters, F features and MS's data series obtained in Sect. 2.2. Once done, for each of the series in MS, we will obtain K clusters, where each cluster represents a subset of the 52 Spanish cities that have similar economic features, or what we call here an "*economic profile*". The clustering is done using our newly-proposed approach called DcGA-Kmeans. It is a two-phased technique that first computes the initial clusters' centroids by solving the clustering as an optimisation problem defined by Eq. (2). Then, using these centroids, the K-means performs the clustering. In the following, we introduce our proposal starting by the K-means then the DcGA (see Algorithm 3).

Algorithm 3. Economic profiling heuristic

---

for all economic data series **do**

  %% **Select K initial Centroids via DcGA** %%

  Random initialisation of the population.

  while maximum number of iterations not reached **do**

    for each grid's node **do**

      Selection via binary-tournament on the Von Neumann neighbourhood.

      Breeding via DPX1 crossover and distance-based mutation.

      Evaluation and synchronous replacement.

    end for

  end while

  %% **Perform Clustering via K-means** %%

  repeat

    Affect each data point to its closest centroid obtained using the DcGA.

    Update the centroids.

  until {No change of data points after centroid update}

  Return the list of cities in each cluster (i.e. profile).

end for

The K-Means Algorithm. The K-means was ranked the 2nd of top-10 data mining algorithms considering that it has a relative simplicity ($\mathcal{O}(NKFT)$), promising efficiency and applicability to several types of data [11]. Still, other advanced clustering techniques exist, but our goal is to demonstrate our proposal's operability even using simple ones. The K-means creates non-overlapping data clusters where each group's intra-similarity is maximised, while all groups' inter-similarity is minimised. Assuming $D = \{X_1, \ldots, X_N\}$ is a set of data points to be clustered, the clustering problem can be formulated using Eq. (2), where W_X is the weight of data point X, N_i is the number of points assigned to the ith cluster C_i, K is the number of clusters set by the user. First, the K-means selects K initial cluster centroids, then each data point is assigned to its closest centroid ($\sum_{X \in C_i} \frac{W_X X}{N_i}$) in terms of squared Euclidean distance. Each data points' collection will constitute a cluster. Later, each centroid is updated by considering the points that have been assigned to it. This process is repeated until no data point changes its cluster. One should keep in mind that several metrics exist for expressing the clustering distance, being the Euclidean one the most widely-used, including in our work, despite its weaknesses [11]. Our aim is to show that a K-means with simple settings can still provide a meaningful profiling.

$$\text{Min} \sum_{i=1}^{K} \sum_{X \in C_i} W_X \|X - \sum_{X \in C_i} \frac{W_X X}{N_i}\|^2 \tag{2}$$

The Distance-Based Cellular Genetic Algorithm. The solution to the clustering problem defined in Eq. (2) is proven to be NP-hard, which motivates the use of heuristic-based techniques to solve it [9]. Cellular genetic algorithms are simple, computationally-affordable and efficient solvers [10]. They use a grid-shaped population (e.g. 2D), where each grid's node is a possible solution to the problem being solved. The interactions between individuals are restricted to a certain neighbourhood (e.g. Von Neumann) (see Fig. 3(b)). The cGA evolves its individuals towards fitter states by applying a series of operators until a stop criterion is reached. Although it is hard to draw the basic cGA's feature since many implementations exist [3], but generally, for each grid's node, the cGA starts by performing I) a selection on the neighbourhood of the processed individual, II) crossover and mutation, III) the produced individual(s) is(are) evaluated using the fitness function of the problem being solved and finally, IV) a replacement decides on the population's components for the next iteration.

Our DcGA is based on the cGA. Each individual, say $\overrightarrow{X}$, is integer-coded and represents a possible configuration of clusters' ID for each city: $\overrightarrow{X} = \{x_1, \ldots, x_N\}$, where N is the number of cities to be clustered and $x_i \in \{1, \ldots, K\}$, where K is the number of clusters to be formed. Figure 3(b) represents a DcGA's individual configuration, where 5 cities are clustered in 4 clusters: the 1st, 2nd, 3rd, both 4th and 5th cities belong to the 4th, 1st, 3rd and 2nd clusters, respectively.

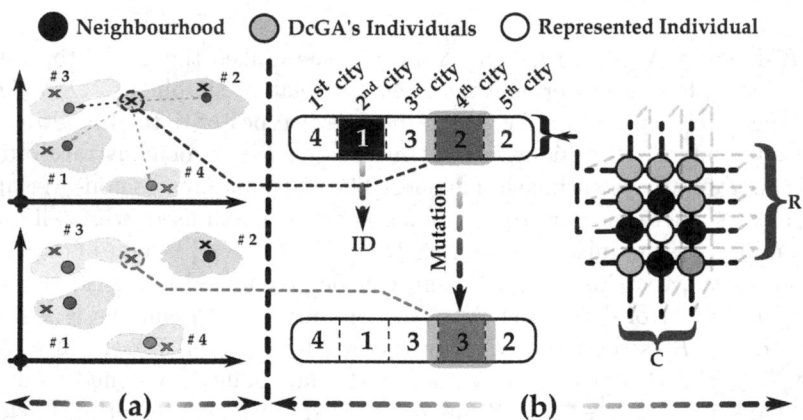

Fig. 3. DcGA's: (a) distance-based mutation and (b) solution representation

The DcGA starts by randomly initialising a grid of R×C individuals. For each individual, the DcGA selects another via a binary tournament on the Von Neumann neighbourhood of the processed node. Then, according to a probability P_c, a well-established variant of two-point crossover called DPX1 [3] is applied on the processed individual and the selected one. The DPX1 results in two new offspring, where the closest one (in terms of sum of absolute value of subtractions) to the best parent is kept. Figure 4 is an example of our implementation of DPX1, where Switch 1 and 2 represent the limits of the alleles' being exchanged between parents.

Fig. 4. The DPX1 crossover

After this, for each element of the produced offspring, our new distance-based mutation is applied according to a probability P_m. The new value of a mutated element is ruled by a dynamic probability P_d which is correlated to the widely-used Euclidean distance between the city's features and the clusters' centroids [11]. The smaller the distance the greater the probability to mutate the city cluster's ID to the one it is having a smaller distance with. Figure 3(a) illustrates how the distance-based mutation works where it represents a city that belongs to the 2nd cluster and is mutated to the 3rd one since it has a closer proximity with it. The mutated offspring is evaluated using the *fitness function* defined by Eq. (2) and compared to the processed individual. If it is better, it will be placed in an auxiliary population, otherwise it will be discarded. Once the synchronous replacement done, the process is performed all over again on the following individual in the grid. The whole process is repeated until all the grid's nodes are browsed. Once done, this marks the end of one DcGA iteration. The DcGA is executed again until the maximum number of iterations is reached.

It is to bear in mind that the new DcGA we propose here has been achieved after several systematic experiments studying several techniques including three types of mutation operators: distance-based (the one we propose here), opposition-based and random mutation. In our work, we emphasise our study on the mutation since previous works (e.g. [5]) have showed its major influence in clustering problems. Four metrics have been used during the comparison: I) sum of the distances between each cluster's samples and its centroid, II) sum of the distances between each cluster's centroid and its farthest sample, III) sum of the distances between each cluster's centroid and the nearest sample from the remaining clusters, IV) sum of the distances between the clusters' centroids. Considering all four metrics, our proposed distance-based cGA has been found to be the best. In addition, the present DcGA's design is made on purpose to ensure its applicability to other datasets and applications. In the following, we provide details of the mutation operators and metrics being studied.

The cGAs' Variants and Clustering Metrics. Our distance-based mutation has been initially compared against random and opposition-based mutations.

The first type of mutation assigns a randomly-chosen cluster ID to a given city when $r \leq P_m$, where r is a random number. The second mutation affects a cluster ID x_i' based on the following formula $(K - x_i + 1)$, where K is the number of profiles to be formed and x_i, x_i' represent the ith city's original and mutated cluster IDs, respectively. Figure 5(a) and (b) illustrate how both mutations act on a configuration with 12 cities that need to be clustered in 4 economic profiles.

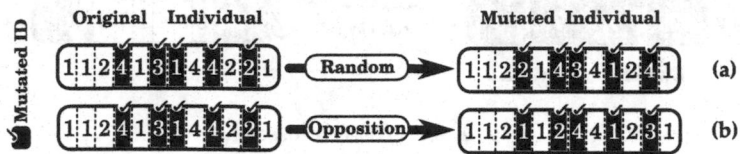

Fig. 5. Mutation: (a) random and (b) opposition-based

To analyse, from different aspects, the quality of the clustering obtained by the three cGAs' variants, we have considered three metrics besides the one described in Eq. (2). The Eqs. (3)–(5) represent, respectively, I) sum of the distances between each cluster's centroid and its farthest sample, II) sum of the distances between the clusters' centroids, III) sum of the distances between each cluster's centroid and the nearest sample from the remaining clusters.

$$\text{Min} \sum_{i=1}^{K} \underset{\forall X \in C_i}{\text{Max}} \, W_X \| X - \sum_{X \in C_i} \frac{W_X X}{N_i} \|^2 \qquad (3)$$

$$\text{Max} \sum_{i=1}^{K} \sum_{j=1, j>i}^{K} \| \sum_{X \in C_i} \frac{W_X X}{N_i} - \sum_{X \in C_j} \frac{W_X X}{N_j} \|^2 \qquad (4)$$

$$\text{Max} \sum_{i=1}^{K} \underset{\forall X \in C_j, \, j=1...K, \, j>i}{\text{Min}} \| W_X X - \sum_{X \in C_i} \frac{W_X X}{N_i} \|^2 \qquad (5)$$

2.4   Economic Prediction

Our prediction module is based on LSTM ANN. Like classical multi-layer perceptron ANN, the LSTM is composed of layers of neurons. The signal propagates through the network to make the prediction. When going into the details, the LSTM are a type of recurrent ANN in which learning signal go in both sideways going in loop through the ANN. The recurrent connections add a memory to the network to harness the ordered nature of the input sequences. Indeed, instead of mapping inputs to outputs, the ANN learns a mapping functions for the inputs over time to an output which unlocks time series for the ANN. Bearing in mind this, the computational unit of an LSTM ANN is called a memory cell or block. The former is composed of weights (input, output and internal states) and gates

(Forget, input and output), which are functions that govern the information flow in the cell. Several types of LSTM ANNs exist (e.g. stacked, convolutional, bidirectional, etc.), where each one has its own features. Therefore, it is hard to draw the standard architecture of an LSTM ANN. Nonetheless, Fig. 6 illustrates the blueprint of a classical stacked LSTM ANN that includes several LSTM layers, which provides a sequence of output rather than one [6].

Fig. 6. Architecture of a stacked LSTM ANN

The original data used by our prediction module is a 180×1 value. This corresponds to the information recorded monthly during 15 years (2003–2017). Then, this sample is normalised to a range of $[0,1]$ using the min-max technique. Once this is done, we set our time-series-like data in samples of the form $\{(t - n), \ldots, t\}$, by shifting one data series value each time. This produces a 176×5 prediction value. We use 70% of the original data (123×5 values) for training our model and the remaining 30% (53×5 values) as testing data. Technically speaking, our model trains using the values recorded from January 2003 to July 2013, while our tests are done on the data recorded from August 2013 to December 2017. Our implementation of LSTM is composed of three layers of 32, 24 and 12 neurons. As a machine learning algorithm, we use Adam version of stochastic gradient descent and the mean squared error as a loss function. We train our model for 50 epochs using a batch size equal to 32.

3 Experimental Results and Analysis

In this section, we present the results obtained by both the profiling and prediction modules of our proposal.

3.1 The *-CGAs' Variants Comparison

Based on offline tuning, the three cGA's variants have been run with a grid of 40×10 nodes, $P_c = 0.5$ and $P_m = 0.2$. The experiments have been repeated over 30 executions, where each has a stopping criterion of 50 iterations. The latter has been chosen since beyond it, we found that the algorithm(s) evolution decreases (or stagnates). Table 1 presents a summary results obtained when comparing the cGA's variants using the distance-based, random and opposition-based mutations. The results consists in how many times a given cGA variant has achieved the best results, in each economic data series, according to the three metrics explained in Sect. 2.3 and by applying a clustering using 1–10 clusters. The ranking takes into account the average of the 30 values obtained in each metric and for each K clusters, where the best scores are highlighted in bold and light-gray. One can note that in 32 out of the 33 data series, our proposed DcGA is the one obtaining the best results. Thus, the former solver is the only one being investigated in the rest of the experiments. For further detailed results, one can access to [4].

Table 1. The scores of cGAs using: distance, opposition and random-based mutations

Series\cGAs	1	2	3	4	5	6	7	8	9	10	11	12	13	14	15	16	17	18	19	20	21	22	23	24	25	26	27	28	29	30	31	32	33
Distance	22	22	22	22	22	23	22	22	23	22	22	22	22	22	22	22	24	22	22	22	22	22	24	25	24	25	22	22	22	22	22	22	22
Opposition	20	8	21	14	8	12	13	12	16	7	14	19	19	19	11	8	8	13	10	13	4	15	13	18	17	17	8	12	11	12	16	20	18
Random	6	18	5	12	18	13	13	14	9	19	12	7	7	7	15	18	16	13	16	13	22	11	11	5	7	6	18	14	15	14	10	6	8

Figure 7 represents the fitness evolution of the cGA's variants when profiling the data series of *men unemployment* using metric n° 1 and 6 profiles. On the other hand, Fig. 8 illustrates the cGAs' fitness evolution when profiling *women unemployment* using metric n° 1 and 10 profiles. The figures have been created using randomly-chosen executions and both show the superiority of our proposal, as well as its continuous and smooth convergence compared to others.

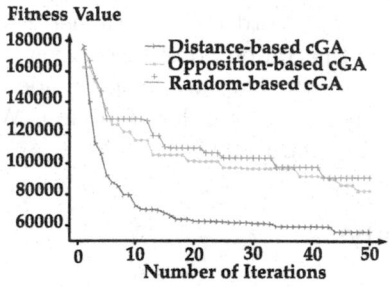

Fig. 7. Men unemployment

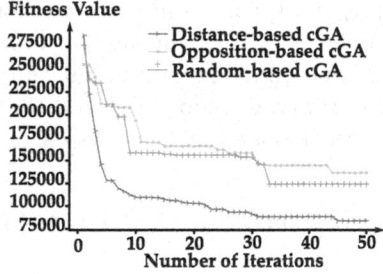

Fig. 8. Women unemployment

It is also worth stating that all our results have been confirmed using both Kruskal-Wallis and a post-hoc tests with a significance level of 0.05.

3.2   Economic Profiling

The implementation and database manipulation have been done using Python 3.8.5 and SQLite 3.27.2, respectively. The two first steps (i.e. data collection, formatting and pre-processing) can be seen directly in the database that we have obtained [4]. Thus, in the following, we will present the results of the remaining phases.

Based on our experiments, we found that using our elbow-based analysis, the ideal number of clusters K to be applied is 3 [4]. Using our feature reduction heuristic [4], we found that when considering 2 months (90 features), in 16 times, it is generating the lowest distance between its clustering and the one obtained using the original 180 features, 3 months (60 features) is producing it 13 times, 4 months (45 features) is giving the lowest distance 7 times, 12 months (15 features) produces it 14 times and the 6 months (30 features) is producing the lowest distance 17 times which is the best one. Thus, we set F to 30 features.

Afterwards, as explained in Sect. 2.2, we reduce the number of economic data series to be considered during the economic profiling. Experiments extracted the following MS list: Activity Percentage of {1} Men and {2} Women. {3} Men Unemployment Percentage. Also, Employment Percentage of {4} Men (I), {5} Women (I), {6} Men (II) and {7} Women (II). Index. of {8} Monthly Variation, {9} Aliments and non alcoholic beverages, {10} Alcoholic beverages and tobacco, {11} Dress and footwear, {12} Health, {13} Transportation, {14} Communications, {15} Leisure and culture, {16} Teaching, {17} Restaurants and Hotels and {18} Other goods and Services. Total companies {19} Without employees, with a workforce {20} From 3 to 5, {21} From 6 to 9, {22} From 10 to 19, {23} From 20 to 49, {24} From 50 to 99, {25} From 100 to 199, {26} From 200 to 499, {27} From 500 to 999, {28} From 1000 to 4999. As it can be seen, the data series reduction has considered only 28 data series instead of the original 33 economic pieces of information that we have selected in Sect. 2.1.

Now, we present the results of the economic profiling described in Sect. 2.3. This has been achieved using $K = 3$ clusters, which corresponds to the number of economic profiles. The number of features has been reduced to 30 features which designates two data values per year. The economic profiling has been done for the 52 Spanish cities and this according to 28 data series. Table 2 presents the results of this phase, where we show in a different colour each economic profile: orange for the 1st profile, yellow for the 2nd one and green for the 3rd. Considering the large amount of data contained in each data series, it is impossible to present further correlations between the original data and the clustering achieved in this section or the centroids coordinates for each data series. For further experimentations on all the presented steps, one can use our profiling approach's code and also our experimentation results available at [4].

As it can be seen in Table 2, our approach could classify the 52 Spanish cities into 3 profiles and this for each of the 28 data series. This substantial decrease in the cases facilitates taking decisions and dealing with abnormal situations (e.g. worldwide pandemic). Indeed, for instance, this allows a smart governance by allowing rulers to establish at most 3 economic strategies (e.g. budget allowance) instead of 52 strategies (one for each city). This shows another keen advantage of our proposal which is that the number of profiles can be set in compliance with the decision-makers' needs and application purposes. Also, even if the profiling is based on economic data, we found that some cities are indirectly clustered considering a common geographical (e.g. overseas cities: Melilla, Ceuta, Santa Cruz de Tenerife), administrative (e.g. Comunidad de Andalucía: Córdoba, Cádiz, Sevilla, Málaga, etc.) or demographical (e.g. Madrid and Barcelona) situation (Fig. 9).

Table 2. Results of the economic profiling

City \ Data Series N°	1	2	3	5	6	8	9	10	11	12	13	14	15	16	17	18	19	20	21	23	24	25	26	27	28	29	30	31
A Coruña	1	3	2	2	3	2	2	2	2	1	1	2	1	2	1	3	1	2	2	3	1	3	2	2	3	1	2	2
Albacete	1	3	3	2	3	2	2	3	1	2	1	1	1	1	1	1	3	2	3	1	3	2	3	1	2	3	1	1
Alicante/Alacant	2	3	3	2	3	2	2	1	1	2	1	3	1	3	1	3	3	2	2	3	1	3	2	2	3	1	1	1
Almería	3	1	1	2	3	2	2	3	1	2	2	2	3	1	3	2	2	2	3	1	3	2	3	1	2	3	1	1
Araba/Álava	3	1	2	3	1	3	3	2	2	2	1	2	1	2	2	3	3	1	3	1	3	2	3	1	2	3	1	1
Asturias	1	2	3	1	2	1	1	2	1	1	1	1	1	2	1	3	1	2	2	3	1	3	2	2	3	1	2	2
Ávila	1	2	3	1	2	1	1	3	1	2	2	2	3	2	1	1	3	1	3	1	3	2	3	1	2	3	1	1
Badajoz	2	2	1	1	2	1	1	1	1	2	2	3	3	3	3	2	2	2	3	1	3	2	3	1	2	3	1	1
Barcelona	3	1	3	3	1	3	3	1	2	1	1	2	1	2	2	1	1	1	1	2	2	1	1	3	1	2	3	3
Bizkaia	2	3	2	2	3	2	2	2	2	1	1	2	1	2	2	3	1	1	2	3	1	3	2	2	3	1	2	2
Burgos	2	3	2	3	3	3	2	2	2	1	1	2	1	3	1	1	1	3	2	3	1	3	2	3	1	2	3	1
Cáceres	1	2	3	1	2	1	1	2	1	2	2	1	1	2	3	3	2	2	3	1	3	1	3	1	2	3	1	1
Cádiz	2	3	1	1	2	1	1	1	1	2	2	1	1	3	3	1	3	2	2	3	1	3	2	2	3	1	1	1
Cantabria	2	3	2	2	3	2	2	1	1	1	1	3	1	2	3	3	1	3	1	3	2	3	1	2	3	1	1	1
Castellón/Castelló	3	1	3	2	3	2	2	1	2	2	2	3	3	3	1	2	2	2	3	1	3	2	3	2	3	1	2	1
Ceuta	3	2	3	2	2	2	1	1	3	3	2	1	2	1	1	3	2	3	3	1	3	2	3	1	2	3	1	1
Ciudad Real	2	2	3	2	2	2	1	3	1	2	1	1	1	2	3	2	2	2	3	1	3	2	3	1	2	3	1	1
Córdoba	2	3	1	1	2	1	1	1	1	1	1	2	1	2	1	2	1	2	3	1	3	2	3	1	2	3	1	1
Cuenca	2	2	3	2	2	2	1	3	1	2	2	1	3	1	3	2	3	2	3	1	3	2	3	1	2	3	1	1
Gipuzkoa	2	3	2	3	1	3	3	2	2	1	2	2	1	1	2	3	1	2	2	1	3	3	2	2	3	1	2	1
Girona	3	1	3	3	1	3	3	2	1	2	1	3	2	1	1	1	3	1	3	1	3	2	2	3	1	1	1	1
Granada	2	3	1	1	2	1	1	1	1	2	2	3	1	1	3	2	2	2	1	3	2	3	1	2	3	1	1	1
Guadalajara	3	1	2	3	1	3	3	1	2	1	1	3	2	2	2	2	3	1	3	2	3	1	2	3	1	1	1	1
Huelva	2	3	1	1	2	1	1	1	1	2	2	1	1	2	3	2	2	2	3	1	3	2	3	1	2	3	1	1
Huesca	2	3	2	3	3	3	2	3	2	1	1	2	1	1	3	2	3	1	3	1	3	2	3	1	2	3	1	1
Illes Balears	3	1	3	3	1	3	3	1	1	1	3	2	1	2	3	3	1	2	3	1	3	2	2	3	1	2	2	2
Jaén	2	2	1	1	2	1	1	1	1	1	1	1	1	1	3	2	3	3	1	3	2	3	1	2	3	1	1	1
La Rioja	3	3	2	3	3	3	2	1	1	1	1	1	1	2	2	2	2	1	3	1	3	2	3	1	2	3	1	1
Las Palmas	3	1	1	2	3	2	2	1	3	3	3	1	2	2	3	1	2	3	2	3	1	3	2	2	3	1	2	1
León	1	2	3	1	2	1	1	3	2	2	1	1	3	2	3	1	1	2	3	1	3	2	3	1	2	3	1	1
Lleida	3	1	2	3	1	3	3	2	1	1	1	1	1	2	1	3	1	3	1	3	2	3	1	2	3	1	1	1
Lugo	1	2	2	1	3	1	2	2	1	1	1	1	1	3	2	2	1	3	3	1	3	2	3	1	2	3	1	1
Madrid	3	1	2	3	1	3	3	1	1	2	1	3	1	2	2	2	3	1	1	2	2	1	1	3	1	2	3	3
Málaga	2	3	1	1	2	1	1	1	1	2	2	1	1	1	3	2	3	1	2	3	1	3	2	2	3	1	1	1
Melilla	3	2	3	2	2	2	1	2	3	3	2	3	2	3	1	2	2	2	3	1	3	2	3	1	2	3	1	1
Murcia	3	3	3	3	3	3	2	1	1	1	3	1	2	1	2	2	2	3	1	3	2	2	3	1	1	2	2	2
Navarra	3	1	2	3	1	3	3	2	1	1	1	2	1	3	2	1	2	1	3	1	3	2	3	1	2	2	2	2
Ourense	1	2	3	1	2	1	1	1	1	1	2	1	1	2	1	2	3	3	1	3	2	3	1	2	3	1	1	1
Palencia	2	2	2	2	2	2	1	2	1	2	2	1	3	3	1	1	1	2	3	1	3	2	3	1	2	3	1	1
Pontevedra	2	3	3	2	3	2	2	3	1	1	1	2	1	2	1	3	1	2	2	3	1	3	2	2	3	1	1	1
Salamanca	1	2	2	2	2	2	1	2	1	2	2	1	3	3	1	2	1	3	1	3	2	3	1	2	3	1	1	1
Santa Cruz de Tenerife	3	1	1	2	3	2	2	1	3	3	3	3	2	1	1	3	2	3	2	3	1	3	2	2	3	1	1	1
Segovia	2	3	2	3	3	3	2	2	2	1	1	2	3	2	2	1	1	1	3	1	3	2	3	1	2	3	1	1
Sevilla	3	3	1	2	2	2	1	1	1	2	2	1	1	3	1	2	3	1	2	3	1	3	2	2	3	1	2	2
Soria	2	3	2	3	3	3	2	2	2	1	1	2	3	1	1	2	1	2	3	1	3	1	3	1	2	3	1	1
Tarragona	3	1	3	3	1	3	3	1	1	1	2	2	3	1	3	1	1	1	2	3	1	3	2	3	1	3	3	1
Teruel	1	2	2	2	3	2	2	3	2	2	2	1	2	3	3	1	2	3	1	3	2	3	1	2	3	1	1	1
Toledo	3	3	3	3	2	3	1	3	1	2	2	1	3	1	1	1	2	3	3	1	3	2	3	1	2	3	1	1
Valencia/València	3	1	3	2	3	2	2	1	2	2	1	3	1	3	2	3	1	2	2	3	1	3	2	2	3	1	2	2
Valladolid	2	3	2	3	3	3	2	3	1	2	1	2	3	1	1	2	1	2	3	1	3	2	3	1	2	3	1	1
Zamora	1	2	3	1	2	1	1	2	2	2	1	2	1	1	1	2	3	2	3	1	3	2	3	1	1	1	1	1
Zaragoza	3	3	2	3	3	3	2	1	1	2	1	1	1	1	1	2	3	1	2	3	1	3	2	2	3	1	2	2

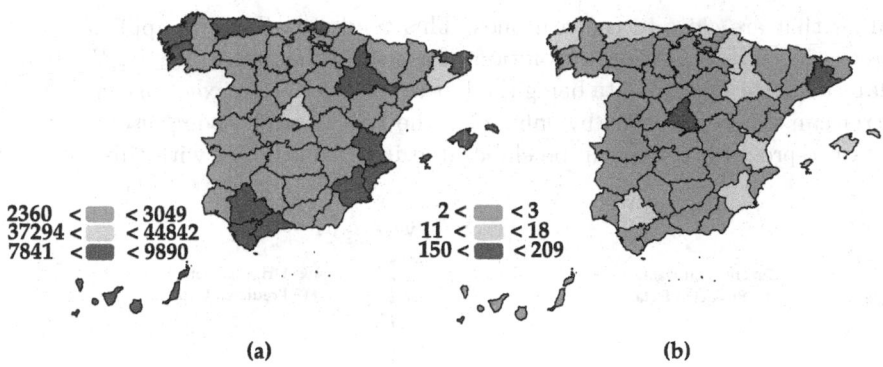

(a)   (b)

Fig. 9. Visual profiling (a) companies with 3–5 and (b) 1000–4999 employees

Figure 10(a) and (b) represent a parallel-coordinates' illustration of the profiles' centroids obtained for two extreme data series representing the companies with **I**) 3 to 5 and **II**) 1000 to 4999 employees, respectively. Data normalisation hs been done using the min-max technique. The features displayed are those selected as described in Sect. 2.2. As it can be seen, our approach could extract distinct centroids, although for hard instances, our approach could provide better profiling by fine-tuning our AI module (e.g. number of runs, iterations, operators, etc.), applying other ML algorithms, etc.

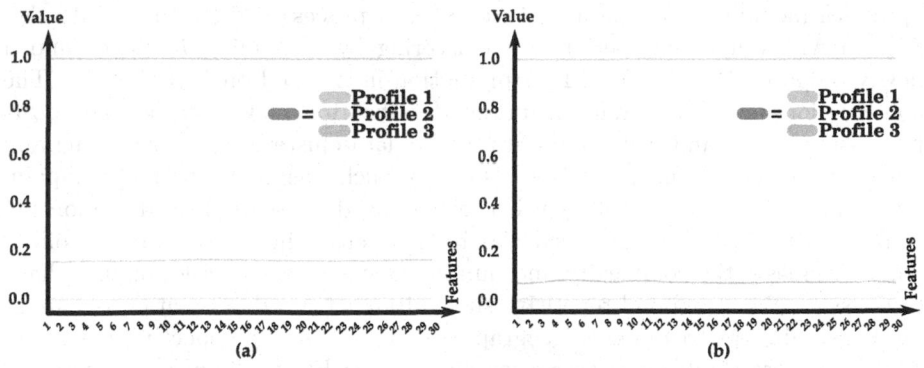

Fig. 10. Parallel coordinates of (a) companies with 3–5 and (b) 1000–4999 employees

3.3   Economic Prediction

Since it is impossible to present the predictions made for all the 52 Spanish cities, we provide here the predictions made for the capital city Madrid for both companies with 3–5 and 1000–4999 employees. Furthermore, one can use our source code to perform further predictions on the rest of the Spanish cities or data series. As it can be seen in Fig. 11(a) and (b), our model is able of predicting values that are close to the real ones. This tends to affirm its applicability for forecasting future unknown evolution of a given economical metric. We believe that due to the type of data being used, it is hard to obtain exact prediction. The latter can still be achieved by enhancing the LSTM model being used. Also, the data pre-processing step can be elaborated in order to cope with this shortfall.

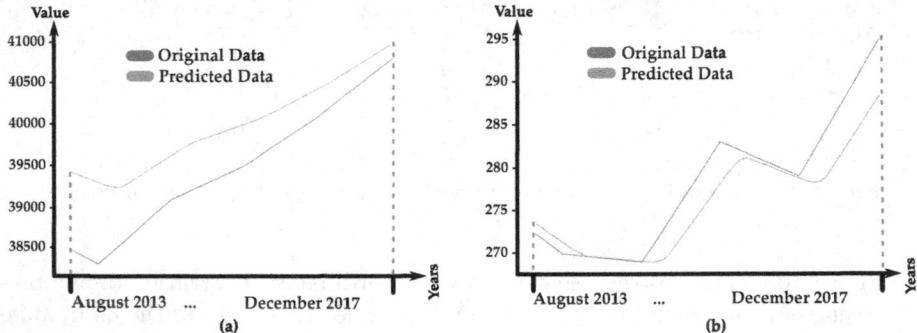

Fig. 11. Prediction of the city of Madrid for companies with (a) 3–5 and (b) 1000–4999 employees

4   Conclusion

We have proposed an automatic nation-wise economic profiling and prediction approach including: I) a data ETL to extract, process and load raw data and II) an AI module that performs a clustering by combining K-means and a newly-proposed DcGA and III) a prediction unit based on LSTM ANN. The contributions of our work range from the identification, collection, pre- and post-processing of new and real-world data to the data-clustering pre-processing and automatic profiling and prediction. Our approach has been applied on Spain, studying all its 52 cities, 33 types of economic data series recorded monthly from 2003 to 2017. Our proposal classified all cities into 3 economic profiles, which decreases the complexity and number of cases that decision-makers have to consider. We also found repetitive similarity patterns due to cities' common-hidden administrative or demo/geographic situation. As a perspective, we intend to design more flexible and user-oriented ML and EC techniques for economic profiling and apply them on larger countries (e.g. China). Also, considering the temporal nature of the economic data, we seek to design a more elaborated profiling by applying multivariate time-series clustering techniques and compare the obtained profiling with the one of classical clustering methods.

Acknowledgements. This work resulted from a stay at the University of Malaga (Spain) using the grant *"stays of researchers with prestigious recognition"* and is also partially funded by the Universidad de Málaga, Consejería de Economía y Conocimiento de la Junta de Andalucía and FEDER under grant number UMA18-FEDERJA-003 (PRECOG); under grant PID 2020-116727RB-I00 (HUmove) funded by MCIN/AEI/10.13039/501100011033; and TAILOR ICT-48 Network (No. 952215) funded by EU Horizon 2020 research and innovation programme. Furthermore, the views expressed are purely those of the writer and may not in any circumstances be regarded as stating an official position of the European Commission.

References

1. Bonaccorso, G.: Mastering Machine Learning Algorithms: Expert Techniques for Implementing Popular Machine Learning Algorithms, Fine-Tuning Your Models, and Understanding How They Work, 2nd edn. Packt Publishing, Birmingham (2020)
2. Chakrabarty, N., Rana, S., Chowdhury, S., Maitra, R.: RBM based joke recommendation system and joke reader segmentation. In: Deka, B., Maji, P., Mitra, S., Bhattacharyya, D.K., Bora, P.K., Pal, S.K. (eds.) PReMI 2019. LNCS, vol. 11942, pp. 229–239. Springer, Cham (2019). https://doi.org/10.1007/978-3-030-34872-4_26
3. Dahi, Z.A., Alba, E.: The grid-to-neighbourhood relationship in cellular GAs: from design to solving complex problems. Soft. Comput. **24**(5), 3569–3589 (2019). https://doi.org/10.1007/s00500-019-04125-w
4. Dahi, Z.A., Luque, G., Alba, E.: Database, source code and results of the proposed machine learning-based approach for economics-tailored applications. https://github.com/Zakaria-Dahi/ML-Economis.git. Accessed 09 Feb 2022
5. El-Shorbagy, M.A., Ayoub, A.Y., Mousa, A.A., El-Desoky, I.M.: An enhanced genetic algorithm with new mutation for cluster analysis. Comput. Stat. **34**(3), 1355–1392 (2019). https://doi.org/10.1007/s00180-019-00871-5
6. Istiake Sunny, M.A., Maswood, M.M.S., Alharbi, A.G.: Deep learning-based stock price prediction using LSTM and bi-directional LSTM model. In: 2020 2nd Novel Intelligent and Leading Emerging Sciences Conference (NILES), pp. 87–92 (2020)
7. Kimball, R., Caserta, J.: The Data Warehouse ETL Toolkit: Practical Techniques for Extracting, Cleaning, Conforming and Delivering Data. Wiley, Hoboken (2011)
8. Ozden, E., Guleryuz, D.: Optimized machine learning algorithms for investigating the relationship between economic development and human capital. Comput. Econ. (2021)
9. Pérez-Ortega, J., Almanza-Ortega, N.N., Vega-Villalobos, A., Pazos-Rangel, R., Zavala-Díaz, C., Martínez-Rebollar, A.: The K-means algorithm evolution. In: Sud, K., Erdogmus, P., Kadry, S. (eds.) Introduction to Data Science and Machine Learning (chap. 5). IntechOpen, Rijeka (2020)
10. Whitley, L.D.: Cellular genetic algorithms. In: Proceedings of the 5th International Conference on Genetic Algorithms, p. 658. Morgan Kaufmann Publishers Inc., San Francisco (1993)
11. Wu, J.: Advances in K-Means Clustering: A Data Mining Thinking. Springer, Heidelberg (2012). https://doi.org/10.1007/978-3-642-29807-3
12. Yamarone, R.: The Trader's Guide to Key Economic Indicators. Bloomberg Financial Series, 3rd edn. Wiley, Hoboken (2012)

Multiobjective Electric Vehicle Charging Station Locations in a City Scale Area: Malaga Study Case

Christian Cintrano$^{(\boxtimes)}$⬡ and Jamal Toutouh⬡

ITIS, University of Málaga, Málaga, Spain
{cintrano,jamal}@lcc.uma.es

Abstract. This article presents a multiobjective variation of the problem of locating electric vehicle charging stations (EVCS) in a city known as the Multiobjective Electric Vehicle Charging Stations Locations (MO-EVCS-L) problem. MO-EVCS-L considers two conflicting objectives: maximizing the quality of service of the charging station network and minimizing the deployment cost when installing different types of charging stations. Two multiobjective metaheuristics are proposed to address MO-EVCS-L: the Non-dominated Sorting Genetic Algorithm, version II (NSGA-II) and the Strength Pareto Evolutionary Algorithm, version 2 (SPEA2). The experimental analysis is performed on a real-world case study defined in Malaga, Spain, and it compares the proposed approaches with a baseline algorithm. Results show that the SPEA2 computes the most competitive solutions, even though both metaheuristics found an accurate set of solutions that provide different trade-offs between the quality of service and the installation costs.

Keywords: Electric vehicles · Infrastructure location · Sustainable mobility · Multiobjective optimization · NSGA-II · SPEA2

1 Introduction

In recent years, sustainability has become a major goal for industry, academia, and society as a whole. Society has steadily moved towards ecological awareness, adapting their lifestyles to promote environmental initiatives, cleaner means of production, and environmentally friendly energy sources. One of the most critical changes in current activities concerns urban mobility, where an effective transition towards inclusive, efficient, and low-carbon means of transport is being experienced [6]. Electric mobility provides citizens with a cleaner and safer way of getting around without gas emissions.

Electric vehicles (i.e., electric cars, scooters, and motorbikes) are powered by an electric motor charged using energy from the electricity grid. They have shown rapid and sudden growth and expansion, as they are preferred mobility options for the younger generation. Consequently, electric vehicles have a relevant socio-economic impact [14]. In addition, electric vehicles have a higher overall efficiency

© Springer Nature Switzerland AG 2022
J. L. Jiménez Laredo et al. (Eds.): EvoApplications 2022, LNCS 13224, pp. 584–600, 2022.
https://doi.org/10.1007/978-3-031-02462-7_37

than combustion engines vehicles. They save (on average) 40% of energy while contributing to reducing gas emissions, when the electricity used to charge the vehicles is obtained from green sources [12].

When considering the deployment of electric vehicles in large cities, a relevant logistical problem arises, similar to other location problems related to public services in the context of smart cities [9,17,22]. One of the most relevant subproblems is the effective and efficient location of charging points for electric vehicles [10]. The main objective of this problem is to provide a good quality of service to citizens while keeping costs reasonable for the city administration.

Different authors have tried to address this siting problem from different perspectives. One of the common ways in the literature is to address the location of charging stations as ILP or MILP problems. The researchers usually used these methods to maximize the economic profits of installing new charging stations [1,2,15], minimizing the total walking distance according to parking patterns estimated using realistic urban data [3], or maximizing coverage to improve demand [11,25]. Some authors have relied on open data to improve the quality of service offered to citizens by taking into consideration the energetic constraints of the area [4,21]. The cited articles work with a mono-objective view of the problem. However, in the real world, the location of charging stations have different objectives.

This article presents a new multiobjective variation of the problem of locating electric vehicle charging stations (EVCS) in a city known as the Multiobjective Electric Vehicle Charging Stations Locations (MO-EVCS-L) problem. Two objectives are considered: maximizing the quality of service of the charging station network to the citizens, and minimizing the deployment and installation cost of the new stations. Both objectives need data about different aspects of a city: locations of neighborhoods, streets, etc., energy data such as types of charging stations or energetic capability of electrical substations, and economic data such as installation costs. To obtain this data, different open data sources were used. This research uses a realistic case study defined in the city of Malaga, Spain.

The main contributions of this article are: a) defining and formulating a new realistic multiobjective problem for locating electric vehicle charging station on a city scale, taking into account the quality of service, power restrictions, and deployment costs; b) proposing two multiobjective metaheuristics to address the proposed problem; c) devising specific evolutionary operators; and d) addressing the problem on a realistic instance defined using real-world data.

The rest of the article is organized as follows: Sect. 2 presents the problem addressed in this study. Section 3 describes the algorithms used in the experimentation as well the operators designed for the problem. The experimental setup and evaluation are reported in Sects. 4 and 5 Finally, Sect. 6 presents the conclusions and formulates the main lines for future work.

2   The Multiobjective Electric Vehicle Charging Stations Location Problem

The problem considered in this paper aims to select the best locations of EVCSs to maximize the quality of service provided to users and simultaneously take into account the infrastructure deployment costs. Different types of EVCS are considered. The kind of EVCSs determines the number of users that can be served per unit of time, the charging time, and the installation costs.

The quality of service is evaluated according to the users that can be served (each EVCS may attend the users that live within a defined *service distance*), the charging time, and the citizens that any charging station does not serve. The deployment cost has two main components: *i)* the infrastructure installation expenses for the required charging equipment and the construction of a new station and *ii)* the cost of connecting the installed station to the power grid.

The two discussed objectives (quality of service and deployment costs) are in conflict, because installing charging stations close to the residences of all tentative clients would require a significant investment, which in turn may not produce in adequate expected revenues for the institutions in charge of the management of the electric vehicle charging system. Thus, in order to assist the decision-makers, the main research outcome of the addressed problem is to provide solutions (i.e., EVCS locations) that properly samples the different trade-offs between these problem objectives.

2.1   Mathematical Formulation

The mathematical formulation of the addressed optimization problem is defined considering the following elements:

- A set $S = \{s_1, \ldots, s_M\}$ of candidate road segments for installing EVCSs. Each road segment s_i can be the location of only one charging station.
- A set $C = \{c_1, \ldots, c_N\}$ of the locations of the tentative users. Nearby locations are grouped in clusters, as usual in the related literature. The number of clients to serve at each cluster c is u_c. The distance from the cluster c to the charging station $s \in S$ is $dc_{c,s}$. A cluster of clients c is served by the charging station located in s if the $dc_{c,s}$ distance is lower or equal to the Ds_s *service distance*, i.e., $dc_{c,s} \leq Ds_s$. $C_s \subseteq C$ represents the set of clusters of clients served by station installed in s, and $NC \subseteq C$ defines the set of clusters not served by any charging station.
- A set $E = \{e_1, \ldots, e_T\}$ of electrical substations that supply the power to the charging stations. Due to the power distribution restrictions, each electrical substation e can serve electricity only to a given subset of of candidate road segments enclosed in a given city area A_e, named electrical substation influence area. In turn, the maximum power allocated for EVCSs, i.e., the electricity distributed by substation e that can be used to feed the electric vehicle charging stations limited by MP_e.

– A set $J = \{j_1, \ldots, j_H\}$ of EVCS types. Each type has its own charging time ct_j, equipment and building cost cc_j, connection to the grid cost cg_j, and required electric power from the electrical substations ep_j. By convention, the model assumes $cc_0 = cg_0 = 0$ and $ct_0 = \infty$ to characterize segments where no charging station is located.

Equations 3–7 describe the proposed optimization model, using the following variables: x_s is and integer variable, $x_s = j_i$ when a charging station of type $j_i \in J$ is installed in segment s, and $x_s = 0$ otherwise; and $y_{e,s}$ is a binary variable, $y_{e,s} = 1$ if the electrical substation e is feeding the charging station located in s and 0 otherwise.

The quality of service provided by the deployed infrastructure is defined in Eq. (3) as the sum of the service provided by each charging station installed in $s \in S$ to the subset C_s of clusters within its service distance minus the number of clients in clusters not served by any charging station NC. NC, defined in Eq. (5), is the complementary set of the set of all clusters served by all charging stations, see Eq. (1). The service provided by the EVCS deployed in s is proportional to the number of citizens in the cluster u_c and inversely proportional to the time required to charge an electric vehicle ct_{x_s}. The quality of service is proposed to be maximized.

$$SC = \bigcup_{s \in S} C_s \quad \forall\, s \in S : x_s \neq 0 \quad (1) \qquad\qquad NC = C \setminus SC \qquad (2)$$

The installation cost of a EVCS considers the sum of the infrastructure cost cc_{x_s} and the cost of connecting the station to its electrical substation, defined in Eq. (4). The budget required to connect the charging station to the electrical substation is proportional to the distance between them $d_{e,s}$ and the cost of wiring cg_{x_s}. The cost is proposed to be minimized.

$$\max \sum_{s \in S} \left(\sum_{c \in C_s} \frac{u_c}{ct_{x_s}} \right) - \sum_{nc \in NC} u_{nc} \qquad (3)$$

$$\min \sum_{c \in C} \sum_{s \in S} (d_{e,s} \cdot cg_{x_s} + cc_{x_s}) \qquad (4)$$

subject to

$$NC = C \setminus SC \qquad\qquad\qquad\qquad\qquad (5)$$

$$d_{c,s} \cdot y_{e,s} \leq De \qquad\qquad \forall\, e \in E,\ s \in S \quad (6)$$

$$\sum_{s \in S} y_{e,s} \cdot ep_{x_s} \leq MP_e \qquad\qquad \forall\, e \in E \qquad (7)$$

Regarding the problem constraints, Eq. (6) imposes that the distance between an electrical substation and any charging station it feds is lower that De, i.e., the charging station in s is in the A_e of the electrical substation e. In turn, the constraint in Eq. (7) guarantees that the total power consumption of all charging stations that are fed by a given electrical substation is lower or equal than MP_e.

3 Algorithms

Two state-of-the-art MOEAs are applied in this study to address MO-EVCS-L *Non-dominated Sorting Genetic Algorithm, version II* (NSGA-II) [8] and *Strength Pareto Evolutionary Algorithm, version 2* (SPEA2) [26]. NSGA-II and SPEA-2 have been successfully applied in many problems in different application areas in smart cities [17,18,20,22,23]. This section presents their main features and describes the specific operators applied in this research.

3.1 NSGA-2

The evolutionary search applied by NSGA-II uses a non-dominated elitist ordering to mitigate the complexity of the dominance check, a crowding technique to keep solutions diversity, and a fitness assignment method that takes into account dominance ranks and crowding distance values. Algorithm 1 presents the pseudo-code of NSGA-II evolving a population P (size N).

Algorithm 1. Pseudo-code of the NSGA-II algorithm

```
 1: t ← 0                                              ▷ generation counter
 2: offspring ← ∅
 3: initialize(P(0))                                  ▷ population initialization
 4: while not stopping_criterion do
 5:     evaluate(P(t))                                ▷ population evaluation
 6:     R ← P(t) ∪ offspring
 7:     fronts ← non-dominated sorting(R))
 8:     P(t+1) ← ∅; i ← 1
 9:     while |P(t+1)| + |fronts(i)| ≤ N do
10:         crowding distance(fronts(i))
11:         P(t+1) ← P(t+1) ∪ fronts(i)
12:         i ← i+1
13:     end while
14:     sorting by distance (fronts(i))
15:     P(t+1) ← P(t+1) ∪ fronts(i)[1:(N - |P(t+1)|)]
16:     selected ← selection(P(t+1))
17:     offspring ← evolutionary operators(selected)
18:     t ← t + 1
19: end while
20: return computed Pareto front
```

3.2 SPEA-2

SPEA2 was an evolution of the SPEA algorithm. SPEA2 is distinct from other MOEAs because it applies the *strength* concept on the fitness computation, which is based on both Pareto dominance and diversity. Thus, the *strength* measures how many solutions dominate (and are dominated by) each candidate solution. In turn, a density estimation is also considered for fitness assignment.

Furthermore, an elite population is defined to store the non-dominated individuals found during the search to apply elitism.

SPEA2 working on a population P (size N) is shown in Algorithm 2. The *elitePop* parameter represents the elite population, which has *eliteSize* size. The most similar individuals are removed by a pruning method to assure that the size of the elite population is always *eliteSize* when the elite population is full.

Algorithm 2. Schema of the SPEA2 algorithm.

1: $t \leftarrow 0$;
2: elitePop $\leftarrow \emptyset$
3: **initialize**($P(0)$)
4: **while** not stopcriterion **do**
5:    **evaluate**($P(t)$)
6:    R $\leftarrow P(t) \cup$ elitePop
7:    **for** $s_i \in$ R **do**
8:       $si_{raw} \leftarrow$ computeRawFitness(s_i,R)
9:       $si_{density} \leftarrow$ computeDensity(s_i,R)
10:      $si_{fitness} \leftarrow si_{raw} + si_{density}$
11:    **end for**
12:    elitePop $\leftarrow$ nonDominated(R)
13:    **if** size(elitePoP) > eliteSize **then**
14:       elitePop $\leftarrow$ removeMostSimilar(elitePop)
15:    **end if**
16:    selected $\leftarrow$ **selection**(R)
17:    offspring $\leftarrow$ **variation operators**(selected)
18:    $t \leftarrow t + 1$
19: **end while**
20: **return** computed Pareto front

3.3   Main Operators

The proposed NSGA-II and SPEA2 for the locating the electric vehicle charging stations include the main following features:

Solution Encoding. Solutions are encoded as a vector of integers in the range $[0,|J|]$. Each position in the vector represent a possible location for the charging station (i.e., indexed by $s_1,...,s_M$), and the corresponding integer value on index s_k represents one of the possible electric vehicle charging type, i.e., $j_i \in J$. The special value '0' is used to represent the situation where no charging station is installed in the segment s_k. Figure 1 presents an example of solution encoding for a sample scenario with eight tentative locations $\{1, ..., 8\}$ and two types of charging stations $\{1, 2\}$.

Initialization. The population is initialized by applying a random procedure that creates feasible solutions. The initialization process iterates over the areas of influence of each electrical power station A_e. For each A_e, it randomly selects

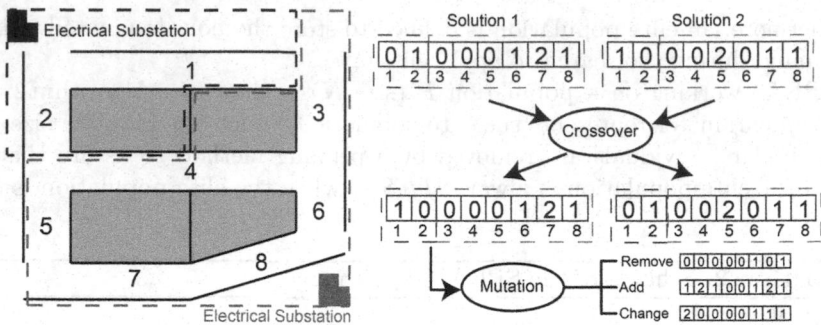

Fig. 1. Example of solution encoding of an scenario with eight possible locations for EVCS and two types of stations, and the evolutionary operators.

a tentative location s_i and adds a randomly chosen EVCS j_k to it. If the power consumption restriction in Eq. 7 is met, it selects another location in A_e to install a new EVCS. This process is repeated while power consumption restriction is fulfilled.

Selection, Replacement, and Fitness Assignment. NSGA-II applies the $(\mu + \lambda)$ evolution model. Tournament selection is applied, with tournament size of two individuals. The tournament criteria is based on dominance, and if the two compared individuals are non-dominated, the selection is made based on crowding distance. Fitness assignment is performed considering Pareto dominance rank and crowding distance values.

Evolutionary Operators. The recombination operator applied is a variation of the standard n-point crossover applied over two selected individuals specifically devised to address MO-EVCS-L named k electrical substation influence area crossover (k-A_eX). That randomly exchange the deployment configuration of the electrical areas of influence of k power stations (A_e defined in Sect. 2). This operator works as follows: given two parents (individuals), k-A_eX randomly selects k A_e and exchanges between parents the information of the selected A_e to create two offspring (see Fig. 1). The mutation operator is based on randomly modifying specific attribute of a randomly selected segment in the A_e of a given individual (i.e., x_s). There are three different potential changes shown in Fig. 1: *i)* if there is a charging station (i.e., $x_s \neq 0$), the charging station can either *removed*; *ii)* if there is no any charging station (i.e., $x_s = 0$), a randomly chosen charging station is selected to *add* it in the represented segment(i.e., x_s is replaced by an integer value uniformly selected in the range $[0, Z-1]$); and *iii)* the values of two different attributes x_s and $x_{s'}$ are *changed* with each other regardless of their values. Te recombination and mutation operators are applied with probability p_C and p_M, respectively. Figure 1 represents both evolutionary operators.

Solution Feasibility. The restriction defined in Eq. 7 may not be met after the application of evolutionary operators, i.e., the total power consumption of all charging stations in A_e could be higher than MP_e. Thus an operator is applied to randomly remove charging stations installed in A_e until the restriction is fulfilled.

4   Experimental Setup

This section summarizes the methodology applied for the experimental analysis of the proposed MOEAs to address MO-EVCS-L.

4.1   Problem Instance

The experimentation is performed over a realistic scenario defined on the city of Malaga, Spain. Around 567,953 citizens spread over 363 neighborhoods live in this city. The road map is composed by 33,550 road segments. Each road can be selected for the placement of a EVCS (see Fig. 2). The city's electric power is supplied by 14 electrical substations. These electrical substations limits the number of stations that we can install in a specific area. When we install a EVCS we can choose between different types of stations according to the different charging speed. In this article, we consider two different types of charging stations to be installed: fast charging stations (type 1) and super-fast charging stations (type 2). Each one have different energy consumption requirements, installation (equipment/building and connection) costs, and also times for fully charging a standard electric vehicle. Table 1 summarizes the main characteristics of both electric vehicle charging station types.

Table 1. Main features of the considered charging stations.

Type (j)	ct_j	ep_j	cc_j	cg_j
Fast (1)	120 min	7.4 kW	13,915€	1.15€
Super-fast (2)	15 min	50.0 kW	39,930€	1.35€

All the information about the city is obtained by open data sources as Open Street Maps [19] or the electrical company itself. With this scenario we test our algorithms in a real urban area.

4.2   Evaluated Metrics

Three relevant multiobjective optimization metrics were considered for results evaluation: generational distance (GD), inverted generational distance (IGD), and relative hypervolume.

GD measures the average distance from the solutions computed by the MOEA to their closest solution in the *Pareto-front* [24]. Let us assume the points found by the MOEA are the objective vector set $Sol = \{s_1, s_2, ..., s_{|Sol|}\}$

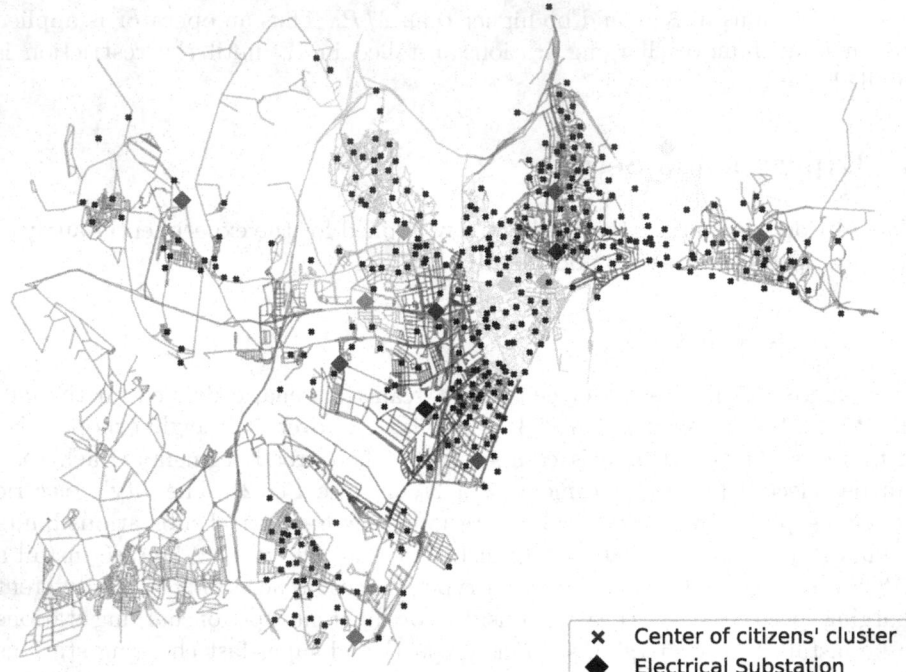

Fig. 2. Citizen' clusters, electrical substations, and road map of Malaga, Spain. Each edge represent a street segment associated with a substation.

and the reference points set (*Pareto-front*) is $P = \{p_1, p_2, ..., p_{|P|}\}$. Then, the GD is computed according to Eq. (8), where d_i represents the distance from s_i to its nearest reference point in P (when $n = 2$ the Euclidian distance is used). In turn, we evaluated the generational distance plus (GD$^+$) proposed by Ishibushi et al. [13]. GD$^+$ is evaluated according to Eq. (9), where for minimization problems the modified distance between s_i and the nearest point in P is computed as $d_i^+ = \max\{s_i - p_i, 0\}$.

$$GD(Sol) = \frac{1}{|Sol|} \cdot \left(\sum_{i=1}^{|Sol|} d_i^n\right)^{\frac{1}{n}} \quad (8) \qquad GD^+(Sol) = \frac{1}{|Sol|} \cdot \left(\sum_{i=1}^{|Sol|} d_i^{+^2}\right)^{\frac{1}{2}} \quad (9)$$

IGD performance indicator inverts the GD and measures the distance from any point in P to the closest point in *Sol* [5]. Equation (10) presents the IGD computation, where $\widehat{d_i}$ represents the distance from p_i to the closest reference solution in *Sol*. Besides, the inverted generational distance plus (IGD$^+$) [13] was considered in the experimental analysis. The IGD$^+$ performance metric is weakly Pareto compliant wheres the original IGD is not. The IGD$^+$ metric es computed as it is shown in Eq. (11), where for minimization problems the modified distance between p_i and the nearest reference point in Sol is computed as $d_i^+ = \max\{s_i - p_i, 0\}$.

$$IGD(Sol) = \frac{1}{|P|} \cdot \left(\sum_{i=1}^{|P|} \hat{d_i}^n\right)^{\frac{1}{n}} \quad (10) \qquad IGD^+(Sol) = \frac{1}{|P|} \cdot \left(\sum_{i=1}^{|P|} \hat{d_i}^{+2}\right)^{\frac{1}{2}} \quad (11)$$

Hypervolume (HV) is the most popular quality indicators to evaluate MOEAs. The HV value of Sol is the volume of the area that is dominated by objective vectors in Sol and bounded by the reference point q as it is shown in Eq. (12), where the function *volume* is the Lebesgue measure [27]. A large HV value indicates that Sol approximates the Pareto front well in terms of both convergence and diversity. The RHV metric is computed as the relative value of HV to the maximal hypervolume of the *Pareto front*.

$$HV(Sol) = volume \left(\bigcup_{s \in Sol} [s_1, q_1] \times ... \times [s_n, q_n]\right) \quad (12)$$

In turn, the solutions provided are evaluated in terms of split ted quality of service. Thus, two metrics are defined: a) sum of service provided by each station, defined in Eq. 13, and b) sum disconnected users, defined in Eq. 14, which represents the number of inhabitants not served by any charging station.

$$QoS = \sum_{s \in S} \left(\sum_{c \in C_s} \frac{u_c}{ct_{x_s}}\right) \quad (13) \qquad\qquad du = \sum_{nc \in NC} u_{nc} \quad (14)$$

4.3   Parameter Settings and Execution Platform

A set of parametric setting experiments were performed to determine the best parameter values for the proposed MOEAs. The parameter setting analysis were made over the proposed scenario. Both MOEAs apply the same initialization, crossover, and mutation operators. The population size ($\#p$) and the maximum number of generations ($\#g$) were calibrated in preliminary experiments. The analysis confirmed that using $\#p = 20$ and $\#g = 500$ provided a good exploration pattern for both MOEAs. In SPEA-2, the size of the elite population was set to 5 individuals, following rules-of-thumb from the related literature [26].

For p_C and p_M, candidate values were $p_C \in \{0.5, 0.7, 0.9\}$ and $p_M \in \{1/14, 4/14, 7/14\}$ (we have 14 zones in or scenario). Each configuration was evaluated over 30 independent executions performed for the proposed MOEAs. The distribution of the relative hypervolume results obtained using each configuration were analyzed by applying the non-parametric Kruskal-Wallis statistical test to determine the configuration that allowed computing the best results. Thus, for NSGA-II, the most competitive results were achieved with $p_C = 0.5$ and $p_M = 1/14$, and for SPEA2, with $p_C = 0.5$ and $p_M = 4/14$.

We perform 30 independent executions of each algorithm. Each one was run on a machine with a Intel Xeon Gold 6240R with two processors, 48 cores at 2.40 GHz and 220 GB. The cluster was managed by HTCondor 8.2.7, which allowed us to perform parallel independent executions to reduce the overall experimentation time. The algorithms are written in python 3.8 using the library DEAP [7] and the source code is available at https://github.com/cintrano/EV-CSL/releases/tag/EvoApps2022.

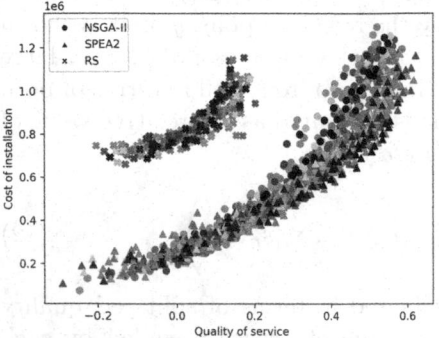
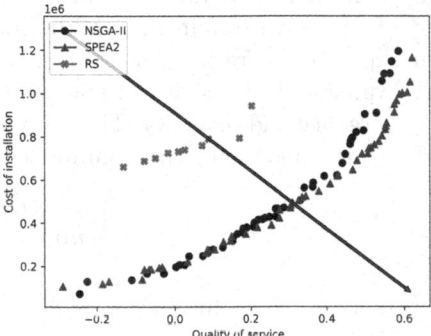

Fig. 3. All computed solutions. (Color figure online)

Fig. 4. Pareto fronts.

4.4   Baseline Method

In order to test the effectiveness of our algorithms, an intelligent Random Search (RS) method to get a baseline of solutions is defined. RS generates feasible solutions using the same constructive method applied to generate the initial population in NSGA-II and SPEA2. The method keeps all non-dominated solutions generated during the process. RS iterates generating new solutions until a stop criteria is reached. In this case, it stops after ruining the maximum execution time of the two MOEAs analyzed here.

5   Experimental Evaluation

This section reports the experimental analysis of the proposed MOEAs to address the real-world case study of MO-EVCS-L.

5.1   Multiobjective Optimization Analysis

Figure 3 shows the non-dominated solutions computed by each independent execution of the evaluated algorithms. The different marker colors indicate the each independent executions. In turn, Fig. 4 illustrates the three Pareto fronts computed by NSGA-II, SPEA2, and RS, i.e., all non-dominated solutions computed considering all the 30 independent executions performed by each method. The arrow indicates the direction of the *best* solutions.

Figure 3 indicates that RS is the least competitive algorithm. The RS set of solutions represent charging covering shorter ranges of quality of service and deployment costs than the MOEAs. All solutions computed by NSGA-II and SPEA2 dominate the RS solutions, i.e., the RS solutions provide less quality of service while requiring higher deployment costs. In turn, NSGA-II and SPEA2 show robustness because the average dispersion of solutions for the same value of each problem objective was below 20% of each one of the independent runs.

Results in Fig. 4 show that the MOEAs are able to compute accurate solutions, properly sample the Pareto front of the problem, and demonstrate the practical applicability of the proposed approach. For deployment costs lower than 0.5×10^6, both methods present solutions with close trade-offs between quality of service and deployment costs. However, for higher installation costs, SPEA2 is able to improve over the solutions computed by NSGA-II, i.e., SPEA2 solutions are able to provide a better quality of service at the same installation costs.

Regarding multiobjective optimization metrics, Table 2 reports relevant statistical values of the evaluated multiobjective metrics (i.e., minimum, maximum, mean, standard deviation, median and IQR) for the evaluated MOEAs.

Table 2. Statistics of multiobjective metrics for the executions of each algorithm.

Algorithm	Metric	Minimum	Mean ± Std	Median	iqr	Maximum
NSGA-II	RHV	0.74	0.82 ± 0.03	0.83	0.05	0.88
	GD	6028.16	9777.69 ± 2569.45	9174.03	2487.18	17151.89
	GD$^+$	0.07	1853.90 ± 2396.43	986.55	3165.70	9132.95
	IGD	18213.75	29024.71 ± 8099.82	27602.76	7603.97	57134.82
	IGD$^+$	0.17	13727.04 ± 9850.17	12702.03	7333.46	44933.12
SPEA2	RHV	**0.83**	**0.89** ± 0.03	**0.89**	0.03	**0.95**
	GD	**3738.85**	**7460.10** ± 1472.98	**7211.45**	1496.41	**10550.35**
	GD$^+$	**0.02**	**66.54** ± 179.47	**0.06**	0.02	**712.32**
	IGD	**13494.63**	**21287.26** ± 5191.87	**20642.87**	6307.04	**37576.88**
	IGD$^+$	**595.11**	**8327.67** ± 6036.19	**7301.21**	7973.87	**26004.91**

According to the results in Table 2, SPEA2 is the most competitive method among the evaluated ones addressing MO-EVCS-L. SPEA2 present lower values for the distance-based metrics, i.e., GD, GD$^+$, IGD, and IGD$^+$ (see in bold in Table 2), which represents that SPEA2 computed solutions are closer to the Pareto front than the NSGA-II ones. The RHV results of SPEA2 are higher than the NSGA-II ones (see in bold in Table 2), which indicates that SPEA2 converged to more competitive solutions than NSGA-II.

Statistical analysis is performed to evaluate the statistical significance of these results. As the distribution of results follow a non-normal distribution,

Kruskal-Wallis statistical test is applied. The results confirmed that SPEA2 outperforms NSGA-II with a confidence higher than 99%, i.e., $p - values \ll 0.001$ for all evaluated metrics. These results imply better convergence towards the Pareto front of the problem and a better coverage of the Pareto space by solutions computed by SPEA2

Table 3 reports the values of the multiobjective metrics evaluated for the Pareto front provided by NSGA-II and SPEA2. Results in Table 3 show that the Pareto front computed by SPEA2 provides the best values for each metric.

Table 3. MOEAs metrics for the Pareto front computed by NSGA-II and SPEA2.

Algorithm	RHV	GD	GD$^+$	IGD	IGD$^+$
NSGA-II	0.947	5790.885	590.648	10131.143	0.031
SPEA2	**0.992**	**1175.294**	**0.009**	**2776.893**	**595.026**

5.2 Computational Time Evaluation

This section discusses the execution time of the evaluated methods. Table 4 shows the relevant statistics of the execution time in minutes of NSGA-II and SPEA2 when addressing the proposed instance of MO-EVCS-L. RS is not included since its stop condition is set as running for the maximum running time of both MOEAs, i.e., 1515 s.

Table 4. NSGA-II and SPEA2 execution times (in seconds).

Algorithm	Minimum	Mean ± sd	Median	iqr	Maximum
NSGA-II	1214.37	1271.17 ± 61.70	1242.04	61.86	1515.38
SPEA2	1214.25	1272.85 ± 49.20	1260.53	47.11	1515.96

Results in Table 4 show that there are no significant differences between the execution time required by the evaluated MOEAs. The execution time is between 20 and 25 min, which entails a low computational cost.

5.3 Comparative Analysis

A few samples of computed solutions are compared in terms of two metrics of quality of service: the sum of the service provided by the stations (QoS) and the disconnected users (du). For a fair comparison, the solutions compared are the ones that require the same deployment cost. Three were selected according to a percentage over the maximum deployment cost computed: 50%, 75%, and 90%. Table 5 reports the results. Besides, it includes the number of stations of each type installed. Figure 5 illustrates the solution with the 75% of the cost.

Table 5. Quality of service metrics for the selected run of each algorithm.

Cost	Algorithm	QoS	du	# stations type = 1	# stations type = 2
50%	NSGA-II	1204666	317563	21	9
	SPEA2	**1413229**	**298201**	19	10
75%	NSGA-II	1072286	280995	39	10
	SPEA2	**1529501**	**275702**	31	13
90%	NSGA-II	1502587	**262766**	43	13
	SPEA2	**1619564**	274091	40	14

According to the results in Table 5, SPEA2 provides the best QoS values for the three evaluated solutions. For the costs of 50% and 75%, SPEA2 leaves fewer users disconnected. However, NSGA-II has fewer citizens that are not served by any charging station for the cost of 90%. Finally, it can be seen that SPEA2 deployments have more EVCS of type 2 than NSGA-II, and NSGA-II installs more EVCS of type 1 than SPEA2.

● **NSGA-II**
● **SPEA2**

Fig. 5. Geographical locations of the solutions computed by NSGA-II and SPEA2 with a 75% of cost.

6   Conclusions and Future Work

This paper presents a multiobjective evolutionary approach to address the problem of locating electric vehicle charging stations in a city, a relevant challenge of the current sustainability and clean mobility concerns.

The multiobjective problem of locating electric vehicle charging stations studied considers two conflicting objectives: quality of service and deployment costs. The problem formulation is more realistic than previous approaches. On the one hand, it considers both kinds of users: citizens served by the stations and a long way to get to the nearest one. On the other hand, it explicitly models real energy supply constraints and deployment costs.

The proposed problem formulation as MO-EVCS-L is more realistic than previous approaches. On the one hand, it considers the two types of users: citizens served by the charging stations and those disconnected from the charging station network (i.e., not attended by any charging station). It is important considering the unserved users because this may make it difficult for these citizens to purchase electric vehicles. On the other hand, it explicitly models real energy supply constraints and deployment costs.

Two variations of MOEAs (NSGA-II and SPEA2) that apply specific evolutionary operators have been proposed to solve MO-EVCS-L. The problem has been solved over a real city-scale scenario, the city of Malaga (Spain). The results obtained show that the SPEA2 is the most competitive approach. However, bot MOEAs provide accurate location plans to assist in making decisions on the location of EVCSs taking into account the quality of service and the cost of installation.

The main lines for future work are related evaluating exact approaches such ILP variations to address this multiobjective optimization problem; devising other operators; applying a race-based method to configure the algorithms, e.g., irace [16]; increasing the realism of the model by considering general citizen's mobility behavior, the location of points of interest (i.e., hospitals, industrial areas, malls, etc.), or the vehicle fleet; and to the definition of real instances over other cities.

Acknowledgements. This research was partially funded by the Universidad de Málaga, Consejería de Economía y Conocimiento de la Junta de Andalucía and FEDER under grant number UMA18-FEDERJA-003 (PRECOG); under grant PID 2020-116727RB-I00 (HUmove) funded by MCIN/AEI/ 10.13039/501100011033; and TAILOR ICT-48 Network (No. 952215) funded by EU Horizon 2020 research and innovation programme.

References

1. Brandstätter, G., Kahr, M., Leitner, M.: Determining optimal locations for charging stations of electric car-sharing systems under stochastic demand. Transp. Res. Part B Methodol. **104**, 17–35 (2017)

2. Çalık, H., Fortz, B.: Location of stations in a one-way electric car sharing system. In: 2017 IEEE Symposium on Computers and Communications (ISCC), pp. 134–139. IEEE (2017)
3. Chen, T.D., Kockelman, K.M., Khan, M.: Locating electric vehicle charging stations: parking-based assignment method for Seattle, Washington. Transp. Res. Rec. **2385**(1), 28–36 (2013)
4. Cintrano, C., Toutouh, J., Alba, E.: Citizen centric optimal electric vehicle charging stations locations in a full city: case of Malaga. In: Alba, E., et al. (eds.) CAEPIA 2021. LNCS (LNAI), vol. 12882, pp. 247–257. Springer, Cham (2021). https://doi.org/10.1007/978-3-030-85713-4_24
5. Coello Coello, C.A., Reyes Sierra, M.: A study of the parallelization of a coevolutionary multi-objective evolutionary algorithm. In: Monroy, R., Arroyo-Figueroa, G., Sucar, L.E., Sossa, H. (eds.) MICAI 2004. LNCS (LNAI), vol. 2972, pp. 688–697. Springer, Heidelberg (2004). https://doi.org/10.1007/978-3-540-24694-7_71
6. Coffman, M., Bernstein, P., Wee, S.: Electric vehicles revisited: a review of factors that affect adoption. Transp. Rev. **37**(1), 79–93 (2016)
7. De Rainville, F.M., Fortin, F.A., Gardner, M.A., Parizeau, M., Gagné, C.: DEAP: enabling nimbler evolutions. SIGEVOlution **6**(2), 17–26 (2014)
8. Deb, K.: Multi-objective Optimization Using Evolutionary Algorithms. Wiley, Hoboken (2001)
9. Fabbiani, E., Nesmachnow, S., Toutouh, J., Tchernykh, A., Avetisyan, A., Radchenko, G.: Analysis of mobility patterns for public transportation and bus stops relocation. Program. Comput. Softw. **44**(6), 508–525 (2018)
10. Falchetta, G., Noussan, M.: Electric vehicle charging network in Europe: an accessibility and deployment trends analysis. Transp. Res. Part D: Transp. Environ. **94**, 102813 (2021)
11. Frade, I., Ribeiro, A., Gonçalves, G., Antunes, A.P.: Optimal location of charging stations for electric vehicles in a neighborhood in Lisbon, Portugal. Transp. Res. Rec. **2252**(1), 91–98 (2011)
12. Hipogrosso, S., Nesmachnow, S.: Analysis of sustainable public transportation and mobility recommendations for montevideo and parque rodó neighborhood. Smart Cities **3**(2), 479–510 (2020)
13. Ishibuchi, H., Masuda, H., Tanigaki, Y., Nojima, Y.: Modified distance calculation in generational distance and inverted generational distance. In: Gaspar-Cunha, A., Henggeler Antunes, C., Coello, C.C. (eds.) EMO 2015. LNCS, vol. 9019, pp. 110–125. Springer, Cham (2015). https://doi.org/10.1007/978-3-319-15892-1_8
14. Kumar, R., Alok, K.: Adoption of electric vehicle: a literature review and prospects for sustainability. J. Clean. Prod. **253**, 119911 (2020)
15. Lin, H., Bian, C., Li, H., Sun, Q., Wennersten, R.: Optimal siting and sizing of public charging stations in urban area. In: 2018 Joint International Conference on Energy, Ecology and Environment (ICEEE 2018) and International Conference on Electric and Intelligent Vehicles (ICEIV 2018), p. 7 (2018)
16. López-Ibáñez, M., Dubois-Lacoste, J., Pérez Cáceres, L., Stützle, T., Birattari, M.: The irace package: iterated racing for automatic algorithm configuration. Oper. Res. Perspect. **3**, 43–58 (2016)
17. Massobrio, R., Toutouh, J., Nesmachnow, S., Alba, E.: Infrastructure deployment in vehicular communication networks using a parallel multiobjective evolutionary algorithm. Int. J. Intell. Syst. **32**(8), 801–829 (2017)
18. Nesmachnow, S., Rossit, D.G., Toutouh, J.: Comparison of multiobjective evolutionary algorithms for prioritized urban waste collection in Montevideo, Uruguay. Electron. Notes Discrete Math. **69**, 93–100 (2018)

19. OpenStreetMap contributors: planet dump retrieved from https://planet.osm.org (2017). https://www.openstreetmap.org
20. Péres, M., Ruiz, G., Nesmachnow, S., Olivera, A.C.: Multiobjective evolutionary optimization of traffic flow and pollution in Montevideo, Uruguay. Appl. Soft Comput. **70**, 472–485 (2018)
21. Risso, C., Cintrano, C., Toutouh, J., Nesmachnow, S.: Exact approach for electric vehicle charging infrastructure location: a real case study in Málaga, Spain. In: Nesmachnow, S., Hernández Callejo, L. (eds.) ICSC-Cities 2021. CCIS, vol. 1555, pp. 42–57. Springer, Cham (2021). https://doi.org/10.1007/978-3-030-96753-6_4
22. Rossit, D.G., Toutouh, J., Nesmachnow, S.: Exact and heuristic approaches for multi-objective garbage accumulation points location in real scenarios. Waste Manage. **105**, 467–481 (2020)
23. Toutouh, J., Rossit, D., Nesmachnow, S.: Soft computing methods for multiobjective location of garbage accumulation points in smart cities. Ann. Math. Artif. Intell. **88**(1), 105–131 (2020)
24. Van Veldhuizen, D.: Multiobjective evolutionary algorithms: classifications, analyses, and new innovations. Air Force Institute of Technology (1999)
25. Wagner, S., Götzinger, M., Neumann, D.: Optimal location of charging stations in smart cities: a points of interest based approach (2013)
26. Zitzler, E., Laumanns, M., Thiele, L.: SPEA2: improving the strength pareto evolutionary algorithm for multiobjective optimization. In: Giannakoglou, K., Tsahalis, D., Périaux, J., Papailiou, K., Fogarty, T. (eds.) Evolutionary Methods for Design Optimization and Control with Applications to Industrial Problems, pp. 95–100 (2001)
27. Zitzler, E., Thiele, L.: Multiobjective optimization using evolutionary algorithms— a comparative case study. In: Eiben, A.E., Bäck, T., Schoenauer, M., Schwefel, H.-P. (eds.) PPSN 1998. LNCS, vol. 1498, pp. 292–301. Springer, Heidelberg (1998). https://doi.org/10.1007/BFb0056872

Resilient Bio-inspired Algorithms

Brain Programming and Its Resilience Using a Real-World Database of a Snowy Plover Shorebird

Roberto Pineda[1], Gustavo Olague[1(✉)], Gerardo Ibarra-Vazquez[2],
Axel Martinez[1], Jonathan Vargas[3], and Isnardo Reducindo[2]

[1] EvoVisión Laboratory, CICESE Research Center, Carretera Ensenada-Tijuana
3918, Zona Playitas, 22860 Ensenada, BC, Mexico
olague@cicese.mx
[2] Facultad de Ingeniería, Universidad Autónoma de San Luis Potosí,
Dr. Manuel Nava 8, Col. Zona Universitaria Poniente,
78290 San Luis Potosí, SLP, Mexico
[3] Coastal Solutions, Cornell Lab of Ornithology, Ithaca, NY, USA

Abstract. Even when deep convolutional neural networks have proven
to be effective at saliency detection, they have a vulnerability that should
not be ignored: they are susceptible to adversarial attacks, making them
highly unreliable. Reliability is an important aspect to consider when
it comes to salient object detection; without it, an attacker can render
the algorithm useless. Brain programming–an evolutionary methodology
for visual problems–is highly resilient and can withstand even the most
intense perturbations. In this work, we perform for the first time a study
that compares the resilience against adversarial attacks and noise perturbations using a real-world database of a shorebird called the Snowy
Plover in a visual attention task. Database images were taken on the field
and even posed a detection challenge due to the nature of the environment and the bird's physical characteristics. By attacking three different
deep convolutional neural networks using adversarial examples from this
database, we prove that they are no match for the brain programming
algorithm when it comes to resilience, suffering significant losses in their
performance. On the other hand, brain programming stands its ground
and sees its performance unaffected. Also, by using images of the Snowy
Plover, we refer to the importance of resilience in real-world issues where
conservation is present. Brain programming is the first highly resilient
evolutionary algorithm used for saliency detection tasks.

Keywords: Adversarial attacks · Salient object detection ·
Robustness · Wildlife conservation · Genetic programming

J. L. Jiménez Laredo et al. (Eds.): EvoApplications 2022, LNCS 13224, pp. 603–618, 2022.
https://doi.org/10.1007/978-3-031-02462-7_38

1 Visual Attention

Within the neuroscience community, it is understood that visual attention is a natural process of the brain, whose functionality is to perceive salient visual characteristics of the most prominent objects in a scene. This cognitive task is strictly necessary for humans due to the impossibility of focusing on two different objects simultaneously. *Visual attention* is an ability that allows a creature, living or artificial, to immediately gaze towards the objects of interest in a visual environment [4]. The objects of interest refer to regions in the image that project from the environment and contain essential information at one moment in time. This selective process that filters visual information from the retina to the visual cortex is described by two visual subsystems, and the two-channel hypothesis for visual perception is one of the most accepted descriptions, which describe the ventral and the dorsal stream. Both subsystems receive the same visual information as input, but the difference resides in the transformations that it suffers on each stream. First, the dorsal stream relates to the task of obtaining a spatial location in the visual processing system, so it is known as the "where" or "how" stream. It begins at the retina, where the information is projected towards the V1 layer of the brain, which is part of the primary visual cortex. This stream moves towards the V2 and V3 layers and continues towards the posterior parietal cortex and adjacent areas. In general, it is widely accepted that the dorsal stream carries out the visual attention process, and the most popular paradigm for this process is the feature integration theory [20]. The first computational model for visual attention was proposed by Koch [9], in which the image is split into multiple dimensions to obtain conspicuity maps, which are then fused into a single representation called saliency map.

On the other hand, the ventral stream is known as the "what" stream because it is usually associated to object recognition and shape representation tasks [16]. From a computational perspective, the ventral stream is considered a hierarchic and biologically inspired information processing system that specializes in object recognition [7].

1.1 Artificial Dorsal Stream

The artificial dorsal stream (ADS) is a model based on the feature integration theory [20], which suggests that attention should be processed serially. The first step in the ADS is the image acquisition stage. To fully understand the practice of the dorsal stream, it is necessary to understand the image as the first input to a visual system, which is defined as the graph of a function. This function sets the grounds to understand the transformation of the physical and geometric properties of the scene.

Definition 1. Image as the graph of a function. *Let f be a function $f : U \subset \mathbb{R}^2 \to \mathbb{R}$. The graph or image I of f is a subset of $\mathbb{R}^3$ that consists of the points $(x, y, f(x,y))$, in which the ordered pair (x, y) is a point in U and $f(x, y)$ is the value at that point. Symbolically, the image $I = \{(x, y, f(x,y)) \in \mathbb{R}^3 | (x, y) \in U\}$.*

In this stage, the system considers digital color images in the RGB color space and their transformations to CMYK and HSV as the algorithm's input values. Therefore, the color image is defined as a set of components for each color space: $I_{color} = \{I_r, I_g, I_b, I_c, I_m, I_y, I_k, I_h, I_s, I_v\}$, where each element corresponds to the color spaces RGB (Red, Green, Blue), CMYK (Cyan, Magenta, Yellow, Black), and HSV (Hue, Saturation, Value). After that, three visual operators are applied to highlight the image characteristics related to the color, orientation, and shape dimensions. These ADS operations are evolved through genetic programming to obtain an optimal set of visual operators (EVO), which are specialized functions used to create a set of visual maps (VM) throughout these three dimensions. A fourth visual map describing intensity is not evolved and is calculated with the average of the RGB color bands. An image pyramid is created and reduced for each visual map until a conspicuity map (CM) is obtained for each of the dimensions previously mentioned. These CMs are integrated through an evolved feature integration function (EFI) into a single saliency map, highlighting the objects of interest in the input image. The process to create the visual maps is described next.

Table 1. Functions and terminals for the ADS.

Functions for EVO_O	Terminals for EVO_O
$A + B$, $A - B$, $A \times B$, A/B, $\lvert A \rvert$, $\lvert A + B \rvert$, $\lvert A - B \rvert$, $\log_2(A)$, $A/2$, A^2, $\sqrt{A}$, $k \times A$, A/k, $A^{1/k}$, A^k, $(1/k) + A$, $A - (1/k)$, $G_{\sigma=1}(A)$, $G_{\sigma=2}(A)$, $D_x(A)$, $D_y(A)$, $round(A)$, $\lfloor A \rfloor$, $\lceil A \rceil$, $inf(A, B)$, $sup(A, B)$, $thr(A)$, $AttenuateBorders(A)$, $ConvGabor(A)$	$I_r, I_g, I_b, I_c, I_m, I_y, I_k, I_h, I_s, I_v$, $D_x(I_{color})$, $D_{xx}(I_{color})$, $D_y(I_{color})$, $D_{yy}(I_{color})$, $D_{xy}(I_{color})$, $AttenuateBorders(I_{color})$, $ConvGabor(I_{color})$,
Functions for EVO_C	Terminals for EVO_C
$A + B$, $A - B$, $A \times B$, A/B, $\log_2(A)$, $exp(A)$, $\lvert A \rvert$, A^2, $\sqrt{A}$, $(A)^c$, $thr(A)$, $round(A)$, $\lfloor A \rfloor$, $\lceil A \rceil$, $k \times A$, A/k, $A^{1/k}$, A^k, $(1/k) + A$, $A - (1/k)$	$I_r, I_g, I_b, I_c, I_m, I_y, I_k, I_h, I_s, I_v, DKL_r$, $DKL_\Phi, DKL_\Theta, Op_{r-g}(I), Op_{b-y}(I)$
Functions for EVO_S	Terminals for EVO_S
$A + B$, $A - B$, $A \times B$, A/B, $\lvert A \rvert$, $k \times A$, A/k, $A^{1/k}$, A^k, $(1/k) + A$, $A - (1/k)$, $round(A)$, $\lfloor A \rfloor$, $\lceil A \rceil$, $A \oplus SE_d$, $A \oplus SE_s$, $A \oplus SE_{dm}$, $A \ominus SE_d$, $A \ominus SE_s$, $A \ominus SE_{dm}$, $Sk(A)$, $Perim(A)$, $A \circledast SE_d$, $A \circledast SE_s$, $A \circledast SE_{dm}$, $T_{hat}(A)$, $B_{hat}(A)$, $A \circledcirc SE_s$, $A \odot SE_s$, $thr(A)$	$I_r, I_g, I_b, I_c, I_m, I_y, I_k, I_h, I_s, I_v$
Functions for EFI	Terminals for EFI
$A + B$, $A - B$, $A \times B$, A/B, $\lvert A \rvert$, $\lvert A + B \rvert$, $\lvert A - B \rvert$, $k \times A$, A/k, $A^{1/k}$, A^k, $(1/k) + A$, $A - (1/k)$, $Hist(A)$, $round(A)$, $\lfloor A \rfloor$, $\lceil A \rceil$, $thr(A)$, $(A)^2$, $\sqrt{A}$, $exp(A)$, $G_{\sigma=1}(A)$, $G_{\sigma=2}(A)$, $D_x(A)$, $D_y(A)$	$CM_d, D_x(CM_d), D_{xx}(CM_d), D_y(CM_d)$, $D_{yy}(CM_d), D_{xy}(CM_d)$

The *Color Visual Map* VM_C allows creating a new image that contains salient information relating to the color dimension of the input image. Here, the image is transformed by the function $EVO_C : I_{color} \rightarrow VM_C$ to highlight the color characteristics. This and the other visual maps use the set of functions and terminals from Table 1 to generate operators for each image dimension. The notation is summarized next: the function's input can be any of the terminals and the output of any functions or a composition among them. I is the input image.

The *Orientation Visual Map* is produced by applying $EVO_O : I_{color} \rightarrow VM_O$. This operator is evolved with genetic programming to optimize the extraction of information about corners and edged inside the input image I or in its color components I_{color}. This way, the visual map values VM_O represent the prominence or characteristics of the orientation dimension.

The *Shape Visual Map* is obtained by the function defined as $EVO_S : I_{color} \rightarrow VM_S$. This operator is evolved to extract interesting features based on the appearance and structure of objects on the scene using mathematical morphology.

The *Intensity Visual Map* is obtained by the simple formula $VM_{Int} = \frac{I_r + I_g + I_b}{3}$, where I_r, I_g y I_b are the red, green, and blue color bands. Resulting in a visual map VM_{Int} that is directly calculated using the pixel values obtained by the camera and that represents the intensity.

Conspicuity Maps. The next step is to compute the conspicuity maps (CM), which are obtained by a center-surround function that is applied to the visual maps to simulate the center-surround receptive fields in the retina, a natural structure that measures the differences in firing rates between the center and the surroundings of the ganglion cells. The goal is to generate a conspicuity map CM per dimension, according to the model proposed by Walther et al. [21]. This process is composed of two parts. First, the compute of the CM is modeled as the difference between fine and coarse scales, which are calculated by a nine-level pyramid $P_d^{\sigma=0} = \{P_d^{\sigma=1}, P_d^{\sigma=2}, P_d^{\sigma=3}, ..., P_d^{\sigma=8}\}$. Each pyramid is calculated from its matching VM_d using a Gaussian blur filter, which results in an image half the size of the input visual map. This process is repeated recursively eight times. Secondly, the pyramid P_d^{σ} is used as the input data to generate six center-surround maps that result from the difference between some of the levels of the pyramid, calculated as:

$$Q_d^j = P_d^{\sigma = \lfloor \frac{j+9}{2} \rfloor + 1} - P_d^{\sigma = \lfloor \frac{j+2}{2} \rfloor + 1} \tag{1}$$

where $j = \{1, 2, ..., 6\}$. The levels P_d^{σ} have different sizes and are downscaled to calculate their differences. Then, each one of these maps is normalized and combined into a single map using across-scale addition to obtain a single CM_d per dimension.

1.2   Feature Integration

The second stage of the ADS combines the output CMs from the previous step to output a single map known as saliency map (SM). The feature integration operation is defined by the mapping $FI : CM_d \rightarrow SM$, where CM_d are the conspicuity maps of the four dimensions $d = \{C, O, S, I\}$. At combining the CMs, the values computed at pixels location represent the prominence of each point in the scene. Then, a spreading algorithm is applied by starting from the pixel with the highest absolute value to define the most prominent region in the image. This prominent region is also known as *proto-object* [17]. It is important to mention that the input image can contain one or more proto-objects. Finally, the map is turned into a binary image, defining the proto-object P composed of n points.

2   Brain Programming

Here we present an evolutionary methodology where multiple programs are evolved while being linked to a complex hierarchical system that imitates an artificial brain [15]. Brain programming (BP) describes a genetic programming strategy where the ADS can be optimized to mimic the specialized areas of the brain through a set of operators. It is focused on functions to extract and combine the relevant information that solves the problem of visual attention. The genetic representation of the ADS consists of four trees linked to an artificial dorsal stream. BP evolves different visual operators VO into evolved visual operators EVO, including the operation in charge of the feature integration stage EFI, resulting in an optimized saliency map (OSM). Each tree has its own set of functions and terminals carefully selected according to the brain's functionality that we want to emulate. These functions and terminals are defined in Table 1.

The first tree of the BP evolutionary process emulates orientation. The orientation operator VO_o is evolved to EVO_O using the set of functions and terminals to highlight corners, edges, and other characteristics related to orientation. The second tree emulates the color dimension by evolving the visual operator VO to EVO_C to reproduce the process of color perception. The third tree codifies the shape dimension by evolving the VO_S operator into EVO_S to highlight the image's appearance and object structure. Finally, the fourth tree emulates how the characteristics are combined to create the optimized saliency map using the EFI functions and terminals, highlighting the characteristics of the object of interest in the scene.

Figure 1 shows the process to evolve these four functions. BP takes the number of generations to evaluate, the population size, and the number of trees to be evolved per individual. Every individual in the initial population is a set of four trees indicated by the *num_trees* parameter. We set the number of generations (*num_gen*) and the number of individuals per populations (*size_pop*) as 30. The initialization method applied to the population is known as the *ramped half-and-half* method [10]. This approach creates half of the initial population with the *grow* method and the other half with the *full* method. The *grow* method produces unbalanced trees allowing branches of different lengths, and the *full* method makes balanced trees, with all branches of the same length. The length of the tree must not exceed the specified maximum to prevent trees grow without control during the evolution. Also, the size of the individuals should not exceed the defined maximum depth to prevent trees from continue growing uncontrollably. To limit the size of any given individual within the population, we apply

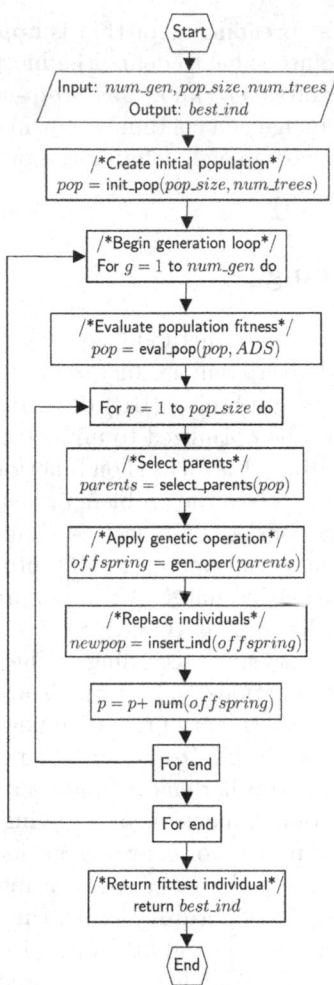

Fig. 1. Flowchart of the brain programming strategy.

two max three depth parameters. The first is *dynamic max depth*, which is a maximum three depth and may not be surpassed by an individual unless its fitness is better than the best solution found so far. When this happens, this parameter is augmented to the three depth of the new fittest individual and reduced to a lower three depth. The second parameter is the *real max depth*, which is limited and no individual should surpass. The termination criterion is defined as the maximum number of generations. Table 2 displays the values applied by BP during the runs.

After the initial population is created and evaluated with a fitness function [4], the next step is to breed the next generation of individuals by the use of the genetic operations of selection, crossover, and mutation. The parents are obtained using a fitness-proportionate selection method implemented with the *roulette-wheel strat-*

Table 2. Main parameters of the BP algorithm.

Parameters	Description
Generations	30
Population size	30 individuals
Initialization	Ramped half & Half
Crossover at chromosome level	0.4
Crossover at gene level	0.4
Mutation at chromosome level	0.1
Mutation at gene level	0.1
Three depth	Dynamic depth selection
Dynamic max depth	7 levels
Real max depth	9 levels
Selection	Roulette-wheel
Elitism	Keep the best individual

egy, which selects the fittest individuals based on probability, and so the crossover and mutation are sequentially applied until a new population is created. The brain programming paradigm reflects nature in a way that it is a process that never ends as any other evolutionary algorithm. Nevertheless, BP differs from the genotype and hierarchical representation and specialized processes such as the Gaussian pyramid and the center-surround process. The purpose is to find an individual with the highest fitness and stop at 30 generations.

3 Adversarial Attacks

In the automatic object recognition world, state-of-the-art has focused on models based on neural networks. These highly powerful learning models can achieve exceptional results in object recognition problems [11] due to the nature of their architectures, which are based on massively parallel non-linear steps [19]. However, they tend to have counter-intuitive properties, one of which involves the lack of robustness, that is, resistance against small perturbations on their input images. Slight modifications on their pixels should not alter the classification task, given that the image is still the same. Szegedy et al. [19] noticed that applying a non-random imperceptible perturbation to a test image made a neural network change its prediction. These perturbations are found when optimizing the input to maximize prediction error, contrary to the purpose of a neural network, which is to minimize the error. These types of images are known as *adversarial examples*. In the present work, we present for the first time that attacks designed for classification tasks transfer to a different task like object detection considering a real-world application. Also, we show that an evolutionary computation methodology, such as brain programming, is highly robust to such attacks.

Another property of adversarial examples involves *transferability*. It is said that an adversarial example designed to be incorrectly classified by a model M_1, is also misclassified by a model M_2 [19]. This property proves that it is possible to generate adversarial examples using one model as the source and attack many other models without knowing their architectures. The action of attacking an object detection model using an adversarial example is known as *adversarial attack*.

The purpose of adversarial attacks is to disturb a neural network detection model or classifier in a testing environment. It is also possible to attack systems that operate in the real world, where the attacker cannot make pixel-by-pixel modifications on an input image [12]. Examples include robots that perceive the world through cameras and sensors, video vigilance systems, and mobile applications for image or sound classification.

This work has another purpose for experimenting with adversarial attacks using a live bird database. Given that the Snowy Plover is classified as a threatened species, it is essential to work to benefit its protection. Adversarial attacks expose the vulnerabilities of many neural networks used for wildlife conservation. These neural networks have been embedded in real-life monitoring systems such as camera traps for animal identification, and counting [14], drones for scanning a habitat [3] and even acoustic recording devices [8]. Many of these applications involve defending against the attacks of individuals performing illegal activities that put wildlife in danger, such as poaching, which has become an increasing problem in today's world and a subject that conservationists desperately need a solution. Other activities that have become a threat to a species wellbeing are the constant increase in human population and their disturbance on the animal's natural habitat, such as beaches where the Snowy Plover and many other species of shorebirds reside. By exposing the vulnerabilities of these neural networks, the science community can focus on improving the resilience of their algorithms to prevent attacks by malicious individuals.

In this work, we generate several types of adversarial examples using a database composed of images of the Snowy Plover for experimentation. These types of adversarial examples are explained in the following sections.

3.1   Fast Gradient Sign Method

Goodfellow et al. [5] proposed a white-box method to attack neural networks by exploiting the gradient, which is the weights correction in the backpropagation process of a neural network to build an adversarial example that maximizes the loss function. The equation for the FGSM is defined as:

$$\eta = \epsilon \; \text{sign} \; (\nabla J(\theta, x, y)) \quad , \tag{2}$$

where θ are the model's parameters, x the input data to the model, y the associated output to x, and $J()$ is the loss function used to train the neural network. This attack computes the gradient over the loss function around the current value of the model parameter with respect to the image and the associated outputs as $\nabla J(\theta, x, y)$. Then, it adjusts the intensity with $\epsilon \; \{0 \leq \epsilon \leq 255\}$ in the direction that maximizes the loss, obtaining an optimal perturbation η. The ϵ value determines the perturbation intensity on the adversarial example generated. The perturbation is nonexistent for a value $\epsilon = 0$, while perturbation is absolute for $\epsilon = 255$ where the input data is virtually lost. Because of this, one must choose an optimal value of ϵ enough to fool a detection model and adequate for the perturbation to remain unnoticeable.

3.2   Adversarial Patch

This method proposed by [2] generates a patch that is highly salient for a neural network. When generating and placing this patch over any location in the input image, neural networks cannot determine their nature and classify them the same way as a salient object. Also, this patch is scene-independent, allowing the attacker to perform a real-world attack without knowing the specifics of any model or how it works; it can just be printed and placed randomly on a real-world scene. The adversarial patch proposed by Brown et al. uses the expectation over transformation framework to train the patch [1]. This patch is trained over various images in the computing process, applying transformations like rotation, translation, and scale and constantly optimizing the patch using gradient descent. The result is a patch that works independently of whatever is in the background.

3.3   Multipixel Attack

The Multipixel attack is inspired by the One-pixel method described by Su et al. [18], in which only one pixel belonging to the input image is modified. This attack is formally defined as an optimization problem with certain limitations. Being $f()$ the model that receives the input data x, $e(x)$ is the vector of pixel modifications on x, the target class adv and the variable d limits the perturbation $e(x)$ until a specified maximum, the goal of adversarial examples is to find an optimal solution $e(x)^*$ for the following equation:

$$\max f_{adv}(x + e(x)) \text{ subject to } ||e(x)||_0 \leq d \tag{3}$$

where $d = 1$ for the one-pixel attack, and $d > 1$ for multipixel attack. However, the required image size for the one-pixel modification to fool a detection algorithm is too small compared to other adversarial example generation methods. A 32×32 pixel image is used in their work, but these dimensions are not big enough to simulate a real-world case. Mainly, detection algorithms embedded into real-world systems capture images with a bigger resolution; that is why the multipixel method aims to extend the number of pixels modified by increasing the value of d. This way, this pixel modification method becomes effective in bigger-sized images; however, the perturbation becomes more noticeable.

3.4   Noise

We also used inserted three types of noise in the input images in order to evaluate the performance of the saliency detection algorithms during perturbations that may arise on the image acquisition stage. The noises used are Gaussian, Salt & Pepper, and Speckle.

4   Experiments and Results

4.1   Image Database

The Snowy Plover database, named the SNPL database, is a set of 250 photographs of the Snowy Plover shorebird, one of the main focuses in this work. Each one of the 250 photographs was taken in the bird's natural habitat, which consists of sandy beaches and mudflats on the pacific coast of Baja California, Mexico. Each photograph was captured in natural environments, with professional DSLR cameras and different weather, exposure, and range conditions. These images show one or more plovers, whose size on the scene varies according to the distance at which the photograph was taken and the focal length. Another essential feature of these images involves the background having a variety of distractors like plants, water bodies, and small objects like shells and debris. Also, many of these images contain occlusions that partially cover the object of interest. Lastly, given that the animal's sandy habitat and the primary color of its feathers have similar color tones, a significant proportion of these images lack high contrast between the background and the object of interest. Because of these properties, the SNPL database is a great instance to study such a real-world problem. The SNPL database was proposed to compare the performance of saliency detection algorithms using a real-world problem, in contrast with other databases containing objects of interest with optimal size and exposure conditions specifically designed for academic purposes or putting a novel algorithm to the test.

The SNPL database was divided into two randomly selected sets; a training set and a validation set. We use a 70:30 split ratio, so 175 images are used for training and 75 for validation. The pixel size of each image is 224×224, a general image size used in a neural network training process, and an acceptable one since brain programming works with all image sizes. Each image was manually segmented to generate the ground-truth binary images, which we used to obtain the precision and recall values for evaluation.

4.2   Adversarial Example Generation

To test the robustness of our brain programming algorithm or the resistance against adversarial attacks, we generated various adversarial examples corresponding with the validation set of our database. The process used to generate them is described in the following sections.

Adversarial Examples with FGSM. We generated six subsets of adversarial examples using the Fast Gradient Sign Method (FGSM) [5] and the PiCANet neural network [13]. Each of the six subsets was generated with a specific epsilon value; thus, the images of each subset hold a different perturbation intensity.

(a) Original images.   (b) Ground-truth.

Fig. 2. Several samples within the SNPL image database.

The epsilon values used are $\epsilon = \{2, 4, 8, 16, 32, 64\}$. Commonly, the perturbation is imperceptible on the values of $\epsilon = \{2, 4\}$, and becomes noticeable starting from $\epsilon = 8$ (Figs. 2, 3).

(a) $\epsilon = 0$ (b) $\epsilon = 2$ (c) $\epsilon = 4$ (d) $\epsilon = 8$

(e) $\epsilon = 16$ (f) $\epsilon = 32$ (g) $\epsilon = 64$

Fig. 3. FGSM perturbation on the SNPL database. The bigger the epsilon value, the stronger the perturbation. The value $\epsilon = 0$ equals to the unperturbed image.

Adversarial Examples with Adversarial Patch. We generated two different patch sizes to evaluate if the size of the patch has any influence on the detection algorithms performance and how much the results vary. We compute these patches using the adversarial patch method [2] and the ResNet-50 neural network [6] to create two subsets of adversarial examples. One contains images with a patch size of 70 × 70 pixels, covering 31% of the original image size. The other subset contains images with a patch size of 50 × 50 pixels, covering 22% of the original image size. The patch on each image was placed on a random location, regardless of where the patch ended up covering–partially or totally–the object of interest.

Adversarial Examples with Multipixel Attack. We generated one subset of adversarial examples using the Multipixel method. For this type of perturbation, we established a value of $d = 10,000$, remembering that d defines the number of modified pixels in the image. This value gave us a subset of adversarial examples where the perturbation, although not inconspicuous, is enough to allow the objects of interest to remain visible and pose a challenge to the neural networks.

Adversarial Examples with Gaussian Noise. We generated one subset of adversarial examples where Gaussian noise was inserted in each image. For each example generated, we defined a standard deviation of $\sigma = 30$, which implements a grain effect with enough magnitude to recognize the objects in the scene.

Adversarial Examples with Salt-Pepper Noise. We generated one subset of adversarial examples using Salt and Pepper noise. This work added noise in color images; thus, the noise does not manifest itself as black and white dots but as colored dots. This process takes each color band's minimum and maximum values and distributes them around the image.

Adversarial Examples with Speckle Noise. In the last subset of adversarial examples, we added Speckle noise. Since this type of noise manifests itself according to the physical conditions of the scene at the moment of creating the image, it was necessary to implement a method to simulate it. To achieve this, we took random values from a normal distribution using the image dimensions as the source, and we multiplied them by a variance of 0.3. This process resulted in a subset of images perturbed with Speckle noise (Fig. 4).

4.3 Results

Here we describe the results gathered from experimenting with each adversarial attack. We split the results among three tables to compare and explain the results that pertain to each perturbation method. The bold values refer to the lowest score that an algorithm obtained among all the adversarial attacks.

FGSM Attack. Table 3 shows the results after attacking the brain programming algorithm (GBVS-BP) and three different neural networks using the FGSM adversarial attack. At first glance, we observe that most of these algorithms

(a) Adv. Patch   (b) Patch (Small)   (c) Multipixel   (d) Gaussian

(e) Salt&Pepper   (f) Speckle

Fig. 4. Different type of perturbations on the SNPL database.

Table 3. Results of the FGSM adversarial attack using the SNPL database. The epsilon ϵ value controls the intensity of the perturbation on the image.

Algorithm	Original	$\epsilon = 2$	$\epsilon = 4$	$\epsilon = 8$	$\epsilon = 16$	$\epsilon = 32$	$\epsilon = 64$
GBVS-BP	52.6	53.31	52.13	50.31	49.09	45.01	42.24
BASNet	64.5	62	59.4	56.1	40.9	17.1	**1.3**
PiCANet	86.04	72.91	63.7	57.64	44.85	9.26	**8.9**
DHSNet	44	43.2	43	42.1	38.6	31.7	**17.8**

do not achieve high performance in the saliency detection task when using this database, except PiCANet. This result is mainly due to the image capturing process, camera quality, and different characteristics of capturing these photographs, like distance from the subject and the many occlusions and distractors in the scene. Even when these three neural networks achieve high scores when testing on different datasets [22] where the objects of interest involve a high degree of saliency, a higher challenge by the proposed database begins to break them apart. This outcome becomes even more noticeable when implementing adversarial attacks. In the case of FGSM, all the neural networks show a gradual decrease in performance beginning at $\epsilon = 2$ and end up plummeting when reaching $\epsilon = 64$. In the case of brain programming, testing the GBVS-BP algorithm with the original images puts it below the BASNet and PiCANet neural networks with a score of 52.6%. However, after implementing an attack with $\epsilon = 16$ it manages to surpass every single neural network when it comes to robustness and ends up with a score way above them after attacking with $\epsilon = 64$. In the end, the BASNet neural network decreased its performance by 98%, PiCANet by 89.6%, and DHSNet by 59.5%. While GBVS-BP only went down by 19.6%.

Table 4. Results of the Multipixel and Adversarial Patch attacks using the SNPL database. Column AdvPatch (Sm) represents a smaller patch.

Algorithm	Original	Multipixel	AdvPatch	AdvPatch(Sm)
GBVS-BP	52.6	47.7	41.26	**39.32**
BASNet	64.5	22.2	21.3	21.5
PiCANet	86.04	13.29	72.61	82.66
DHSNet	44	36.1	34.8	30

Adversarial Patch and Multipixel Attack. Table 4 shows the results using the multipixel and adversarial patch attack. In the case of the multipixel attack, we noticed a significant decrease in scores among the three neural networks, being PiCANet falls from 86.04% to 13.29% with this attack alone, while BASNet suffers almost the same consequences. Again, due to the nature of the SNPL database, these neural networks cannot maintain the necessary robustness to keep achieving decent detection. The GBVS-BP algorithm only shows a slight decrease in performance, demonstrating a high level of robustness. In the case of the adversarial patch, PiCANet manages to be surprisingly robust, decreasing 86.04% to 72.06% at the standard size patch and 82.66% at the minor sized patch, achieving a high degree of detection even when a patch inserts itself in the scene as another type of salient object. Lastly, the brain programming algorithm manages to stay with a score of around 40% at the moment of both patch attacks, overcoming the BASNet and DHSNet neural networks.

Table 5. Results of the adversarial attacks using Gaussian, Salt & Pepper and Speckle noise in the SNPL database.

Algorithm	Original	Gaussian	Salt & Pepper	Speckle
GBVS-BP	52.6	48.4	45.74	44.98
BASNet	64.5	35	34.6	12.8
PiCANet	86.04	32.79	26.93	7.37
DHSNet	44	38.9	38.2	33.6

Noise Attack. Table 5 shows the results after attacking using noise. Noise attacks continue to prove that neural networks are not immune to them. In all cases, the brain programming algorithm surpasses the three neural networks when it comes to robustness. The Speckle noise affects it the most, although not significantly. BASNet and PiCANet suffer great declines in their performance, being once more the Speckle noise the one that destroys them almost completely (BASNet suffers a total of 80% decrease and PiCANet a 91% decrease). DHSNet maintains certain robustness, but not enough to surpass the brain programming algorithm.

5   Conclusions and Discussion

This work presented the brain programming evolutionary algorithm and its resilience against adversarial attacks. Even when state-of-the-art neural networks have proven to be highly powerful at saliency detection, they cannot keep up when fed with small perturbations on their input images, also known as adversarial examples. On the other hand, brain programming can withstand even the most intense perturbations, even when the scenes on the input images possess small objects of interest and several challenges like distractors and occlusions. As we show in the experiments, the three types of adversarial attack methods (FGSM, Adversarial Patch, Multipixel Attack) and the three types of noise (Gaussian, Salt & Pepper, Speckle) were not able to break the brain programming algorithm but surpassed the performance of the tested neural networks at the highest levels of perturbation. Also, we used what we called a real-world database for experimentation, and we emphasized the importance of robustness when working with these types of images that real-life systems use.

Nowadays, adversarial attacks are a hot topic because deep learning is susceptible to manipulation by an attacker. This work shows that BP exhibits a robustness property against adversarial attacks. This property is a breakthrough due to its resilience to these perturbations. The attacks that we consider in this study were adapted from image classification research to the problem of salient object detection. Such attacks are real threats to the security of computer vision systems. This work points towards a new research avenue where evolutionary computing methodologies can edge over deep learning approaches. Also, we foresee the combination of BP with other computer vision and deep learning approaches to achieve robustness and high performance. In summary, this work points towards secure solutions of salient object detection algorithms, where resilience to these attacks must be considered from now on.

References

1. Athalye, A., Engstrom, L., Ilyas, A., Kwok, K.: Synthesizing robust adversarial examples (2018)
2. Brown, T.B., Mané, D., Roy, A., Abadi, M., Gilmer, J.: Adversarial patch (2018)
3. Doull, K., Chalmers, C., Fergus, P., Longmore, S., Piel, A., Wich, S.: An evaluation of the factors affecting 'poacher' detection with drones and the efficacy of machine-learning for detection. Sensors 21(12), 4074 (2021). https://doi.org/10.3390/s21124074
4. Dozal, L., Olague, G., Clemente, E., Hernandez, D.: Brain programming for the evolution of an artificial dorsal stream. Cogn. Comput. 6(3), 528–557 (2014)

5. Goodfellow, I.J., Shlens, J., Szegedy, C.: Explaining and harnessing adversarial examples (2015)
6. He, K., Zhang, X., Ren, S., Sun, J.: Deep residual learning for image recognition (2015)
7. Hubel, D., Wiesel, T.: Receptive fields of single Neurones in the cat's striate cortex. J. Physiol. **148**(3), 574–591 (1953)
8. Kahl, S., Wood, C.M., Eibl, M., Klinck, H.: BirdNet: a deep learning solution for avian diversity monitoring. Ecol. Inform. **61**, 101236 (2021). https://doi.org/10.1016/j.ecoinf.2021.101236, https://www.sciencedirect.com/science/article/pii/S1574954121000273
9. Koch, C., Ullman, S.: Shifts in selective visual attention: towards the underlying neural circuitry. Humam Neurobiol. **4**, 219–227 (1985)
10. Koza, J.: Genetic Programming: On the Programming of Computers by Means of Natural Selection, 1st edn. A Bradford Book, Cambridge (1992)
11. Krizhevsky, A., Sutskever, I., Hinton, G.: ImageNet classification with deep convolutional neural networks, pp. 1106–1114 (2012)
12. Kurakin, A., Goodfellow, I.J., Bengio, S.: Adversarial examples in the physical world (2016)
13. Liu, N., Han, J., Yang, M.H.: PicaNet: learning pixel-wise contextual attention for saliency detection. In: Proceedings of the IEEE Conference on Computer Vision and Pattern Recognition (2018)
14. Norouzzadeh, M., et al.: Automatically identifying, counting, and describing wild animals in camera-trap images with deep learning. Proc. Natl. Acad. Sci. United States Am. **115**(25), E5716–E5725 (2018). https://doi.org/10.1073/pnas.1719367115
15. Olague, G., Clemente, E., Dozal, L., Hernandez, D.: Evolving an artificial visual cortex for object recognition with brain programming. In: Schuetze, O. et al. (eds.) EVOLVE - A Bridge between Probability, Set Oriented Numerics, and Evolutionary Computation III. Studies in Computational Intelligence, vol. 500, pp. 97–119. Springer, Heidelberg (2014). https://doi.org/10.1007/978-3-319-01460-9_5
16. Oram, M., Perrett, D.: Modeling visual recognition from neurobiological constraints. Neural Netw. **7**(6), 945–972 (1994)
17. Rensink, R.: Seeing, sensing and scrutinizing. Vis. Res. **40**(10–12), 1469–1487 (2000)
18. Su, J., Vargas, D.V., Sakurai, K.: One pixel attack for fooling deep neural networks. IEEE Trans. Evol. Comput. **23**(5), 828–841 (2019). https://doi.org/10.1109/tevc.2019.2890858, http://dx.doi.org/10.1109/TEVC.2019.2890858
19. Szegedy, C., et al.: Intriguing properties of neural networks (2014)
20. Treisman, A., Gelade, G.: A feature-integration theory of attention. Cogn. Psychol. **12**(1), 97–136 (1980)
21. Walther, D., Koch, C.: Modeling attention to salient proto-objects. Neural Netw. **19**(9), 1395–1407 (2006)
22. Wang, W., Lai, Q., Fu, H., Shen, J., Ling, H., Yang, R.: Salient object detection in the deep learning era: an in-depth survey. IEEE Trans. Patt. Anal. Mach. Intell. (2021)

Resilient Bioinspired Algorithms: A Computer System Design Perspective

Carlos Cotta[1,2]([✉])(iD) and Gustavo Olague[3](iD)

[1] Dept. Lenguajes y Ciencias de la Computación, ETSI Informática,
Campus de Teatinos, Universidad de Málaga, 29071 Málaga, Spain
ccottap@lcc.uma.es
[2] ITIS Software, Universidad de Málaga, Málaga, Spain
[3] EvoVisión Laboratory, CICESE Research Center,
Carretera Ensenada -Tijuana 3918, Col. Playitas, 22860 Ensenada, B.C., Mexico
olague@cicese.mx

Abstract. Resilience can be defined as a system's capability for returning to normal operation after having suffered a disruption. This notion is of the foremost interest in many areas, in particular engineering. We argue in this position paper that is a crucial property for bioinspired optimization algorithms as well. Following a computer system perspective, we correlate some of the defining requirements for attaining resilient systems to issues, features, and mechanisms of these techniques. It is shown that bioinspired algorithms do not only exhibit a notorious built-in resilience, but that their plasticity also allows accommodating components that may boost it in different ways. We also provide some relevant research directions in this area.

Keywords: Resilience · Bioinspired optimization · Robustness · Computer systems

1 Introduction

Stemming from the Latin word *resilire* (to jump back, or to rebound), dictionaries commonly define resilience as (1) the ability of something to return to its original shape after it has been pulled, stretched, pressed, bent, etc. and (2) the ability to become strong, healthy, or successful again after something bad happens. While the first definition is more in line with a literal Material Science interpretation, the second one has a more figurative sense that nevertheless seems more appropriate within a computational context: it captures the ability of a computer system to deliver again its functionality after a disruptive event takes place. While we shall revisit the meaning of resilience later on, let us note here in that it is commonly the case within such a computational context that resilience is identified as a synonym for fault tolerance and safety at

Carlos Cotta acknowledges support by Universidad de Málaga, Campus de Excelencia Internacional Andalucía Tech.

© Springer Nature Switzerland AG 2022
J. L. Jiménez Laredo et al. (Eds.): EvoApplications 2022, LNCS 13224, pp. 619–631, 2022.
https://doi.org/10.1007/978-3-031-02462-7_39

critical applications. However, notice that fault tolerance does not necessarily implies bouncing back to normal operation, and can simply entail a well-defined behavior after a fault [6]. Hence, we can argue that resilience goes beyond fault tolerance, or at least that it has its own particularities, some significant overlap with fault tolerance notwithstanding.

Resilience turns out to be a fundamental feature of technological systems. Our daily life is notoriously dependent on the availability and proper functioning of many networks, computing infrastructures, and most importantly on numerous algorithms running on them. Bioinspired optimization methods are no exception to this, since not only they can be deployed on irregular, dynamic computational environments [8] but they also may have to face challenges of diverse nature during their regular operation (dynamic environments, uncertain objectives, byzantine faults, etc.). It is therefore of the foremost interest to analyze these techniques, the challenges they have to cope with, and the way they can be appropriately designed from a resilience viewpoint. This position is defended in this work, in which we try to draw some rough lines to map the ground using some lessons from other engineering fields, as well as identifying some challenges in pursue of resilience properties. The rest of this paper is organized as follows: we firstly discuss some general issues about resilience at large (Sect. 2.1), and dive into the engineering perspective on this property, and the requirements to achieve it (Sect. 2.2); then, we proceed to discuss these requirements in the context of bioinspired algorithms, i.e., what they typically entail and/or how they are often approached (Sect. 3); we close the paper with an outlook and a sketch of some challenges we believe are important in this area (Sect. 4).

2   Background

2.1   What Is Resilience?

There are many different definitions of resilience [22]. The United Nations defined it in the General Assembly Resolution 71/276 as *"the ability of a system, community or society exposed to hazards to resist, absorb, accommodate, adapt to, transform and recover from the effects of a hazard in a timely and efficient manner, including through the preservation and restoration of its essential basic structures and functions through risk management,"* cf. [49]. In line with this definition, the potential sources for sudden, disruptive events are numerous, and they can have natural or anthropogenic causes: natural disasters, malicious human activity, health crises, economic meltdowns, and so on [55]. If we consider a physical, a computational, or a technosocial system [51], the specificities of the disruptive events can be different. However, the bottom-line remains: they expose vulnerabilities in the corresponding systems (and in a meta-level, so can they also do in higher systems that use the former). In response and –most importantly– in anticipation of this, it is necessary to foster resilience.

Building resilience allows reverting to normal conditions after a shock. Conversely, a lack of resilience prevents the restoration of these conditions. In this sense, it is essential to distinguish between resilience and the related notion of *robustness*: while the former refers to the capacity to withstand shocks dynamically, the latter sometimes denote the ability to resist shocks without adapting [7]. For example, it is possible to make a system robust by endowing it with redundant components to ensure continuous operation even if some of them fail. On the other hand, a system could be resilient by reconfiguring some components to keep delivering the required functionality after a shock (perhaps, having to endure a transient period of degraded performance until the reconfiguration is effective). Of course, these two possibilities are not mutually exclusive and can be ideally combined cost-effectively.

Resilience is essential to ensure *sustainability* (and conversely, absence of resilience results in unsustainability and adverse feedback loops after a crisis). Quite counterintuitively, sustainability can also be related to risk-exposure, in a phenomenon known as the volatility paradox [25], whereby systems with a low systemic risk build-up increasingly fragile, ultimately undermining sustainability; much like the immune system requires being confronted with pathogens to build up defenses, a system that endures crises at a higher frequency will develop resilience over time. Furthermore, disruptions (temporary shocks notwithstanding) are the catalysts of growth and innovation, which are essential for sustained progress instead of stagnation (biologists and evolutionary computation practitioners will identify a common theme here).

2.2   Resilience from an Engineering Perspective

Beyond the general definition of resilience provided in the previous subsection, it is possible to include more specific definitions withing the context of Engineering and more particularly in the context of Information, and Communications Technology. This will pave the way for attaining a better characterization of resilience in the domain of bioinspired optimization techniques. To this end, we will also indicate anthropogenic issues and malicious activities that generate a lack of trustworthiness in popular deep-learning methodologies. Also, we will provide examples in robotic technologies and machine learning industries for the interested reader.

According to [9], a resilient system acting within time, environmental, and operating conditions is that which is ready to perform its intended function, guaranteeing the absence of improper system alterations with the ability to anticipate and accommodate changes while executing and conducting servicing and inspection so that in case of a fault, quick restoration to a specified working condition must be achieved, or otherwise discontinue of the operation in a safe way. Figure 1 shows different attributes and measures of resilience relevant to this goal. Let us briefly discuss these:

- *Reliability* is a measure of the extent to which a system can provide a continued service up to a certain time. It is therefore particularly relevant to safety critical system in which service discontinuation is not an acceptable possibility.
- *Security* encloses a subset of attributes, including integrity, maintainability, and availability.

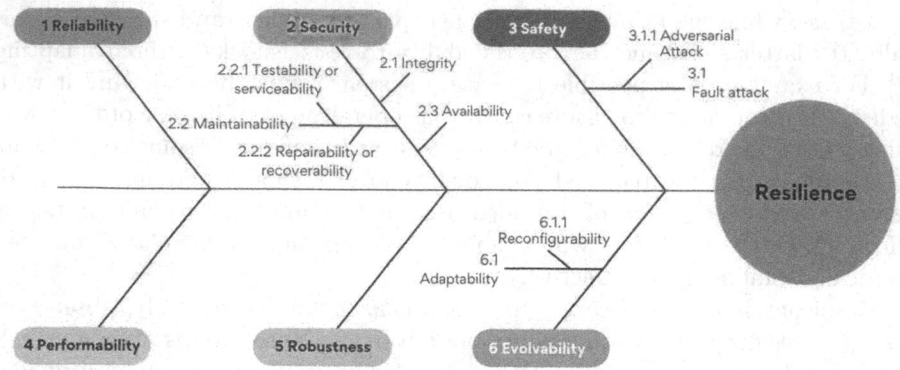

Fig. 1. Resilience is a term with multiple aspects such as reliability, security, safety, performability, robustness, and evolvabiliy.

- *Integrity* refers to absence of improper system states (be these states physical or –as it is usually the case in algorithms– logical). Keeping integrity thus refers to preventing improper alterations to the state.
- *Maintainability* captures the extent to which a system that has been damaged/compromised can be repaired.
- *Availability* refers to the readiness for service. It could be defined as a measure of the how often a certain system is functional.
- *Safety* amounts to the system's ability to avoid catastrophic failures, understood as any failure that causes damage to others systems and/or compromises the safety of these. It is thus a measure of the fail-safe capabilities of the system.
- *Performability* measures the extent to which a system performs above specific functioning requirements (be these, speed, accuracy, resource consumption, etc.). It is therefore an indicator of interest for systems whose performance can be determined in a quantitative way, as it is the case of bioinspired optimization algorithms.
- *Robustness* captures whether the system can deliver correct service conditions beyond the typical domain of operation, and without fundamental changes to the original system (cf. Sect. 2.1).
- *Evolvability* measures the extent to which the system can perform changes on itself, be it decreasing its level of performance or reliability for a specific time range to compensate for faults or during exceptional circumstances (graceful degradation) or by adapting any aspect of its functioning in order to ensure appropriate (or even improved) performance. It is therefore related to the notions of *elasticity* and *adaptability*. Specifically, engineers consider that a resilient system must have the ability to be adaptable (which is to be understood as the ability to evolve while executing; Therefore, adaptability is a subset of evolvability and may require anticipating changes prior to the resulting damage, or simply taking actions reacting to such changes).

As we may observe, resilience encompasses important attributes and measures that people use across science and engineering. Such concepts help to conceptualize different aspects needed to explain resilience. For example, resiliency naturally appears in robotics and machine learning in connection to malevolent external actions. Regarding the former, malicious attacks represent a challenging issue and preventing security vulnerabilities due to human factors is a significant subject aiming to implement and maintain effective countermeasures [53]. As to the latter, security is indeed an open issue since the technology is susceptible to adversarial attacks from hackers [1]. From the viewpoint of evolutionary computation we argue that human and data modeling are two aspects that need higher attention from the research community with interest in the development of bioinspired resilient systems.

3   Biuinspired Algorithms as Resilient Systems

Having laid out a general view of resilience in the previous section, as well as the requirements to attain this property in a computer system, let us discuss how these apply to the particular case of bioinspired optimization algorithms. Of course, this particular algorithmic paradigm has its own specifities, which render some of the issues defined before as only tangentially applicable to these techniques. This fact notwithstanding, the core requirements for resilience are relevant in this domain and can be actually quantified, as we shall see. Indeed, it is possible to group these requirements in a natural way into a number of feature sets which are described next.

3.1   Integrity and Safety

While integrity and safety may appear to be in principle some of the least applicable resilience features in this case, they actually characterize an important issue that has been extensively studied in this domain. These features can be seen as natural sides of a common issue in bioinspired optimization techniques: the normal operation of the system should not result in damage in its own state (let alone catastrophic damage), even accounting for potential external factors. Leaving aside the latter for a moment, it turns out there is indeed an internal operational factor that can cause damage to the algorithm state, namely convergence to suboptimal regions of the search space. This is typically due to the presence of deceptive features [52] in the search landscape, whereby the algorithm is led to local optima which may be in some cases far from the actual global optimum [24]. As mentioned before, the issue of premature convergence and deceptiveness has been one of the major topics studied in the literature, both from theoretical and experimental perspectives. The use of populations is widely regarded as one of the primary safeguards of bioinspired optimization techniques with respect to deceptive local optima, though their ability to escape from these will greatly depend on the effective maintenance of diversity. Fortunately, there are numerous mechanisms whereby diversity can be promoted, either proactively (e.g., use of non-panmictic populations [18], ad-hoc operators [11,16], etc. – see also [31]) or reactively (e.g., random immigrants [48], triggered hypermutation [29], etc. – cf. Sect. 3.2). This also means that it is possible to capitalize on the knowledge available on convergence metrics (see, e.g., [12]) in order to measure the integrity $I(t)$ of the algorithm.

If external factors come into play, integrity and safety become more prominent features. Consider for example the case of volunteer computing (VC) networks. The existence of malicious agents who operate within these networks providing false results (i.e., *cheaters*) has been long documented [42]. Research suggests that distributed evolutionary algorithms running in this kind of hostile environments can indeed tolerate some degree of cheating, and would theoretically converge to the optimum given enough time [30]. Needless to say, these malicious agents can exert some other pernicious influence on distributed applications, but most of these are either implementation-dependent (e.g., overflowing buffers, injecting code, etc.) or can be better dealt with by other resilience requirements (e.g., crashes – see Sect. 3.3).

3.2   Evolvability and Adaptability

Evolvability and adaptability are flagship features of bioinspired optimization techniques, and surely one of their *raisons d'être*. Indeed, parameter adaptation is deeply rooted as a core principle of some bioinspired computation flavors (e.g., evolution strategies), and was taxomized well before the turn of the century [21]. Of course, adaptation (and self-adaptation) does not limit to parameter control in bioinspired methods. As a matter of fact, it can be found in other components such as population structures [15], or the definition of variation operators (e.g., local search mechanisms in memetic algorithms [44]), just to name a few. This is not surprising, since one of the keystones in practical (meta)heuristic problem-solving is the fact that tuning the optimization technique to the problem under consideration is paramount for achieving top performance, and that transferring a part of this tuning/customization effort from the human designer to the algorithm itself –i.e., by endowing it with smart mechanisms to self-adapt to the problem– has been a long pursued goal in the field of metaheuristics [10]. To a large extent, this is something that lies precisely at the root of the notion of memetic computing [32] (which is understood as the harmonic coordination of complex computational structures composed of interacting modules –memes– for problem solving, whose representation is stored and manipulated by the algorithm itself) and hyperheuristics [14] (which comprise heuristic techniques for intelligently selecting or generating a suitable heuristic for a given situation).

The previous examples underpin the amenability of bioinspired optimization methods for accommodating (self-)adaptive components and effectively taking advantage of them. From a broader perspective, these components are responsible for endowing the algorithm with self-⋆ properties, namely any property whereby a certain system can exert advanced control on its own functioning and/or structure [4], ultimately contributing to the resilience of the former.

Such properties may include self-organizing, self-healing, self-configuring or self-scaling among many others [5], and they turn out to be essential to cope with some major environmental disruptions as shown next. Before proceeding to that, let us note that evolvability is multi-faceted, and therefore quantifying $E(t)$ may depend on the particular feature subject to study (see, e.g., [45,46]).

3.3 Performability, Recoverability and Robustness

Performability and recoverability arguably capture the quintessential features required to develop resilience in this context. As anticipated before, it is very common that from a computational perspective resilience is equated to (or at least strongly connected to) fault tolerance. There is a large truth in this connection, at least to the extent that fault tolerance is understood as the ability to keep delivering the expected performance in the presence of failures (maybe tautologically so). Then again, fault tolerance can be more broadly assumed to mean well-defined behavior in the presence of failures [6]. Furthermore, even if we assume that delivering uninterrupted service is what defines a fault-tolerant system, an argument could be done as to whether this is achieved by means of some built-in robustness (i.e., the system is capable to withstand failures without needing to adapt or change its behavior), or by developing resilience (i.e., the system recovers its performance after a transient degradation phase, by means of some internal adaptation or reconfiguration), cf. Sect. 2.1. As we shall see, bioinspired optimization methods can achieve fault tolerance under either interpretation, although in one case they may simply rely on its intrinsic architecture, whereas in the other the inclusion of appropriate mechanisms may be required.

Focusing on evolutionary algorithms in particular, basic fault-tolerance has been analyzed from different perspectives. It has been established that the use of populations provides some intrinsic redundancy, whereby moderate losses of individuals do not result in major performance degradation in master-slave panmictic models [27], fine-grained (cellular) decentralized models [26], and coarse-grained (island-based) decentralized models [20], and this robustness can be enhanced via standard fault-tolerance mechanisms such as checkpointing [33]. A more interesting perspective from the resilience viewpoint can be attained by endowing the algorithm with self-* properties, as mentioned in Sect. 3.2. Relevant properties in this context are self-scaling (i.e., exerting internal reconfiguration in response to changes in the computational substrate) [34] and self-healing (i.e., performing actions to correct any damage infringed by external disruptions) [35]. Performability is approached from a different angle in *brain programming* [39], a symbolic paradigm that uses the power of genetic programming combined with neuroscientific modeling and that is aimed at purposive vision. Therein, generality is a designed property where models are constantly trained in one problem and tested on a different problem, not only changing the dataset but the whole visual task [38].

The bottom-line is here that –as mentioned in Sect. 3.2– these methods can naturally accommodate self-adaptive and reactive components that enable sensibly responding to failures in a resilient way. Furthermore, the very quantitative nature in which the performance of bioinspired optimization methods can be measured leads in turn to an amenable quantification of performability $P(t)$ at time t, by comparing to regular undegraded performance values, or by determining the maximum level at which the algorithm yielded acceptable performance up to time t (i.e., $P(t) = \max(L \mid \forall t' \leqslant t : \psi(t') \geqslant L)$, where $\psi(t)$ is the algorithm's performance at time t).

3.4   Reliability and Availability

Reliability and availability have in this particular context a significant overlap with the features previously discussed. If service continuity (understood as the effective fulfilment of the optimization purposes of the specific bioinspired method considered) is pursued in the presence of computational failures, we would be in the scenarios depicted in previous section. The previous setting is not the only possible one in order to assess reliability. In fact, there is a very interesting and relevant subfield of research that deals with dynamic optimization, namely the use of these methods in scenarios in which the optimization target changes along time [2]. Needless to say, this poses great challenges to any optimization technique, and require the use of appropriate mechanisms (such as using archives of previous solutions, diversity-preservation policies, and control mechanisms to anticipate, detect, or react to changes in the optimization target – see [28,54]) in order to be able to provide trustworthy operation in this kind of environments. From a different perspective, reliability aspects also appear in vision metrology systems (where thousands of simulation evaluations using complex nonlinear least-squares analysis are required, often relying in surrogate models) [37] and in the analysis of corner extraction [36].

In either case, it must be noted that reliability is often a property used to characterize systems where failures are unacceptable (e.g., safety critical systems) [9], which means that in this particular bioinspired optimization context adequate thresholds would have to be defined to characterize when the transient degraded performance does not render the optimization service discontinued. From a more quantitative point of view, the reliability $R(t)$ could be here defined as the probability of the algorithm delivering acceptable functionality at time t, which could be approximated as the proportion of runs in which acceptable behavior is observed. Likewise, the availability $A(t)$ could be approximated as the fraction of the time in which the algorithm delivered correct performance.

3.5   Sustainability

Sustainability refers to the ability to maintain a trend or a process in the long run. It is a concept that is frequently brought up in connection to human activities and the impact that these have on the environment and the toll they exert on future resource availability. As indicated in Sect. 2.1, sustainability is one of the natural consequences of resilient operation. In this sense, sustainability is not just an application goal of bioinspired methods (and any other AI method, for that matter), but also an operational requirement of these techniques. Unsurprisingly, AI methods have been identified as having a significant carbon footprint [47]. While a standard of measurement is still absent for quantifying energy consumption and carbon emission in the life cycle of AI methods [13], it is clear that monitoring the emission level of these techniques, and prioritizing energetically efficient computational platforms and algorithms is crucial. This has led to the notions of *red AI* and *green AI* [43]. The former refers to AI research that seeks to obtain state-of-the-art results through the use of massive computational power, and particularly applies to scenarios in which the computational effort scales at a substantially larger pace than the gains obtained in the results [19], whereas green AI research provides novel results without increasing computational cost or even reducing it. This is a topic that is gaining momentum in the context of machine-learning, but for which some seminal works notwithstanding, e.g., [3,17], requires further analysis in the context of bioinspired optimization.

4   Outlook and Challenges

Resilience in engineering and computer science is a well-established research area, and we found a rich connection with bioinspired approaches in most aspects typically studied in other domains. We identified five research axes, namely (i) integrity and safety, (ii) evolvability and adaptability, (iii) performability and recoverability, (iv) reliability and availability, and (v) sustainability. We give a first account of the kind of problems researchers study in each compound set of attributes. An important lesson learned is that bioinspired methods have a great deal of intrinsic resilience and –most importantly– are flexible enough to admit being augmented with components to boost resilience in their different aspects.

We envision that future research will not just focus on studying the resilience of bioinspired optimization methods, but should also exploit resilience as a stimulus for optimization. In this sense, it must be noted that exposure to disruptions is very often the catalyst for breakthroughs. This has been observed in many systems, including bioinspired methods (e.g., see [34,50]), and lies in the spirit of some long-known approaches such as competitive coevolution [41] and, more recently, adversarial attacks (e.g., [23,40]). The latter have a longer history in the area of computer vision and machine learning, and are bound to have an important impact in bioinspired optimization as well. Needless to say, these strategies are targeted to disrupt evolutionary equilibrium, hence exerting a continuous and directed selective pressure, but analogous strategies aimed to strategically attack other aspects of the algorithm are not inconceivable, and would provide the playground for the evolution of resilience, and indirectly for the improvement of the underlying optimization process. Another line of research which we envision will gain momentum in the near future is the development of green bioinspired optimization methods, following the trend of the machine-learning community.

References

1. Akhtar, N., Mian, A.: Threat of adversarial attacks on deep learning in computer vision: a survey. IEEE Access **6**, 14410–14430 (2018)
2. Alba, E., Nakib, A., Siarry, P. (eds.): Metaheuristics for Dynamic Optimization, Studies in Computational Intelligence, vol. 433. Springer, Berlin (2013)
3. Diaz Álvarez, J., et al.: A fuzzy rule-based system to predict energy consumption of genetic programming algorithms. Comput. Sci. Inf. Syst. **15**(3), 635–654 (2018)
4. Babaoglu, O., Jelasity, M., Montresor, A., Fetzer, C., Leonardi, S., van Moorsel, A., van Steen, M. (eds.): SELF-STAR 2004. LNCS, vol. 3460. Springer, Heidelberg (2005). https://doi.org/10.1007/b136551
5. Berns, A., Ghosh, S.: Dissecting self-⋆ properties. In: Third IEEE International Conference on Self-Adaptive and Self-Organizing Systems - SASO 2009, pp. 10–19. IEEE Press, San Francisco, CA (2009)
6. Bouvry, P., et al.: Resilience within ultrascale computing system: challenges and opportunities from Nesus project. Supercomput. Front. Innov. **2**(2), 46–63 (2015)
7. Brunnermeier, M.: The Resilient Society. Endeavor Literary Press, Colorado Springs (2021)
8. Camacho, D., et al.: From ephemeral computing to deep bioinspired algorithms: new trends and applications. Future Generation Comput. Syst. **88**, 735–746 (2018)
9. Castano, V., Schagaev, I.: Resilient Computer System Design. Springer, Cham (2015). https://doi.org/10.1007/978-3-319-15069-7
10. Cotta, C., Sevaux, M., Sörensen, K. (eds.): Adaptive and Multilevel Metaheuristics, Studies in Computational Intelligence, vol. 136. Springer-Verlag, Berlin Heidelberg (2008)
11. Cotta, C., Troya, J.: Using dynastic exploring recombination to promote diversity in genetic search. In: Schoenauer, M., et al. (eds.) Parallel Problem Solving From Nature VI. Lecture Notes in Computer Science, vol. 1917, pp. 325–334. Springer-Verlag, Paris (2000)
12. Derrac, J., García, S., Hui, S., Suganthan, P.N., Herrera, F.: Analyzing convergence performance of evolutionary algorithms: a statistical approach. Inf. Sci. **289**, 41–58 (2014)
13. Dhar, P.: The carbon impact of artificial intelligence. Nat. Mach. Intell. **2**, 423–425 (2020)
14. Drake, J.H., Kheiri, A., Özcan, E., Burke, E.K.: Recent advances in selection hyper-heuristics. Eur. J. Oper. Res. **285**(2), 405–428 (2020)
15. Fernandes, C.M., Rosa, A.C., Laredo, J.L., Merelo, J., Cotta, C.: Dynamic models of partially connected topologies for population-based metaheuristics. In: 2018 IEEE Congress on Evolutionary Computation (CEC), pp. 1–8. IEEE, Rio de Janeiro, Brazil (2018)
16. Fernandes, C., Laredo, J., Rosa, A., Merelo, J.: The sandpile mutation genetic algorithm: an investigation on the working mechanisms of a diversity-oriented and self-organized mutation operator for non-stationary functions. Appl. Intell. **39**(2), 279–306 (2013)
17. de Vega, F.F., Chávez, F., Díaz, J., García, J.A., Castillo, P.A., Merelo, J.J., Cotta, C.: A cross-platform assessment of energy consumption in evolutionary algorithms. In: Handl, J., Hart, E., Lewis, P.R., López-Ibáñez, M., Ochoa, G., Paechter, B. (eds.) PPSN 2016. LNCS, vol. 9921, pp. 548–557. Springer, Cham (2016). https://doi.org/10.1007/978-3-319-45823-6_51

18. Harada, T., Alba, E.: Parallel genetic algorithms: a useful survey. ACM Comput. Surv. **53**(4), 86:1–86:39 (2020)
19. Henderson, P., Hu, J., Romoff, J., Brunskill, E., Jurafsky, D., Pineau, J.: Towards the systematic reporting of the energy and carbon footprints of machine learning. J. Mach. Learn. Res. **21**(248), 1–43 (2020)
20. Hidalgo, J., Lanchares, J., Fernández de Vega, F., Lombraña, D.: Is the island model fault tolerant? In: Genetic and Evolutionary Computation - GECCO 2007, pp. 2737–2744. ACM, New York (2007)
21. Hinterding, R., Michalewicz, Z., Eiben, A.: Adaptation in evolutionary computation: a survey. In: Fourth IEEE Conference on Evolutionary Computation, pp. 65–69. IEEE Press, Piscataway, New Jersey (1997)
22. Huq, M., Sarker, M., Prasad, R., et al.: Resilience for disaster management: opportunities and challenges. In: Alam, G., et al. (eds.) Climate Vulnerability and Resilience in the Global South, pp. 425–442. Springer, Cham (2021)
23. Ibarra-Vazquez, G., Olague, G., Chan Ley, M., Puente, C., Soubervielle-Montalvo, C.: Brain programming is immune to adversarial attacks: towards accurate and robust image classification using symbolic learning. Swarm Evol. Comput. (2022, to appear)
24. Jones, T., Forrest, S.: Fitness distance correlation as a measure of problem difficulty for genetic algorithms. In: Eshelman, L.J. (ed.) Proceedings of the Sixth International Conference on Genetic Algorithms, pp. 184–192. Morgan Kaufmann (1995)
25. Kubitza, C.: Tackling the volatility paradox: spillover persistence and systemic risk. ECONtribute Discussion Paper No. 079 (2021). https://doi.org/10.2139/ssrn.2858763
26. Laredo, J., Castillo, P., Mora, A., Merelo, J., Fernandes, C.: Resilience to churn of a peer-to-peer evolutionary algorithm. Int. J. High Performance Syst. Architecture **1**(4), 260–268 (2008)
27. Lombraña González, D., Jiménez Laredo, J., Fernández de Vega, F., Guervós, J.M.: Characterizing fault-tolerance in evolutionary algorithms. In: Parallel Architectures and Bioinspired Algorithms. Springer-Verlag (2012). https://doi.org/10.1007/978-3-642-28789-3_4
28. Mavrovouniotis, M., Li, C., Yang, S.: A survey of swarm intelligence for dynamic optimization: algorithms and applications. Swarm Evol. Comput. **33**, 1–17 (2017)
29. Morrison, R., De Jong, K.: Triggered hypermutation revisited. In: Proceedings of the 2000 Congress on Evolutionary Computation, vol. 2, pp. 1025–1032. IEEE (2000)
30. Muszyński, J., Varrette, S., Bouvry, P., Seredyński, F., Khan, S.U.: Convergence analysis of evolutionary algorithms in the presence of crash-faults and cheaters. Comput. Math. Appl. **64**(12), 3805–3819 (2012)
31. Neri, F.: Diversity management in memetic algorithms. In: Neri, F., Cotta, C., Moscato, P. (eds.) Handbook of Memetic Algorithms, Studies in Computational Intelligence, vol. 379, pp. 155–167. Springer-Verlag, Berlin Heidelberg (2012)
32. Neri, F., Cotta, C.: Memetic algorithms and memetic computing optimization: a literature review. Swarm Evol. Comput. **2**, 1–14 (2012)
33. Nogueras, R., Cotta, C.: Studying fault-tolerance in island-based evolutionary and multimemetic algorithms. J. Grid Comput. **13**, 351–374 (2015)
34. Nogueras, R., Cotta, C.: Studying self-balancing strategies in island-based multimemetic algorithms. J. Comput. Appl. Math. **293**, 180–191 (2016)

35. Nogueras, R., Cotta, C.: Self-healing strategies for memetic algorithms in unstable and ephemeral computational environments. Nat. Comput. **16**(2), 189–200 (2016). https://doi.org/10.1007/s11047-016-9560-7
36. Olague, G., Hernandez, B., Dunn, E.: Hybrid evolutionary ridge regression approach for high-accurate corner extraction. In: 2003 IEEE Computer Society Conference on Computer Vision and Pattern Recognition, 2003. Proceedings, vol. 1, pp. I-I (2003)
37. Olague, G.: Automated photogrammetric network design using genetic algorithms. Photogrammetric Eng. Remote Sens. **68**(5), 423–431 (2002)
38. Olague, G., Clemente, E., Hernández, D.E., Barrera, A., Chan-Ley, M., Bakshi, S.: Artificial visual cortex and random search for object categorization. IEEE Access **7**, 54054–54072 (2019). https://doi.org/10.1109/ACCESS.2019.2912792
39. Olague, G., Hernández, D.E., Clemente, E., Chan-Ley, M.: Evolving head tracking routines with brain programming. IEEE Access **6**, 26254–26270 (2018)
40. Qiu, H., Custode, L.L., Iacca, G.: Black-box adversarial attacks using evolution strategies. In: Proceedings of the Genetic and Evolutionary Computation Conference Companion, pp. 1827–1833. GECCO 2021, Association for Computing Machinery, New York, NY, USA (2021)
41. Rosin, C.D., Belew, R.K.: New Methods for Competitive Coevolution. Evol. Comput. **5**(1), 1–29 (1997)
42. Sarmenta, L.F.: Sabotage-tolerance mechanisms for volunteer computing systems. Future Generation Comput. Syst. **18**(4), 561–572 (2002)
43. Schwartz, R., Dodge, J., Smith, N., Etzioni, O.: Green AI. Commun. ACM **63**(12), 54–63 (2020)
44. Smith, J.E.: Self-adaptative and coevolving memetic algorithms. In: Neri, F., Cotta, C., Moscato, P. (eds.) Handbook of Memetic Algorithms, Studies in Computational Intelligence, vol. 379, pp. 167–188. Springer, Berlin (2012). https://doi.org/10.1007/978-3-642-23247-3_11
45. Smith, T., Husbands, P., O'Shea, M.: Fitness landscapes and evolvability. Evol. Comput. **10**(1), 1–34 (2002)
46. Soria-Alcaraz, J.A., Espinal, A., Sotelo-Figueroa, M.A.: Evolvability metric estimation by a parallel perceptron for on-line selection hyper-heuristics. IEEE Access **5**, 7055–7063 (2017)
47. Strubell, E., Ganesh, A., McCallum, A.: Energy and policy considerations for modern deep learning research. Proc. AAAI Conf. Artif. Intell. **34**(09), 13693–13696 (2020)
48. Tinós, R., Yang, S.: A self-organizing random immigrants genetic algorithm for dynamic optimization problems. Genetic Program. Evol. Mach. **8**, 255–286 (2007)
49. United Nations High-level Political Forum on Sustainable Development (HLPF): 2018 HPLF thematic review: Transformation towards sustainable and resilient societies - building resilience (2018), sustainabledevelopment.un.org/hlpf/2018
50. de Vega, F.F., Cantú-Paz, E., López, J.I., Manzano, T.: Saving resources with plagues in genetic algorithms. In: Yao, X., et al. (eds.) PPSN 2004. LNCS, vol. 3242, pp. 272–281. Springer, Heidelberg (2004). https://doi.org/10.1007/978-3-540-30217-9_28
51. Vespignani, A.: Predicting the behavior of techno-social systems. Science **325**, 425–428 (2009)
52. Whitley, D.: Mk landscapes, NK landscapes, MAX-KSAT: a proof that the only challenging problems are deceptive. In: Silva, S., Esparcia-Alcázar, A.I. (eds.) Proceedings of the 2015 Annual Conference on Genetic and Evolutionary Computation, pp. 927–934. ACM, New York, NY (2015)

53. Yaacoub, J.P.A., Noura, H.N., Salman, O., Chehab, A.: Robotics cyber security: vulnerabilities, attacks, countermeasures, and recommendations. Int. J. Inf. Secur. (2021)
54. Yazdani, D., Cheng, R., Yazdani, D., Branke, J., Jin, Y., Yao, X.: A survey of evolutionary continuous dynamic optimization over two decades-part A. IEEE Trans. Evol. Comput. **25**(4), 630–650 (2021)
55. Zobel, C., Khansa, L.: Characterizing multi-event disaster resilience. Comput. Oper. Res. **42**, 83–94 (2014)

Evolutionary Robotics

Seeking Specialization Through Novelty in Distributed Online Collective Robotics

Amine Boumaza[(✉)]

Université de Lorraine, CNRS, Inria, LORIA, 54000 Nancy, France
amine.boumaza@loria.fr

Abstract. Online Embodied Evolution is a distributed learning method for collective heterogeneous robotic swarms, in which evolution is carried out in a decentralized manner. In this work, we address the problem of promoting reproductive isolation, a feature that has been identified as crucial in situations where behavioral specialization is desired. We hypothesize that one way to allow a swarm of robots to specialize on different tasks is through the promotion of diversity. Our contribution is twofold, we describe a method that allows a swarm of heterogeneous agents evolving online to maintain a high degree of diversity in behavioral space in which selection is based on originality. We also introduce a behavioral distance measure that compares behaviors in the same conditions to provide reliable measurements in online distributed situations. We test the hypothesis on a concurrent foraging task and the experiments show that diversity is indeed preserved and, that different behaviors emerge in the swarm; suggesting the emergence of reproductive isolation. Finally, we employ different analysis tools from computational biology that further support this claim.

Keywords: Online evolutionary robotics · Behavioral diversity · Task specialization · Phylogeny

1 Introduction

Embodied evolutionary robotics (EER), aims to design collective behaviors for a swarm of heterogeneous agents evolving online [8]. In a nutshell, these are algorithms in which evolution is carried out in a decentralized manner, where each agent, typically a mobile robot, runs an EA on board and exchanges genetic material with other agents when they meet. Selection and variation are performed locally by the agent. As such, there is no central process that governs evolution, in contrast with traditional evolutionary robotics (ER). These algorithms have been successfully applied in different contexts and have been shown to be robust in open-ended environments [2]. However, due to their distributed nature and the fact that they operate online, some problems remain challenging for EER. For example, evolving task specialization or division of labor has been very challenging and remains an open problem.

© Springer Nature Switzerland AG 2022
J. L. Jiménez Laredo et al. (Eds.): EvoApplications 2022, LNCS 13224, pp. 635–650, 2022.
https://doi.org/10.1007/978-3-031-02462-7_40

It has been shown that evolving specialized behaviors requires reproductive isolation and tailored selection operators [16]. The fact that agents can spread their genetic material to any other agent does not help the emergence and the conservation of specialized behaviors. One way to solve this issue is to limit mating encounters trough geographical isolation (*Allopatric speciation*) where contact between agents from different regions is limited [20]. However, forcing geographical isolation limits the range of problems and environments that the swarm can tackle. One of the objectives of online EER is long-term adaptation in open-ended environments; having such a requirement can be a big limitation.

In the present work, our goal is to propose a new hypothesis that could facilitate behavior specialization. Our main argument can be stated as follows: if reproductive isolation is a requirement to behavior specialization, can this isolation be enforced in other ways than geographically? For instance, can we favor this isolation during the reproduction phase by choosing the "right" genetic material? This type of isolation can be considered as an instance of *Sympatric speciation*, akin to what exists in nature where reproductive isolation evolves within a population sharing the same environment, allowing the divergence into different species [15]. There exist in nature some evidence that behavioral separation can be a plausible mechanism for promoting *Sympatric speciation*. We propose here to explore the idea of promoting isolation by promoting behavioral diversity.

Ever since the introduction of novelty search [14], there has been a large body of work on the idea of searching for originality; disposing of objectives. It opened the way to many clever and powerful algorithms in evolutionary robotics (ER) and later lead to quality diversity (QD) algorithms which elegantly combine the search for novel and fit solutions [17,21].

Although very efficient in traditional ER, their application in Online collective robotics raises few challenges. For instance, they rely on centralized archives that are incrementally filled with newly discovered behaviors, whereas EER are decentralized and emphasize local information exchange and no history maintenance. Furthermore, the property of EER that is probably the greatest challenge, is the fact that they operate online. Individual behaviors resulting from the interactions of the agents with their environment or other agents, cannot be predicted and cannot be reproduced, which make their comparison difficult.

The search for novelty and the promotion of diversity, have been previously introduced in collective robotics. For instance, [9] applied novelty search on a swarm of agents on an aggregation task. In that instance, the authors used NEAT [22] a centralized off-line algorithm and all agents shared the same controller (homogeneous agents). It showed that novelty can improve exploration for solutions in swarm robotics. In the case of online collective robotics, [12] introduced a decentralized instance of MAP-Elites [17] where agents exchange locally behavioral maps. That work showed that behavioral diversity can be possible without reproductive isolation, however the algorithm needs task specific behavioral descriptors. Furthermore, measuring the behavior of solutions on agents in specific situations may not generalize well to other situations due to the dynamics of the environment.

1.1 Objectives

As said previously, evolving specialized behaviors requires reproductive isolation. In some cases, choosing specially crafted selection operators, and fitness assignment schemes can also help for the emergence of specialization. For instance, [11] showed that using a "market" mechanism that favors the least prevalent behavior by artificially increasing its fitness value, allows the algorithm to balance between two behaviors. One can even argue that the "market" mechanism creates some level of isolation, as it devalues behaviors based on their frequency in the population and then selection pressure limits their spread. The central theme of this paper follows a similar general idea.

We argue that promoting diversity, as it is done in Novelty Search [14] for example, can increase reproductive isolation and thus favors the emergence of specialization. To this end, we propose a selection scheme that promotes diversity by selecting solutions based on their originality in behavioral space. The selection scheme respects the online and distributed nature of EER, it operates and uses generic behavioral descriptors measures locally on the agents. The experiments we present aim at verifying the following hypotheses claiming that:

H1. it is possible to promote diversity with the proposed selection scheme and,
H2. promoting diversity favor reproduction isolation.

In the following, we start by describing the $(\mu, 1)$-ONLINE EEA and the different selection schemes we used. We then describe our experimental procedure along with the tested scenarios and finally, conclude with a discussion of the results.

2 Methods

2.1 The $(\mu, 1)$-ONLINE Embodied EA

The main inspiration of the $(\mu, 1)$-ONLINE EEA is the original version of minimally Environment-driven Distributed EA (mEDEA) [3] to which with we add a selection operator that we will describe later (Algorithm 1). The algorithm considers a swarm of λ mobile agents a^j with $j = 1, \ldots, \lambda$ each executing a neuro-controller whose parameters are $\mathbf{x}^j$ (the active genome). Each agent maintains a list L^j, initially empty, in which it stores other genomes that it receives from other agents.

At each time step $t < t_{\max}$, an agent executes its active controller and broadcasts its genome within a limited range. In parallel, it listens for genomes originating from other agents, and when a genome is received (a mating event), it is stored in the agent's list L^j (its local population). This procedure is executed in parallel on all agents during $t_{\max}$ steps, the evaluation period or one generation.

During the evaluation phase, fitness values are assigned to the individual agents based on their own performance with regard to the given task and continuously updated (line 12). The fitness values are transmitted along the genome and stored on the receiving end and if an agent receives an already seen genome,

---

Algorithm 1: $(\mu, 1)$-ONLINE EEA.

```
 1  for  1 ≤ j ≤ λ in parallel do
 2        x^j ← random()
 3        a^j is active
 4        σ^j ← σ₀

 5  repeat
 6        for  1 ≤ j ≤ λ in parallel do
 7              t ← 0, f^j ← 0, L^j ← ∅
 8              while  t < t_max do
 9                    t ← t + 1
10                    if a^j is active then
11                          execute(x^j)
12                          update(f^j)
13                          if t > τ t_max then
14                                broadcast(x^j, f^j, σ^j)

15                    L^j ← L^j ∪ listen()
16              if L^j ≠ ∅ then
17                    x̄ ← select(L^j)
18                    σ ← adapt(σ^x̄)
19                    x^j ← mutate(x̄, σ)
20              else a^j is not active

21  until termination condition met
```

---

it updates the genome's fitness. To further ensure that fitness values are accurate, a *maturation age* is required of the agent before broadcasting [13]. This is set as $\tau\, t_{\max}$ with $\tau < 1$.

At the end of a generation, agents select a genome $\bar{\mathbf{x}} \in L^j$, and replace their active genome with a mutated copy of the selected one. Their list is then emptied and a new generation begins. In this work, the genome is a vector $\mathbf{x} \in \mathbb{R}^N$, which represents the weights of the neuro-controller. Only the weights undergo evolution (fixed-topology).

Each genome has its own mutation step-size σ whose initial values is σ_0, and when broadcasting, this value is also sent along with the genome and its fitness. On the receiving end, when a genome is selected from the local list it is mutated using its σ value. Step-sizes are adapted before the mutation takes place, where each agent chooses with probability (0.5) to either increase or a decrease the step-size. This rule increases the likelihood that the most adapted value between the two, will survive and spread in the swarm. The update rule is defined as:

$$\sigma = \begin{cases} \min\left(\sigma(1+\gamma),\ \sigma^+\right) & \text{if increase} \\ \max\left(\sigma(1-\gamma),\ \sigma^-\right) & \text{if decrease} \end{cases}$$

where σ^+ and σ^- are the allowable upper and lower bounds and γ some positive constant (Table 1). Mutation is Gaussian using the updated step-size:

$$\mathbf{x}^j := \bar{\mathbf{x}} + \sigma^2 \times \mathcal{N}(1,0). \tag{1}$$

In the event where an agent had not mating opportunities and finishes its evaluation period with an empty list $L^j = \emptyset$, it becomes inactive; a state during which the agent is motionless and which can last multiple generations. During this period, the inactive agent continues to listen for incoming genomes from other agents passing by, and once $L^j \neq \emptyset$ the agent becomes active again at the beginning of the next generation.

The number of genomes the agents collects $\mu^j = |L^j|$ ($0 \leq \mu^j \leq \lambda$) is conditioned by its mating encounters. Since the communication range is limited, agents that travel long distances will increase their probability of mating. We should note that mating encounters allow agents to spread their active genome and to collect new genetic material. The algorithm stores only one copy of the same genome[1] if it is received more than once.

The main difference between mEDEA and Algorithm 1 is that we do not consider a listening phase. Agents broadcast if they are active and listen all the time, whereas in mEDEA, agents must be in a specific listening state to record incoming genetic material.

2.2 Seeking New Behaviors

In novelty base search [14], the goal of the evolutionary algorithm is to discover solution that exhibit new unseen behaviors. These solutions are recorded in an archive which prevents the algorithm to "rediscover" similar solutions later. New solutions are compared to ones in the archive, and if novel enough they extend the archive. In our context, the notion of novelty, is defined differently. Since agents empty their genome list after selecting a new active genome, we do not have an archive. The selection scheme we propose selects solutions at the agent level by choosing the solutions that have the most different behavior from other solutions in the local list. The open question now is how do we measure behavior? Traditionally, selection based on behavioral distances requires the definition of a so-called behavioral descriptor; a set of features that capture the behavior of solutions, in a given environment. These descriptors allow to compute a distance between behaviors which is the basis for selection. These behaviors descriptors fall generally in two categories: 1) task related, requiring expert knowledge about the task or 2) generic descriptors, that can be applied on a large class of problems. This latter category often relies on sensor and actuator data to characterize behaviors [10,19].

In addition to the lack of an archive, we believe that measuring the behavior during the lifetime of the agent online is not a reliable basis for comparison. Because agents live specific experiences depending on their encounters, it is difficult to generalize a measurement outside from the situation that it was measured

---

[1] The term same is here used in the sense "originating from the same agent".

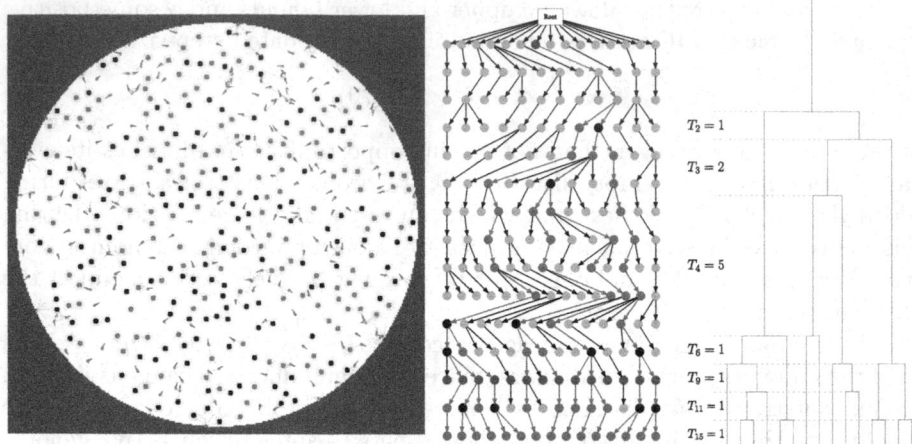

Fig. 1. A screenshot of the simulator (left), black and red circles represent items to be collected, the rest are agents. An example of a phylogenetic tree for 15 agents and 14 generations (middle) and the corresponding coalescence tree (right). Gray nodes represent active genomes, black nodes represent coalescence events and the green node the MRCA of the surviving genomes. The generation at which the takeover event occurs is where all active genomes are colored in blue (all descendants of a single initial genome, also colored in blue.) The coalescence tree is a different representation of the sub-tree that is colored in red. (Color figure online)

in. Consider for example, a genome that realized a given behavior during its lifetime, the same genome on a different vehicle would most certainly exhibit a different behavior. It would have lived a different experience, crossing the path of different agents and obstacles at different times. If we measure its behavior on its first experience, it will not be a reliable measure on which we could base a behavioral distance. Furthermore, since genomes get only one chance to be evaluated, reevaluations in different conditions is not possible.

To overcome these limitations, we propose to estimate the behavior of the genomes regardless of their agents' experience. If the goal is to identify the most different behavior in the agent's selection pool, we propose to simulate behaviors and measure their difference on fake sensory data and use these measurements as a selection criterion. The important step here is to present all genomes in the selection pool the same "fake" sensory data, as if they experienced the same situation, and compare their "would be" response to that fake situation. We use the terms "fake" and "would be" to emphasize the fact that the sensory data is not measured by agents moving in the environment, but randomly generated and, the responses are not executed by the agents, they are just the outputs of the neuro-controller.

To be more specific, let us consider one agent of the swarm, let it be a. We note $L^a = \{x_1 \ldots x_\mu\}$ the set of genomes it collected during its last lifetime. Our goal is to identify $\bar{x} \in L^a$ that has the most different behavior.

At the selection step, agent a creates the set $\mathcal{I}$ consisting of K "fake" input vectors such as $\mathcal{I} = \{\mathbf{I}_k | \mathbf{I}_k \sim \text{Uniform}(0,1)^{|s|}\}$, where $|s|$ is the number of sensors and $[0, 1]$ is the range of values they can take. Let $g : \mathbb{R}^N \times [0, 1]^{|s|} \to [-1, 1]^{|e|}$ such as, $\mathbf{O}_{i,k} = g(\mathbf{x}_i, \mathbf{I}_k)$ be the function that computes the "would be" output of a neuro-controller with weights $\mathbf{x}_i$ on "fake" input $\mathbf{I}_k$. We define the sequence $\mathcal{B}_{\mathbf{x}_i} = (\mathbf{O}_{i,1}, \mathbf{O}_{i,2}, \ldots, \mathbf{O}_{i,K})$ as the behavior of genome $\mathbf{x}_i$, note that $\mathcal{B}_{\mathbf{x}_i}$ contains $K \times |e|$ individual output values and is considered as a vector. The genome with the most different behavior is defined as the one whose behavior vector has the largest average distance to all others or:

$$\bar{\mathbf{x}} = \underset{\mathbf{x} \in I^a}{\arg \max} \left(\frac{1}{|I^a|} \sum_{y \in L^a} \| \mathcal{B}_{\mathbf{x}} - \mathcal{B}_{\mathbf{y}} \| \right).$$

At each generation, each agent generates its own set $\mathcal{I}$, present it to all genomes in its list and then selects its next active genome using the above distance. In the following, we name this selection scheme (BS) for behavioral selection.

Objective Based Selection. As a baseline for the experiments, we use a traditional fitness based selection scheme. Different fitness-based selection operators have been applied in EER contexts and it has been shown in multiple instances that their level of selection pressure is correlated with the performance of the swarm on the task [7]. Here, we chose two objective base selection schemes: on the one hand fitness proportionate selection (FPS) whose benefits on behavior specialization have been discussed in [4] and on the other hand, and elitist selection (ES).

Multi-objective Selection. Having defined the above selection schemes, here we follow the ideas of [18] and introduce an instance of "diversity selection" in which selection is based on multi-objective criteria using fitness and novelty (MOBS). In this case, the set of Pareto-optimal solutions[2], with respect to fitness and novelty, in the agent's local population constitutes the selection pool. The agent then selects a random genome from the front. Here the novelty objective is the same as defined above.

3   Experiments

3.1   Simulation

We considered a concurrent foraging task in which agents have to collect items placed randomly in the environment (Fig. 1). There exist two types of items (red and black) and agents are rewarded one unit of fitness every time they collect

---

[2] Recall that the Pareto set is the set of all non-dominated solutions and that, x dominates y imply that x in is not worse than y with respect to all objectives and x is strictly better than y with respect to at least one objective.

Table 1. Simulation parameters.

Arena diam.	1000 pix.	λ	200
Nb. items	300	t_{max}	2000 tics
Sens./com. range	16 pix.	g_{max}	500 generations
Agent/item diam.	6 pix.	σ^0	0.25
Max trs. vel.	2 pix./tic	(σ^-, σ^+)	(0.01, 0.5)
Max rot. vel.	30°/tic	γ	0.35
Init. weights	$[-2, 2]$	τ	0.2
Nb. runs	64	Nb. fake inputs	128

one item regardless of its type. This task has been extensively studied in the context of task specialization [11,12]. Collecting one type of items or the other is considered two distinct tasks (although similar in nature), because agents perceive the types of items trough different sensors.

The experiments were performed on the Roborobo simulator [5] an environment that allows to run experiments on large swarms of agents[3]. In this simulator, agents are e-puck like mobile robots with limited range sensors and two differential drive wheels. Sensors are placed at 12 locations around the agent's body (7 facing the front, 3 facing the back and two facing each side). Agents can perceive 4 different things in the environment: obstacles, other agents and the 2 types of items, and can move using 2 motors. All sensor values are in $[0, 1]$ and actuator values are in $[-1, 1]$. To collect items, agents must bump onto them, and when collected, the items reappear at some random location. There is always the same number of items in the environment.

The neuro-controller we consider here is a simple feed-forward perceptron with a hyperbolic tangent activation function. The genome encodes the weights ([1 bias neuron + 12 sensors × (2 types of items + 1 agent + 1 obstacle)] × 2 outputs = 98 weights) of this controller as a real vector.

All instances of the algorithm[4] were run on the exact same environment, with the same conditions and runs were repeated 64 times. All the parameters of the experiments are summarized in Table 1.

3.2 Measures

We present here the measures that were used in the experiment. To estimate the quality of the foraging, we define the swarm fitness at generation g as $\hat{f}(g) = \sum_{j=1}^{\lambda} f^j(g)$ where $f^j(g)$ is the fitness of agent j at generation g.

Assessing Behavioral Diversity. To measure the behavioral diversity of the swarm, we use two behavioral descriptors: the ratios of red to black items collected and the max distances traveled. The former measures the diversity in

---

[3] Roborobo3 at commit `f108c030f51a991e8fabd92aaaecb87d5ad7032a`.
[4] The code can be downloaded at: https://gitlab.inria.fr/boumaza/public-code.

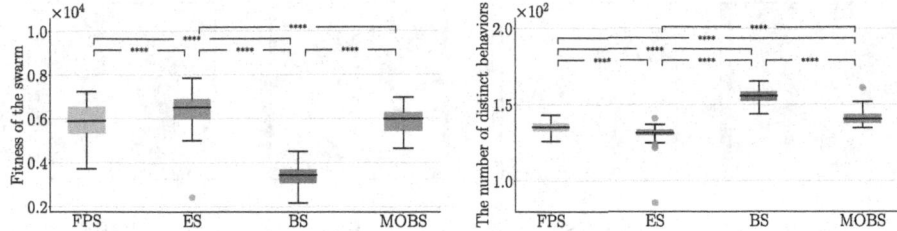

Fig. 2. The fitness of the swarm (left) and the number of distinct behaviors (right). Box-plots of 64 values (runs) where individual values in each are averages over all generation. Statistical significance indicated by the number of asterisks (n asterisks indicate a p-value $< 10^{-n}$).

termo of items collected, and the latter in terms of exploration range. This range is measured as the maximum distance the agent had during its lifetime from its starting position[5]. These behavioral descriptors are discretized into a two-dimensional map on which the agents of the swarm are binned; i.e. each cell of the map counts the number of agents that expressed the behavior represented by the cell's coordinates. The larger the number of occupied cells in the map, the larger the number of distinct behaviors in the swarm. These types of behavioral measures were also used in [12]. They are not considered during the selection step of the algorithm, they are computed off-line, after the simulation.

Assessing Diversity Using Phylogeny. One way to measure if the population maintains its diversity throughout time is to inspect the population dynamics. For that, we take a gene perspective and study their genealogy independently of the agents. At each generation of Algorithm 1, we record the descendants of the genes and construct a *phylogenetic tree*. Nodes in this tree represent active genomes (those that get selected on agents) and edges represent parenthood relations (child nodes are one mutation away from their parents). Edges relate genomes regardless of their vehicles; i.e., parents and offspring may have existed on different agents.

All the initial genomes (the ones created randomly) are children of a "root" node[6]. Each node, that is neither the root nor an initial genome, has at most one parent, and can have at most λ children (the number of agents). Finally, the depth of the tree is at most g_{max}, and at each level there are at most λ nodes. Since inactive agents do not have active genomes, they are not present in the tree. A simplified phylogenetic tree with 15 initial genes over 14 generations is shown in the center of Fig. 1. From a phylogenetic tree, we extract the corresponding *coalescence tree* which synthesize the genealogy of the last surviving genomes (the population at the last generation). This tree is constructed from the last

---

[5] The exploration range is bounded by the diameter of environment.
[6] This allows to construct a connected graph to ease the analysis and does not affect the results.

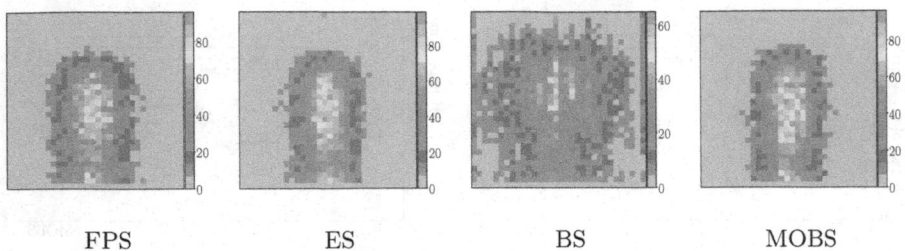

FPS   ES   BS   MOBS

Fig. 3. Behavioral Maps with 32×32 bins. x-axis represent the item type ratio (leftmost bins represent behaviors that collect only black items, those that collect equally are in the middle, and those that collect only red are on the right) the y-axis represent the maximum traveled distance (zero distance at the topmost, maximum distance at the bottom). Maps aggregate behaviors from the last 10 generations. (Color figure online)

generation up to the first where, at each step lineages are merged whenever two or more genomes share the same parent (a coalescence event).

The most recent common ancestor (MRCA) is the youngest genome from which all the end-survivors descend. Given a phylogenetic tree, it can be identified going back in time from the lowest leaves up following their ancestors until we find the sole ancestor which can either be, a regular node or the root node. The time of the most recent common ancestor (TMRCA) is the height of the MRCA (12 on Fig. 1). This time indicates the genetic closeness, since it is related to the number of generations and thus the number of mutations between the MRCA and a current population. The larger it is, the farthest apart are the genomes in the population, the smaller it is, the more its descendants are related. Another measure we use, is the branch lengths of the coalescence tree. It gives a sense of how much history genomes share. Genomes would share the least history if they come from a common ancestor far back in time and then evolved along distinct lineages. Furthermore, the less history genomes share, the more reproductive isolation is present. Assuming there are $r \leq \lambda$ survivors at the last generation, the branch length is defined as:

$$L = \sum_{k=2}^{r} k \cdot T_k, \qquad (2)$$

where T_k is the number of generations there were k distinct lineages in the tree (Fig. 1, right).

Finally, we define the takeover time as the generation at which all active gnomes in the population descend from only one initial genome [1,6]. This can be seen on Fig. 1 where the takeover event happens at the 12th generation. Related to this, we also measure the survival rate of the initial population throughout the generations, which is defined as the proportion of the initial genomes that have offspring at depth d.

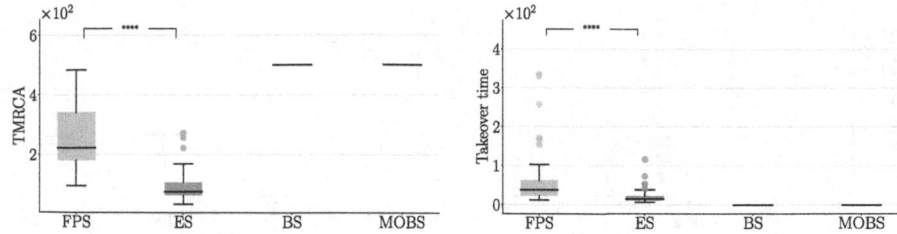

Fig. 4. TMRCA (left) and Takeover time (right). Flat boxes indicate the event did not occur. Box-plots of 64 values (runs) where individual values in each are averages over all generation. Statistical significance indicated by the number of asterisks (n asterisks indicate a p-value $< 10^{-n}$).

Statistical Significance. To compare all selection schemes, each instance is executed 64 times, and we compare the above measures using median values and percentiles. Furthermore, we perform a pairwise comparison using a Mann-Whitney U test with the null hypothesis being *"the samples of the results of both instances are drawn from the same distribution"*. We establish that instance "A" outperforms instance "B" on a given measure, if the median value for instance "A" is better than the one for "B" and there is significant statistical difference between the distributions. The level of confidence at which we reject the null hypothesis is indicated by the number of asterisks on the figures, when applicable.

4   Results and Discussion

We begin our analysis by considering the population fitness on all instances (Fig. 2, left). Recall that the fitness of one agent is the number of items it collects during its lifetime regardless of their color. As expected, the instance with the highest selection pressure (ES) evolves a swarm that collects the most items. If we look at the fitness proportionate selection (FPS), it comes in second, gathering slightly fewer items (the difference is statistically significant). This result is not new, for the collection task, several authors reported that increasing the selection pressure increases the fitness, see for example [7]. The instance that performed the worst is the one that disregards fitness (BS) which is also expected. In the middle comes the multi-objective selection schemes (MOBS) that performs as good as fitness proportionate selection in terms of the task.

Does Behavioral-Selection Promote Diversity? We start by looking at the number of distinct behaviors in the swarm (Fig. 2, right). Here we see that behavioral-selection (BS) play its role and the number of distinct behaviors in this case is significantly higher than in the other instances. On the other hand, ES is the instance that create the least diversity. Here again MOBS comes in the middle.

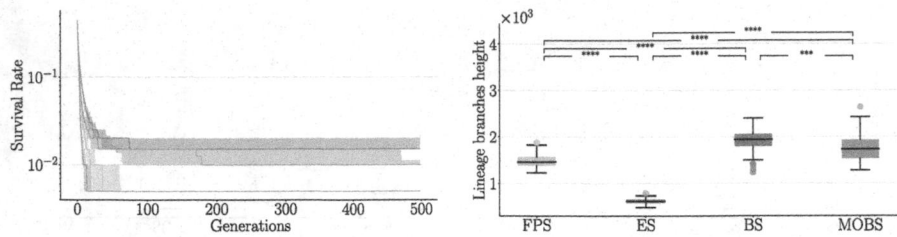

Fig. 5. Survival rate (left) and branch length (right). Curve on the left represent median (solid line) and the range between the 25th and the 75th percentiles. Their colors follow the same scheme as the box-plots on the right. Statistical significance indicated by the number of asterisks (n asterisks indicate a p-value $< 10^{-n}$). (Color figure online)

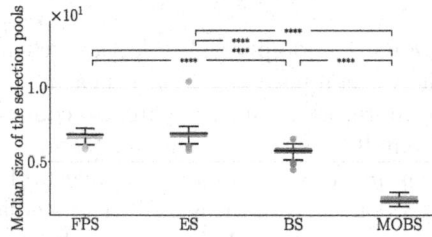

Fig. 6. The size of the selection pool. Box-plots of 64 values (runs) where individual values in each are averages over all generation.

The number of distinct behaviors and their characterization is better viewed on the behavioral maps (Fig. 3). Here we see that the behavioral-selection scheme creates diverse behaviors on all the spectrum. We also notice that there are two groups of agents that only collect items of one type (leftmost or rightmost bins). Some agents specialize in collecting red items and other in collecting black items.

On the other hand, the behavioral maps of the other instances (ES, FPS, MOBS) do not show specialized behaviors. They are mostly clustered in the middle (ratio of $\frac{1}{2}$); behavioral specialization did not occur. Furthermore, in all instances, there is a group of agents that have a large exploration range (bottom of the map). This is the algorithm overcoming environmental pressure: controllers that travel long distances increase their chances of mating and survival.

Does Behavioral-Selection Promote Isolation? To answer this question, we compare the instances on the basis of the coalescence measures we described above. If we inspect the TMRCA (Fig. 4, left), it is very clear that the most elitist instance (ES) has the lowest TMRCA. At each generation, only the few (locally) most fit solutions survive, reducing the genetic pool at the swarm level. FPS on the other hand allows (with a small probability) less fit solutions to survive which delays the appearance of the common ancestor. On the other hand, in all diversity-selection instances, there was no MRCA! This is indicated by the flat box-plot at the value 500 (the total number of generation). This reflects the

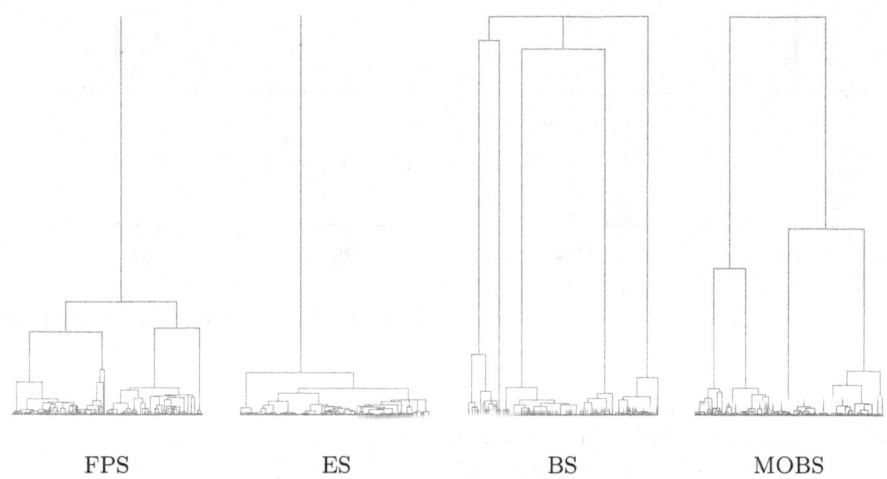

FPS   ES   BS   MOBS

Fig. 7. Coalescence trees of typical runs for each selection instance.

fact that in those instances, there were distinct lineages that coexisted since the start of the simulation that never coalesced. Said differently, there were subsets of genomes that never shared any genetic material, which we may qualify as niches. This is an indication that reproductive isolation occurred.

How many niches where there? To answer this question, we can inspect the survival rate curves (Fig. 5, left). The curves represent the rate, out of λ, of the initial solutions that have decedents at a given generation. We can see that for ES and FPS, the rate drops rapidly to $\frac{1}{\lambda}$. However for the diversity-selection instances, the drop is less rapid and converges between $\frac{2}{\lambda}$ and $\frac{5}{\lambda}$; i.e., between 2 or 5 niches. The rate at which the survival curves drops, indicates the rate at which the population becomes more homogeneous. This can also be seen if we look at the takeover-time (Fig. 4, right). In the case of ES, the population is quickly taken-over by one of the initial solutions and in the case of FPS, the takeover happens a bit later. In the case of diversity-selection the takeover event never happened and, incidentally, this observation can be drawn directly from the lack of MRCA above. Finally, if we look at the branch length (Fig. 5, right), we notice that the higher values are those for BS and MOBS indicating that the end-survivors of these instances share the least genetic history. To illustrate the above results, Fig. 7 shows coalescence trees from typical runs of all the instances.

We were puzzled at the poor performance in terms of diversity of the multi-objective instance. The phylogenetic measures indicate that some level of isolation occurred, however it may not be a direct result of behavioral diversity, this is at least not reflected in the behavioral maps. These algorithms have been reported to perform well on different settings, how come it is not the case here? We believe that the answer to this question lies in the size of the local populations collected by the agents. Indeed, these populations contain few solutions

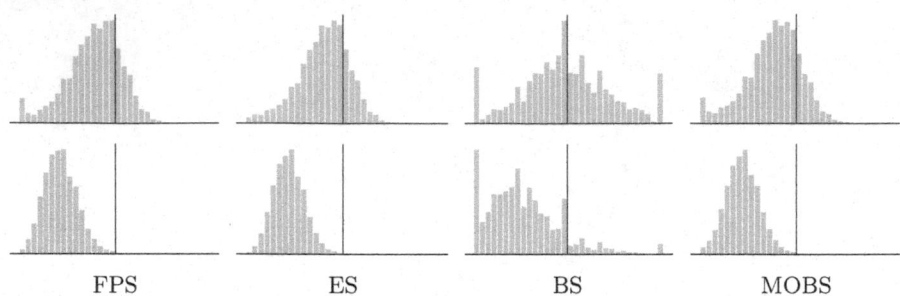

FPS   ES   BS   MOBS

Fig. 8. Histogram of red items ratios across the population. Each bar counts the number of agents collecting at that ratio. The range $[0, 1]$ is clustered in 32 bins. The vertical black line at value 0.5 indicates that equal amounts of black and red items are collected. The histogram aggregate counts from the final 10 generations. The top row is the case where black items reward more, and bottom row in the case were black items are more abundant. (Color figure online)

compared to the swarm size, and when the Pareto set is extracted, the selection pool becomes even smaller (Fig. 6). Sampling randomly the next solution from such a small set does not leave enough room for creativity. Furthermore, since both the fittest and the most different solutions in the local population, are part of the Pareto set[7], on very small samples, the algorithm behaves as if it were a mix of both ES and BS.

Finally, fitness proportionate selection has been reported as very important in promoting behavioral specialization [4]. Why was it not the case here? The answer at this stage is still speculative and need further investigations. We believe that it is probably the case i.e., FPS favors behavioral specialization. However, for it to do so, may require reproductive isolation. In our simulation, there was none since our hypothesis was that it emerges from diversity. In [4], reproductive isolation was enforced geographically.

4.1   Increasing the Pressure from the Environment

In the following, we briefly discuss cases where we modified the environment to create a bias toward favoring one of the two item types. Our goal is to investigate if diversity selection can cope with the added environmental pressure. On the first environment, we modified the reward system, where before, collecting items brings one unit of fitness regardless of their type, now black items reward more, 10 times more. On the second environment, we modified the proportions of each item type, now 75% of items are black and the rest are red. In both these modified environments, we expect the swarm will "prefer" to collect more black items since they are either more abundant thus easily found, or more rewarding. This was the case and can be seen on Fig. 8 which presents the distribution of red item ratios across the population. In all instances with a task selection pressure

---

[7] They are both strictly better than the others on one objective.

(FPS, ES, MOBS) the swarm tends to favor black items and no agent specialized on red (no bars on the right). However, the behavioral-selection instance (BS) is, to some extent, able to cope with the added environmental pressure since some agents specialize on only red items.

5 Conclusions

It has been shown that in order to evolve specialized behaviors, reproductive isolation must be present. This isolation, can either be enforced geographically (Allopatric speciation) in the environment or through some other mechanism that limits reproduction between species sharing the same environment (Sympatric speciation). We proposed the idea of favoring isolation of the latter kind through the use of diversity-selection schemes and tested the hypothesis on a concurrent foraging task. The results suggest that it is the case, when diversity is enforced, a proportion of the population specialized on one of the two possible tasks. In order to measure behaviors reliably, we introduced a procedure that estimates behavioral distances on board the agents. This procedure compares the outputs of candidate controllers on the same reference input data. To verify if indeed isolation occurred, we used phylogenetic measures borrowed from coalescence theory. We also briefly investigate the resilience of the selection schemes in situations with added environmental pressure and the results were very promising.

The next steps are to test if the conclusions hold on different tasks and to further investigate the multi-objective selection method, as we believe it could be a good candidate to promote diversity and quality in EER settings. However, the issue of the archive or lack of archive, needs to be addressed. One possibility to increase the selection pool, could be by extending the lifetime or the number of agents which will increase the number of mating encounters. Furthermore, extending the behavioral comparison to neural architecture other than feed-forward networks can also be interesting to tackle more challenging tasks.

References

1. Bäck, T.: Selective pressure in evolutionary algorithms: a characterization of selection mechanisms. In: Michalewicz, Z., Schaffer, J.D., Schwefel, H.P., Fogel, D.B., Kitano, H. (eds.) Proceedings of the First IEEE International Conference on Evolutionary Computation, pp. 57–62 (1994)
2. Bredeche, N., Haasdijk, E., Prieto, A.: Embodied evolution in collective robotics: a review. Front. Robot. AI **5**, 12 (2018)
3. Bredeche, N., Montanier, J.-M.: Environment-driven embodied evolution in a population of autonomous agents. In: Schaefer, R., Cotta, C., KoLodziej, J., Rudolph, G. (eds.) PPSN 2010. LNCS, vol. 6239, pp. 290–299. Springer, Heidelberg (2010). https://doi.org/10.1007/978-3-642-15871-1_30
4. Bredeche, N., Montanier, J.M., Carrignon, S.: Benefits of proportionate selection in embodied evolution: a case study with behavioural specialization (2017)

5. Bredeche, N., Montanier, J.M., Weel, B., Haasdijk, E.: Roborobo! A fast robot simulator for swarm and collective robotics. CoRR abs/1304.2888 (2013)
6. E. Goldberg, D., Deb, K.: A comparative analysis of selection schemes used in genetic algorithms. Found. Gen. Alg. 1, 69–93 (1991). Elsevier. https://doi.org/10.1016/B978-0-08-050684-5.50008-2
7. Fernández Pèrez, I.N., Boumaza, A., Charpillet, F.: Comparison of selection methods in on-line distributed evolutionary robotics. In: Proceedings of Alife 2014, pp. 282–289. MIT Press, New York (2014)
8. Ficici, S., Watson, R., Pollack, J.: Embodied evolution: a response to challenges in evolutionary robotics. In: Proceedings of the 8th European Workshop on Learning Robots (1999)
9. Gomes, J., Urbano, P., Christensen, A.L.: Introducing novelty search in evolutionary swarm robotics. In: Dorigo, M., et al. (eds.) ANTS 2012. LNCS, vol. 7461, pp. 85–96. Springer, Heidelberg (2012). https://doi.org/10.1007/978-3-642-32650-9_8
10. Gomez, F.J.: Sustaining diversity using behavioral information distance. In: Proceedings of the 11th Annual Conference on Genetic and Evolutionary Computation, pp. 113–120. Association for Computing Machinery, New York (2009). https://doi.org/10.1145/1569901.1569918
11. Haasdijk, E., Bredeche, N., Eiben, A.E.: Combining environment-driven adaptation and task-driven optimisation in evolutionary robotics. PLoS ONE 9(6), e98466 (2014). https://doi.org/10.1371/journal.pone.0098466
12. Hart, E., Steyven, A.S.W., Paechter, B.: Evolution of a functionally diverse swarm via a novel decentralised quality-diversity algorithm. In: Proceedings of the Genetic and Evolutionary Computation Conference, Kyoto, Japan, pp. 101–108. ACM, , 15–19 July 2018. https://doi.org/10.1145/3205455.3205481
13. Karafotias, G., Haasdijk, E., Eiben, A.E.: An algorithm for distributed on-line, onboard evolutionary robotics. In: Proceedings of GECCO 2011, pp. 171–178. ACM (2011)
14. Lehman, J., Stanley, K.O.: Abandoning objectives: evolution through the search for novelty alone. Evol. Comput. 19(2), 189–223 (2011). https://doi.org/10.1162/EVCO_a_00025
15. Smith, J.M.: Sympatric speciation. Am. Naturalist 100(916), 637–650 (1966)
16. Montanier, J.M., Carrignon, S., Bredeche, N.: Behavioral specialization in embodied evolutionary robotics: why so difficult? Front. Robot. AI 3, 38 (2016)
17. Mouret, J., Clune, J.: Illuminating search spaces by mapping elites. CoRR abs/1504.04909 (2015). http://arxiv.org/abs/1504.04909
18. Mouret, J.B., Doncieux, S.: Overcoming the bootstrap problem in evolutionary robotics using behavioral diversity. In: Eleventh Conference on Congress on Evolutionary Computation (CEC 2009), pp. 1161–1168. IEEE Press, Trondheim (2009). https://hal.archives-ouvertes.fr/hal-00473147
19. Mouret, J.B., Doncieux, S.: Encouraging behavioral diversity in evolutionary robotics: an empirical study. Evol. Comput. 20(1), 91–133 (2012)
20. Prieto, A., Becerra, J.A., Bellas, F., Duro, R.J.: Open-ended evolution as a means to self-organize heterogeneous multi-robot systems in real time. Robot. Auton. Syst. 58(12), 1282–1291 (2010)
21. Pugh, J.K., Soros, L.B., Stanley, K.O.: Quality diversity: a new frontier for evolutionary computation. Front. Robot. AI 3, 40 (2016)
22. Stanley, K.O., Miikkulainen, R.: Evolving neural networks through augmenting topologies. Evol. Comput. 10(2), 99–127 (2002)

Open-Ended Search for Environments and Adapted Agents Using MAP-Elites

Emma Stensby Norstein[1(✉)], Kai Olav Ellefsen[1], and Kyrre Glette[1,2]

[1] Department of Informatics, University of Oslo, Oslo, Norway
emmaste@ifi.uio.no
[2] RITMO, University of Oslo, Oslo, Norway

Abstract. Creatures in the real world constantly encounter new and diverse challenges they have never seen before. They will often need to adapt to some of these tasks and solve them in order to survive. This almost endless world of novel challenges is not as common in virtual environments, where artificially evolving agents often have a limited set of tasks to solve. An exception to this is the field of open-endedness where the goal is to create unbounded exploration of interesting artefacts. We want to move one step closer to creating simulated environments similar to the diverse real world, where agents can both find solvable tasks, and adapt to them. Through the use of MAP-Elites we create a structured repertoire, a map, of terrains and virtual creatures that locomote through them. By using novelty as a dimension in the grid, the map can continuously develop to encourage exploration of new environments. The agents must adapt to the environments found, but can also search for environments within each cell of the grid to find the one that best fits their set of skills. Our approach combines the structure of MAP-Elites, which can allow the virtual creatures to use adjacent cells as stepping stones to solve increasingly difficult environments, with open-ended innovation. This leads to a search that is unbounded, but still has a clear structure. We find that while handcrafted bounded dimensions for the map lead to quicker exploration of a large set of environments, both the bounded and unbounded approach manage to solve a diverse set of terrains.

Keywords: Evolutionary algorithms · Virtual creatures · Environments · Map-Elites · Open-endedness · Modular robots

1 Introduction

Virtual creatures that *learn* locomotion skills have attracted significant research interest. Even so there has not been much research that combines the optimisation of the controller, morphology and environment. All three of these components play an important role in determining the behaviour of an agent, but much of the research in this field focuses on either the environment [21,23,24] or the morphology [7,22]. The research that focuses on both morphology and environment often uses a limited set of environments [1,15,25].

© Springer Nature Switzerland AG 2022
J. L. Jiménez Laredo et al. (Eds.): EvoApplications 2022, LNCS 13224, pp. 651–666, 2022.
https://doi.org/10.1007/978-3-031-02462-7_41

When evolving the morphology and controller simultaneously [14], and when evolving to solve a difficult task directly [2], it is common to become stuck in a local optima. Environmental variation could potentially alleviate some of the difficulty by providing stepping stones to more difficult environments, and by introducing environments that require different morphologies to be solved.

One work that considers the optimisation of both agents and environments is the Paired Open-Ended Trailblazer (POET) [24]. In the field of open-endedness the goal is not to find a single solution, but to find many interesting solutions [19]. In POET agents are optimised to solve environments, at the same time as the environments are optimised for giving the agents new challenges. Constraining the evolving environments by criteria relating to the agent fitness ensures environments that are neither too difficult nor too hard. This creates a push towards novel but solvable environments, leading to an open-ended stream of new tasks.

As mentioned, open-ended algorithms aim to explore as many interesting solutions as possible. Since interestingness is difficult to define [20] and optimise for, it is common to take inspiration from novelty search algorithms to instead create solutions that are as different as possible from what has previously been found. The hope is often that finding solutions that are different from each other will make it more likely to find the interesting ones. In novelty search [12] an archive of previously found solutions is kept, and new solutions are compared to the archive in order to look for solutions that are different from what is already found. This generates a diverse set of solutions. However, in order to make the diverse set of found solutions useful we may also want the solutions to have high quality. This leads us to a family of algorithms called Quality-Diversity algorithms [6,13], that aim to balance search for novelty with optimisation, to create an archive of solutions that are both diverse and solve their task efficiently. This class of algorithms is often used to ensure that the phenotypic search space is covered, to avoid getting stuck in local optima, while still optimising to solve a set objective.

A popular quality diversity algorithm is MAP-Elites [16]. Map-Elites has been used to optimise both the controller and morphology of robots [3,17]. In MAP-Elites evolution takes place in an archive that is shaped as a grid, where each cell in the grid can hold one solution. MAP-Elites aims to fill the grid while also performing an elitist search for the best candidate within each cell. The position of a candidate within the grid is determined by a set of feature descriptors, called the behaviour dimensions, each relating to one dimension of the grid. These behaviour dimensions are normally set up by the researchers depending on the solution types they want to explore.

The behaviour dimensions can be difficult to design. In order to avoid creating them by hand, methods for automatic definition of behaviour dimensions have been proposed. AURORA [8] uses an autoencoder [11], that has been trained on a set of found candidates, to encode the candidates into a shorter feature vector. This vector determines the placement of the candidates in the MAP-Elites grid. As more candidates are discovered the autoencoder gradually learns a better representation of the search space, allowing the space of possible solutions to be mapped without user-defined descriptors.

Another method for automatically creating behavioural dimensions was presented by Gaier et al. [10], who attempt to use MAP-Elites to reconstruct an image generated by a Compositional pattern producing network (CPPN) [18]. Like AURORA, this method also uses autoencoders. However, instead of using the encoded feature vector produced by the autoncoder as behaviour dimensions, it instead uses the mean square error between a candidate and the attempted reconstruction of the candidate by the autoencoder. This error value indicates how novel or unexpected the solution is: If the autoencoder cannot reconstruct it, it cannot have seen many similar solutions in its training data. This behavior descriptor thereby becomes a measure of how original a solution is, allowing the exploration of solutions with different degree of familiarity. The autoencoder is retrained at intervals, causing the dimensions to shift to encourage exploration of images different from what is in the map. By keeping solutions with different levels of familiarity, the map gradually builds a record of the most notable images previously explored. As the second behaviour dimension the number of nodes in the CPPN is used.

Our approach takes inspiration from the Paired Open-Ended Trailblazer (POET) [23,24], which explores pairs of environments and agents solving them, with the goal of endlessly innovating to create ever-more challenging environments and agents solving them. However, unlike POET, our approach attempts a structured exploration of new solutions, aiming to fill up a grid-shaped repertoire of environments, by using the MAP-Elites [16] algorithm (Fig. 1). With this grid structure solutions in adjacent squares can be used as stepping stones for the agents to solve increasingly difficult environments. Our *static* approach uses handcrafted features of the environment as map dimensions. However, to explore open-ended generation of environments we also take inspiration from Gaier [10], and test the use of autoencoder error as a *dynamic* map dimension. We found that while both the static and dynamic approach managed to solve a diverse set of environments, the static bounded dimensions led to larger exploration of environments.

Our contributions are twofold. 1) We explore the possibility of creating a structured repertoire of tasks and agents with MAP-Elites, and show that this approach is capable of generating a diverse set of terrains and virtual creatures that manage to walk through them. In this preliminary work we limit the tasks to locomotion on different terrains. While the terrains are unbounded, the task of locomotion is not. To truly achieve our goal of unlimited tasks we will in the future have to evolve not only the terrain but also the tasks that the agents solve. 2) We test the use of an autoencoder as a behaviour dimension in the map to allow for unbounded innovation.

2   Methods

Together the body, brain and environment determines the behaviour of a virtual creature. We evolve all these three components with the goal of finding diverse terrains and agents that walk through them. Inspired by MAP-Elites [16] we

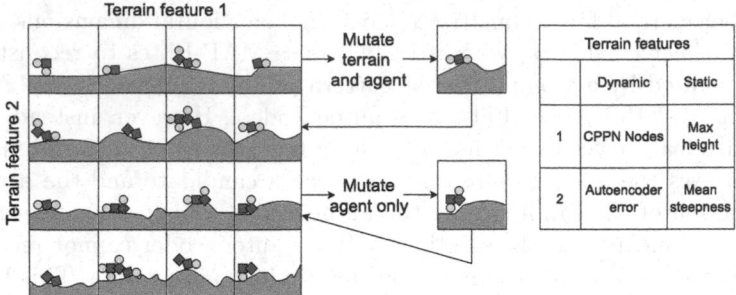

Fig. 1. Each cell in the map can hold a pair of one terrain and one virtual creature. In each iteration pairs from some cells are chosen to be mutated. The virtual creature is always mutated, while the environment is only mutated for some of the chosen pairs. The mutated pairs are inserted into the map if their fitness is higher than the fitness of the pair already in the cell they belong to.

optimise within an archive structured as a 2d grid[1]. Each cell in the grid holds a pair of one agent and one environment. If multiple pairs have been found for a single cell only the pair with the highest fitness score is kept. Terrain features of the environment determines the placement of the pair within the grid. We compare two variants of our approach. The first uses handcrafted grid dimensions, and will be referred to as the static approach. The second uses a combination of handcrafted and automatically defined dimensions, and will be referred to as the dynamic approach. As the testbed for our algorithm we will use a simulation environment created by Veenstra et al.[2] [22], where 2D modular virtual creatures move through a course, attempting to reach the end. This environment is convenient as it is not computationally heavy, and because it allows changing all three components that we are interested in evolving: Terrain, morphology and controller.

2.1 Simulation Environment

We test our approach using an OpenAI Gym [4] simulator for 2D modular virtual creatures [22], the simulator uses the Box2D physics engine [5]. The creatures consist of circles and rectangles, and can be represented as trees. The root module of a creature is always a rectangle. Every rectangle module can connect to up to three new modules. Circle modules cannot connect to any new modules, so all circle modules will be leaf nodes of the virtual creature tree. In addition to their shape the modules have parameters for size and the angle at which they are connected to their parent.

The virtual creature moves across a 2d terrain, which is 220 units long. The first 20 units of the terrain is a startpad, which is always flat. The environment is

---

[1] Source code is available at https://github.com/EmmaStensby/environment-map.
[2] https://github.com/FrankVeenstra/gym_rem2D.

defined by specifying the height of the terrain at each unit. A creature's fitness is defined as the number of units its root module has progressed along the terrain. The creatures are simulated for up to 2000 time steps, after this the simulation is stopped to ensure that the time spent to evaluate a single individual is not too long. A vertical line moves after the simulated creature at a speed of 0.02 units per time step. If the line reaches the creature the simulation will end, quickly eliminating individuals that do not move.

2.2   Environment Encoding

Like in POET-Enhanced [23], terrains are generated by a compositional pattern producing network [18] (CPPN). The initialisation and mutation parameters of the CPPN are the same as those used in POET-enhanced, as we wished to use a method for terrain generation already established in the literature. 200 values evenly distributed between 0 and 1 are evaluated by the CPPN to create a vector containing the height of the 200 units of the terrain.

2.3   Agent Encoding

Table 1. Encoding of a module and controller within the bitstring that encodes a modular virtual creature.

Module (rectangle)	12 bits	Module (circle)	8 bits	Controller	12 bits
Bit 0–3	Width	Bit 0–3	Radius	Bit 0–3	Amplitude
Bit 4–7	Height	Bit 4–7	Angle	Bit 4–7	Period
Bit 8–11	Angle	–	–	Bit 8–11	Phase

An agent is encoded as a bitstring, which can be decoded into a tree structure representing both a 2d modular virtual creature and its controller. Representing the agent as a bitstring eases the design of mutation operators. The agents were mutated by flipping bits with a probability of 0.05. We use a decentralised controller where each module is controlled by a sine wave. The bitstring has a length of 288 bits. The first 48 bits are decoded into four rectangle modules, the next 32 bits are decoded into four circle modules, the next 96 bits are decoded into eight controllers. The controller for each module produces a sine wave that controls the angle of a module in relation to its parent module. How the bits relate to the parameters of the modules and controllers is summarised in Table 1. The last 112 bits are decoded into a tree where each node contains one of the modules and one of the controllers previously defined. The tree is generated by performing the following steps:

1. Add all available connection points for modules to a list.
2. Pass through the bitstring until a 1 is reached. For every 0 passed remove one connection point from the top of the list.

3. The next 6 bits are decoded into two numbers between 0 and 7, which decide
   which of the eight modules and which of the eight controllers are to be used
   at the connection point now at the top of the list.
4. Add all new connections to the list, and repeat from step 2.

These steps continue until the end of the string is reached, or the list of connection points is empty.

2.4 Environment-Agent MAP-Elites

Our algorithm keeps an archive shaped as a 2d map, the map has 25 by 25 cells.
Figure 2 shows how the maps may look during runtime. Each cell in the map can
hold one environment and one agent. The environment and agent form a pair, and
the fitness of the pair is determined the agent's fitness in the environment. The
placement of a pair in the map is determined by the map's behaviour dimensions.

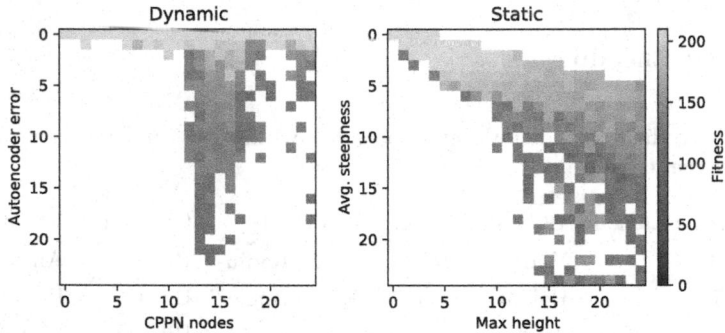

Fig. 2. Example maps from a dynamic and static run. These maps are the archives
used by the MAP-Elites algorithm. The color of each cell represents the fitness of the
pair in that position of the grid. The maps are taken from the runs with median *average
map fitness* (see Fig. 4) out of all 29 runs. (Color figure online)

Before a run is started the maps are bootstrapped with initial solutions. 500
random pairs of environments and agents are created and placed in their respective cells in the grid. The 500 initial environments are generated by mutating
a flat environment. The initial environments will then be spread across a small
area of the map. If there are several pairs that belong in the same cell the one
with the highest fitness is kept, and the rest are discarded.

Next the MAP-Elites algorithm is applied. One iteration of the algorithm
consists of the following three steps:

Select 500 random pairs from the grid, the same pair can be selected multiple
times.
Mutate the agent in all pairs. Mutate the environment with a probability of 0.2.
Insert the mutated pairs into the grid.

When attempting to insert the mutated pairs into the map they are first evaluated to determine their fitness. Next the cell that they belong to is found. If the cell is empty they are inserted as long as their fitness is above a threshold of 100. If the cell is occupied the pair with the highest fitness will be kept, while the other is discarded.

2.5   Behaviour Dimensions

We compare two different ways of defining the behaviour dimensions of the map, the *static* approach and the *dynamic* approach. The static approach has the maximum height of the terrain as the first dimension, and the average steepness as the second dimension. The dynamic approach has the number of nodes in the CPPN, which generates the terrain, as the first dimension. The second dimension is the reproduction error of an autoencoder. Since the autoencoder is retrained at regular intervals, this dimension changes as the map is filled. The second dimension in both the static and the dynamic approach is scaled by a constant to ensure the map holds reasonable environments. This is necessary because it is difficult to create environments that have very high values for average steepness and autoencoder error. The second dimension for the static approach is scaled by 50, while the second dimension for the dynamic approach is scaled by 5.

The autoencoder used for the dynamic approach has an input layer with 200 nodes, three hidden layers with respectively 64, 32 and 64 nodes, and an output layer with 200 nodes. It is trained with the adam optimiser, and the loss function is the mean squared error. It is bootstrapped by training on 500 randomly generated simple terrains at the start of each run, and is retrained every 100 iterations. When it is retrained it is trained on all environments currently in the map. The reproduction error, used as the novelty measure for the behaviour dimension, is defined as the mean absolute error between the terrain and the reproduced terrain from the autoencoder.

3   Recording Data

3.1   Reference Maps

In addition to the map used as the archive for our algorithm we also record the pairs found in a separate map with higher resolution. All explored solutions may be recorded in the reference map, regardless of whether they were placed in the MAP-Elites archive. The reference map has 100 by 100 cells and is used to compare the solutions found by the dynamic approach to those found by the static approach. The behaviour dimensions for the reference maps are the same as the dimensions for the static map, except for the resolution. The static approach therefore has an advantage when filling the reference map as it has access to almost the same dimensions during runtime.

3.2   Found and Solved Environments

We also record explored solutions in two lists. These two archives hold respectively found and solved environments. Each time a new pair is explored is added to the archive of found environments, as long as there is no environment already in the archive that is too similar to it. An environment is regarded as too similar if there is an environment in the archive to which it has an absolute error of less than 25. If the pair has a fitness above 200 it is also added to the archive of solved environments. The archive for solved environments has a lower threshold for absolute error at 2.5.

3.3   Environment Difficulty

We analyse some of our results by approximating the environment difficulty. The difficulty is measured by discretising the steepness of the terrain into several categories. Next each hill in the terrain is localised. A hill is defined as a continuous section of terrain units where all units belong to the same steepness category. The hills are then assigned a value based on their length and steepness category, see Table 2. The difficulty of the terrain is the sum of the values for all its hills.

Table 2. Difficulty values for hills.

		Steepness						
		<−2.4	<−0.24	<−0.024	−0.024 to 0.024	>0.024	>0.24	>2.4
Units	1–3	−3	−2	−1	0	2	4	6
	4–8	−4	−3	−2	0	4	6	8
	>8	−5	−4	−3	0	6	8	10

4   Results

To select parameters 200 trials with random parameters were performed for each of the two approaches. The parameters were handpicked based on the results of the trials. Next we performed 29 runs of each of our two approaches on 16 cores for 16 h. The number of iterations completed within this time varied between the runs. However, the average number of iterations was 1364 which corresponds to evaluating 682 000 individuals.

4.1   Reference Maps and Performance

In Fig. 3 we see the reference maps for the two approaches. We can see that the dynamic approach has explored a significantly smaller part of the map than the static approach. This is expected as the static approach optimises directly to fill dimensions very similar to those of the reference map. Both of the approaches has high fitness in environments in the top left corner of the map, and decreasing fitness towards the bottom right. In the top right corner there is an area where all found pairs have 0 fitness. In this area it is not possible to create solvable environments due to the terrain features required by the map dimensions. These environments have high maximum terrain height, but low average steepness. This is only possible to achieve by setting the first step in the terrain to a high value, which creates a tall wall immediately after the startpad.

Figure 4 shows statistics about the performance of the 29 runs. The static approach performs better than the dynamic approach in both coverage of the reference map, and average map fitness. The average map fitness is the average fitness per square in the reference map. The two approaches performed similarly for average fitness of the found solutions and the total number of solved environments.

In Fig. 5 we can see how the coverage of the MAP-Elites archives develops for both approaches. This graph excludes some runs that completed very few iterations and is therefore only meant to illustrate the effect of the autoencoder training on the behaviour dimensions. For the dynamic approach we can see the effect of the autoencoder training every 100 iterations. As the behaviour dimensions change, some pairs that were previously in separate cells end up in the same cell, and the coverage drops slightly.

4.2 Analysis of Found Environments

In Fig. 6 we can see how the solved environments are distributed with regards to difficulty. The distribution of the solved environments is slightly different for the two approaches. Although the two approaches has solved approximately the same number of environments, the dynamic approach seems to have solved more simple environments, while the static approach has solved quite a few difficult environments.

Figure 7 displays some examples of solved environments. The environments have been scaled to highlight terrain features. We can see qualitatively that the algorithm produces various different terrains and agents.

In Fig. 8 we quantitatively analyse all explored environments with regards to difficulty. We can clearly see that the static approach has explored significantly more environments than the dynamic approach.

5   Discussion

We have explored the possibility of creating a structured repertoire of environments, and agents solving them, using MAP-Elites. We expected the grid structure in MAP-Elites to aid the agents in exploring and solving increasingly difficult environments by using adjacent squares in the map as stepping stones. The reference maps in Fig. 3 showed us that our approach is indeed capable of filling a map with environments and agents. The environments seem to be distributed as expected within the map, with the easy environments in the top left

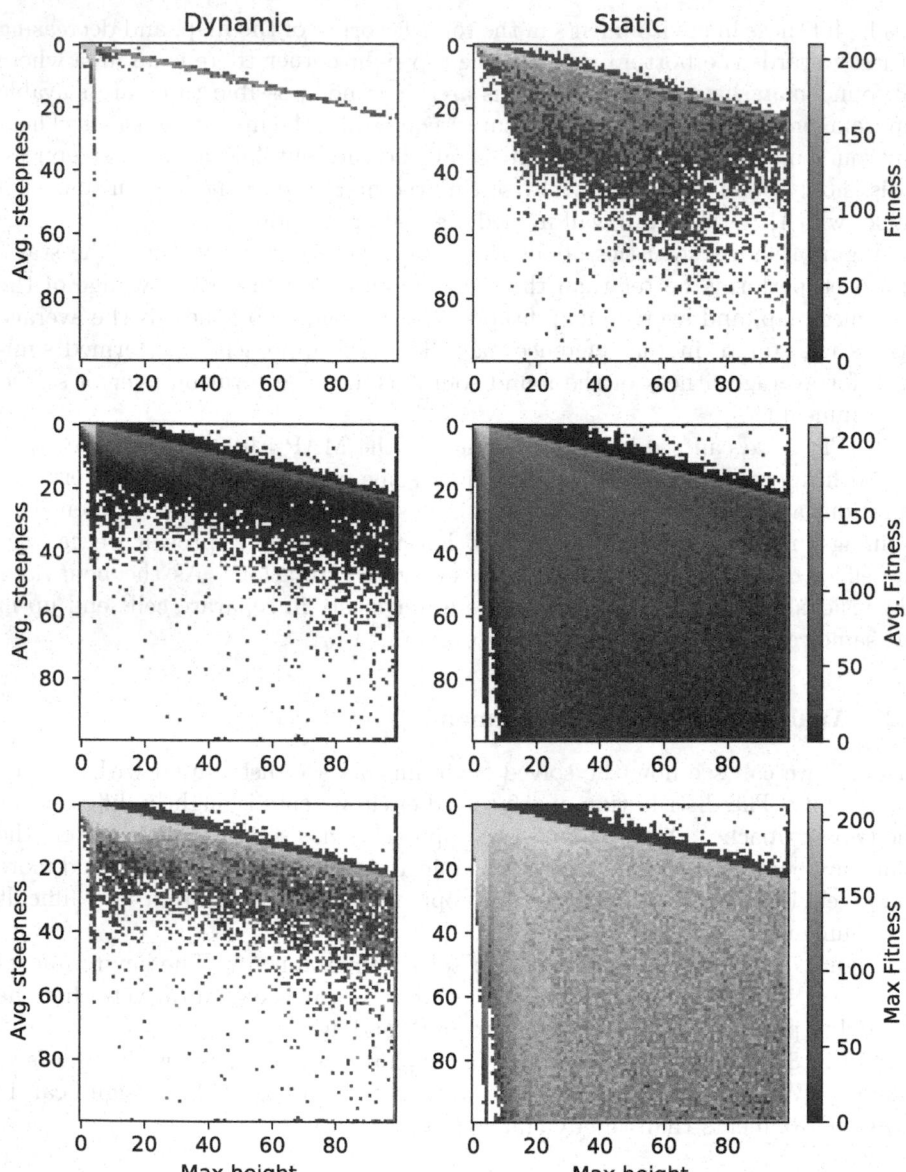

Fig. 3. Reference maps record the progress of the two approaches. The color of each square represents the fitness of the pair found in that cell. The top row shows the reference maps from single example runs. The example maps are taken from the runs with median *average map fitness* out of all 29 runs. The second row shows the mean fitness, and the bottom row shows the maximum fitness over all 29 runs. (Color figure online)

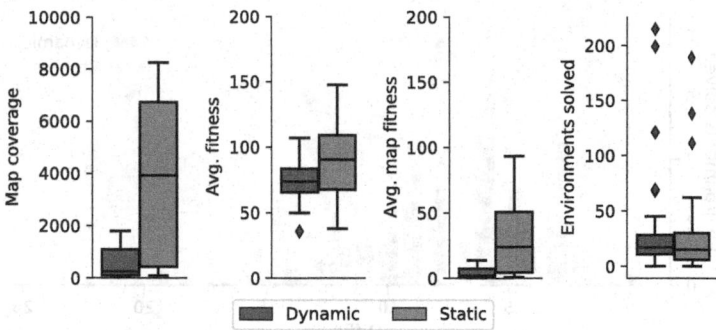

Fig. 4. From left to right we see 1) the coverage of the reference maps, 2) the average fitness of the pairs present in the reference maps 3) the sum of the fitness of all individuals in the reference maps divided by the number of squares, and 4) the number of environments solved.

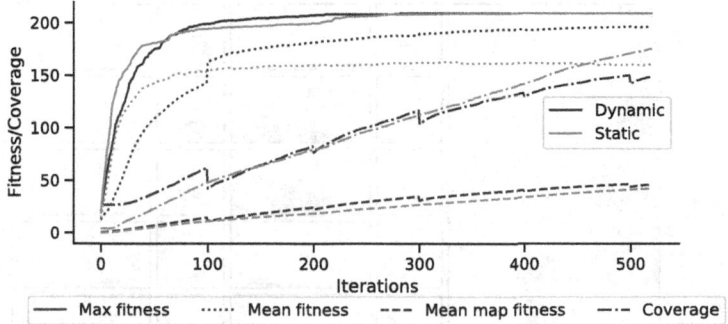

Fig. 5. This graph shows 1) maximum fitness, 1) mean fitness of found pairs, 3) average map fitness and 4) map coverage over time. These are measured on the maps used as archives by MAP-Elites during runtime (see Fig. 2). The graphs show the mean over all 29 runs. Note that this figure excludes some runs that completed very few iterations and is therefore meant only to illustrate the effect of the autoencoder training on the behaviour dimensions.

corner, and difficult environments in the bottom right corner. This is verified by the fitness found gradually decreasing towards the bottom right. We can see qualitatively in Fig. 7 that the approach seems capable of solving diverse environments, as the randomly drawn solved environments are quite different from each other. Although no conclusions can be drawn from the few environments plotted, they seem to get increasingly bumpy as the difficulty increases.

The dynamic approach filled significantly less of the reference maps in Fig. 3 than the static approach. This was an expected result as the static approach optimises directly to explore the features of the reference map, while the dynamic approach optimises for a different novelty measure. Figure 8 showed a quantitative analysis of the environments found, and confirms that the static approach

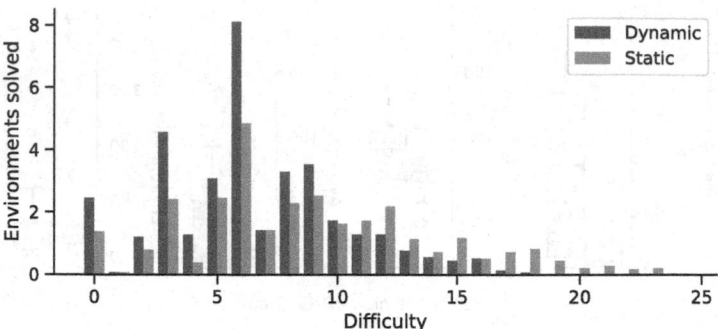

Fig. 6. This histogram shows the distribution in difficulty of the solved environments. Each bar shows the mean number of environments solved for that difficulty over all 29 runs.

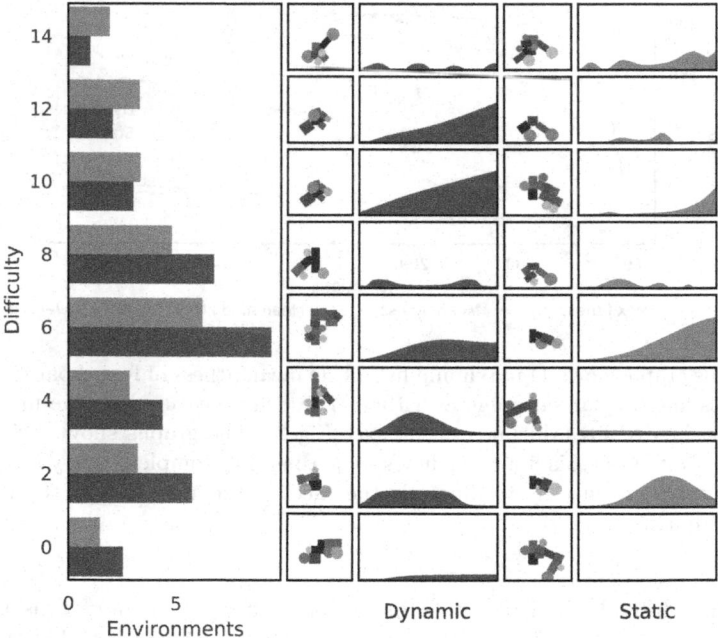

Fig. 7. The histogram on the left shows the same data as the histogram in Fig. 6. However, the number of bins have been halved by combining every two bins. On the right a pair from each bin is shown. The pairs are drawn randomly from their respective bins. Note that the virtual creatures shown are larger than their actual size compared to the environments, and that the y axis for the environments has been scaled to highlight terrain features. However, all the displayed environments are scaled equally so they can be compared to each other.

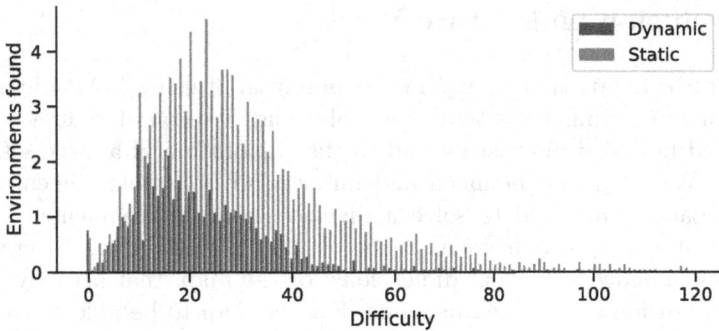

Fig. 8. This histogram shows the distribution in difficulty of the found, but not necessarily solved, environments. Each bar shows the mean number of environments found for that difficulty over all 29 runs.

explored a larger diversity of environments. While handcrafted behaviour dimensions may be difficult to create in some cases, they performed better than the automatically defined dimensions in our case. However, in other, more complex, domains where handcrafted dimensions may be more difficult to create, the automatically defined dimensions from the autoencoder can be an alternative that is more general and can be applied to most types of environments.

Another benefit of the dynamic approach is that it could in theory continue exploring new environments for longer than the static approach. The static approach would likely stagnate once all squares in its map have been filled with high fitness pairs that are difficult to replace, while the dynamic approach could keep retraining the autoencoder and change its dimensions. We did not have the opportunity to see whether such an effect would appear in our experiments, as we did not run the experiment for long enough for this to happen. A main limitation for such continuous exploration is the capability of the autoencoder. A requirement for the dynamic approach to keep endlessly exploring the available environments in more detail, is that the autoencoder is capable of storing information about all found environments and discern new environments from these. This becomes increasingly difficult as more environments are discovered, and further experiments may be necessary for the autoencoder to be able to do its part.

We do not only want to *generate* diverse terrains, we also want to *solve* them. We discovered that the static and dynamic approach had solved approximately the same number of environments (Fig. 6), despite the difference in the number of found environments (Fig. 8). This may indicate that although the dynamic approach has explored less of the environment search space, it may have explored the top left section of the maps more thoroughly, leading to the increased number of solved easy environments.

6   Conclusion and Future Work

This work was an attempt to explore the potential of using MAP-Elites to generate both interesting tasks and their solutions. We found that a map with handcrafted bounded dimensions lead to the exploration of a large set of environments. We compared bounded and unbounded behaviour dimensions, and both approaches managed to solve a diverse set of environments. The main limitation of our approach seems to be that it is challenging to create general and open-ended behaviour dimensions for the map, that actually allow for endless generation of new environments. For the map to be able to continually develop novel tasks, the autoencoder used to describe the novelty must be able to represent many previously found environments, and meaningfully discern them from new environments. It would be interesting to further explore the choice of behaviour dimensions. Either by finding out what properties are necessary for the autoencoder to perform well, even as the amount of data it trains on becomes very large, or by exploring alternative behaviour dimensions. An approach like AURORA [8] could be explored as an alternative to our approach, as its feature extraction may lead to an interesting structure to the repertoire.

Another direction could be to explore possible uses of an already generated repertoire of environments and solutions. While generating novel environments automatically is interesting in itself, it could be even more interesting if the skills stored in the repertoire could somehow be leveraged to quickly adapt to new never before seen environments. In this case it would be interesting to either design the behaviour dimensions of the map so that they correlate with the skills necessary to solve the environments, or use methods for adapting through trial and error [9].

Other extensions to our current work could include improving the efficiency of the search within the map, for example by introducing crossover or other mechanisms that create interaction throughout the map. The efficiency should also be compared with other existing methods that generate terrains and their solutions, such as the Paired Open-Ended Trailblazer [23], or to quality-diversity methods with unstructured repertoires, such as Novelty Search with Local Competition [13].

Acknowledgments. This work was partially supported by the Research Council of Norway through its Centres of Excellence scheme, project number 262762. The simulations were performed on resources provided by UNINETT Sigma2 - the National Infrastructure for High Performance Computing and Data Storage in Norway. Thank you to Frank Veenstra for support using the 2D simulator for modular robots.

References

1. Auerbach, J.E., Bongard, J.C.: Environmental influence on the evolution of morphological complexity in machines. PLoS Comput. Biol. **10**(1), e1003399 (2014)

2. Bongard, J.C.: Morphological and environmental scaffolding synergize when evolving robot controllers: artificial life/robotics/evolvable hardware. In: Proceedings of the 13th Annual Conference on Genetic and Evolutionary Computation. In: GECCO 2011, pp. 179–186. Association for Computing Machinery, Dublin (2011)
3. Bossens, D.M., Mouret, J.-B., Tarapore, D.: Learning behaviour-performance maps with meta-evolution. In: Proceedings of the 2020 Genetic and Evolutionary Computation Conference, pp. 49–57 (2020)
4. Brockman, G., et al.: OpenAI Gym (2016)
5. Catto, E.: Box2D (2019)
6. Chatzilygeroudis, K., Cully, A., Vassiliades, V., Mouret, J.-B.: Quality-diversity optimization: a novel branch of stochastic optimization. In: Pardalos, P.M., Rasskazova, V., Vrahatis, M.N. (eds.) Black Box Optimization, Machine Learning, and No-Free Lunch Theorems. SOIA, vol. 170, pp. 109–135. Springer, Cham (2021). https://doi.org/10.1007/978-3-030-66515-9_4
7. Cheney, N., et al.: Scalable co-optimization of morphology and control in embodied machines. J. Roy. Soc. Interface 15(143), 20170937 (2018)
8. Cully, A.: Autonomous skill discovery with quality-diversity and unsupervised descriptors. In: Proceedings of the Genetic and Evolutionary Computation Conference. GECCO 2019, pp. 81–89. Association for Computing Machinery, Prague (2019)
9. Cully, A., et al.: Robots that can adapt like animals. Nature 521(7553), 503–507 (2015)
10. Gaier, A., Asteroth, A., Mouret, J.-B.: Are quality diversity algorithms better at generating stepping stones than objective-based search?. In: Proceedings of the Genetic and Evolutionary Computation Conference Companion. GECCO 2019, pp. 115–116. Association for Computing Machinery, Prague (2019)
11. Hinton, G.E., Salakhutdinov, R.R.: Reducing the dimensionality of data with neural networks. In: Science 313(5786), 504–507 (2006)
12. Lehman, J., Stanley, K.O.: Abandoning objectives: evolution through the search for novelty alone. Evol. Comput. 19(2), 189–223 (2011)
13. Lehman, J., Stanley, K.O.: Evolving a diversity of virtual creatures through novelty search and local competition. In: Proceedings of the 13th Annual Conference on Genetic and Evolutionary Computation. GECCO 2011, pp. 211–218. Association for Computing Machinery, Dublin (2011)
14. Lipson, H., et al.: On the difficulty of co-optimizing morphology and control in evolved virtual creatures. In: Artificial Life Conference Proceedings 13, pp. 226–233. MIT Press (2016)
15. Miras, K., Ferrante, E., Eiben, A.E.: Environmental influences on evolvable robots. PloS ONE 15(5), e0233848 (2020)
16. Mouret, J.-B., Clune, J.: Illuminating search spaces by mapping elites. arXiv preprint arXiv:1504.04909 (2015)
17. Nordmoen, J., et al.: MAP-elites enables powerful stepping stones and diversity for modular robotics. Front. Robot. AI 8, 56 (2021)
18. Stanley, K.O.: Compositional pattern producing networks: a novel abstraction of development. Genet. Program. Evolv. Mach. 8(2), 131–162 (2007)
19. Stanley, K.O.: Why open-endedness matters. Artif. Life 25(3), 232–235 (2019)
20. Taylor, T., et al.: Open-ended evolution: perspectives from the OEE workshop in York. Artif. Life 22(3), 408–423 (2016)
21. Open Ended Learning Team, et al.: Open-ended learning leads to generally capable agents. arXiv preprint arXiv:2107.12808 (2021)

22. Veenstra, F., Glette, K.: How different encodings affect performance and diversification when evolving the morphology and control of 2D virtual creatures. In: Artificial Life Conference Proceedings, vol. 32, pp. 592–601 (2020)
23. Wang, R., et al.: Enhanced POET: open-ended reinforcement learning through unbounded invention of learning challenges and their solutions. In: International Conference on Machine Learning, pp. 9940–9951. PMLR (2020)
24. Wang, R., et al.: POET: open-ended coevolution of environments and their optimized solutions. In: Proceedings of the Genetic and Evolutionary Computation Conference. GECCO 2019, pp. 142–151. Association for Computing Machinery, Prague (2019)
25. Zhao, A., et al.: RoboGrammar: graph grammar for terrain-optimized robot design. ACM Trans. Graph. (TOG) **39**(6), 1–16 (2020)

Out of Time: On the Constrains that Evolution in Hardware Faces When Evolving Modular Robots

Rodrigo Moreno[(✉)] and Andres Faiña

IT University of Copenhagen, Copenhagen, Denmark
rodr@itu.dk

Abstract. With the recent advances of modular robots and low-cost manipulators, the evolution of robots, including morphologies and controllers, has become possible to perform in a physical setup without using any simulators. In this scenario, the evolution cannot be parallelized and the wall time becomes a scarce resource that should be used wisely. This paper analyses different algorithms by using the wall time as a stopping criterion for evolution, and it takes into account that wall time depends on the evaluation time plus the time to assemble and disassemble robots before and after an evaluation. The experiments have been performed in simulation, but the evaluation and assembly time have been carefully modelled from previous hardware experiments. Results suggest that (i) genetic algorithms are severely penalized, (ii) genetic algorithms can be improved by performing several evaluations of controllers for each morphology, and that (iii) evolutionary strategies that can chain several evaluations of robots with close morphologies can outperform other evolutionary algorithms. This finding is not surprising, but to the best of our knowledge previous attempts to evolve modular robots in hardware have not employed evolutionary strategies.

Keywords: Modular robots · Evolutionary robotics · Morphological evolution · Evolution in hardware

1 Introduction

Autonomous robots help automate a lot of different tasks, but they must be designed in a suitable way to be successful. In order to obtain robot designs adapted to a specific task, several authors have proposed to evolve the morphology and controller of a robot at the same time [16,27]. These works perform the evolutionary process mostly in simulation as several morphological variations are evaluated in a short time. However, this advantage comes with an important drawback: the reality gap.

Due to simplified physics and bad modelled features, simulators are not capable of accurately simulating the real world. Therefore, evolution exploits these artifacts and produces robot designs that do not perform well in the real world. There are several approaches to reduce the effect of the reality gap. Some of

© Springer Nature Switzerland AG 2022
J. L. Jiménez Laredo et al. (Eds.): EvoApplications 2022, LNCS 13224, pp. 667–682, 2022.
https://doi.org/10.1007/978-3-031-02462-7_42

them apply different strategies during the evolution [13] or generate a population of diverse solutions with the hope that some will work as expected in reality [6]. However, it is also possible to avoid simulation and perform evolution directly in hardware. This last approach avoids the reality gap, but faces important challenges.

Varying the morphology of real robots requires a considerable amount of time which could make evolution in hardware unfeasible, specially if each morphology is built from scratch. Current solutions to solving this problem include (1) using specific robotic platforms that can change the length of their limbs and other parts [25], (2) using modular robot platforms [2,19] and (3) combining modular robot systems with 3D printed parts [1,9]. While 3D printing can create robots with almost any shape, it increases the building time and produces non-reusable parts. On the other end, robots that can only change the size of their limbs, although highly reusable, severely restrict the morphological space they can reach. Thus, there is a trade-off on the morphological search space and the reusability of the hardware employed that needs to be balanced when evolving morphologies in hardware [18].

The aforementioned systems make evolution of morphologies in hardware feasible. However, most hardware evolution works ignore the fact that real robot evaluation cannot be parallelized. Evolutionary algorithms are population-based algorithms, and therefore can be parallelized in software easily. However, evolution of morphologies in hardware requires custom and specialized setups that makes having more than one evaluation setup very costly. To the best of our knowledge, all attempts to evolve robotic morphologies in hardware have used only one platform to perform the robot evaluation. Furthermore, the space in these setups is usually reduced and different robots are built with the same basic components, thus it is also not feasible to build and store a population of assembled robots to evaluate them in a future time. This means that each robot needs to be built, evaluated with a controller, and disassembled.

Taking into account the time to build and subsequently disassemble a robot is many times larger than the time required to evaluate it, we propose to use the wall time of the evolution, i.e. the time measured by a wall clock as the evaluations are performed in the real world, as the stop criterion for an evolutionary algorithm rather than the number of evaluations or generations. In this paper, we explore how this strict stop condition affects different strategies that use time more effectively than a traditional strategy for evolving control and morphology of a robot. Specifically, we compare a basic genetic algorithm (GA), used as a baseline, the same GA but testing each morphology with 5 different controllers, and the Edhmor system [7]. The effect of changing the building speed per module on the algorithm results is also analyzed. The experiments have been done in a simulator, but all parameters were chosen to be as close to a physical evaluation as possible.

The next section describes related work that give a higher chance for the controller of a robot to be optimized when optimizing morphology and control with evolutionary algorithms.

2   Related Work

Motivated by the fact that morphological changes are destructive [15], several works have suggested to adapt the controller for each morphology in evolutionary robotics. Chocron proposed a nested genetic algorithm for evolving modular manipulators, where the outer algorithm was in charge of the morphological evolution and the inner one obtained a suitable controller for each manipulator [5]. To reduce the number of evaluations needed in the controller adaptation, other authors have investigated a Lamarckian type inheritance, where each robot adapts its controller, based on the controllers inherited from its parents, and passes the optimized controller to its offspring [11]. Similarly, a recent article by Goff et al. proposes keeping an archive of controllers as an inheritance mechanism [8]. Different learning methods for optimizing controllers for different morphologies of modular robots are evaluated in [14], but there is no evolution of morphologies.

All these works try to adapt or learn a controller for each morphology rather than use a joint evolution of morphology and control. Furthermore, all use the number of generations or evaluations to stop the evolution and do not consider the building time of the robots. In addition, most works use an enormous budget of evaluations which is not available when evolving in hardware. In this paper, rather than focusing on optimizing controllers as morphologies change per se, we look at evolution from the perspective of the wall time and how to balance the time spent evaluating new controllers for changing morphologies and the time spent assembling and disassembling morphologies with this strict stop condition.

3   Materials and Methods

This section describes the three main aspects that we use for our experiments: The Emerge modules that are used for building the robots, the three different methods used to evolve the morphologies and controllers, and the calculation of the wall clock time.

3.1   Emerge Modules

In this paper, we use the Emerge (Easy Modular Embodied Robot Generator) modular robot[1], which is an open source robotic module designed to be easy to build, maintain, and modify [19]. The mating magnetic connectors of the modules allow easy assembly and disassembly of robotic morphologies in seconds either by a human operator or by a robotic manipulator [17]. In addition, magnetic connectors make assembled robots robust against collisions as they can break apart without damaging the modules in case of a collision or if an excessive torque is applied to a connector.

---

[1] More information about the Emerge robot can be found at https://sites.google.com/view/emergemodular.

Fig. 1. Emerge Modular Robot: the magnetic connections allows a quick assembly of the modules to build a robot, which is useful to evolve morphologies and controllers in reality. An evolved morphology with the base module in the center and several basic modules connected is shown: (right) simulation and (left) reality.

Each module has one servo motor and four connection faces, one of them is connected to the bottom end of the motor and the other three are connected to a bracket, forming a U shape, which is actuated by the shaft of the motor. Connectors in all faces are built with a 3D printed layer and a printed circuit board (PCB) layer. Spring pins are soldered to the PCB layers, which allow module faces to share power and communications. Additionally, a base module with eight connection faces is used as a starting module to build the robots. The base provides a battery and a centralized controller that sends commands to the motors through the motor communication bus while also being able to communicate with an external computer. Both modules are shown in Fig. 1. The basic features of the Emerge modules are described in Table 1. A more detailed description of the Emerge modules can be found in [19].

While the evolution can be performed in a physical setup, we have chosen to carry out the experiments of this paper in simulation to speed up the process and fine tune the algorithms for future hardware runs. Thus, we employ the CoppeliaSim simulator [26] in which the Emerge modules have been already modelled. All parameters of the simulation are set to replicate the physical modules and connections between modules break when facing high torques and forces.

3.2   Algorithms

The following three algorithms are tested in this paper with the wall time stopping condition: A genetic algorithm, a genetic algorithm with additional controller evaluations, and the Edhmor algorithm. All of them were implemented using the Java Evolutionary Algorithm Framework (JEAF) [3].

The three algorithms use a tree encoding representation for their individuals that represents the morphology and the controller of a robot. The nodes of the graph represent the modules, and the edges represent the connection between modules. Each node contains the module type, which is fixed in this study (the root node is always a base module and the rest are individual basic Emerge modules) and the parameters of the module controller. An edge contains the face of the parent node where the child is attached and the orientation of the

child (rotation of 0 or 90°). While the modules can be rotated around the center
of a connector in multiples of 90°, we only take into account two rotations as the
same behaviour can be achieved by adding π to the phase offset of the controller.

Genetic Algorithm (GA). A genetic algorithm is used to evolve robot morphologies and controllers from the Emerge modules and its results are used as
a base to compare the results obtained with the other algorithms. Similar to
the methods employed in [23], the genetic algorithm selects parents for crossover
using a simple random tournament (with a tournament size of 2) and has two
different mutation probability parameters: one for the morphology of an individual and another for each parameter of the controller. The crossover is performed
by selecting a random node of both parents (without considering the root node)
and swapping their downstream branches. The morphological mutation operator selects one of these operations, each with a 1/3 probability: add a node to
any random module with a free face in the robot, change the orientation of a
module and the face where it connects to its parent, and delete a random node
and its children. All robots go through a mutation of their controller, where
each parameter can be mutated with the probability specified. If a mutation
occurs, the new value is obtained by adding a Gaussian noise $\mathcal{N}(0, 0.2)$, which is
scaled by the range of the parameter, to the old value. If the mutated parameter
falls outside a prespecified range, a bounce-back function is used to restore the
parameter to its bounds (a circular bounce-back function in case of the phase
offset parameter) [22].

A genetic algorithm is expected to take longer to cycle through generations
and make a less efficient use of wall clock time when evolving real robot morphologies and controllers as each evaluation encompasses an assembly and disassembly step. All individuals are assembled, tested, and disassembled, even if they
have similar or the same morphology. Figure 2 shows the evaluation sequence
when evolving robot morphologies and controllers in a genetic algorithm.

Genetic Algorithm with Additional Controller Evaluations (GA-ACE).
As the basic GA spends most of the evolution time in assembling and disassembling
robots, we have also introduced a modified version of the GA that uses the evolution time more effectively. For each robot built, the modified genetic algorithm
executes 5 additional evaluations in which only the controller is changed and thus
it can also take advantage of already assembled robots to evaluate and optimize
their controllers. This kind of evolutionary algorithm is expected to obtain better
individuals than a standard genetic algorithm as it can perform more evaluations
on the same wall clock time at the expense of slightly reducing the number of different morphologies tested. The 5 additional controller evaluations are obtained
by mutating the controller with the same parameters used for the standard GA.

Edhmor. We have selected as a third algorithm the Evolutionary Designer of
Heterogeneous Modular Robots (Edhmor) [7], which is a custom evolutionary

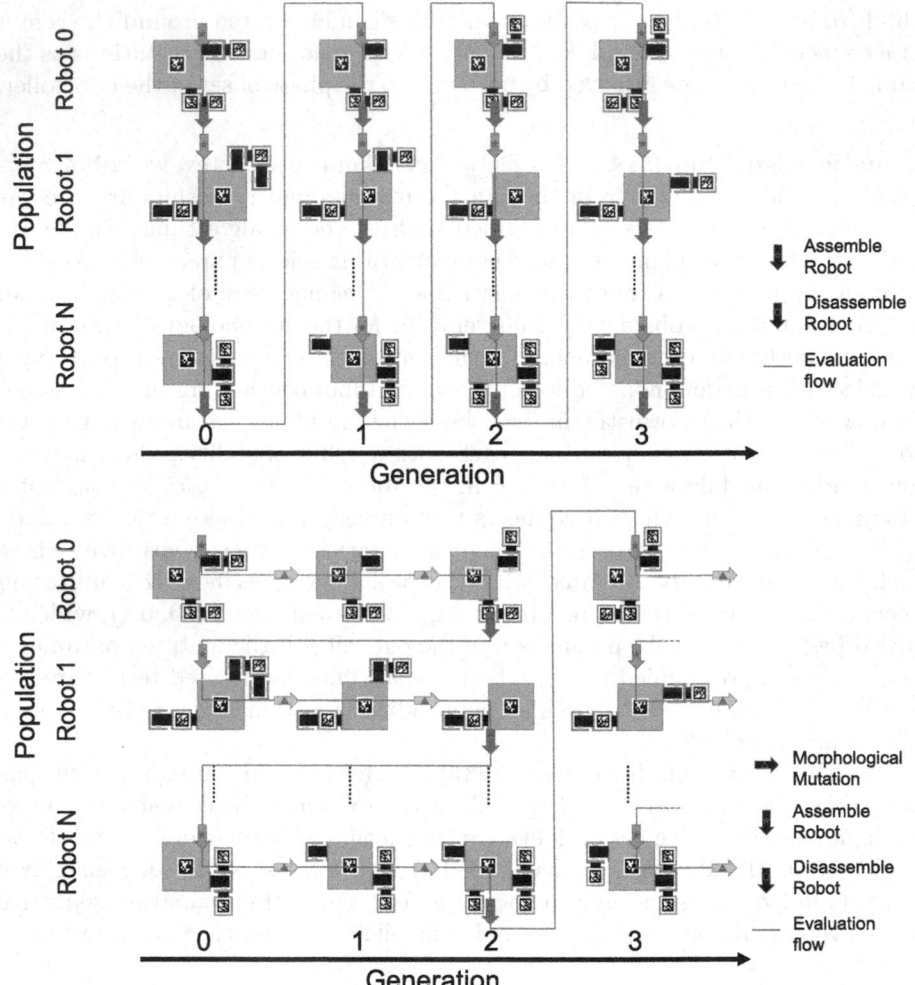

Fig. 2. Evaluation sequence of individual robots when evolving robot morphologies and controllers using a genetic algorithm (top) and the Edhmor algorithm (bottom).

strategy to evolve modular robots. Similar to other works [4], the Edhmor system has a simple mechanism to force and protect innovations: It forces morphological innovations by adding modules to the robots in a growing phase, which is followed by other phases where the morphology and controller adapt to the newly introduced modules.

Table 1. Simulation experiments parameters

Parameter	Edhmor	GA	GA-ACE
Max wall clock time (s)	172800 (48 h)		
Population	20		
Repetitions	40		
Module assembly time (s/module)	20,40		
Morphology mutation probability	N/A		0.1
Controller mutation probability	0.2		0.1
Selection tournament size	N/A		2
Settling time (s)	6		
Evaluation time (s)	38		
Module motor torque (Nm)	1.8		
Module actuator range (radians)	$[-\frac{\pi}{2}, \frac{\pi}{2}]$		
Simulation time step (ms)	50		
Physics engine time step (ms)	5		
Force sensor torque threshold (Nm)	1		

We summarize the Edhmor algorithm and specify the tuned parameters that are used in this paper to reduce the number of evaluations and make evolution in hardware feasible. For a more detailed explanation of Edhmor, we refer the reader to [7]. After generating a random population, the following algorithm phases are applied in a loop until the stop criterion is met:

1. *Growing phase* (2 iterations): Adds one child module to a random module of the robot. The orientation and the connection faces between the new module and its parent are generated randomly. The new module is also tested in two additional positions by changing the orientation of the module and where it is connected to its parent. The best of the three robots with the newly introduced module replaces the original robot, even if it is fitness is lower than the fitness of its parent.
2. *Morphological adaptation phase* (2 iterations): A module which is not the root node is selected randomly from the robot. Three new robots are created by mutating the module connection (the parent attaching face and the module orientation). The best of the three robots only replaces the parent if its fitness is better compared to its parent fitness.
3. *Control adaptation phase* (1 iteration): The controller of the robot is mutated six times to generate six new different controllers. The best of the six robots only replaces the parent if the fitness is better compared to that of its parent. The controller mutation operator is the same as in the GA algorithm, but the probability of mutation has been increased, see Table 1.

4. *Pruning phase* (1 iteration): Generates several morphological mutations by iterating over all the modules of the robot (except the root node) and removing them from the robot with their children. If a robot has M modules, this produces $M - 1$ morphological mutations. The best of the pruned robots replaces the parent if its fitness is better compared to that of its parent.
5. *Replacement phase* (1 iteration, immediately run after the pruning phase): Removes the worst 4 individuals and replaces them with 4 individuals produced after applying a symmetry mutation (randomly selecting a limb attached to the root base node and making its reflection through the XZ or YZ planes). Half of the individuals used to produce the mutation are the 2 best robots in the population and the others are chosen randomly. If a symmetry mutation is not possible, a random robot is created instead.

In contrast to previous algorithms, Edhmor keeps the robots assembled between morphological mutations of the same robot across different generations or phases, making only small changes each time (Fig. 2). As Edhmor only needs the fitness of the other robots of the population at the replacement phase, it can apply 6 generations (2 module additions, 2 morphological adaptations, 1 controller adaptation, and 1 prune phase) to the same robot and then change to the next robot of the population (see Fig. 2). Edhmor is thus expected to be able to perform more robot evaluations in the same amount of wall clock time as it does not use as much time assembling and disassembling robots as a genetic algorithm would take.

3.3   Wall Time Calculation

We have used a simplified model to calculate the assembly time for each robot based on the time that it takes to assemble one module (MAT), and the robot evaluation time (EVALT). When a population is created and evaluated, the wall time is increased for each robot as shown in Eq. (1).

$$wallTime = \sum_{r=1}^{r=20} (Modules_r * \text{MAT} + \text{EVALT}) \tag{1}$$

Where *Modules* means the number of modules of a specific robot in the population. In the GA and GA-ACE algorithms, *wallTime* is increased at each generation by Eq. (2).

$$wallTime = \sum_{r=1}^{r=20} (\text{Modules}_r * \text{MAT} * 2 + \text{EVALT} * (\text{ACE} + 1)) \tag{2}$$

Where ACE is the number of additional controller evaluations (0 for the GA). Notice that this equation takes into account the assembly and disassembly time of the robots.

In Edhmor, the time of each generation depends on the phase that is being performed and is calculated based on Eqs. (3) to (6).

$$wallTime_{grow/morph} = \sum_{r=1}^{r=20} (\text{MAT} * (\text{VAR}_r + 1) + \text{EVALT} * \text{VAR}_r) \qquad (3)$$

$$wallTime_{controller} = \sum_{r=1}^{r=20} (\text{EVALT} * \text{VAR}_r) \qquad (4)$$

$$wallTime_{prune} = \sum_{r=1}^{r=20} ((Modules_r - \min N) * \text{MAT}) + \text{EVALT} * \text{VAR}_r) \qquad (5)$$

$$wallTime_{replace} = \sum_{r=1}^{r=20} (Modules_r * \text{MAT})) \qquad (6)$$

Where VAR represents the number of robot variations produced in each phase, and N is the number of modules, thus $\min N$ is the minimum number of modules allowed.

4   Experimental Setup

Using the three algorithms, robots are evolved for a locomotion task. In this paper, the controller is kept very simple and each module generates an oscillatory movement where the only parameter that the evolution can adjust is the phase offset. The angle of each joint is controlled by Eq. (7).

$$angle = \frac{\pi}{2} \cdot sin(2 \cdot t + \varphi) \qquad (7)$$

Where φ, the phase offset ($[0, 2\pi)$), is encoded in each individual's chromosome and t is the simulation time. Individual robot solutions are tested by placing them in the center of a simulated flat surface environment and allowing them to move for about 38 s. Simulation is carried out in the CoppeliaSim simulator. The fitness is calculated as the final position of the base module measured in a straight line in the (x, y) plane from the starting position of the robot, as in Eq. (8). The first six seconds of the simulation are not taken into account to discard transient movements of the robot (settling time).

$$Fitness = d((x_{t=38}, y_{t=38}), (x_{t=6}, y_{t=6})) \qquad (8)$$

Each method of evolution is run repeatedly 40 times. In the case of the genetic algorithms, a mutation probability of 0.1 is used for both morphological and controller mutations. In all methods used, robots can have a maximum of 16 modules and a minimum of 3 modules. The population is composed of 20 individuals. Robots start always with the flat base module described in Sect. 3.1 as the root node. The initial population is composed of random robots generated

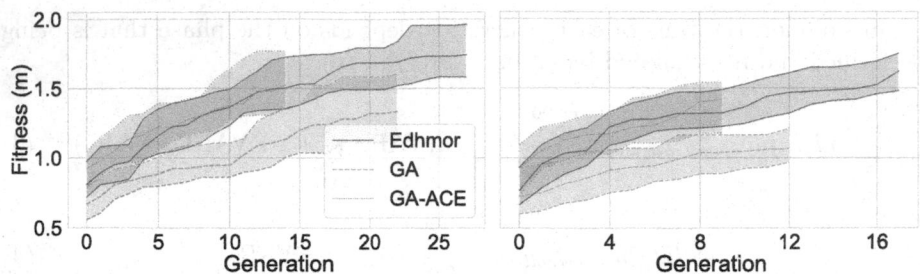

Fig. 3. Best individual fitness vs generation in 40 runs of evolving morphology and control with the Edhmor system, a standard genetic algorithm (GA) and a genetic algorithm modified to allow additional controller evaluations (GA-ACE), and allowing (left) 20 s and (right) 40 s of assembly per module. Center line shows the median of each group and the shaded area shows the interquartile range (IQR).

with a maximum of 8 modules attached and a minimum of 3. The assembly time per module is established as a constant and used to register the wall clock time in each algorithm. All algorithms are tested first with a 20 s assembly time per module in the morphology and then with a 40 s assembly time per module (20 s per module is approximately the time that takes to assemble a module manually as reported in [19]). On average, a complete robot has 10 modules and therefore the full assembly time is 200 s in the 20 s per module case when starting to build the robot from scratch. Evolution is stopped after 172800 s (48 h) of simulated wall-clock time have passed.

5   Results

Evolution of the best individual fitness over time is often presented as a graph of fitness vs the number of generations that the algorithm performs in the allotted time. Generations represent the cycles of selection, reproduction, and replacement, which in some cases have a direct relation to the wall clock time spent. This style of graph is shown for all evolutionary algorithms used in Fig. 3 and shows the median and interquartile range (IQR) in the case of both using 20 s of assembly time per module and 40s of assembly time per module respectively.

It can be seen that the three evolutionary methods do not achieve the same number of generations. This is a direct consequence of limiting the wall clock time and of the way in which all three evaluate their individuals. It can also be observed that, in the case of using 40 s to assemble each module in a morphology, all methods achieve a fewer number of generations as more time is spent assembling and disassembling robots. In both figures, the GA-ACE method achieves better fitness than the standard GA, even when completing fewer generations.

A more interesting graph can be drawn by changing the X-axis variable from generations to the actual wall clock time spent. This gives us a better picture of how the fitness would change in a real evolutionary run with assembly and

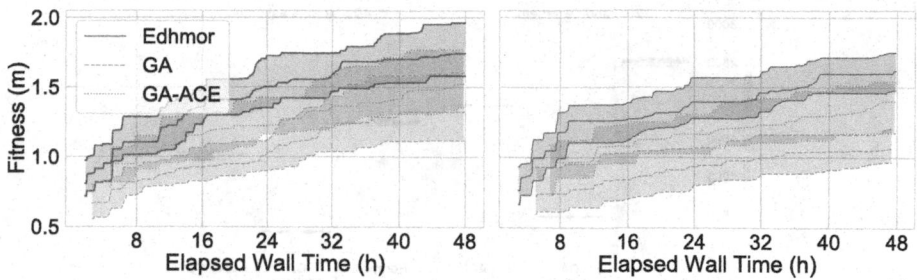

Fig. 4. Best individual fitness vs wall clock time (in hours) in 40 runs of evolving morphology and control with the Edhmor system, a standard genetic algorithm (GA) and a genetic algorithm modified to allow additional controller evaluations (GA-ACE), and allowing (left) 20 s and (right) 40 s of assembly per module. Center line shows the median of each group and the shaded area shows the interquartile range (IQR).

disassembly processes. The wall clock time is measured inside each of the evolutionary algorithms as the simulated robots are evaluated (Sect. 3.3), however the exact times can be registered at different moments in different runs due to the different morphologies that appear. As a consequence, and to be able to compare between runs and evolutionary methods, the missing values between runs are interpolated with a linear interpolation. After this process ends, the missing tail values are filled by repeating the closest value.

Figure 4 shows the best fitness in 40 runs of all three evolutionary methods against wall clock time for 20 s of assembly time per module and 40 s of assembly time per module, respectively. Again, the median and interquartile range (IQR) are shown for each group. These figures show that the Edhmor system produces the best individual fitness in the time allotted and at almost all times. The GA-ACE best individuals follow the Edhmor ones and the standard genetic algorithm individuals are the worst performing.

These figures show that going from 20 s to 40 s of assembly time per module increases the separation between the different algorithm groups. A Kruskall-Wallis test showed that there is a statistically significant difference between the fitness of the best individuals of each evolutionary method at the end of all runs, for a 5% significance level (20 s: $p < 0.0001$, 40 s: $p < 0.0001$). A post hoc Dunn test with Bonferroni correction showed that all groups have a statistically significant difference with each other (20 s: all $p < 0.022$, 40 s: all $p < 0.005$).

The better performance achieved by the Edhmor system can be attributed to performing a higher number of fitness evaluations than the other two as was expected. Figure 5 shows the final number of fitness evaluations for each method in 40 runs using 20 s and 40 s of assembly time per module, respectively. Again, as more time is used performing assembly and disassembly processes, runs with 40s of assembly time per module perform fewer fitness evaluations.

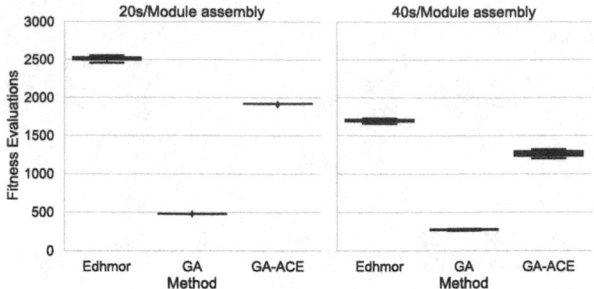

Fig. 5. Final number of fitness evaluations performed by each of the evolutionary methods used when using (left) 20 s of assembly time per module and (right) 40 s of assembly time per module.

6   Discussion

In this paper, we have investigated the effect of using the wall time as a stopping condition, and taking into account the assembly and disassembly time, when evolving the morphology and controller of robots built using a modular robot platform. Under this strict stopping condition, it was shown that the three algorithms analyzed achieve a different final number of generations and fitness evaluations (Figs. 3 and 5). This is due mainly to how the robot evaluations are organized in each of the three algorithms. Additionally, results show that the time it takes for assembly and disassembly of one module in each morphology directly affects the number of generations and fitness evaluations achieved, which is expected as this time cannot be used to evaluate robots and takes a sizable part of the assigned wall time.

The Edhmor and GA-ACE approaches are shown to produce individuals with better fitness than the standard GA (Fig. 4). In the case of GA-ACE, the increase in fitness is because of the extra evaluations used for optimizing the controller, which coincides with what is observed in other studies that perform controller optimization between morphological changes in evolutionary algorithms [5,8,11]. In fact, this algorithm is the closest evaluated in this paper to an evolutionary algorithm with Lamarckian features as the controller is optimized for a morphology and then transmitted to the robot offspring in the following generation. The main difference is that in our approach the controller mutations are randomly generated and there is no specific algorithm for optimizing the controller apart from the ongoing outer evolutionary algorithm. We could test the influence of using an specific algorithm for optimizing the controller in future work, but we should highlight the minimal budget allowed for controller evaluation (only 6 controller evaluations are tested for each morphology). The emphasis in controller optimization also allows GA-ACE to achieve a higher number of fitness evaluations than the normal GA (Fig. 5), as it does not disassemble and assemble robots as often, and could point to evolutionary algorithms with Lamarckian mechanisms being more efficient when time is limited.

In the case of Edhmor, the increase in fitness can be attributed to two related reasons: First, the Edhmor evaluation flow, keeping the a similar morphology assembled for six generations, allows it to perform a higher number of fitness evaluations in the same time than the other two algorithms (Fig. 2). And second, this evaluation flow keeps making small changes to the morphology of the robot and allows the controllers to adapt to these small changes before removing modules and replacing the worst robots, which intermixes morphology and control changes in a more granular way.

Another advantage of the Edhmor algorithm when working under limited time constrains is that it protects innovations in the morphology of robots similar to the mechanisms presented in [4], as described in Sect. 3.2. This innovation protection can be found also in other evolutionary algorithms as is the case of MAP-Elites [20]. In the MAP Elites case, innovation protection is achieved by maintaining a diverse population using a grid of desired features. New morphological changes can be stored in one space of the grid and will only be replaced if a better performing robot with similar features is found, which can be the same robot morphology with a better controller. However, depending on the features selected for the grid similar robots can imply big morphological changes thus the performance of MAP-Elites under wall time constrains could vary widely. Nevertheless, MAP-Elites can be used as a first step in simulation to produce a diverse set of robot morphologies and controllers that can later be used as the seed population of a hardware evolution.

Regarding the values selected for assembling one module (20 and 40 s), we believe that they are conservative. 20 s is approximately the time reported in [19] for manual assembly. In [2], a robot of three modules was built by a manipulator in 210 s (70 s per module). In the ARE project, it takes around three minutes to assemble a 3D printed part with three different modules (45 s per component) [10], but excluding the time used for 3D printing. Even in the case where only the length of the limbs is modified without changing the structure of the robot, the change could last up to 90 s [25].

While the results of the paper are not surprising, it is important to notice that few attempts of evolution of robots in hardware have focused on improving the evaluation flow. To the best of our knowledge, none of them have employed evolutionary strategies as a way to reduce the wall time of the evolution [2,8,11,24]. Time is a scarce resource in hardware evolution and algorithms that evaluate similar morphologies consecutively can improve this bottleneck.

Finally, we would like to point out that in this particular case where the time for evolution is severely limited, a technique that can take advantage of the robot evaluation time to improve the controller of a robot would be highly desirable. An example of such a technique is reinforcement learning [12], which works by updating a controller policy based on a reward obtained each time the robot performs an action in the environment. Although of episodic nature, reinforcement learning can be setup to optimize an average reward at each time step of the robot evaluation and thus would be able to optimize the controller of a robot on this smaller time scale. Learning and evolution have been combined in the past for tackling different tasks and to study the relation between these two processes in reality [21]. A balance would still be needed between the tight time budget that would be allowed for learning, and the time needed to change between morphologies.

7   Conclusions and Future Work

In this work, we have studied the effect of using the wall time as a stop condition when evolving the morphology and control of robots built with the Emerge modules. Results showed that the evolutionary algorithms tested were severely constrained by the wall time stop condition. Thus, for evolutionary algorithms to be effective in real hardware tasks, the algorithms must take careful consideration in their use of time. Specifically, they should take into account factors that affect the amount of time that can effectively be used to evaluate robots such as damaged robots, resetting robots to their initial position or building time. In this paper, we have focused on the time necessary for assembly and disassembly robots, which the results showed to directly affect the number of generations and evaluations performed by each algorithm. Therefore, algorithms must be carefully designed to use time efficiently when performing evolutionary robotics experiments in reality, something that is often not considered.

In particular, evolutionary strategies can be more efficient as they can generate several offspring per parent and evaluate them consecutively. In the Edhmor case, small variations in the morphology performed by the morphological mutation phases of the algorithm keep the majority of the robot assembled between changes and as a consequence do not use as much time for assembly and disassembly as the other two algorithms. This allows the Edhmor algorithm to perform the highest number of fitness evaluations overall, and combined with the local search performed by these small variations in morphology and control, also allows it to find the best individuals in the end. From this, it can be concluded that strategies that keep making morphological changes while keeping assembly times at a minimum hold a high potential of obtaining high fitness results when on time constrains.

Future work includes testing the results found in this paper in the real Emerge platform. Furthermore, we would like to explore how the diversity of solutions is affected by the wall time constrain more in detail by creating a pool of diverse solutions in simulation using the MAP-Elites algorithm and using these solutions to seed the Edhmor algorithm. Finally, we would like to test whether using reinforcement learning to optimize controllers when testing different morphologies is effective under a tight evaluation time budget.

References

1. Auerbach, J., et al.: RoboGen: robot generation through artificial evolution. In: Proceedings of the Fourteenth International Conference on the Synthesis and Simulation of Living Systems, Artificial Life, pp. 136–137. The MIT Press, New York (2014). https://doi.org/10.7551/978-0-262-32621-6-ch022

2. Brodbeck, L., Hauser, S., Iida, F.: Morphological evolution of physical robots through model-free phenotype development. PLoS ONE **10**(6), 1–17 (2015). https://doi.org/10.1371/journal.pone.0128444
3. Caamaño, P., Tedín, R., Paz-Lopez, A., Becerra, J.A.: JEAF: a Java evolutionary algorithm framework. In: IEEE Congress on Evolutionary Computation, pp. 1–8. IEEE (2010)
4. Cheney, N., Bongard, J., SunSpiral, V., Lipson, H.: Scalable co-optimization of morphology and control in embodied machines. J. Roy. Soc. Interface **15**(143), 20170937 (2018)
5. Chocron, O.: Evolutionary design of modular robotic arms. Robotica **26**(3), 323–330 (2008). https://doi.org/10.1017/S0263574707003931
6. Cully, A., Clune, J., Tarapore, D., Mouret, J.B.: Robots that can adapt like animals. Nature **521**(7553), 503–507 (2015). https://doi.org/10.1038/nature14422
7. Faíña, A., Bellas, F., López-Peña, F., Duro, R.J.: EDHMoR: evolutionary designer of heterogeneous modular robots. Eng. Appl. Artif. Intell. **26**(10), 2408–2423 (2013). https://doi.org/10.1016/j.engappai.2013.09.009
8. Goff, L.K.L., et al.: Morpho-evolution with learning using a controller archive as an inheritance mechanism. arXiv:2104.04269 [cs], September 2021
9. Hale, M.F., et al.: Hardware design for autonomous robot evolution. In: 2020 IEEE Symposium Series on Computational Intelligence (SSCI), pp. 2140–2147, December 2020. https://doi.org/10.1109/SSCI47803.2020.9308204
10. Hale, M.F., et al.: The ARE robot fabricator: how to (re)produce robots that can evolve in the real world. In: The 2019 Conference on Artificial Life, pp. 95–102. MIT Press, Cambridge (2019). https://doi.org/10.1162/isal_a_00147.xml
11. Jelisavcic, M., Glette, K., Haasdijk, E., Eiben, A.E.: Lamarckian evolution of simulated modular robots. Front. Robot. AI **6**, 9 (2019). https://doi.org/10.3389/frobt.2019.00009
12. Kober, J., Bagnell, J.A., Peters, J.: Reinforcement learning in robotics: a survey. Int. J. Robot. Res. **32**(11), 1238–1274 (2013). https://doi.org/10.1177/0278364913495721
13. Koos, S., Mouret, J.B., Doncieux, S.: The transferability approach: crossing the reality gap in evolutionary robotics. IEEE Trans. Evol. Comput. **17**(1), 122–145 (2013). https://doi.org/10.1109/TEVC.2012.2185849
14. Lan, G., De Carlo, M., van Diggelen, F., Tomczak, J.M., Roijers, D.M., Eiben, A.E.: Learning directed locomotion in modular robots with evolvable morphologies. Appl. Soft Comput. **111** (2021). https://doi.org/10.1016/j.asoc.2021.107688
15. Lipson, H., Sunspiral, V., Bongard, J., Cheney, N.: On the difficulty of co-optimizing morphology and control in evolved virtual creatures. In: Artificial Life Conference Proceedings 13, pp. 226–233. MIT Press (2016)
16. Marbach, D., Ijspeert, A.J.: Online optimization of modular robot locomotion. In: IEEE International Conference Mechatronics and Automation, vol. 1, pp. 248–253. IEEE (2005)
17. Moreno, R., et al.: Automated reconfiguration of modular robots using robot manipulators. In: 2018 IEEE Symposium Series on Computational Intelligence (SSCI), pp. 884–891, November 2018. https://doi.org/10.1109/SSCI.2018.8628628
18. Moreno, R., Faina, A.: Reusability vs morphological space in physical robot evolution. In: Proceedings of the 2020 Genetic and Evolutionary Computation Conference Companion, pp. 1389–1391 (2020)
19. Moreno, R., Faiña, A.: EMERGE modular robot: a tool for fast deployment of evolved robots. Front. Robot. AI **8**, 198 (2021). https://doi.org/10.3389/frobt.2021.699814

20. Mouret, J.B., Clune, J.: Illuminating search spaces by mapping elites. arXiv:1504.04909 [cs, q-bio], April 2015
21. Nolfi, S., Bongard, J., Husbands, P., Floreano, D.: Evolutionary robotics. In: Siciliano, B., Khatib, O. (eds.) Springer Handbook of Robotics, pp. 2035–2068. Springer, Cham (2016). https://doi.org/10.1007/978-3-319-32552-1_76
22. Nordmoen, J., Nygaard, T.F., Samuelsen, E., Glette, K.: On restricting real-valued genotypes in evolutionary algorithms. In: Castillo, P.A., Jiménez Laredo, J.L. (eds.) EvoApplications 2021. LNCS, vol. 12694, pp. 3–16. Springer, Cham (2021). https://doi.org/10.1007/978-3-030-72699-7_1
23. Nordmoen, J., Veenstra, F., Ellefsen, K.O., Glette, K.: MAP-elites enables powerful stepping stones and diversity for modular robotics. Front. Robot. AI **8**, 56 (2021). https://doi.org/10.3389/frobt.2021.639173
24. Nygaard, T.F., Martin, C.P., Samuelsen, E., Torresen, J., Glette, K.: Real-world evolution adapts robot morphology and control to hardware limitations. In: Proceedings of the Genetic and Evolutionary Computation Conference. GECCO 2018, pp. 125–132. Association for Computing Machinery, New York, July 2018. https://doi.org/10.1145/3205455.3205567
25. Nygaard, T.F., Martin, C.P., Torresen, J., Glette, K., Howard, D.: Real-world embodied AI through a morphologically adaptive quadruped robot. Nat. Mach. Intell. **3**(5), 410–419 (2021). https://doi.org/10.1038/s42256-021-00320-3
26. Rohmer, E., Singh, S.P.N., Freese, M.: V-REP: a versatile and scalable robot simulation framework. In: IROS 2013, pp. 1321–1326. IEEE, Tokyo, November 2013. https://doi.org/10.1109/IROS.2013.6696520
27. Sims, K.: Evolving virtual creatures. In: Proceedings of the 21st Annual Conference on Computer Graphics and Interactive Techniques - SIGGRAPH 1994, pp. 15–22. ACM Press, New York (1994). https://doi.org/10.1145/192161.192167

Analysis of Evolutionary Computation Methods: Theory, Empirics, and Real-world Applications

Neuroevolution Trajectory Networks of the Behaviour Space

Stefano Sarti[✉][iD], Jason Adair[iD], and Gabriela Ochoa[iD]

University of Stirling, Stirling, Scotland, UK
{stefano.sarti,jason.adair,gabriela.ochoa}@stir.ac.uk

Abstract. A network-based modelling technique, search trajectory networks (STNs), has recently helped to understand the dynamics of neuroevolution algorithms such as NEAT. Modelling and visualising variants of NEAT made it possible to analyse the dynamics of search operators. Thus far, this analysis was applied directly to the NEAT genotype space composed of neural network topologies and weights. Here, we extend this work, by illuminating instead the behavioural space, which is available when the evolved neural networks control the behaviour of agents. Recent interest in *behaviour characterisation* highlights the need for divergent search strategies. Quality-diversity and Novelty search are examples of divergent search, but their dynamics are not yet well understood. In this article, we examine the idiosyncrasies of three neuroevolution variants: novelty, random and objective search operating as usual on the genotypic search space, but analysed in the behavioural space. Results show that novelty is a successful divergent search strategy. However, its abilities to produce diverse solutions are not always consistent. Our visual analysis highlights interesting relationships between topological complexity and behavioural diversity which may pave the way for new characterisations and search strategies.

Keywords: Search trajectory networks · Behavioural space · NEAT · Novelty search · Divergent search

1 Introduction

Hard optimisation problems may require more than following a fitness gradient – this is often due to intrinsic characteristics of deceptive domains. In neuroevolution, search based on fitness alone could become stagnant, due to local optima regions with strong attraction basins. An algorithm offering a solution to this, in the context of neuroevolution, is *Novelty Search*, where a search strategy seeks to discover behaviours which are *novel* and *distant* from those previously explored. Through this, remarkable results were achieved; highlighting that behavioural diversity can enhance exploration to solve difficult neuroevolution tasks.

Sarti and Ochoa [22] showed that modelling and visualising the dynamics of NEAT variants using search trajectory networks [16,17] can bring new under-

© Springer Nature Switzerland AG 2022
J. L. Jiménez Laredo et al. (Eds.): EvoApplications 2022, LNCS 13224, pp. 685–703, 2022.
https://doi.org/10.1007/978-3-031-02462-7_43

standing into neuroevolution. Here we propose Neuroevolution Trajectory Networks (NTNs), an adaptation of STNs to the neuroevolution domain, to characterise the dynamics of algorithms in the space of the agent's behaviours. In contrast to previous work [22], where STNs were applied directly on the complex NEAT genomes, here we model the trajectories of the best performing agents in the behaviour space. Given the complexity and redundancy of NEAT genomes, mapping and partitioning the search space as required for STN analysis, results in large network models with sporadic search convergence (frequently visited nodes attracting the search process). The key idea in this article is, instead, to use a much simpler and smaller space, the behavioural space, which in the context of agents solving 2D mazes is simply the end Cartesian (x, y) coordinates of the simulated agents. Encoding (x, y) coordinates as nodes in NTN models (rather than encoding NEAT genomes), is a much simpler task, already tested for STNs in continuous domains [16,17], and provides much more compact and thus easy to analyse models.

We aim to visualise and analyse the derived NTN models from solving medium and hard maze problems (as proposed in [7]) using three NEAT search strategies: novelty, objective and random search. The main contributions of this article are to:

- Provide an analysis tool, NTNs, to examine divergent search and behavioural characterisation in neuroevolution
- Assess the relationships between topological structure, behavioural diversity, and divergent search

This paper is organised as follows. Section 2 details related lines of previous work. Section 3 describes our experimental methodology. Section 4 presents the performance and behavioural diversity analysis of the studied search strategies. Section 5 describes the creation and analysis of the NTN models. Finally, Sect. 6 summarises our main findings and suggestions for future work.

2   Related Work

Recent neuroevolution research has focused on divergent optimisation with a specific interest in increasing diversity [20], moving away from the pursuit of objective and convergent optimisation. The rational for this shift in focus may be attributed to observations in evolutionary biology [25] - it is suggested that high diversity and unique niches result in higher fitness and survival qualities.

This has led to considerable research in the identification of mechanisms that allow for high quality and high diversity, often referred to as *quality/diversity* algorithms [1,4,5,9]. One of the first exemplar of these algorithms was proposed by Lehman and Stanley [8,10], where a *novelty* metric was used to force divergent search and increase behavioural diversity. This method was devised to overcome strong regions of local optima in a maze navigation domain with a highly deceiving structure. An extension of this algorithm is *novelty search with local competition* (NSLC) in which the true principals of quality-diversity were incorporated [9]. NSLC uses multi-objective NSGA-II to evaluate solutions, seeking high-quality behavioural diversity.

Behaviour Characterisation. A comprehensive evaluation of this class of algorithms is provided in [20] in which the best performing QD algorithm was in fact NSLC. The authors go on to define *behavioural characterisation* (BC) as a convention by which behaviours are categorised and classified. This classification can either be aligned or unaligned to the notion of *quality* determined by the assessment domain. In their research, nothing in between these two characterisations is examined. In this, the authors find that not all behaviours are equally important but they highlight that a BC aligned to the end goal leads to favorable diversity and high exploitative qualities to reach the domain goal. Although, the counterargument, as seen previously in [9,10,13] is that the evolution of diversity should originate from BC not directly related to the notion of quality, as these non-directed explorations would lead to divergent ways to discover wider varieties of optimal capabilities [12]. In this research, we intend to detect these behavioural characterisations and offer a tool for their evaluation, with particular attention to topological complexities. Topological complexities often dictate the behaviour of an evolved agent. Here, we elucidate what roles, if any, topologies play in BC and diversification.

Search Trajectory Networks (STNs). STNs are data-driven, graph-based models of search trajectories where nodes represent a given state of the search process and edges represent search progression between consecutive states. Once a system is modelled as a graph (network) powerful analytical and visualisation tools from the field of complex networks can be applied [15]. STNs were initially proposed to characterise differential evolution and particle swarm optimisation for several classical continuous optimisation benchmark functions [17]. STN analysis was later extended to cover not only population-based algorithms but also stochastic local search methods, and both continuous and combinatorial optimisation problems [16]. Recently, these models have been applied to the Cyclic Bandwidth Sum problem in [14], which further corroborates the usefulness of this modelling technique and its applicability to different domains.

3   Experimental Methodology

As a benchmark, we use the classic 2D maze navigation domain outlined in [8]. The task involves an agent (robot) controlled by a neural network navigating a maze from a starting point to an end point, for a fixed number of time steps. The agent's physiognomy is presented in Fig. 1. The agent has six rangefinder sensors that indicate the distance to the nearest obstacle. These are rays (represented as arrows in Fig. 1) originating from the body of the agent, which detect obstacles that are in close proximity, returning the computed distance to such obstacles. The four pie-slice radar sensors are known as the field of view (FOV) that act as a compass, pointing the agent towards the goal (maze exit point). When the line from the goal location to the centre of the robot falls within these (FOV degrees are specified in Fig. 1) the specific radar sensor becomes activated. The activation of the sensors are returned as inputs for the maze-navigating agent to compute behaviours and stored to represent the state of the agent at each simulation time steps.

The outputs computed by the ANN represent the actions that the agent can perform. There are two actuators (actions) which relate to forces that either rotate and/or propel the agent's body. These correspond to changes in linear and/or angular velocity.

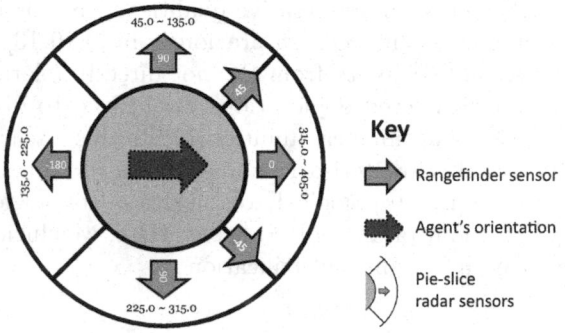

Fig. 1. Physiognomy of the maze navigating agent. The agent has six rangefinder sensors for obstacles detection and four pie-slice radar sensors acting as a compass to detect the goal orientation. Pie-slice labels indicate the degree range of the compass, and arrow labels indicate the rangefinder sensors positions, both in reference to the agent's orientation. Illustration adapted from [19]

The maze domain is relevant for testing novelty search as it has a deceptive fitness landscape. The fitness function used in [8] is formulated on the proximity of the agent to the goal at the end of any given maze navigation simulation. The navigation is made difficult as the maze present walls that form "culs-de-sac". These dead ends that lead close to the goal are local optima to which an objective-based algorithm is likely to converge to. These are especially accentuated in the hard maze map (see Fig. 2b); traps to the search progress are highlighted in red. We used the two maps designed in [8], described below.

Medium Maze (Low Deception). Figure 2a shows the map for the medium maze. This configuration is of low to medium difficulty. The map presents areas of low deception that can be circumvented by the agent without major difficulty. The path from the starting point (yellow square) to the goal (green square) is reasonably linear with a lower chance, as compared to the hard map, for the agent to get trapped in between walls.

Hard Maze (High Deception). Figure 2b illustrates the hard maze configuration. This map is harder as the placement of the walls generate local minima (red shaded circles) capable of trapping the search progress of agents traversing the maze. These areas of high deception are what most challenges the neuroevolution search strategies.

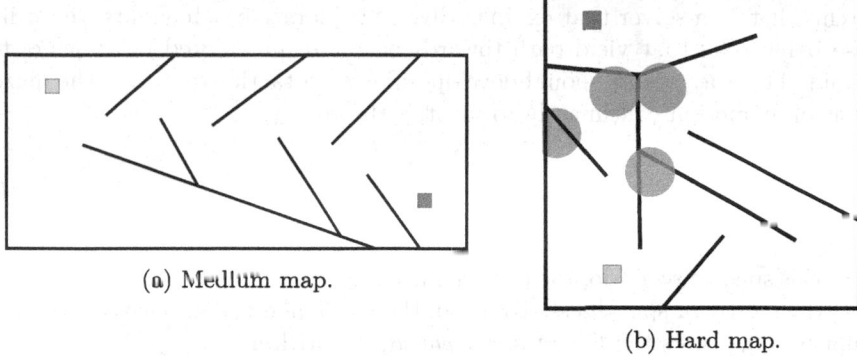

(a) Medium map.

(b) Hard map.

Fig. 2. Maze navigation maps. In both maps, the yellow square represents the starting position of the simulation and the green square represents the goal (exit point). In the hard maze, the landscape local optima are highlighted in red. Image adapted from [19] (Color figure online)

3.1 NEAT Variants

Previous work [2, 21–24, 26] indicates that the use of crossover in NEAT does not improve the algorithm's performance. Therefore, we deemed the use of crossover unnecessary and possibly counterproductive for this experiment. We use the original NEAT-Python implementation [19], but switch off the recombination operator. The search strategies compared, outlined below, are the same ones used in [8].

Fitness-Based Search. Standard evolutionary algorithms use a fitness function to guide the search process. In the maze domain, high fitness is achieved based on an agent's proximity to the goal at the end of the navigation task (trial time steps).

$$\mathcal{L} = \sqrt{\sum_{i=1}^{2} (a_i - b_i)^2} \tag{1}$$

Equation 1 is used to calculate the Euclidean distance between the agent's simulated location with respect to the goal (exit point of the maze). $\mathcal{L}$ represents the specific root-mean-squared error function used for the proximity evaluation. Where a denotes the position of the agent at the end of the simulation and b the location of the maze exit (expressed as 2 dimensional coordinates). The result of this loss function is subtracted from the initial distance of the agent to the end goal and subsequently normalised to produce the fitness value.

Novelty Search. This strategy, as the authors describe, is somewhat counter-intuitive [8]; in order to yield successful results in deceptive domains the objective must be abandoned. The idea is to define a function which uses the *novelty* of the agents' behaviours as a metric of performance. Novelty, specific to NEAT, can either be *structural* (novelty of the ANNs topologies) or *behavioural* (novelty of the ANNs explorative behaviours). The goal is to achieve an effective diverging search. That is, a solver that exhibits diverse explorative behaviours, rewarding those behaviours that yield path towards new and unexplored locations of the domain. This way agents should develop unforeseen tactics to escape the maze's basins of attractions, ultimately to identify the goal.

$$\text{dist}(x, \mu) = \frac{1}{n} \sum_{j=n}^{n} |x_j - \mu_j| \tag{2}$$

In this specific scenario, the performance of the neurocontrollers is calculated using the metric of sparseness. To do so, the implementation, derived from [19], similarly to [8], uses the *k-nearest neighbours* algorithm.

In Eq. 2, $\text{dist}(x, \mu)$ is the novelty score denoting the behavioural difference between two agents, computed as the distance between the two trajectory vectors (one vector per agent; x and μ). Trajectory vectors, which are traced by agents, are comprised of bi-dimensional maze coordinates of size n. x_j and μ_j are the values of the compared vectors (x and μ) at position j. To simplify the calculation, in this implementation, only the agent's trial end coordinates ($j = n$) are considered as the coordinates of interest. This way we can determine the final position of the agent and therefore the distance to the goal.

Random Search. Random search is the simplest of the search strategies. It assigns continuous random values derived from a pseudo-random number generator to the evaluated genomes. This strategy is merely used as control to explore the effect of divergent search, and it is expected not to yield good performance.

3.2   Parameters

Table 1 outlines the parameters values used in our experiments, we emulate the values used in [8]. All parameters, with the exception of the solver time steps, are identical for both maze maps. Similarly, the parameter values are the same for the three algorithm variants. The k parameter in the k-nearest neighbours algorithm, is required for the sparseness calculation, only necessary in novelty search. The solver time steps had to be increased for the hard maze, as for this specific implementation [19] our tests have shown that 400 time steps was not a sufficient allowance to reach the goal in this map.

The coefficients $c1$, $c2$ and $c3$ are all NEAT specific parameters. Standard default values have been used for these. For NTNs modelling on all variants, the partition of the behavioural space is achieved by rounding off to $1e - 0$ the x, y spacial coordinates, and the fitness evaluations to $1e - 2$.

Table 1. NEAT parameter values used. The k parameter (k-nearest neighbours) is relevant only for the novelty search strategy.

Parameter	Value
Population size	250
Maximum generations	1,000
Solver time steps (medium maze)	400
Solver time steps (hard maze)	600
Solution fitness value	1.00
Fitness threshold	0.95
Bias range	$[-30, 30]$
Weight range	$[00, 00]$
$c1$	1
$c2$	1
$c3$	3
Probability add link	0.1
Probability add node	0.005
k (k-nearest neighbours)	16

4 Results

4.1 Analysis Setup

Performance is measured in terms of evaluations required to reach a solution (efficiency) and the quality of solutions at the end of the evaluation cycle (efficacy). The former also highlights the success rate of each strategy. Metrics are computed over 30 independent runs. Out of these 30 runs for each search strategy, 10 were selected for the NTNs construction and analysis, in order to have models of manageable size, yet producing a comprehensive picture of the spectrum of search behaviours observed (Sect. 5.1 details how these were selected).

In our experiments we consider a solution to be achieved, if the fitness reaches a value of 1.00. The maximum number of iterations is set to 1,000 which corresponds to approximately 250,000 evaluations in total (with a population of 250 genomes). The number of maze simulator solving steps is set to 400 for the medium maze and 600 for the hard maze as this requires substantially more exploration time to find the solution (see Sect. 3.2).

4.2 Performance Analysis

It is known from previous work [8,10] that novelty search outperforms both objective and random search on deceptive domains. For completeness, we reproduce some of those results. Figure 4 illustrates the performance of the three search strategies averaged over 30 runs. We clearly observe that both objective and novelty search are able to reach a fitness of 1.0 and discover a behaviour that solves the maze in Fig. 3a. On the other hand, random search does not produce successful solvers and significantly under-performs compared to the other variants, as expected. The main observable difference between objective and novelty search is that objective search on average requires more iterations/evaluations to find solvers. This can be gathered from the distribution of evaluations of the 30 runs shown in Fig. 3c.

In the hard maze we observe that novelty search is the only strategy capable of finding solving behaviours; however, higher average fitness does not signify a better quality of solution. Differently from [10] which achieved 97% success rate (39 out of 40), in our approach, out of the 30 runs the success rate was only found to be 16.6% (5 out of 30). As we observe in Fig. 3d the spread of the evaluations necessary to reach a solution is very wide, showing a rather unstable system.

4.3   Behavioural Diversity Analysis

Amongst the search strategies analysed, it is known that both novelty and random search are divergent strategies, as opposed to convergent strategies such as objective search. From [8,10], it is known that objective search is subject to deception due to the greediness of its search. Despite being a divergent strategy, random search does not offer valuable solutions to the problems. The only succeeding variant in the highly deceptive problem (hard maze) is novelty. Therefore, this section focus on the finer details of what makes each strategy different. We assess these using metrics available from the NEAT algorithm such as topologies and behaviours. As previously discussed in Sect. 2, a correct behavioural characterisation as shown in [20] is fundamental for a successful algorithm that explores solutions with the aim of diversifying. BC, as the name suggests, primarily concerns actual behaviours derived from evolved agents. Another way of setting novelty search is to use the phenotypical structures of evolved agents as the diversity characterisation [19]. To the best of our knowledge this type of setting and related examination has not been attempted before.

Topological Complexity and Behavioural Diversity. We take into consideration two metrics for the diversity analysis: *topological complexities* and *distinct behaviours*. From the algorithm we can record the topological information in terms of number of neurons and number of connection. As our analysis is aimed at neuroevolutionary dynamics, we only consider aspects of the topologies that are evolved. Therefore, we augment this metric by summing the number of neurons and connections together, to then subtract those *input* and *output* neurons, as they are not evolved but preset parameters. To perform a fair comparative analysis we normalise this derived complexity metric to be in the range (0–1). The examined literature was unable to offer an alternative metric to this, which would be suitable for our experiment. As these shallow neurocontrollers are formed by neurons and connections which are expressed in the same genetic code, we summarise them as a unique value determining topological complexity. This is also outlined in the documentation of the algorithm used [11].

Behavioural diversity is also a metric obtained from the actions of an agent. Several consecutive action will trace a path of locations in the Cartesian space of the maze domain. The dimensionality of this actions vector can vary; therefore, to simplify this, as it is used in this implementation (Sect. 3.1), we consider a behaviour solely as the ending x and y coordinates. Other behaviour vectors configurations, capable of capturing more information may be possible, but in this approach, we deemed the chosen one to be sufficient for the purpose of this examination.

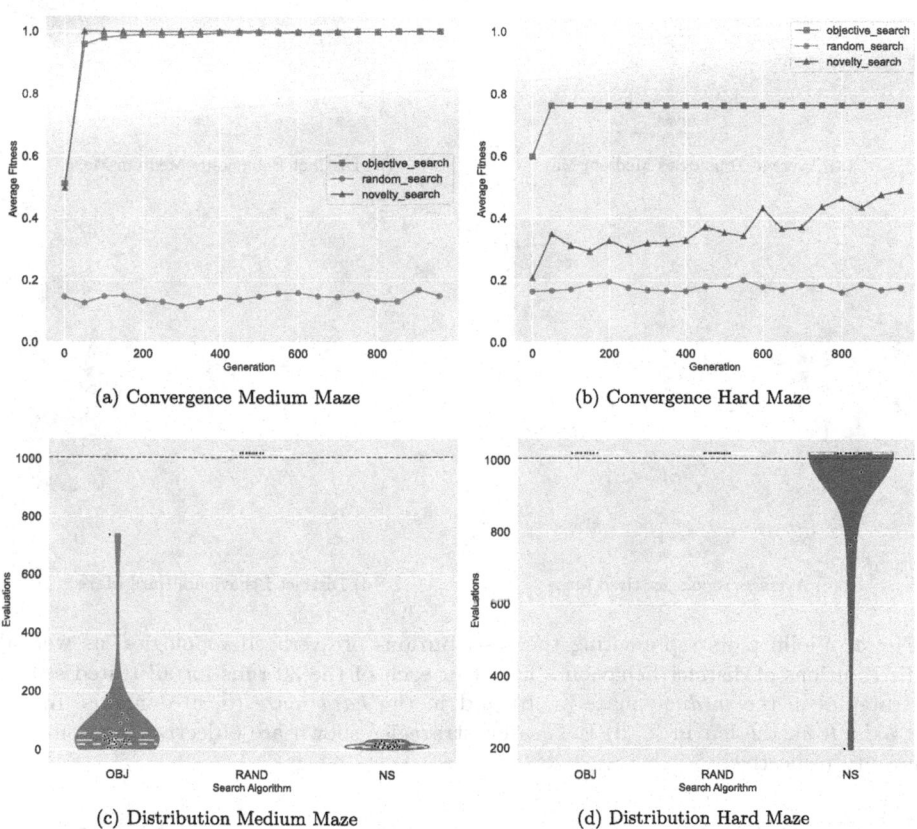

(a) Convergence Medium Maze

(b) Convergence Hard Maze

(c) Distribution Medium Maze

(d) Distribution Hard Maze

Fig. 3. Convergence plot (top) representing the averaged fitness performance over 30 runs and violin plots (bottom) showing distributions of successful vs failed runs (failed runs shown above the dotted line) for all tested search strategies in the *medium* maze (a, c) and in the *hard* maze (b, d) domains. In the x axis, from the left in (c, d) the search strategies shown are objective, random and novelty respectively.

In order to make sense of these coordinates and to discretise the space of possible behaviours we pre-process these values by reducing their precision. This is a common technique and a requirement of STNs analysis.

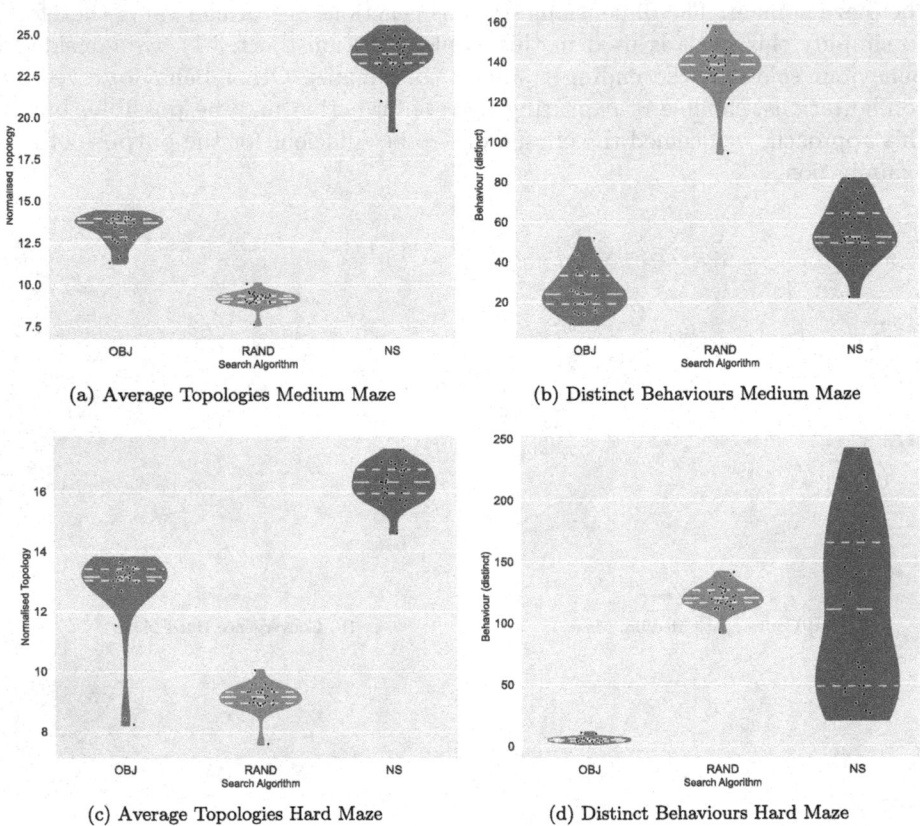

(a) Average Topologies Medium Maze

(b) Distinct Behaviours Medium Maze

(c) Average Topologies Hard Maze

(d) Distinct Behaviours Hard Maze

Fig. 4. Violin plots representing the distributions of averaged topologies, as well as distributions of distinct behaviours found in each of the 30 runs for all tested search strategies in the *medium* maze (a, b) and in the *hard* maze (d, c) domains. In the *x* axis, from the left in (c, d) the search strategies shown are objective, random and novelty respectively.

Figure 4b and d show the distributions of the normalised topologies for each of the 30 runs for the medium and hard maze respectively. The normalised topology values are derived as an average of all the best topologies from all the iterations for each of the runs. This way we are proposing a planarisation of time, flattening this dimension, to capture all the topologies seen during the run. Similarly, this practice is done in NTNs modelling. Figure 4b and d present the number of distinct behaviours recorded in each run for both mazes respectively. These distributions are a metric to assess the diversity of solution and we argue they provide a sense of how divergent a search strategy can be.

The main insight observable from these plots, which will be used to further validate the NTNs analysis are:

- Novelty search has on average more complex topological structures of the genomes than random and objective search. The spread of topologies in random search is narrow and they are notably less complex.
- Looking at random search, we observe high and consistent diversity values (narrow spread). As simpler topologies appear capable of creating more diverse behaviours.
- Random search is not an effective strategy, due to the poor fitness performance. Judged in terms of diversity, this strategy produces more consistent results.
- Novelty search distribution of distinct behaviours presents high variance. This signals an instability of this strategy and high variation between low and high diversity generation.

5 Neuroevolution Trajectory Networks

NTNs are STNs applied to neuroevolution dynamics. The model definitions of STNs are applicable to NTNs and can be found in [16], we reproduce these below for completeness and extend these to include some exceptions, such as a model variation named compressed NTNs (CNTNs). This compressed model, originally inspired form an idea applied to local optima networks [18], is devised to deal with search spaces that present large amounts of neutrality. Neutrality refers to adjacent portions of the search space with the same fitness. Modelling neutrality is relevant for NEAT, as it is well known that there are many ways to set neural network weights which may instantiate a similar behaviour [25]. Compressing equal edges will provide clearer visualisations.

To define a network model, we need to specify their nodes and edges. The relevant definitions are given below.

Representative Solution. A solution to the problem (in this study, an evolved behaviour) at a given iteration is one that represents the status of the search process in the space of behaviours. Since NEAT is a population-based algorithm, the solution with best fitness in the population at a given iteration is chosen as the representative solution.

Location. A non-empty subset of solutions that results from extracting behaviours as x and y coordinates. Each solution in the search space is mapped to one location. Several similar solutions are generally mapped to the same location, as the locations represent a partition of the behaviour search space.

Search Trajectory. Given a sequence of representative solutions in the order in which they are encountered during evolution. A search trajectory is a sequence, mapping solutions to corresponding location.

Node. A location in a search trajectory. The set of nodes is denoted by N.

Edges. Edges are directed and connect two consecutive locations in the search trajectory. Edges are weighted with the number of times a transition between two given nodes occurred during the NTN modelling process. The set of edges is denoted by E.

Neuroevolution Trajectory Network (NTN). A directed graph NTN = (N, E), with node set N, and edge set E as defined above.

Compressed Node. A node that aggregates a set of connected nodes (a connected component) in the NTN with the same fitness value. The set of compressed nodes, is denoted by CN.

Compressed Edges. After the compression (supra), there are no edges between nodes with the same fitness. The set of edges belonging to the compressed nodes are aggregated and their weights summed. This is the compressed edges set, CE.

Compressed NTN. A directed graph CNTN = (CN, CE), where nodes are the compressed nodes CN and edges the compressed edge set CE.

5.1   Sampling and Model Construction

Models are constructed from a data log derived by running the studied search strategies. Specifically, a list of steps connecting two adjacent representative solutions in the search process is recorded. Solutions (evolved agents in the case of NEAT) are represented by their behavioural signature; their final x and y maze coordinates. Each stored search step is formed by two consecutive representative solutions being linked; these transitions become the edges of the network model. Each representative solution stores other attributes for the analysis; such as fitness value and topological complexity. A log file is populated with steps from multiple selected runs. Following this, a post-processing stage maps solutions to locations, and models the network object as per the definitions above.

To extract the models, 10 out of the 30 independent runs were selected for each search strategy on both maze problems. We ranked the runs and selected top 3, the bottom 3, and 4 intermediate runs in therms of fitness performance. The idea was to select a representative sample of the 30 runs; as generating models for all the runs would have made the plots too complex to extract and visually perceive meaningful features.

Network Metrics and Visualisation. Once a system is modelled as a graph or network, a variety of metrics can be computed; such as the degree distribution, length of paths, community structure, and centrality of nodes to name a few [15]. For simplicity, we selected seven network metrics to bring insight into the behaviour of the search variants studied. These metrics are summarised in Table 2.

Visual assessments enable us to appreciate structural features which may be difficult to infer from the network metrics alone. The commonly used node-edge diagrams that we have deployed assign nodes to points in the 2-dimensional Euclidean space, and connect adjacent nodes by lines. For directed graphs, arrowheads are used to indicate the direction of connections. Vertices are drawn here using basic geometrical shapes. Vertices properties such as size, colour and shapes are used to convey more information. In our approach we decorate the network with a palette gradient to signify the complexity of the evolved topologies, size is the strength of incoming edges. The graph visualisations in this paper were produced with the igraph library [3] of the R programming language. For the plots we used a *force-directed* layout algorithms [6], which strive to satisfy some generally accepted criteria, such as distributing the nodes evenly on the plane, minimising the number of edges crossing and keeping edges lengths approximately uniform.

Table 2. Description of network metrics.

nodes	Number of nodes	Metric detailing diversity of solutions
solutions	Number of nodes that reach the desired fitness target	Number of behaviours achieving the maze goal
w-edges	Number of worsening edges	Indicative of the amount of non-greedy exploration
n-paths	Number of shortest paths from start nodes to solution nodes in the CNTN	Indicative of the effectiveness and solutions' reachability. (not expressed if solution not identified)
p-length	Average lengths of the shortest paths from start nodes to solution nodes in the CNTN	Indicative of efficiency of explorations. (not expressed if solution not identified)
complexity	Average topological complexity	Indicative of complexity in relation to divergent and convergent searches respectively
in-strength	Average strength of vertices' incoming connections	Indicative of the extent to which some nodes attract the search process. (calculated as a normalised average)

5.2   Results and Discussion

Figure 5 shows the CNTNs for the two maze maps. Each individual figure depicts the 10 selected runs of each search strategy. These illustrations and their relative metrics of Table 3 helped to reveal the following main observations.

- The networks of objective search highlight the greedy, convergent behaviour of this search strategy (see *w-edges* in metrics Table 3). In the medium maze the best solution is easily achieved with some convergence to common behaviours. In the hard maze, the behaviour search process rapidly gets attracted to a common sub-optimal behaviour. A range of simple to complex topologies reach the goal in the medium maze. In the hard maze, all agents that get stuck in a local optimum are of high topological complexities (dark red).
- The networks of random search are, in both domains, maintaining levels of high diversity, with attracting behaviours. In both cases, with this search strategy there seems to be common attracting behaviours at the end of the trajectories. The large value in *w-edges* of the metrics Table 3 as well as *in-strength* are visible in the network and appear as an oscillating behaviour between improving and worsening behaviours. This highlights the truly random, yet diverse characteristics of this divergent search. Here the agents' complexities appears to be mid to low (blue).
- The networks of novelty search in the medium maze present very similar levels of node convergence, same locations are re-visited several times close in the trajectories, generating a higher *w-edges* value. Paths to solution are similar in terms of average length (*p-length*) but novelty being a divergent search, can be slightly more explorative. Due to this nature we can also observe similar levels of topological complexities as random search.
- An interesting feature can be discerned from this visualised model (Fig. 5f). In the centre, a distinctive rim is drawn, composed of nodes that have similar size (convergence), the nodes inside this circle present lesser diversity (high convergence values) and they appear to be more topologically complex. The nodes outside this circle visually appear simpler, more diverse, and are capable of greater exploration towards reaching the solving behaviour.
- We assume that this conformation (supra) signifies a breakthrough from higher complexity, low diversity of nodes attracted by local minima, to simpler networks, leading to higher diversity, leading to more successful explorations.

Table 3 shows the values of the network metrics described in Table 2 for the two mazes and the 3 search strategies.

Table 3. Network metrics.

	Medium maze			Hard maze		
	Objective	Random	Novelty	Objective	Random	Novelty
nodes	66	653	67	25	536	701
solutions	1	0	1	0	0	1
w-edges	0	0.460	0.293	0	0.421	0.456
n-path	10	0	10	0	0	9
p-length	5.7	N/A	5.9	N/A	N/A	8.555
complexity	0.360	0.155	0.177	0.339	0.135	0.226
in-strength	0.015	0.015	0.016	0.070	0.022	0.004

As behaviours in this case are characterised by Cartesian x and y coordinates, in Fig. 6 we propose an illustration of the neuroevolution trajectories in both geographical representation of the mazes. For simplicity we only represent the best run found for each search strategy. This will provide a visual appreciation of the diversity of each variant and the behavioural optimisation and topological diversity necessary to fully explore and achieve each domain's goal. In this instances, the size of the nodes are a proxy for the normalised topological complexity of each solution.

From these visualisation we can appreciate the following salient features:

- Looking at the behaviours of random search and their topologies, we are able to further corroborate that simpler topologies present higher explorative capabilities as they diverge away from the starting point of the maze.
- In novelty search, specifically on the hard maze, the breakthrough seen in the NTN model of Fig. 5f, is somehow resembling the changes in topological complexities happening from the maze's diagonal upward channel. When topologies diverge from this deceptive area they appear to become topologically less complex.

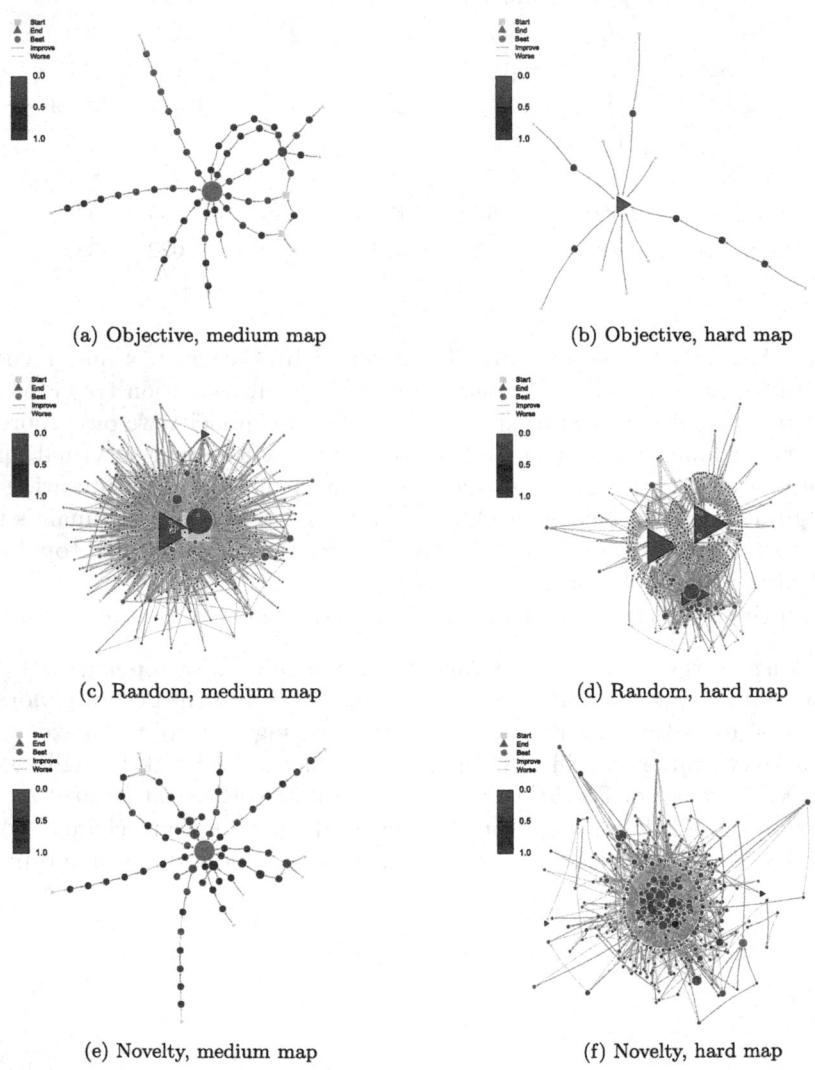

(a) Objective, medium map

(b) Objective, hard map

(c) Random, medium map

(d) Random, hard map

(e) Novelty, medium map

(f) Novelty, hard map

Fig. 5. CNTNs for objective, random and novelty search on the two maze maps. Each sub-figure presents the network model composed of 10 runs for a single search strategy. The size of nodes is proportional to the number of locations in the compressed nodes. Nodes colors are a gradient that signifies the complexity of the network from (0–1], and the shapes specify key search stages.

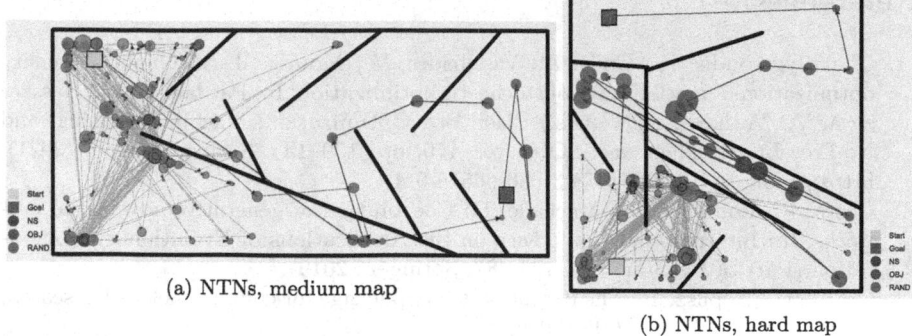

(a) NTNs, medium map

(b) NTNs, hard map

Fig. 6. Visualisation of the best NTN for each search strategy in the medium and hard maze respectively. Size of node signifies topological complexity.

6   Conclusion and Future Work

We propose a visual assessment of the behavioural characterisation of three search strategies deployed in NEAT, tested in medium to highly deceptive maze domains. We adapted an emerging visualisation technique STNs, to the neuroevolution realm: NTNs. We used NTNs to model the evolutionary search process as seen in the behavioural space. Our analysis highlights important characteristics and relationships between behavioural diversity and topological complexities. Although random search is a divergent strategy that does not produce good results, using it in our analysis helped to highlight salient characteristics. Amongst others: topologies that are less complex tend to create increasingly distinct and diverse behaviours with traits of high exploration capabilities, which can eventually lead to successful solutions. This phenomena was also visible by examining the conformation of the best NTNs of novelty search in the hard maze domain geographical representation.

Our examination provides the inceptive insight that alternatives between a closely aligned and non-aligned BC may exist. *Transitive BC* may be an appropriate way to describe this. That is, the focus of the search strategy should be placed on a BC that is indirectly related to divergence and exploration, in the pursuit of quality-diversity. We propose that this indirect search mechanism could further enhance the performance of divergent search strategies. Future work will analyse this claim of *transitive behaviour characterisation* by deploying it as a search technique and testing its validity, while extending its generality by exploiting different domains.

References

1. Chatzilygeroudis, K., Cully, A., Vassiliades, V., Mouret, J.-B.: Quality-diversity optimization: a novel branch of stochastic optimization. In: Pardalos, P.M., Rasskazova, V., Vrahatis, M.N. (eds.) Black Box Optimization, Machine Learning, and No-Free Lunch Theorems. SOIA, vol. 170, pp. 109–135. Springer, Cham (2021). https://doi.org/10.1007/978-3-030-66515-9_4
2. Costa, V., Lourenço, N., Machado, P.: Coevolution of generative adversarial networks. In: International Conference on the Applications of Evolutionary Computation (Part of EvoStar). pp. 473–487. Springer (2019)
3. Csardi, G., Nepusz, T.: The iGraph software package for complex network research. Int. J. Complex Syst. **1695** (2006)
4. Cully, A., Demiris, Y.: Quality and diversity optimization: a unifying modular framework. IEEE Trans. Evol. Comput. **22**(2), 245–259 (2018)
5. Doncieux, S., Laflaquière, A., Coninx, A.: Novelty search: a theoretical perspective. In: Proceedings of the Genetic and Evolutionary Computation Conference, pp. 99–106 (2019)
6. Kamada, T., Kawai, S.: An algorithm for drawing general undirected graphs. Inf. Process. Lett. **31**, 7–15 (1989)
7. Lehman, J., Stanley, K.O.: Exploiting open-endedness to solve problems through the search for novelty. In: ALIFE Xi, pp. 329–336 (2008)
8. Lehman, J., Stanley, K.O.: Abandoning objectives: evolution through the search for novelty alone. Evol. Comput. **19**(2), 189–222 (2011). https://doi.org/10.1162/EVCO_a_00025
9. Lehman, J., Stanley, K.O.: Evolving a diversity of virtual creatures through novelty search and local comp. In: GECCO, pp. 211–218 (2011)
10. Lehman, J., Stanley, K.O., et al.: Exploiting open-endedness to solve problems through the search for novelty (2008)
11. McIntyre, A., Kallada, M., Miguel, C.G., da Silva, C.F.: NEAT-Python. https://github.com/CodeReclaimers/neat-python
12. Meyerson, E., Lehman, J., Miikkulainen, R.: Learning behavior characterizations for novelty search, pp. 149–156 (2016). https://doi.org/10.1145/2908812.2908929
13. Mouret, J.B., Doncieux, S.: Encouraging behavioral diversity in evolutionary robotics: an empirical study. Evol. Comput. **20**(1), 91–133 (2012)
14. Narvaez-Teran, V., Ochoa, G., Rodriguez-Tello, E.: Search trajectory networks applied to the cyclic bandwidth sum problem. IEEE Access **9**, 1–1 (2021). https://doi.org/10.1109/access.2021.3126015
15. Newman, M.E.J.: Networks: An Introduction. Oxford University Press, Oxford (2010)
16. Ochoa, G., Malan, K.M., Blum, C.: Search trajectory networks: a tool for analysing and visualising the behaviour of metaheuristics. Appl. Soft Comput. **109**, 107492 (2021). https://doi.org/10.1016/j.asoc.2021.107492
17. Ochoa, G., Malan, K.M., Blum, C.: Search trajectory networks of population-based algorithms in continuous spaces. In: Castillo, P.A., Jiménez Laredo, J.L., Fernández de Vega, F. (eds.) EvoApplications 2020. LNCS, vol. 12104, pp. 70–85. Springer, Cham (2020). https://doi.org/10.1007/978-3-030-43722-0_5
18. Ochoa, G., Veerapen, N., Daolio, F., Tomassini, M.: Understanding phase transitions with local optima networks: number partitioning as a case study. In: Hu, B., López-Ibáñez, M. (eds.) EvoCOP 2017. LNCS, vol. 10197, pp. 233–248. Springer, Cham (2017). https://doi.org/10.1007/978-3-319-55453-2_16

19. Omelianenko, I.: Hands-On Neuroevolution with Python. Packt Publishing Limited, Birmingham (2019)
20. Pugh, J.K., Soros, L.B., Stanley, K.O.: Quality diversity: a new frontier for evolutionary computation. Front. Robot. AI **3**(JUL), 1–17 (2016). https://doi.org/10.3389/frobt.2016.00040
21. Real, E., et al.: Large-scale evolution of image classifiers. In: International Conference on Machine Learning, pp. 2902–2911. PMLR (2017)
22. Sarti, S., Ochoa, G.: A NEAT visualisation of neuroevolution trajectories. In: Castillo, P.A., Jiménez Laredo, J.L. (eds.) EvoApplications 2021. LNCS, vol. 12694, pp. 714–728. Springer, Cham (2021). https://doi.org/10.1007/978-3-030-72699-7_45
23. Siebel, N.T., Sommer, G.: Evolutionary reinforcement learning of artificial neural networks. Int. J. Hybrid Intell. Syst. **4**(3), 171–183 (2007)
24. Silva, F., Correia, L., Christensen, A.L.: Evolutionary online behaviour learning and adaptation in real robots. Roy. Soc. Open Sci. **4**(7) (2017). https://doi.org/10.1098/rsos.160938
25. Stanley, K.O., Clune, J., Lehman, J., Miikkulainen, R.: Designing neural networks through neuroevolution. Nat. Mach. Intell. **2**, 24–35 (2019)
26. Stanley, K.O., Miikkulainen, R.: Competitive coevolution through evolutionary complexification. J. Artif. Intell. Res. **21**, 63–100 (2004)

Parameter Tuning for the $(1 + (\lambda, \lambda))$ Genetic Algorithm Using Landscape Analysis and Machine Learning

Maxim Pikalov[iD] and Vladimir Mironovich[(⊠)][iD]

ITMO University, 49 Kronverkskiy pr., Saint Petersburg 197101, Russia
mironovich.vladimir@gmail.com

Abstract. The choice of parameter values in evolutionary algorithms greatly affects their performance. Many popular parameter tuning techniques are limited by the tuning budget for finding a good set of parameter values. Recently, we proposed an approach to parameter tuning that uses fitness landscape analysis and machine learning to recommend good parameter values for problem instances based on their landscape features. Using fitness landscape features allows to identify similar problems and use parameter tuning data obtained on benchmark problems, significantly reducing the tuning budget requirements.

In this paper, we present our study of the landscape-aware parameter tuning approach for the $(1 + (\lambda, \lambda))$ genetic algorithm. We evaluate the performance of the algorithm tuned by this approach on the linear integer weights problem and the MAX-3SAT problem, in addition to the W-model problem used for the collection of training data. Our results suggest that the proposed approach allows to make meaningful parameter choices and shows good performance without high fitness evaluation budget requirements.

Keywords: Parameter tuning · Landscape analysis · Black-box optimization

1 Introduction

Choosing an appropriate set of parameter values for an evolutionary algorithm is essential for their practical application [13]. The choice of bad parameters may result in significant performance loss or even make it impossible to find the optimum for the given optimization problem. Thus, there exist multiple methods for ensuring good parameter choices, most commonly referred to as *parameter tuning* and *parameter control* [7].

Parameter tuning [6] tries to find good parameters for the algorithms before the optimization run is made. Popular automated techniques, such as irace [14] and SMAC [9] are based on the evaluation of different parameter combinations on the set of problem instances in order to find the best set of parameters. These

© Springer Nature Switzerland AG 2022
J. L. Jiménez Laredo et al. (Eds.): EvoApplications 2022, LNCS 13224, pp. 704–720, 2022.
https://doi.org/10.1007/978-3-031-02462-7_44

parameters are not changed in the optimization process. Unfortunately, using such techniques is usually not possible in a budget-constrained environment.

The good parameter values can also change during the optimization run. In parameter control, the parameters are chosen during the optimization run according to certain established rules or methods [11]. While this approach can be used in more cases, especially with tight budget restrictions where parameter tuning cannot be used, it is still a lot harder to choose good parameters without prior training on the problem instances.

Fitness landscape analysis [16] is an actively developed tool for studying optimization problems and their properties. Extracting information from the problem instances can help not only improve our understanding of the problem we solve but also make it possible to find connections between different optimization problems with similar landscape features. *Exploratory landscape analysis* [15] further considers fitness landscape features extracted from intermediate solutions and their fitness values. As a result, it is possible to extract information about the fitness function and its properties even in the black-box environment.

Fitness landscape analysis has found various applications, mostly in continuous optimization domain [17]. Successful applications of landscape analysis for automated selection of algorithms with machine learning [12] and dynamic algorithm selection [10] show that there may be great potential for using the fitness landscape analysis for parameter choice in optimization algorithms.

Recently, we proposed an approach for automatic parameter tuning for evolutionary algorithms that uses fitness landscape analysis and machine learning techniques [18]. The proposed approach aims to use the ability of fitness landscape analysis tools to provide information on problem instances and to show similarities between different problem instances. Using the training data collected over different instances of model problems with various fitness landscape features, we are able to skip the training process which is usually required in parameter tuning techniques. Instead, we consider that problems with similar landscape features require similar parameter values. As a result, we only need to evaluate a small number of solutions in order to compute the landscape features and present the recommended parameter values. Successful transfer of knowledge about suitable parameters from one problem to another using fitness landscape analysis may help improve the current methods for parameter choice in evolutionary algorithms.

In this paper, we present our study of the proposed approach on the $(1 + (\lambda, \lambda))$ genetic algorithm (GA) [4] with four static parameters. Following the initial explorations presented in [18] we evaluate the proposed approach on two additional problems: the linear integer weights problem and MAX-3SAT problem [8] and summarize the results on the W-model problem that was used for the collection of the training data. Furthermore, we analyze the parameter choices given by the neural network and try to understand if the given parameter recommendations can be explained by underlying landscape changes in problems.

The results suggest that the proposed landscape-aware method for parameter tuning provides good values for the parameters of the $(1 + (\lambda, \lambda))$ genetic algorithm. Specifically, the algorithm with the parameters suggested by the neural network usually ranks better than alternative versions of the algorithm considered in this study. Additionally, we observe that the $(1 + (\lambda, \lambda))$ GA with parameters suggested by our approach significantly outperforms the algorithm with the best parameter values in the training data. This suggests that the machine learning algorithm actually trains on the data set to provide meaningful parameter choices compared to just suggesting the best-observed solution over all data. Finally, we observe successful transfer of knowledge on good parameter choices between different problems, albeit with possibly similar fitness landscape features.

Overall, our results seem to indicate that there is a significant potential in applying fitness landscape analysis and machine learning tools to parameter tuning. In the future, we plan to extend this approach to parameter control, i.e. reevaluating the landscape features and changing parameters accordingly during the optimization process. Additionally, there is a lot to be done in regards to the proposed parameter tuning method itself. We are planning to consider more difficult and different (in terms of landscapes) optimization problems, with a slight focus on permutation-based problems, where the landscape analysis tools are less developed and would require more studies. It would also be important to further study how different sampling strategies used for evaluation of landscape features affect our results. Finally, it would be great to develop a simple tool for parameter tuning in evolutionary algorithms that uses our current and future results.

2   Preliminaries

2.1 $(1 + (\lambda, \lambda))$ Genetic Algorithm

The $(1 + (\lambda, \lambda))$ genetic algorithm, proposed in [4], is a two-phase genetic algorithm that was shown to be quite efficient on several theoretical and practical problems. The algorithm tries to explore the search space by using high mutation rates in the first phase. In the second phase, it applies a crossover mechanism to counteract the possible negative effects of high mutation rates on individuals. Parameter choices in the $(1 + (\lambda, \lambda))$ GA significantly affect its performance and motivates research on its generalization and, in particular, parameter tuning [2].

In this work, we consider the $(1 + (\lambda, \lambda))$ genetic algorithm with four parameters $\{\lambda_1, \lambda_2, k, c\}$, similar to the one studied in [2]. The algorithm's pseudocode is listed in Algorithm 1 and it works as follows:

- during the first phase of each iteration, the algorithm creates λ_1 mutant individuals by applying the mutation operator with mutation rate k/n, where n is the problem size;
- the best mutant individual is selected to crossover with the parent individual;

Algorithm 1. The $(1 + (\lambda, \lambda))$ GA with parameters $\lambda_1, \lambda_2, k, c$

1: $n \leftarrow$ problem size
2: $\lambda_1 \leftarrow$ mutation phase population size
3: $\lambda_2 \leftarrow$ crossover phase population size
4: $k \leftarrow$ mutation coefficient
5: $c \leftarrow$ crossover probability
6: Initialize: $x \leftarrow$ uniformly from $\{0,1\}^n$
7: **for** $t \leftarrow 1, 2, 3, \ldots$ **do**
8: $\quad \ell \sim \mathcal{B}(n, k/n)$
9: $\quad$ **for** $i \in 1, 2, \ldots, \lambda_1$ **do** $\triangleright$ Phase 1: Mutation
10: $\quad\quad x^{(i)} \leftarrow$ flip ℓ uniformly chosen bits in x
11: $\quad$ **end for**
12: $\quad x' \leftarrow$ uniformly from $\{x^{(j)} \mid f(x^{(j)}) = \max\{f(x^{(i)})\}\}$
13: $\quad$ **for** $i \in 1, 2, \ldots, \lambda_2$ **do** $\triangleright$ Phase 2: Crossover
14: $\quad\quad$ **for** $j \in 1, 2, \ldots, n$ **do**
15: $\quad\quad\quad y_j^{(i)} \leftarrow x'_j$ with probability c, otherwise x_j
16: $\quad\quad$ **end for**
17: $\quad$ **end for**
18: $\quad y \leftarrow$ uniformly from $\{y^{(j)} \mid f(y^{(j)}) = \max\{f(y^{(i)})\}\}$
19: $\quad$ **if** $f(y) \geq f(x)$ **then** $\triangleright$ Selection
20: $\quad\quad x \leftarrow y$
21: $\quad$ **end if**
22: **end for**

- in the second phase, λ_2 individuals are created with the crossover operator that takes bits from the mutant individual with probability c;
- the best crossover individual replaces the parent if it is at least as good;
- the process repeats until the optimum is found.

For experiments presented in this paper we used a modified generic implementation of the $(1 + (\lambda, \lambda))$ genetic algorithm that was first presented in [1]. The implementation of the algorithms and the benchmark problems used for evaluation of the proposed approach is available on GitHub[1].

2.2 Benchmark Problems

In this paper, we use three benchmark problems for the development and evaluation of the proposed approach: W-model problem, Linear Integer Weights problem and MAX-3SAT problem.

The W-model problem [19] is a configurable benchmark optimization problem based on the well-known OneMax problem: finding a hidden optimum based on the number of matching bits in a binary string.

W-model introduces several tunable layers that affect the fitness landscape of the underlying problem and thus makes it suitable for various benchmarking tasks. The tunable layers of the W-model problem are as follows:

---

[1] https://github.com/vmironovich/generic-onell.

- *neutrality*, which introduces areas with equal fitness values;
- *epistasis*, which introduces dependency between different individual genes;
- *ruggedness*, which introduces so-called mountains and valleys into the fitness landscape;
- *dummy*, which introduces variables that do not affect the fitness value of the individual.

W-model allows us to create different problem instances with varying fitness landscape properties, which is essential to collecting a performance data set required in our approach. For this study, we used the C++ implementation of the W-model problem from the IOHProfiler [5].

Linear Integer Weights problem, or simply, LININT_W is the problem of finding the largest value of a linear function with positive integer weights with a constraint on the maximum value of the weight W. In other words, LININT_W is the problem of maximizing a function of n Boolean variables:

$$\text{LININT}_W = \sum_{i=1}^{n} w_i \cdot x_i, 1 \leq w_i \leq W$$

LININT_W problem is similar to the OneMax problem (notably, OneMax is equivalent to LININT_1), but allows us to modify the correlation between the fitness value and the distance to the optimal value, which decreases with the increase of W. Thus we use the LININT_W problem for evaluation of our approach on problems with different landscape features.

Finally, we use the implementation of the *MAX-3SAT* problem [8] from the generic $(1 + (\lambda, \lambda))$ GA repository. This problem is studied in evolutionary computation both theoretically and practically, and it was shown that (1+1) EA and $(1 + (\lambda, \lambda))$ GA show runtimes similar to those they show on the OneMax problem.

The main task in the Boolean satisfiability problem in the maximization setting is to find such a bit string $x = \{x_i\}, i = 1, \ldots, n$, in order to maximize the number of satisfied expressions in a boolean formula in the CNF (conjunctive normal form) with three variables in each clause.

2.3   Collecting Training Data

The training dataset consists of average runtimes (the number of fitness evaluations required to find the optimum) of the parameterized $(1 + (\lambda, \lambda))$ GA presented in Sect. 2.1 with different parameter values on different instances of the W-Model problem.

For the $(1 + (\lambda, \lambda))$ GA we consider the following parameters:

- $\lambda_1 \in [2, 3, 4, \ldots, 9]$;
- $\lambda_2 \in [2, 3, 4, \ldots, 9]$;
- $k \in [1, 3, 5, 7, 9]$;
- $c \in [0.01, 0.03, 0.05, 0.07, 0.09]$.

We computed the average runtime of $(1 + (\lambda, \lambda))$ GA with all possible combinations of these parameters on the instances of W-Model problem with the following parameters:

- $n \in [8, 16, 32, 64, 128]$;
- dummy $\in [1, 1 - 1/n, 1 - 2/n, ..., 0.8]$;
- neutrality $\in [0, 1, 2, ..., n \times 0.2]$;
- epistasis $\in [0, 2, 3]$.

For each combination of $(1 + (\lambda, \lambda))$ GA parameters $\{\lambda_1, \lambda_2, k, c\}$ and W-model problem parameters $\{n, \text{dummy}, \text{neutrality}, \text{epistasis}\}$ we run the $(1 + (\lambda, \lambda))$ genetic algorithm 25 times. We then choose the set of parameters with the lowest mean runtime as the best set of parameters for the particular problem instance.

For each W-model problem instance from the performance dataset we used FLACCO to compute 25 separate vectors of 35 fitness landscape features. This significantly increases the amount of data we can use for the training of the neural network while also taking into the consideration the randomness of computed features. Each feature vector is calculated from 100 random individuals which should be taken into account when considering the budget of the algorithm. For now, we do not use the evaluated individuals in the algorithm, although it may be useful to initialize the optimization process from the best individual observed during the evaluation of features.

Combining the features with performance data results in a dataset of best $(1 + (\lambda, \lambda))$ GA parameters for certain fitness landscape features computed by FLACCO and problem size n: $\{n, \text{features}\} \Rightarrow \{\lambda_1, \lambda_2, k, c\}$. We can then use this data to train the neural network that can suggest certain parameter values for a given set of landscape features.

2.4   Training the Neural Network

We've trained a FNN (feedforward neural network) with two hidden layers with 32 and 16 nodes, respectively. Other architectures were also considered: networks with a number of hidden layers from 1 to 4, in which there were from 4 to 32 neurons, the test and validation results of which turned out to be worse than the final one.

Various optimization algorithms such as Adam, SGD (Stochastic Gradient Descent) and BFGS with Bounded Memory (Broyden-Fletcher-Goldfarb-Shanno) were considered. The Adam algorithm showed the smallest mean square error, so it was chosen for the final model training. ReLU (Rectified Linear Unit) with a regularization parameter equal to 10^{-4} was used as an activation function, while the learning rate was constant with an initial value of 10^{-3}. 20% of the entire dataset has been set aside for verification. The neural network training process stopped when the test result did not improve by at least 10^{-4} for ten consecutive epochs.

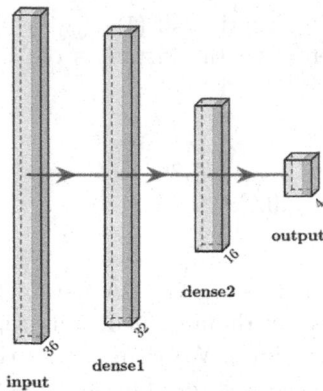

Fig. 1. Final model architecture

The final neural network architecture used for the experiments is shown in Fig. 1. We provide the computed fitness landscape features for the problem instance as an input and get the recommended parameter values for the $(1 + (\lambda, \lambda))$ GA as an output. The source code for the resulting neural network used in this study is available on GitHub[2].

3   Experiments

In this section, we present our empirical findings on the efficiency of the proposed approach. We evaluate the running time performance of the algorithms, i.e. the number of fitness function evaluations required to find an optimum solution.

For each considered benchmark problem, we compare the performance of the algorithm with suggested parameters, which we call the $(1 + (\lambda, \lambda))$ GA tuned, with following $(1 + (\lambda, \lambda))$ GA variations available in the generic $(1 + (\lambda, \lambda))$ GA repository:

- $(1 + (\lambda, \lambda)), \lambda = 4$: $(1 + (\lambda, \lambda))$ GA with default parameter choices $\lambda_2 = \lambda_1, k = \lambda_1, c = 1/\lambda_1$ for $\lambda_1 = 4$;
- $(1 + (\lambda, \lambda)), \lambda \leq 2 \log n$ and $(1 + (\lambda, \lambda)), \lambda \leq n$: $(1 + (\lambda, \lambda))$ GA with dynamic λ tuned according to the 1/5-th rule [3] with the listed upper bound on λ;
- $(1 + (\lambda, \lambda)), \lambda \sim pow(2.5)$ the $(1 + (\lambda, \lambda))$ GA with λ sampled from power-law distribution with $\beta = 2.5$;

We also consider the set of parameters in the training data set, which shows the best overall performance among all parameter choices. The parameter values are $\lambda_1 = 2, \lambda_2 = 2, k = 8, c = 0.02$, and we denote the $(1 + (\lambda, \lambda))$ GA with such parameters as $(1 + (\lambda, \lambda))$, single best. For the W-model problem, we also evaluated the $(1+1)$ evolutionary algorithm and random local search (RLS) as an additional baseline for algorithm comparison.

---

[2] https://github.com/engich/1ll-param-model.

Table 1. Recommended parameters for the W-model problem

Configuration				$(1 + (\lambda, \lambda))$ GA parameters			
n	dum	neu	epi	λ_1	λ_2	c	k
All-default				4	4	0.25	4
All-singlebest				2	2	0.02	8
256	0.8	50	1	4	2	0.071	9
256	0.95	0	0	3	5	0.052	1
256	1	0	3	3	3	0.034	3
512	0.85	0	2	5	7	0.048	2
512	0.90	50	1	5	4	0.012	4
512	0.99	25	2	4	5	0.069	9

3.1 W-Model Problem

We initially evaluated the performance of the proposed approach on the W-model instances with problem sizes $n = 256$ and $n = 512$. Note that for the training data we used only W-model instances with the population size of $n = 128$. Increasing the value of n also increases the number of different values in the dummy layer and increases the upper bound for the neutrality layer.

Examples of recommended parameters for W-model configurations are shown in Table 1. We compare them to two other sets of parameters: ALL-DEFAULT is the $(1 + (\lambda, \lambda))$ GA with static parameters $\lambda_2 = \lambda_1, k = \lambda_1, c = 1/\lambda_1$ with $\lambda_1 = 4$, recommended by the authors of the algorithm. ALL-SINGLEBEST— $(1 + (\lambda, \lambda))$ GA with parameters that showed on average the best performance among all the considered parameter sets on W-model configurations from the model training dataset. These parameter values do not change on different instances of the problem. Each W-model configuration is defined by its parameters: n—dimension, dum, neu, epi—parameters of dummy, neutrality, and epistasis layers respectively. For each instance of the W-model problem, we present the parameters recommended by the neural network.

As can be seen from Table 1, the model suggests different $(1 + (\lambda, \lambda))$ GA parameters for different W-model configurations. Notably, we can see that the crossover and mutation population sizes are increased along with n, although no more obvious dependencies can be seen from this data.

The resulting performance metrics are shown in Table 2 for $n = 256$ and Table 3 for $n = 512$. For each algorithm, we provide its average rank among all the considered algorithms $Mean_r$, the median of the rank $Median_r$, the standard deviation of the rank Std_r, and the average difference in the number of function evaluations $Mean_d$ by the algorithm in comparison with tuned $(1 + (\lambda, \lambda))$ GA, excluding evaluations for collecting a sample for landscape analysis. Negative difference values indicate that a particular algorithm performs better than tuned $(1 + (\lambda, \lambda))$ GA, while positive values indicate that the algorithm performs worse. Figures 2 and 3 show the boxplots for the ranks of the considered algorithms over considered problem instances.

Table 2. Comparison of algorithms on the W-model problem for $n = 256$

Algorithm	$Mean_r$	$Median_r$	Std_r	$Mean_d$
RLS	2.49	2	1.25	−45.45
$(1 + (\lambda, \lambda))$, tuned	**3.25**	3	1.32	0
(1+1) EA	3.75	3	1.26	68.17
$(1 + (\lambda, \lambda))$, single best	3.81	4	1.69	84.33
$(1 + (\lambda, \lambda)), power - law\ \lambda$	4.44	5	1.39	126.44
$(1 + (\lambda, \lambda)), \lambda = 4$	6.43	6	1.15	306.69
$(1 + (\lambda, \lambda)), \lambda \leq n$	6.66	7	1.44	419.24
$(1 + (\lambda, \lambda)), \lambda \leq 2\ln n$	6.87	7	1.36	467.93

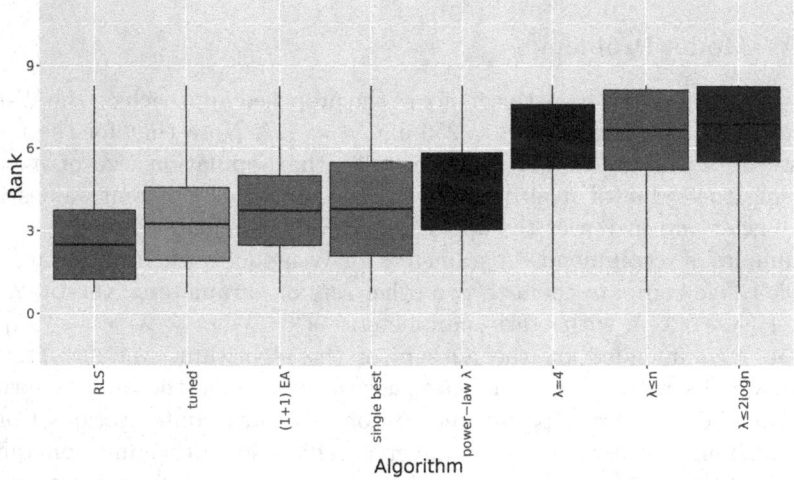

Fig. 2. Algorithms' ranks on the W-model problem with $n = 256$

Table 3. Comparison of algorithms on the W-model problem for $n = 512$

Algorithm	$Mean_r$	$Median_r$	Std_r	$Mean_d$
RLS	2.93	3	1.83	−154.48
$(1 + (\lambda, \lambda))$, tuned	**3.13**	3	1.52	0
$(1 + (\lambda, \lambda)), \lambda \leq n$	3.40	3	1.91	204.54
$(1 + (\lambda, \lambda)), \lambda \leq 2\log n$	3.86	4	1.58	247.22
$(1 + (\lambda, \lambda)), power - law\ \lambda$	4.53	5	1.20	399.10
$(1 + (\lambda, \lambda)), \lambda = 4$	5.09	5	2.14	706.25
(1+1) EA	6.32	6	1.35	886.95
$(1 + (\lambda, \lambda))$, single best	6.56	7	1.48	1272.10

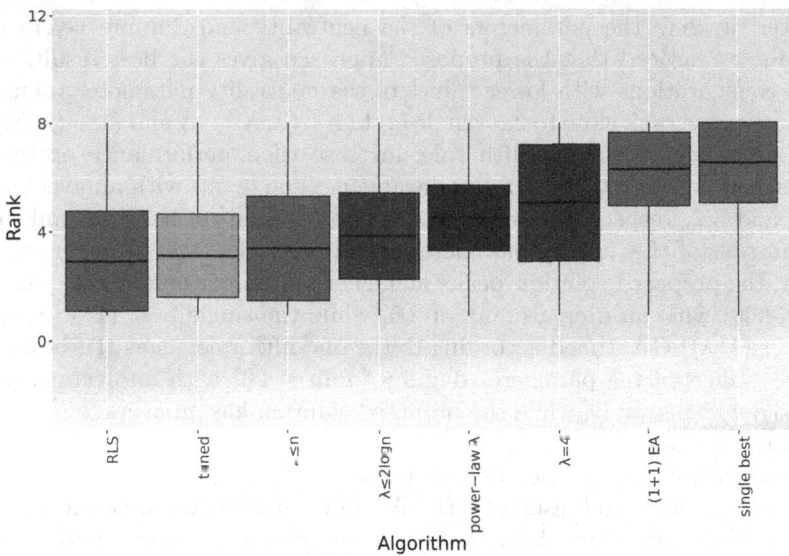

Fig. 3. Algorithms' ranks on W-model problem with $n = 512$

From the results, we can see that on average the tuned $(1 + (\lambda, \lambda))$ GA is the best out of all considered variations of the $(1 + (\lambda, \lambda))$ GA. In most cases, the performance difference is greater than the number of evaluations required for the evaluation of fitness landscape features and it increases with problem size. The versions of the $(1 + (\lambda, \lambda))$ GA with dynamic parameter setting improve their performance with the increased problem size, which may mean that they either need time to find the good parameter value or benefit greatly from changing λ in the different stages of optimization. Still, the tuned $(1 + (\lambda, \lambda))$ GA ranks better than them which suggests that it is able to choose good parameter values before the optimization run.

Additionally, there is a significant difference in the performance of the tuned $(1 + (\lambda, \lambda))$ GA and single best $(1 + (\lambda, \lambda))$ GA. For $n = 256$ they are ranked similarly which may suggest that the neural network just chooses the best set of parameters in the training data. This is proven wrong for $n = 512$, where the single best algorithm shows the worst performance out of all, while the tuned $(1 + (\lambda, \lambda))$ GA still shows great performance. This observation makes us believe that the neural network is capable of learning parameter choices dependent on the landscape features.

While the tables show the average results grouped by problem dimension, specific ranges of W-model parameters were also studied in order to understand the advantages and disadvantages of the trained model and the proposed method. Dividing the ranges of parameters into intervals, a comparative analysis of the performance of the considered algorithms was carried out on the corresponding configurations of the W-model problems based on the average ranks and the average difference in the number of function evaluations.

After dividing the parameters of the neutrality and dummy layers into 5 intervals, we noticed that the proposed approach gives the best results on W-model configurations with lower values of the neutrality parameter (neu ≤ 40) with an average rank close to 2.5. single best $(1 + (\lambda, \lambda))$ GA and $(1 + (\lambda, \lambda))$ GA, tuned according to the one-fifth rule, increase their performance on intervals with higher values of the neutrality parameter (neu ≥ 50) with an average rank of 3.8 and 3.7, respectively, while the proposed solution has the rank of 3.5. The intervals of the dummy parameter of the W-model layer show the opposite results: the proposed solution peaks at higher values of the dummy parameter (dum ≥ 30) with an average rank of 2.5, while the single best $(1 + (\lambda, \lambda))$ GA and $(1 + (\lambda, \lambda))$ GA, tuned according to the one-fifth rule, show the best results at lower values of the parameter dummy (dum ≤ 20) with an average rank of 3.8 and 3.9 respectively, while the proposed solution has an average rank of 3.4.

Finally, we notice that the $(1 + (\lambda, \lambda))$ algorithm is the closest in performance to the algorithms that do not use the parameter tuning, that is, RLS and $(1 + 1)$ EA, which show good results for the W-model problem on dimension $n = 256$. However, these algorithms begin to noticeably decrease in performance with an increase in the dimension of the problem, and the RLS algorithm is not able to find the optimum on several instances of the W-model problem. Perhaps these results are explained by the fact that the parameters suggested by the proposed method for the $(1 + (\lambda, \lambda))$ algorithm allow it to repeat the behaviour of these simple algorithms, thereby solving specific instances of the problem faster, on which $(1 + (\lambda, \lambda))$ GA with its typical parameter values is inferior in performance to the RLS and $(1 + 1)$ EA algorithms.

3.2   Linear Integer Weights Problem

We evaluated the proposed approach on the LININT$_W$ problem. This problem was not used for the collection of training data, thus it represents a harder challenge for our approach. The weights w_i in the experiments were chosen randomly and independently. In total, about 1000 random configurations were considered with the parameter $W \in [2, 3, 4, 5]$ and the dimension of the problem $n \in [256, 512]$.

Examples of suggested parameters for LININT$_W$ problem instances are shown in Table 4. Similarly to the table for the W-model problem, ALL-DEFAULT is $(1 + (\lambda, \lambda))$ GA with recommended parameters for $\lambda_1 = 4$, and ALL-SINGLEBEST is the $(1 + (\lambda, \lambda))$ GA with parameters that showed the best performance on average. Each LININT$_W$ configuration is defined by parameters: n—dimension, w—maximum weight value, s—seed of the random number generator used to generate the weights.

Notably, the recommended parameter values for different instances of the LININT$_W$ problem differ a lot less than in the case of the W-model problem. It should be noted that with an increase of the dimension of the problem, the values of the parameters λ_1, λ_2 increase, while c and k remain small, which significantly differs from the best parameters on average from the training dataset. Another observation is that for large dimensions n of the problem, larger values of the parameters λ are chosen more often.

Table 4. Recommended parameters for the LININT_W problem

Configurations			$(1 + (\lambda, \lambda))$ GA parameters			
n	w	Seed	λ_1	λ_2	c	k
All-default			4	4	0.25	4
All-singlebest			2	2	0.02	8
256	2.0	266342530426951530	3	4	0.071	1
256	5.0	2148839832026378154	3	3	0.009	1
256	3.0	4842151723410319667	4	4	0.023	1
512	2.0	3242964031826183473	4	5	0.012	2
512	4.0	2963942577492853385	4	8	0.017	2
512	5.0	3963578705995886099	5	7	0.015	1

Table 5. Comparison of algorithms on the LININTW problem

Algorithm	$Mean_r$	$Median_r$	Std_r	$Mean_d$
$(1 + (\lambda, \lambda))$, tuned	**2.38**	2	1.12	0
$(1 + (\lambda, \lambda))$, $\lambda \leq 2\log n$	2.45	2	1.16	47.14
$(1 + (\lambda, \lambda))$, $\lambda \leq n$	2.59	2	1.18	54.32
$(1 + (\lambda, \lambda))$, $\lambda = 4$	3.92	4	1.24	123.47
$(1 + (\lambda, \lambda))$, single best	4.75	5	1.19	164.90
$(1 + (\lambda, \lambda))$, power-law λ	4.83	5	1.42	286.21

The experimental results for the LININT_W problem are presented in Table 5. Figure 4 shows the boxplots for the ranks of the algorithms on considered instances of the LININT_W problem.

Analyzing the results, we can notice that on average the tuned $(1 + (\lambda, \lambda))$ GA algorithm turns out to be more efficient than other $(1 + (\lambda, \lambda))$ GA variations. On the one hand, the results are similar to the results for the W-model problem, since the proposed method has the lowest average rank among all $(1 + (\lambda, \lambda))$ GA variations, on the other hand, the results are more stable. This may be explained by the fact that different LININT_W instances still share the same landscape features, while in the case of the W-model problem changing layer variables may result in a significant change to the fitness landscape.

The results were analyzed for different values of the maximum weight parameter w, while the average rank of the proposed method for larger values of w turned out to be lower than the average rank for smaller w. For instance, for $w = 2$, the average rank of the proposed method is 2.40, and for $w = 5$ the methods ranks better at 2.31.

3.3   MAX-3SAT Problem

Finally, we evaluated the proposed approach on the MAX-3SAT problem, which also was not used for the training dataset. Expressions of the Boolean formula in the experiments were generated randomly and independently. In total, about 1000 random configurations with the dimension of the problem $n \in [256, 512]$ were considered.

Examples of parameters for the MAX-3SAT problem instances are shown in Table 6. Similarly to the table for the W-model problem, ALL-DEFAULT is $(1 + (\lambda, \lambda))$ GA with recommended parameters for $\lambda_1 = 4$, and ALL-SINGLEBEST is the $(1 + (\lambda, \lambda))$ GA with parameters that showed the best performance on

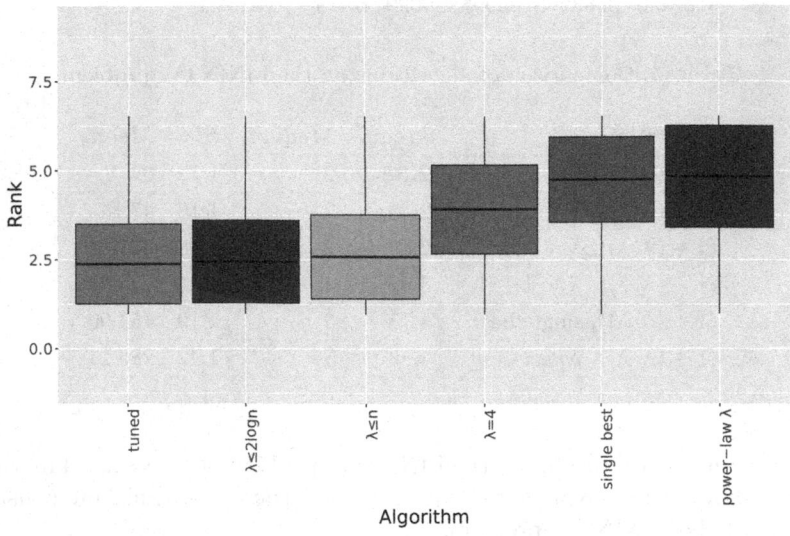

Fig. 4. Algorithms' ranks on the LININT$_W$ problem

Table 6. Recommended parameters for the MAX-3SAT problem

Configuration		$(1 + (\lambda, \lambda))$ GA parameters			
n	seed	λ_1	λ_2	c	k
All-default		4	4	0.25	4
All-singlebest		2	2	0.02	8
256	5109391727087492123	4	5	0.009	1
256	4834392486058466841	3	4	0.018	2
256	7667734744654931535	4	5	0.011	2
512	5709887852471747307	4	6	0.047	1
512	2775983206201735441	3	5	0.015	3
512	3361070628007961843	4	6	0.033	1

average. Each MAX-3SAT configuration is defined by parameters: n—dimension, s—generator element of the random number generator used to generate Boolean formula expressions.

Similarly to the LININT$_W$ problem, there is not much difference in recommended parameter values over different problem instances of the MAX-3SAT problem. Indeed, W-model layers introduce drastic changes to the fitness landscape, while most of the MAX-3SAT and LININT$_W$ instances are supposed to be similar in terms of their landscape features. For the MAX-3SAT problem the trained model suggests for the parameter λ_2 to be larger, and c and k to be smaller than the standard or the best on average parameters. The value of $\lambda_1 = 3$ or 4 is a fairly stable recommendation, which may be due to the size n of the problem.

Table 7. Comparison of algorithms on the MAX-3SAT problem

Algorithm	Mean$_r$	Median$_r$	Std$_r$	Mean$_d$
$(1 + (\lambda, \lambda))$, tuned	**2.45**	2	1.10	0
$(1 + (\lambda, \lambda))$, $\lambda \le 2\log n$	2.57	2	1.21	68.22
$(1 + (\lambda, \lambda))$, $\lambda \le n$	2.79	3	1.29	104.54
$(1 + (\lambda, \lambda))$, $\lambda = 4$	3.82	4	1.21	209.10
$(1 + (\lambda, \lambda))$, single best	4.90	5	0.98	299.50
$(1 + (\lambda, \lambda))$, power-law λ	5.07	5	1.32	355.52

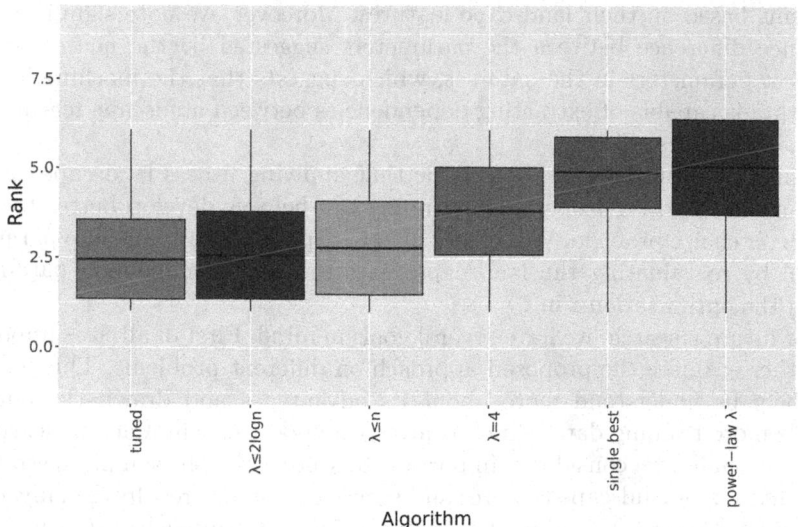

Fig. 5. Algorithms' ranks on the MAX-3SAT problem

The results for the MAX-3SAT problem are presented in Table 7. Figure 5 shows the boxplots for the ranks of the algorithms on considered instances of the MAX-3SAT problem.

Similarly to the other results, the proposed tuned $(1 + (\lambda, \lambda))$ GA algorithm is more efficient than other $(1 + (\lambda, \lambda))$ GA variations. In contrast to the W-model problem, more stable results and smaller spreads of ranks are observed. This may be once again explained by the fact that the MAX-3SAT instances with different configurations, although they have different characteristics of the landscape, are still more similar to each other than different W-model configurations.

4   Conclusion and Future Work

In this paper, we study an approach for automatic parameter tuning using fitness landscape analysis and machine learning. This approach aims to use the fitness landscape analysis to transfer knowledge about good parameters from problem instances evaluated previously to new problem instances with similar landscape features. As a result, this allows us to skip the training step, commonly present in tuning approaches, by using data obtained on problems with various fitness landscape features.

We evaluated the proposed approach on the $(1+(\lambda, \lambda))$ genetic algorithm with four parameters and several optimization problems. We collected performance data for the algorithm on different instances of the W-model problem with various fitness landscape features and trained a neural network on this data. The network is capable to suggest good parameter values for different optimization problems based on their landscape features. Moreover, we note significant performance difference between the parameters suggested by the neural network and best parameters in the data set, which suggests that the machine learning algorithm is capable of extracting dependencies between landscape features and parameter choices.

From our results, we are optimistic that applying fitness landscape analysis and machine learning to parameter tuning can help us develop better tools for parameter choice problem. We further plan to expand this approach to parameter control by reevaluating the landscape features and recommended parameters during the optimization run.

For future research, we have several goals in mind. First of all, it is important to further evaluate the proposed approach on different problems. This may not only help us understand more about its advantages and drawbacks but also provide more training data, which is always a good thing in machine learning.

Additionally, we consider it important to study different sampling strategies for evaluation of landscape features and their effect on our results. Finally, developing a simple tool for parameter tuning using our approach would make our results easily available to others as well as make it much easier to collect the performance data of different algorithms, which is a significant requirement for the efficiency of our approach.

Acknowledgments. The reported study was funded by RFBR and CNRS, project number 20-51-15009.

References

1. Bassin, A., Buzdalov, M.: The $(1 + (\lambda, \lambda))$ genetic algorithm for permutations. In: Proceedings of Genetic and Evolutionary Computation Conference Companion, pp. 1669–1677. ACM (2020)
2. Dang, N., Doerr, C.: Hyper-parameter tuning for the $(1 + (\lambda, \lambda))$ GA. In: Proceedings of Genetic and Evolutionary Computation Conference, pp. 889–897 (2019)
3. Doerr, B., Doerr, C.: Optimal parameter choices through self-adjustment: applying the 1/5-th rule in discrete settings. In: Proceedings of Genetic and Evolutionary Computation Conference, pp. 1335–1342 (2015)
4. Doerr, B., Doerr, C., Ebel, F.: From black-box complexity to designing new genetic algorithms. Theor. Comput. Sci. **567**, 87–104 (2015)
5. Doerr, C., Wang, H., Ye, F., van Rijn, S., Bäck, T.: IOHprofiler: a benchmarking and profiling tool for iterative optimization heuristics (2018). https://arxiv.org/abs/1810.05281, IOHprofiler is available at https://github.com/IOHprofiler
6. Eiben, A.E., Smit, S.K.: Parameter tuning for configuring and analyzing evolutionary algorithms. Swarm Evol. Comput. **1**(1), 19–31 (2011)
7. Eiben, Á.E., Michalewicz, Z., Schoenauer, M., Smith, J.E.: Parameter control in evolutionary algorithms. In: Lobo, F.G., Lima, C.F., Michalewicz, Z. (eds.) Parameter Setting in Evolutionary Algorithms. SCI, vol. 54, pp. 19–46. Springer, Heidelberg (2007). https://doi.org/10.1007/978-3-540-69432-8_2
8. Garey, M.R., Johnson, D.S.: Computers and Intractability: A Guide to the Theory of NP-Completeness. W. H. Freeman & Co., New York (1979)
9. Hutter, F., Hoos, H.H., Leyton-Brown, K.: Sequential model-based optimization for general algorithm configuration. In: Coello, C.A.C. (ed.) LION 2011. LNCS, vol. 6683, pp. 507–523. Springer, Heidelberg (2011). https://doi.org/10.1007/978-3-642-25566-3_40
10. Janković, A., Doerr, C.: Adaptive landscape analysis. In: Proceedings of the Genetic and Evolutionary Computation Conference Companion, pp. 2032–2035 (2019)
11. Karafotias, G., Hoogendoorn, M., Eiben, Á.E.: Parameter control in evolutionary algorithms: trends and challenges. IEEE Trans. Evol. Comput. **19**(2), 167–187 (2015)
12. Kerschke, P., Trautmann, H.: Automated algorithm selection on continuous black-box problems by combining exploratory landscape analysis and machine learning. Evol. Comput. **27**(1), 99–127 (2019)
13. Lobo, F.G., Lima, C.F., Michalewicz, Z. (eds.): Parameter Setting in Evolutionary Algorithms. SCI, vol. 54. Springer, Heidelberg (2007). https://doi.org/10.1007/978-3-540-69432-8
14. López-Ibáñez, M., Dubois-Lacoste, J., Cáceres, L.P., Stützle, T., Birattari, M.: The irace package: iterated racing for automatic algorithm configuration. Oper. Res. Perspect. **3**, 43–58 (2016)
15. Mersmann, O., Bischl, B., Trautmann, H., Preuss, M., Weihs, C., Rudolph, G.: Exploratory landscape analysis. In: Proceedings of the 13th Annual Conference on Genetic and Evolutionary Computation, pp. 829–836 (2011)

16. Mersmann, O., Preuss, M., Trautmann, H.: Benchmarking evolutionary algorithms: towards exploratory landscape analysis. In: Schaefer, R., Cotta, C., Kołodziej, J., Rudolph, G. (eds.) PPSN 2010. LNCS, vol. 6238, pp. 73–82. Springer, Heidelberg (2010). https://doi.org/10.1007/978-3-642-15844-5_8
17. Ochoa, G., Malan, K.: Recent advances in fitness landscape analysis. In: Proceedings of the Genetic and Evolutionary Computation Conference Companion. GECCO 2019, pp. 1077–1094. Association for Computing Machinery, New York (2019). https://doi.org/10.1145/3319619.3323383
18. Pikalov, M., Mironovich, V.: Automated parameter choice with exploratory landscape analysis and machine learning. In: Proceedings of the Genetic and Evolutionary Computation Conference Companion. GECCO 2021, pp. 1982–1985 (2021). https://doi.org/10.1145/3449726.3463213
19. Weise, T., Wu, Z.: Difficult features of combinatorial optimization problems and the tunable W-Model benchmark problem for simulating them. In: Proceedings of Genetic and Evolutionary Computation Conference Companion, pp. 1769–1776 (2018)

Towards a Principled Learning Rate Adaptation for Natural Evolution Strategies

Masahiro Nomura[✉] and Isao Ono

Tokyo Institute of Technology, Meguro City, Japan
nomura.m.ad@m.titech.ac.jp, isao@c.titech.ac.jp

Abstract. Natural Evolution Strategies (NES) is a promising framework for black-box continuous optimization problems. NES optimizes the parameters of a probability distribution based on the estimated natural gradient, and one of the key parameters affecting the performance is the learning rate. We argue that from the viewpoint of the natural gradient method, *the learning rate should be determined according to the estimation accuracy of the natural gradient*. To do so, we propose a new learning rate adaptation mechanism for NES. The proposed mechanism makes it possible to set a high learning rate for problems that are relatively easy to optimize, which results in speeding up the search. On the other hand, in problems that are difficult to optimize (e.g., multimodal functions), the proposed mechanism makes it possible to set a conservative learning rate when the estimation accuracy of the natural gradient seems to be low, which results in the robust and stable search. The experimental evaluations on unimodal and multimodal functions demonstrate that the proposed mechanism works properly depending on a search situation and is effective over the existing method, i.e., using the fixed learning rate.

Keywords: Natural evolution strategies · Learning rate adaptation · Natural gradient

1   Introduction

Natural Evolution Strategies (NES) [8,21–23] is a promising framework for black-box continuous optimization problems. Instead of directly seeking the optimal solution x^*, NES optimizes the parameter θ of a probability distribution $p(x|\theta)$. The expectation of the objective function $f(x)$ over the solution space $J(\theta) = \int f(x)p(x|\theta)dx$ is minimized by repeatedly updating the parameter of the probability distribution based on the estimated natural gradient [3]. In this study, we focus on NES using a multivariate normal distribution as the probability distribution. The natural gradient plays an important role in evolution strategies and randomized algorithms: For example, the rank-μ update [11,12] of CMA-ES [9] can also be regarded as using the estimated value of the natural gradient [2]. Information Geometric Optimization [18], which is a generalized

© Springer Nature Switzerland AG 2022
J. L. Jiménez Laredo et al. (Eds.): EvoApplications 2022, LNCS 13224, pp. 721–737, 2022.
https://doi.org/10.1007/978-3-031-02462-7_45

framework of NES and the rank-μ update of CMA-ES, has been actively studied in recent years [1, 6, 19].

As with other evolution strategies, one of the critical parameters in NES is a learning rate for the parameter of the probability distribution. If the learning rate is too high, the parameter update will be unstable and the performance will deteriorate. On the other hand, if the learning rate is too low, the speed of approaching the optimal solution will be slow, resulting in poor performance. Therefore, setting an appropriate learning rate is essential for maximizing the performance of NES.

There are a few studies on learning rate adaptation of NES. DX-NES proposed by Fukushima et al. switches the learning rate based on the norm of the evolution path which accumulates the movement of the *normalized* mean vector [7]. The effectiveness of switching the learning rate in DX-NES is demonstrated empirically. In fact, the recently proposed DX-NES variants [16, 17], which also employ switching of the learning rate, show promising performance on unconstrained and implicitly constrained black-box optimization problems. Another learning rate adaptation method based on maximum likelihood estimation is proposed in the literature of the CMA-ES [13].

In this paper, we propose a new learning rate adaptation mechanism in view of the natural gradient method; our work is based on a principle that *the learning rate of the natural gradient method should depend on its estimation accuracy.* To measure the estimation accuracy of the natural gradient, we calculate the movement of Kullback-Leibler (KL) divergence, which was first introduced in the population size adaptation of the CMA-ES [14, 15]. We extend the notion to the learning rate adaptation in NES.

The aim of this study is to understand the behavior of NES with the learning rate adaptation mechanism, rather than developing a method that achieves state-of-the-art performance, as this is the first work of learning rate adaptation based on the estimation accuracy of the natural gradient in NES. To that end, we decide to incorporate the learning rate adaptation into xNES [8], which is a simple and promising variant of NES.

The rest of this paper is organized as follows. In Sect. 2, we describe xNES algorithm. In Sect. 3, we propose a learning rate adaptation mechanism based on the estimation accuracy of the natural gradient. In Sect. 4, we experiment on unimodal and multimodal benchmark problems to investigate the effect of the learning rate adaptation mechanism. Section 5 concludes with summary and future direction of this work.

2   xNES

xNES [8] uses a multivariate normal distribution $\mathcal{N}(m, \sigma^2 BB^\top)$ as the probability distribution. Here, $m \in \mathbb{R}^d$ is the mean vector, $\sigma \in \mathbb{R}_{>0}$ is the step-size, and $B \in \mathbb{R}^{d \times d}$ is the normalization transformation matrix where $\det(B) = 1$. The update of xNES is performed by using an estimated natural gradient in the parameter space of the multivariate normal distribution.

xNES first initializes the parameters m, σ, and B. Then, the following steps are repeated until a stopping criterion is met.

Step 1. Sampling and Sorting
For $i \in \{1, \cdots, \lambda\}$, sample λ solutions $x_i \sim \mathcal{N}(m, \sigma^2 BB^\top)$ as follows. Generate d-dimensional standard normal vectors $z_i \sim \mathcal{N}(0, I)$ and compute $x_i = m + \sigma B z_i$. Evaluate the generated solutions on the objective function and obtain their objective values. Then, sort the solutions according to their evaluation values.

Step 2. Estimating Natural Gradient
Estimate the natural gradient based on the sorted solutions as follows:

$$G_\delta = \sum_{i=1}^{\lambda} w_i z_i, G_M = \sum_{i=1}^{\lambda} w_i (z_i z_i^\top - I),$$
$$G_\sigma = \mathrm{Tr}(G_M)/d, G_B = G_M - G_\sigma \cdot I,$$

where w_i is the weight function

$$w_i = \frac{\max\left(0, \ln\left(\frac{\lambda}{2} + 1\right) - \ln(i)\right)}{\sum_{j=1}^{\lambda} \max\left(0, \ln\left(\frac{\lambda}{2} + 1\right) - \ln(i)\right)} - \frac{1}{\lambda}.$$

The weight function holds $\sum_{i=1}^{\lambda} w_i = 0$. Note that xNES uses the weight function instead of using raw evaluation values. This technique is called *fitness shaping*, and it improves the robustness of the algorithm due to the invariance for the monotone transformation of the objective function and enables linear convergence [6].

Step 3. Updating Parameters
Based on the estimated natural gradient, update the parameters of the multivariate normal distributions as follows:

$$m \leftarrow m + \eta_m \sigma B G_\delta,$$
$$\sigma \leftarrow \sigma \cdot \exp(\eta_\sigma/2 \cdot G_\sigma),$$
$$B \leftarrow B \cdot \exp(\eta_B/2 \cdot G_B),$$

where η_m, η_σ, and η_B are the learning rates for updating m, σ, and B, respectively. These learning rates above have default values [8]; $\eta_m = 1$ and $\eta_\sigma = \eta_B = \frac{3}{5} \cdot \frac{(3 + \log(d))}{d\sqrt{d}}$.

3 Learning Rate Adaptation

3.1 Motivation

While default values of the learning rates are presented in xNES, Fukushima et al. have pointed out that the default values are too conservative in a certain situation and there is much room for improvement [7]. However, simply increasing the learning rate causes performance degradation in problems where it is difficult to estimate the natural gradient. It is thus important to adapt the learning rate according to the search situation in order to maximize the performance of xNES.

In this work, we try to adapt the learning rates η_σ and η_B. That is, we focus on only the learning rates related to the covariance matrix. We fix $\eta_m = 1$, which is the default value presented in [8] and widely used in the literature of the CMA-ES [10,12] as well.

To this end, we introduce a learning rate adaptation mechanism that dynamically adapts the learning rates based on the estimation accuracy of the natural gradient. To quantify the estimation accuracy of the natural gradient, we introduce an evolution path in the *parameter space* [15], which accumulates successive parameter movements. The length of the evolution path in the parameter space, described in detail in Sect. 3.2, is used to measure the accuracy of the natural gradient. We believe that, if the length of the evolution path is larger than its expectation under a random function, the accuracy is high, as the tendency of parameter update can be captured. On the contrary, we believe that, if the length of the evolution path is close to its expectation under a random function, the estimation is dominated by noise, and the accuracy is low.

In this study, we consider an evolution path in the parameter space of only the covariance matrix, not the mean vector, because the learning rate for the mean vector is fixed. This is different from existing studies that use the evolution path in the parameter space [14,15]. We will investigate the behavior of the evolution path in Sect. 4.

3.2   Evolution Path for Covariance Matrix

In this work, we introduce an evolution path in the parameter space of the covariance matrix to quantify the estimation accuracy of the natural gradient. We use a modification of the evolution path proposed in [15], which considers both the mean vector and the covariance matrix. Let $\Sigma := \sigma^2 B B^\top$. The covariance movement matrix $\delta\Sigma^{(t+1)}$ is defined to capture the movement of the covariance matrix from iteration t to $t + 1$, which is updated as

$$\delta\Sigma^{(t+1)} = (\sigma^{(t+1)})^2 B^{(t+1)} B^{(t+1)\top} - (\sigma^{(t)})^2 B^{(t)} B^{(t)\top}. \tag{1}$$

We then define the evolution path in the parameter space of the covariance matrix.

$$p_\Sigma^{(t+1)} = (1 - \beta)p_\Sigma^{(t)} + \sqrt{\beta(2 - \beta)}\frac{\mathcal{I}_{\Sigma^{(t)}}^{\frac{1}{2}}\delta\Sigma^{(t+1)}}{\mathbb{E}[\|\mathcal{I}_{\Sigma^{(t)}}^{\frac{1}{2}}\delta\Sigma^{(t+1)}\|^2]^{\frac{1}{2}}}, \tag{2}$$

where β is a cumulation factor of the evolution path and $\mathcal{I}_{\Sigma^{(t)}}$ is the Fisher information matrix of the covariance matrix of the multivariate normal distribution. The expectation $\mathbb{E}[\cdot]$ is taken under a random function $f(x) = \epsilon$, where ϵ is independently drawn from the identical distribution for each evaluation. We use the approximation of $\mathbb{E}[\|\mathcal{I}_{\Sigma^{(t)}}^{\frac{1}{2}} \delta \Sigma^{(t+1)}\|^2]^{\frac{1}{2}}$, which will be derived in Sect. 3.4.

Using the result from Eq. (21) and Appendix B in [14], we define the *length of the evolution path* p_Σ, which represents the movement of the KL divergence in the parameter space of the covariance matrix, as follows:

$$l_\theta^{(t+1)} := \frac{\text{Tr}\left(\left(p_\Sigma^{(t+1)}\right)^2\right)}{2}. \tag{3}$$

Although we do not consider the movement of the KL divergence in the parameter space of the mean vector, we use θ as a notation for the parameter space of the probability distribution.

Under a random function, the length of the evolution path approaches 1 as the iteration t increases. Therefore, comparing the length of the evolution path with *the normalization factor* $\gamma_\theta^{(t+1)}$ which is updated as

$$\gamma_\theta^{(t+1)} = (1 - \beta)^2 \gamma_\theta^{(t)} + \beta(2 - \beta), \tag{4}$$

we can obtain the estimation of the accuracy of the parameter update. The initial parameter $\gamma_\theta^{(0)}$ is set to $\gamma_\theta^{(0)} = 0$.

3.3   Updating Learning Rate

In this section, we give a procedure for the learning rate adaptation. As described, we argue that the learning rate should depend on the estimation accuracy of the natural gradient. When the accuracy is high, the learning rate should be increased, and when the accuracy is low, the learning rate should be decreased. The learning rate adaptation is performed as follows:

$$\eta_\sigma^{(t+1)} = \eta_\sigma^{(t)} \exp\left(\beta_\sigma\left(\frac{l_\theta^{(t+1)}}{\alpha_\sigma} - \gamma_\theta^{(t+1)}\right)\right), \tag{5}$$

$$\eta_B^{(t+1)} = \eta_B^{(t)} \exp\left(\beta_B\left(\frac{l_\theta^{(t+1)}}{\alpha_B} - \gamma_\theta^{(t+1)}\right)\right), \tag{6}$$

where $\alpha_\sigma, \alpha_B, \beta_\sigma$, and β_B are pre-defined hyperparameters. It is possible to set different hyperparameters for η_σ and η_B, respectively, if needed. In this study, we employ the same value for easier interpretation, i.e., $\alpha_\sigma = \alpha_B$ and $\beta_\sigma = \beta_B$.

We clip the learning rates to prevent them from being updated to unexpected ranges by the following equations:

$$\eta_\sigma^{(t+1)} \leftarrow \text{clip}(\eta_\sigma^{(t+1)}, \eta_\sigma^{\min}, \eta_\sigma^{\max}), \tag{7}$$

$$\eta_B^{(t+1)} \leftarrow \text{clip}(\eta_B^{(t+1)}, \eta_B^{\min}, \eta_B^{\max}), \tag{8}$$

where $\eta_\sigma^{\max}$ and $\eta_\sigma^{\min}$ are the maximum and the minimum values of the learning rate for step-size σ, respectively. Similarly, $\eta_B^{\max}$ and $\eta_B^{\min}$ are the maximum and the minimum values of the learning rate for the normalized transformation matrix B, respectively. The clip function is defined as $\mathrm{clip}(u, a, b) := \min(\max(u, a), b)$.

To prevent extrapolation in the update of the parameter, we set the maximum value of the learning rates to 1, i.e., $\eta_\sigma^{\max} = \eta_B^{\max} = 1$. Also, the minimum value of the learning rates is set to the default value of xNES, as it is pointed out that the setting of the learning rates in xNES is often too conservative [7]. Therefore, we use the values recommended in [8] for $\eta_\sigma^{\min}$ and $\eta_B^{\min}$, i.e., $\eta_\sigma^{\min} = \eta_B^{\min} = \frac{3}{5} \cdot \frac{(3+\log(d))}{d\sqrt{d}}$.

3.4 Approximation of $\mathbb{E}[\|\mathcal{I}_{\Sigma^{(t)}}^{\frac{1}{2}} \delta \Sigma^{(t+1)}\|^2]^{\frac{1}{2}}$

In this section, we derive an approximation of $\mathbb{E}[\|\mathcal{I}_{\Sigma^{(t)}}^{\frac{1}{2}} \delta \Sigma^{(t+1)}\|^2]^{\frac{1}{2}}$, which represents a change of the KL divergence in terms of the covariance matrix.

Let $C^{(t)} = B^{(t)} B^{(t)\top}$, $\delta \Sigma = \sigma^{(t+1)2} C^{(t+1)} - \sigma^{(t)2} C^{(t)}$, $\Sigma^{-1} = \sigma^{(t)-2} C^{(t)-1}$, $\delta\sigma = \sigma^{(t+1)}/\sigma^{(t)}$, and $\delta C = C^{(t+1)} - C^{(t)}$. To derive the approximation, we use the Slepian-Bangs formula [5,20] and obtain $\|\mathcal{I}_{\Sigma^{(t)}}^{\frac{1}{2}} \delta \Sigma\|^2 = \delta \Sigma^\top \mathcal{I}_{\Sigma^{(t)}} \delta \Sigma = \frac{1}{2} \mathrm{Tr}\left(\delta \Sigma \Sigma^{-1} \delta \Sigma \Sigma^{-1}\right)$. We will derive the expectation of this equation. From the result provided by Nishida and Akimoto [15], we can obtain $\mathbb{E}[\mathrm{Tr}\left(\delta \Sigma \Sigma^{-1} \delta \Sigma \Sigma^{-1}\right)] = \mathbb{E}[\delta\sigma^4]\mathrm{Tr}\left(\mathbb{E}\left[\left(C^{(t)-1/2} \cdot \delta C \cdot C^{(t)-1/2}\right)^2\right]\right) + d(\mathbb{E}[\delta\sigma^4] - 2\mathbb{E}[\delta\sigma^2])$. We then need to derive the approximation of $\mathbb{E}[\delta\sigma^4], \mathbb{E}[\delta\sigma^2]$, and $\mathrm{Tr}\left(\mathbb{E}\left[\left(C^{(t)-1/2} \cdot \delta C \cdot C^{(t)-1/2}\right)^2\right]\right)$.

Derivation of $\mathbb{E}[\delta\sigma^a]$ $(a = 2, 4)$:

The update equation of step-size in xNES can be rewritten as

$$\sigma^{(t+1)} = \sigma^{(t)} \cdot \exp\left(\frac{\eta_\sigma}{2d}\left(\sum_{j=1}^{d}\sum_{i=1}^{\lambda} w_i \left([z_i]_j^2 - 1\right)\right)\right).$$

Then, by the second order Taylor expansion, for any $a \in \mathbb{R}$,

$$\mathbb{E}[\delta\sigma^a] = \mathbb{E}\left[\exp\left(a\frac{\eta_\sigma}{2d}\left(\sum_{j=1}^{d}\sum_{i=1}^{\lambda} w_i \left([z_i]_j^2 - 1\right)\right)\right)\right]$$

$$\approx 1 + a \cdot \frac{\eta_\sigma \lambda}{2}\mathbb{E}[w_i([z_i]_j^2 - 1)] + \frac{a^2}{2} \cdot \left(\frac{\eta_\sigma}{2d}\right)^2 \mathbb{E}\left[\left(\sum_{j=1}^{d}\sum_{i=1}^{\lambda} w_i \left([z_i]_j^2 - 1\right)\right)^2\right].$$

We will thus calculate the expectations in the above equation. First, from $\mathbb{E}[[z_i]^2] = \mathbb{V}[[z_i]] + \mathbb{E}[[z_i]] = 1$, $\mathbb{E}[w_i([z_i]_j^2 - 1)] = 0$. Next, noting that $\mathbb{V}[z^2] = 2$ and the independence,

$$\mathbb{E}\left[\left(\sum_{j=1}^{d}\sum_{i=1}^{\lambda} w_i\left([z_i]_j^2 - 1\right)\right)^2\right] = \left(\underbrace{\mathbb{E}\left[\sum_{j=1}^{d}\sum_{i=1}^{\lambda} w_i\left([z_i]_j^2 - 1\right)\right]}_{=0}\right)^2 + \mathbb{V}\left[\sum_{j=1}^{d}\sum_{i=1}^{\lambda} w_i\left([z_i]_j^2 - 1\right)\right]$$

$$= \sum_{j=1}^{d}\sum_{i=1}^{\lambda} w_i^2 \mathbb{V}[[z_i]_j^2] = 2d/\mu_w,$$

where $\mu_w = \sum_{i=1}^{\lambda} 1/w_i^2$. By combining these results, $\mathbb{E}[\delta\sigma^a] \approx 1 + \frac{a^2\eta_\sigma^2}{4d\mu_w}$. We thus obtain

$$\mathbb{E}[\delta\sigma^4] \approx 1 + \frac{4\eta_\sigma^2}{d\mu_w}, \mathbb{E}[\delta\sigma^2] \approx 1 + \frac{\eta_\sigma^2}{d\mu_w},$$

$$\mathbb{E}[\delta\sigma^4] - 2\mathbb{E}[\delta\sigma^2] \approx \frac{2\eta_\sigma^2}{d\mu_w} - 1.$$

Derivation of $\mathrm{Tr}\left(\mathbb{E}\left[\left(C^{(t)-1/2} \cdot \delta C \cdot C^{(t)-1/2}\right)^2\right]\right)$:

Let $\Delta = \mathrm{Tr}\left(\sum_{i=1}^{\lambda} w_i(z_i z_i^\top - I)\right)$. The first order Taylor expansion of C in xNES can be obtained as

$$C^{(t+1)} \approx C^{(t)} + \eta_B C^{(t)1/2}\left(\Delta - \frac{\mathrm{Tr}(\Delta)}{d}I\right)C^{(t)1/2}.$$

From $\delta C \approx \eta_B C^{(t)1/2}\left(\Delta - \frac{\mathrm{Tr}(\Delta)}{d}I\right)C^{(t)1/2}$, $C^{(t)-1/2}\delta C C^{(t)-1/2} \approx \eta_B\left(\Delta - \frac{\mathrm{Tr}(\Delta)}{d}I\right)$,

$$\mathrm{Tr}\left(\mathbb{E}\left[\left(C^{(t)-1/2} \cdot \delta C \cdot C^{(t)-1/2}\right)^2\right]\right)$$

$$\approx \mathrm{Tr}\left(\mathbb{E}\left[\left(\eta_B\left(\Delta - \frac{\mathrm{Tr}(\Delta)}{d}I\right)\right)^2\right]\right)$$

$$= \mathbb{E}\left[\mathrm{Tr}\left(\eta_B\left(\Delta - \frac{\mathrm{Tr}(\Delta)}{d}I\right)\right)^2\right] = \eta_B^2\left\{\mathbb{E}\left[\mathrm{Tr}(\Delta^2)\right] - \frac{1}{d}\mathbb{E}[\mathrm{Tr}(\Delta)^2]\right\}.$$

Derivation of $\mathbb{E}\left[\mathrm{Tr}(\Delta^2)\right]$ **and** $\mathbb{E}[\mathrm{Tr}(\Delta)^2]$**:**

$$
\mathbb{E}\left[\mathrm{Tr}(\Delta^2)\right] = \mathbb{E}\left[\mathrm{Tr}\left(\sum_{i=1}^{\lambda}\sum_{j=1}^{\lambda} w_i w_j (z_i z_i^\top - I)(z_j z_j^\top - I)\right)\right]
$$

$$
= \sum_{i=1}^{\lambda}\sum_{j=1}^{\lambda} w_i w_j \mathbb{E}\left[\mathrm{Tr}\left((z_i z_i^\top - I)(z_j z_j^\top - I)\right)\right]
$$

$$
= \sum_{i=1}^{\lambda}\sum_{j=1}^{\lambda} w_i w_j \mathbb{E}\left[\mathrm{Tr}\left(z_i z_i^\top z_j z_j^\top\right) - \mathrm{Tr}(z_i z_i^\top) - \mathrm{Tr}(z_j z_j^\top) + \mathrm{Tr}(I)\right]
$$

$$
= \sum_{i=1}^{\lambda}\sum_{j=1}^{\lambda} w_i w_j (\mathbb{E}[(z_i^\top z_j)^2] - \mathbb{E}[z_i^\top z_i] - \mathbb{E}[z_j^\top z_j] + d).
$$

Note that $\mathbb{E}[z_i^\top z_i] = d$. Then, $\forall i, j \geq 1 (i \neq j)$, $\mathbb{E}[(z_i^\top z_j)^2] = d$, $\mathbb{E}[(z_i^\top z_i)^2] = d^2 + 2d$. Therefore,

$$
\mathbb{E}\left[\mathrm{Tr}(\Delta^2)\right] = \sum_{i=1}^{\lambda} w_i^2 (d^2 + 2d - d - d + d) + \sum_{i,j:i\neq j} w_i w_j \underbrace{(d - d - d + d)}_{=0}
$$

$$
= \sum_{i=1}^{\lambda} w_i^2 (d^2 + d) = (d^2 + d)/\mu_w.
$$

We then derive $\mathbb{E}\left[\mathrm{Tr}(\Delta)^2\right]$. First, $\Delta = \mathrm{Tr}\left(\sum_{i=1}^{\lambda} w_i (z_i z_i^\top - I)\right)$ is written as

$$
\mathrm{Tr}\left(\sum_{i=1}^{\lambda} w_i (z_i z_i^\top - I)\right) = \sum_{i=1}^{\lambda} w_i (\mathrm{Tr}(z_i z_i^\top) - d)
$$

$$
= \sum_{i=1}^{\lambda} w_i \mathrm{Tr}(z_i z_i^\top)
$$

$$
= \sum_{i=1}^{\lambda} w_i \|z_i\|^2.
$$

In the second line, we used $\sum_{i=1}^{\lambda} w_i = 0$. Then, $\mathbb{E}[\mathrm{Tr}(\Delta)^2] = \mathbb{E}[\sum_{i=1}^{\lambda} w_i^2 \|z_i\|^4 + \sum_{i,j:i \neq j} w_i w_j \mathbb{E}[\|z_i\|^2] \mathbb{E}[\|z_j\|^2]] = (d^2 + 2d)/\mu_w - d^2/\mu_w = 2d/\mu_w$. We used $\sum_{i,j:i \neq j} w_i w_j = \sum_{i,j} w_i w_j - \sum_{i=1}^{\lambda} w_i^2 = -1/\mu_w$ due to $\sum_{i=1}^{\lambda} w_i = 0$.

By combining these results,

$$\mathrm{Tr}\left(\mathbb{E}\left[\left(C^{(t)-1/2} \cdot \delta C \cdot C^{(t)-1/2}\right)^2\right]\right) = \frac{\eta_B^2}{\mu_w}(d^2 + d - 2).$$

Approximation Result:

From the results above,

$$\mathbb{E}[\mathrm{Tr}\left(\delta \Sigma \Sigma^{-1} \delta \Sigma \Sigma^{-1}\right)]$$

$$= \underbrace{\mathbb{E}[\delta\sigma^4]}_{\approx 1 + \frac{4\eta_\sigma^2}{d\mu_w}} \underbrace{\mathrm{Tr}\left(\mathbb{E}\left[\left(C^{(t)-1/2} \cdot \delta C \cdot C^{(t)-1/2}\right)^2\right]\right)}_{\approx \eta_B^2(d^2+d-2)/\mu_w} + d\underbrace{(\mathbb{E}[\delta\sigma^4] - 2\mathbb{E}[\delta\sigma^2])}_{\approx \frac{2\eta_\sigma^2}{d\mu_w} - 1} + d$$

$$\approx \frac{1}{\mu_w}\left\{\eta_B^2\left(1 + \frac{4\eta_\sigma^2}{d\mu_w}\right)(d^2 + d - 2) + 2\eta_\sigma^2\right\}.$$

Therefore,

$$\mathbb{E}[\|\mathcal{I}_{\Sigma^{(t)}}^{\frac{1}{2}} \delta \Sigma^{(t+1)}\|^2] \approx \frac{1}{\mu_w}\left\{\frac{\eta_B^2}{2}\left(1 + \frac{4\eta_\sigma^2}{d\mu_w}\right)(d^2 + d - 2) + \eta_\sigma^2\right\}. \qquad (9)$$

We recalculate this approximation every iteration because it depends on the dynamically changing learning rates, η_σ and η_B.

3.5 Overall Procedure

The overall procedure of xNES with the proposed learning rate adaptation mechanism is shown in Algrithm 1. The parameters $\eta_\sigma^{\mathrm{def}}$ and η_B^{def} are the recommended setting of the learning rates in [8], and O is the zero matrix. The procedures in line 3–14 are the same as xNES. In line 15, the covariance movement matrix is updated. In line 16, the expectation of the length of the evolution path under a random function is approximated by using Eq. (9). In line 17, the evolution path in the parameter space of the covariance matrix is updated. In line 18, the length of the evolution path is calculated. In line 19, the normalization factor for the evolution path is updated. In line 20–21, the learning rates for the step-size and the normalization transformation matrix are updated with clipping.

Algorithm 1. xNES with the learning rate adaptation.

Input: $m^{(0)} \in \mathbb{R}^d, \sigma^{(0)} \in \mathbb{R}_+, B^{(0)}\mathbb{R}^{d \times d}, \lambda \in \mathbb{N}$

Input: $\alpha_\sigma, \alpha_B, \beta_\sigma, \beta_B, \eta_\sigma^{\min}, \eta_B^{\min}, \eta_\sigma^{\max}, \eta_B^{\max}$

1: $t = 0, p_\Sigma^{(0)} = O, \gamma_\theta^{(0)} = 0, \eta_\sigma^{(0)} = \eta_\sigma^{\text{def}}, \eta_B^{(0)} = \eta_B^{\text{def}}, \eta_m = 1$

2: **while** stopping criterion not met **do**

3:      **for** $i \in \{1, \cdots, \lambda\}$ **do**

4:          $z_i \sim \mathcal{N}(0, I)$

5:          $x_i = m^{(t)} + \sigma^{(t)}B^{(t)}z_i$

6:      **end for**

7:      Evaluate the solutions and sort $\{(z_i, x_i)\}$

8:      $G_\delta = \sum_{i=1}^\lambda w_i z_i$

9:      $G_M = \sum_{i=1}^\lambda w_i(z_i z_i^\top - I)$

10:     $G_\sigma = \text{Tr}(G_M)/d$

11:     $G_B = G_M - G_\sigma \cdot I$

12:     $m^{(t+1)} = m^{(t)} + \eta_m \sigma^{(t)} B^{(t)} G_\delta$

13:     $\sigma^{(t+1)} = \sigma^{(t)} \cdot \exp(\eta_\sigma^{(t)}/2 \cdot G_\sigma)$

14:     $B^{(t+1)} = B^{(t)} \cdot \exp(\eta_B^{(t)}/2 \cdot G_B)$

15:     $\delta\Sigma^{(t+1)} = (\sigma^{(t+1)})^2 B^{(t+1)} B^{(t+1)\top} - (\sigma^{(t)})^2 B^{(t)} B^{(t)\top}$

16:     Approximate $\mathbb{E}[\|\mathcal{I}_{\Sigma^{(t)}}^{\frac{1}{2}} \delta\Sigma^{(t+1)}\|^2]^{\frac{1}{2}}$ by Eq. (9)

17:     $p_\Sigma^{(t+1)} = (1 - \beta)p_\Sigma^{(t)} + \sqrt{\beta(2 - \beta)} \dfrac{\mathcal{I}_{\Sigma^{(t)}}^{\frac{1}{2}} \delta\Sigma^{(t+1)}}{\mathbb{E}[\|\mathcal{I}_{\Sigma^{(t)}}^{\frac{1}{2}} \delta\Sigma^{(t+1)}\|^2]^{\frac{1}{2}}}$

18:     $l_\theta^{(t+1)} = \text{Tr}(p_\Sigma^{(t+1)2})/2$

19:     $\gamma_\theta^{(t+1)} = (1 - \beta)^2 \gamma_\theta^{(t)} + \beta(2 - \beta)$

20:     $\eta_\sigma^{(t+1)} = \text{clip}\left(\eta_\sigma^{(t)} \exp\left(\beta_\sigma\left(\dfrac{l_\theta^{(t+1)}}{\alpha_\sigma} - \gamma_\theta^{(t+1)}\right)\right), \eta_\sigma^{\min}, \eta_\sigma^{\max}\right)$

21:     $\eta_B^{(t+1)} = \text{clip}\left(\eta_B^{(t)} \exp\left(\beta_B\left(\dfrac{l_\theta^{(t+1)}}{\alpha_B} - \gamma_\theta^{(t+1)}\right)\right), \eta_B^{\min}, \eta_B^{\max}\right)$

22:     $t \leftarrow t + 1$

23: **end while**

Table 1. Definition of benchmark problems.

Definition
$f_{\text{Sphere}}(x) = \sum_{i=1}^d x_i^2$
$f_{\text{Ellipsoid}}(x) = \sum_{i=1}^d (1000^{\frac{i-1}{d-1}} x_i)^2$
$f_{\text{Rastrigin}}(x) = 10d + \sum_{i=1}^d (x_i^2 - 10\cos(2\pi x_i))$
$f_{\text{Bohachevsky}}(x) = \sum_{i=1}^{d-1}(x_i^2 + 2x_{i+1}^2 - 0.3\cos(3\pi x_i) - 0.4\cos(4\pi x_{i+1}) + 0.7)$

4 Experiments

In this section, we investigate the following research questions (RQs).

RQ1. When the learning rate is fixed, how does the evolution path in Eq. (2) of xNES behave on unimodal and multimodal functions?

RQ2. How is the learning rate adapted in xNES with the proposed learning rate adaptation mechanism?

RQ3. Does xNES with the proposed learning rate adaptation mechanism achieve better performance than xNES with fixed learning rates?

We first describe the experimental setups in Sect. 4.1. In Sect. 4.2, we investigate the behavior of the evolution path in xNES with a fixed learning rate (RQ1). We then investigate the behavior of the evolution path and the learning rate in xNES with the *adaptive* learning rate mechanism (RQ2) in Sect. 4.3. Finally, we compare the performance of xNES with the proposed adaptive learning rate mechanism and that with fixed learning rates (RQ3) in Sect. 4.4. The code for running the proposed method is available at **https://github.com/nomuramasahir0/xnes-adaptive-lr**.

4.1   Experimental Setups

Table 1 shows the definition of benchmark problems used in the experiment. We employ two unimodal functions (Sphere and Ellipsoid) and two multimodal functions (Rastrigin and Bohachevsky). While the Rastrigin function has strong multimodality, the Bohachevsky function has relatively weak multimodality. In this experiment, we set the dimension to $d = 10$. The initial parameters are set to $m^0 = [3, \cdots, 3], \sigma^{(0)} = 2.0, B^{(0)} = I$ in the Sphere, Ellipsoid, and Rastrigin functions, and $m^0 = [8, \cdots, 8], \sigma^{(0)} = 7.0, B^{(0)} = I$ in the Bohachevsky function.

The hyperparameters for the proposed learning rate adaptation mechanism are: $\eta_\sigma^{\max} = \eta_B^{\max} = 1, \eta_\sigma^{\min} = \eta_B^{\min} = \frac{3}{5} \cdot \frac{(3+\log(d))}{d\sqrt{d}}$, as described in Sect. 3.3. And we set $\alpha_\sigma = \alpha_B = 1.3$ and $\beta = \beta_\sigma = \beta_B = 0.2$ based on our preliminary experiments. η_σ^{def} and η_B^{def} are set to their default values, i.e., $\eta_\sigma^{\text{def}} = \eta_B^{\text{def}} = \frac{3}{5} \cdot \frac{(3+\log(d))}{d\sqrt{d}}$.

4.2   Evolution Path with Fixed Learning Rate

Figure 1 shows a typical behavior of the best evaluation value $f(x_{\text{best}})$ and the length of the evolution path l_θ of xNES with a fixed learning rate on the bench-

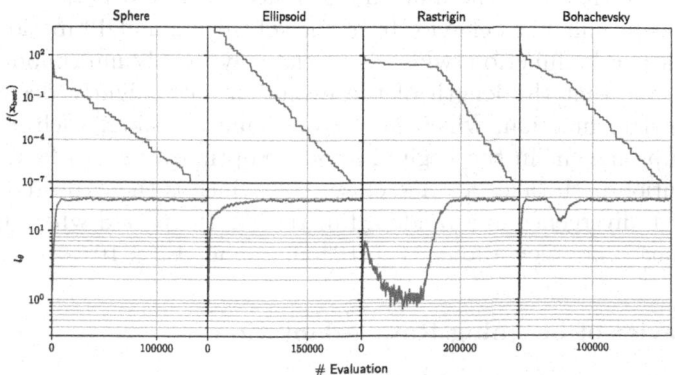

Fig. 1. Typical behavior of xNES with a fixed learning rate on the 10-dimensional benchmark problems. The horizontal axis represents the number of evaluations. The vertical axes represent the best evaluation value $f(x_{\text{best}})$ and the length of the evolution path l_θ, respectively.

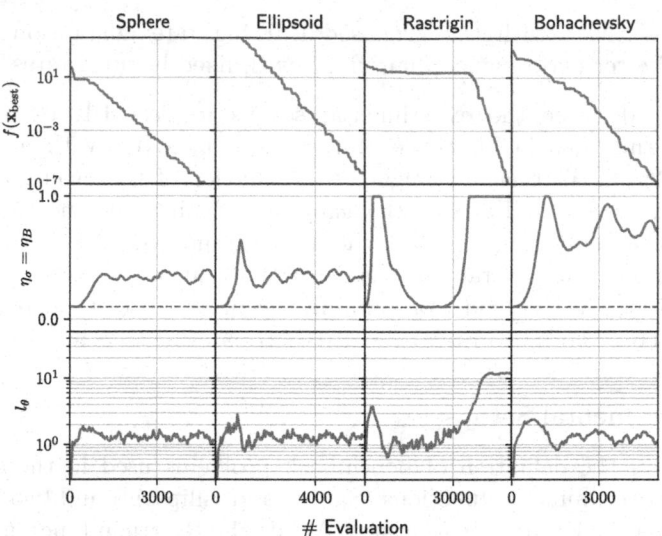

Fig. 2. Typical behavior in xNES with the proposed learning rate adaptation mechanism on the 10-dimensional benchmark problems. The green dotted line in the learning rate graphs indicates the default value. The horizontal axis represents the number of evaluations. The vertical axes represent the best evaluation value $f(x_{\text{best}})$, the learning rates η_σ and η_B, and the length of the evolution path l_θ, respectively. (Color figure online)

mark problems. We use the default learning rate and set the population size $\lambda = 400$ to obtain a reliable estimation of the evolution path.

In the result of the Sphere and Ellipsoid functions where $f(x_{\text{best}})$ is improved quickly, the length of the evolution path $l(\theta)$ becomes long (>1). We believe this is because the estimation accuracy of the natural gradient should be high in relatively easy objective functions (e.g., unimodal functions).

On the other hand, in the multimodal functions where $f(x_{\text{best}})$ may not be improved easily, different behavior from that of the unimodal functions appears. In the Bohachevsky function, which is a relatively weakly multimodal function, we can observe that the length of the evolution path slightly decreases once. In the Rastrigin function, which has strong multimodality, such a decreasing behavior is prominent in the beginning of the optimization. In fact, the length of the evolution path takes a value close to 1, which is the expected amount of change in KL divergence in a random function, in the period where the number of evaluations is between about 0.5×10^5 and about 1.2×10^5.

4.3   Behavior of Learning Rate Adaptation

A typical behavior of xNES with the proposed learning rate adaptation mechanism is depicted in Fig. 2. In addition to the learning rates η_σ and η_B, the corresponding objective function value $f(x_{\text{best}})$ and the length of the evolution path l_θ are also shown. We employ $\lambda = 30$ for the Sphere and Ellipsoid functions, $\lambda = 300$ for the Rastrigin function, $\lambda = 50$ for the Bohachevsky function, respectively. It is observed in each function that the learning rates also increase when the length of the evolution path increases.

To investigate the effect of the setting of the population size, we conduct an experiment with $\lambda = 10, 20, 30, 40$, and 50 on the 10-dimensional Sphere function. Figure 3 shows the result of the experiment. In $\lambda = 10$, the length of the evolution path l_θ does not increase and the learning rates, η_σ and η_B, are then not changed at all. We think that this is because the accuracy of the parameter update is low under the setting of a small population size. We observe that, as λ is increased, the length of the evolution path increases, and, as a result, the learning rate also increases. The result suggests that the proposed mechanism can adapt the learning rate appropriately, measuring the estimation accuracy of natural gradient. This dynamic learning rate adaptation depending on the population size is an advantage over DX-NES [7], which statically injects the population size into the setting of the learning rate.

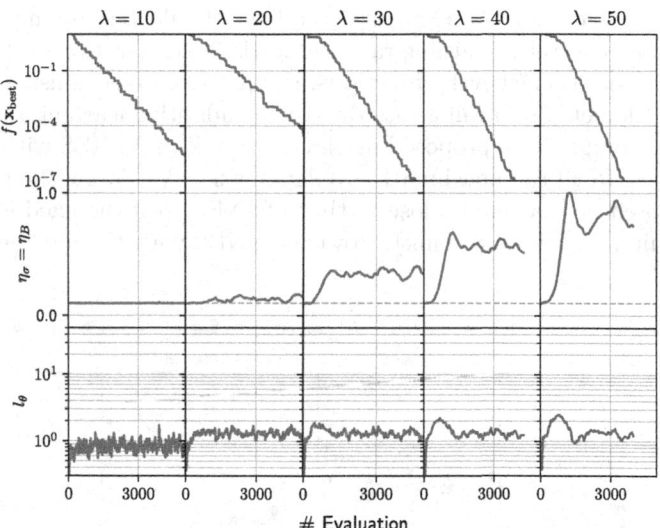

Fig. 3. Typical behavior in xNES with the proposed learning rate adaptation mechanism on the 10-dimensional sphere function. The experiment is performed with the population size $\lambda = 10, 20, 30, 40$, and 50. The green dotted line in the learning rate graphs indicates the default value. (Color figure online)

4.4 Fixed Learning Rate vs. Adaptative Learning Rate

To check the effectiveness of the proposed mechanism, we compare the performance of xNES with the proposed learning rate adaptation mechanism and that of xNES with fixed learning rates (= the default value ×1, 2, 4, 6, 8, and 10). The performance metrics is the average number of evaluations until $f(x_{\text{best}})$ reaches a target function value over successful trials divided by the success rate [4]. The target function value is set to 10^{-8}. A trial is successful if the target function value is found. We set the maximum number of evaluations to 5×10^5. For the Sphere function and the Ellipsoid function, we employ the population size $\lambda = 10, 20, 30, 40$, and 50. Note that the recommended value of the population size presented in [8] is included, i.e., $4 + \lfloor 3 \ln(10) \rfloor = 10$. For the Rastrigin function, we employ the population size $\lambda = 200, 250, 300, 350$, and 400. For the Bohachevsky function, we employ the population size $\lambda = 30, 40, 50, 60$, and 70. We perform 50 trials to calculate the performance metrics for the Sphere and Ellipsoid functions. We perform 200 trials to calculate it for the Rastrigin and Bohachevsky functions.

Figure 4 shows the result of the experiment. We first compare the proposed mechanism (red) and xNES with the default learning rate (blue). In the Sphere and Ellipsoid functions, when $\lambda = 10$, the performance is almost the same, which is consistent in Sect. 4.3. As λ increases, the proposed mechanism shows better performance than xNES with the default learning rate. This is because that the estimation accuracy of the natural gradient should become high when λ is large, which increases the learning rate and leads to acceleration of the search. In the Rastrigin and Bohachevsky functions, the proposed mechanism outperforms xNES with the default learning rate due to the adaptive learning rate.

Next, we compare the proposed mechanism (red) and xNES with other fixed learning rates. In all the benchmark problems, when λ is large, the performance of the proposed mechanism is close to that of xNES with the fixed learning rate of the default value times 8 (pink). However, xNES with the fixed learning rate

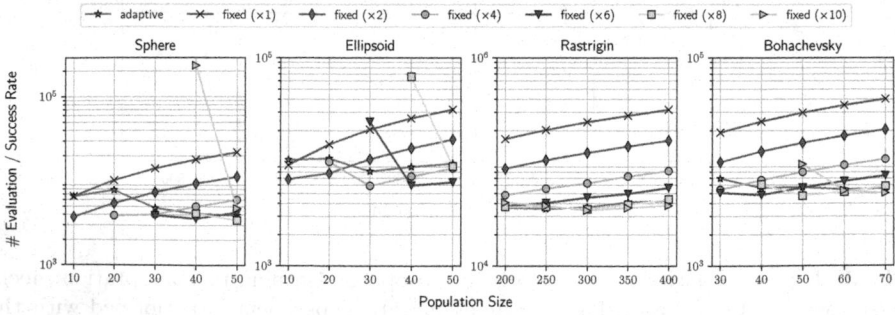

Fig. 4. Performance comparison of xNES with the proposed learning rate adaptation mechanism (red) and xNES with the fixed learning rates (blue, green, yellow, purple, pink, and cyan) on 10-dimensional benchmark problems. The horizontal axis represents the population size. The vertical axis represents the average number of evaluations divided by the success rate, which is the smaller, the better it is. Note that, if no successful trials exist at a population size, nothing is plotted at the population size. (Color figure online)

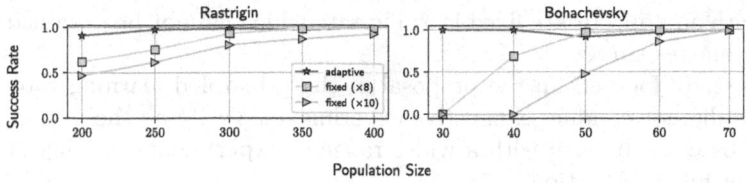

Fig. 5. Success rate of xNES with the proposed learning rate adaptation method (red), xNES with the fixed learning rate of the default value times 8 (pink), and xNES with the fixed learning rate of the default value times 10 (cyan) in the multimodal functions. (Color figure online)

of the default value times 8 fails to find the optimum in the Sphere and Ellipsoid functions with small population sizes ($\lambda = 10, 20$, and 30) because the learning rate is too high. On the other hand, the proposed mechanism does not increase the learning rate so much by measuring the estimation accuracy of the natural gradient when the population size is small, which enables stable search.

From the result in the multimodal functions, we can observe that the proposed mechanism is competitive with xNES with high learning rates when the population size is large. In particular, in the Rastrigin function, the proposed mechanism and xNES with the fixed learning rate of the default value times 8 and 10 achieve almost the same performance in terms of the average number of evaluations of successful trials divided by the success rate. This means that the number of evaluations required to find the optimum is about the same if an appropriate restart is performed when the optimum is failed to find. Figure 5 shows the success rate of the proposed mechanism (red), xNES with the fixed learning rate of the default value times 8 (pink), and xNES with the fixed learning rate of the default value times 10 (cyan). While these methods are competitive when the population size is large, xNES with the fixed learning rates are more likely to fail when the population size is small. This result suggests that the proposed mechanism is more robust than xNES with a fixed learning rate. A higher success rate in the proposed mechanism is also practically beneficial, as it is often difficult to implement an appropriate restart strategy.

5 Conclusion

In this paper, we proposed a novel learning rate adaptation mechanism for NES. The proposed mechanism adapts the learning rate based on the estimation accuracy of the natural gradient, which is inspired by the population size adaptation mechanism of the CMA-ES [14,15]. We introduced the evolution path in the parameter space of the covariance matrix. Based on the length of the evolution path, we update the learning rates related to the covariance matrix. The numerical experiments using unimodal and multimodal benchmark functions demonstrated that the proposed mechanism can appropriately adapt the learning rates depending on the estimation accuracy of the natural gradient. Additionally, xNES with the proposed mechanism achieved comparable performance as xNES with an appropriate fixed learning rate which cannot be obtained *without* prior parameter surveys.

This study focused on the proposal of the principled learning rate adaptation, and did not conduct exhaustive experiments. Verifying the performance of the proposed mechanism with a wider range of experimental setting is thus an important future direction.

Acknowledgement. The authors thank anonymous reviewers for their helpful comments. This work was partially supported by JSPS KAKENHI Grant Number JP20K11986.

References

1. Akimoto, Y., Auger, A., Hansen, N.: Comparison-Based Natural Gradient Optimization in High Dimension. In: Proceedings of the 2014 Annual Conference on Genetic and Evolutionary Computation. pp. 373–380. ACM (2014)
2. Akimoto, Y., Nagata, Y., Ono, I., Kobayashi, S.: Bidirectional relation between CMA evolution strategies and natural evolution strategies. In: Schaefer, R., Cotta, C., Kołodziej, J., Rudolph, G. (eds.) PPSN 2010. LNCS, vol. 6238, pp. 154–163. Springer, Heidelberg (2010). https://doi.org/10.1007/978-3-642-15844-5_16
3. Amari, S.I., Douglas, S.C.: Why natural gradient? In: Proceedings of the 1998 IEEE International Conference on Acoustics, Speech and Signal Processing, ICASSP 1998 (Cat. No. 98CH36181), vol. 2, pp. 1213–1216. IEEE (1998)
4. Auger, A., Hansen, N.: A restart CMA evolution strategy with increasing population size. In: 2005 IEEE Congress on Evolutionary Computation, vol. 2, pp. 1769–1776. IEEE (2005)
5. BANGS II, W.J.: Array Processing with Generalized Beam-Formers. Yale University, New Haven (1971)
6. Beyer, H.G.: Convergence analysis of evolutionary algorithms that are based on the paradigm of information geometry. Evol. Comput. **22**(4), 679–709 (2014)
7. Fukushima, N., Nagata, Y., Kobayashi, S., Ono, I.: Proposal of distance-weighted exponential natural evolution strategies. In: 2011 IEEE Congress of Evolutionary Computation (CEC), pp. 164–171. IEEE (2011)
8. Glasmachers, T., Schaul, T., Yi, S., Wierstra, D., Schmidhuber, J.: Exponential natural evolution strategies. In: Proceedings of the 12th Annual Conference on Genetic and Evolutionary Computation, pp. 393–400 (2010)
9. Hansen, N.: The CMA evolution strategy: a comparing review. In: Lozano, J.A., Larrañaga, P., Inza, I., Bengoetxea, E. (eds.) Towards a New Evolutionary Computation. SFSC, vol. 192, pp. 75–102. Springer, Heidelberg (2006). https://doi.org/10.1007/3-540-32494-1_4
10. Hansen, N.: The CMA evolution strategy: a tutorial. arXiv preprint arXiv:1604.00772 (2016)
11. Hansen, N., Kern, S.: Evaluating the CMA evolution strategy on multimodal test functions. In: Yao, X., et al. (eds.) PPSN 2004. LNCS, vol. 3242, pp. 282–291. Springer, Heidelberg (2004). https://doi.org/10.1007/978-3-540-30217-9_29
12. Hansen, N., Müller, S.D., Koumoutsakos, P.: Reducing the time complexity of the derandomized evolution strategy with covariance matrix adaptation (CMA-ES). Evol. Comput. **11**(1), 1–18 (2003)

13. Loshchilov, I., Schoenauer, M., Sebag, M., Hansen, N.: Maximum likelihood-based online adaptation of hyper-parameters in CMA-ES. In: Bartz-Beielstein, T., Branke, J., Filipič, B., Smith, J. (eds.) PPSN 2014. LNCS, vol. 8672, pp. 70–79. Springer, Cham (2014). https://doi.org/10.1007/978-3-319-10762-2_7

14. Nishida, K., Akimoto, Y.: Population size adaptation for the CMA-ES based on the estimation accuracy of the natural gradient. In: Proceedings of the Genetic and Evolutionary Computation Conference 2016, pp. 237–244 (2016)

15. Nishida, K., Akimoto, Y.: PSA-CMA-ES: CMA-ES with population size adaptation. In: Proceedings of the Genetic and Evolutionary Computation Conference, pp. 865–872 (2018)

16. Nomura, M., Ono, I.: Natural evolution strategy for unconstrained and implicitly constrained problems with ridge structure. In: 2021 IEEE Symposium Series on Computational Intelligence (SSCI), pp. 1–7. IEEE (2021)

17. Nomura, M., Sakai, N., Fukushima, N., Ono, I.: Distance-weighted exponential natural evolution strategy for implicitly constrained black-box function optimization. In: 2021 IEEE Congress on Evolutionary Computation (CEC), pp. 1099–1106. IEEE (2021)

18. Ollivier, Y., Arnold, L., Auger, A., Hansen, N.: Information-geometric optimization algorithms: a unifying picture via invariance principles. J. Mach. Learn. Res. 18(1), 564–628 (2017)

19. Otwinowski, J., LaMont, C.H., Nourmohammad, A.: Information-geometric optimization with natural selection. Entropy 22(9), 967 (2020)

20. Slepian, D.: Estimation of signal parameters in the presence of noise. Trans. IRE Prof. Group Inf. Theory 3(3), 68–89 (1954)

21. Sun, Y., Wierstra, D., Schaul, T., Schmidhuber, J.: Efficient natural evolution strategies. In: Proceedings of the 11th Annual Conference on Genetic and Evolutionary Computation, pp. 539–546. ACM (2009)

22. Wierstra, D., Schaul, T., Glasmachers, T., Sun, Y., Peters, J., Schmidhuber, J.: Natural evolution strategies. J. Mach. Learn. Res. 15(1), 949–980 (2014)

23. Yi, S., Wierstra, D., Schaul, T., Schmidhuber, J.: Stochastic search using the natural gradient. In: Proceedings of the 26th Annual International Conference on Machine Learning, pp. 1161–1168. ACM (2009)

Convergence of Anisotropic Consensus-Based Optimization in Mean-Field Law

Massimo Fornasier[1,2] , Timo Klock[3] , and Konstantin Riedl[1(✉)]

[1] Department of Mathematics, Technical University of Munich, Munich, Germany
{massimo.fornasier,konstantin.riedl}@ma.tum.de
[2] Munich Data Science Institute, Munich, Germany
[3] Department of Numerical Analysis and Scientific Computing, Simula Research Laboratory, Oslo, Norway
timo@simula.no

Abstract. In this paper we study anisotropic consensus-based optimization (CBO), a population-based metaheuristic derivative-free optimization method capable of globally minimizing nonconvex and nonsmooth functions in high dimensions. CBO is based on stochastic swarm intelligence, and inspired by consensus dynamics and opinion formation. Compared to other metaheuristic algorithms like Particle Swarm Optimization, CBO is of a simpler nature and therefore more amenable to theoretical analysis. By adapting a recently established proof technique, we show that anisotropic CBO converges globally with a dimension-independent rate for a rich class of objective functions under minimal assumptions on the initialization of the method. Moreover, the proof technique reveals that CBO performs a convexification of the optimization problem as the number of particles goes to infinity, thus providing an insight into the internal CBO mechanisms responsible for the success of the method. To motivate anisotropic CBO from a practical perspective, we further test the method on a complicated high-dimensional benchmark problem, which is well understood in the machine learning literature.

Keywords: High-dimensional global optimization · Metaheuristics · Consensus-based optimization · Mean-field limit · Anisotropy

1 Introduction

Several problems arising throughout all quantitative disciplines are concerned with the global unconstrained optimization of a problem-dependent objective function $\mathcal{E} : \mathbb{R}^d \to \mathbb{R}$ and the search for the associated minimizing argument

$$v^* = \arg\min_{v \in \mathbb{R}^d} \mathcal{E}(v),$$

which is assumed to exist and be unique in what follows. Because of nowadays data deluge such optimization problems are usually high-dimensional. In machine

© Springer Nature Switzerland AG 2022
J. L. Jiménez Laredo et al. (Eds.): EvoApplications 2022, LNCS 13224, pp. 738–754, 2022.
https://doi.org/10.1007/978-3-031-02462-7_46

learning, for instance, one is interested in finding the optimal parameters of a neural network (NN) to accomplish various tasks, such as clustering, classification, and regression. The availability of huge amounts of training data for various real-world applications allows practitioners to work with models involving a large number of trainable parameters aiming for a high expressivity and accuracy of the trained model. This makes the resulting optimization process a high-dimensional problem. Since typical model architectures consist of many layers with a large amount of neurons, and include nonlinear and potentially nonsmooth activation functions, the training process is in general a high-dimensional nonconvex optimization problem and therefore a particularly hard task.

Metaheuristics have a long history as state-of-the-art methods when it comes to tackling hard optimization problems. Inspired by self-organization and collective behavior in nature or human society, such as the swarming of flocks of birds or schools of fish [3], or opinion formation [20], they orchestrate an interplay between locally confined procedures and global strategies, randomness and deterministic decisions, to ensure a robust search for the global minimizer. Some prominent examples are Random Search [19], Evolutionary Programming [7], Genetic Algorithms [11], Ant Colony Optimization [6], Particle Swarm Optimization [14] and Simulated Annealing [1].

CBO follows those guiding principles, but is of much simpler nature and more amenable to theoretical analysis. The method uses N particles $V^1, \ldots, V^N$, which are initialized independently according to some law $\rho_0 \in \mathcal{P}(\mathbb{R}^d)$, to explore the domain and to form a global consensus about the minimizer v^* as time passes. For parameters $\alpha, \lambda, \sigma > 0$ the dynamics of each particle is given by

$$dV_t^i = -\lambda \left(V_t^i - v_\alpha(\widehat{\rho}_t^N) \right) dt + \sigma D\big(V_t^i - v_\alpha(\widehat{\rho}_t^N)\big) dB_t^i, \tag{1}$$

where $\widehat{\rho}_t^N$ denotes the empirical measure of the particles. The first term in (1) is a drift term dragging the respective particle towards the momentaneous consensus point, a weighted average of the particles' positions, computed as

$$v_\alpha(\widehat{\rho}_t^N) := \int v \frac{\omega_\alpha(v)}{\|\omega_\alpha\|_{L_1(\widehat{\rho}_t^N)}} \, d\widehat{\rho}_t^N(v), \quad \text{with} \quad \omega_\alpha(v) := \exp(-\alpha\mathcal{E}(v))$$

and motivated by the fact that $v_\alpha(\widehat{\rho}_t^N) \approx \arg\min_{i=1,\ldots,N} \mathcal{E}(V_t^i)$ for $\alpha \gg 1$ if the arg min is unique. To feature the exploration of the energy landscape of $\mathcal{E}$, the second term in (1) is a diffusion injecting randomness into the dynamics through independent standard Brownian motions $((B_t^i)_{t\geq 0})_{i=1,\ldots,N}$. The two commonly studied diffusion types are isotropic [2,9,18] and anisotropic [4] diffusion with

$$D\big(V_t^i - v_\alpha(\widehat{\rho}_t^N)\big) = \begin{cases} \left\| V_t^i - v_\alpha(\widehat{\rho}_t^N) \right\|_2 \mathrm{Id}, & \text{for isotropic diffusion,} \\ \mathrm{diag}\left(V_t^i - v_\alpha(\widehat{\rho}_t^N) \right), & \text{for anisotropic diffusion,} \end{cases}$$

where $\mathrm{Id} \in \mathbb{R}^{d \times d}$ is the identity matrix and $\mathrm{diag} : \mathbb{R}^d \to \mathbb{R}^{d \times d}$ the operator mapping a vector onto a diagonal matrix with the vector as its diagonal. The term's scaling encourages in particular particles far from $v_\alpha(\widehat{\rho}_t^N)$ to explore larger regions. The coordinate-dependent scaling of anisotropic diffusion has proven to be particularly beneficial for high-dimensional optimization problems by yielding dimension-independent convergence rates (see Fig. 1) and therefore improving both computational complexity and success probability of the algorithm [4,8].

A theoretical convergence analysis of the CBO dynamics is possible either on the microscopic level (1) or by analyzing the macroscopic behavior of the particle density through a mean-field limit. In the large particle limit a particle is not influenced by individual particles but only by the average behavior of all particles. As shown in [12], the empirical random particle measure $\widehat{\rho}^N$ converges in law to the deterministic particle density $\rho \in \mathcal{C}([0,T], \mathcal{P}(\mathbb{R}^d))$, which weakly (see Definition 1) satisfies the non-linear Fokker-Planck equation

$$\partial_t \rho_t = \lambda \mathrm{div}\big((v - v_\alpha(\rho_t))\, \rho_t\big) + \frac{\sigma^2}{2} \sum_{k=1}^{d} \partial_{kk} \Big(D(v - v_\alpha(\rho_t))_{kk}^2\, \rho_t\Big). \qquad (2)$$

A quantitative analysis of the convergence rate remains, on non-compact domains, an open problem, see, e.g., [9, Remark 2]. Analyzing a mean-field limit such as (2) allows for establishing strong qualitative theoretical guarantees about CBO methods, paving the way to understand the internal mechanisms at play.

Prior Arts. The original CBO work [18] proposes the dynamics (1) with isotropic diffusion, which is analyzed in the mean-field sense in [2]. Under a stringent well-preparedness condition about ρ_0 and $\mathcal{C}^2$ regularity of $\mathcal{E}$ the authors show consensus formation of the particles at some $\widetilde{v}$ close to v^* by first establishing exponential decay of the variance $\mathrm{Var}\,(\rho_t)$ and consecutively showing $\widetilde{v} \approx v^*$ as a consequence of the Laplace principle [17]. This analysis is extended to the anisotropic case in [4].

Motivated by the surprising phenomenon that, on average, individual particles of the CBO dynamics follow the gradient flow of $v \mapsto \|v - v^*\|_2^2$, see [9, Figure 1b], the authors of [9] develop a novel proof technique for showing global convergence of isotropic CBO in mean-field law to v^* under minimal assumptions. Following [9, Definition 1], we speak of convergence in mean-field law to v^* for the interacting particle dynamics (1), if the solution ρ_t to the associated mean-field limit dynamics (2) converges narrowly to the Dirac delta δ_{v^*} at v^* for $t \to \infty$. The proof is based on showing an exponential decay of the energy functional $\mathcal{V} : \mathcal{P}(\mathbb{R}^d) \to \mathbb{R}_{\geq 0}$, given by

$$\mathcal{V}(\rho_t) := \frac{1}{2} \int \|v - v^*\|_2^2\, d\rho_t(v). \qquad (3)$$

This simultaneously ensures consensus formation and convergence of ρ_t to δ_{v^*}.

Contribution. In view of the effectiveness and efficiency of CBO methods with anisotropic diffusion for high-dimensional optimization problems, a thorough understanding is of considerable interest. As we illustrate in Fig. 1, anisotropic

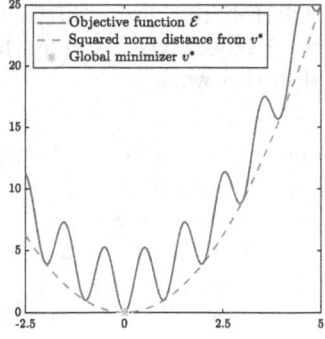
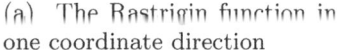
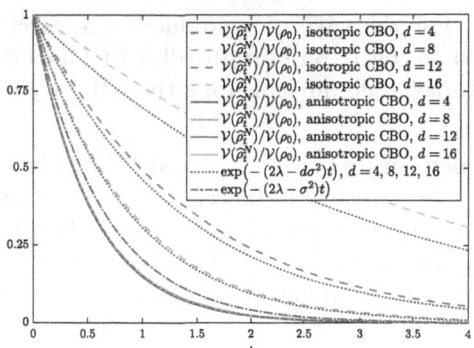

(a) The Rastrigin function in
one coordinate direction

(b) Evolution of $\mathcal{V}(\widehat{\rho}_t^N)$ for isotropic and anisotropic CBO for different dimensions

Fig. 1. A demonstration of the benefit of using anisotropic diffusion in CBO. For the Rastrigin function $\mathcal{E}(v) = \sum_{k=1}^{d} v_k^2 + \frac{5}{2}(1 - \cos(2\pi v_k))$ with $v^* = 0$ and spurious local minima (see (a)), we evolve the discretized system of isotropic and anisotropic CBO using $N = 320000$ particles, discrete time step size $\Delta t = 0.01$ and $\alpha = 10^{15}$, $\lambda = 1$, and $\sigma = 0.32$ for different dimensions $d \in \{4, 8, 12, 16\}$. We observe in (b) that the convergence rate of the energy functional $\mathcal{V}(\widehat{\rho}_t^N)$ for isotropic CBO (dashed lines) is affected by the ambient dimension d, whereas anisotropic CBO (solid lines) converges independently from d with rate $(2\lambda - \sigma^2)$.

CBO [4] converges with a dimension-independent rate as opposed to isotropic CBO [2,9,18], making it a particularly interesting choice for problems in high-dimensional spaces, e.g., from signal processing and machine learning applications. In this work we extend the analysis of [9] from isotropic CBO to CBO with anisotropic diffusion. More precisely, we show global convergence of the anisotropic CBO dynamics in mean-field law to the global minimizer v^* under minimal assumptions about the initial measure ρ_0 and for a rich class of objectives $\mathcal{E}$. Furthermore, utilizing some tweaks in the implementation of anisotropic CBO, such as a random mini-batch idea and a cooling strategy of the parameters as proposed in [4,10], we show that CBO performs well, in fact, almost on par with state-of-the-art gradient-based methods, on a long-studied machine learning benchmark in 2000 dimensions, despite using just 100 particles and no gradient information. This encourages the use and further investigation of CBO as a training algorithm for challenging machine learning tasks.

Organization. In Sect. 2 we first recall details about the well-posedness of the mean-field dynamics (2) in the case of anisotropic diffusion before we present the main theoretical result about the convergence of anisotropic CBO in mean-field law. The proof follows in Sect. 3. Section 4 illustrates the practicability of the method on a benchmark problem and Sect. 5 concludes the paper.

For the sake of reproducible research, in the GitHub repository https://github.com/KonstantinRiedl/CBOGlobalConvergenceAnalysis we provide the Matlab code implementing the CBO algorithm used in this work.

Notation. $B_r^\infty(u)$ is a closed ℓ^∞ ball in $\mathbb{R}^d$ with center u and radius r. For the space of continuous functions $f : X \to Y$ we write $\mathcal{C}(X,Y)$, with $X \subset \mathbb{R}^n$, $n \in \mathbb{N}$, and a suitable topological space Y. For $X \subset \mathbb{R}^n$ open and for $Y = \mathbb{R}^m$, $m \in \mathbb{N}$, the function space $\mathcal{C}_c^k(X,Y)$ contains functions $f \in \mathcal{C}(X,Y)$ that are k-times continuously differentiable and compactly supported. Y is omitted if $Y = \mathbb{R}$.

The objects of study are laws of stochastic processes, $\rho \in \mathcal{C}([0,T], \mathcal{P}(\mathbb{R}^d))$, where $\mathcal{P}(\mathbb{R}^d)$ contains all Borel probability measures over $\mathbb{R}^d$. $\rho_t \in \mathcal{P}(\mathbb{R}^d)$ is a snapshot of such law at time t and ϱ some fixed distribution. Measures $\varrho \in \mathcal{P}(\mathbb{R}^d)$ with finite p-th moment are collected in $\mathcal{P}_p(\mathbb{R}^d)$. For any $1 \le p < \infty$, W_p denotes the Wasserstein-p distance. $\mathbb{E}(\varrho)$ is the expectation of a probability measure ϱ.

2   Global Convergence in Mean-Field Law

In this section we first recite a well-posedness result about the Fokker-Planck Eq. (2) and then present the main result about global convergence.

2.1   Definition of Weak Solutions and Well-Posedness

We begin by defining weak solutions of the Fokker-Planck Eq. (2).

Definition 1. *Let* $\rho_0 \in \mathcal{P}(\mathbb{R}^d)$, $T > 0$. *We say* $\rho \in \mathcal{C}([0,T], \mathcal{P}(\mathbb{R}^d))$ *satisfies the Fokker-Planck Eq. (2) with initial condition* ρ_0 *in the weak sense in the time interval* $[0,T]$, *if we have for all* $\phi \in \mathcal{C}_c^\infty(\mathbb{R}^d)$ *and all* $t \in (0,T)$

$$
\begin{aligned}
\frac{d}{dt} \int \phi(v)\, d\rho_t(v) = &-\lambda \int \sum_{k=1}^{d} (v - v_\alpha(\rho_t))_k \partial_k \phi(v)\, d\rho_t(v) \\
&+ \frac{\sigma^2}{2} \int \sum_{k=1}^{d} D\big(V_t^i - v_\alpha(\widehat{\rho}_t^N)\big)_{kk}^2 \partial_{kk}^2 \phi(v)\, d\rho_t(v)
\end{aligned}
\tag{4}
$$

and $\lim_{t\to 0} \rho_t = \rho_0$ *pointwise.*

In what follows the case of CBO with anisotropic diffusion is considered, i.e., $D(V_t^i - v_\alpha(\widehat{\rho}_t^N)) = \mathrm{diag}\,(V_t^i - v_\alpha(\widehat{\rho}_t^N))$ in Eqs. (1), (2) and (4).

Analogously to the well-posedness results [2, Theorems 3.1, 3.2] for CBO with isotropic diffusion, we can obtain well-posedness of (2) for anisotropic CBO.

Theorem 1. *Let* $T > 0$, $\rho_0 \in \mathcal{P}_4(\mathbb{R}^d)$ *and consider* $\mathcal{E} : \mathbb{R}^d \to \mathbb{R}$ *with* $\underline{\mathcal{E}} := \mathcal{E}(v^*) > -\infty$, *which, for some constants* $C_1, C_2 > 0$, *satisfies*

$$
|\mathcal{E}(v) - \mathcal{E}(w)| \le C_1(\|v\|_2 + \|w\|_2)\, \|v - w\|_2, \quad \text{for all } v, w \in \mathbb{R}^d,
$$
$$
\mathcal{E}(v) - \underline{\mathcal{E}} \le C_2\big(1 + \|v\|_2^2\big), \quad \text{for all } v \in \mathbb{R}^d.
$$

If in addition, either $\sup_{v \in \mathbb{R}^d} \mathcal{E}(v) < \infty$, *or, for some* $C_3, C_4 > 0$, $\mathcal{E}$ *satisfies*

$$\mathcal{E}(v) - \underline{\mathcal{E}} \geq C_3 \|v\|_2^2, \quad \text{for all } \|v\|_2 \geq C_4,$$

then there exists a law $\rho \in C([0,T], \mathcal{P}_4(\mathbb{R}^d))$ *weakly satisfying Eq.* (2).

Proof. The proof is based on the Leray-Schauder fixed point theorem and uses the same arguments as the ones provided for [2, Theorems 3.1, 3.2]. ∎

Remark 1. As discussed in [9, Remark 7], the proof of Theorem 1 justifies an extension of the test function space $C_c^\infty(\mathbb{R}^d)$ in Definition 1 to

$$C_*^2(\mathbb{R}^d) := \left\{ \phi \in C^2(\mathbb{R}^d) : |\partial_k \phi(v)| \leq c(1 + |v_k|) \text{ and } \|\partial_{kk}^2 \phi\|_\infty < \infty \right.$$
$$\left. \text{for all } k \in \{1, \ldots, d\} \text{ and some constant } c > 0 \right\}.$$

2.2 Main Results

We now present the main result about global convergence in mean-field law for objective functions that satisfy the following conditions.

Definition 2 (Assumptions). *We consider functions* $\mathcal{E} \in C(\mathbb{R}^d)$, *for which*

A1 there exists $v^* \in \mathbb{R}^d$ *such that* $\mathcal{E}(v^*) = \inf_{v \in \mathbb{R}^d} \mathcal{E}(v) =: \underline{\mathcal{E}}$, *and*
A2 there exist $\mathcal{E}_\infty, R_0, \eta > 0$, *and* $\nu \in (0, \infty)$ *such that*

$$\|v - v^*\|_\infty \leq \frac{1}{\eta}(\mathcal{E}(v) - \underline{\mathcal{E}})^\nu \quad \text{for all } v \in B_{R_0}^\infty(v^*), \tag{5}$$

$$\mathcal{E}_\infty < \mathcal{E}(v) - \underline{\mathcal{E}} \quad \text{for all } v \in \left(B_{R_0}^\infty(v^*)\right)^c. \tag{6}$$

Assumption A2 can be regarded as a tractability condition of the energy landscape around the minimizer and in the farfield. Equation (5) requires the local coercivity of $\mathcal{E}$, whereas (6) prevents that $\mathcal{E}(v) \approx \underline{\mathcal{E}}$ far away from v^*.

Definition 2 covers a wide range of function classes, including for instance the Rastrigin function, see Fig. 1a, and objectives related to various machine learning tasks, see, e.g., [10].

Theorem 2. *Let* $\mathcal{E}$ *be as in Definition 2. Moreover, let* $\rho_0 \in \mathcal{P}_4(\mathbb{R}^d)$ *be such that*

$$\rho_0(B_r^\infty(v^*)) > 0 \quad \text{for all } r > 0. \tag{7}$$

Define $\mathcal{V}(\rho_t) := 1/2 \int \|v - v^*\|_2^2 d\rho_t(v)$. *Fix any* $\varepsilon \in (0, \mathcal{V}(\rho_0))$ *and* $\tau \in (0,1)$, *parameters* $\lambda, \sigma > 0$ *with* $2\lambda > \sigma^2$, *and the time horizon*

$$T^* := \frac{1}{(1-\tau)(2\lambda - \sigma^2)} \log\left(\frac{\mathcal{V}(\rho_0)}{\varepsilon}\right). \tag{8}$$

Then there exists $\alpha_0 > 0$, *which depends (among problem dependent quantities) on* ε *and* τ, *such that for all* $\alpha > \alpha_0$, *if* $\rho \in C([0, T^*], \mathcal{P}_4(\mathbb{R}^d))$ *is a weak solution* *to the Fokker-Planck Eq.* (2) *on the time interval* $[0, T^*]$ *with initial condition* ρ_0*, we have* $\min_{t \in [0, T^*]} \mathcal{V}(t) \leq \varepsilon$*. Furthermore, until* $\mathcal{V}(\rho_t)$ *reaches the prescribed accuracy* ε*, we have the exponential decay*

$$\mathcal{V}(\rho_t) \leq \mathcal{V}(\rho_0) \exp\left(-(1 - \tau)\left(2\lambda - \sigma^2\right) t\right)$$

and, up to a constant, the same behavior for $W_2^2(\rho_t, \delta_{v^*})$*.*

The rate of convergence $(2\lambda - \sigma^2)$ obtained in Theorem 2 is confirmed numerically by the experiments depicted in Fig. 1. We emphasize the dimension-independent convergence of CBO with anisotropic diffusion, contrasting the dimension-dependent rate $(2\lambda - d\sigma^2)$ of isotropic CBO, cf. [9, Theorem 12].

3   Proof of Theorem 2

This section provides the proof details for Theorem 2, starting with a sketch in Sect. 3.1. Sections 3.2–3.4 present statements, which are needed in the proof and may be of independent interest. Section 3.5 completes the proof.

Remark 2. Without loss of generality we assume $\underline{\mathcal{E}} = 0$ throughout the proof.

3.1   Proof Sketch

The main idea is to show that $\mathcal{V}(\rho_t)$ satisfies the differential inequality

$$\frac{d}{dt}\mathcal{V}(\rho_t) \leq -(1 - \tau)\left(2\lambda - \sigma^2\right)\mathcal{V}(\rho_t) \tag{9}$$

until $\mathcal{V}(\rho_T) \leq \varepsilon$. The first step towards (9) is to derive a differential inequality for $\mathcal{V}(\rho_t)$ using the dynamics of ρ, which is done in Lemma 1. In order to control the appearing quantity $\|v_\alpha(\rho_t) - v^*\|_2$, we establish a quantitative Laplace principle. Namely, under the inverse continuity property A2, Proposition 1 shows

$$\|v_\alpha(\rho_t) - v^*\|_2 \lesssim \ell(r) + \frac{\sqrt{d}\exp(-\alpha r)}{\rho_t(B_r^\infty(v^*))}, \quad \text{for sufficiently small } r > 0,$$

where ℓ is decreasing with $\ell(r) \to 0^+$ as $r \to 0$. Thus, $\|v_\alpha(\rho_t) - v^*\|_2$ can be made arbitrarily small by suitable choices of $r \ll 1$ and $\alpha \gg 1$, as long as we can guarantee $\rho_t(B_r^\infty(v^*)) > 0$ for all $r > 0$ and at all times $t \in [0, T]$. The latter requires non-zero initial mass $\rho_0(B_r^\infty(v^*))$ as well as an active Brownian motion, as made rigorous in Proposition 2.

3.2   Evolution of the Mean-Field Limit

We now derive the evolution inequality of the energy functional $\mathcal{V}(\rho_t)$.

Lemma 1. *Let* $\mathcal{E} : \mathbb{R}^d \to \mathbb{R}$, *and fix* $\alpha, \lambda, \sigma > 0$. *Moreover, let* $T > 0$ *and let* $\rho \in \mathcal{C}([0, T], \mathcal{P}_4(\mathbb{R}^d))$ *be a weak solution to Eq. (2). Then* $\mathcal{V}(\rho_t)$ *satisfies*

$$\frac{d}{dt}\mathcal{V}(\rho_t) \leq - \left(2\lambda - \sigma^2\right)\mathcal{V}(\rho_t) + \sqrt{2}\left(\lambda + \sigma^2\right)\sqrt{\mathcal{V}(\rho_t)}\,\|v_\alpha(\rho_t) - v^*\|_2$$

$$+ \frac{\sigma^2}{2}\|v_\alpha(\rho_t) - v^*\|_2^2.$$

Proof. Noting that $\phi(v) = 1/2\,\|v - v^*\|_2^2$ is in $\mathcal{C}_*^2(\mathbb{R}^d)$ and recalling that ρ satisfies the identity (4) for all test functions in $\mathcal{C}_*^2(\mathbb{R}^d)$, see Remark 1, we obtain

$$\frac{d}{dt}\mathcal{V}(\rho_t) = -\lambda \int \langle v - v^*, v - v_\alpha(\rho_t)\rangle\,d\rho_t(v) + \frac{\sigma^2}{2}\int \|v - v_\alpha(\rho_t)\|_2^2\,d\rho_t(v),$$

where we used $\partial_k\phi(v) = (v - v^*)_k$ and $\partial_{kk}^2\phi(v) = 1$ for all $k \in \{1, \ldots, d\}$. Following the steps taken in [9, Lemma 14] yields the statement. $\square$

3.3   Quantitative Laplace Principle

The Laplace principle asserts that $-\log(\|\omega_\alpha\|_{L_1(\varrho)})/\alpha \to \underline{\mathcal{E}}$ as $\alpha \to \infty$ as long as the global minimizer v^* is in the support of ϱ. Under the assumption of the inverse continuity property this can be used to qualitatively characterize the proximity of $v_\alpha(\varrho)$ to the global minimizer v^*. However, as it neither allows to quantify this proximity nor gives a suggestion on how to choose α to reach a certain approximation quality, we introduced a quantitative version in [9, Proposition 17], which we now adapt suitably to satisfy the anisotropic setting.

Proposition 1. *Let* $\underline{\mathcal{E}} = 0$, $\varrho \in \mathcal{P}(\mathbb{R}^d)$ *and fix* $\alpha > 0$. *For any* $r > 0$ *we define* $\mathcal{E}_r := \sup_{v \in B_r^\infty(v^*)}\mathcal{E}(v)$. *Then, under the inverse continuity property A2, for any* $r \in (0, R_0]$ *and* $q > 0$ *such that* $q + \mathcal{E}_r \leq \mathcal{E}_\infty$, *we have*

$$\|v_\alpha(\varrho) - v^*\|_2 \leq \frac{\sqrt{d}(q + \mathcal{E}_r)^\nu}{\eta} + \frac{\sqrt{d}\exp(-\alpha q)}{\varrho(B_r^\infty(v^*))}\int \|v - v^*\|_2\,d\varrho(v).$$

Proof. Following the lines of the proof of [9, Proposition 17] but replacing all ℓ^2 balls and norms by ℓ^∞ balls and norms, respectively, we obtain

$$\|v_\alpha(\varrho) - v^*\|_\infty \leq \frac{(q + \mathcal{E}_r)^\nu}{\eta} + \frac{\exp(-\alpha q)}{\varrho(B_r^\infty(v^*))}\int \|v - v^*\|_\infty\,d\varrho(v).$$

The statement now follows noting that $\|\cdot\|_\infty \leq \|\cdot\|_2 \leq \sqrt{d}\,\|\cdot\|_\infty$. $\square$

3.4 A Lower Bound for the Probability Mass Around v^*

In this section we provide a lower bound for the probability mass of $\rho_t(B_r^\infty(v^*))$, where $r > 0$ is a small radius. This is achieved by defining a mollifier ϕ_r so that $\rho_t(B_r^\infty(v^*)) \geq \int \phi_r(v)\, d\rho_t(v)$ and studying the evolution of the right-hand side.

Lemma 2. *For $r > 0$ we define the mollifier $\phi_r : \mathbb{R}^d \to \mathbb{R}$ by*

$$\phi_r(v) := \begin{cases} \prod_{k=1}^d \exp\left(1 - \frac{r^2}{r^2 - (v-v^*)_k^2}\right), & \text{if } \|v - v^*\|_\infty < r, \\ 0, & \text{else.} \end{cases} \tag{10}$$

We have $\mathrm{Im}(\phi_r) = [0,1]$, $\mathrm{supp}(\phi_r) = B_r^\infty(v^*)$, $\phi_r \in C_c^\infty(\mathbb{R}^d)$ and

$$\partial_k \phi_r(v) = -2r^2 \frac{(v-v^*)_k}{\left(r^2 - (v-v^*)_k^2\right)^2} \phi_r(v),$$

$$\partial_{kk}^2 \phi_r(v) = 2r^2 \left(\frac{2\left(2(v-v^*)_k^2 - r^2\right)(v-v^*)_k^2 - \left(r^2 - (v-v^*)_k^2\right)^2}{\left(r^2 - (v-v^*)_k^2\right)^4} \right) \phi_r(v).$$

Proof. ϕ_r is a tensor product of classical well-studied mollifiers. □

Proposition 2. *Let $T > 0$, $r > 0$, and fix parameters $\alpha, \lambda, \sigma > 0$. Assume $\rho \in C([0,T], \mathcal{P}(\mathbb{R}^d))$ weakly solves the Fokker-Planck Eq. (2) in the sense of Definition 1 with initial condition $\rho_0 \in \mathcal{P}(\mathbb{R}^d)$ and for $t \in [0,T]$. Furthermore, denote $B := \sup_{t \in [0,T]} \|v_\alpha(\rho_t) - v^*\|_\infty$. Then, for all $t \in [0,T]$ we have*

$$\rho_t\left(B_r^\infty(v^*)\right) \geq \left(\int \phi_r(v)\, d\rho_0(v)\right) \exp(-qt),$$

$$for \quad q := 2d \max\left\{ \frac{\lambda(cr + B\sqrt{c})}{(1-c)^2 r} + \frac{\sigma^2(cr^2 + B^2)(2c+1)}{(1-c)^4 r^2}, \frac{2\lambda^2}{(2c-1)\sigma^2} \right\}, \tag{11}$$

where $c \in (1/2, 1)$ can be any constant that satisfies $(1-c)^2 \leq (2c-1)c$.

Remark 3. In order to ensure a finite decay rate $q < \infty$ in Proposition 2 it is crucial to have a non-vanishing diffusion $\sigma > 0$.

Proof (Proposition 2). By the properties of the mollifier in Lemma 2 we have

$$\rho_t\left(B_r^\infty(v^*)\right) \geq \int \phi_r(v)\, d\rho_t(v).$$

Our strategy is to derive a lower bound for the right-hand side of this inequality. Using the weak solution property of ρ and the fact that $\phi_r \in C_c^\infty(\mathbb{R}^d)$, we obtain

$$\frac{d}{dt} \int \phi_r(v)\, d\rho_t(v) = \sum_{k=1}^d \int \left(T_{1k}(v) + T_{2k}(v)\right) d\rho_t(v) \tag{12}$$

$$\text{with} \quad T_{1k}(v) := -\lambda\, (v - v_\alpha(\rho_t))_k\, \partial_k \phi_r(v)$$

$$\text{and} \quad T_{2k}(v) := \frac{\sigma^2}{2}\, (v - v_\alpha(\rho_t))_k^2\, \partial_{kk}^2 \phi_r(v)$$

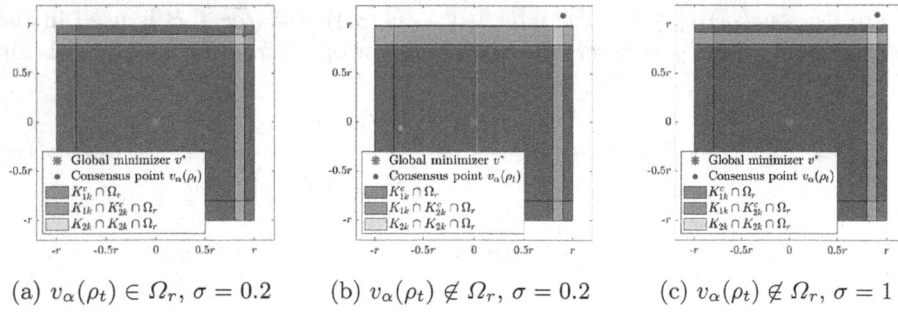

(a) $v_\alpha(\rho_t) \in \Omega_r,\ \sigma = 0.2$ (b) $v_\alpha(\rho_t) \notin \Omega_r,\ \sigma = 0.2$ (c) $v_\alpha(\rho_t) \notin \Omega_r,\ \sigma = 1$

Fig. 2. Visualization of the decomposition of Ω_r as in (15) for different positions of $v_\alpha(\rho_t)$ and values of σ.

for $k \in \{1, \ldots, d\}$. We now aim for showing $T_{1k}(v) + T_{2k}(v) \geq -q\phi_r(v)$ uniformly on $\mathbb{R}^d$ individually for each k and for $q > 0$ as in the statement. Since the mollifier ϕ_r and its derivatives vanish outside of $\Omega_r := \{v \in \mathbb{R}^d : \|v - v^*\|_\infty < r\}$ we restrict our attention to the open ℓ^∞-ball Ω_r. To achieve the lower bound over Ω_r, we introduce for each $k \in \{1, \ldots, d\}$ the subsets

$$K_{1k} := \left\{v \in \mathbb{R}^d : |(v - v^*)_k| > \sqrt{c}r\right\} \tag{13}$$

and

$$K_{2k} := \left\{v \in \mathbb{R}^d : -\lambda\,(v - v_\alpha(\rho_t))_k\,(v - v^*)_k \left(r^2 - (v - v^*)_k^2\right)^2 \right.$$
$$\left. > \tilde{c}r^2 \frac{\sigma^2}{2}(v - v_\alpha(\rho_t))_k^2\,(v - v^*)_k^2\right\}, \tag{14}$$

where $\tilde{c} := 2c - 1 \in (0, 1)$. For fixed k we now decompose Ω_r according to

$$\Omega_r = (K_{1k}^c \cap \Omega_r) \cup (K_{1k} \cap K_{2k}^c \cap \Omega_r) \cup (K_{1k} \cap K_{2k} \cap \Omega_r), \tag{15}$$

which is illustrated in Fig. 2 for different positions of $v_\alpha(\rho_t)$ and values of σ. In the following we treat each of these three subsets separately.

Subset $K_{1k}^c \cap \Omega_r$: We have $|(v - v^*)_k| \leq \sqrt{c}r$ for each $v \in K_{1k}^c$, which can be used to independently derive lower bounds for both terms T_{1k} and T_{2k}. For T_{1k}, we insert the expression for $\partial_k\phi_r$ from Lemma 2 to get

$$T_{1k}(v) = 2r^2\lambda\,(v - v_\alpha(\rho_t))_k \frac{(v - v^*)_k}{\left(r^2 - (v - v^*)_k^2\right)^2}\phi_r(v)$$

$$\geq -2r^2\lambda \frac{|(v - v_\alpha(\rho_t))_k||(v - v^*)_k|}{\left(r^2 - (v - v^*)_k^2\right)^2}\phi_r(v) \geq -\frac{2\lambda(\sqrt{c}r + B)\sqrt{c}}{(1 - c)^2r}\phi_r(v)$$

$$=: -q_1\phi_r(v),$$

where $|(v - v_\alpha(\rho_t))_k| \leq |(v - v^*)_k| + |(v^* - v_\alpha(\rho_t))_k| \leq \sqrt{c}r + B$ is used in the last inequality. For T_2 we insert the expression for $\partial^2_{kk}\phi_r$ from Lemma 2 to obtain

$$T_{2k}(v) = \sigma^2 r^2 \left(v - v_\alpha(\rho_t)\right)^2_k \frac{2\left(2\left(v - v^*\right)^2_k - r^2\right)\left(v - v^*\right)^2_k - \left(r^2 - \left(v - v^*\right)^2_k\right)^2}{\left(r^2 - \left(v - v^*\right)^2_k\right)^4} \phi_r(v)$$

$$\geq -\frac{2\sigma^2(cr^2 + B^2)(2c+1)}{(1-c)^4 r^2}\phi_r(v) =: -q_2\phi_r(v),$$

where the last inequality uses $(v - v_\alpha(\rho_t))^2_k \leq (\sqrt{c}r + B)^2 \leq 2(cr^2 + B^2)$.

Subset $K_{1k} \cap K^c_{2k} \cap \Omega_r$: As $v \in K_{1k}$ we have $|(v - v^*)_k|_2 > \sqrt{c}r$. We observe that $T_{1k}(v) + T_{2k}(v) \geq 0$ for all v in this subset whenever

$$\left(-\lambda\left(v - v_\alpha(\rho_t)\right)_k\left(v - v^*\right)_k + \frac{\sigma^2}{2}\left(v - v_\alpha(\rho_t)\right)^2_k\right)\left(r^2 - \left(v - v^*\right)^2_k\right)^2$$
$$\leq \sigma^2\left(v - v_\alpha(\rho_t)\right)^2_k\left(2\left(v - v^*\right)^2_k - r^2\right)\left(v - v^*\right)^2_k. \tag{16}$$

The first term on the left-hand side in (16) can be bounded from above exploiting that $v \in K^c_{2k}$ and by using the relation $\tilde{c} = 2c - 1$. More precisely, we have

$$-\lambda(v - v_\alpha(\rho_t))_k(v - v^*)_k\left(r^2 - (v - v^*)^2_k\right)^2 \leq \tilde{c}r^2\frac{\sigma^2}{2}(v - v_\alpha(\rho_t))^2_k(v - v^*)^2_k$$
$$= (2c-1)r^2\frac{\sigma^2}{2}(v - v_\alpha(\rho_t))^2_k(v - v^*)^2_k \leq \left(2\left(v - v^*\right)^2_k - r^2\right)\frac{\sigma^2}{2}(v - v_\alpha(\rho_t))^2_k(v - v^*)^2_k,$$

where the last inequality follows since $v \in K_{1k}$. For the second term on the left-hand side in (16) we can use $(1 - c)^2 \leq (2c - 1)c$ as per assumption, to get

$$\frac{\sigma^2}{2}\left(v - v_\alpha(\rho_t)\right)^2_k\left(r^2 - (v - v^*)^2_k\right)^2 \leq \frac{\sigma^2}{2}\left(v - v_\alpha(\rho_t)\right)^2_k(1 - c)^2 r^4$$
$$\leq \frac{\sigma^2}{2}(v - v_\alpha(\rho_t))^2_k(2c-1)r^2 cr^2 \leq \frac{\sigma^2}{2}(v - v_\alpha(\rho_t))^2_k\left(2\left(v - v^*\right)^2_k - r^2\right)(v - v^*)^2_k.$$

Hence, (16) holds and we have $T_{1k}(v) + T_{2k}(v) \geq 0$ uniformly on this subset.

Subset $K_{1k} \cap K_{2k} \cap \Omega_r$: As $v \in K_{1k}$ we have $|(v - v^*)_k|_2 > \sqrt{c}r$. We first note that $T_{1k}(v) = 0$ whenever $\sigma^2\left(v - v_\alpha(\rho_t)\right)^2_k = 0$, provided $\sigma > 0$, so nothing needs to be done if $v_k = (v_\alpha(\rho_t))_k$. Otherwise, if $\sigma^2\left(v - v_\alpha(\rho_t)\right)^2_k > 0$, we exploit $v \in K_{2k}$ to get

$$\frac{(v - v_\alpha(\rho_t))_k(v - v^*)_k}{\left(r^2 - (v - v^*)^2_k\right)^2} \geq \frac{-|(v - v_\alpha(\rho_t))_k||(v - v^*)_k|}{\left(r^2 - (v - v^*)^2_k\right)^2}$$
$$> \frac{2\lambda(v - v_\alpha(\rho_t))_k(v - v^*)_k}{\tilde{c}r^2\sigma^2|(v - v_\alpha(\rho_t))_k||(v - v^*)_k|} \geq -\frac{2\lambda}{\tilde{c}r^2\sigma^2}.$$

Using this, T_{1k} can be bounded from below by

$$T_{1k}(v) = 2r^2\lambda \, (v - v_\alpha(\rho_t))_k \, \frac{(v - v^*)_k}{\left(r^2 - (v - v^*)_k^2\right)^2} \phi_r(v) \geq -\frac{4\lambda^2}{\tilde{c}\sigma^2}\phi_r(v) =: -q_3\phi_r(v).$$

For T_{2k}, the nonnegativity of $\sigma^2\,(v - v_\alpha(\rho_t))_k^2$ implies $T_{2k}(v) \geq 0$, whenever

$$2\left(2\,(v - v^*)_k^2 - r^2\right)(v - v^*)_k^2 \geq \left(r^2 - (v - v^*)_k^2\right)^2.$$

This holds for $v \in K_{1k}$, if $2(2c - 1)c \geq (1 - c)^2$ as implied by the assumption.

Concluding the Proof: Using the evolution of ϕ_r as in (12) and the individual decompositions of Ω_r for the terms $T_{1k} + T_{2k}$, we now get

$$\frac{d}{dt}\int \phi_r(v)\,d\rho_t(v) = \sum_{k=1}^{d}\Bigg(\int_{K_{1k}\cap K_{2k}^c\cap\Omega_r}\underbrace{(T_{1k}(v) + T_{2k}(v))}_{\geq 0}\,d\rho_t(v)$$

$$+ \int_{K_{1k}\cap K_{2k}\cap\Omega_r}\underbrace{(T_{1k}(v) + T_{2k}(v))}_{\geq -q_3\phi_r(v)}\,d\rho_t(v) + \int_{K_{1k}^c\cap\Omega_r}\underbrace{(T_{1k}(v) + T_{2k}(v))}_{\geq -(q_1+q_2)\phi_r(v)}\,d\rho_t(v)\Bigg)$$

$$\geq -d\max\{q_1 + q_2, q_3\}\int \phi_r(v)\,d\rho_t(v) = -q\int\phi_r(v)\,d\rho_t(v).$$

An application of Grönwall's inequality concludes the proof.   □

3.5   Proof of Theorem 2

We now have all necessary tools to conclude the global convergence proof.

Proof (Theorem 2). Lemma 1 provides a bound for the time derivative of $\mathcal{V}(\rho_t)$, given by

$$\frac{d}{dt}\mathcal{V}(\rho_t) \leq -\left(2\lambda - \sigma^2\right)\mathcal{V}(\rho_t) + \sqrt{2}\left(\lambda + \sigma^2\right)\sqrt{\mathcal{V}(\rho_t)}\,\|v_\alpha(\rho_t) - v^*\|_2$$

$$+ \frac{\sigma^2}{2}\|v_\alpha(\rho_t) - v^*\|_2^2.$$

$$\text{(17)}$$

Now we define the time $T \geq 0$ by

$$T := \sup\left\{t \geq 0 : \mathcal{V}(\rho_{t'}) > \varepsilon \text{ and } \|v_\alpha(\rho_{t'}) - v^*\|_2 < C(t') \text{ for all } t' \in [0, t]\right\}, \quad (18)$$

where

$$C(t) := \min\left\{\frac{\tau}{2}\frac{(2\lambda - \sigma^2)}{\sqrt{2}\,(\lambda + \sigma^2)}, \sqrt{\tau\frac{(2\lambda - \sigma^2)}{\sigma^2}}\right\}\sqrt{\mathcal{V}(\rho_t)}.$$

Combining (17) with (18), we have by construction for all $t \in (0, T)$

$$\frac{d}{dt} \mathcal{V}(\rho_t) \leq -(1 - \tau)\left(2\lambda - \sigma^2\right) \mathcal{V}(\rho_t).$$

Grönwall's inequality implies the upper bound

$$\mathcal{V}(\rho_t) \leq \mathcal{V}(\rho_0) \exp\left(-(1 - \tau)\left(2\lambda - \sigma^2\right) t\right), \quad \text{for } t \in [0, T]. \tag{19}$$

Accordingly, we note that $\mathcal{V}(\rho_t)$ is a decreasing function in t, which implies the decay of the auxiliary function $C(t)$ as well. Hence, we may bound

$$\max_{t \in [0,T]} \|v_\alpha(\rho_t) - v^*\|_2 \leq \max_{t \in [0,T]} C(t) \leq C(0), \tag{20}$$

$$\max_{t \in [0,T]} \int \|v - v^*\|_2 \, d\rho_t(v) \leq \max_{t \in [0,T]} \sqrt{2\mathcal{V}(\rho_t)} \leq \sqrt{2\mathcal{V}(\rho_0)}. \tag{21}$$

To conclude that $\mathcal{V}(\rho_T) \leq \varepsilon$, it now remains to check three different cases.

Case $T \geq T^*$: If $T \geq T^*$, we can use the definition of T^* in (8) and the time-evolution bound of $\mathcal{V}(\rho_t)$ in (19) to conclude that $\mathcal{V}(\rho_{T^*}) \leq \varepsilon$. Hence, by definition of T in (18), we find $\mathcal{V}(\rho_T) = \varepsilon$ and $T = T^*$.

Case $T < T^*$ and $\mathcal{V}(\rho_T) = \varepsilon$: Nothing needs to be discussed in this case.

Case $T < T^*$, $\mathcal{V}(\rho_T) > \varepsilon$, and $\|v_\alpha(\rho_T) - v^*\|_2 \geq C(T)$: We shall show that there exists $\alpha_0 > 0$ so that for any $\alpha \geq \alpha_0$ we have $\|v_\alpha(\rho_T) - v^*\|_2 < C(T)$, a contradiction, which proves that the case never occurs. To do so, we define

$$q := \min\left\{\left(\eta C(T)/(2\sqrt{d})\right)^{1/\nu}, \mathcal{E}_\infty\right\}/2 \quad \text{and} \quad r := \max_{s \in [0, R_0]}\left\{\max_{v \in B_s^\infty(v^*)} \mathcal{E}(v) \leq q\right\}.$$

By construction, $r \leq R_0$ and $q + \mathcal{E}_r = q + \sup_{v \in B_r^\infty(v^*)} \mathcal{E}(v) \leq 2q \leq \mathcal{E}_\infty$. Furthermore, we note that $q > 0$ since $C(T) > 0$. By continuity of $\mathcal{E}$ there exists $s_q > 0$ such that $\mathcal{E}(v) \leq q$ for all $v \in B_{s_q}^\infty(v^*)$, thus yielding also $r > 0$. Therefore, we can apply Proposition 1 with q and r as above together with (21) to get

$$\|v_\alpha(\rho_T) - v^*\|_2 \leq \frac{1}{2} C(T) + \frac{\sqrt{d} \exp\left(-\alpha q\right)}{\rho_T(B_r^\infty(v^*))} \sqrt{2\mathcal{V}(\rho_0)}. \tag{22}$$

Furthermore, by (20) we have $\max_{t \in [0,T]} \|v_\alpha(\rho_t) - v^*\|_2 \leq C(0)$, implying that all assumptions of Proposition 2 hold. Therefore, there exists $a > 0$ so that

$$\rho_T(B_r^\infty(v^*)) \geq \int \phi_r(v) \, d\rho_0(v) \exp(-aT) > 0,$$

where we used (7) for bounding the initial mass ρ_0, and the fact that ϕ_r is bounded from below on $B_{cr}^\infty(v^*)$ for any $c < 1$. Then, by using any $\alpha > \alpha_0$ with

$$\alpha_0 = \frac{1}{2q}\left(\log d - 2 \log \rho_T(B_r^\infty(v^*)) + \log\left(\frac{\mathcal{V}(\rho_0)}{\mathcal{V}(\rho_T)}\right) + 2 \log\left(\frac{\lambda + \sigma^2}{\tau\left(2\lambda - \sigma^2\right)}\right)\right),$$

(22) is strictly smaller than $C(T)$, giving the desired contradiction. $\square$

4   A Machine Learning Example

In this section, we showcase the practicability of the implementation of anisotropic CBO as described in [4, Algorithm 2.1] for problems appearing in machine learning by training a shallow and a convolutional NN (CNN) classifier for the MNIST dataset of handwritten digits [16]. Let us emphasize that it is not our aim to challenge the state of the art for this task by employing the most sophisticated model or intricate data preprocessing. We merely believe that this is a well-understood, complex, high-dimensional benchmark to demonstrate that CBO achieves good results already with limited computational capacities.

Let us now describe the NN architectures used in our numerical experiment, see also Fig. 3. Each input image is represented by a matrix of dimension 28×28 with entries valued between 0 and 1 depending on the grayscale of the respective pixel. For the shallow neural net (see Fig. 3a) the image is first reshaped to a vector $x \in \mathbb{R}^{728}$ before being passed through a dense layer of the form $\text{ReLU}(Wx+b)$ with trainable weight matrix $W \in \mathbb{R}^{10 \times 728}$ and bias vector $b \in \mathbb{R}^{10}$. The CNN (see Fig. 3b) has learnable kernels and its architecture is similar to the one of the LeNet-1, cf. [15, Section III.C.7]. In both networks a batch normalization step is included after each ReLU activation, which entails a considerably faster training process. Moreover, in the final layers a softmax activation function is applied so that the output can be interpreted as a probability distribution over the digits. In total, the number of unknowns to be trained in case of the shallow NN is 7850, which compares to 2112 free parameters for the CNN. We denote the parameters of the NN by θ and its forward pass by f_θ.

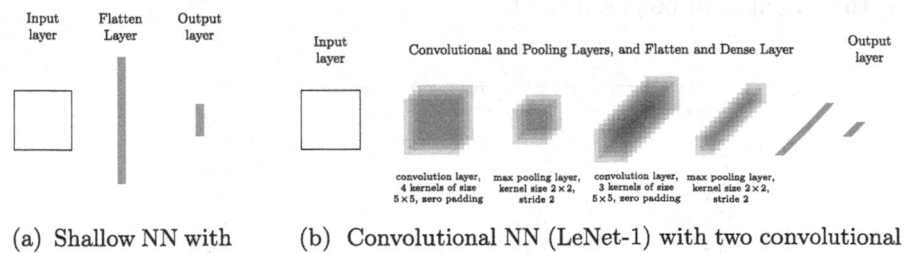

(a) Shallow NN with one dense layer

(b) Convolutional NN (LeNet-1) with two convolutional and two pooling layers, and one dense layer

Fig. 3. Architectures of the NNs used in the experiments of Sect. 4.

As a loss function during training we use the categorical crossentropy loss $\ell(\widehat{y}, y) = -\sum_{k=0}^{9} y_k \log(\widehat{y}_k)$ with $\widehat{y} = f_\theta(x)$ denoting the output of the NN for a training sample (x, y) consisting of image and label. This gives rise to the objective function $\mathcal{E}(\theta) = \frac{1}{M} \sum_{m=1}^{M} \ell(f_\theta(x^m), y^m)$, where $(x^m, y^m)_{m=1}^{M}$ denote the M training samples. When evaluating the performance of the NN we determine the accuracy on a test set by counting the number of successful predictions.

The used implementation of anisotropic CBO combines ideas presented in [10, Section 2.2] with the algorithm proposed in [4]. More precisely, it employs random mini-batch ideas when evaluating the objective function $\mathcal{E}$ and when computing the consensus point v_α, meaning that $\mathcal{E}$ is only evaluated on a random subset of size $n_\mathcal{E}$ of the training dataset and v_α is only computed from a random subset of size n_N of all particles. While this reduces the computational complexity, it simultaneously increases the stochasticity, which enhances the ability to escape from local optima. Furthermore, inspired by Simulated Annealing, a cooling strategy for the parameters α and σ is used as well as a variance-based particle reduction technique similar to ideas from Genetic Algorithms. More specifically, α is multiplied by 2 after each epoch, while the diffusion parameter σ follows the schedule $\sigma_{epoch} = \sigma_0/\log_2(epoch + 2)$. For our experiments we choose the parameters $\lambda = 1$, $\sigma_0 = \sqrt{0.4}$ and $\alpha_{initial} = 50$, and discrete time step size $\Delta t = 0.1$ for training both the shallow and the convolutional NN. We use $N = 100$ particles, which are initialized according to $\mathcal{N}\big((0,\ldots,0)^T, \mathrm{Id}\big)$. The mini-batch sizes are $n_\mathcal{E} = 60$ and $n_N = 10$ and despite v_α being computed only on a basis of n_N particles, all N particles are updated in every time step, referred to as the full update in [4]. We emphasize that hyperparameters have not been tuned extensively.

In Fig. 4 we report the results of our experiment. While achieving a test accuracy of almost 90% for the shallow NN, we obtain around 97% accuracy with the CNN. For comparison, when trained with backpropagation with finely tuned parameters, a comparable CNN achieves 98.3% accuracy, cf. [15, Figure 9]. In view of these results, CBO can be regarded as a successful optimizer for machine learning tasks, which performs comparably to the state of the art. At the same time it is worth highlighting that CBO is extremely versatile and customizable, does not require gradient information or substantial hyperparameter tuning and has the potential to be parallelized.

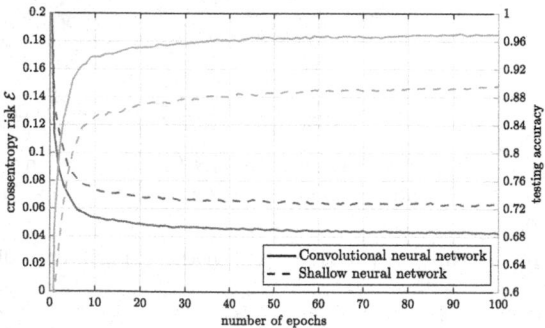

Fig. 4. Comparison of the performances of a shallow (dashed lines) and convolutional (solid lines) NN with architectures as described in Figs. 3a and b, when trained with a discretized version of the anisotropic CBO dynamics (1). Depicted are the accuracies on a test dataset (orange lines) and the values of the objective function $\mathcal{E}$ (blue lines), which was chosen to be the categorical crossentropy loss on a random sample of the training set of size 10000. (Color figure online)

5 Conclusion

In this paper we establish the global convergence of anisotropic consensus-based optimization (CBO) to the global minimizer in mean-field law with dimension-independent convergence rate by adapting the proof technique developed in [9]. It is based on the insight that the dynamics of individual particles follow, on average, the gradient flow dynamics of the map $v \mapsto \|v - v^*\|_2^2$. Furthermore, by utilizing the implementation of anisotropic CBO suggested in [4], we demonstrate the practicability of the method by training the well-known LeNet-1 on the MNIST data set, achieving around 97% accuracy after few epochs with just 100 particles.

In subsequent work we plan to extend our theoretical understanding of CBO to the finite particle regime, and aim to provide extensive numerical studies. We also intend to use this approach to explain the mean-field law convergence behavior of other metaheuristics such as Particle Swarm Optimization, see, e.g., [5,13].

Acknowledgements. MF acknowledges the support of the DFG Project "Identification of Energies from Observations of Evolutions" and the DFG SPP 1962 "Nonsmooth and Complementarity-Based Distributed Parameter Systems: Simulation and Hierarchical Optimization". KR acknowledges the financial support from the Technical University of Munich – Institute for Ethics in Artificial Intelligence (IEAI).

References

1. Aarts, E., Korst, J.: Simulated Annealing and Boltzmann Machines. A Stochastic Approach to Combinatorial Optimization and Neural Computing. Wiley-Interscience Series in Discrete Mathematics and Optimization. Wiley, Hoboken (1989)
2. Carrillo, J.A., Choi, Y.P., Totzeck, C., Tse, O.: An analytical framework for consensus-based global optimization method. Math. Models Methods Appl. Sci. **28**(6), 1037–1066 (2018)
3. Carrillo, J.A., Fornasier, M., Toscani, G., Vecil, F.: Particle, kinetic, and hydrodynamic models of swarming. In: Mathematical Modeling of Collective Behavior in Socio-Economic and Life Sciences, pp. 297–336. Model. Simul. Sci. Eng. Technol., Birkhäuser Boston, Boston (2010)
4. Carrillo, J.A., Jin, S., Li, L., Zhu, Y.: A consensus-based global optimization method for high dimensional machine learning problems. ESAIM Control Optim. Calc. Var. **27**(suppl.), Paper No. S5, 22 (2021)
5. Cipriani, C., Huang, H., Qiu, J.: Zero-inertia limit: from particle swarm optimization to consensus based optimization. arXiv:2104.06939 (2021)
6. Dorigo, M., Blum, C.: Ant colony optimization theory: a survey. Theor. Comput. Sci. **344**(2–3), 243–278 (2005)
7. Fogel, D.B.: Evolutionary Computation: Toward a New Philosophy of Machine Intelligence, 2nd edn. IEEE Press, Piscataway (2000). https://ieeexplore.ieee.org/book/5237910
8. Fornasier, M., Huang, H., Pareschi, L., Sünnen, P.: Anisotropic diffusion in consensus-based optimization on the sphere. arXiv:2104.00420 (2021)

9. Fornasier, M., Klock, T., Riedl, K.: Consensus-based optimization methods converge globally in mean-field law. arXiv:2103.15130 (2021)
10. Fornasier, M., Pareschi, L., Huang, H., Sünnen, P.: Consensus-based optimization on the sphere: convergence to global minimizers and machine learning. J. Mach. Learn. Res. **22**(237), 1–55 (2021)
11. Holland, J.H.: Adaptation in Natural and Artificial Systems. An Introductory Analysis with Applications to Biology, Control, and Artificial Intelligence. University of Michigan Press, Ann Arbor (1975)
12. Huang, H., Qiu, J.: On the mean-field limit for the consensus-based optimization. arXiv:2105.12919 (2021)
13. Huang, H., Qiu, J., Riedl, K.: On the global convergence of particle swarm optimization methods. arXiv:2201.12460 (2022)
14. Kennedy, J., Eberhart, R.: Particle swarm optimization. In: Proceedings of ICNN 1995 - International Conference on Neural Networks, vol. 4, pp. 1942–1948. IEEE (1995)
15. LeCun, Y., Bottou, L., Bengio, Y., Haffner, P.: Gradient-based learning applied to document recognition. Proc. IEEE **86**(11), 2278–2324 (1998)
16. LeCun, Y., Cortes, C., Burges, C.: MNIST handwritten digit database (2010). http://yann.lecun.com/exdb/mnist/
17. Miller, P.D.: Applied Asymptotic Analysis. Graduate Studies in Mathematics, vol. 75. American Mathematical Society, Providence (2006)
18. Pinnau, R., Totzeck, C., Tse, O., Martin, S.: A consensus-based model for global optimization and its mean-field limit. Math. Models Methods Appl. Sci. **27**(1), 183–204 (2017)
19. Rastrigin, L.: The convergence of the random search method in the extremal control of a many parameter system. Autom. Remote Control **24**, 1337–1342 (1963)
20. Toscani, G.: Kinetic models of opinion formation. Commun. Math. Sci. **4**(3), 481–496 (2006)

Author Index

Printed in the United States
by Baker & Taylor Publisher Services

Printed in the United States
by Baker & Taylor Publisher Services